Biology of Termites: A Modern Synthesis

David Edward Bignell · Yves Roisin · Nathan Lo
Editors

Biology of Termites: A Modern Synthesis

Foreword by Bernard J. Crespi

Springer

Editors
Prof. David Edward Bignell
School of Biological and Chemical Sciences
Queen Mary University of London
Mile End Rd.
E1 4NS London
United Kingdom
d.bignell@qmul.ac.uk

Prof. Yves Roisin
Behavioural and Evolutionary Ecology
Université Libre de Bruxelles
Av. F.D. Roosevelt 50
1050 Bruxelles
Belgium
yroisin@ulb.ac.be

Dr. Nathan Lo
School of Biological Sciences
University of Sydney
Macleay Bldg. A12
2006 Sydney New South Wales
Australia
Nathan@usyd.edu.au

ISBN 978-90-481-3976-7 e-ISBN 978-90-481-3977-4
DOI 10.1007/978-90-481-3977-4
Springer Dordrecht Heidelberg London New York

Library of Congress Control Number: 2010937355

Cover illustration: imagoes, soldiers, workers and larvae of *Syntermes territus* Emerson. Photo: Yves Roisin

Printed on acid-free paper

Springer is part of Springer Science+Business Media (www.springer.com)

While preparing the final proofs of this volume, we learned with great sadness of the death of Professor Charles Noirot on 23 September 2010, at the age of 88. Noirot was the greatest termite biologist of his generation and arguably the greatest of any generation, past or present. No part of the field escaped his influence, not least since his publications at one time or another covered every conceivable facet of the subject, though most notably that of post-embryonic development and its regulation (Pasteels 2001), but principally because he refined the lexicon and set standards for the observation, documentation and analysis of termite societies and associations which we all follow today. Our world of termitology is the less for his passing, but hugely the greater for his life and work. We dedicate this book to his memory.

London, UK
Bruxelles, Belgium
Sydney, Australia

David Bignell
Yves Roisin
Nathan Lo

Reference

Pasteels JM (2001) Charles Noirot: an obligatory reference. Insect Soc 48:185–186

Foreword

A greater and longer childhood: the comparative biology of termites.

When I was a young graduate student, Richard Alexander instilled in me the importance of cutting one's teeth on a taxonomic group – be it crickets, thrips, poison-dart frogs, barnacles or prairie dogs – and letting the deep, unresolved biological questions well up from the vast diversity within even the most humble clade. Especially among insects, a fascination emerges from systematic-through-genomic familiarity, and even a love of small, idiosyncratic creatures can develop as they lead us through convoluted pathways and tunnels of discovery.

Social insects hold a special place in the pantheons of biophilia, given the close parallels of their lives with ours. Who among us does not admit the abiding centrality of family life – the mutualistic peace entwined with strife and conflicts, first when we are children and later with bairns of our own? Of all social insects, termites may fit most closely with the human condition, as parents, alloparents, builders, soldiers, and biochemical-genomic engineers, ensconced in extended nuclear families that expand and heroically defend their homes. But perhaps most of all, as children.

Benjamin Franklin penned the phrase, "our whole life is but a greater and longer childhood", referring to ourselves, but equally applicable to termite evolution since the first cockroach helped its mother and father rear a younger sibling. A greatly-extended childhood has indeed been seen as key to modern human evolution (Hrdy 2009; Konner 2010), and it was certainly crucial to unlocking the potential of termites to dominate the globe. For both humans and isopterans, early development in a relatively-safe, nourishing niche potentiated such a heterochronic yet plastic shift, and led ultimately to a broad swath of ecological, morphological and behavioral specializations, as well as expansion of family conflicts and confluences of interest beyond those of almost all other creatures.

This book draws the vast diversity of termite adaptations together in the framework of a range of greater unresolved biological questions, from genetics through ecology and social behavior to macroevolution, that termites, as such-special insects, can help us answer. As such, the book brings into sharp focus the tremendous usefulness of termites as research systems, and the close interplay between systematics, evolutionary biology, ecology, and other, more-proximate and mechanistic disciplines. Each chapter celebrates in rigorous up to date detail, some vital facet of termite biology: their taxonomy, phylogeny, genetics, symbioses, physiology, morphology, pheromones, ecology, behavior, and status as pests, all linked by the unique characteristics, and the extended or lifelong childhoods, that define the group. As such, this book is indispensible to any serious student of Isoptera, as a source of facts, ideas, and syntheses upon which to build our future understanding. Indeed, the advances in termite biology since the last comprehensive compendium in 2000 are spectacular – from endogenous cellulases, facultatively-asexual queens, to establishment of firm links between ecology, life history and social structure that permit robust comparisons with other taxa.

And with humans, of course. The origin and form of human childhood can be seen as an evolutionary battleground of parent versus beloved-offspring strategies and countering moves: children selected to delay the age of puberty and accrue marginally-more investment, mothers selected to shorten age at weaning, reduce inter-birth intervals, and become more queen-like than any other primate (Haig 2010). Among termites, analysing evolutionary trajectories of life-history timing and inter-family interactions awaits fine-scale, dovetailed phylogenetic and behavioral-ecological studies, that are likely to further astonish and please human termitophiles – in part due to their implications for all areas of isopteran biology. This book provides the scaffold for building, conceptually and empirically, across all such disciplines and questions.

For many students of termites, the book comes at a special time, after we have lost intellectual parents of social-insect biology, Ross Crozier and William Hamilton, and must ourselves build on their ideas. *Biology of Termites: a Modern Synthesis* should serve as inspiration and foundation for new discoveries that would make them proud – and fascinated by the ongoing complexities of children and social life under soil and bark.

Vancouver, Canada

Bernard J. Crespi

References

Haig D (2010) Transfers and transitions: parent-offspring conflict, genomic imprinting, and the evolution of human life history. Proc Natl Acad Sci USA 107(Suppl 1):1731–1735

Hrdy S (2009) Mothers and others: the evolutionary origins of mutual understanding. Harvard University Press, Cambridge, MA

Konner M (2010) The evolution of childhood. Harvard University Press, Cambridge, MA

Preface

The initiative for this new volume was made in 2007 by Springer and came a few years after their merger (in 2003) with the Kluwer organisation, which had published the successful preceding compilation *Termites: evolution, sociality, symbioses, ecology* (Abe T, Bignell DE, Higashi M (eds) (2000) Kluwer Academic Publishers, Dordrecht, 466p). The Kluwer book had been the first attempt to draw together the many different threads of academic termite science since the three monumental volumes of P-P Grassé's *Termitologia* (Masson, Paris, 1983, 1984, 1985), and the first academic research compendium in English since the two classic volumes of *Biology of Termites* edited by Krishna K and Weesner FM (Academic Press, New York, NY, 1969, 1970).

The purposes of the present book are (1) to provide extended coverage to areas of termite science which have advanced spectacularly since 2000, these being largely fields in which extensive molecular data have now become available (phylogeny and biogeography, immune defence, population genetics, caste determination, digestive biochemistry and intestinal microbiology) (2) to reflect new concepts of eusociality in termites and to present the emerging views of social evolution within the Isoptera, now that this group has been confirmed as a monophyletic lineage (3) to provide points of comparison with ants and some other social Hymenoptera, on social organisation and co-evolution with symbionts (4) to provide consolidating reviews in selected areas where the literature has continued to grow (cladistic taxonomy, mound architecture and function, intestinal morphology and function, community ecology and pest status for wood in service) and (5) to present reviews that were notably absent from the 2000 Kluwer book (chemical ecology and termites as pests of agriculture). Our expectation is that the new book will be regarded as the standard reference work for about a decade, and will help termites to be seen as good models for fundamental research in developmental biology, microbial ecology and social evolution, as well as major players in the ecology of the biosphere, with impact in many areas of the human economy.

We made a deliberate decision not to commission updates in certain areas that were covered in 2000. These were palaeontology, foraging, intracellular symbiosis, symbiosis with protists (sensu stricto), energy metabolism, population ecology, termites and soils, and greenhouse gas production. In some cases these topics are now subsumed under other headings, so that new material will still be reported, and in other cases our feeling is that the existing literature (including reviews published elsewhere in the last 10 years) is sufficient. We remind readers that termite biology still rests on the large body of older descriptive material of an essentially timeless nature (for example covering morphology, anatomy, behaviour, nest architecture and economic damages) and that in these respects the volumes by Krishna and Weesner and by Grassé are in no sense outdated. As before, we debated whether applied termite biology (termite control) should have a place in an academic book: the outcome is two pleasingly complementary chapters (18 and 19) dealing, respectively, with termites in agriculture and with invasive species which defy the high endemicity otherwise shown by isopterans. Both chapters explore contemporary concepts and contain extensive bibliographies.

Although we have now passed the 10th anniversary of their untimely deaths, we still need to acknowledge that the main modern initiative to draw leading termite biologists together to identify the major advances and issues in the field was that of Takuya Abe and Masahiko Higashi, so in that sense the new book is still very much their legacy. We can also acknowledge the influence of the late Ross Crozier in advancing the science of social insect genetics, the fruit of which is much in evidence in this volume. In commissioning authors for the book, the editors have tried to balance contributions between the major countries where termite research is supported and carried out: thus the UK, USA, France, Germany, Belgium, The Netherlands, Portugal, Japan and Australia are all represented. Authors from developing countries where termites are important organisms are still missing, but this absence will surely not persist for another decade. A taxonomic index, with authorities and information on synonymy, will be published separately at a future date.

London, UK — David E. Bignell
Bruxelles, Belgium — Yves Roisin
Sydney, Australia — Nathan Lo
27 May 2010

Contents

Contributors

Duur K. Aanen Laboratory of Genetics, Wageningen University, 6708 PB Wageningen, The Netherlands, duur.aanen@wur.nl

David E. Bignell School of Biological and Chemical Sciences, Queen Mary, University of London, London E1 4NS, UK; Institute for Tropical Biology and Conservation, Universiti Malaysia Sabah, Kota Kinabalu, Sabah, Malaysia, d.bignell@qmul.ac.uk

Christian Bordereau Développement et Communication Chimique chez les Insectes, Université de Bourgogne, F-21000 Dijon, France, christian.bordereau@u-bourgogne.fr

Andreas Brune Department of Biogeochemistry, Max Planck Institute for Terrestrial Microbiology, 35043 Marburg, Germany, brune@mpi-marburg.mpg.de

Mark S. Bulmer Department of Biological Sciences, Towson University, MD 21252, USA, mbulmer@towson.edu

Paul Eggleton Termite Research Group and Soil Biodiversity Programme, Entomology Department, Natural History Museum, London SW7 5BD, UK, P.eggleton@nhm.ac.uk

Theodore A. Evans CSIRO Division of Entomology, Canberra ACT 2600, Australia, theo.evans@csiro.au

Kenneth J. Howard Department of Entomology, University of Maryland, College Park, MD 20742, USA, howark2@sage.edu

Claudia Husseneder Department of Entomology, LSU Agricultural Center, Baton Rouge, LA 70803, USA, chusseneder@agcenter.lsu.edu

David T. Jones Soil Biodiversity Group, Department of Entomology, Natural History Museum, London SW7 5BD, UK, david.jones@nhm.ac.uk

Judith Korb Behavioural Biology, University of Osnabrueck, D-49076 Osnabrueck, Germany, judith.korb@biologie.uni-osnabrueck.de

Nathan Lo Behaviour and Genetics of Social Insects Laboratory, School of Biological Sciences, University of Sydney, Sydney, NSW 2006, Australia, nathan@usyd.edu.au

Kenji Matsuura Laboratory of Insect Ecology, Graduate School of Environmental Science, Okayama University, Okayama 700-8530, Japan, kenjijpn@cc.okayama-u.ac.jp

Toru Miura Laboratory of Ecological Genetics, Graduate School of Environmental Science, Hokkaido University, Sapporo 060-0810, Japan, miu@ess.hokudai.ac.jp

Christine A. Nalepa Entomology Department, North Carolina State University, Raleigh, NC 27695-7613, USA, christine_nalepa@ncsu.edu

Tânia Nobre Laboratory of Genetics, Wageningen University, 6708 PB Wageningen, The Netherlands, tania.mesquitanobre@wur.nl

Moriya Ohkuma Microbe Division (Japan Collection of Microorganisms), RIKEN BioResource Center, Saitama 351-0198, Japan, mohkuma@riken.jp

Jacques M. Pasteels Evolutionary Biology and Ecology, Université Libre de Bruxelles, B-1050 Bruxelles, Belgium, jmpastee@ulb.ac.be

Yves Roisin Evolutionary Biology and Ecology, Faculté des Sciences, Université Libre de Bruxelles, B-1050 Brussels, Belgium, yroisin@ulb.ac.be

Rebeca B. Rosengaus Department of Biology, Northeastern University, Boston, MA 02115, USA, r.rosengaus@neu.edu

Corinne Rouland-Lefèvre Unité de recherche Biodiversité et fonctionnement du sol, Institut de recherche pour le développement, 93143 Bondy Cedex, France, Corinne.Rouland-lefevre@bondy.ird.fr

Michael E. Scharf Entomology and Nematology Department, University of Florida, Gainsville, FL 32611-0620, USA, mescharf@ufl.edu

Barbara L. Thorne Department of Entomology, University of Maryland, College Park, MD 20742, USA, bthorne@umd.edu

Gaku Tokuda Tropical Biosphere Research Center, COMB, University of the Ryukyus, Nishihara 903-0213, Japan, tokuda@comb.u-ryukyu.ac.jp

James F.A. Traniello Department of Biology, Boston University, Boston, MA 02215, USA, jft@bio.bu.edu

Edward L. Vargo Department of Entomology, North Carolina State University, Raleigh, NC 27695, USA, ed_vargo@ncsu.edu

Hirofumi Watanabe National Institute of Agrobiological Sciences, Tsukuba 305-8634, Japan, hinabe@affrc.go.jp

Chapter 1
An Introduction to Termites: Biology, Taxonomy and Functional Morphology

Paul Eggleton

Abstract Termites are fully social insects, with an extraordinary range of morphological forms. It is now clearly established that they are a very specialised form of cockroach, with far more complex social systems than other cockroaches, and with a far wider range of diets. Termites all live in colonies, with reproductives (kings, queens, and nymphs), soldiers and "helpers" (true workers and also immature stages that assist within the colony to some extent). Termite morphological and anatomical adaptations are caste-specific, with structures evolving independently in reproductives (to allow dispersal, pair bonding and fecundity), workers (foraging and feeding, tending and feeding of immatures, nest construction) and soldiers (only defence). The modifications seen in termite societies are similar to those found in the somatic parts of multicellular organisms, leading to the idea that a termite colony is best thought of as a single organism (or, more controversially, a "superorganism"). The structures that termites build, the mounds and nests, might also be defined as part of this organism. Mounds and nests contribute greatly to the well-being of termite colonies by providing shelter, fortifications and climate control. Overall, termites have amongst the most complex social, anatomical and structural adaptations of any animal.

1.1 Introduction

Most people are aware of termites, either directly or indirectly. Throughout the tropics everyone knows that termites are voracious eaters of houses and crops, while a smaller number understand that they also have a role in improving soil quality. In the US they cause more economic damage than fire and flood combined, predominantly by feeding on structural timbers. In contrast, here in the UK, where there are no termites, they are mostly known by anecdote and through an iconic sequence in David

P. Eggleton (✉)
Termite Research Group and Soil Biodiversity Programme, Entomology Department, Natural History Museum, London, SW7 5BD, UK
e-mail: P.eggleton@nhm.ac.uk

D.E. Bignell et al. (eds.), *Biology of Termites: A Modern Synthesis*,
DOI 10.1007/978-90-481-3977-4_1, © Springer Science+Business Media B.V. 2011

Attenborough's *Life on Earth* when he crawled inside a termite mound. In this chapter I will introduce termites, by briefly discussing their biology, classification and functional morphology. I set the scene for later chapters by presenting background information and placing termites in a wider context. Termites are eusocial cockroaches. They live in complex societies that can be modelled as "superorganisms" – where the individuals form part of a larger self-regulating entity (Wilson 1992). They feed on dead plant material at different stages of decomposition (Donovan et al. 2001). They live predominantly in tropical regions, where they are by far the most important decomposer animals (Davies et al. 2003).

1.2 Diversity, Taxonomy, Classification

Termites are easy to distinguish from all other insect groups, as they have very distinctive morphologies in all their castes. However, the phylogenetic position of termites has been long debated. The majority view is that termites form part of the order Dictyoptera (along with cockroaches and mantids), and that within that order they are phylogenetically nested within the cockroaches (Inward et al. 2007a; Legendre et al. 2008; Ware et al. 2008). They are close to the Blattidae, and their sister group is the woodroach, *Cryptocercus*. The details of these relationships are discussed in Chapter 2. Some researchers have argued that this means that an order-level classification for termites is unwarranted (Eggleton et al. 2007; Inward et al. 2007a), but others (Lo et al. 2007; Engel et al. 2009) have rejected this idea, predominantly for reasons of nomenclatural stability. In this chapter I simply use the word "termite" as it unambiguously describes the same set of species. A related problem is the use of the term "cockroach-like" as it is tautological if used to describe features of a group that is actually within the cockroaches. Here I will use it to describe a similarity to the common ancestor of termites and *Cryptocercus*, or more simply, as similarity to "non-termite cockroaches". There is a direct analogy between this problem and that concerning the group left when dinosaurs are defined without the birds, leaving the "non-avian dinosaurs".

There are approximately 2,600 described species of termites (Kambhampati and Eggleton 2000) with perhaps 500–1,000 species still left to describe. The most recent higher classification (Engel et al. 2009) splits those species into nine extant families and proposes two fossil ones, with one "family" of uncertain status. Generally, the basal families are more cockroach-like and the apical ones more specialised: this can be particularly seen in the Mastotermitidae and the "Termopsidae". All groups are fully eusocial, although some dry-wood-nesting termites (Kalotermitidae) may not have functionally active workers.

The Mastotermitidae, with just a single extant species restricted to northern Australia, are acknowledged to be the oldest family, with many cockroach-like features (e.g. wing venation, retention of an ootheca, presence of the endosymbiont *Blattabacterium*). However, they have true workers and nest away from their food. Two slightly more recent groups, the 'Termopsidae' and the Kalotermitidae both nest and feed in single pieces of dead wood: the termopsids in wet logs in temperate rain forests, and the kalotermitids in dry wood in tropical rainforest tree canopies, isolated islands and semi-deserts. These groups do not have true workers, although they always have soldiers. A specialised group within or close to the termopsids, the Hodotermitidae, are grass-feeders across semi-arid regions of the Old World, and have well developed nests built away from their food. The Rhinotermitidae are widespread, being the only family that extends significantly into subtropical and warm temperate regions, for example in North America and Europe (Weesner 1965). They have true workers and mostly build their nests away from their food. The Termitidae make up the bulk of extant species and are dominant in tropical regions, particularly rain forests and savannas.

Other families have been proposed in the past – in particular the Indotermitidae, which are clearly specialised Apicotermitinae. The Serritermitidae, still recognised in the most recent classification (Engel et al. 2009), is probably better placed as a subfamily of the Rhinotermitidae. The present family-level classification is still likely to change in the future as two families, the Rhinotermitidae and the "Termopsidae" (split up by Engel into a number of families), are probably not natural groupings.

The ranks between family and generic levels are generally poorly worked out. The Mastotermitinae has only one species. The Termopsidae have three fairly disparate subfamilies (Termopsinae, Stolotermitinae, Porotermitinae). The Kalotermitidae have no accepted subfamily classification. The Rhinotermitidae have subfamilies that mostly contain a single or two genera, with the exception of the Rhinotermitinae. Within the Termitidae seven subfamilies were most recently recognised by Engel et al. (2009): Macrotermitinae, Sphaerotermitinae, Foraminitermitinae, Apicotermitinae, Syntermitinae, Termitinae, and Nasutitermitinae. However, this subfamily-level classification still remains somewhat unsatisfactory and, despite recent improvements, needs a complete revision. This has been known for some time (e.g. see Kambhampati and Eggleton 2000; Inward et al. 2007b), particularly with respect to the Termitinae, which are clearly a paraphyletic group, and the Syntermitinae, which are probably recognised at too high a taxonomic rank. The most important barrier to a reclassification within the Termitidae is the difficulty of resolving relationships between taxa at the most apical part of the termite tree (Inward et al. 2007b).

1.3 The Colony

A termite colony is a family of termite individuals all living together. It generally has an inanimate and an animate part. The animate part is the individuals living within the colony; the inanimate part is the structures built by individuals within which they live. Sometimes the inanimate part of the colony is just a few tunnels, but often it is a very extensive and sophisticated structure.

The animate part of a colony has immatures and typically three main adult castes: reproductives (queens, kings, and alates), workers, and soldiers. The queen is generally the only egg-laying individual in the colony. The king is her consort and his only task appears to be to mate with her regularly (Korb 2008). The alates are winged reproductives preparing to leave the nest in order to swarm, to pair and to start new colonies. Workers, on the other hand, never leave the nest except to forage for food. They are the mainstay of the colony, and their roles are numerous. They forage for food and water, build and repair colony structures, and tend the immatures, alates, the king and the queen. Soldiers have only one job: to defend the colony, and particularly the queen and the king. This simple description of caste structure is complicated by some species that have no soldiers and others that have no workers. All termites species have one or the other, however, and so all termite species are eusocial, because they have atleast one sterile caste that is pre-determined during the immature stages (Boomsma 2009).

The life cycle is similar in all termites. Colonies produce winged reproductives ("alates"), often at the start of the rainy season in drier or seasonal habitats, but all the year round in wetter, aseasonal habitats (Martius et al. 1996). These reproductives land on the ground or on a piece of dead wood and pair up: one male with one female. The pair then found a colony, either in the soil or in dead wood. They mate and produce workers (or pseudergates) that begin to tend young, to build colony structures and to forage for food. Slightly later in the colony's development soldiers are produced. When the colony has reached maturity alates are produced again, and the cycle continues.

1.4 The Colony as (Super)organism

An individual in a termite colony is not really like a standard solitary insect. If you separate it from the colony it will die. Even the alates, which can survive for some time apart from the nest, need to pair and form a new colony to survive. The main reason for this obligate association with the colony is that each caste lacks some element that is present in a solitary insect. Workers and soldiers have no reproductive tract; soldiers and reproductives cannot feed themselves; workers and reproductives generally cannot defend themselves effectively; soldiers and workers cannot disperse. These functions, reproduction, feeding, defence and dispersal, are all, by necessity, combined in a solitary insect. The question then arises – to what degree is an individual termite a real biological individual? Using an analogy with a human body, we generally do not think of our organs as individual organisms, so why should we think the same of individual termites? This reasoning by analogy leads to the idea that a termite colony is the individual, and so to the concept of the superorganism.

A superorganism is defined as a collection of agents that can act in concert to produce phenomena governed by the collective (Holldobler and Wilson 2009), although, as we will see, this definition fits the concept of an organism equally well. The superorganism concept is, of course, not restricted to termites: all other fully eusocial insects have colonies that can be defined in the same way. The key evolutionary point in superorganisms is that it is the breeding entity, the colony, that is selected. This leads to some interesting conceptual possibilities, because the way that termites interact with the environment looks very different if we deal with the colony as the "individual" rather than dealing with a single termite as the "individual". It seems eminently sensible to treat the colony as the individual organism, as the parts of the colony do not have all the necessary properties of an organism. Organisms must persist and they must reproduce (in fact, they must persist long enough *to* reproduce). The elements that make up the colony do not individually have these two properties. None can persist and reproduce on their own without the resources of the colony. Alates, which seem closest in overall form and function to solitary cockroaches, will die rapidly without the assistance of the first workers produced at the inception of the colony. This way of thinking of a colony tells against the term "superorganism" as, in fact, the colony is simply an organism constructed at a higher level of organisation.

One of the first naturalists to realise how similar a termite mound is to a much larger single animal was Eugene Marais, an Afrikaner who wrote a fascinating book (*The Soul of the White Ant*[1]) emphasising the similarity between termite colonies and humans. He likened the roles of individual parts of a colony to the functions of the human body, and although the idea was a little contrived and fanciful, it points the way, I believe, to a generally valid way of understanding the functional biology of termites – by considering the functions of the colony and how they are achieved.

When discussing caste structure and function it seems useful to define some generally accepted terms. Individuals in a colony are either *immature* or *mature*, depending on whether they can develop further. *Castes* are the morphologically distinct, task-specific groups of individuals within a colony. In a termite colony these comprise *reproductives* (winged reproductives: alates, queens, kings), *workers*, *soldiers* and immatures. Immatures are known as either *larvae* (if they have no wing buds) or *nymphs* (if they have wing buds). Soldier may have up to five worker *morphs* (e.g. in *Psammotermes*). More commonly there may be major and minor soldiers (in many fungus-growing termites) and occasionally a third morph (e.g. *Acanthotermes*, *Velocitermes*). Major and minor workers are very common throughout the group, particularly in the Macrotermitinae and Nasutitermitinae. In all termites, nymphs, larvae, workers and soldiers are all juveniles, as they retain their prothoracic glands (Noirot and Pasteels 1987).

The colony organism divides its functions through its inanimate and animate parts, as follows:

1. Reproduction and dispersal (alates, queens, king)
2. Construction, feeding and tending (workers):
3. Active defence (soldiers)
4. Protection, homeostasis, fortification (nest, mound)

[1] Marais's work first appeared in a series of magazine articles in the early 1920s, and was published as a book (*Die Siel van die Mier*) in 1937, a year after his death. Modern editions in English have been published by Penguin Books, and most recently by New York University Press and by Osiran Books. The text is widely available. A similar work (*The Life of the White Ant*) produced in 1927 by Maurice Maeterlinck (George Allen and Unwin Ltd) is now considered to plagiarise Marais's ideas.

1.5 Reproduction and Dispersal: Alates

As with all animals, termites need to disperse, mate and establish new homes. The alates do all these tasks. They are produced within the colony either continuously or periodically. Generally if they are produced seasonally they develop to maturity just before the rainy season and they all fly off at the same time in great swarms. The large number of alates produced is probably an anti-predator herd-like defence (Jeschke and Tollrian 2007). These flying termites provide food for many other animals; in one study 200 vertebrate predators in 31 species attacked alates from a *Macrotermes subhyalinus* colony, during four separate nuptial flights (Dial and Vaughan 1987).

The alates fly for a variable time and then land on the ground to search for a mate. The flight presumably ensures that there is a greater probability of finding a mate from another, less closely related, nest (but see Chapter 12 by Vargo and Husseneder, this volume). Once a mate has been found most termites run around as a pair looking for a suitable nesting site (Mitchell 2007). One individual (usually the male) grabs on the end of the abdomen of an individual of the opposite sex and they run together (*tandem running*). In a few species individuals of one sex shed their wings and then the pair takes off again and flies on briefly. In many ecosystems the ground is only soft enough to dig into during the rainy season, which partially explains the flight timing. Once the pair has dug a small chamber they mate and the queen (as she is now) lays eggs that give rise to workers. Soldiers are only produced later.

Alates, are, in sense, closest to solitary insects. They generally look like long, thin cockroaches, with slender wings in which generally the two pairs of wings look essentially the same (Fig. 1.1). Alates interact with the outside environment

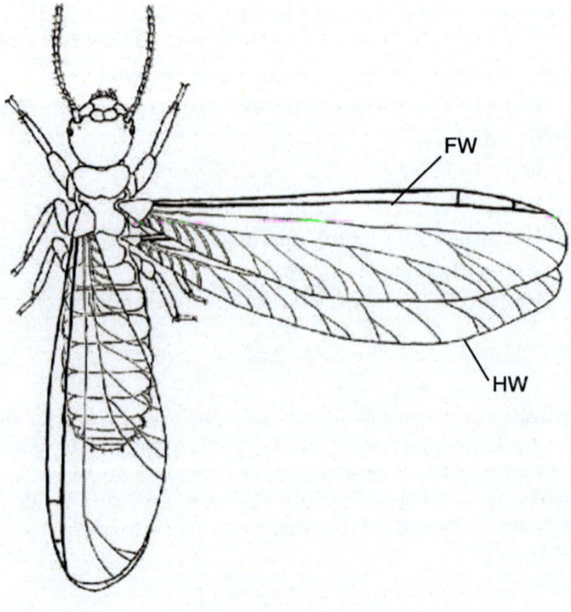

Fig. 1.1 Alate structure (illustrated by *Tenuirostritermes*), FW = forewing, HW = hindwing from Weesner (1970)

much more than the other castes and we can use them to set out the basic body plan of termites because they look more similar to other closely related insect groups. Like most insects a termite alate is split into three body regions – a head, a thorax and an abdomen. There is, however, considerable variation in key structures across the termite families (Table 1.1). The following functional morphology descriptions generally follow the far more detailed descriptions of Weesner (1970), and employ his terminology. I have used forward/front for anterior and behind/back for posterior. I have concentrated on describing only the major features of termite morphology and anatomy, focusing on the structures that are most important functionally.

1.5.1 Head (Figs. 1.1 and 1.2)

All termite alates have *eyes* and one pair of *lateral ocelli* on the dorsal (top) surface: eyes are obviously necessary for dispersal and mate recognition (Fig. 1.2). Termites do not have a median ocellus and in some species the lateral ocelli are also absent. The rest of the dorsal part of the head is arranged as in most orthopteroid insects with a *labrum* at the front end, followed by a *clypeus*, subdivided into a *postclypeus* and *anteclypeus*. The labrum is generally small and tongue-shaped in all alates. Behind the clypeus, and separated from the clypeus by the *epistomal suture* is the main part of the head capsule, with the *frons* at the front and the *epicranium* behind. The boundary between the frons and the epicranium is not well defined. In some termites (Table 1.1) there is a pit in the middle-front of the frons, called the *fontanelle*. The fontanelle is apparently non functional in alates, but has a defensive function in soldiers, as it is the end of a gland that produces anti-predator chemicals (Santos and Costa-Leonard 2006). On either side and to the front of the frons are the paired *antennae*. These are long thin, unbranched and unclubbed, structures in termites, with 11–33 segments. As in all insects they have a number of sensory functions. The segment closest to the head is known as the *scape*, the segment next to it is the *pedicel*, which is nearly always shorter than the scape. All the segments beyond the scape are known collectively as the *flagellum.* From the side the eyes are below the frons on both sides. Below the eyes is the *gena* or "cheek". The mandibles can be clearly seen in front of the antennae.

The structures on the underside of the head are rather complex (Fig. 1.2). The mouthparts have three components: the *maxillae,* the *labium,* and the *mandibles*. The maxillae and labium both carry palps for food sensing and handling. They are generally very similar in all termites and do not merit further discussion here. Alate mandibles are generally simplified versions of those found in the workers (see below). Alates do not appear to have any mandibular defences against predators.

1.5.2 Thorax (Fig. 1.1)

As with all insects, the thorax is divided into three segments: *prothorax*, *mesothorax*, and *metathorax*. Each segment has a pair of legs. The mesothorax and metathorax also have a pair of wings. The legs and wings are anchored to plates along the thorax. I will not describe these in detail as they have a broadly similar function in most insects. The mesothorax and metathorax have these plates well developed,

Table 1.1 Main morphological characters of termites families. Wing venation: SC, subcosta; R, radius; RS, radial sector; M, media; Cu, cubitus; A, anal; F, forewing; H, hindwing. In the first four families a short subcostal vein is sometimes present between the costal margin and the radial sector. The anal vein is sometimes present as a stub in the same families. Does not include fossil groups (see Engel et al. 2009)

Family	Mastotermitidae	Termopsidae	Hodotermitidae	Kalotermitidae	Rhinotermitidae, Serritermitidae	Termitidae
Postclypeus	Undivided	Undivided	Undivided	Undivided	Divided, keeled	Divided
Fontanelle	Absent	Absent	Absent	Absent	Present	Present
Pronotum	Flat	Flat	Flat	Flat	Flat	Saddle-shaped
Wing venation	SC, R, RS, M, Cu, A	SC, R, RS, M, Cu	SC, R, RS, M, Cu	SC, R, RS, M, Cu	[SC+R], RS, M, Cu	[SC+R], RS, M, Cu
Wing scale	F>>H	F>>H	F>>H	F>>H	F>>H[a]	F~H
Tarsal segments	5	5/4	5/4	4	4/3	4/3
Malpighian tubules	12–15	8	8	8	8	4 (2)[b]
Flagellates present in gut	Yes	Yes	Yes	Yes	Yes	No
Feeding groups	I	I	I	I	I	II, III, IV
Workers	Yes[c]	No	Yes	No	Yes[d]	Yes

[a]F ~ H in *Psammotermes*.
[b]Two Malpighian tubules in *Labiotermes*.
[c]Development pathway simpler than in other groups.
[d]No true workers in *Glossotermes* and *Prorhinotermes*.

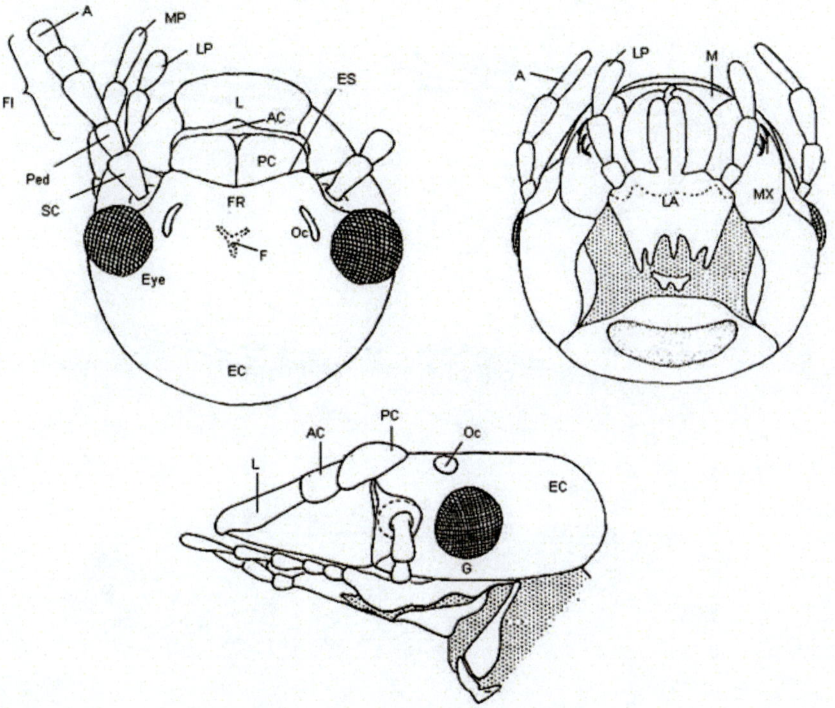

Fig. 1.2 Head of *Tenuirostritermes* (labels incomplete), from Weesner (1970). Head structures: A, antenna; SC, scape; Ped, pedicel; Fl, flagellun; MP, maxillary palps; LP, labial palps; AC, anteclypeus; PC, postclypeus; ES, epistomal suture; FR, frons; Oc, ocellus; EP, epicranium; M, mandible; MX, maxilla; LA, labium; G, gena

but the prothorax, which has no wings, has much smaller plates. The plates on the top (dorsal) surface of the thorax are well developed and vary somewhat across the termites. These are the *pronotum*, *mesonotum* and *metanotum*. The *pronotum* is usually simple and shield-like, but can be variable. The pronotum is saddle-shaped in all Termitidae, but flat in all the other families (Table 1.1).

1.5.3 Wings (Figs. 1.1 and 1.3)

Termite alates are generally poor flyers: their technique is to launch themselves into the air and fly in a random direction. Their main anti-predator defence seems to be the very large numbers of alates that are released from the colony simultaneously. Larger termites generally fly further than smaller ones, with very large alates being strong but poorly directed fliers (Weesner 1965). Alates fly to get clear of the parental colony, to find places with a suitable mate, and to find a suitable place to found a colony. Here there is a clear difference between those termites that are one-piece nesters, who must find a suitable nest site within their feeding substrate, and separate-piece nesters, where the colony can be founded in the soil at some distance from the feeding substrate.

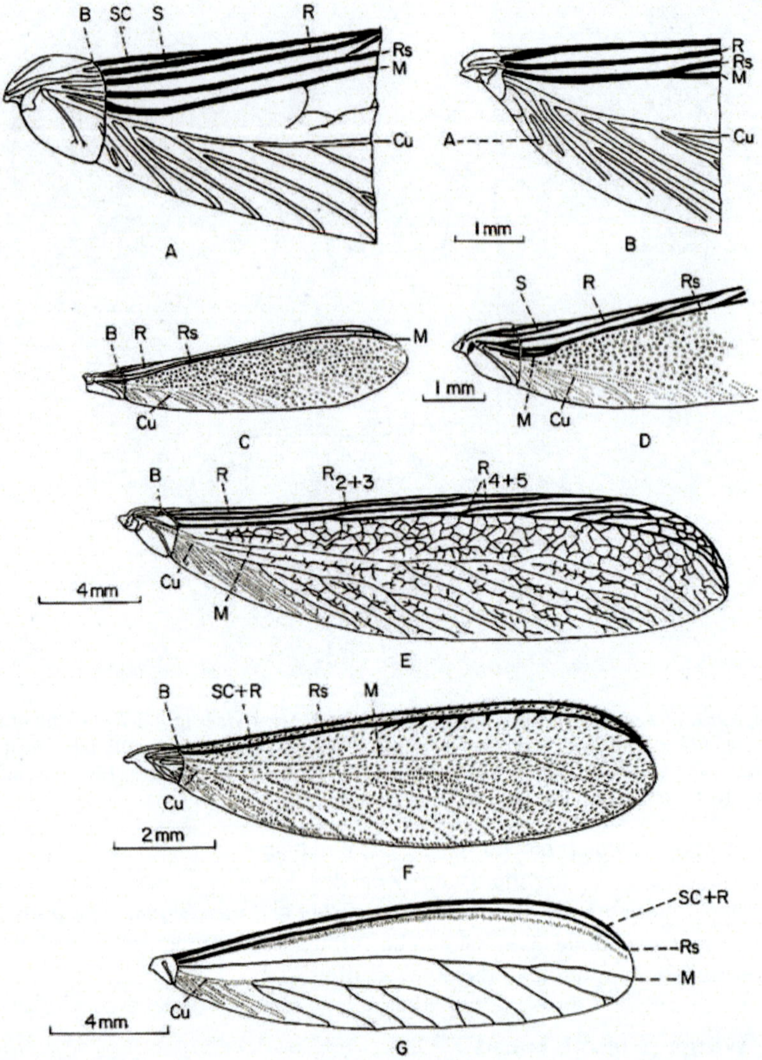

Fig. 1.3 Wing structure in different groups of termites. Wing venation. **a** Forewing of *Neotermes malatensis*; **b** hind wing of same; **c** forewing of *Glyptotermes chapmani*; **d** forewing of *Rugitermes athertoni*; **e** forewing of *Zootermopsis angusticollis*; **f** forewing of *Coptotermes pacificus*; **g** forewing of *Nasutitermes pictus*. A, anal vein; B, basal or humeral suture; Cu, cubitus vein; M, media vein; R, R_{2+3}, radius veins; Rs, R_{4+5}, radial sector; SC, costal margin; S, subcosta, from Weesner (1970)

Termite fore wings and hind wings are very similar, hence the traditional scientific name for the termite order, the Isoptera (Greek: iso = same, ptera = wing). The wings are held parallel with the body at rest and at right angles when flying. The basal part of the wing is covered by a wing scale. The end of this scale is where the wings are shed after landing, there is a suture (the *humeral suture*) here, which allows the wings to be shed automatically, in all families except the Mastotermitidae, where the wing is chewed off above the scale. Wings are clearly useless underground. The fore wing scale is always larger than the hind wing scale, but the relative sizes differ between families (Table 1.1).

Termite families differ most profoundly in the venation in the wings. There has been a progressive simplifying of the venation in more phylogentically apical groups (Table 1.1), so that the Termitidae have the simplest wings, and the Mastotermitidae have the most complex (Fig. 1.3). The Mastotermitidae also have a unique cockroach-like structure, the anal lobe, a part of the wing lost in all other termites. Historically, wing venation has been hard to describe, with many different systems proposed. In termites, Emerson's (1965) system is usually used. The following notes are for a temite, like *Mastotermes darwinensis,* which has the full complement of veins. The front of the wing has a heavily sclerotised false vein (false because it does not develop from a tracheal tube), the *costal margin*. The precosta and costa veins found in some other insects are absent. The next vein inwards is the short *subcosta*. Following that are three *radius* veins, R_1, R_2 and R_3, which may be separate or fused in any combination. The next vein is the *radial sector*, which reaches to the wing, or close to it, and may split into several sub-veins. Next to this vein, in the middle of the wing, is the *cubitus*, which branches as it goes down the wing, ending on the wing margin. The final set of veins, the *anal veins*, is only well developed in the Mastotermitidae. The functional significance of all these structures is unclear, as all termites seem to be about as bad as each other at flying. The loss of individual veins may represent a general body size reduction across the termite phylogenetic tree.

1.5.4 Legs (Fig. 1.4)

Termites are reliant on their legs for most of their movement – termite alates fly only briefly. The legs are fairly constant across castes, although soldier legs can be heavier and more conspicuously armed. Termite legs follow the standard pattern for insects. Starting closest to the body these are the *coxa*, *trochanter*, *femur*, *tibia* and *tarsus*. In termites the coxa of the second and third pair of legs is divided by a deep suture into the *meron* and the coxa proper. The trochanter is short, the femur relatively large, and the tibia relatively long and thin. The tarsus has a variable number of short joints (Table 1.1) followed by a long terminal joint with a large claw. The tibia also has a variable number of *tibial spurs* at its far end. In some termites there is an *arolium*, between the claws. This sticky pad-like structure is absent in most termites, probably because they do not generally have to climb up smooth surfaces (Crosland et al. 2005).

1.5.5 Abdomen

The abdomen has ten segments, consisting of upper and lower plates: the *tergites* and the *sternites* respectively. Nine of the ten tergites are wide and substantial, while

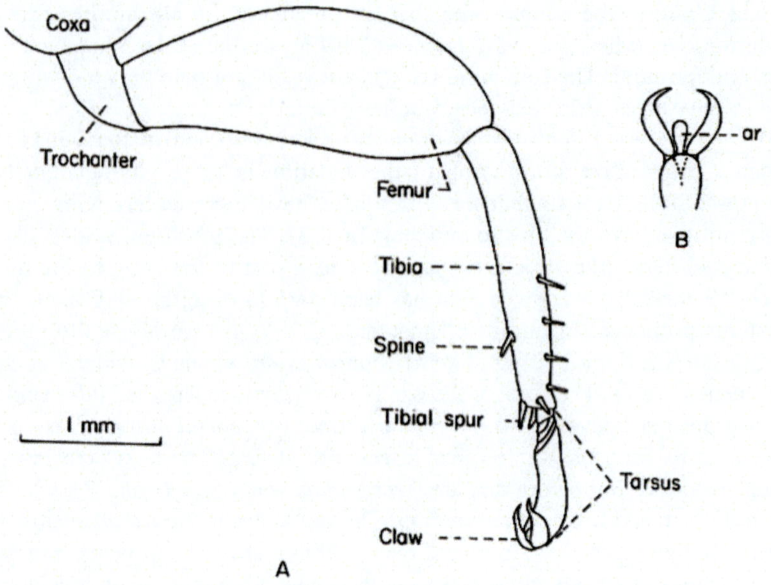

Fig. 1.4 Leg structure in *Zootermopsis angusticollis*, from Weesner (1970)

the tenth (the *epiproct*) is elongated and tapering. The tergites are identical in males and females. The first sternite is small or absent. The second through to sixth sternite are broader than long and similar in males and females. The seventh sternite of the female alate (the *hypogynium*) is large, often completely covering sternites eight and nine, which are modified. These modifications of the sternites do not occur in males and this is one of the most reliable ways to sex alates. Mature males generally have *styli* (unjointed processes) at the ends of the ninth sternite, but these vary widely across species. Mature females lack these structures. The tenth sternite (the *paraproct*) is similar in both sexes. It is divided in the middle, forming a pair of plates, one on each side of the body. The sides of the paraprocts have segmented *cerci,* usually with two joints. Both the cerci and styli are sensory structures, their gradual loss in termite evolutionary history suggest that they may not be very useful inside a colony or when foraging.

The internal reproductive system of termites is greatly simplified when compared with other cockroaches. The male has no intromittent organ, and in most species the sperm are immotile and aflagellate. The striking exception to this is in the Mastotermitidae, where the sperm are multiflagellate but appear to have limited motility (Riparbelli et al. 2009). This pattern is repeated in the females, where the genital structures are again simple, except in the Mastotermitidae, which have a clearly defined ovipositor, homologous with those found in cockroaches (Nalepa and Lenz 2000). The mechanism of mating is very poorly studied, but it is known that the ovarioles are very well developed in the Termitidae, where some queens become extraordinarily distended (*physogastric*) and lay many thousands of eggs during a lifetime. The lack of motile sperm, intromittent organs, and the general lack of sexual dimorphism is unsurprising in a group where monogamous lifetime pair bonds are the norm and where, therefore, sperm competition does not occur (Morrow 2004).

Termite eggs are unadorned, with smooth surfaces. They are laid singly in all species except *Mastotermes*, which lays eggs in an ootheca-like structure, similar to that found in other cockroaches, but simplified (Nalepa and Lenz 2000). The eggs are always laid within the colony, so the protective role of the ootheca is unnecessary.

1.6 Worker Morphology

The head, thorax and abdomen of workers is essentially similar to those in the alates, except for the absence of wings and any genital structures. Workers, however, have much more strongly developed mandibles, reinforced with generally small amounts of zinc and manganese (Cribb et al. 2008). In the Kalotermitidae, the mandibles are strongly reinforced with zinc to allow them to break into the very strong dry dead wood on which they feed. Almost all worker termites are blind, as they lack compound eyes. The few exceptions are all early branching groups, some of which, but not all, forage above ground. However, there are also many blind surface foragers so the connection between above ground foraging and eyes may be mostly a phylogenetic signal.

1.7 Construction, Feeding and Tending

Workers do almost all the work within the colony, and they live and work predominantly within the colony. Even when foraging, most workers are protected under sheeting or runways. Only a relatively few grass, microepiphyte and litter-feeders forage unprotected on the surface. In those cases the foraging morphs are conspicuously more sclerotised than the non-foraging morphs. Workers are sterile, they never reproduce. However, in many species workers can develop into supplementary reproductives if the primary reproductives die (Roisin 2001, see also Chapter 9 by Miura and Scharf and Chapter 10 by Matsuura, this volume). Not all termite species have true workers. In the Kalotermitidae and the Termopsidae there are no true workers, and the immatures may not help very much (Korb 2008). In all these cases the termites nest and feed in dead wood and there is no requirement for nest building or foraging. It is not clear how much tending these groups do, possibly, in some cases none. However, even in this case all the species are eusocial, as they have soldiers that never reproduce.

1.7.1 Feeding

Only workers forage. Termites feed on dead plant material at all stages of decomposition ("humification", see Donovan et al. 2000; Hyodo et al. 2008). This includes, in order of humification: microepiphytes, living stems and roots, dry grass, dead leaves, dead wood, very decayed dead wood plastered with soil, humus and (apparently mineral) soil. Feeding preferences vary between species and higher taxa and can be classified according to their position on the humification gradient leading to four feeding groups (Donovan et al. 2001; see Chapter 14 by Bignell, this volume). Group I feeds on dead wood and grass and have relatively simple guts. Group II feeds on wood, grass, leaf litter and microepiphytes and have more complex guts. Group III feeds on humus (i.e. soil-like material with recognisable plant material in it). Group IV feeds on soil (i.e. soil-like material with a high proportion of silica and no recognisable plant material).

Termite worker-imago mandibles are very variable in the number of their marginal teeth, but functionally they seem to fit into two groups: (a) grinding (milling), and (b) pounding (pestle and mortar) (Donovan et al. 2001; Fig. 1.5). The molar plates are heavily ridged in the grinding type, with the left molar plate concave and the right convex. As the mandibles rub against each other, the molar plates grind up the plant material, often dead wood, which is a very fibrous material. In the pounding type the polar plate is convex in the left mandible and concave in the right mandible and both mandibles have no ridges. These two structures act

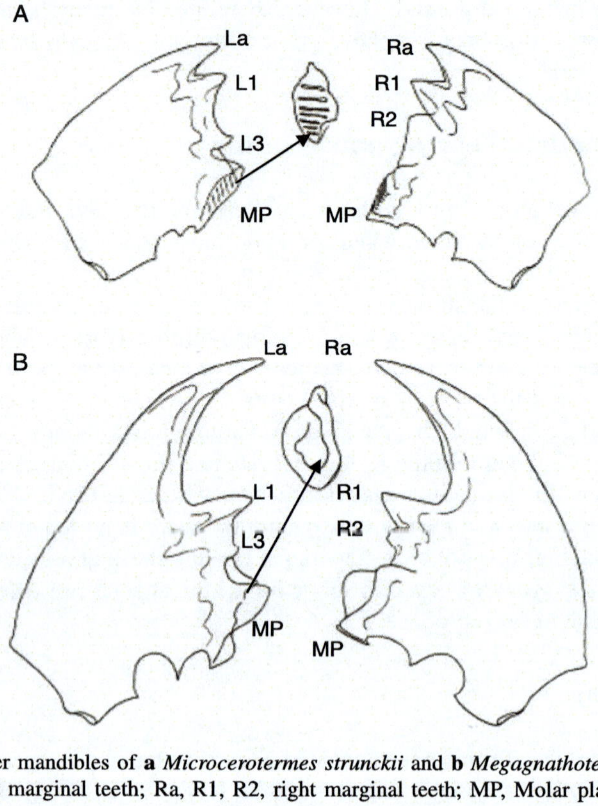

Fig. 1.5 Worker mandibles of **a** *Microcerotermes strunckii* and **b** *Megagnathotermes notandus*. La, L1, L3, left marginal teeth; Ra, R1, R2, right marginal teeth; MP, Molar plate, from Sands (1998)

like a pestle and mortar, pounding the feeding substrate, which is usually soil or humus. Correlated with this molar plate structure is the size of the first apical tooth. In the grinding type, the tooth is short, and the distance between the apical and first marginal teeth is also short. The apical and first marginal teeth seem to be used to tear off fibres of wood. In the pounding type the first apical tooth is long and the distance between the apical and first marginal teeth is also long. In this case the apical and first marginal teeth seem to be used to cut out a relatively large volume of soil. These mandible adaptations are associated not only with diet but also with gut structure.

Termite guts are very complicated structures. Early branching families are very like closely related cockroaches but later branching families are strongly modified. The general pattern is for the cockroach-like guts to have a generalist detrivore type structure, while the more derived guts switch to a more complex hindgut fermentation structure. Insect guts are split into three distinct parts: *foregut* (stomatodeum), *midgut* (mesenteron) and *hindgut* (proctodeum). The food is ingested and passes into the foregut, which contains the *crop* (stomach) and then the *proventriculus* (gizzard), a grinding organ. From the proventriculus the food enters the midgut, which is the primary source of enzymes in most insects. The midgut has *midgut caeca* which are enzyme production and enzyme-product absorption sites. At the junction between the midgut and the hindgut are the *Malpighian tubules,* which have an excretory function. This junction also has a *procotodeal valve* in many insects. The partly digested food then passes into the hindgut, which is of variable length and structure, but usually contains a rectum and an anus. The hindgut is often short in insects that have no significant microbial symbioses.

In the least derived termite guts (and in the closely related wood-feeding cockroach, *Cryptocercus*), all of which feed on wood or grass, the structure is essentially as described above (Fig. 1.6a). The proventriculus is well developed and assists in fragmenting the food. The midgut is relatively extensive, and often has mid-gut caeca, which are points of absorption for fluid resulting from enzymatic breakdown. The hindgut is relatively small, with proctodeal segment 3 (or paunch: an enlarged sac at the beginning of the hindgut) the best developed part. In all of the termite families except the Termitidae, the paunch is packed with flagellates. This form of gut represents a combination of a generalist detrivore-type (as in cockroaches, e.g. *Periplaneta*) and a hindgut fermentation system, where the flagellates ferment the partly-digested food under anaerobic conditions. Most textbooks have generally attributed the digestion of dead wood to the hindgut flagellates. However, we now know that all studied wood- and grass-feeding termites produce their own cellulases, and in addition that many wood-feeding termites in the Termitidae digest wood efficiently without any flagellates (see Chapter 3 by Lo et al., Chapter 14 by Bignell, and Chapter 16 by Brune and Ohkuma, this volume). Overall digestion is therefore divided between mid-gut enzyme production and hindgut microbial fermentation.

Nearly all the important variations from this basic intestinal model are found in the Termitidae. In that family alone the hindgut flagellates have been lost and this appears to have accelerated the evolution of physiological and anatomical

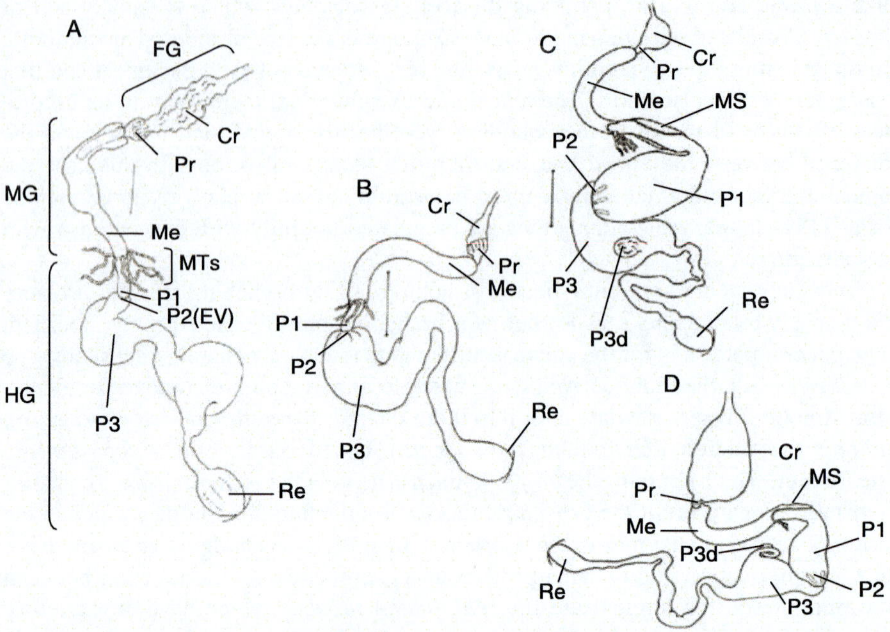

Fig. 1.6 Gut figures from Sands (1998), showing the variable elaborations of the hindgut, unravelled for illustration. Cr, crop; EV, enteric valve; FG, foregut; HG, hindgut; MG, midgut; Me, mesenteron; MS, mixed segment; MTs, Malpighian tubules; P1, P2, P3, proctodaeal segments; P3d, diverticulum; Re, rectum. **a**, *Hodotermes* (Hodotermitidae), feeding-group I, grass-feeder; **b**, *Coptotermes* (Rhinotermitidae), feeding-group I, wood-feeder; **c**, *Cubitermes* (*Cubitermes*-group Termitidae), Feeding-group IV, soil-feeder; **d**, *Ophiotermes* (*Cubitermes*-group Termitidae), Feeding-group IV, mound-feeder

innovations. The main changes are the simplification of the proventriculus, the loss of the midgut caeca, the development of a *mixed segment*, a reduction in the number of Malpighian tubules, the development of a sclerotised *enteric valve* between the first and third parts of the hindgut, and the extensive (and very variable) development and compartmentalisation of the hindgut (Fig. 1.6c, d). All of these changes represent feeding either more efficiently on an existing food (wood, grass) or on entirely new substrates (humus, soil). In both cases selection seems to have produced hindgut fermentation systems of greater sophistication. A system of numbering hindgut sections in the Termitidae (which are commonly also separated compartments) was devised by Holmgren (1909) and was most recently reviewed by Noirot (2001). It provides, with just a few variations in the published literature, a consistent scheme of homology by allocating the enteric valve to its own section (P2); the section anterior to the valve is thus the P1, and those posterior to it are therefore P3, P4 and P5 (the last being the rectum).

The most elaborated guts in termites are those of the *Cubitermes*-group termitids that feed on soil with no discernible plant materia in it. They show all of the above innovations (Fig. 1.6d, e). In *Cubitermes,* a typical member of the group, the mandibles are of the pounding type, and the homogenised soil passes through the mouth to a small proventriculus, which is poorly sclerotised and seemingly does not alter the ingested soil. The midgut is short and tubular and overlaps with the (embryologically proctodaeal) P1 for about half its length, to produce the mixed segment. This structure is unique to termites and appears to be a fluid exchange site associated with excretion and a preliminary microbial processing of the food (see Chapter 14 by Bignell and Chapter 16 by Brune and Ohkuma, this volume). The Malpighian tubules attach at the junction between the midgut and the hindgut and are clustered over the mixed segment to form a *Malpighian knot*. Beyond the mixed segment the P1 expands greatly to form a large sac. At the posterior end of this sac are the heavily sclerotised ridges of the enteric valve (Fig. 1.6). This structure is highly variable across the Termitidae and its function is somewhat obscure. In soil-feeders it probably helps to separate clay particles, which have abundant soil organic matter (SOM) associated with them, from silica (sand) particles, which are inert. The enteric valve may ensure that clay particles stay in the hindgut longer than silica particles (Donovan 2002). Peristaltic contractions may assist this process. Beyond the P1 is another very large sac, the P3, which in the Cubitermes-group has a distinctive *P3 diverticulum* (which expands and contracts as coarse fractions of soil flux in and out of it). The mixed segment, P3 and P4 are packed with prokaryotic microbes (see Chapter 14 by Bignell, Chapter 15 by Ohkuma and Brune, and Chapter 16 by Brune and Ohkuma, this volume). The whole gut allows highly refractory organic material in soils to be digested, probably, in part, by releasing proteins, peptides and amino acids immobilised as SOM (Ji and Brune 2005, 2006). The most extreme guts are those of the *Cubitermes*-group mound feeders (the *Ophiotermes*-group), that feed on the already digested faeces of other *Cubitermes*-group species in the mound matrix. These have enormously enlarged crops and salivary glands, and curiously twisted P3 diverticula (Fig. 1.6). The large crop is probably developed to cope with the very large amount of mound material that they must ingest in order to extract anything digestible from the mounds.

Between the two extremes of gut structure there are numerous forms, which generally fall in an intermediate position on the termite phylogenetic tree (see Chapter 2 by Lo and Eggleton, this volume). The fungus-growing termites (Macrotermitinae) have retained the a rather phylogenetically-basal gut structure, probably because much of their forage is digested by the mutualistic fungus.

1.7.2 Nursing

The immatures of termites are small white, unsclerotised and essentially helpless. They have to be fed by the workers/helpers. The workers assimilate the food that they eat and re-secrete nutrients in a form palatable to the immatures. In the non-termitid familes, which all have gut flagellates, this is closely connected with passing the flagellates from workers to larvae, and occurs by *proctodeal trophallaxis*. The immatures are fed by secretions from the anus, which contain the symbionts and woody particles. This method of feeding does not occur in the termitids, where the flagellates have been lost. *Stomodeal trophallaxis*, where immatures are fed from glands in the head through the mouth is found in all termites.

Grooming is important in all groups, particularly in order to remove potentially parasitic microbial populations. The relative amount of time that workers spent grooming is probably related to parasite load in the nesting and feeding substrates (e.g. termopsids in wet wood probably have a higher fungal load than kalotermitids in dry wood).

1.8 Active Defence: Soldiers

The soldiers in a colony have only one function – to defend the colony. They generally have large, highly sclerotised heads and powerful, highly modified, mandibles. They also often have chemical defences (Prestwich 1984). Soldiers were the first sterile caste to evolve (Thorne et al. 2003). It seems probable that the most important termite predators are ants, and much of the variation in soldier morphology is in response to ant predation pressure. Vertebrate predation may also be very important, but there is little evidence that these predators generally kill entire colonies. Generally vertebrate termitophages can wound a colony, but only ants can destroy it (Leal and Oliveira 1995).

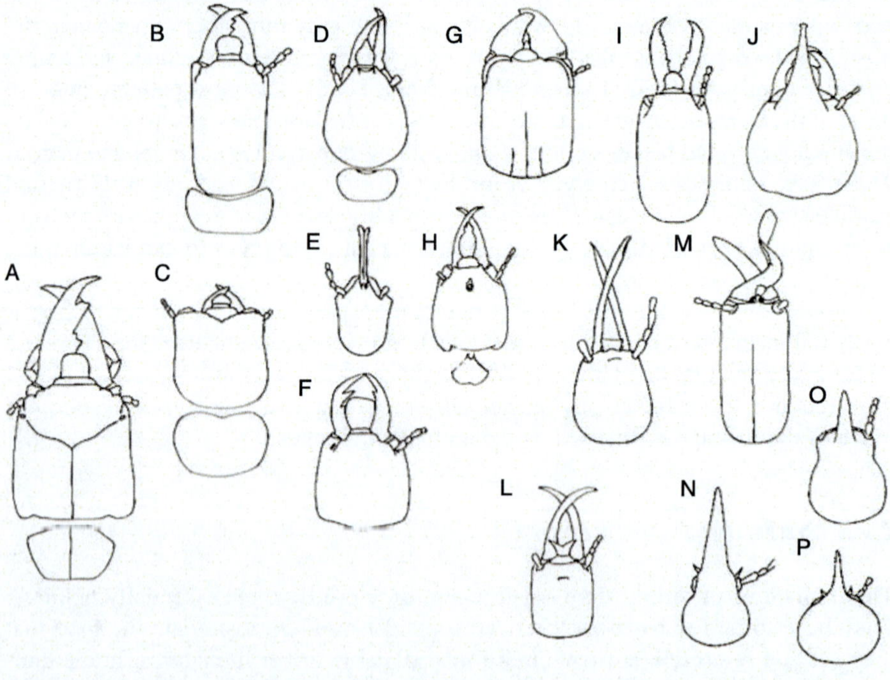

Fig. 1.7 Plan view of soldier head capsules. **a**, *Archotermopsis wroughtoni*; **b**, *Rugitermes bicolor*; **c**, *Cryptotermes verruculosus*; **d**, *Coptotermes sjostedti*; **e**, *Rhinotermes hispidus* (minor soldier); **f**, *Rhinotermes hispidus* (major soldier); **g**, *Jugositermes tuberculatus*; **h**, *Acanthotermes acanthothorax* (minor soldier); **i**, *Microcerotermes fuscotibialis*; **j**, *Armitermes grandidens*; **k**, *Promirotermes orthocopes*; **l**, *Procubitermes niapuensis*; **m**, *Pericapritermes urgens*; **n**, *Angularitermes nasutissimus*; **o**, *Coarctotermes suffuscus*; **p**, *Nasutitermes octopilis*, from Weesner (1970)

Soldiers show the greatest of variation of any caste, not only between species but also within species. Many species have several morphs and sometimes the morphs have very divergent structures, due to very different methods of defence. At the generic and species level soldier morphology is the most important source of taxonomic characters. These variations can be seen almost entirely in the head capsules of soldiers (Fig. 1.7). The labrum, for example, is highly variable: it can be: tongue shaped (basal condition, Figs. 1.7a–d, g, i), bilobed (Termitinae, Fig. 1.7 l), asymmetrical (Termitinae, Fig. 1.7m), and guttered (Rhinotermitinae, Fig. 1.7e–f). The postmentum is usually enlarged, often with process or node. Soldier mandibles are also highly variable – sometimes reduced or vestigial (Rhinotermitinae, Termitidae: Nasutitermitinae, Figs. 1.7e–f, n–p). The fontanelle can be absent (basal condition, Fig. 1.7a–c), highly enlarged (Coptotermitinae, Fig. 1.7d), or at the end of a variable length process (Syntermitinae, Nasutitermitinae, Figs. 1.7j, n–p). All these variations clearly relate to the way the soldiers repel predators. For example, in those cases where soldiers block tunnels (*phragmosis*) the soldier head width has been shown to be under stabilising succession (Matsuura 2002; Roux et al. 2009).

Soldier morphology, however, cannot be separated from the colony structure, the nature of the feeding substrate, and the energetic cost of producing soldiers for the colony. These all interact to produce four main defence strategies: *strong-point*, *counter-attack*, *labyrinth*, and *trail-guarding*.

a. Strong-point defence is where soldiers retreat to defensible points within the nest (e.g. soldier heads as in Fig. 1.7a, c, l).
b. Counter-attack defence is where soldiers come out of the nest to swamp the enemy (e.g. soldier heads as in Fig. 1.7e–f, j).
c. Labyrinth defence is where the workers build a complex set of tunnels, most of which are not occupied (as in *Cephalotermes*, where soldiers are very rare). This means that any ant predators entering the next will have very long search times before they encounter any prey.
d. Trail-guarding is where a trail of foraging workers are defended by actively guarding soldiers (e.g. soldier heads as in Fig. 1.7o). The most striking example of this is found in *Hospitalitermes*, where workers scrape micro-epiphytes from the bark of standing trees and nasutes soldiers stand at the edge of the foraging parties and are highly effective at deterring ants from attacking (Miura and Matsumoto 1998).

Both the type and number of soldiers depends on the defence strategy used. Strong-point defenders have large heads, and in some species they have ridges or crests on the head to secure the head within the tunnels. As strong points are necessarily limited, and such soldiers are presumably expensive to construct, there tends to be low soldier:worker ratios in such colonies. In contrast, counter-attack relies on having very large numbers of soldiers, and so the soldiers must be relatively cheap to produce. In a labyrinth nest the onus is on the workers to produce the complex nest structures and there is obviously a cost involved in producing such extensive structures.

Table 1.2 Morphological characters of termitid clades. Inner and outer refer to the position of the posterior mesenteric extension forming the mixed segment, and are with respect to the edge of the intestinal coil in situ (see Sands 1998). Feeding groups are those of Donovan et al. (2001). IIf refers to feeding-group II fungus-growing termites

	Macrotermitinae	Foraminitermitinae, Sphaerotermitinae	Apicotermitinae	Termitinae, nasutitermitinae
Mixed segment	No	No	Inner	Outer
Feeding groups	IIf	II, III	III, IV	II, III, IV
Malpighian tubule attachment	At midgut-hindgut junction	At midgut-hindgut junction	To posterior midgut	At midgut-hindgut junction
Malpighian knot	No	Pseudo-	True	True

Prestwich (1984) classified termite soldiers according to their physical and chemical defences. Recent attempts to improve on the classification using more sophisticated methods of analysing shapes have only been partially successful (e.g. Scholtz et al. 2008), although they do support the general value of the scheme. All non-termitids have fully functional soldiers. However, within the termitids there are two clades (possibly even three) where soldiers have been lost – the *Anoplotermes*-group, and the *Protohamitermes*- group (Inward et al. 2007b). A number, but not all, of these are autothysic. *Autothysis* is the self-destructive rupturing of the abdomen when in contact with a predator ant to extrude the sticky, and possibly poisonous, guts (Sands 1982). The mechanism, presumably designed to work in tunnels causes the predator to stick to the autothysic worker, and with repeated encounters the accumulated bodies will eventually block the tunnel and repel the attack (Table 1.2).

1.9 Protection, Stability, Fortification: Nests and Mounds

Termite colonies live inside a *nest*, and if the nest is covered with or made from earth to create a structure protruding above the soil surface then it becomes a *mound*. Mounds and nests protect against the environment, keep the internal climate stable, and fortify against predators (a range of nest types is shown in Fig. 1.8). However, not all termite colony structures fulfil all these functions. In fact, there are probably more termites that produce amorphous underground nests than those that produce complex multi-function nests or mounds (e.g. see Eggleton et al. 1996).

Abe and co-workers developed a nesting classification for termites dependent on the position of the nesting and feeding substrates (e.g. see Higashi et al. 1992). *One-piece nesters* nest and feed in the same substrate (usually wood). *Intermediate-nesters* nest and feed in the same substrate but can forage into that substrate nearby. *Separate-piece nesters* nest in one substrate and forage out into another. It is only the separate-piece nesters that habitually need material from outside the colony area to construct their nests.

The most complex colony constructions are found in the Termitidae, Rhinotermitidae and Hodotermitidae – groups which all have true workers and most of which are separate-piece nesters. Where there are no true workers (e.g. in Termopsidae, Kalotermitidae, *Prorhinotermes*) colony constructions tend to be simpler. The most complex nest building of all is found in the Macrotermitinae (the fungus-growing termites), exemplified by the huge soil mounds built by species in the genus *Macrotermes*. Mounds of equal size but lower internal complexity are built by the Australian termitid *Nasutitermes triodiae*. A mature living *Macrotermes* mound is arguably the most complex colonial organism known in nature. A case can be made for all the castes, the mound, and the fungus, all being part of a single organism. If so it is also probably the only organism that can completely decompose dead plant material, including all its lignocellulose, leaving virtually no residue (Ohkuma 2003).

Termites build their nests with faeces. This makes perfect sense as their substrates are relatively inert to pathogens, are cheap to produce, and are generally good structural materials. There are two sorts of building material: partly digested plant matter, which produces *carton nests*, and soil, which produces *soil nests and mounds*. Not all of these nests are conspicuous, as many colony centres are found underground, particularly in tropical rain forests (Eggleton et al. 1996). The soldierless Apicotermitinae, for example, generally build no obvious nest structures and appear to live in a set of amorphous tunnels. In non-termitid mound builders, the carton nest is built inside a soil mound (e.g. in many Australian *Coptotermes*) but in the termitids soil mounds usually have no carton material inside. The fungus comb of the Macrotermitinae may well be homologous with the carton nests of *Coptotermes*, with the carton material being used as food for the mutualistic fungus (Donovan et al. 2001). A substantial minority of termites, particularly the single-piece nesting kalotermitids and termopsids, nest in dead wood and live within the tunnels created by their feeding. Although these groups do produce faecal structures, they do not make carton nests.

1.9.1 Protection and Stability

The main environmental challenges for colonies are rainfall, heat and gas exchange. These problems tend to become greater with colony size and complexity. Within rain forests temperatures are buffered, but rainfall can be torrential and continuous. The erosion of mounds then becomes a very serious problem. However, carton mounds in tropical rain forests rarely have anti-rain protection, probably because carton is a more rain-resistant material than soil. Mounds in the soil, however, often have "drip tip"-like structures that allow the efficient flow of water over and around them.

Temperature fluctuations outside rain forests are a much bigger problem: isolated termites can survive only in a very narrow range of temperatures. Most termites use the thermal properties of their nesting substrate to aid temperature control. In hot dry areas this means nesting in the soil, which is an extremely thermally buffered environment (Turner and Soar 2008). Termite nesting within logs may also be reasonably well temperature-buffered. Large colonies in dry, hot areas have significant temperature control problems, and this is where architecturally-complex mounds have a role. This is particularly true in Australia, Africa and, to a lesser extent, South America in the genera *Coptotermes*, *Nasutitermes*, *Macrotermes*, *Cornitermes* and *Syntermes*. *Macrotermes* on the African plains build the most complex known non-human architectural structures. In such habitats, mound structure may often be a trade off between thermoregulation and gas exchange (Korb 2003). For example, Turner and Soar (2008) have shown that, in at least some species, big mounds in savannas act like lungs. The old model of mound ventilation (the "thermosiphon") proposed by (Luscher 1951) now seems not to fit the available data (Korb and Linsenmair 2000)

1.9.2 Fortification

The need for fortification from predators is very clear, as many animals feed on termites (Das and Coe 1994; De Visser et al. 2008). The nest/mound structure is just as important in defence as the soldier morphology and behaviour. Many termite nests are made of carton and are not particularly strong. Carton nests are usually defended using counter-attack strategies. Many mounds, however, are made up of faecal soil, and these are usually defended as strong-points. In *Cubitermes*, for example, the cells within the soil mound are connected by short tunnels or apertures that are just the right diameter for soldiers to block them (Perna et al. 2008). In the most sophisticated nests (e.g. within the Macrotermitinae), the colony has a "queen cell", a highly fortified central chamber in which the queen and king are defended as a last resort.

There seems to have been a long history of predation of termite mounds and nests. The earliest plausible termite-feeding animal was the Jurassic mammal *Fruitafossor*, which had very similar adaptations to extant termite-eating mammals, such as aardvarks and anteaters (Luo and Wible 2005). At least one group of dinosaurs, the alvarezsuarids, seem to have been specialised termitophages, with fore limbs adapted to break into termite colonies in dead wood (Longrich and Currie 2009). Modern termite-eating mammals are very common, particularly in the tropics. Those that break directly into mounds include aardvarks, anteaters, pangolins, many species of bears, apes (particularly chimpanzees), numbats, many lizards, amphibians etc. The Aardwolf appears to feed only on grass-harvesting termites of the genus *Trinervitermes* (Koepfli et al. 2006), which are common in African savannahs. Chimpanzees that feed on termite mounds use a range of tools (Suzuki et al. 1995) to extract termites, and *Australopithecus* appears to have used bone tools in a similar way (Backwell and d'Errico 2001). However, the primary predators of termites are undoubtedly ants (see above), although in some cases this predator-prey relationship is complicated by ants that are *inquilines* (i.e. live inside termite colonies). These may provide some protection from other raiding ants, although overall, completely ant-free nests seem to be the healthiest. Many other species live inside termite nests, as commensals, predators, parasites, or mutualists (Jaffe

Fig. 1.8 Epigeal termite mounds, illustrated by **a**, *Lacessititermes* sp. (SE Asia); **b**, *Cubitermes severus* (West Africa); **c**, *Procubitermes arboricola* (West/Central Africa); **d**, *Macrotermes bellicosus* (East Africa); **e**, *Amitermes laurensis* (Australia). All photos from the Termite Research Group photo collection, NHM

et al. 1995). In addition, mounds can be important refugia to protect animals from fire (Yarnell et al. 2008) or just to give protection and shelter in otherwise hostile environments. Mounds are not just ecologically important when they are occupied. Abandoned termite mounds can be a source of important nutrients for large mammals, particularly in nutrient poor areas (Ruggiero and Fay 1994), and their eventual erosion is a normal part of longer-term local nutrient cycling (Holt and Lepage 2000) (Fig. 1.8).

1.10 Conclusions

Termite colonies are organism-like entities with strong division of labour between different subcomponents. This leads to a wide range of anatomical and morphological adaptations within each species, and therefore, within each individual termite mound. Comparable variation between is only seen in other colonial organisms, particularly other social insects (ants, corbiculate bees, vespids). However, none can match the complexity of anatomical specialisation, body form and colony structure of termites.

References

Backwell LR, d'Errico F (2001) Evidence of termite foraging by Swartkrans early hominids. Proc Natl Acad Sci U S A 98:1358–1363

Boomsma JJ (2009) Lifetime monogamy and the evolution of eusociality. Philos Trans R Soc Lond B Biol Sci 364:3191–3207

Cribb BW, Stewart A, Huang H, et al (2008) Unique zinc mass in mandibles separates drywood termites from other groups of termites. Naturwissenschaften 95:433–441

Crosland, MWJ, Su, N-Y, Scheffrahn, RH (2005) Arolia in termites (Isoptera): functional significance and evolutionary loss. Insectes Soc, 52:63–66

Das I, Coe M (1994) Dental morphology and diet in anuran amphibians from South India. J Zool 233:417–427

Davies RG, Eggleton P, Jones DT, et al (2003) Evolution of termite functional diversity: analysis and synthesis of local ecological and regional influences on local species richness. J Biogeogr 30:847–877

De Visser SN, Freymann BP, Schnyder H (2008) Trophic interactions among invertebrates in termitaria in the African savanna: a stable isotope approach. Ecol Entomol 33:758–764

Dial KP, Vaughan TA (1987) Opportunistic predation on alate termites in Kenya. Biotropica 19:185–187

Donovan SE (2002) A morphological study of the enteric valves of the Afrotropical Apicotermitinae (Isoptera: Termitidae). J Nat Hist 36:1823–1840

Donovan SE, Eggleton P, Bignell DE (2001) Gut content analysis and a new feeding group classification of termites. Ecol Entomol 26:356–366

Donovan SE, Jones DT, Sands WA, Eggleton P (2000) The morphological phylogenetics of termites (Isoptera). Biol J Linn Soc 70:467–513

Eggleton P, Beccaloni G, Inward D (2007) Save Isoptera: a comment on Inward et al. – response to Lo et al. Biol Lett 3:564–565

Eggleton P, Bignell DE, Sands WA, et al (1996) The diversity, abundance and biomass of termites under differing levels of disturbance in the Mbalmayo Forest Reserve, southern Cameroon. Philos Trans R Soc Lond B Biol Sci 351:51–68

Emerson AE (1965) A review of the Mastotermitidae (Isoptera), including a new fossil genus from Brazil. Am Mus Novit 2236:1–46

Engel MS, Grimaldi DA, Krishna K (2009) Termites (Isoptera): their phylogeny, classification, and rise to ecological dominance. Am Mus Novit 3650:1–27

Higashi M, Abe T, Burns TP (1992) Carbon-nitrogen balance and termite ecology. Proc R Soc Lond B Biol Sci 249:303–308

Holldobler B, Wilson EO (2009) The superorganism: the beauty, elegance, and strangeness of insect societies. W. W. Norton, New York, NY and London, 522 pp

Holmgren N (1909) Termitenstudien I. Anatomische Untersuchungen. Klg Svenska Vetenskapsakad Handl 44:1–215

Holt JA, Lepage M (2000) Termites and soil properties. In: Abe T, Bignell DE, Higashi M (eds) Termites: evolution, sociality, symbioses, ecology. Kluwer Academic Publisher, Dordrecht, pp 389–407

Hyodo F, Tayasu L, Konaté S, et al (2008) Gradual enrichment of ^{15}N with humification in a below-ground food web: relationship between ^{15}N and diet age determioned using ^{14}C. Funct Ecol 22:516–522

Inward D, Beccaloni G, Eggleton P (2007a) Death of an order: a comprehensive molecular phylogenetic study confirms that termites are eusocial cockroaches. Biol Lett 3:331–335

Inward DJG, Vogler P, Eggleton P (2007b) A comprehensive phylogenetic analysis of termites (Isoptera) illuminates key aspects of their evolutionary biology. Mol Phylogenet Evol 44: 953–967

Jaffe K, Ramos C, Issa S (1995) Trophic interactions between ants and termites that share common cests. Ann Entomol Soc Am 88:328–333

Jeschke JM, Tollrian R (2007) Prey swarming: which predators become confused and why? Anim Behav 74:387–393

Ji R, Brune A (2005) Digestion of peptidic residues in humic substances by an alkali-stable and humic-acid tolerant proteolytic activity in the gut of soil-feeding termites. Soil Biol Biochem 37:1648–1655

Ji R, Brune A (2006) Nitrogen mineralization, ammonia accumulation, and emission of gaseous NH_3 by soil-feeding termites. Biogeochemistry 78:267–283

Kambhampati S, Eggleton P (2000) Taxonomy and phylogrny of termites. In: Abe T, Bignell DE, Higashi M (eds) Termites: evolution, sociality, symbioses, ecology. Kluwer Academic Publishers, Dordrecht, pp 1–23

Koepfli KP, Jenks SM, Eizirik E, et al (2006) Molecular systematics of the Hyaenidae: relationships of a relictual lineage resolved by a molecular supermatrix. Mol Phylogenet Evol 38:603–620

Korb J (2003) Thermoregulation and ventilation of termite mounds. Naturwissenschaften 90: 212–219

Korb J (2008) Termites, hemimetabolous diploid white ants? Front Zool 5:15

Korb J, Linsenmair KE (2000) Ventilation of termite mounds: new results require a new model. Behav Ecol 11:486–494

Leal IR, Oliveira PS (1995) Behavioral ecology of theneotropical termite hunting ant *Pachycondyla (=Termitopone) marginata* – colony founding, group-raiding and migratory patterns. Behav Ecol Sociobiol 37:373–383

Legendre F, Whiting MF, Bordereau C, et al (2008) The phylogeny of termites (Dictyoptera: Isoptera) based on mitochondrial and nuclear markers: implications for the evolution of the worker and pseudergate castes, and foraging behaviors. Mol Phylogenet Evol 48:615–627

Lo N, Engel MS, Cameron S, et al (2007) Save Isoptera: a comment on Inward et al. Biol Lett 3:562–563

Longrich NR, Currie PJ (2009) *Albertonykus borealis*, a new alvarezsaur (Dinosauria: Theropoda) from the Early Maastrichtian of Alberta, Canada: implications for the systematics and ecology of the Alvarezsauridae. Cretaceous Res 30:239–252

Luo ZX, Wible JR (2005) A late Jurassic digging mammal and early mammalian diversification. Science 308:103–107

Luscher M (1951) Air-conditioned nests. Sci Am 205:138–145

Martius C, Bandeira AG, da Silva Medeiros LG (1996) Variation in termite alate swarming in rain forests of central Amazonia. Ecotropica 2:1–11

Matsuura K (2002) Colony-level stabilization of soldier head width for head-plug defense in the termite *Reticulitermes speratus* (Isoptera: Rhinotermitidae). Behav Ecol Sociobiol 51:172–179

Mitchell JD (2007) Swarming and pairing in the fungus-growing termite, *Macrotermes natalensis* (Haviland) (Isoptera: Macrotermitinae). Afr Entomol 15:153–160

Miura T, Matsumoto T (1998) Foraging organization of the open-air processional lichen-feeding termite *Hospitalitermes* (Isoptera, termitidae) in Borneo. Insectes Soc 45:17–32

Morrow EH (2004) How the sperm lost its tail: the evolution of aflagellate sperm. Biol Rev 79: 795–814

Nalepa CA, Lenz M (2000) The ootheca of *Mastotermes darwiniensis* Froggatt (Isoptera: Mastotermitidae): homology with cockroach oothecae. Proc R Soc Lond B Biol Sci 267: 1809–1813

Noirot C (2001) The gut of termites (Isoptera). Comparative anatomy, systematics, phylogeny. II. Higher termites (Termitidae). Ann Soc Entomol Fr 37:431–471

Noirot C, Pasteels JM (1987) Ontogenic development and evolution of the worker caste in termites. Experientia 43:851–860

Ohkuma M (2003) Termite symbiotic systems: efficient biorecycling of lignocellulose. Appl Microbiol Biotechnol 61:1–9

Perna A, Jost C, Couturier E (2008) The structure of gallery networks in the nests of termite *Cubitermes* spp. revealed by X-ray tomography. Naturwissenschaften 95:877–884

Prestwich GD (1984) Defense-mechanisms of termites. Annu Rev Entomol 29:201–232

Riparbelli MG, Callaini G, Mercati D, et al (2009) Centrioles to basal bodies in the spermiogenesis of *Mastotermes darwiniensis* (Insecta, Isoptera). Cell Motil Cytoskeleton 66:248–259
Roisin Y (2001) Caste sex ratios, sex linkage, and reproductive strategies in termites. Insectes Soc 48:224–230
Roux EA, Roux M, Korb J (2009) Selection on defensive traits in a sterile caste – caste evolution: a mechanism to overcome life-history trade-offs? Evol Dev 11:80–87
Ruggiero RG, Fay FM (1994) Utilization of termitarium soils by elephants and its ecological implications. Afri J Ecol 32:222–232
Sands WA (1982) Agonistic behavior of African soldierless Apicotermitinae (Isoptera, Termitidae). Sociobiology 7:61–72
Sands WA (1998) The identification of worker castes of termite genera from soil of African and the Middle East. CAB International, Wallingford, CT
Santos CA, Costa-Leonard AM (2006) Anatomy of the frontal gland and ultramorphology of the frontal tube in the soldier caste of species of Nasutitermitinae (Isoptera, Termitidae). Microsc Res Tech 69:913–918
Scholtz OI, Macleod N, Eggleton P (2008) Termite soldier defence strategies: a reassessment of Prestwich's classification and an examination of the evolution of defence morphology using extended eigenshape analyses of head morphology. Zool J Linn Soc Lond 153:631–650
Suzuki S, Kuroda S, Nishihara T (1995) Tool-set for termite-fishing by chimpanzees in the Ndoki Forest, Congo. Behaviour 132:219–235
Thorne BL, Breisch NL, Muscedere ML (2003) Evolution of eusociality and the soldier caste in termites: influence of intraspecific competition and accelerated inheritance. Proc Natl Acad Sci U S A 100:12808–12813
Turner JS, Soar RM (2008) Beyond biomimicry. What termites can tell us about realizing the living building. Proceedings of the First International Conference on Industrialized, Intelligent Construction (I3CON) 1: 1–18
Ware JL, Litman J, Klass KD, Spearman LA (2008) Relationships among the major lineages of Dictyoptera: the effect of outgroup selection on dictyopteran tree topology. Syst Entomol 33:429–450
Weesner F (1965) The termites of the United States. The National Pest Control Association, Elizabeth, NJ, 70 pp
Weesner F (1970) External anatomy. In: Krishna K, Weesner F (eds) Biology of termites, vol I. Academic Press, New York, NY, pp 1–23
Wilson EO (1992) The effects of complex social-life on evolution and biodiversity. Oikos 63:13–18
Yarnell RW, Metcalfe DJ, Dunstone N, et al (2008) The impact of fire on habitat use by the short-snouted elephant shrew (*Elephantulus brachyrhynchus*) in North West Province, South Africa. Afr Zool 43:45–52

Chapter 2
Termite Phylogenetics and Co-cladogenesis with Symbionts

Nathan Lo and Paul Eggleton

Abstract Termites are key decomposer insects in numerous ecosystems in the tropics and beyond, and their unique social systems provide a major counterpoint to those of hymenopteran social insects. Our knowledge of the phylogenetics and systematics of the group have traditionally lagged behind those of other important insect groups, but significant progress has now been made. Here we review recent phylogenetic studies of relationships both among termites, and between termites, cockroaches and mantids. We also discuss studies of co-cladogenesis between termites and two groups of symbionts: cellulolytic hindgut flagellates, and *Blattabacterium.* A consensus has emerged that the sister-group of termites is the wood-feeding cockroach genus *Cryptocercus,* and that the digestion of wood by the common ancestor of these two groups was aided by cellulolytic hindgut flagellates. The basal phylogenetic position of *Mastotermes darwiniensis* among termites has been confirmed, however agreement on the phylogenetic positions of members of the Kalotermitidae, Termopsidae and Hodotermitidae has yet to be reached. Relationships between and within the Rhinotermitidae and Serritermitidae also remain to be settled. Key lineages of the major family Termitidae, however, are now fairly well established.

2.1 Introduction

Termites are the most important decomposer insects in tropical forest, tropical savanna, and desert ecosystems. Their social systems are unique, and they provide an important evolutionary contrast with analogous hymenopteran systems. In addition, some termites are often important timber and crop pests. It is therefore surprising to find the systematics of the group are so poorly developed. In particular there has been no review of termite phylogenetic hypotheses since Eggleton (2001)

N. Lo (✉)
Behaviour and Genetics of Social Insects Laboratory, School of Biological Sciences, University of Sydney, Sydney, NSW 2006, Australia
e-mail: nathan@usyd.edu.au

D.E. Bignell et al. (eds.), *Biology of Termites: A Modern Synthesis*,
DOI 10.1007/978-90-481-3977-4_2, © Springer Science+Business Media B.V. 2011

even though there have been a number of recent comprehensive phylogenetic papers. After many decades where vague hypotheses and poor taxon sampling were typical (reviewed by Eggleton 2001), we can finally put forward corroborated hypotheses of relationships among a properly representative set of taxa.

In this chapter we begin by examining the position of termites among the Dictyoptera, and describe the co-cladogenesis of two groups of symbionts with their termite and cockroach hosts: the intracellular symbiont *Blattabacterium cuenoti* (known simply as *Blattabacterium*), and intestinal parabasalid and oxymonad cellulolytic flagellates. We then review the sources of phylogenetic data (character sets) for termites themselves, and how they have been used to construct phylogenetic trees and construct classifications. This sets the stage for an examination of family-group relationships within the termites, and relationships within the largest termite family, the Termitidae. Co-cladogenesis between termites and their cellulolytic flagellates and fungal symbionts are also discussed. In the case of relationships among termites, we deal primarily with papers published since 2000, as existing reviews of termite phylogenetics (Kambhampati and Eggleton 2000; Eggleton 2001) cover the preceding period.

2.2 Phylogenetic Origins of the Termites

> The evidence is overwhelming that this roach is either the ancestor of termites, or is closely related to the ancestor, which is extinct. (Cleveland et al. 1934, on *Cryptocercus*)

Together with the appearance of sexual reproduction and multicellularity, the emergence of eusocial organisms is considered to be one of the major transitions in evolution (Maynard Smith and Szathmary 1997). How termites evolved is therefore a subject of outstanding interest, particularly because of the many differences between them and hymenopteran eusocial insects (i.e. all ants, and many species of wasps and bees). These differences include a diplo-diploid chromosome system, and hemimetaboly (vs haplodiploidy and holometaboly in hymenopteran social insects). In the case of bees and wasps, the study of solitary and facultatively social species has enhanced our understanding of eusociality in these groups. There are, however, no such solitary or primitively social species of termites: all are eusocial. Nonetheless, knowledge of the closest living relatives of termites, and their biology, has the potential to shed light on how termite eusociality evolved.

2.2.1 Pre-cladistic Studies of Termite Relatives

Speculations on the affinities of the termites date back to the mid-1800s, and proposed sister groups included the Neuroptera (lace-wings), Embioptera (web-spinners), Psocoptera (booklice), and Dermaptera (earwigs) (reviewed in Imms 1920). Hagen suggested in 1868 there was a close relationship between termites and cockroaches, and this view became dominant in the early twentieth century, on the basis of various similarities between members of the two groups, including the presence of an ootheca in *Mastotermes darwiniensis*, and the structure of the ovipositor (Crampton 1920; Imms 1920; Walker 1922; Crampton 1923; Hill 1926); see also Nalepa and Lenz (2000). A similarly close relationship between mantids and cockroaches also became recognized, on the basis of wing venation characters, ovipositor characters, and the production of oothecae (Walker 1922).

Further support for a close relationship between termites and cockroaches came from Jucci's (1932, 1952) and Koch's (1938) work, which demonstrated the presence of the specialized intracellular symbiont *Blattabacterium* in the fat body of *M. darwiniensis*. This bacterium had previously been found only from the fat-bodies of cockroaches (reviewed in Buchner 1965). Cleveland et al. (1934) performed the first examination of the intestinal symbionts of the wood-feeding cockroach *Cryptocercus punctulatus*, and found that they harboured a number of flagellate genera also found in the guts of termites, but nowhere else in nature. This led him to propose a close relationship between the two taxa.

Crampton (1938) proposed the superorder Panisoptera for the termites, cockroaches and mantids. Subsequent studies on the head structure and ovipositor of termites, cockroaches and mantids supported the view that they form a monophyletic group (Hudson 1945, 1947; Marks and Lawson 1962). Nonetheless, many textbooks of the era, and those in subsequent decades, grouped cockroaches and the mantids into the order Dictyoptera, and termites into the separate order Isoptera (e.g. Imms 1957; Gillott 1995). McKittrick (1965) recognized similarities in the proventriculus and female genitalia between *M. darwiniensis* and *C. punctulatus*, and suggested termites and cockroaches were more closely related to each other than was generally appreciated at the time. She recommended that termites be placed as a third suborder within Dictyoptera. The phylogenetic trees she presented, however, contained only cockroaches, and did not indicate the position of termites in relation to cockroaches (McKittrick 1964).

2.2.2 Cladistic Analyses

Relationships among the termites, cockroaches, and mantids, and particularly the placement of termites and *Cryptocercus,* were the subject of dispute from the 1970s until the turn of the twenty-first century. A likely contributing factor to this controversy is the lack of development of many morphological characters in termites, relative to cockroaches. The presence of markedly juvenile characters in termites, described by Nalepa and Bandi (2000) as "paedomorphosis", is found not only in the sterile castes, but also in the imagoes. Paedomorphic characters include the penis and ovipositor, which are rudimentary or absent, the male accessory glands, which are poorly differentiated, and tarsal and antennal segment numbers, which are fewer. Such reductions and losses result in fewer characters for cladistic analysis, and since they are also known to occur in some cockroach lineages, it can be difficult to determine whether characters are pleisomorphic or paedomorphic.

The two main hypotheses concerning the sister group of termites have been: (1) a sister group relationship between mantids and all cockroaches, with termites as the most basal branch (the view found in many textbooks, and proposed by Boudreaux

1979); (2) termites and *Cryptocercus* as sister groups, and either nested within the cockroaches, or sister group to them, with mantids basal (formally proposed by Hennig 1981). The first of these two hypotheses was supported in a cladistic analysis of 70 morphological, developmental and behavioural characters by Thorne and Carpenter (1992). Thorne (1991) had previously proposed an ancient horizontal transfer event to explain the presence of similar parabasalid and oxymonad flagellates in termites and *Cryptocercus*, but this was disputed by Nalepa (1991). A re-analysis of Thorne and Carpenter's data, including modification of polarity assumptions, found support for termites and cockroaches as sister groups (Kristensen 1995). A number of studies based on DNA sequence data (and in one case, combined DNA and morphological data) nevertheless provide support for the [cockroaches+mantids] + [termites] hypothesis (DeSalle et al. 1992; Kambhampati 1995; Wheeler et al. 2001). However, these DNA-based studies suffered either from inadequate taxon sampling (i.e. absence of representatives from each of the recognized cockroach families), or the exclusive use of mitochondrial DNA, which due its relatively fast rate of evolution, may be unsuitable for resolving cladogenesis events that took place at least 130 million years ago (when the first fossils for termites, mantids, and modern cockroaches appear).

Klass (1995) performed a comparative morphological analysis of Dictyopteran representatives, revising previous descriptions and carefully considering polarity via comparisons with a variety of outgroups. He focussed particularly on male genitalia and the proventriculus. In these analyses, he found support for termites as sister group to *Cryptocercus*, nested within the cockroaches, with mantids basal (hypothesis 2). A subsequent formal cladistic analysis recovered the same relationships (Klass and Meier 2006). Lo et al. (2000) also found significant support for this relationship based on analyses of the relatively slow-evolving 18S gene, and one mitochondrial gene (COII), from representatives of each of the main cockroach lineages, as well as termites and mantids. An analysis of endogenous cellulase genes (present only in termites and cockroaches) gave the same result (Chapter 3 by Lo et al., this volume). A number of subsequent studies have included additional genes and taxa, but have consistently recovered a termite-*Cryptocercus* clade (Deitz et al. 2003; Lo et al. 2003; Kjer 2004; Terry and Whiting 2005; Inward et al. 2007a; Lo et al. 2007a; Pellens et al. 2007; Ware et al. 2008; Legendre et al. 2008). There now appears to be overwhelming evidence for this grouping. The possibility remains that an as-yet-unexamined cockroach is closer to termites than *Cryptocercus*, but is considered unlikely (Eggleton 2001).

The morphological evidence for the phylogenetic position of *Cryptocercus* has long been disputed between Klass (see above), and Grandcolas (Grandcolas 1994, 1996; Gäde et al. 1997), whose studies suggest that *Cryptocercus* is deeply nested in the cockroach family Polyphagidae. Numerous challenges to the morphological work of Grandcolas are presented in detail by Klass (2001). Moreover, Grandcolas (1996) used Isoptera as an outgroup taxon for his cladistic analysis of Blattaria, and thus could not test the relationship between termites and *Cryptocercus*.

2.3 Co-cladogenesis Between Cockroaches, Termites, and Their Symbionts

Co-cladogenesis refers to the parallel process of speciation between host and symbiont, such that phylogenetic trees of each partner are equivalent, or approximately so. Parabasalid and oxymonad hindgut flagellates, and the intracellular symbiont *Blattabacterium,* have been shown to be transmitted from host parent to offspring, and thus have the potential to exhibit co-cladogenesis with their hosts. Here we provide a very brief overview of the biology of these symbionts, and describe the results of recent molecular phylogenetic analyses on them.

2.3.1 Blattabacterium

Blattabacterium cuenoti is a gram-negative intracellular symbiont within the Cytophaga-Flavobacterium-Bacteriodes phylum of bacteria (Bandi et al. 1994). It has been found in all cockroaches examined to date, including representatives of five of the main families, with one exception (Nocticolidae) (Lo et al. 2007a). In termites, it is found in *Mastotermes darwiniensis*, but no other species. The symbionts reside in specialized cells of the fat-body, called bacteriocytes, or mycetocytes. Their primarily role is thought to be the recycling of urates (nitrogenous waste products) into usable nitrogen for the host (reviewed in Cochran (1985)), although genome sequencing studies have shown that no recognizable pathways for uricolysis are present in the genomes of these bacteria (Sabree et al. 2009). During oogenesis, the bacteria are transmitted from the fat body to oocytes via a highly co-ordinated process (Sacchi et al. 1998). If this transmission is strictly vertical (with no horizontal (infectious) transmission between cockroaches and termites), genes in the symbionts can act as phylogenetic markers for their hosts. Comparisons of molecular phylogenetic trees for host and symbiont have found that they are indeed congruent, with no evidence for horizontal transfer (Bandi et al. 1995; Clark et al. 2001; Lo et al. 2003) (Fig. 2.1).

2.3.2 Cellulolytic Flagellates

With the exception of the Termitidae, members of all other termite families and *Cryptocercus* harbour unique genera of parabasalid and oxymonad flagellates, found nowhere else in nature (Cleveland et al. 1934; Inoue et al. 2000). The flagellates, which are passed through the generations by proctodeal trophallaxis, contribute to the digestion of cellulose, and their hosts are dependent upon them for survival (Chapter 15 by Ohkuma and Brune, this volume). As described above, two hypotheses concerning how these insects acquired their flagellates have been proposed. The

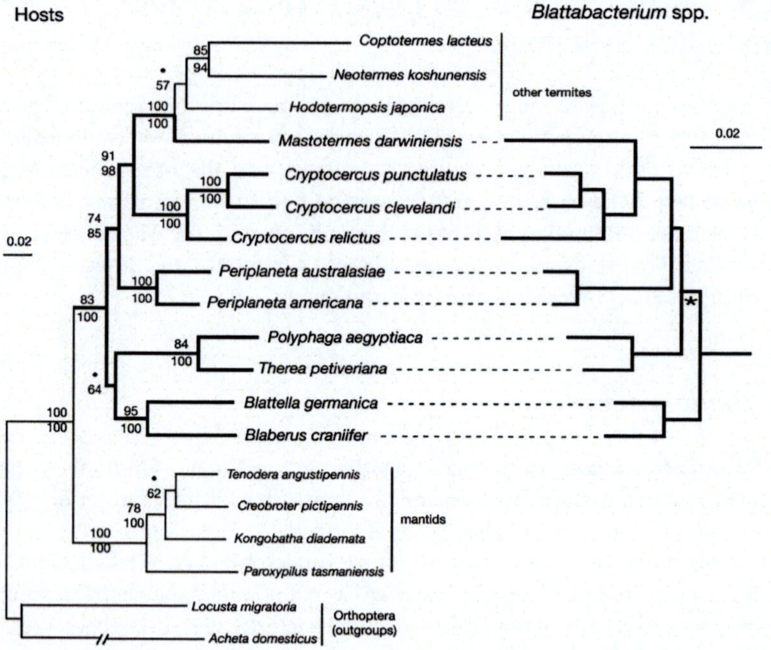

Fig. 2.1 Phylogeny of dictyopteran species and a comparison with the phylogeny of *Blattabacterium* spp. The host topology was obtained from a maximum parsimony (MP) analysis of 18S rDNA sequences combined with mitochondrial COII, 12S rDNA, and 16S rDNA sequences. *Bold lines* indicate those dictyopteran taxa that harbor *Blattabacterium* spp. The topology of *Blattabacterium* spp. was obtained from MP analysis of the 16S rRNA gene. The *asterisk* indicates the only node in the topology that was in disagreement with that based on the host phylogeny; comparisons of host and symbiont trees showed they were not significantly different. Reprinted, with permission, from Lo et al. (2003)

first is that horizontal transfer occurred between *Cryptocercus* or a similar cockroach no longer extant and the ancestors of extant termites (Thorne 1990). The second is that the flagellates were present in an ancestor of the two lineages, and have been passed down in a primarily vertical fashion (i.e. parent to offspring) since then (Nalepa 1991). These hypotheses were tested in two recent molecular phylogenetic studies of members of two taxonomic groups of flagellates present in both termites and *Cryptocercus punctulatus* (members of the genus *Trichonympha,* and members of the genera *Euconympha* and *Teranympha*) (Ikeda-Ohtsubo and Brune 2009; Ohkuma et al. 2009). For each taxonomic group, representatives from termite species clustered together, and were the sister group to representatives present in *Cryptocercus punctulatus* (Fig. 2.2; see also Chapter 15 by Ohkuma and Brune, this volume). These results support the second hypothesis described above. If horizontal transfer had occurred (i.e. hypothesis 1), one would expect, for example, either the *Trichonympha* from *Cryptocercus* to be nested within those from the termites, or vice versa, rather than separated.

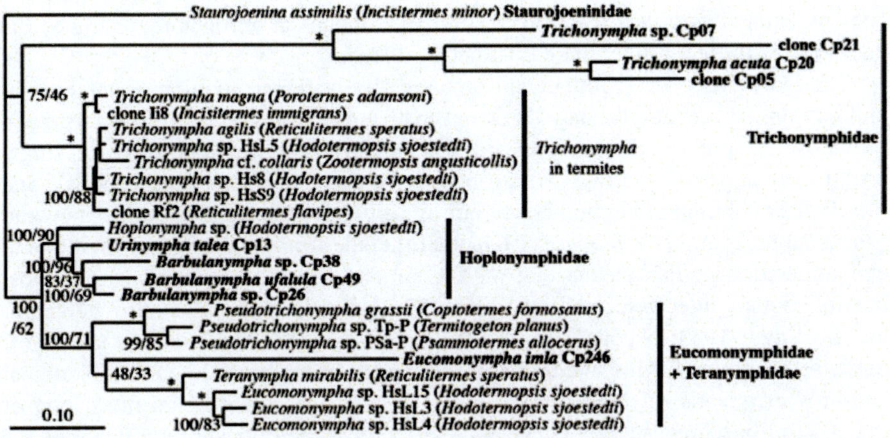

Fig. 2.2 Phylogeny of flagellates from the parabasalid order Trichonymphida present in *Cryptocercus punctulatus* (*bold*) and termites (host names shown in *brackets*), based on maximum likelihood analysis of 1,142 base pairs of the 18S rRNA gene. *Trichonympha* spp. from various termite species cluster together, as do those from *C. punctulatus*, suggesting vertical transfer from a common ancestor, as opposed to horizontal transfer. A similar pattern is found for members of the genera *Eucomonympha* and *Teranympha*. Reprinted, with permission from Ohkuma et al. (2009)

2.4 Fossil History and Key Events Leading to the Origin of Termites

Cockroach-like fossils have been found in late Carboniferous deposits and are ~310 million years old (MYO) (Laurentiaux 1951). The key characteristic setting these fossils apart from extant cockroaches is the presence of a long, external ovipositor. This structure becomes reduced in size over time, until it becomes very short or internalised, as is found in extant taxa (Thorne and Carpenter 1992). Such fossils, which lack an ovipositor, and which fit into extant families, appear in early Cretaceous deposits (~130 million years old). The earliest termite fossils (~135 million years old) come from a similar epoch (Thorne et al. 2000; Engel et al. 2009), as do those for mantids (Grimaldi 1997). These data suggest that the stem group for Dictyoptera arose around the time of the late Jurassic, and radiated soon afterwards.

The agreement between phylogenetic trees for *Blattabacterium* and their termite and cockroach hosts suggests that the ancestor of all extant cockroaches (with the possible exception of Nocticolidae) was infected with this bacterium. Co-cladogenesis between these two partners has thus been occurring for at least 130 million years. An analysis of the rate of evolution of the 16S rRNA gene of *Blattabacterium*, based on the first appearance of termites, supports the notion that extant cockroach lineages (again with the possible exception of Nocticolidae) diverged not earlier than the late Jurassic. *Blattabacterium* was apparently lost in the ancestor of the lineage leading to all termites other than *M. darwiniensis*, or, less parsimoniously, multiple times in later termite lineages. This loss may have been associated with the evolution of alternative means of nitrogen recycling by the ancestors of these termite lineages (Lo et al. 2003).

Other key events in termite evolution were, firstly, the acquisition of parabasalid and oxymonad cellulolytic flagellates in the ancestor of [*Cryptocercus* + termites], secondly the presumed transition from semelparous reproduction (i.e. a single brood over a pair's lifetime) to iteroparous reproduction (multiple broods), and thirdly the evolution of helping behaviour by siblings in the nest. The consistent and robust support for a [*Cryptocercus* + Isoptera] clade from numerous morphological and molecular studies means that all the similarities between the two are most parsimoniously interpreted as homologies. The biology of *Cryptocercus* (Seelinger and Seelinger 1983; Nalepa 1984, 1988a, b; Klass et al. 2008) is thus useful for inferring scenarios for the evolution of eusociality in termites although it is almost certainly erroneous to think of *Cryptocercus* as the ancestor of termites, one of the possibilities suggested by Cleveland et al. (1934). Significant modifications are likely to have occurred in both lineages since their divergence from a common, wood-feeding cockroach ancestor. One example is the loss of wings in the lineage leading to *Cryptocercus*.

2.5 Taxonomic Implications of the Phylogenetic Position of Termites

The sister group relationship between *Cryptocercus* and termites has potentially major taxonomic consequences, particularly for the rank of termites within the Dictyoptera classification. Given the modern prevailing view, it may no longer be appropriate to consider termites an order within the Dictyoptera, because they are nested within another order, the Blattaria. A taxonomic rank lower than an order may be more suitable. Inward et al. (2007a) proposed that this rank should be family, making termites the Termitidae. However, there was considerable opposition to this from others, both termite systematists and termite biologists (Lo et al. 2007b), and at present the proposed compromise is that the termites are to be treated as an epifamily of cockroaches. (An epifamily is intermediate in rank between family and superfamily). The new epifamily name of the termites is therefore proposed to be Termitoidae (Eggleton et al. 2007). However, this suggestion may not be adopted by many systematists, and we expect to see the informal group name "Isoptera" being used extensively for some time to come.

2.6 Termite Phylogeny: Morphological Character Sets

2.6.1 Alate Morphology

Alate morphology holds a particularly important place in termite systematics as most termite fossils are alates (e.g. see Engel et al. 2009). In particular, wing morphology is of great importance, as many fossils are just wing impressions. For extant species, the alates are of less importance because the largest termite family, the Termitidae, have wings that vary very little. Within lower termites, however, alate characters are important at all taxonomic levels, and wing venation and wing structure characters are useful.

2.6.2 Soldier Morphology

The morphology of the soldier head capsule and mandibles are very variable and taxonomically distinctive (see Chapter 1, this volume, by Eggleton). These structures have always been the mainstay of taxonomic work, providing the main set of characters for the delineation of genera and species. However, they have rarely been used in phylogenetic studies and the most recent work by Donovan et al. (2000) and Scholtz et al. (2008) show that as markers soldier characters suffer from incongruence with other character sets, probably due to convergence in defence strategies. This can be seen clearly in two morphological forms: the nasus, a forward projection of the head with a reservoir opening externally from the frontal gland and which has probably evolved twice, and asymmetrical snapping mandibles that can be estimated to have evolved from two to five times in trees which are not derived from soldier character data. In addition, the soldier caste has been lost at least twice (in the Apicotermitinae and the Protohamitermitinae). Inward et al. (2007b) came to the conclusion that soldier characters were too homoplasious to use in phylogenetic analyses.

The temptation in the past has been to assume transition series where none may actually be present. This is clearly seen, for example, in the phylogenetic trees of Ahmad (1950), which have a plausible transitional series structure for both the Nasutitermitinae and the Termitinae. In the Nasutitermitinae this consists of a basal genus, *Syntermes*, with a very poorly developed nasus and strongly developed mandibles, through to intermediate forms with strongly developed nasuses and moderately strongly developed mandibles, and finally to forms with nasuses without mandibles (Fig. 2.3). Similarly, in the Termitinae there are species with symmetrical biting mandibles, through to snapping symmetrical mandibles, and to forms with an increasing degree of asymmetry. In both of these extremely plausible cases, however, the group suggested by the transition series are polyphyletic in molecular trees (Inward et al. 2007b).

2.6.3 Worker-Imago Mandible Morphology

There is a very similar problem with worker mandibles. Again, the mandibles are very variable, with considerable taxonomic value at the generic level and below. However, the configuration of the mandibles is strongly influenced by feeding substrate (see Chapter 1 by Eggleton, this volume), and again their value as phylogenetic markers is probably relatively low. This is perhaps surprising given that one of the first major attempts to examine termite phylogenetics (Ahmad 1950) used worker mandible characters. His findings have not been corroborated by subsequent analyses. However, Ahmad does recognises groupings in

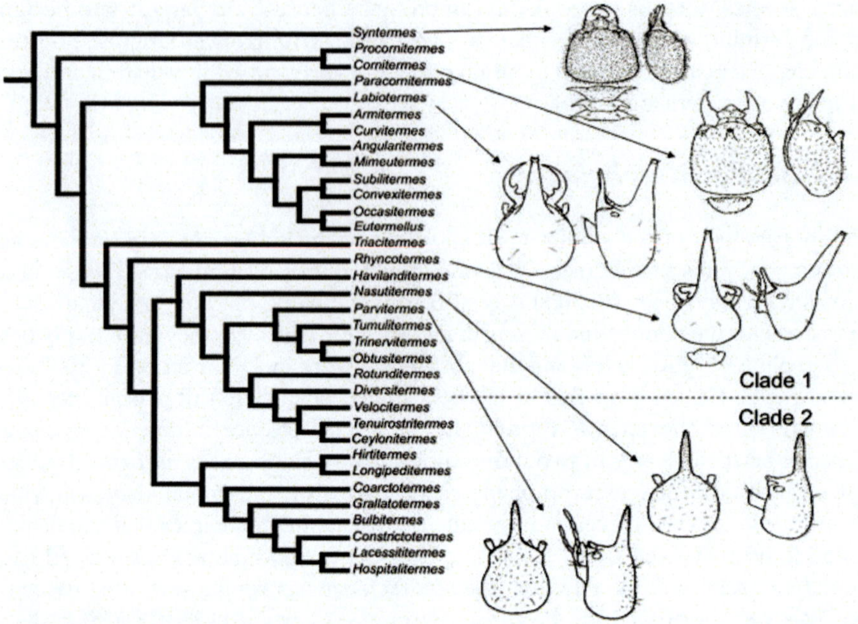

Fig. 2.3 Plausible transition series for the two clades that made up the old Nasutitermitinae. However, in the most recent phylogenetic trees Clade 1 is nested within the *Amitermes*-group and clade 2 is nested within the Nasutitermitinae *sensu stricto*. Figures from Constantino (1995), with permission

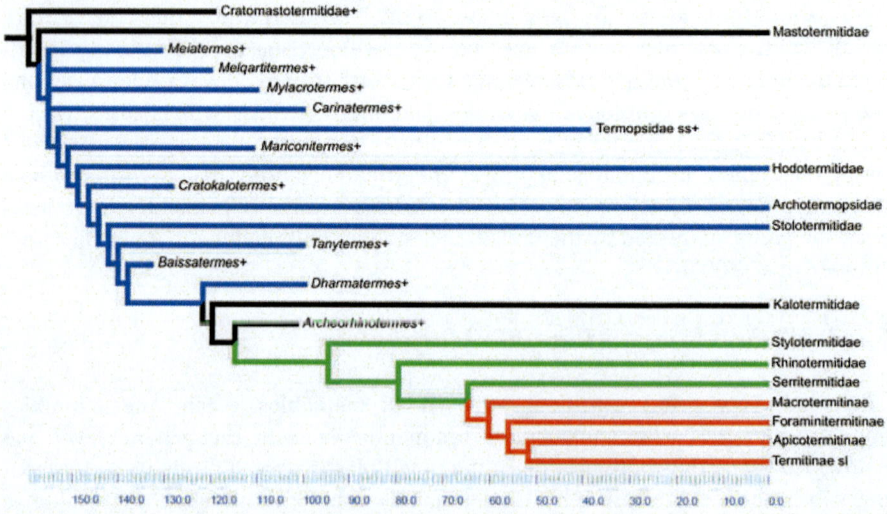

Fig. 2.4 Phylogenetic tree redrawn from Engel et al. (2009) as a chronogram. *Blue* branches, Termopsidae *sensu lato; green* branches Rhinotermitidae *sensu lato; red* branches, Termitidae. Taxa suffixed with a '+' are extinct. Scale bar is millions of years ago

his paper that have been recovered by other phylogenetic analyses: he placed *Neocapritermes* and *Planicapritermes* together, and recognised the anomalous positions of *Protohamitermes*, *Cephalotermes* and *Cylindrotermes*. There is, therefore, just a hint that mandibles may have some use after all, perhaps in the most apical taxa. However, this would require the untangling of homologies: a job that must await a more robust baseline classification.

2.6.4 Worker Gut Anatomy

The basic structure of the termite gut conforms to the general insect plan of foregut, midgut and hindgut. However, the configuration of the gut differs considerably in different groups (see Chapter 1, this volume, by Eggleton). The gut of worker termites is an extremely complicated coiled three dimensional structure, especially within the termitid soil-feeders. The important parts of the gut for taxonomic work are described by Noirot (1995, 2001) in his two modern synoptic anatomical works: we will not repeat details here except when they define a particular clade. In brief, however, the principal character sets relate to: the torsion of the gut, the relative length of the hindgut and midgut, the shape and structure of the mixed segment – an area of the gut where the mesenteron and proctodeum overlap, the number and attachment arrangement of the Malpighian tubules, and the size and shape of the first and third segments of the proctodeum (P1 and P3). The enteric valve seating arrangement (the enteric valve being the variably sclerotised structure that sits between the P1 and the P3), has been used extensively in termite taxonomy (e.g. Sands 1972). However, as with many other variable termite characters, the enteric valve is too complex to be of much use above the generic level, because homologies are very hard to assess. For functional comments on gut anatomy, see Chapters 14 by Bignell and 16 by Brune and Ohkuma (this volume).

2.6.5 Other Character Sets

Although Cleveland recognised the value of flagellate protists in linking *Cryptocercus* and the stem-group termites, the phylogenetic signal from the flagellates is messy (see below). *Stolotermes* (Termopsidae) shares unique flagellate groups with *Cryptocercus,* as does *Paraneotermes* (Kalotermitidae). In contrast, *Mastotermes*, the sister group of all other termites, does not. However, the absence of flagellates is a good synapomorphy for the Termitidae.

2.7 Phylogenetic and Taxonomic Relationships Among Termites

Termites are generally classified into seven families, which are assumed to be major clades (Table 2.1). However, until recently, there has been very little evidence

Table 2.1 Extant families of termites

Family	Food	Habitat	Regions	Number of spp
Mastotermitidae	w	Dry forest/savanna	N Australia	1
Hodotermitidae	g	Arid grasslands	Africa, Palaearctic	15
Termopsidae	ww	Temperate rain forest	Bipolar	
Kalotermitidae	dw	Tropical tree canopy, oceanic islands, deserts, coasts	Cosmopolitan	~400
Rhinotermitidae	w	Rain forest edges and gaps, temperate woodland, oceanic islands, mangroves, deserts	Pantropical, also in sub-tropics, warm temperate	~300
Serritermitidae	nests ?	Cerrado	S America	~2
Termitidae	diverse	Tropical rain forests, savannas, semi-deserts	Pantropical	~ 2,000

w, wood; g, grass; ww, wet wood; dw, dry wood.

presented to support the monophyly of each of these clades. Prior to 2000, the evidence for the relationships among the "families" (i.e. large clades) of termites was built on studies of a very few species. These species were assumed to be exemplars for clades, but the monophyly of the clades had not been established. Where a more extensive set of species were analysed, they were either from a limited geographical range (Ohkuma et al. 2004), or their morphology alone was studied (Donovan et al. 2000). This work is reviewed by Eggleton (2001).

Since 2005, three far more comprehensive studies have been published; two molecular (Inward et al. 2007b; Legendre et al. 2008) and one morphological, but including comprehensive fossil data (Engel et al. 2009). We will refer to them as the Inward, Legendre and Engel studies, respectively. Engel built a new phylogenetic framework and classification from a consideration of both extant and fossil species and is therefore the most taxonomically complete account. Our discussion of higher taxonomic relationships is based on that new framework (Figs. 2.4 and 2.5), and on comparisons with both Inward and Legendre, particularly where the reconstructions differ (Table 2.2). One of the main obstacles to progress identified by Eggleton (2001), that the monophyly of the families had not been proven, still applies in some parts of the tree, but we now have a lot more information.

In order to assist in our understanding of the patterns in the available data we have run a combined analysis on the Engel morphological data and the mitochondrial COII data from Inward (the most taxon-comprehensive molecular dataset). We added COII data to the Engel data at the generic level and ran a rapid TNT search (Goloboff et al. 2008), using all search parameters, 500 random addition sequences. This search gave us 150 trees of length 2,336, and we display the trees as a 50% majority rule tree (Fig. 2.6). We use this tree to indicate the ambiguities that still remain in our understanding of termite relationships.

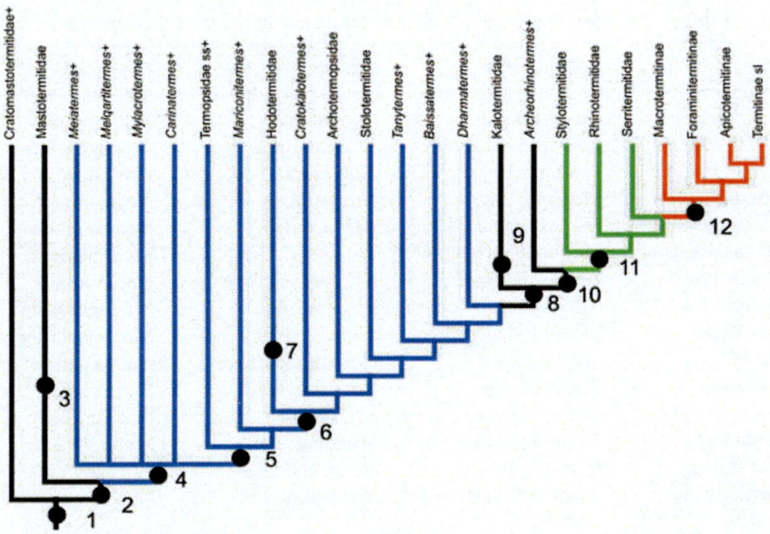

Fig. 2.5 Key synapomorphies from Fig. 2.4. See Table 2.3 for details of individual synapomorphies. Extinct clades are indicated by a '+'

2.7.1 Termites with Flagellates (The So-Called "Lower Termites")

The earliest branching termite lineage appears to be the Cratomastotermitidae, represented by the single fossil species, *Cratomastotermes wolfschwenningeri* (Engel et al. 2009). This termite, known only from amber, has many plesiomorphic wing venation characters (i.e. is very "cockroach-like"). Among living termites, there is now general agreement that the Mastotermitidae is sister to all other extant termites (see also Thompson et al. 2000a). This is confirmed in all three of the recent studies. The cockroach-like elements of its biology (anal lobe in hindwing, wings non-dehiscent, production of oothecae, presence of *Blattabacterium*) act as clear markers of its position.

Traditionally, the next branching group of termites after the Mastotermitidae has been known as the Termopsidae *sensu lato*, sometimes as a single family and sometimes split into two families (wood-nesting and -feeding Termopsidae, and ground-nesting grass-harvesting Hodotermitidae), but generally thought to be one (or two) monophyletic groups. The validity of this assumption is challenged by the fossil evidence, where the family splits into a large grade of species (Fig. 2.4), many of which were described originally as Hodotermitidae. The name-bearing genus of the family (*Termopsis*, only known from fossils) is, in the Engel tree, in a clade (the Termopsidae senus stricto) separate from the other genera that were contained within the larger family. The extant species appear to form three distinct clades (formally recognised by Engel), as the families (Hodotermitidae, Archotermopsidae and Stolotermitidae). The molecular evidence is ambiguous. The Inward and Legendre trees both have the Stolotermitidae and the Hodotermitidae as monophyletic, but not

Table 2.2 Clades used in Fig. 2.4 and their status in the three key analyses

Clade	Engel	Inward	Legendre
Cratomastotermes†	One exemplar	Not sampled	Not sampled
Mastotermitidae	Monophyletic	One exemplar	One exemplar
Meiatermes grade†	Paraphyletic	Not sampled	Not sampled
Termopsidae ss†	Monophyletic	Not sampled	Not sampled
Mariconitermes	One exemplar	Not sampled	Not sampled
Hodotermitidae ss	Monophyletic	Monophyletic	Monophyletic
Cratokalotermes†	One exemplar	Not sampled	Not sampled
Archotermitidae	Paraphyletic	Paraphyletic	Monophyletic
Stolotermitidae	Monophyletic	Monophyletic	Monophyletic
Tanytermes†	One exemplar	Not sampled	Not sampled
Baissatermes†	One exemplar	Not sampled	Not sampled
Dharmatermes†	One exemplar	Not sampled	Not sampled
Kalotermitidae	Monophyletic	Monophyletic	Monophyletic
Stylotermitidae	Monophyletic	Not sampled	Not sampled
Archeorhinotermes†	Sister to rhinos + termitids	Not sampled	Not sampled
Rhinotermitidae	Monophyletic	Paraphyletic	Paraphyletic
Serritermitidae	Sister to termitids	Within rhinos	Sister to rhinos + termitids
Termitidae	Monophyletic	Monophyletic	Monophyletic
Macrotermitinae	One exemplar	Monophyletic	Monophyletic
Sphaerotermitinae	Not sampled	One exemplar	One exemplar, most basal termitid
Foraminitermitinae	Not sampled	Monophyletic	One exemplar
Apicotermitinae	Paraphyletic	Monophyletic	Not sampled
Syntermitinae	Paraphyletic	Monophyletic	Monophyletic
Termitinae/ Nasutitermitinae	Paraphyletic	Monophyletic	Monophyletic
Cubitermes-group	Not sampled	Polyphyletic	One exemplar
Amitermes-group	Paraphyletic	Monophyletic	Not sampled
Termes-group	Not sampled	Monophyletic	Monophyletic
Mirocapritermes-group	One exemplar	Monophyletic	Not sampled
Protohamitermes-group	Not sampled	Monophyletic	Not sampled
Nasutitermitinae	Monophyletic	Polyphyletic	Paraphyletic

the Archotermopsidae (because *Hodotermopsis*, an archotermopsid, is sister to the hodotermitids in the Inward tree and sister to the stolotermitids in the Legendre tree). Clearly, *Hodotermopsis* needs to be studied in more detail. In the [Engel+COII] tree, Archotermopsidae becomes monophyletic with *Hodotermopsis* nested within that group.

All three analyses agree on the monophyly of the Kalotermitidae (see also Thompson et al. 2000b), but differ in their position relative to the Termopsidae sensu lato. Inward and Engel place the termopsids more basal than the kalotermitids, Legendre places the termopsids more apical. The balance of evidence seems to

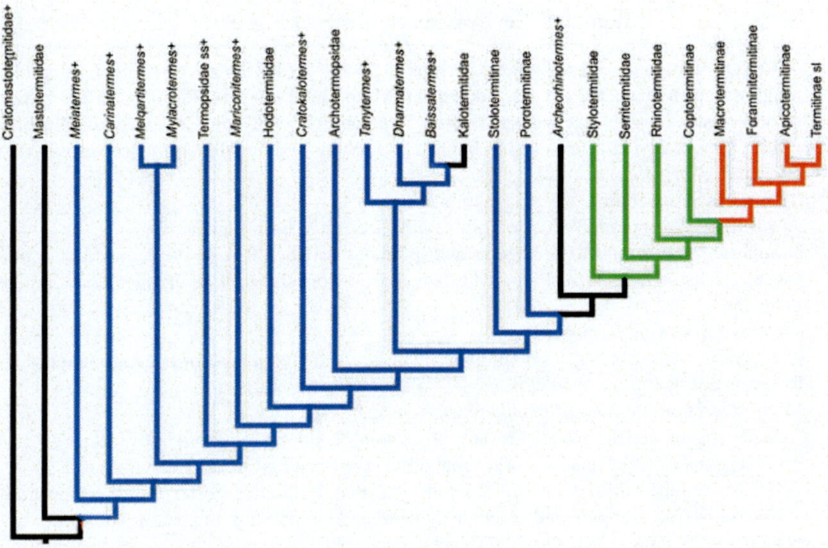

Fig. 2.6 Combined Engel/COII tree, obtained by parsimony (see text). Colour code as in Fig. 2.3

suggest a more basal placement of the "extant termopsids", in so far as that means anything phylogenetically. This position is also found in the [Engel+Inward] trees.

The monophyly of the Rhinotermitidae is disputed: the group is paraphyletic in all three analyses, and in an additional study which focussed on the family (Lo et al. 2004). Engel went as far as to split the group up into two families, the Stylotermitidae and the Rhinotermitidae. *Stylotermes*, the only genus within the Stylotermitidae, is a wood-feeding termite found in drier parts of Asia, which was not, unfortunately, included in either the Inward or Legendre studies. It has not been possible, therefore, to provide molecular corroboration for Engel's phylogenetic hypothesis.

The Serritermitidae is another problematic group. Defined as a separate subfamily predominantly due to the unique morphology of the soldiers, its status as a separate family has been debated (Lo et al. 2004; Inward et al. 2007b). The phylogenetic evidence is confused. Legendre places *Serritermes* as the sister to the [rhinotermitids + termitids] (see also Thompson et al. 2000a). Inward places the genus within a paraphyletic "rhinotermitid" grade. Two studies indicate that *Glossotermes* is the sister group of *Serritermes*, which together fall morphologically within the Rhinotermitidae (Lo et al. 2004; Cancello and DeSouza 2005). However, in Engel *Serritermes* is sister to the termitids. In the combined [Engel+COII] tree, *Serritermes* is again placed within a "rhinotermitid" grade.

The relationships among the presently recognised rhinotermitid subfamilies is very unclear – particularly between the Heterotermitinae and the Coptotermitinae, where the latter group may well be nested within the former group, with

Table 2.3 Key synapomorphies from Fig. 2.5

1	– Hindgut with ventral torsion, anal lobe of hindwing present, ootheca formed, midgut long, midgut with caeca, P1 short, dozens of Malpighian tubules, *Blattobacterium* present, sternal glands on segment 3, 4 and 5, tarsi 5-segmented, ocelli present (no median ocellus), postclypeus undivided (unique parabasalid and oxymonad flagellate protists), cross veins connecting longitudinal veins.
2	– No cross veins connecting longitudinal veins.
3	– Multiflagellate sperm, reduction of muscle sheath just before gizzard, compartmentalisation of P3, huge armed sphincter between P3 and P4, reduced midgut caeca, gizzard with columns I and II reduced, true workers but with distinct developmental pathway.
4	– Forewing Rs simple in scale.
5	– Occipital sulcus absent, hind wing basal suture mechanism present but rudimentary ['termopsids']; *Blattobacterium* absent, no anal lobe, no ootheca formed, 10 or fewer Malpighian tubules, sternal glands present on segments 4 and 5.
6	– Ocelli absent, sternal glands present on segment 4, subsidiary tooth between apical and first marginal of right imago-worker mandible (assumed lost in Hodotermitidae).
7	– Greatly reduced midgut caeca, P2 strongly armed, P3 dilated posteriorly [hodotermitids].
8	– Forewing Cua elongate and extensively developed, extending to apex or subapex, posteriorly along one half-wing, with six or more branches; hindgut without ventral torsion (some exceptions in Kalotermitidae), no midgut caeca, eight or fewer Malpighian tubules, rectal gland armature reduced, sternal glands present on segment 5, tarsi 4-segmented.
9	– [Reduced muscle sheath at end of crop, midgut ring shape, caecum-like paunch], wing membrane nodulose or pimplate [kalotermitids].
10	– C+Sr+R and Rs extremely close and simple and parallel, fontanelle present, [Archaeorhinotermes onwards].
11	– Postclypeus divided by a longitudinal furrow, distinct from anteclypeus, often raised above frons, crop musculature greatly reduced, stomodeal valve cuticle with a thin striation, amotile aflagellate sperm.
12	– No flagellates, forewing bases the same size as hindwing bases, no sclerotised tooth on columns I and II of gizzard, four (rarely two) Malpighian tubules, no limiting groove between P1 and P2 [hindgut with dorsal torsion].

Heterotermes more closely related to *Coptotermes* than *Reticulitermes* (Lo et al. 2004). Other subfamilies are represented by single genera, and may therefore be rather isolated phylogenetically.

In summary, the general order of major clades within at least the extant termites is now moving towards a consensus. Most recent studies agree on the order (Mastotermitidae ("Hodotermitidae/Termopsidae" (Kalotermitidae (Rhinotermitidae (Termitidae))))). However, the "termopsids" are not monophyletic, and the rhinotermitids may not be. Better taxon sampling is required in both groups. In addition, the "termopsid" grade is made up predominantly of fossils, many of which cannot be placed in an extant family. The subfamily classification of both "termopsids" and "rhinotermitids" is very provisional, particularly because so many of the subfamilies contain a single genus, probably on long branches due to large numbers of extinctions. The classification of both groups requires a substantial research effort in both molecules and morphology. The Engel classification must be considered provisional.

2.7.2 Co-cladogenesis of Cellulolytic Flagellates Among Termites

Kitade (2004) examined the distribution of different oxymonad and hypermastigid flagellates among the different families of termites. He found that within a given family, members tended to harbour the same types of flagellates. This was particularly the case for Rhinotermitidae and Kalotermitidae. However, this correspondence did not translate to deeper phylogenetic relationships among families, and he found cases where horizontal transfer must have occurred between the ancestors of extant termites. For example, the flagellate fauna of *Reticulitermes* and *Hodotermposis* were found to be similar, whereas the fauna of *Reticulitermes* and other rhinotermitids are quite different. Since *Reticulitermes* and *Hodotermposis* are from different families, horizontal transfer between their ancestors is likely to have occurred. This can be explained by the presence of each of these genera in temperate regions of East Asia, and their tendency to inhabit identical environments (i.e. rotting logs). Thus, although no evidence has been found for horizontal transfer between the ancestors of *Cryptocercus* and termites (see above), the same cannot be said for transfer among different kinds of termites.

A more recent study of the flagellate genus *Pseudotrichonympha* in termites of the family Rhinotermitidae (Fig. 2.7) has confirmed the initial hypotheses of Kitade (Noda et al. 2007). Phylogenetic trees for the hosts and symbionts were inferred from DNA sequences, and shown to display a high level of congruence, with the exception of the genus *Reticulitermes*. Since flagellate fauna are transmitted by proctodeal trophallaxis, rather than directly in the eggs, it is expected that their phylogenetic relationships will be relatively less congruent with host relationships, when compared with transovarially transmitted symbionts. While the results of Noda et al. (2007) do confirm this notion, they also reveal that co-cladogenesis

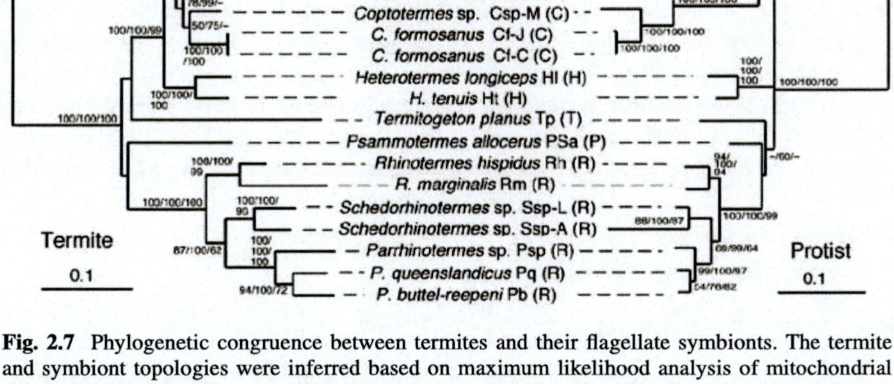

Fig. 2.7 Phylogenetic congruence between termites and their flagellate symbionts. The termite and symbiont topologies were inferred based on maximum likelihood analysis of mitochondrial 16S rRNA and COII genes (termites) and the 18S rRNA genes (symbionts). Reprinted, with permission from Noda et al. (2007)

of gut-symbionts with their hosts can occur over remarkably long time scales (i.e. the evolution of an entire insect family).

2.7.3 Relationships Within Termitidae (The So-Called "Higher Termites")

Worker gut structure provides the main basis for the existing Termitidae classification and ideas about termitid relationships. However, the classification is not really well worked out, as many of the subfamilies are not adequately diagnosed. The main features used in the present classification are the number, position and arrangement of the Malpighian tubules (Table 2.3, within termitids only). These were proposed by Sands (1972) and the only substantial amendment to that scheme is Engel and Krishna (2004), who erected or re-erected three new subfamilies (Sphaerotermitinae, Foraminitermitinae and Syntermitinae).

Ideas concerning relationships within the Termitidae have been confused by existing ideas about clades within the group. This is expressed in the classifications that have been proposed over time. Sands (1972) proposed four subfamilies: Macrotermitinae, Apicotermitinae, Termitinae and Nasutitermitinae. This arrangement led researchers to believe erroneously that these subfamilies were monophyletic and so to undertake phylogenetic analyses using those subfamilies which themselves were therefore inappropriate exemplar taxa. Since Sands (1972) several new subfamilies have been proposed (Engel et al. 2009), the Foraminitermitinae, Sphaerotermitinae, and Syntermitinae, predominantly based on the findings of Noirot (2001) concerning gut anatomy.

The Inward analysis is the only study that deals with all parts of the termitids. Its results are conclusive at higher levels but inconclusive at the very tip of the termitid tree. The relationship (Macrotermitinae ((Foramanitermitinae, Sphaerotermitinae) (Apicotermitinae) (Termitinae + Nasutitermitinae))) is well supported, particularly by gut anatomical characters (Fig. 2.8). The problem lies, however, in clarifying where the parts of the Termitinae + Nasutitermitinae should be placed. It seems clear, looking at the levels of the Bremer support in that part of the tree, that very rapid evolution at the base of Termitinae/Nasutitermitinae has occurred, which makes sorting out relationships very difficult. Clearly a reclassification will have to proceed with some caution. We can, however, recognise some sub-clades within the Termitinae + Nasutitermitinae (Fig. 2.8). These are reasonably strongly supported in the Inward analysis but the relationships among them are unresolved.

2.7.4 Co-cladogenesis Between Fungus-Growing Termites and Termitomyces

Fungus-growing termites (Macrotermitinae) are acknowledged to be the sister group of all the other termitids and are unique in having a mutualistic relationship with a basidomycete fungus, *Termitomyces*. The exact details of this mutualism are discussed in Chapter 8 by Nobre et al., this volume, but, in brief, the fungus lives inside termite colonies, on a diet of partly termite-digested dead plant material, and

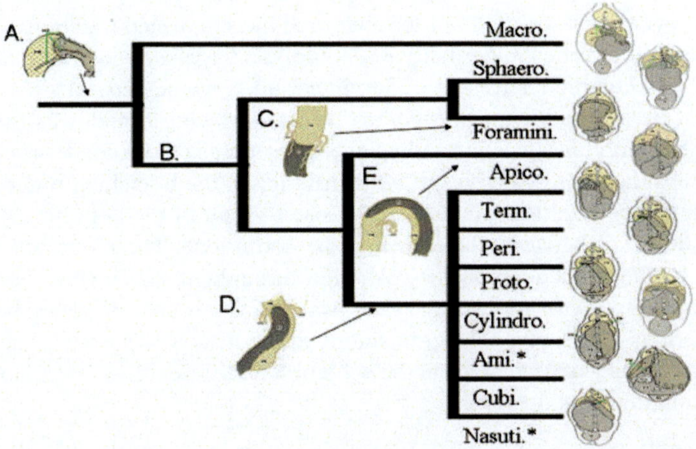

Fig. 2.8 Termitid tree, based on Inward et al. (2007b). **a** Malpighian tubules reduced to four, evenly spaced around distal edge of mesenteron. **b** Fungal mutualists. **c** Pseudomalpighian knot. **d** Mixed segment on inner edge of gut. **e** Mixed segement on outer edge of gut. Macro. = Macrotermitinae, Sphaero. = Sphaerotermitinae, Foramini. = Foraminitermitinae, Apico. = Apicotermitinae, Term. = *Termes*-group, Peri. = *Pericapritermes*-group, Proto. = *Protohamitermes*-group, Cylindro. = *Cylindrotermes*-group, Ami. = *Amitermes*-group, Cubi. = *Cubitermes*-group, Nasuti. = Nasutitermitinae. Note that the Syntermitinae are nested within the *Amitermes*-group. Figures from Sands (1998), used with permission

provide food for the termite colony. This close relationship has led to a number of studies on the phylogenetics of the termite and the fungus, and to studies of co-cladogenesis. Phylogenetic analyses of the termite and fungus data (Aanen et al. 2002; Aanen and Eggleton 2005) show that the mutualism evolved only once, in Africa, and has colonised Asia and Madagascar, but no other continents.

The termite and fungus phylogenetic trees have been shown to be broadly congruent (Aanen et al. 2002), but multiple host shifting has occurred, particularly at lower taxonomic levels. This is likely to be due to the generally horizontal transmission mode from one colony to the next. In most Macrotermitinae the workers that develop first have to pick up the spores to start the fungus growing in the termite colony. This means that if the preferred fungal strain is not found, then another less-favoured strain can be taken to back to the nest. This mode of transition is found in all termites except for *Microtermes* and *Macrotermes bellicosus*, where transmission is vertical, the alates carrying mycelium in their guts. Horizontal transmission may constrain long distance dispersal in the Macrotermitinae, as shown in Madagascar, where only *Microtermes* (with vertical transmission) has managed to disperse across the Mozambique Channel (Nobre et al. 2010).

2.8 Conclusions

The last decade has seen the application of molecular phylogenetic techniques, as well as morphology-based cladistic analyses, to increasingly larger datasets of termites, their cockroach relatives, and their symbionts. This has led to an end to the debate over the sister-group of termites, and has confirmed traditional hypotheses, including the vertical inheritance of cellulolytic flagellates and *Blattabacterium* from the last common ancestor of termites and *Cryptocercus, Mastotermes* as the earliest branching extant lineage, and fungus-growing termites as the earliest branching termitid lineage. Outstanding questions include the second earliest branch (i.e. Termopsidae representatives vs Kalotermitidae), the branching order of members of the Rhinotermitidae (including the sister group of the Termitidae), and the exact relationships between clades within the Termitidae. These questions may be answered by the application of low-cost, high-throughput sequencing and genomic techniques, which will increase the data available by orders of magnitude. Once those questions are resolved, key transitions in the history of termites, such as the appearance of the worker caste, and the evolution of different feeding strategies, can be more confidently addressed.

References

Aanen DK, Eggleton P (2005) Fungus-growing termites originated in African rain forest. Curr Biol 15:851–855

Aanen DK, Eggleton P, Rouland-Lefevre C et al (2002) The evolution of fungus-growing termites and their mutualistic fungal symbionts. Proc Natl Acad Sci U S A 99:14887–14892

Ahmad M (1950) The phylogeny of termite genera based on imago-worker mandibles. Bull Am Mus Nat Hist 95:37–86

Bandi C, Damiani G, Magrassi L et al (1994) *Flavobacteria* as intracellular symbionts in cockroaches. Proc R Soc Lond B 257:43–48

Bandi C, Sironi M, Damiani G et al (1995) The establishment of intracellular symbiosis in an ancestor of cockroaches and termites. Proc R Soc Lond B 259:293–299

Boudreaux HB (1979) Arthropod phylogeny with special reference to insects. John Wiley and Sons, New York, NY

Buchner P (1965) Endosymbiosis of animals with plant microorganisms. Interscience, New York, NY

Cancello EM, DeSouza O (2005) A new species of *Glossotermes* (Isoptera): reappraisal of the generic status with transfer from the Rhinotermitidae to the Serritermitidae. Sociobiology 45:31–51

Clark JW, Hossain S, Burnside CA, Kambhampati S (2001) Coevolution between a cockroach and its bacterial endosymbiont: a biogeographical perspective. Proc R Soc Lond B 268: 393–398

Cleveland LR, Hall SR, Saunders EP, Collier J (1934) The wood-feeding roach *Cryptocercus*, its protozoa, and the symbiosis between protozoa and roach. Mem Am Acad Sci 17: 185–342

Cochran DG (1985) Nitrogen excretion in cockroaches. Annu Rev Entomol 10:29–39

Constantino R (1995) Revision of the neotropical termite genus *Syntermes* Holmgren (Isoptera: Termitidae). Univ Kans Sci Bull 55:455–518

Crampton GC (1920) The terminal abdominal structures of the primitive Australian termite *Mastotermes darwiniensis* Froggatt. Trans R Entomol Soc Lond 1920:137–145

Crampton GC (1923) A comparison of the terminal abdominal structures of an adult alate female of the primitive termite *Mastotermes darwiniensis* with those of the roach *Periplaneta americana*. Bull Brooklyn Entomol Soc 18:85–93

Crampton GC (1938) The interrelationships and lines of descent of living insects. Psyche 45: 165–181

Deitz LL, Nalepa CA, Klass KD (2003) Phylogeny of the Dictyoptera re-examined (Insecta). Entomol Abh 1:69–91

DeSalle R, Gatesy J, Wheeler W, Grimaldi D (1992) DNA sequences from a fossil termite in Oligo-Miocene amber and their phylogenetic implications. Science 257:1933–1936

Donovan SE, Jones DT, Sands WA, Eggleton P (2000) Morphological phylogenetics of termites (Isoptera). Biol J Linn Soc 70:467–513

Eggleton P (2001) Termites and trees: a review of recent advances in termite phylogenetics. Insectes Soc 48:187–193

Eggleton P, Beccaloni G, Inward D (2007) Save Isoptera: a comment on Inward et al – response to Lo et al. Biol Lett 3:564–565

Engel MS, Grimaldi DA, Krishna K (2009) Termites (Isoptera): their phylogeny, classification, and rise to ecological dominance. Am Mus Novit 3650:1–27

Engel MS, Krishna K (2004) Family-group names for termites (Isoptera). Am Mus Novit 3432:1–9

Gäde G, Grandcolas P, Kellner R (1997) Structural data on hypertrehalosaemic neuropeptides from *Cryptocercus punctulatus* and *Therea petiveriana*: how do they fit into the phylogeny of cockroaches? Proc R Soc Lond B 264:763–768

Gillott C (1995) Entomology. Plenum Press, New York, NY

Goloboff PA, Farris JS, Nixon KC (2008) TNT, a free program for phylogenetic analysis. Cladistics 24:774–786

Grandcolas P (1994) Phylogenetic systematics of the subfamily Polyphaginae, with the assignment of *Cryptocercus* Scudder, 1862 to this taxon (Blattaria, Blaberoidea, Polyphagidae). Syst Entomol 19:145–158

Grandcolas P (1996) The phylogeny of cockroach families: a cladistic appraisal of morpho-anatomical data. Can J Zool 74:508–527

Grimaldi D (1997) A fossil mantis (Insecta: Mantodea) in Cretaceous amber of New Jersey, with comments on the early history of the Dictyoptera. Am Mus Novit 3204:1–11

Hennig W (1981) Insect phylogeny. John Wiley & Sons, New York, NY

Hill GF (1926) Australian termites (Isoptera) Notes on *Stolotermes*, *Calotermes* and *Coptotermes*, with descriptions of new species. Proc R Soc Vic 38:192–214

Hudson GB (1945) A study of the tentorium in some orthopteroid Hexapoda. J Entomol Soc South Afr 8:71–90

Hudson GB (1947) Studies in the comparative anatomy and systematic importance of the hexapod tentorium II Dermaptera, Embioptera, and Isoptera. J Entomol Soc South Afr 9: 99–108

Ikeda-Ohtsubo W, Brune A (2009) Cospeciation of termite gut flagellates and their bacterial endosymbionts: *Trichonympha* species and *Candidatus* Endomicrobium trichonymphae. Mol Ecol 18:332–342

Imms AD (1920) On the structure and biology of *Archotermopsis*, together with descriptions of new species of intestinal protozoa, and general observations on the Isoptera. Philos Trans R Soc Lond B 209:75–180

Imms AD (1957) A general textbook of entomology. Methuen, London

Inoue T, Kitade O, Yoshimura T, Yamaoka I (2000) Symbiotic associations with protists. In: Abe T, Bignell DE, Higashi M (eds) Termites: evolution, sociality, symbiosis, ecology. Kluwer Academic Publishers, Dordrecht, pp 275–288

Inward D, Beccaloni G, Eggleton P (2007a) Death of an order: a comprehensive molecular phylogenetic study confirms that termites are eusocial cockroaches. Biol Lett 3:331–335

Inward DJ, Vogler AP, Eggleton PE (2007b) A comprehensive phylogenetic analysis of termites (Isoptera) illuminates key aspects of their evolutionary biology. Mol Phylogenet Evol 44: 953–967

Jucci C (1932) Sulla presenza di batteriociti nel tessuto adiposo dei Termitidi. Boll Zool; Atti XI Congr Intern Zool Padova 1930 Arch Zool Ital 16:1422–1429

Jucci C (1952) Symbiosis and phylogenesis in the Isoptera. Nature 169:837

Kambhampati S (1995) A phylogeny of cockroaches and related insects based on DNA sequence of mitochondrial ribosomal RNA genes. Proc Natl Acad Sci U S A 92:2017–2020

Kambhampati S, Eggleton P (2000) Taxonomy and phylogeny of termites. In: Abe T, Bignell DE, Higashi M (eds) Termites: evolution, sociality, symbiosis, ecology. Kluwer Academic Publishers, Dordrecht, pp 1–23

Kitade O (2004) Comparison of symbiotic flagellate faunae between termites and a wood-feeding cockroach of the genus *Cryptocercus*. Microbes Environ 19:215–220

Kjer KM (2004) Aligned 18S and insect phylogeny. Syst Biol 53:506–514

Klass K-D (1995) Die Phylogeny der Dictyoptera PhD thesis, Ludwig Maximilians Universität

Klass K-D (2001) Morphological evidence on blattarian phylogeny: "phylogenetic histories and stories" (Insecta, Dictyoptera). Dtsch Entomol Z 48:223–265

Klass K-D, Meier R (2006) A phylogenetic analysis of Dictyoptera (Insecta) based on morphological characters. Entomol Abh 63:3–50

Klass K-D, Nalepa CA, Lo N (2008) How useful are the wood-feeding cockroaches *Cryptocercus* and *Perisphaeria boleiriana* as models of termite evolution (Insecta: Dictyoptera). Mol Phylogenet Evol 46:809–817

Koch A (1938) Symbiosestudien, 3: Die intrazellulare symbiose von *Mastotermes darwiniensis* Froggatt. Z Morph Okol Tiere 34:384–609

Kristensen NP (1995) Forty years' insect phylogenetic systematics. Zool Beitr 36:83–124

Laurentiaux D (1951) Le problème des blattes paléozoiques a ovipositeur externe. Ann Paleontol 37:187–194

Legendre F, Whiting MF, Bordereau C et al (2008) The phylogeny of termites (Dictyoptera: Isoptera) based on mitochondrial and nuclear markers: implications for the evolution of the worker and pseudergate castes, and foraging behaviors. Mol Phylogenet Evol 48:615–627

Lo N, Bandi C, Watanabe H, Beninati T (2003) Evidence for co-cladogenesis between diverse dictyopteran lineages and their intracellular endosymbionts. Mol Biol Evol 20:907–913

Lo N, Beninati T, Stone F et al (2007a) Cockroaches that lack *Blattabacterium* endosymbionts: the phylogenetically divergent genus *Nocticola*. Biol Lett 3:327–330

Lo N, Engel MS, Cameron S et al (2007b) Save Isoptera: a comment on Inward et al. Biol Lett 3:562–563

Lo N, Kitade O, Miura T et al (2004) Molecular phylogeny of the Rhinotermitidae. Insectes Soc 51:365–371

Lo N, Tokuda G, Watanabe H et al (2000) Evidence from multiple gene sequences indicates that termites evolved from wood-feeding cockroaches. Curr Biol 10:801–804

Marks EP, Lawson FA (1962) A comparative study of the Dictyoptera ovipositor. J Morphol 111:139–171

Maynard Smith J, Szathmary E (1997) The major transitions in evolution. Oxford University Press, Oxford

McKittrick FA (1964) Evolutionary studies of cockroaches. Mem Cornell Univ Agr Exp Stn 389:1–197

McKittrick FA (1965) A contribution to the understanding of cockroach – termite affinities. Ann Entomol Soc Am 58:18–22

Nalepa CA (1984) Colony composition, protozoan transfer and some life-history characteristics of the woodroach *Cryptocercus punctulatus* Scudder (Dictyoptera, Cryptocercidae). Behav Ecol Sociobiol 14:273–279

Nalepa CA (1988a) Reproduction in the woodroach *Cryptocercus punctulatus* Scudder (Dictyoptera, Cryptocercidae) – mating, oviposition, and hatch. Ann Entomol Soc Am 81:637–641

Nalepa CA (1988b) Cost of parental care in the woodroach *Cryptocercus punctulatus* Scudder (Dictyoptera, Cryptocercidae). Behav Ecol Sociobiol 23:135–140

Nalepa CA (1991) Ancestral transfer of symbionts between cockroaches and termites: an unlikely scenario. Proc R Soc Lond B 246:185–189

Nalepa CA, Bandi C (2000) Characterising the ancestors: paedomorphosis and termite evolution. In: Abe T, Bignell DE, Higashi M (eds) Termites: evolution, sociality, symbiosis, ecology. Kluwer Academic Publishers, Dordrecht, pp 53–76

Nalepa CA, Lenz M (2000) The ootheca of *Mastotermes darwiniensis* Froggatt (Isoptera: Mastotermitidae): homology with cockroach oothecae. Proc Biol Sci 267: 1809–1813

Nobre T, Eggleton PE, Aanen DK (2010) Vertical transmission as the key to the colonization of Madagascar by fungus-growing termites? Proc R Soc Lond B 277:359–365

Noda S, Kitade O, Inoue T et al (2007) Cospeciation in the triplex symbiosis of termite gut protists (*Pseudotrichonympha* spp), their hosts, and their bacterial endosymbionts. Mol Ecol 16: 1257–1266

Noirot C (1995) The gut of termites (Isoptera) – comparative anatomy, systematics, phylogeny. 1 Lower termites. Ann Soc Entomol Fr 31:197–226

Noirot C (2001) The gut of termites (Isoptera) comparative anatomy, systematics, phylogeny II – Higher termites (Termitidae). Ann Soc Entomol Fr 37:431–471

Ohkuma M, Noda S, Hongoh Y et al (2009) Inheritance and diversification of symbiotic trichonymphid flagellates from a common ancestor of termites and the cockroach *Cryptocercus*. Proc R Soc Lond B 276:239–245

Ohkuma M, Yuzawa H, Amornsak W et al (2004) Molecular phylogeny of Asian termites (Isoptera) of the families Termitidae and Rhinotermitidae based on mitochondrial COII sequences. Mol Phylogenet Evol 31:701–710

Pellens R, D'Haese C, Belles X et al (2007) The evolutionary transition from subsocial to eusocial behaviour in Dictyoptera: phylogenetic evidence for modification of the "shift-in-dependent-care" hypothesis with a new subsocial cockroach. Mol Phylogenet Evol 43:616–626

Sabree ZL, Kambhampati S, Moran NA (2009) Nitrogen recycling and nutritional provisioning by *Blattabacterium*, the cockroach endosymbiont. Proc Natl Acad Sci U S A 106: 19521–19526

Sacchi L, Nalepa CA, Bigliardi E et al (1998) Some aspects of intracellular symbiosis during embryo development of *Mastotermes darwiniensis* (Isoptera: Mastotermitidae). Parassitologia 40:309–316

Sands WA (1972) The soldierless termites of Africa (Isoptera: Termitidae). Bull Br Mus Nat Hist (Entomol) Suppl 18:1–244

Sands WA (1998) The identification of worker castes of termite genera from soils of Africa and the Middle East, CAB International, UK.

Scholtz OI, Macleod N, Eggleton P (2008) Termite soldier defence strategies: a reassessment of Prestwich's classification and an examination of the evolution of defence morphology using extended eigenshape analyses of head morphology. Zool J Linn Soc 153:631–650

Seelinger G, Seelinger U (1983) On the social organization, alarm and fighting in the primitive cockroach *Cryptocercus punctulatus* Scudder. Z Tierpsychol 61:315–333

Terry MD, Whiting MF (2005) Mantophasmatodea and phylogeny of the lower neopterous insects. Cladistics 21:240–257

Thompson GJ, Kitade O, Lo N, Crozier RH (2000a) On the origin of termite workers: weighing up the phylogenetic evidence. J Evol Biol 17:217–220

Thompson GJ, Miller LR, Lenz M, Crozier RH (2000b) Phylogenetic analysis and trait evolution in Australian lineages of drywood termites (Isoptera, Kalotermitidae). Mol Phylogenet Evol 17:419–429

Thorne BL (1990) A case for ancestral transfer of symbionts between cockroaches and termites. Proc R Soc Lond B 241:37–41

Thorne BL (1991) Ancestral transfer of symbionts between cockroaches and termites: an alternative hypothesis. Proc R Soc Lond B 246:191–195

Thorne BL, Carpenter JM (1992) Phylogeny of the Dictyoptera. Syst Entomol 17:253–268

Thorne BL, Grimaldi DA, Krishna K (2000) Early fossil history of the termites. In: Abe T, Bignell DE, Higashi M (eds) Termites: evolution, sociality, symbiosis, ecology. Kluwer Academic Publishers, Dordrecht, pp 77–94

Walker EM (1922) The terminal structures of orthopteroid insects: a phylogenetic study II The terminal abdominal structures of the male. Ann Entomol Soc Am 15:1–87

Ware JL, Litman J, Klass K-D, Spearman LA (2008) Relationships among the major lineages of Dictyoptera: the effect of outgroup selection on dictyopteran tree topology. Syst Entomol 33:429–450

Wheeler WC, Whiting MF, Wheeler QD, Carpenter JM (2001) The phylogeny of the extant hexapod orders. Cladistics 17:113–169

Chapter 3
Evolution and Function of Endogenous Termite Cellulases

Nathan Lo, Gaku Tokuda, and Hirofumi Watanabe

Abstract In the past decade, a century of debate over whether termites produce their own cellulases has been resolved by the application of molecular genetic techniques. Cellulase genes were present in ancient bilaterian animals, and have been passed down to termites and many other invertebrate lineages over several hundred million years. Termites contain multiple endoglucanase gene copies, all of which come from glycosyl hydrolase family 9, but the roles of the different gene copies are not yet clear. Enzyme assays and RNAi experiments indicate that endogenous cellulases play a key role in termite metabolism. The overall contribution of these enzymes in members of the Termitidae (which lack cellulolytic flagellates) appears to be greater than in members of other families. A major shift in the site of expression of endoglucanases and β-glucosidases from the salivary glands to the midgut has occurred in some members of the speciose family Termitidae. Investigations into the roles of different members of the termite colony in digesting cellulose have begun, and have revealed major variations in the level of expression, including differences between different sized workers. In fungus-growers and soil-feeders, endogenous cellulases appear to be of relatively minor importance, but have nonetheless been retained in the genome.

3.1 Introduction

One of the reasons plants dominate the earth is the recalcitrant nature of their cell walls, in which long cellulose polymers bond with each other to form highly ordered crystalline microfibrils. Together with hemicellulose and lignin, these microfibrils form an insoluble matrix, termed lignocellulose, that is highly resistant to mechanical and enzymatic attack. Microorganisms are chiefly responsible for the degradation of lignocellulose, but some animals have also evolved the ability to

N. Lo (✉)
Behaviour and Genetics of Social Insects Laboratory, School of Biological Sciences, The University of Sydney, Sydney, NSW 2006, Australia
e-mail: nathan@usyd.edu.au

D.E. Bignell et al. (eds.), *Biology of Termites: A Modern Synthesis*,
DOI 10.1007/978-90-481-3977-4_3,

derive their energy needs from this substrate. The most well-known of these are the termites, which succeed in digesting up to 99% of the cellulose present in the plant material they ingest (Esenther and Kirk 1974; Wood 1978). This relatively rare ability has attracted much interest from biologists. During the twentieth century, it was widely believed that termites, and indeed all cellulose-digesting animals, were completely reliant on microbial symbionts to break down cellulose. This view has changed in recent years with the discovery of cellulase genes in the genomes of termites and many other animals. Efficient cellulose digestion in many termite species appears to require both endogenous and intestinal microbial cellulases. Here we briefly review the history of the study of termite cellulases, then focus on endogenous cellulase research during the last decade. We discuss evolutionary aspects of cellulase genes, and the functional significance of their products. Reviews of the symbionts involved in cellulose digestion are provided in Chapter 15 by Ohkuma and Brune and Chapter 16 by Brune and Ohkuma, this volume.

3.2 Cellulose and Cellulases

Cellulose, a polymer of β-1,4-linked glucose, is the most abundant organic compound in the biosphere. The biology of cellulose and the enzymes that degrade it have been reviewed at length elsewhere (Bayer et al. 1998; Béguin and Aubert 1994); here we provide a brief overview. Most cellulose is produced by higher plants and algae, but also by certain amoebae, slime moulds, fungi, marine invertebrates, and bacteria (Richmond 1991). In plants, cellulose is the primary component of cell walls (30–50%), in which it occurs predominantly as long, rigid microfibrils. These microfibrils are composed of numerous linear cellulose molecules oriented in parallel, and extensively connected by hydrogen bonds and van der Waals forces. Throughout their length, they show varying degrees of crystallinity: some regions of the microfibrils are more disordered than others, and are termed "amorphous" or "semi-crystalline" cellulose. The greater part of the cellulose produced by plants is in a crystalline form. The insoluble and structurally complex nature of this substrate makes it particularly resistant to degradation (Tomme et al. 1995).

3.2.1 Cellulase Components

Enzymes that effect hydrolysis of cellulose into glucose are known as cellulases. The three main types are endo-β-1,4-glucanases (EGs; EC 3.2.1.4), cellobiohydrolases (CBHs; EC 3.2.1.91; a form of "exo-glucanase") and β-glucosidases (BGs; EC 3.2.1.21), which convert the short-chain sugars into glucose. EGs hydrolyse internal glycosidic bonds in a random fashion at the surface of the semi-crystalline/amorphous regions of the substrate, while CBHs hydrolyse cellobiose units from either the reducing or non-reducing end of a cellulose chain (Fig. 3.1). Structural features of EGs include either shallow grooves on their surfaces, or deeper

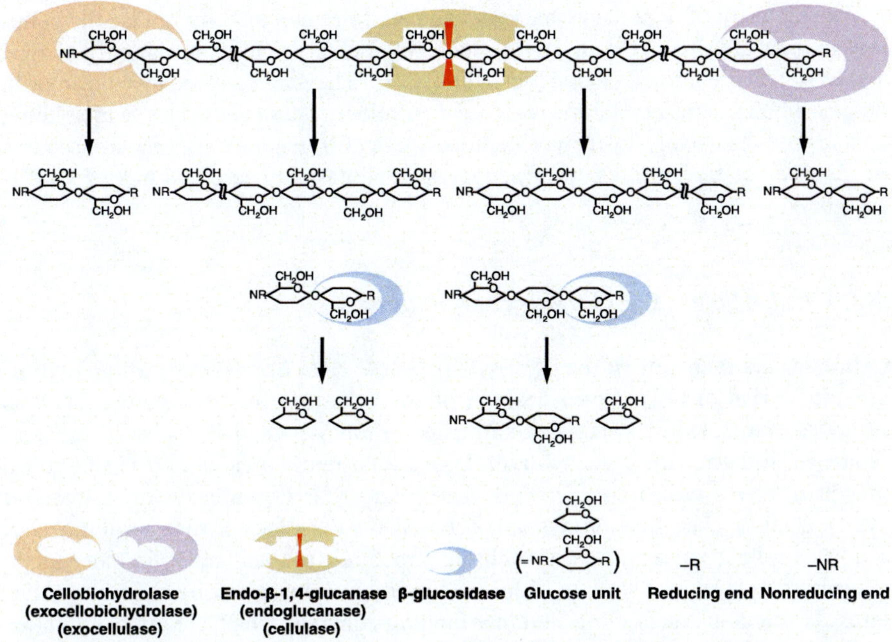

Fig. 3.1 Hydrolysis of cellulose by the three main cellulase types. Reproduced, with permission, from Watanabe and Tokuda (2010)

clefts so they can bind anywhere along the cellulose chain, while those of CBHs include a tunnel-like active site that permits binding only to the end of a cellulose chain (Davies and Henrissat 1995). EGs are generally more common than CBHs in nature. EGs have been found in bacteria, fungi, protists, plants and animals, while CBHs appear to be restricted to bacteria, fungi and protists. BGs are a common cellulase component.

3.2.2 Complete vs. Incomplete Cellulases

Some species of fungi and bacteria (e.g. white and brown rot fungi, *Clostridium*, rumen bacteria) are able to exhaustively digest crystalline cellulose in pure culture, and are said to have "complete" or "true" cellulases. Complete cellulases typically consist of both an EG and a CBH, along with a BG. EGs and CBHs work synergistically: the former producing "nicks" in the semi-crystalline regions of the microfibrils, which provide sites of attack for the latter. Once the CBHs begin hydrolysing the polymer and reducing its length, additional semi-crystalline sites become available to the EGs, and the cycle can continue to completion (Tomme et al. 1995). Some enzymes are known to have both EG and CBH activity, which give them a special niche in the synergistic process described above (Sakon et al. 1997).

The majority of organisms that produce cellulases can only hydrolyse the cellulose in their diets to a certain extent. Such cellulases typically consist only of EGs and BGs, and are termed "incomplete cellulases". They are believed to act primarily on amorphous, semi-crystalline regions of cellulose. These cellulases, whilst being unable to exhaustively hydrolyse cellulose, can still generate significant amounts of glucose for their producers. Endogenous cellulases of termites belong to this category.

3.2.3 Cellulases Are Glycolsyl Hydrolases

Cellulases are members of the glycosyl-hydrolase class of enzymes, which hydrolyse the myriad of polysaccharides and oligosaccharides found in nature, such as cellulose, starch, chitin, xylan, laminarin, and cellobiose. On the basis of amino-acid sequence similarity, the glycosyl-hydrolases are currently divided into 115 families, of which over a dozen contain EGs, CBHs and BGs (www.cazy.org) (Henrissat 1991). Such a classification system is based on the concept that similarities in sequence reflect conservation of both the structural fold and catalytic mechanism (Henrissat and Davies 1997). In addition to catalytic domains, many microbial cellulases also contain one or more cellulose binding domains, which are typically linked to the catalytic domain by peptides rich in proline, threonine and serine. Cellulose binding domains, which evolved independently from catalytic domains, maintain a high concentration of enzyme near the generally insoluble substrate.

For efficient degradation of cellulose, hemicellulose polymers present on the surface of cellulose microfibrils need to be removed. Hemicelluloses are a diverse group of carbohydrate substrates which are hydrolysed by an equally diverse group of glycosyl hydrolases, including xylanase, mannanase, and arabinosidase. Although some hemicellulases are produced by symbionts present in the hindguts of termites, there is only limited evidence for the presence of endogenous hemicellulases in termites.

3.3 A Brief History of Termite Cellulase Research

The study of cellulose digestion in termites dates back to the early twentieth century, when researchers such as Buscalioni and Comes (1910) and Imms (1920) identified the presence of wood particles in flagellate protists present in termite hindguts. They hypothesized that the protists were mutualists (i.e. providing a benefit to the host). Interestingly, the first researchers to view the protists in the late nineteenth century, such as Leidy (1881) and Grassi and Sandias (1893), had considered these protists to be parasites, or commensals. In a famous defaunation experiment Cleveland (1923) eventually showed that termites (and subsequently wood-roaches; Cleveland et al. 1934) were unable to survive on wood without the aid of the protists, and established that the relationship was one of obligate mutualism. Parallel work on ruminants established that rumen microorganisms were vital in the cellulose digestion process (Hungate 1947). In the case of termites, the debate over the nature of the symbiosis with protists resulted in a strong focus on the ability of flagellates to digest cellulose. The potential contribution by endogenously produced termite enzymes received little attention. As a result, the belief that termites, and indeed animals in general, were completely dependent on symbionts to effect cellulose hydrolysis became widespread.

Evidence for the endogenous production of cellulases by a number of animals, including members of the phyla Mollusca, Annelida, Brachiopoda, Echinodermata and Arthropoda, has been found in various studies throughout the twentieth century and even before (e.g. Biedermann and Moritz 1898; Ray and Julian 1952; Yokoe and Yasumasu 1964; Scrivener et al. 1989). In the case of termites, removal of protists via tetracycline or heat treatment did not drastically reduce the amount of cellulase activity present in some parts of the gut (Yokoe 1964). In other cases, activity was found in regions of the gut (e.g. the midgut) that had few or no detectable symbiotic microorganisms present (Boyle and Mitchell 1978; Schulz et al. 1986; Scrivener et al. 1989). The authors of these studies concluded that the cellulase was endogenous, but the results were not fully accepted by others, since the possibility of contamination from a microbial source, albeit unlikely, had not been completely ruled out (e.g. Zachary and Colwell 1979). In a classic experiment, individual thysanuran (*Ctenolepisma lineata*) eggs were raised under aseptic conditions, and were able to survive on ^{14}C-cellulose (Lasker and Giese 1956), strongly indicating the production of endogenous cellulases. However, it was pointed out that impurities are known to exist in ^{14}C-cellulose as a consequence of its preparation, which may have supported the growth of the insects without the need for a cellulase (Martin 1991). Thus, the dogma that all cellulose digesting animals relied exclusively on gut symbionts persisted and found its way into the majority of biology and biochemistry textbooks of this period (Slaytor 1992). This view was perhaps reinforced by the "acquired enzyme" hypothesis, a proposal supported by enzymatic assays, that cellulose-digestion in organisms such as fungus-growing termites and siricid wasps was dependent on cellulases from ingested microbes (Martin 1982; Martin and Martin 1978). In the termites the assays showed that endogenous enzymes were produced but the argument was made that they were only effective by acting synergistically with enzymes acquired from fungal symbionts (Martin and Martin 1978).

3.4 Discovery of Endogenous Cellulase Genes and Their Evolutionary Origins

In the mid 1990s, the first purifications of endogenous cellulase enzymes from the wood-feeding cockroach *Panesthia cribrata,* and the termites *Reticulitermes speratus* and *Nasutitermes takasagoensis*, were performed using column chromatography (Scrivener and Slaytor 1994; Tokuda et al. 1997; Watanabe et al. 1997). Biochemical assays on the purified proteins showed they were EGs, with some activity against crystalline cellulose (Scrivener and Slaytor 1994; Watanabe et al. 1997). In the case of *R. speratus*, pure protein was used to raise antiserum, which was then used to screen a recombinant phage complementary DNA (cDNA) library. A partial sequence was then identified, and the complete cDNA was obtained by adaptor-mediated PCR of cDNA derived from salivary gland tissue (Watanabe et al. 1998). The endogenous nature of the gene encoding this cDNA, named *RsEG,* was confirmed by Southern Blotting and PCR of a genomic fragment (using DNA extracted from *R. speratus* heads). The genomic fragment was shown to contain a typical eukaryotic intron.

The predicted RsEG protein is 448 amino acids long, and shares 52, 48, and 42%, respectively, with the catalytic domains of selected bacterial, slime mould, and plant members of the glysosyl hydrolase family 9 (from *Cellulomonas fimi*, *Dictyostelium discodeum*, and *Phaseolus vulgaris*), two of which are shown in Fig. 3.2. The amino acid residues thought to be involved in catalysis and substrate-binding in all GH9 members are conserved in RsEG. A conspicuous feature of RsEG is the absence of a cellulose-binding domain. The presence of a single (catalytic) domain in RsEG makes it similar to EGs from plants.

RsEG was one of the first two animal cellulase genes to be cloned. Endogenous EG genes from plant parasitic nematodes were isolated at a similar time (Smant et al. 1998). A key question arising from the discovery of these EGs concerned their evolutionary origin. The nematode EGs were from a completely different glycosyl hydrolase family (GH5) to RsEG. This ruled out a common evolutionary origin for

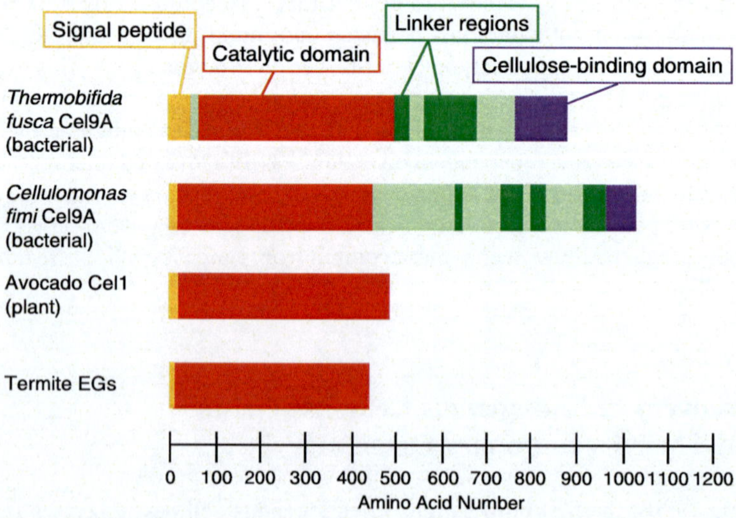

Fig. 3.2 Comparison of termite EG with other microbial and plant EGs. Microbial EGs are usually composed of multiple domains (i.e. catalytic, linker and cellulose-binding domains), while termite cellulases consist of a single catalytic domain

termite and nematode EGs. It was hypothesized that the nematode EG had arisen from a horizontal gene transfer event from a bacterium, though no firm evidence for this is available. Was horizontal gene transfer, perhaps from an ancient hindgut microbe, responsible for the presence of EGs in termites? Soon after the cloning of RsEG, the entire ~13 kb genomic sequence (including the promoter site) of a GH9 EG from the higher termite *Na. takasagoensis* was determined, and found to contain 9 introns (Tokuda et al. 1999) (Fig. 3.3). The discovery of endogenous GH9 genes from an abalone (*Haliotus discus*; Suzuki et al. 2003), and a sea-squirt (*Ciona intestinalis*; Dehal et al. 2002) followed. A comparison of intron positions in these EG genes revealed that a number of them were identical among the three genes (Lo et al. 2003). This was highly suggestive that a gene with introns in the very same positions was present in the ancestor of all bilaterian animals. This hypothesis is now strongly supported: GH9 genes have been discovered in diverse animals, from earthworms, to echinoderms, to honeybees (Davison and Blaxter 2005). Contrary to the traditional belief that endogenous animal cellulases are non-existent or rare, GH9 EGs are quite common animal enzymes, having been inherited over hundreds of millions of years from a common ancestor (Lo et al. 2003). GH9 genes have, however, been lost in many other lineages over evolutionary time, as evidenced by the absence of cellulase genes in the genomes of numerous animals, including all vertebrates, *Drosophila melanogaster*, the silkworm *Bombyx mori*, and the nematode *Caenorhabditis elegans*, to name but a few. Endogenous GH5 and GH45 genes have also been discovered in various animals, including beetles, molluscs and nematodes (Davison and Blaxter 2005). Whether these genes were present in an ancient metazoan ancestor, or whether they are more recent horizontal acquisitions, is not yet clear.

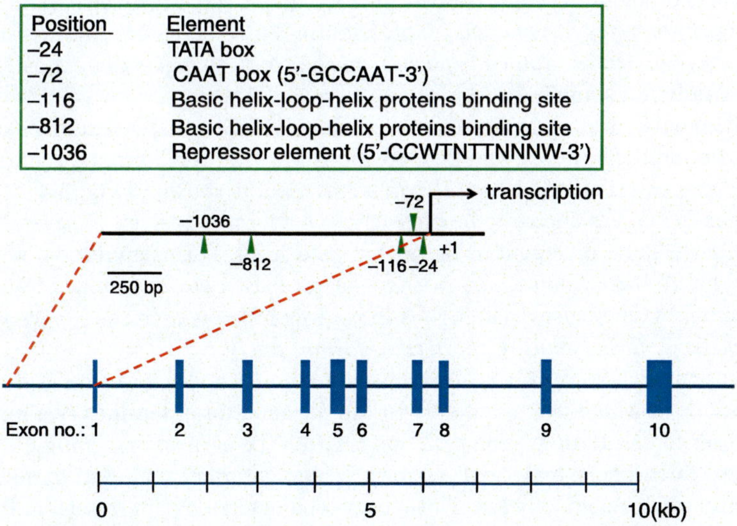

Fig. 3.3 Genomic organization of *NtEG*. The gene is comprised of 10 exons interrupted by 9 introns. 5′-upstream sequence contains some putative typical eukaryotic *cis*-elements

3.5 Endogenous Endoglucanase and β-Glucosidase Copy Number and Expression in Termites

3.5.1 Endoglucanases

Since the cloning of *RsEG* and *NtEG*, GH9 EG cDNA sequences have been cloned from a wider range of termites. The sequencing of a number of full EG cDNAs has revealed the presence of a single catalytic domain, consistent with the structure of other animal GH9 sequences (Nakashima et al. 2002). Some GH9 EGs have been sequenced only partially, using PCR primers designed in highly conserved regions present in termite, cockroach and crustacean EGs (such primers are expected to amplify any GH9 genes that are transcribed). Cloning of the amplified products, followed by sequencing of multiple clones, has revealed the diversity of EGs within each termite species. No evidence for alternative splicing of EG genes has yet been found. Based on protein purification and EST studies, there is no evidence that cellulases from GH families present in other insects (GH5, GH45) are produced by termites.

To examine the diversity and expression site of EGs, Tokuda et al. (2004) performed PCR on representatives of five of the seven termite families. Amplification of cDNAs derived from four different tissues (salivary glands, foregut, midgut, hindgut) of the digestive system of mature workers was attempted. The salivary glands were the only site of endogenous cellulase expression in 4 of the 5 termite families examined. The salivary glands are also the site of EG expression in cockroaches (Lo et al. 2000), although other gut regions have not been examined extensively. The pattern is, however, different in the phylogenetically apical Termitidae. Although EG expression occurs only in the salivary glands of the fungus grower *Odontotermes formosanus*, expression in the wood feeder *Na. takasagoensis* and the soil feeder *Sinocapritermes mushae* has changed dramatically, now occurring only in the midgut.

A phylogenetic tree showing the relationships of EGs in representatives of each of the main termite families, and their relationship with EGs from the cockroaches *Cryptocercus clevelandi* and *Periplaneta americana*, is shown in Fig. 3.4. Multiple paralogous EG copies appear to be present in most termite species. In several cases, these EGs are quite divergent at the amino acid level. For example, the phylogenetically basal *Mastotermes darwiniensis* has at least two EGs, with 86% amino acid identity. In other cases, paralogous EGs show little divergence (*e.g. Neotermes koshunensis, O. formosanus, Coptotermes formosanus*).

The common ancestor of *Cryptocercus* and termites appears to have had at least two EGs, one of which was apparently lost in the ancestor of termites. We infer this by the absence of a termite gene that clusters with the earliest branching *Cr. clevelandi* gene. Such a gene may, however, be revealed in further studies of termite EGs, particularly phylogenetically basal species. A characteristic pattern seen in Fig. 3.4 is that EG copies from one species tend to cluster with one another, rather than with EGs from other species. One explanation is that the ancestors of each of the families represented had a single EG, which was later duplicated in more recently evolved lineages. The exception to this pattern of EG copies from a species clustering together is found in the phylogenetically apical Termitidae, where EGs from *Na. takasagoensis* and *Sinocapritermes mushae* respectively cluster together rather than with their paralogs. Relationships among the EG clusters from the ten species shown are largely congruent with the currently accepted phylogenetic relationships of termites (see Chapter 2 by Lo and Eggleton, this volume), although EGs from relatively few termites have been examined thus far.

An alternative explanation for the pattern in Fig. 3.4 is that ancestors of each of the families represented had multiple EGs, which then underwent gene conversion, resulting in them becoming more similar to each other than to those from other species. Gene conversion occurs when two paralogs recombine, and the exchange of genetic material is non-reciprocal (i.e. only one paralog donates DNA, and itself remains the same). Preliminary studies of gene conversion in cockroach and termite

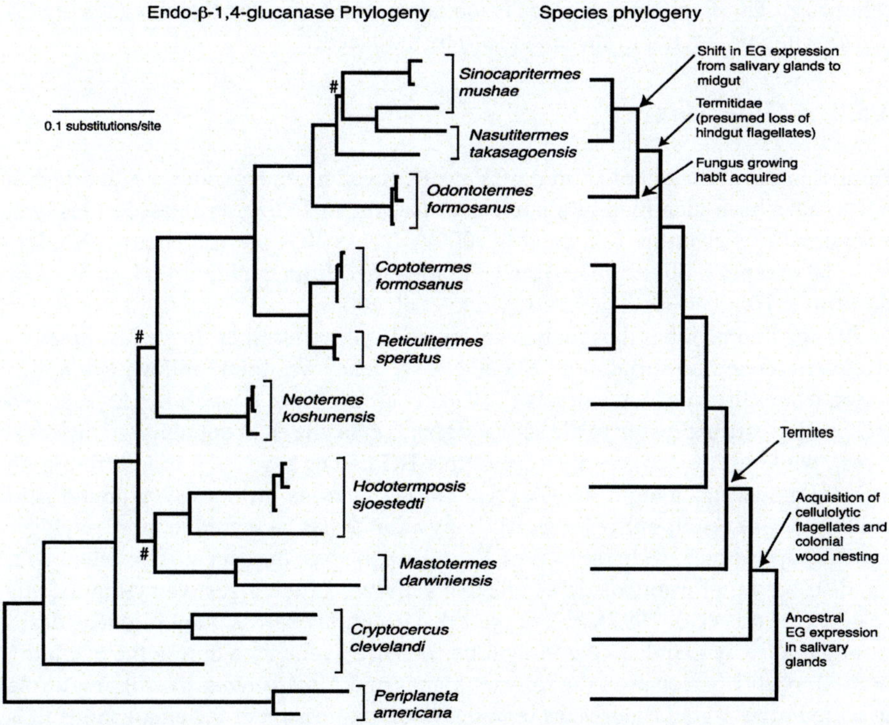

Fig. 3.4 Phylogeny of duplicate EG genes from eight termites, and the cockroaches *Cryptocercus clevelandi* and *Periplaneta americana*, compared with a phylogeny for these species. Multiple EG cDNAs were simultaneously amplified from an individual of each species (with the exception of *Coptotermes formosanus*) using degenerate primers (see text), then cloned and sequenced (~1,000 bp in total). A maximum likelihood tree of the EG cDNA sequences is shown on the *left*. Bootstrap support values were universally high across the topology (>75%) with the exception of nodes marked with a *hash* (#). For further details, see Tokuda et al. (2004)

EGs found no evidence for this phenomenon, with the possible exception of EGs from *M. darwiniensis* (Lo et al. 2000). The fact that EGs tend to mirror phylogenetic relationships suggests that gene conversion is not a key process in their evolution. If it were, one might expect less congruence between EG and species trees.

The maintenance and fixation of duplicated gene copies are believed to result in three primary outcomes (Hahn 2009). Firstly, "neofunctionalization", where one of the duplicates takes on a new function; secondly, "subfunctionalization", where multiple functions of the original gene are divided between the duplicates; thirdly, "conservation", where both duplicates retain the same function. It is not yet understood which of these three outcomes is relevant to paralogous termite EGs. In *R. speratus*, the biochemical properties of two purified EGs with very similar sequences (98% amino acid identity) have been investigated. These enzymes, presumed to represent *RsEG* and its paralog *RsEG2*, have very similar levels of activity against carboxymethylcellulose, but they differ in their activity against microcrystalline cellulose, and in their ability to hydrolyse the short sugars cellotetraose and cellotriose. The significance of this is unknown. It may be that the two gene copies permit a higher amount of the enzyme to be produced.

3.5.2 β-Glucosidase

Parallel to the discovery of animal EGs, genes encoding endogenous β-glucosidases (BGs) have been identified in a number of insects, including termites, beetles, and silkworms (reviewed in Tokuda et al. 2009). These BGs are affiliated with GH1, with the exception of one gene from GH3 identified from a midgut EST analysis of the termite *Hodotermopsis sjostedti* (Yuki et al. 2008).

BG copy number has thus far been examined in three termites. In *Ne. koshunensis* (Kalotermitidae), no variation in BG cDNA sequence was found following amplification from salivary gland extracts (Tokuda et al. 2002). In *R. speratus*, at least two BG paralogs are expressed in the salivary glands (Tokuda et al. unpublished observations), while in *Na. takasagoensis*, multiple BG copies have been found from both salivary gland and midgut cDNA preparations, with as little as 89% amino acid identity. These results suggest that BG copy number has increased in more phylogenetically apical species. In the case of *Na. takasagoensis*, the data are consistent with the distribution of β-glucosidases enzyme activities in the digestive system of this termite (Tokuda et al. 1997), and suggest that a shift in expression of β-glucosidases from the salivary glands to the midgut has occurred sometime during the evolution of the Termitidae. However, the salivary glands of *Na. takasagoensis* still play a role in secretion of β-glucosidases, as opposed to the expression of the endogenous EGs in this species, which are secreted exclusively in the midgut.

3.6 Functional Significance of Endogenous Cellulases

The genetic diversity of endogenous termite cellulase systems, which consist of a few EG paralogs from GH9 (Fig. 3.4), and one or a few BGs from GH1, is far less than that found in some cellulolytic microbes, which usually contain multiple

EGs, CBHs and BGs. For example, the filamentous fungus *Trichoderma reesei* produces at least eight cellulase components from seven different GH families, while the anaerobic bacterium *Clostridium cellulovorans* produces eight components from three different GH families (Tamaru et al. 2000; Tomme et al. 1995). Although endogenous termite cellulases are unable to exhaustively degrade cellulose, they do appear to play a significant role in providing energy for those termites that feed on wood. A recent study showed that inhibition of mRNA expression of an endogenous EG in *R. flavipes* by administration of dsRNA resulted in substantial mortality of the termites in a dose-dependent manner (Zhou et al. 2008), supporting an indispensable role of endogenous cellulases in termite physiology.

The amount of glucose produced by gut extracts from commercial microcrystalline cellulose under laboratory conditions has been examined in various wood-feeding termites (Tokuda et al. 2005) (Glucose production measures the combined activities of EGs and BGs). Among four termite species with flagellates, the percentages of total gut activity from endogenous cellulases were 12, 29, 60 and 39% (the remainder of the activity was from gut flagellates, present in the hindgut) (Fig. 3.5). In the case of *Na. takasagoensis* and *Nasutitermes walkeri*, which lack flagellates, but do contain bacteria in its hindgut, endogenous cellulases account for up to 60% of all activity against microcrystalline cellulose (Tokuda and Watanabe 2007). Notably, the amount of glucose produced by endogenous enzymes is likely to account for all, or most, of the CO_2 respired by each of these six wood-feeding termites (Tokuda et al. 2005).

It is generally thought that EGs can only act in the amorphous regions of cellulose. Presumably, some regions of commercially produced microcrystalline cellulose are less crystalline than others, and it is these regions that are attacked

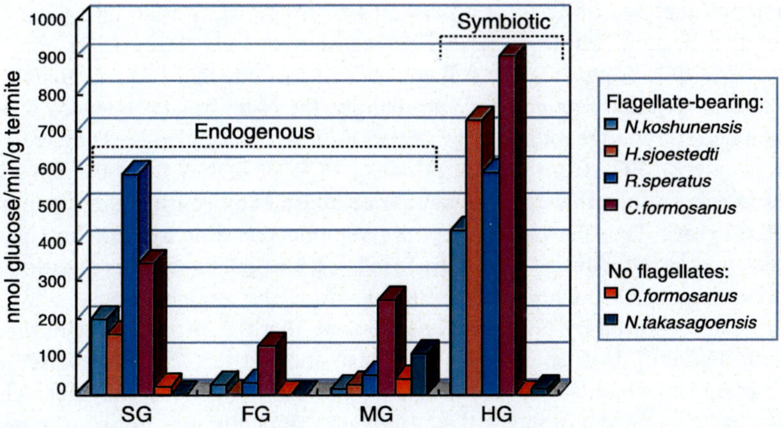

Fig. 3.5 Production of glucose from microcrystalline cellulose using crude extracts of the salivary glands and various gut regions of lower and higher termites. Salivary cellulase activities are endogenous, whereas the hindgut activities are derived from symbiotic microorganisms

by termite cellulases in vitro to produce glucose under laboratory conditions. Under natural conditions, the grinding action of termite jaws and the gizzard (i.e. gastric mills) is likely to increase the number of amorphous regions of cellulose, which can then be attacked by EGs. For example, in the case of *Co. formosanus*, the ingested wood fragments are broken down into small particles less than 10 μm in size in the foregut (Fujita et al. 2010). As an aside, the equivalent process in commercial cellulosic biomass conversion systems accounts for up 40% of the processing cost (Fujita et al. 2010).

The amount of cellulase activity against microcrystalline cellulose among the flagellate-harbouring species *H. sjostedti*, *Ne. koshunensis*, and *R. speratus* drops dramatically between the salivary glands and the foregut (Tokuda et al. 2005). Since the foregut consists of nonsecretory tissue (and is thus unable to produce proteinases), this apparent drop requires an explanation. One possibility is that salivary cellulases are expelled in the saliva along with phagostimulant signals (Reinhard and Kaib 2001a, 2001b), antibiotics (Lamberty et al. 2001), and proteinaceous pheromones (Matsuura et al. 2009). Such cellulases may soften the surface of wood to help termites gnaw. Relatively low cellulase activity in the midguts of these species may be explained by secreted proteinases which hydrolyse salivary cellulases (Fujita et al. 2001). An exception is the midgut of *Co. formosanus*, which has almost half of the total amount of endogenous activity in the gut (Fig. 3.5). Thus, the cellulase in this species retains activity in the intestine after being secreted by the salivary glands and contributes significantly to cellulose digestion in the midgut (Fujita et al. 2010).

The amount of endogenous cellulase activity reaching the hindgut can be difficult to determine due to the presence of microbial activity in this region. Defaunation of the symbiotic protists from *Ne. koshunensis* reduced the hindgut EG activity to approximately 20% of the untreated termites (Sugio et al. 2006), suggesting that some endogenous EGs may reach this region. On the other hand, zymogram and immunoblot analyses of *C. formosanus* showed no evidence that the salivary EGs reached the hindgut (Fujita et al. 2010; Nakashima et al. 2002). Li et al. (2003) isolated two EGs from symbiotic flagellates in the gut of *M. darwiniensis*, and performed N-terminal sequencing. Surprisingly, the N-terminal sequences matched the sequences of endogenous EGs produced in the salivary glands. They were also able to sequence EGs from GH45 following PCR of cDNA from the flagellates, though they were not able to demonstrate cellulase activity from these symbiont-derived enzymes. They concluded that the symbionts rely directly upon host-derived enzymes, and are becoming "metabolic fossils". This conclusion was questioned by Watanabe et al. (2006), who purified enzymes from the hindgut protists of *M. darwiniensis*, and showed by N-terminal sequencing that the enzymes match those of symbiont-derived GH45 sequences. They also showed that the overall EG activity of hindgut symbionts was about half of that of endogenous enzymes. Thus it appears that hindgut symbionts still contribute significantly to cellulose digestion in this species.

In the wood-feeder *Na. takasagoensis*, endogenous cellulase activity matches the pattern of EG and BG gene expression, being present in the midgut rather than the salivary glands. The fungus-grower *Odontotermes formosanus* does not appear to have enough cellulase activity to account for the amount of CO_2 it respires. Its endogenous cellulase activity is less than 20% of that found in wood-feeding termites (Fig. 3.5), and does not appear to make a major contribution to the carbon budget of this termite. Indeed, stable isotope experiments show that the majority of carbon in this termite's tissues comes from ingestion of its fungal symbiont (see Chapter 8 by Nobre et al. and Chapter 16 by Brune and Ohkuma, this volume). The soil-feeding species *Sinocapritermes mushae* (not shown in Fig. 3.5, but see Tokuda et al. 2004) also has very low endogenous cellulase activity. This is perhaps to be expected, given the relatively low amount of cellulose present in humus.

In addition to its function in converting cellobiose to glucose, a novel function for BG has recently been demonstrated. Matsuura et al. (2009) studied the mechanism of recognition by termites of fungal sclerotia, known as "termite balls", present in nests of *R. speratus*. They showed that, in addition to lysozyme, termites use β-glucosidase as an egg-recognition pheromone. The fungus mimics termite eggs by producing BG, which causes termite balls to be tended in the same manner as termite eggs. The sequence of BGs from the fungus has not yet been determined, but it is presumably from the same GH family (i.e. GH1) of the β-glucosidases of *R. speratus*. Egg-tending behaviour was also observed when heat-inactivated β-glucosidase was applied, suggesting that the phenomenon is not directly related to its enzymatic activity nor hydrolysates. A conserved GH1 amino acid sequence may bind to a receptor on a tactile organ on the termites to induce the response.

3.7 Caste Specific Production of Cellulase Genes

Studies on termite cellulases in the twentieth century tended to focus on activity in the worker caste. In recent years some reports on relative expression among different castes have appeared. Using quantitative PCR, Scharf et al. (2005) examined relative amounts of transcription of endogenous EGs in workers, soldiers, pre-soldiers, nymphs, alates, and supplementary reproductives of *Reticulitermes flavipes*. Workers had only slightly more EG transcript abundance than nymphs, about 10 times that of alates, and over 300 times the transcript abundance in presoldiers, soldiers and supplementary reproductives. Although not confirmed, transcript abundance was presumably correlated with the amount of EG activity.

Stomodeal trophallaxis of food is known to occur from workers to soldiers and nymphs, which have reduced capacity to feed by themselves (Qiu-Ying et al. 2008). Whether relatively high transcription of EG in nymphs compared with workers is associated with further processing of trophallactic food, or is used to digest wood they consume themselves, is unclear. The very low amount of EG transcription of soldiers suggests that they do not require EGs to further digest the food provided to them by workers. The same appears to be the case for supplementary reproductives. Alates are known to rely primarily on fat stores during the initial stages of colony initiation, but may supplement their energy needs by consuming cellulose in developing nests.

Fujita et al. (2008) investigated caste-specific cellulase expression in workers and soldiers of *H. sjostedti* and *Na. takasagoensis*, by enzyme assays of EG and BG, and quantitative PCR of EG cDNA. In the case of *H. sjostedti,* pseudergates (which are the main helping/working caste in this species) had much higher EG activity than soldiers in the saliva, but activity levels were relatively low in both castes in other gut regions. This pattern was matched by the pattern of EG transcript abundance in each of the castes. BG activity was also much higher in the saliva of pseudergates than soldiers, however activity in the hindguts (presumably derived from protists) was higher in soldiers than in pseudergates. Similar activity patterns were also observed in another lower termite, *Ne. koshunensis*, except that the hindgut BG activity was higher in the workers than in the soldiers (Sugio et al. 2006).

In *Na. takasagoensis*, activities in three different types of workers (minor, major, and medium) were compared with those in soldiers. EG activity, which was restricted to the midgut in all cases, was highest in minor workers, followed by medium workers, and then major workers. Activity in soldiers was similar to that in major workers. Since major workers are the main foraging caste, it was proposed that food is transferred to minor workers for additional processing before transfer to other castes. EG transcript abundance was not strictly correlated with enzyme activity: relative levels in minor workers, medium workers, and soldiers were similar, and twice that of major workers. The reason for this is not clear, but may be due to post-transcriptional processing of EGs. BG activity did not strictly follow the pattern of EG activity: although it was highest in minor workers, it was lowest in medium workers, and at an intermediate level in major workers and soldiers (in all cases present in the midgut). The authors suggested this may reflect the transfer of food between different worker forms. Small amounts of BG were also present in the salivary glands of minor and major workers, which may be associated with egg-marking, as for *R. speratus*.

3.8 Conclusions

The application of molecular genetic techniques to the study of termite cellulases in the past decade has resolved a century of debate over whether these insects produce their own cellulases. Phylogenetic comparisons of termite endoglucanase genes with those of other animals have shown that they are ancient animal genes, present in a large number of invertebrates. All termites thus far examined contain multiple endoglucanase gene copies, which, in some cases, are quite genetically divergent. The roles of the different cellulase gene products are not yet understood. Based on in vitro enzyme assays, as well as preliminary gene-knockdown experiments, endogenous cellulases appear to play an important role in providing glucose for termites, though they are not capable of exhaustive cellulose hydrolysis due to the absence of a cellobiohydrolase. The relative importance of termite vs symbiont cellulases, the latter of which do contain CBHs, needs further investigation. During the evolution of the speciose Termitidae family, both EG and BG expression has shifted from the salivary glands to the midgut. The functional signficance of this remains to be elucidated. In two successful but phylogenetically separated feeding-groups of termitids, the fungus-growers and the soil-feeders, cellulase activity levels are far lower than those found in wood-feeders, reflecting the role of the fungal symbionts in the former case, and the low levels of cellulose in the food in the latter (Chapter 8 by Nobre et al. and Chapter 14 by Bignell, this volume).

References

Bayer EA, Chanzy H, Lamed R, Shoham Y (1998) Cellulose, cellulases and cellulosomes. Curr Opin Struct Biol 8:548–557

Béguin P, Aubert J (1994) The biological degradation of cellulose. FEMS Microbiol Rev 13:25–58

Biedermann W, Moritz P (1898) Beiträge zur vergeleichenden Physiologie der Verdauung. II. Uver ein celluloselödendes Enzyme im Leversecret der Schnecke (*Helix pomatia*). Pflüdg Arch Gem 73:219–287

Boyle PJ, Mitchell R (1978) Absence of microorganisms in crustacean digestive tracts. Science 200:1157–1159

Buscalioni L, Comes S (1910) La digestione delle membrane vegatali per opera dei conternuti nell 'intestina dei termitidi e il problema, della simbiosi'. Atti Accad Gioenia Sci Nat Catania 3:1–16

Cleveland LR (1923) Symbiosis between termites and their intestinal protozoa. Proc Natl Acad Sci U S A 9:424–428

Cleveland LR, Hall SR, Saunders EP, Collier J (1934) The wood-feeding roach *Cryptocercus*, its protozoa, and the symbiosis between protozoa and roach. Mem Am Acad Sci 17:185–342

Davies G, Henrissat B (1995) Structures and mechanisms of glycosyl hydrolases. Structure 3: 853–859

Davison A, Blaxter M (2005) Ancient origin of glycosyl hydrolase family 9 cellulase genes. Mol Biol Evol 22:1273–1284

Dehal P, Satou Y, Campbell RK et al (2002) The draft genome of *Ciona intestinalis*: insights into chordate and vertebrate origins. Science 298:2157–2167

Esenther GR, Kirk TK (1974) Catabolism of aspen sapwood in *Reticulitermes flavipes*. Ann Entomol Soc Am 67:989–991

Fujita A, Hojo M, Aoyagi T et al (2010) Details of the digestive system in the midgut of *Coptotermes formosanus* Shiraki. J Wood Sci 56:222–226

Fujita A, Miura T, Matsumoto T (2008) Differences in cellulose digestive systems among castes in two termite lineages. Physiol Entomol 33:73–82

Fujita A, Shimizu I, Abe T (2001) Distribution of lysozyme and protease, and amino acid concentration in the guts of a wood-feeding termite, *Reticulitermes speratus* (Kolbe): possible digestion of symbiont bacteria transferred by trophallaxis. Physiol Entomol 26:116–123

Grassi B, Sandias A (1893) The constitution and development of the society of termites: observations on their habits; with appendices on the parasitic protozoa of Termitidae, and on the Embiidae. Q J Microsc Sci 39:245–322

Hahn MW (2009) Distinguishing among evolutionary models for the maintenance of gene duplicates. J Hered 100:605–617

Henrissat B (1991) A classification of glycosyl hydrolases based on amino acid sequence similarities. Biochem J 280:309–316

Henrissat B, Davies G (1997) Structural and sequence-based classification of glycoside hydrolases. Curr Opin Struct Biol 7:637–644

Hungate RE (1947) Studies on cellulose fermentation: III. The culture and isolation for cellulose-decomposing bacteria from the rumen of cattle. J Bacteriol 53:631–645

Imms AD (1920) On the structure and biology of *Archotermopsis*, together with descriptions of new species of intestinal protozoa, and general observations on the Isoptera. Phil Trans R Soc Lond B 209:75–180

Lamberty M, Zachary D, Lanot R et al (2001) Insect immunity. Constitutive expression of a cysteine-rich antifungal and a linear antibacterial peptide in a termite insect. J Biol Chem 276:4085–4092

Lasker R, Giese AC (1956) Cellulose digestion in the silverfish *Ctenolepisma lineata*. J Exp Biol 33:542–553

Leidy J (1881) The parasites of the termites. J Acad Nat Sci Philadelphia III 8:425–447

Li L, Frohlich J, Pfeiffer P, Konig H (2003) Termite gut symbiotic archaezoa are becoming living metabolic fossils. Eukaryot Cell 2:1091–1098

Lo N, Tokuda G, Watanabe H et al (2000) Evidence from multiple gene sequences indicates that termites evolved from wood-feeding cockroaches. Curr Biol 10:801–804

Lo N, Watanabe H, Sugimura M (2003) Evidence for the presence of a cellulase gene in the last common ancestor of bilaterian animals. Proc R Soc Lond B 270:S69–S72

Martin MM (1982) The role of ingested enzymes in the digestive processes of insects. In: Anderson JM, Rayner ADM, Walton DWH (eds) Invertebrate-microbial interactions. Cambridge University Press, Cambridge, pp 155–172

Martin MM (1991) The evolution of cellulose digestion in insects. Phil Trans R Soc Lond B 333:281–288

Martin MM, Martin JS (1978) Cellulose digestion in the midgut of the fungus-growing termite *Macrotermes natalensis*: the role of acquired digestive enzymes. Science 199: 1453–1455

Matsuura K, Yashiro T, Shimizu K et al (2009) Cuckoo fungus mimics termite eggs by producing the cellulose-digesting enzyme β-glucosidase. Curr Biol 19:30–36

Nakashima K, Watanabe H, Saitoh H et al (2002) Dual cellulose-digesting system of the wood-feeding termite, *Coptotermes formosanus* Shiraki. Insect Biochem Mol Biol 32: 777–784

Qiu-Ying H, Wei-Ping W, Rang-Yu M, Chao-Liang L (2008) Studies on feeding and trophallaxis in the subterranean termite *Odontotermes formosanus* using rubidium chloride. Entomol Exp Appl 129:210–215

Ray DL, Julian JR (1952) Occurrence of cellulase in *Limnoria*. Nature 169:32–33

Reinhard J, Kaib M (2001a) Food exploitation in termites: indication for a general feeding stimulating signal in labial gland secretion of Isoptera. J Chem Ecol 27:189–201

Reinhard J, Kaib M (2001b) Thin-layer chromatography assessing feeding stimulation by labial gland secretion compared to synthetic chemicals in the subterranean termite *Reticulitermes santonensis*. J Chem Ecol 27:175–187

Richmond PA (1991) Occurrence and functions of native cellulose. In: Haigler CH, Weimer PJ (eds) Biosynthesis and biodegradation of cellulose. Marcel Dekker Inc, New York, NY, pp 5–24

Sakon J, Irwin D, Wilson DB, Karplus PA (1997) Structure and mechanism of endo/exocellulase E4 from *Thermomonospora fusca*. Nat Struct Biol 4:810–818

Scharf ME, Wu-Scharf D, Zhou X et al (2005) Gene expression profiles among immature and adult reproductive castes of the termite *Reticulitermes flavipes*. Insect Mol Biol 14:31–44

Schulz MW, Slaytor M, Hogan ME, O'Brien RW (1986) Components of cellulase from the higher termite, *Nasutitermes walkeri*. Insect Biochem 16:929–932

Scrivener AM, Slaytor M (1994) Properties of the endogenous cellulase from *Panesthia cribrata* Saussure and purification of major endo-β-1,4-glucanase components. Insect Biochem Mol Biol 24:223–231

Scrivener AM, Slaytor M, Rose HA (1989) Symbiont-independent digestion of cellulose and starch in *Panesthia cribrata* Saussure, an Australian wood-eating cockroach. J Insect Physiol 35: 935–941

Slaytor M (1992) Cellulose digestion in termites and cockroaches: what role do symbionts play? Comp Biochem Physiol 103B:775–784

Smant G, Stokkermans J, Yan YT et al (1998) Endogenous cellulases in animals: isolation of beta-1,4-endoglucanase genes from two species of plant-parasitic cyst nematodes. Proc Natl Acad Sci U S A 95:4906–4911

Sugio K, Simojo K, Isozaki J et al (2006) Distribution of cellulase activities in the salivary glands and the guts of pseudoworkers and soldiers of the drywood-feeding termite *Neotermes koshunensis* (Shiraki) and the effect of defaunation. Jpn J Appl Entomol Zool 50:1–6

Suzuki K, Ojima T, Nishita K (2003) Purification and cDNA cloning of a cellulase from abalone *Haliotis discus hannai*. Eur J Biochem 270:771–778

Tamaru Y, Karita S, Ibrahim A et al (2000) A large gene cluster for the *Clostridium cellulovorans* cellulosome. J Bacteriol 182:5906–5910

Tokuda G, Lo N, Watanabe H et al (1999) Metazoan cellulase genes from termites: intron/exon structures and sites of expression. Biochim Biophys Acta 1447:146–159

Tokuda G, Lo N, Watanabe H et al (2004) Major alteration of the expression site of endogenous cellulases in members of an apical termite lineage. Mol Ecol 13:3219–3228

Tokuda G, Lo N, Watanabe H (2005) Marked variations in patterns of cellulase activity against crystalline- versus carboxymethyl-cellulose in the digestive systems of diverse, wood-feeding termites. Physiol Entomol 30:372–380

Tokuda G, Miyagi M, Makiya H et al (2009) Digestive β-glucosidases from the wood-feeding higher termite, *Nasutitermes takasagoensis*: intestinal distribution, molecular characterization, and alteration in sites of expression. Insect Biochem Mol Biol 39:931–937

Tokuda G, Saito H, Watanabe H (2002) A digestive beta-glucosidase from the salivary glands of the termite, *Neotermes koshunensis* (Shiraki): distribution, characterization and isolation of its precursor cDNA by 5′- and 3′-RACE amplifications with degenerate primers. Insect Biochem Mol Biol 32:1681–1689

Tokuda G, Watanabe H (2007) Hidden cellulases in termites: revision of an old hypothesis. Biol Lett 3:336–339

Tokuda G, Watanabe H, Matsumoto T, Noda H (1997) Cellulose digestion in the wood-eating higher termite, *Nasutitermes takasagoensis* (Shiraki): distribution of cellulases and properties of endo-β-1,4-glucanase. Zool Sci 14:83–93

Tomme P, Warren RA, Gilkes NR (1995) Cellulose hydrolysis by bacteria and fungi. Adv Micro Physiol 37:1–81

Watanabe H, Nakamura M, Tokuda G et al (1997) Site of secretion and properties of endogenous endo-beta-1,4-glucanase components from *Reticulitermes speratus* (Kolbe), a Japanese subterranean termite. Insect Biochem Mol Biol 27:305–313

Watanabe H, Noda H, Tokuda G, Lo N (1998) A cellulase gene of termite origin. Nature 394: 330–331

Watanabe H, Takase A, Tokuda G et al (2006) Symbiotic 'Archaezoa' of the primitive termite *Mastotermes darwiniensis* still play a role in cellulase production. Eukaryot Cell 5:1571–1576

Watanabe H, Tokuda G (2010) Cellulolytic systems in insects. Annu Rev Entomol 55:609–632

Wood TG (1978) Feed and feeding habits of termites. In: Brian MV (ed) Production ecology of ants and termites. Cambridge University Press, New York, NY, pp 55–80

Yokoe Y (1964) Cellulose activity in the termite, *Leucotermes speratus*, with new evidence in support of a cellulase produced by the termite itself. Sci Pap Coll Gen Educ Univ Tokyo 14:115–120

Yokoe Y, Yasumasu I (1964) The distribution of cellulase in invertebrates. Comp Biochem Physiol 13:323–328

Yuki M, Moriya S, Inoue T, Kudo T (2008) Transcriptome analysis of the digestive organs of *Hodotermopsis sjostedti*, a lower termite that hosts mutualistic microorganisms in its hindgut. Zool Sci 25:401–406

Zachary A, Colwell RR (1979) Gut-associated microflora of *Limnoria tripunctata* in marine creosote-treated wood pilings. Nature 282:716–717

Zhou X, Wheeler MM, Oi FM, Scharf ME (2008) RNA interference in the termite *Reticulitermes flavipes* through ingestion of double-stranded RNA. Insect Biochem Mol Biol 38:805–815

Chapter 4
Altricial Development in Wood-Feeding Cockroaches: The Key Antecedent of Termite Eusociality

Christine A. Nalepa

Abstract Altricial development, the production of vulnerable, dependent neonates, is regularly associated with high levels of parental care in both vertebrates and invertebrates and has evolved in at least two phylogenetically distant, biparental wood-feeding cockroach genera: *Cryptocercus* and *Salganea*. Here I make the case that altricial development was the first deviation from standard hemimetabolous development in the biparental termite ancestor, played a fundamental role in the origin of the termite lineage, and can account for many of the novel traits and functions that constitute eusociality in extant Isoptera. Altricial neonates of subsocial wood feeding cockroaches are small, fragile, incapable of feeding themselves, and costly to rear. However, the type of care these juveniles require does not necessarily have to originate from a parent; consequently other members of the social group have the opportunity to undertake brood care obligations. Altricial development therefore allows for the transition to alloparental care, and has a great deal of explanatory power when the evolution of termite eusociality is framed as a hierarchical two-stage process. Stage 1 alloparental care both increased reproductive output of adults and prolonged the development of young helpers. Combined, this changed the nature of the family: it was composed of a greater number of small, vulnerable juveniles. At the same time however, parents were increasingly specializing in reproduction at the expense of their defensive parental duties. Stage 1 alloparental care, then, necessitated Stage 2, the development of a soldier caste, but it also allowed it. Soldiers are a nutritionally dependent caste, so alloparental feeding would have to be part of the behavioral repertoire before soldiers with specialized defensive mandibles could appear. As eusociality progressed, altricial development formed the operational basis for the prolongation of juvenile morphology, physiology and behavior into later ontogeny, accounting for many aspects of termite life history neglected by current hypotheses. Altricial development reinforces the importance of biparental care and hemimetabolous development in the termite ancestor and is a clear indication that the social environment can significantly influence developmental ontogeny

C.A. Nalepa (✉)
Entomology Department, North Carolina State University, Raleigh, NC 27695-7613, USA
e-mail: christine_nalepa@ncsu.edu

D.E. Bignell et al. (eds.), *Biology of Termites: A Modern Synthesis*,
DOI 10.1007/978-90-481-3977-4_4,

in the lineage. The evolution of termite eusociality is a prime example of how evolution uses existing components in the origin of new systems.

4.1 Introduction

> Evolution does not produce novelties from scratch. It works on what already exists (Jacob 1977).

The classic view regarding termite origins is that they either branched directly from cockroaches, or that cockroaches and termites are sister groups (e.g. Cleveland et al. 1934: p. 321). Although this view has been challenged, in recent years there has been overwhelming evidence that termites are a monophyletic clade embedded within cockroaches, and sister group to the cockroach genus *Cryptocercus* (Lo et al. 2000; Inward et al. 2007a, and see Fig. 1 and Table 2 in Ware et al. 2008 for a summary of proposed phylogenetic trees for the Dictyoptera since 1991). This placement is strongly supported, and reflected in phylogenies based on morphology, molecular characters of the insects, phylogeny of bacteroids in the fat body (Lo et al. 2003) and, more recently, the phylogeny of hindgut protozoa (Ohkuma et al. 2009) and their associated bacterial ectosymbionts (Carpenter et al. 2009; Ikeda-Ohtsubo and Brune 2009). Although it is still unsettled as to how these findings affect the taxonomic nomenclature of the insects (Lo et al. 2007; Eggleton et al. 2007), the fundamental results are clear: termites are cockroaches. There are two broad implications of these findings. The first is that we now have an explicit framework, a phylogenetic context that allows us to infer the direction of evolutionary change: an oviparous lineage of wood feeding cockroaches closely related to *Cryptocercus* gave rise to termites. That alone significantly narrows potential evolutionary scenarios because it makes *Cryptocercus* the logical model for inferring character states in the termite ancestor. Of key importance are those characters that may have been instrumental in facilitating the transition from *Cryptocercus*-like biparental family groups, to eusocial colonies with morphological castes and division of labor. The best analyses of adaptive radiations are typically those between sister taxa where one lineage exhibits a key innovation that is absent in the sister lineage (e.g., Mitter et al. 1988; Guyer and Slowinski 1993).

The second implication of these studies is that the phylogenetic split of Isoptera from the remaining Dictyoptera is synonymous with the evolution of termite eusociality. Eusociality is a shared-derived character of Isoptera, and was undoubtedly the key innovation, the evolutionary novelty that allowed termites to phylogenetically pinwheel away from their blattarian relatives. As such, established colonies of extant termites are far removed from their origins. Even the most basal termites have apomorphic development, physiology, behavior, and life history, and are thus in most cases poor models of the initial stages of eusociality. Conditions responsible for the evolution of eusociality in termites are unlikely to be the same as those currently maintaining and driving it. The role of soldiers in contemporary colonies, for example, probably has little relevance in identifying selective pressures associated with the origination of the caste (cf. Thorne et al. 2003; Roux and Korb 2004). Similarly, scenarios involving secondary reproductives in all their forms are flawed because the ability to quickly differentiate functional gonads is a hallmark of established eusociality; it is one developmental option in the extraordinarily plastic caste system currently in place (Roisin 1999). When the reproductive pair is removed from a colony of *Kalotermes flavicollis*, for example, new candidates begin to develop as their replacements within 24 h (Lüscher 1953). If extant wood-feeding cockroaches are any indication, however, progress toward reproductive maturity in juveniles of a termite ancestor would be more realistically measured in years. Although cockroach ontogeny has a great deal of plasticity (Nalepa 1994), it is canalized relative to that of extant termites, with physiological limitations on how rapidly development responds to relevant cues. The termite ancestor likely had lengthy developmental periods (Table 4.1) eventually reaching a fairly large adult size, as do all studied wood-feeding cockroaches (Maekawa et al. 2008). Termite evolutionary scenarios that involve "hopeful reproductives" are therefore less convincing when temporal limitations are considered (see also Roisin 1999); adult *Cryptocercus* rarely, if ever, overlap with their adult offspring (Nalepa, unpublished). It is increasingly clear that we must examine the problem from the cockroach side of the social divide and allow blattarian biology to lead theory.

4.1.1 Evolution of Development

An obvious but uncommon starting point is an examination of dictyopteran developmental systems. The extraordinary flexibility of termite ontogeny (Noirot 1985a, b; Roisin 2000) did not arise de novo; it had to evolve within the limitations and opportunities set by the developmental system of blattarian ancestors (Nalepa 1994 and references therein). This simple idea, that termite polyphenism originated from straightforward, linear hemimetabolous development, is too often overlooked

Table 4.1 Interrelated life history and other characters in a termite ancestor, as exemplified by *Cryptocercus*. Combined, all were crucial antecedents to the evolution of termite eusociality

Character	References
Oviparity	Maekawa et al. (2008)
Biparental care	Maekawa et al. (2008), present chapter
Adult longevity	Bell et al. (2007)
Altricial offspring	Present chapter
Social monogamy	Nalepa and Jones (1991)
Protective log nest	Many authors (e.g., Klass et al. 2008)
Hatching asynchrony	Klass et al. (2008)
Functional semelparity	Nalepa (1988)
Food sharing behaviors	Nalepa (1994)
Nitrogen deficient wood diet	Many authors (e.g., Nalepa 1994)
Hemimetabolous development	Many authors (e.g., Kennedy 1947)
Multiple symbiotic relationships	Nalepa (1994), Nalepa et al. (2001)
Extended developmental periods	Nalepa (1994)

in studies of termite caste evolution (Bordereau 1985, Noirot 1985a, Noirot and Pasteels 1987), and is ripe for re-examination (Nalepa and Bandi 2000, Nalepa 2010). A core principle of evolutionary developmental biology (evo-devo) is the recognition that developmental systems evolve (e.g. McKinney and Gittleman 1995, Hall 1998, True and Haag 2001), and of all eusocial insect taxa, termites probably exhibit the strongest deviation from the developmental trajectory exhibited by their ancestors.

It has been long recognized (e.g. Kennedy 1947), that hemimetabolous development is a distinguishing life history trait in termites (Table 4.1), but the wide-ranging implications of this basic observation have been widely ignored. One corollary is that the very definition of eusociality is biased against termites, because a typically listed criterion is that *adults* of two or more generations co-exist in the same nest (e.g. Hölldobler and Wilson 2008, p. 9). It depends on the definition of "adult" that is used, but if it means terminal-stage, somatically and reproductively mature, fully metamorphosed individuals, then termites do not meet the standard. Overlap of generations in the Isoptera is not the temporal overlap of adult generations, but the overlap of the adult reproductive pair with a succession of developmentally arrested juveniles (Noirot 1982, 1985a). While it is regularly noted that a hemimetabolous development allows these immatures to work, it is less often considered that juvenile stages also have physiology, development, and behavior that differs from that of adults (reviewed in Nalepa 1994, Nalepa and Bandi 2000, Nalepa 2010). Consequently, termites are often uneasily wedged into Hymenoptera-dominated eusociality paradigms. One example is that battles for reproductive dominance were unlikely to have been prominent in the origin of isopteran eusociality. We need to be cautious in taking an overly adultocentric view of evolution (Minelli 2003, 2009); nowhere more so than with termites, where juveniles predominate in colonies.

4.1.2 Evolution of Termite Eusociality: A Two Stage Process

The social and developmental complexity that currently characterizes Isoptera did not spring up fully developed. There was a logical, hierarchical order to the appearance of individual features that currently distinguish the group. Roisin (1999) indicated three major transitions took place during the evolution of termite eusociality from a subsocial ancestor: (1) the evolution of helping behavior (alloparental care)[1], (2) the evolution of the soldier caste, and (3) the evolution of the worker caste. Here I suggest that these three events occurred in exactly the order that Roisin listed them, with each stage dependent on the one that preceded it. Early stages produced novel morphologies, physiologies, and behaviors that served as the raw material for additional, perhaps more powerful forms of selection at later stages. True eusociality made its appearance at Stage 2, when soldiers, the first sterile or

[1]Alloparental care (Stage 1), as used here, is called cooperative breeding by Roisin (1999). I use the terms alloparent and helper interchangeably in this chapter, to mean a juvenile that performs parental tasks but retains the ability to develop to reproductive maturity.

near sterile morphological caste appeared (Roisin 1999). Except for the secondary loss of soldiers in one sub-family, all termites satisfy the condition of eusociality not by the existence of sterile workers, but by the existence of sterile soldiers (Noirot 1985a, Horiuchi et al. 2002). Stage 3, the appearance of permanently sterile workers, either evolved once with some secondary reversions, or did not occur in all lineages (Noirot and Pasteels 1987, 1988; Higashi et al. 1991; Thompson et al. 2000; Roisin 2006, Inward et al. 2007b, among others). The appearance of sterile workers likely occurred in an ancestor that had already met the basic criteria of eusociality (via soldiers), and so falls outside the scope of this chapter.

Prior to the Stage 1 shift to alloparental care, the ancestor was characterized by a unique suite of interdependent life history traits (Andersson 1984, Nalepa 1991, 1994) (Table 4.1). This chapter focusses on just one of these, altriciality, a specific developmental precondition in a subsocial termite ancestor, as exemplified by *Cryptocercus* and some other wood-feeding cockroaches. I suggest that altricial development was the first deviation from standard hemimetabolous development in termite evolution (Nalepa 2010), and a key antecedent to Stage 1, the transition to alloparental care. I then detail how the initial Stage 1 behavioral shift could initiate an evolutionary cascade that results in eusociality at Stage 2, the appearance of a soldier caste.

4.2 Altricial Development

There are obvious gradients in developmental maturity and relative size of new hatchlings, ranging from immature and helpless (altricial) to complete independence from adult support (precocial). This continuum of developmental patterns is well known in vertebrates, and provides a useful basis for discussing patterns of postnatal growth. It has been studied in placental mammals (e.g., Walser 1977, Derrickson 1992), marsupials (e.g. Smith 2001, Shaw and Renfree 2006), and amphibians (e.g., O'Reilly et al. 1998; Kupfer et al. 2006), but is best characterized in birds. Nice (1962) divided the avian altricial-precocial spectrum into categories based on a combination of specific morphological and behavioral characters, determined at the time of hatch: open or closed eyes, presence or absence of feathers, hatchling mobility, and the nature of parental care. Altricial birds are blind, naked, confined to a nest, and incapable of self-care. Precocial birds have open eyes and are covered in feathers; within a few hours of hatch they are mobile and can forage independently (Case 1978, Ricklefs 1983, Starck 1993, Gill 1995).

The degree of development at hatch is also known to vary in a variety of invertebrates, including insects (Alexander 1990, Tallamy 2005). Furthermore it has been proposed that, with only minor adjustments, the categories used for birds by Nice (1962) also may be used to classify neonates in cockroaches: eye development, the thickness and pigmentation of the cuticle, mobility, and the presence and type of parental care (Nalepa and Bell 1997, Bell et al. 2007). The vast majority of cockroaches are precocial. They emerge from the egg with well developed eyes, their cuticle becomes hardened and pigmented within a few hours after hatch, and they

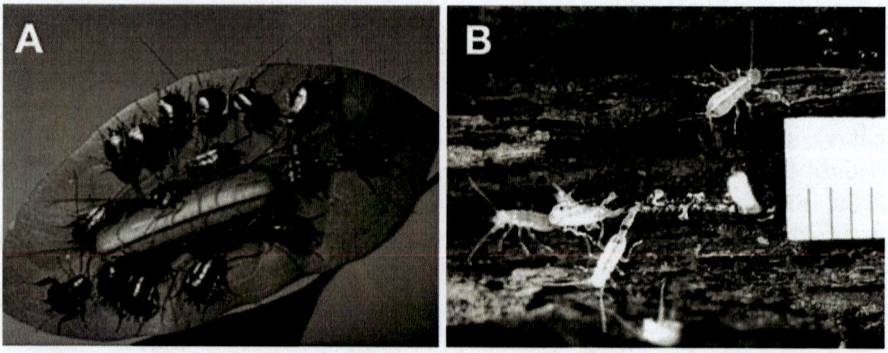

Fig. 4.1 **a** Unidentified neonate cockroaches newly hatched from an ootheca attached to a leaf, Bukit Timah, Singapore. Photo courtesy of Edward S. Ross. **b** *Cryptocercus punctulatus* hatching from an egg case embedded in wood, Mountain Lake Biological Station, Virginia (mm scale)

forage independently for food (Fig. 4.1a). In the few cockroaches with extensive parental care, however, young cockroaches are altricial, evident in their blindness, fragile exoskeleton, and reliance on adults for nourishment.

4.2.1 Altricial Development in Wood Feeding Cockroaches

Cryptocercus represents the extreme of altricial development among cockroaches (Fig. 4.1b) (Nalepa and Bell 1997, Nalepa et al. 2008). First instars lack compound eyes; a small number of pigmented ocelli become visible in the second instar (Huber 1976). The cuticle is pale and thin, with internal organs clearly visible through the dorsal surface of the abdomen; it does not become significantly pigmented until nymphs are nearly half grown. Young nymphs are dependent on adults for food, because hindgut symbionts are not fully established until the third instar (Nalepa 1990). Young *Cryptocercus* are mobile, but typically remain in close proximity to a parent within the confines of their wood nest. Although the evidence suggests that altricial development in this genus may be directly associated with the parental care given to neonates, it is not possible to do a comparative analysis with a closely related non-parental taxon. *Cryptocercus* is the only known oviparous cockroach that lives in biparental family groups (Table 4.1), and its exact relationship to other cockroaches is still in some dispute. Nonetheless, a good example of the correlation of altricial development with nutritionally based biparental care has been described in a subfamily of ovoviviparous wood-feeding cockroaches remote from the *Cryptocercus*/termite clade.

Within the subfamily Panesthiinae (Blaberidae), the morphology, parental behavior, social structure, and ecology of cockroaches in the genus *Salganea* parallel these same characters in *Cryptocercus* to a remarkable degree (Maekawa et al. 2008). Like *Cryptocercus*, *Salganea* lives in a system of galleries and chambers chewed into rotten wood, and exhibits long term, biparental care that includes the defense and feeding of young nymphs. In *Salganea taiwanensis,* young nymphs are fed via

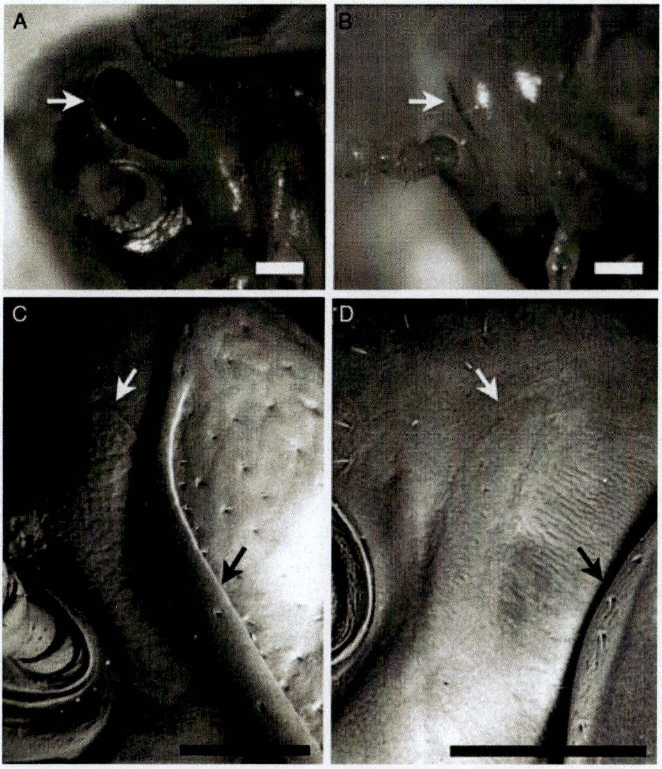

Fig. 4.2 Development of the eye (*white arrows*) in species of first-instar wood feeding cockroaches (Panesthiinae) with different social structure. Light photomicrograph (**a**) and scanning electron micrograph (**c**) of gregarious *Panesthia a. spadica* from Yakushima (*scale bars*, 500 μm); light photomicrograph (**b**) and scanning electron micrograph (**d**) of subsocial *Salganea taiwanensis* from Tokunoshima (*scale bars*, 250 μm). Note the difference between species in the thickness of the pronotal edge (*black arrows*). From Nalepa et al. (2008)

stomodeal trophallaxis. They periodically cling to the mouthparts of a parent and take liquids via oral feeding (Maekawa et al. 2008; Fig. 4.2). Neonates recognize their parents, and if removed from parental care in *Salganea esakii* and *S. taiwanensis,* they suffer high mortality rates (Obata 1988). The closely related wood-feeding genus *Panesthia*, however, typically lives in aggregations and, unlike *Salganea*, solitary small nymphs are often found in the field (Obata 1988, Maekawa 1997). Both *Salganea* and *Panesthia* clearly fall within the same subfamily and had a fairly recent common ancestor (Maekawa et al. 2003). Both live in and feed on rotting wood, and may even be collected from the same log (e.g. *S. esakii* and *P. a. spadica*; Nalepa et al. 2008). We can therefore assume that food quality and exposure to predators and pathogens are comparable. Adults in both taxa are large, dark brown to black in color, and heavily sclerotized. The most notable difference lies in their social structure: *Panesthia* is gregarious, while *Salganea* lives in biparental family groups. A comparative analysis of the morphology of first instars in these two genera indicates that the development of neonates clearly reflects this difference in social structure (Nalepa et al. 2008). Neonates of *Salganea* have a thin, pale cuticle through which the gut and fat body are visible, and they undergo an ontogenetic color change as they progress toward adulthood. First instars of *Panesthia*, however, hatch with a cuticle that is sturdy and pigmented, sometimes as deeply as adults. The eyes of *Salganea* neonates are significantly smaller in area than those of *Panesthia*, relative to conspecific adults, and have fewer, smaller ommatidia (Fig. 4.2). The evidence suggests that in wood-feeding cockroaches, biparental care that includes progressive provisioning can result in a shift in the morphological characters of neonates and an increased reliance on parental solicitude.

Comparison of *Cryptocercus* and *Salganea* indicates that the former takes altricial development a step further in that first instars are smaller in size, relative to their respective adults. In three species of *Cryptocercus*, the relative head capsule width of first instars is about 15–20% of that of adults, while that of *Salganea* is 24–32%, similar to that of the gregarious *Panesthia* (Nalepa et al. 2008). *Cryptocercus* neonates average just 0.06% of their final adult dry weight (Nalepa and Mullins 1992), and their small size is associated with the production of small eggs by the mother. The length of the terminal oocyte in *C. punctulatus* is 5.6% of adult length, contrasting with the 9.3–19.2% exhibited by eleven other species of oviparous cockroaches (Nalepa and Mullins 2009).

Table 4.2 is a summary comparison of developmental characteristics of young wood-feeding cockroaches in the genera *Cryptocercus*, *Salganea* and *Panesthia*.

Table 4.2 Summary comparison of traits relevant to developmental characteristics of young wood feeding cockroaches. *Salganea* and *Panesthia* are closely related within the subfamily Panesthiinae (Blaberidae). *Cryptocercus* (Cryptocercidae) is phylogenetically distant from the Blaberidae. After Nalepa et al. (2008)

	Cryptocercus	*Salganea*	*Panesthia*
Social structure	Biparental family	Biparental family	Multi-adult aggregation
Parental feeding	Yes	Yes[a]	Unlikely
1st instar dependence[b]	+++	++	–
1st instar cuticle	Pale, soft, transparent	Pale, soft, transparent	Dark, robust, opaque
1st instar eyes	Absent	4.7% of number of adult ommatidia[c]	11.3% of number of adult ommatidia[d]
1st instar size[e]	Head-capsule width is 15–20% that of adult	Head-capsule width is 24–32% that of adult	Head-capsule width is 27–28% that of adult

[a]Observed in *S. taiwanensis*.
[b]As measured by mortality when removed from adult care: (+++) complete mortality; (++) high mortality, with developmental delays in survivors; (–) insignificant mortality.
[c]In *S. esakii*; eyes also reduced in other species.
[d]In *P.a. spadica*.
[e]See Nalepa et al. (2008) for data.

Based on these characters, Nalepa et al. (2008) considered *Panesthia* neonates as precocial, and *Cryptocercus* as altricial. *Salganea* was considered semi-altricial, because while first instars exhibit cuticular and ocular reduction, they are not completely dependent on adult care (some that are isolated survive, although their development suffers) and because their size is comparable to that of *Panesthia*. The evidence indicates that, as in vertebrates, parental care in wood-feeding cockroaches alters the physical and social environment of neonates and, as a consequence, can change their developmental ontogeny (Daly and Wilson 1978, Ricklefs 1983, Clutton-Brock and Godfray 1991, Kölliker 2005). Furthermore, altricial development in both vertebrates and invertebrates includes dependence for food and a reduction in sensory and integumentary structures, demonstrating that similar selective pressures working independently can result in similar developmental shifts (Nalepa et al. 2008). The convergence across animal taxa is remarkable, given the disparity in evolutionary starting points.

4.2.2 Food and Protection

Convergent developmental changes such as these are significant observations, as they indicate preferred evolutionary pathways to particular developmental ends (Raff 1996, Farmer 2000), and are generally associated with parental care that occurs in a specific environmental context. Altricial offspring are favored whenever (1) the nest site is a safe place to complete development, and/or (2) food is difficult to obtain (Case 1978, Burley and Johnson 2002). In wood-feeding cockroaches, a rotting log acts as both food and nest, and meets both criteria. The type of logs used by *Cryptocercus* and *Salganea* (see Klass et al. 2008) are protected sites that buffer residents from fluctuations in temperature and humidity, shield them from the elements, conceal them, mechanically protect them, and limit the type of predators to which they are exposed. The relatively recalcitrant wood of these logs may take decades to degrade, thereby assuring a stable base for social interactions. Juveniles are additionally safeguarded by parental vigilance in both genera. Adults are large, well armored, and aggressive in defending their galleries from intruders (Seelinger and Seelinger 1983, Obata 1988, Park and Choe 2003).

The primary selective pressure shaping the altricial-precocial developmental spectrum is the nature of the food source (Imms 1931, Nice 1962, Case 1978, Ricklefs 1983, O'Connor 1984; Clutton-Brock 1991). If food is readily available, both to females for egg production and to the young for growth, precocial development is possible. If, however, food is difficult to obtain or metabolize, altricial development is favored and neonates are dependent on others for the transfer of resources; indeed, the word altricial is derived from the Latin *altor* (sustainer or nourisher; Walser 1977). Wood-feeding cockroaches are surrounded by their food source, but it is not easily utilized. Wood is physically tough, very low in nitrogen, and in *Cryptocercus*, is digested in collaboration with numerous and varied prokaryotes as well as diverse, specific flagellates in the hindgut (e.g., Cleveland et al. 1934; Ohkuma et al. 2009). Until this symbiotic community is fully established in the neonate hindgut, young nymphs are metabolically dependent on parental gut fluids and fecal pellets. It is notable that protozoa in the parental hindgut fluids are large in size, exceedingly numerous, and may themselves be serving as source of proteinaceous food for neonates (Grassé and Noirot 1945, Grassé 1952, Machida et al. 2001; Nalepa et al. 2001).

Salganea and *Panesthia,* like all other cockroaches investigated, produce endogenous cellulase (Maekawa et al. 2008) and have hindguts that are fermentation tanks filled with a diverse anaerobic microbiota, including ciliates, amoebae, flagellates, and a heterogeneous prokaryotic assemblage (Kidder 1937, Maekawa et al. 2008). This gut microbiota is likely established in neonates via coprophagy, a behavior that can occur in either a subsocial or gregarious social setting (Nalepa et al. 2001). There is no evidence to suggest that *Salganea* or *Panesthia* have an obligate relationship with any of the microbial taxa in their hindgut. Nonetheless, adults of *Salganea* feed young nymphs on oral fluids, suggesting that stomodeal trophallaxis in these cockroaches may serve primarily as nutrient transfer.

4.2.3 Evolution of Dependence

The evolution of altricial development is analogous to the loss of characters in caves: there is a rapid decay of "unused" characters, particularly in energy/nutrient-limited environments (e.g. Fong et al. 1995; Hall and Colegrave 2008). Unless the genes involved play other roles, they will be released from selection and undergo mutational degradation (Raff 1996). Thus, if a neonate cockroach is fed and protected, there is a loss of selection for a pigmented, sclerotized cuticle, for well-developed sense organs, and for digestive efficiency. Shimada and Maekawa (2008) recently published compelling evidence for the last of these: they found that in wood-feeding Panesthiinae, there is a correlation between the ability to digest cellulose and dependence on parentally provided food. The expression of genes that code for endogenous cellulase (endo-ß-1,4-glucanase) in first instars of the subsocial species *Salganea esakii* is greatly lower, relative to adults, than in all other developmental stages assayed. In the closely related gregarious species *Panesthia angustipennis*, expression levels in first instars were similar to those of third instars, and much higher relative to adults. These results are based on gene expression, not genotypic change; they nonetheless illustrate that there is relaxed selection on individual digestive physiology in first instars of *Salganea* and support the hypothesis that, under the right conditions, neonates that are fed become more dependent for food (Wesolowski 1994).

4.2.4 Parental Care as Selective Environment

A key point is that parents are an environmental component in the development of offspring traits. Adults change both the physical and social setting in which their progeny develop, and this parentally altered environment selects for ontogenetic characteristics in the offspring (Ricklefs 1983, Wcislo 1989, Clutton-Brock 1991, Wolf and Brodie 1998, Kölliker 2005). All further evolution then occurs within a selective context that includes parental care, and parental characters co-evolve with ontogenetic shifts in neonates (Daly and Wilson 1978, Wesolowski 1994). Once offspring cannot develop normally without parental support, parental care is obligatory, regression to a former state impossible, and the question "why care at all" becomes immaterial (Wesolowski 1994).

4.3 Altricial Offspring: Necessary Precedent to Eusociality

The ontogenetic shift that results in altricial offspring is a common evolutionary pattern associated with parental care in birds and mammals, and is now recognized in the termite sister group *Cryptocercus*, as well as more distantly related cockroaches and the orthopteran *Anurogryllus* (Alexander 1990). In the context of a specific social structure (subsociality), the underlying genetic architecture of these hemimetabolous insects allows for an evolutionary change in the direction of increased dependence at hatch. It is therefore reasonable to assume that this specific ontogenetic pathway also evolved in termite ancestors during the subsocial phase of their evolutionary history, and was a compulsory precondition in allowing the termite ancestor to progress toward eusociality.

4.3.1 Raising Altricial Offspring: Economics

There are two core reasons why altricial development was a key precondition in a subsocial termite ancestor. The first reason is the economics of parental care required in these cases: it is both depreciable and costly. In altricial mammals, parental care is dominated by lactation, in birds it is dominated by provisioning young at the nest, and in known hemimetabolous insects with altricial offspring, neonates are fed on parental secretions (cockroaches) or trophic eggs (*Anurogryllus*). When a parent supplies the metabolic requirements of offspring, it is considered depreciable care, because the benefit to individuals declines with brood size. What a mother gives to one offspring cannot be given to another (Clutton-Brock 1991). In field collections of *Cryptocercus punctulatus* at the end of the natal year, nymphs from small families are developmentally more advanced than nymphs in large families (Nalepa and Mullins 2009), providing support for the idea that parental care is depreciable in this species; the slower development in large families likely reflects juvenile competition for parentally provided resources. Because parents must meet the metabolic demands of altricial neonates, such offspring are also more costly than are precocial young (Case 1978, May and Rubenstein 1984). In birds, the cost of feeding hatchlings typically exceeds that of egg production or gestation by a substantial margin. Consequently, reproduction in altricial taxa is limited largely by post-hatching parental care rather than the energetics of egg production (Ricklefs 1974, Clutton-Brock 1991). *Cryptocercus* fits this syndrome as well. If eggs are removed from pairs of *Cryptocercus*, about half of the pairs will reproduce during the following reproductive season. If parental care is allowed prior to brood removal, however, just 12–18% produce eggs in the subsequent reproductive period (Nalepa 1988).

4.3.2 Raising Altricial Offspring: Temporal Dimension

The second reason why altricial development was a key antecedent in a termite ancestor is that such development changes the timing of parental investment in relation to hatch. Consider a distant ancestor of the *Cryptocercus*-termite clade with no parental care and precocial neonates. The female deposits her ootheca, then leaves (Fig. 4.1a). Her costs of nourishing offspring consist solely of the yolk she deposited in oocytes; after emerging from the egg, first instars forage independently (Fig. 4.3a). The peak metabolic expense for the mother occurs prior to hatch (Case

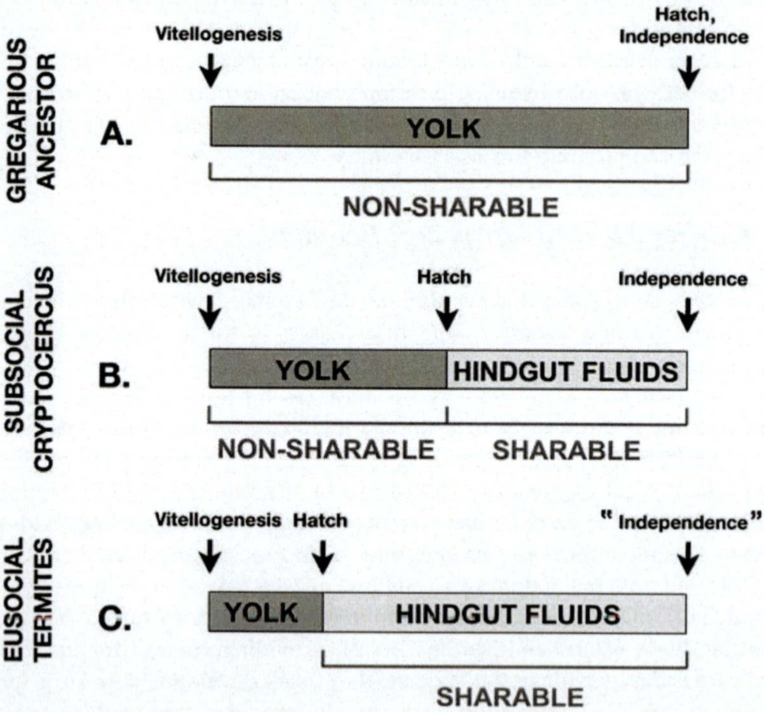

Fig. 4.3 Shift in timing and type of nutritional investment in progeny in relation to hatch during evolution of the termite lineage. **a** Gregarious cockroach ancestor. The cost of nourishing offspring to independence consists primarily of yolk deposited in oocytes (non-sharable investment). **b** Subsocial ancestor, as typified by *Cryptocercus*. Female decreases non-sharable investment by reducing egg size; altricial hatchlings require continued feeding on hindgut fluids (sharable investment) prior to nutritional independence. **c** Extant termites. Non-sharable investment in minimized, sharable investment is maximized

1978). An essential point is that a female has to yolk the eggs herself – a male may offset some of the burden by providing her with nutrients during courtship or mating, but the physiological burden of egg production is solely with the female – it is a non-sharable parental investment. In a biparental family with altricial offspring, as typified by *Cryptocercus*, the female avoids a large metabolic cost prior to hatch by reducing the size of the eggs (Fig. 4.3b). After hatch, however, the tiny offspring require continued parental investment because they cannot feed on their own; hatch is not synonymous with independence. Provisioning the young on gut fluids takes place outside the mother's body, and importantly, the behavior requires no specialized morphological or physiological adaptations. There is nothing stage- or sex-specific about hindgut fluids, as far as we know, so the male or any other member of the social group is able to feed newly hatched dependent young (e.g. Machida et al. 2001). It is parental investment that is not necessarily restricted to a parent. Altricial development gives other members of the social group the opportunity to assume responsibility for some of the duties and allows a parent to share the cost of producing independent offspring.

In *Cryptocercus*, it is presumed that the contribution of the male allowed the female to reduce investment in eggs, as in many other altricial taxa (Zeveloff and Boyce 1982, Blackburn and Evans 1986, Alexander 1990). Biparental care is found in the majority of socially monogamous birds that produce altricial young (Zeveloff and Boyce 1982, Clutton-Brock 1991, Temrin and Tullberg 1995, Burley and Johnson 2002, Cockburn 2004, Thomas and Székely 2005, Olson et al. 2008). Among wood-feeding cockroaches, altricial or semi-altricial development is known only in two taxa that live in biparental families, underscoring the association of social monogamy with altricial development (Table 4.1). Lameere (in Wheeler 1928) recognized that same association during the incipient, subsocial stage of termites: "The young on hatching are unable to obtain sufficient nourishment by attacking so refractory a substance as wood; they are therefore fed by the mother and the great advantage of the father's cooperation in this activity is obvious".

4.3.3 Stage 1: The Transition To Alloparental Care

The importance of altricial development in an oviparous subsocial termite ancestor, as typified by *Cryptocercus*, is that (1) neonates have to be helpless before other members of the social group can help, (2) helping is of a type not necessarily restricted to a reproductively competent female, and (3) post-hatch parental investment is costly, so that other members of the social group can have a substantial impact by helping. This last-named attribute is the basis for the trophic shift hypothesis for Stage 1 eusociality in termites (Nalepa 1988, 1994; Bell et al. 2007). It proposes that a single behavioral change, the assumption of brood care duties by the oldest offspring in the family, was the precipitating event eventually leading to all distinguishing characteristics of termite eusociality. The hypothesis is grounded in data indicating that post-hatch care of neonates is expensive in both *Cryptocercus* and in young termite colonies (e.g. Johnston and Wheeler 2005). After an initial egg laying period in both taxa, reproduction ceases while the adults feed and otherwise care for their offspring (Nalepa 1994). The shift from parental to alloparental care in the ancestor of termites transferred these costs to juvenile care givers, and is furthermore mirrored in the colony ontogeny of extant Isoptera when the oldest offspring in the initial cohort after colony foundation assume responsibility for feeding and otherwise caring for their younger siblings (Nalepa 1988). This single behavioral change is therefore a pathway for making the transition between adaptive points, and accounts with great parsimony for the initial stages of termite eusociality (Nalepa 1988, 1994).

4.3.3.1 Consequences for Parents

An evolutionary shift in responsibility for brood care in the termite ancestor had potentially striking consequences for both parents and alloparents. The adult female, now relieved of the costliest component of her parental investment, could once again channel her reserves into vitellogenesis. Resumption of egg laying in incipient termite colonies is known to be roughly coincident with the assumption of alloparental duties by the first helpers (Buchli 1950, Sands 1965, Nutting 1969). In some birds with altricial young rapid re-nesting is similarly facilitated by division of labor – helpers continue to feed the fledged young while the breeding pair recuperates, renests, and begins another reproductive cycle (Langen 2000). It should be noted in this regard that although *Cryptocercus* typically has just a single reproductive period in the field, brood removal studies indicate that females retain at least some capacity for further oogenesis (Nalepa 1988).

4.3.3.2 Consequences for Alloparents

Because they were immature, the consequences incurred by the first alloparents assuming the cost of feeding younger dependents would be expressed in terms of development, not reproduction (Nalepa 1994). The resultant downsizing of their energy/nutrient budget would halt progressive development; helpers would begin undergoing stationary and regressive molts, without losing their ontogenetic potential (Noirot 1982, Noirot and Pasteels 1987). Their development would stall, just as it does in starved juvenile cockroaches (Nalepa 1994). In extant incipient termite colonies these alloparental "wet nurses" are individuals that had recently become nutritionally self-sufficient, i.e. their hindgut symbionts had just become fully established. This is very early in their ontogenetic development, at the third or fourth instar. In insects in general and cockroaches in particular, the biggest impact of a shift in nutritional status is on early instars, which have small reserves, a high metabolism, and nutritional requirements that differ from those of adults (Scriber and Slansky 1981, Rollo 1986). The concept that helping behavior can have a physiological cost in alloparents is not novel (Heinsohn and Legge 1999). It has been documented in a variety of vertebrates, including birds, fish, and mammals (Taborsky 1984, Heinsohn and Cockburn 1994, Clutton-Brock et al. 1998).

If the original shift to alloparental care in termites is reflected in extant incipient colonies, then the relative youth of the ancestral alloparents is a key observation. First, these initial helpers were a long way from reproductive maturity. The growth of wood-feeding insects is typically very slow (Calow and Townsend 1981, Scriber and Slansky 1981), and wood-feeding cockroaches exhibit several characters typically associated with prolonged development (Danks 1992): they are relatively basal insects that are hemimetabolous and live in a stable habitat, utilizing food of low quality. All Cryptocercidae and wood-feeding Panesthiinae studied to date are slow growing, long-lived cockroaches (Table 4.1) (Rugg and Rose 1990, Bell et al. 2007; Maekawa et al. 2008); development takes up to 7 years in *Cryptocercus clevelandi*. The ancestral alloparents would therefore lack a functional reproductive system, and would not begin developing one for several more years. At the initiation of Stage 1 termite eusociality, these juvenile alloparents were incapable of independent reproduction, either inside or outside the nest. The strong asymmetry in both size and reproductive ability of parents and alloparents would make issues such as controlling helper reproduction or competition for reproductive dominance irrelevant.

Second, the assumption of parental costs at an early developmental stage means that the ontogeny of alloparents was suspended at a point when they still exhibited many altricial characters, most notably a thin, non-pigmented cuticle and a small body size. In *Cryptocercus punctulatus* the body size of third instars (as indicated by head capsule width) is comparable to the relative size of first instars in other wood feeding cockroaches (Nalepa and Mullins 2009), and their internal organs are clearly visible through the transparent, unpigmented cuticle of the abdomen. The ancestral alloparents then, were developmentally suspended at a stage that was small, unsclerotized, and vulnerable. They would be unlikely to leave the protected environment of the nest.

Third, the assumption of parenting costs at a young age could have long-term consequences additional to the immediate effects on progressive development. A change in nutritional status when the architecture and *modus operandi* of the body is becoming established can have a profound impact in determining the life history trajectory of an organism (Metcalf and Monaghan 2001, Rombough et al. 2006). It was very likely, then, that the juvenile physiology of these first developmentally stalled alloparents supplied the raw material for the profound plasticity of ontogenetic trajectories that eventually evolved in the lineage (Nalepa 2010). Young postembryonic cockroaches are unable to undergo metamorphosis (Kikukawa and Tobe 1986), because in early instars metamorphosis does not have the survival value that it does in later developmental stages (Staal 1986). Young cockroach alloparents whose development was arrested in such an indeterminate state would have offered a novel developmental template on which natural selection could act (Nalepa 2010).

4.3.3.3 Origins of Alloparental Behavior

Although the evolution of alloparental behavior is still debated on many levels, it can be defined simply: it is modified parental care. Helpers begin exhibiting parental behavior before they are themselves parents (Nalepa 1994: p. 88; Hunt 2007: p. 189). The age at which provisioning normally commences is predisplaced, with the primary difference being the context in which the behavioral pattern occurs (Jamieson 1989, Craig and Jamieson 1990). This behavioral heterochrony, however, is not necessarily the direct product of selection. Even though a given biological process currently produces a certain benefit, it may not be the property for which it was originally selected (Williams 1966). In birds, the intense parental care associated with altricial lineages may predispose individuals to alloparental care because such behavior is a genetically correlated response to strong selection for parental care in the ancestor. If such a correlated character allowed for a new optimum it would then be maintained by stabilizing selection. The functional significance of alloparental behavior, then, may be irrelevant with respect to the evolutionary origin of the behavior (West-Eberhard 1987, Jamieson 1989, Price and Langen 1992, Ligon and Burt 2004)[2]. One precondition is the close proximity of non-breeders and dependent young. Alloparental care is normally not expressed in lineages where individuals have no close contact with young that are not their own (Ligon and Burt 2004).

As hypothesized for birds, the Stage 1 shift to alloparental care in the termite ancestor was likely associated with strong selection for parental care in the ancestor. Potential alloparents were furthermore in direct contact with younger, dependent siblings because of their confined, log-based lifestyle (Table 4.1). Tellingly, juveniles are reported to exhibit adult behaviors in at least one orthopteran that lives in a nest and has altricial development. In the burrowing cricket *Anurogryllus*, young developmental stages exhibit behaviors that appear only in the late nymphal or adult stages of other crickets. These include transportation of food particles, and burrow excavation by the removal and transport of substratum particles (West and Alexander 1963, Alexander 1990). The studies on *Anurogryllus* are pivotal in illustrating that in hemimetabolous taxa with altricial offspring, adult behavior in juveniles may appear regardless of any evident selective pressures or associated fitness benefits. Nonetheless, the type of adult behavior exhibited by alloparents in the termite ancestor, provisioning with body fluids, had strong potential for influencing family fitness.

4.3.3.4 Consequences for the Family

The essential result of the behavioral shift from parental to alloparental care is that family size started increasing. Developmental delays of juvenile alloparents combined with the renewed egg-laying ability of the mother resulted in an accumulating workforce of staggered developmental stages of juveniles queuing to reach developmental maturity. Initially all alloparents probably exhibited a typical, albeit protracted, cockroach-like hemimetabolous development. Each helper eventually

[2]There are, however, numerous alternate hypotheses offered to explain alloparental behavior in vertebrates (e.g., Heinsohn et al. 1990; Brown and Vleck 1998).

became reproductive and dispersed, and the alloparental family structure was maintained until either the reproductive capacity of the long-lived adults or their log resource declined. At the same time that family size was growing in this Stage 1 ancestor, family members were becoming increasingly vulnerable because many were stalled in an early developmental stage. It is obvious that in extant termites, the typical morphology of the majority of colony members is that of a juvenile *Cryptocercus* (Cleveland et al. 1934), and two key features that characterize such a morphotype are a thin cuticle and a small body size. The ancestral alloparents were thus physically vulnerable (Hadley 1984), but as early stage juveniles, they were also small in size. There are compelling reasons to believe that the evolutionary trend would continue to progress in the direction of a smaller, more fragile morphotype.

A critical factor that selects for small, vulnerable neonates is a predictable workforce to provide assistance. If alloparental care occurs consistently, a female can reduce her investment in reproduction, and allocate the conserved resources to other aspects of life history, such as future breeding attempts (Russell et al. 2007). Once Stage 1 eusociality was well-established in the termite ancestor, hatching as a reference point for parental investment could shift even further (Fig. 4.3c), resulting in an evolutionary cascade in the direction of tighter social integration and smaller body size. The production of smaller eggs allows a female to generate a larger number for a given resource base. Neonates become even more altricial, requiring a greater investment from alloparents, thus compromising the developmental progress of helpers even further. The division of labor between parents and alloparents becomes exacerbated, with parents becoming more specialized for reproduction, and the alloparents becoming more specialized for brood care and the non-reproductive chores of the family. Stasis in alloparental growth and development in combination with decreased egg size eventually resulted in a smaller body size, relative to most cockroaches, in all stages of the ancestor. Once social behavior progressed to the stage where alloparents were reliably feeding their parents as well as their younger siblings, selection for the accumulation of large metabolic reserves in the adults (as in *Cryptocercus, Salganea*) was relaxed. Adult termites currently exhibit numerous juvenile traits when compared to adult cockroaches (Matsuda 1979, Nalepa and Bandi 2000). This suggests that eventually the development of termite adults was also truncated: the ancestors began reaching reproductive maturity at a smaller, somatically earlier stage of ontogeny. Termites therefore may be described as paedomorphic (Nalepa and Bandi 2000), a term denoting descendant species that resemble earlier ontogenetic stages of ancestral species (Reilly 1994).

An additional characteristic of extant termites is the minute size and complete helplessness of the first few instars, supporting the idea that parents currently invest minimal resources into the production of individual eggs. In most termite species, the super-altricial neonates cannot hatch without assistance, and are kept in nurseries where they are tended by siblings. They are pale, transparent, blind, carried from place to place, and fed on digestive fluids, as their unsclerotized mandibles and incomplete gut fauna render them incapable of feeding on the standard lignocellulose diet (Grassi and Sandias 1897, Grassé and Noirot 1945,

Buchli 1958, Noirot 1985b, Raina et al. 2004, among others). The first few instars are markedly embryonic in nature, and although hemimetabolous, they are typically called larvae, in acknowledgement of their dependent status (discussed in Nalepa 2010).

4.3.4 Stage 2: Soldiers

As described above, the Stage 1 family faced a novel selective environment where parents were increasingly devoting their time and resources to egg laying rather than to patrolling and defense. The first helpers with slightly larger jaws and some degree of aggression could provide an immediate fitness advantage, not only by protecting soft-bodied siblings, but by protecting the adults. As the parents became increasingly specialized for reproduction, it was in the family's best interest to shield them from the risks associated with defending family members. An increase in mandible size in some of the ancestral Stage 1 helpers would not be a staggering developmental leap; it would depend largely on the same processes employed during the development of "standard" sized jaws (see Prestwich 1984, Scholtz et al. 2008). Moreover, it is possible that the developmental stasis that characterized the first alloparents also predisposed some of them to the development of larger mandibles. Seamans and Woodruff (1939) experimentally prolonged the developmental period of the cockroach *Blattella germanica* by starving them, and found the sclerotized parts of the body became slightly larger than is characteristic of a normally fed insect.

As the jaws of proto-soldiers became more specialized for defense, they would be less suitable for masticating wood. Extant termite soldiers are unable to feed themselves, and rely on siblings for feeding via trophallaxis (Haverty 1977, Koshikawa et al. 2002; Buczkowski et al. 2007). It therefore follows that the trophallactic feeding of nestmates had to be in place before soldiers with fully developed defensive mandibles could evolve. The nutritional dependency of soldiers reinforces the hypothesis that termite evolution was a hierarchical, stepwise process: Stage 1 (alloparental care) had to be established before Stage 2 (soldier evolution) could occur.

Additional support for a sequential two stage evolutionary process comes from a fundamental difference in the nature of the tasks performed by alloparents and soldiers. In *Cryptocercus*, brood care and family defense are adult behaviors that fall into the "parental care" behavioral syndrome. The large, well-armored adults both feed and defend their young offspring (Seelinger and Seelinger 1983, Nalepa 1994, Park and Choe 2003). The evolution of termite eusociality can be thought of as the sequential delegation of these parental obligations to the older offspring. Family defense, however, is non-depreciable care (Clutton-Brock 1991). Parental vigilance typically benefits the entire brood; depriving one nymph of protection is unlikely to increase the share received by another nymph. In contrast to relieving adults of feeding duties (= depreciable care), relieving adults of defensive duties would not allow parents to increase reproductive output, particularly in the nitrogen stressed environment typical of wood feeders. This is additional support for the idea that Stage 1 alloparental care must have preceded the Stage 2 evolution of soldiers.

4.4 Altricial Development Becomes the Norm

In the majority of extant termite species most colony members are blind, have a thin unpigmented cuticle, and are physiologically dependent on nestmates. These are the very properties that define altricial development in young wood-feeding cockroaches (Nalepa et al. 2008). The resemblance of termites to altricial juvenile cockroaches also includes their small size and their bodily proportions, including a short pronotum, leaving the head exposed. Cleveland et al. (1934) and Huber (1976) both noted the resemblance of early instars of *Cryptocercus* to larger termite species, with the major difference being the more rapid movement and longer antennae of *Cryptocercus*. These observations suggest that termite developmental characteristics and the morphologies they produce originated via the truncation of standard cockroach-like hemimetabolous development. This developmental shift influenced not only soldier evolution, but also an entire repertoire of epiphenomena associated with arrest in an altricial morphology and physiology.

4.4.1 Vulnerability

The physical vulnerability that placed the termite ancestor at increased risk of interspecific predation (and thus may have selected for the evolution of a soldier caste) also increased the risk of intraspecific predation. In cockroaches hatchlings, freshly molted nymphs, and the weak or wounded are the most frequent victims of cannibalism (Bell et al. 2007). In extant termites cannibalism plays a role in caste control, population regulation, and nutrient dynamics (LaFage and Nutting 1978), and likewise may be rooted in the fragile altricial morphology of a juvenile ancestor.

A thin, unmelanized cuticle would also increase vulnerability to pathogens in the humid interior environment of a rotting log. The cuticle as a physical barrier depends on a combination of thickness, tensile strength, and degree of hardening by sclerotization. Melanin, together with phenoloxidase, is an essential component of the invertebrate immune system; thus cuticular color and immune response are not independent because both are associated with melanin production (Hopkins and Kramer 1992, Hajek and St. Leger 1994, Gillespie et al. 1997; Sugumaran 2002, Armitage and Siva-Jothy 2005). Termites currently have sophisticated, multi-tiered strategies for dealing with pathogens (e.g. Traniello et al. 2002), a notable component of which is behavioral resistance via allogrooming (e.g. Rosengaus et al. 1998; Yanagawa and Shimizu 2007). In *Cryptocercus*, allogrooming occurs with highest frequency in the young, altricial instars (Seelinger and Seelinger 1983). This suggests that some constituents of termite disease management, including allogrooming and other sanitary behaviors, may stem from an ancestral insubstantial, unpigmented integument. A thin cuticle also may be more sensitive to fluctuations in levels of environmental moisture, and consequently influential in the development of mechanisms for controlling individual water loss via cuticular hydrocarbon components (Woodrow et al. 2000) and for behaviorally maintaining nest homeostasis (Noirot and Darlington 2000). A small body size further prioritizes hydration management due to the higher surface area/volume ratio (Bonner 2006).

The altricial morphotype is fundamental to explaining isopteran evolution, but was a viable option only because ancestral termites continued to experience the same conditions that allowed for altricial development in neonates of extant wood-feeding cockroaches: they obtained food from relatives (trophallaxis) and they lived in a protected environment (nest, soldiers).

4.4.2 Developmental Plasticity and Alternative Phenotypes

In extant termites the ontogeny of individuals is extraordinarily flexible; organs can develop independently, and molt, growth, and differentiation are disassociated (Noirot and Pasteels 1987). It is reasonable to suggest that this developmental complexity was another consequence of juvenilization in their ancestors. First, young cockroaches are exceptionally sensitive to the social environment ("group effects" of Nalepa 1994, Bell et al. 2007), and it is likely to have been the social environment, rather than the external environment, that was primary source of stimuli shaping developmental trajectories during early termite evolution (Nalepa and Bandi 2000). Second, growth rates and the hormonal control of development of early instar cockroaches differ from those in later instars (Tanaka 1981, Kikukawa and Tobe 1986), most notably in the metamorphic incompetence of the early instars. Third, many animals experience a point during progressive development when phenotypic plasticity is diminished or lost. One mechanism of recovering phenotypic flexibility is via prolongation of the earlier, more malleable ontogenetic stages (e.g. Balon 1981, Hensley 1993, Hentschel 1999, Hatle et al. 2004). Fourth, changes initiated earlier in development have greater potential to generate novelties than changes initiated later in development, and often result in new phenotypes, together with associated life histories (Hall and Miyake 1995, Balon 2002, Raff 2008).

4.5 Conclusions

Stage 1 alloparental behavior was the first step toward full termite eusociality because it increased the reproductive output of the adults, resulted in suspended development of alloparents, and both necessitated and allowed the development of nutritionally dependent soldiers. No dictyopteran species currently exhibits a Stage 1 alloparental lifestyle, although every extant termite passes through a Stage 1 phase during colony ontogeny. Rapid evolutionary progression to Stage 2 may have happened once only, as there is general consensus that termite soldiers are monophyletic (e.g. Noirot 1985a). A point of no return (*sensu* Hölldobler and Wilson 2005) was reached when alloparents began losing their cellulolytic hindgut protozoans during molt. It is clear from the extensive work of L.R. Cleveland that the shift in hormonal titers that accompanied the prolonged juvenile development of alloparents would also cause the death of their gut protozoans during the host molting period (reviewed in Nalepa 1994). Anything that alters the chronological time required to reach a defined point of development also alters the timing of underlying physiological regulatory systems (Spicer and Burggren 2003). The periodic loss of symbionts in alloparents would remove the option of living independently because it mandates access to the hindgut fluids of a nestmate. It was the parentally provided social environment that allowed for the evolution of altricial development in *Cryptocercus*, *Salganea*, and the subsocial stage of the termite ancestor. Altricial development in turn allowed the switch from parental to alloparental care during the initial phase of termite eusociality. Eventually, the juvenile of the ancestor became the stabilized morphological state of the descendent. This transition hypothesis accounts for most key characters of extant termites, and establishes how an ancestral phenotype may be modified to produce novel functional forms – an important but neglected part of evolutionary biology (West-Eberhard 2005).

> We may be surprised by how much new is possible through modifications of the familiar (Moczek 2008).

Acknowledgments I thank Jim Hunt and Kiyoto Maekawa for discussion and feedback. Ideas in this paper were presented in 1996 at the 20th International Congress of Entomology (Florence, Italy) and in 2007 at the 55th Annual Meeting of the Entomological Society of America (San Diego, CA).

References

Alexander RM (1990) How did humans evolve? Reflections on the uniquely unique species. Univ Michigan Mus Zool Spec Publ 1:1–38

Andersson M (1984) The evolution of eusociality. Annu Rev Ecol Syst 15:165–189

Armitage SAO, Siva-Jothy MT (2005) Immune function responds to selection for cuticular color in *Tenebrio molitor*. Heredity 94:650–656

Balon EK (1981) Saltatory processes and altricial to precocial forms in the ontogeny of fishes. Am Zool 21:573–596

Balon EK (2002) Epigenetic processes, when *natura non facit saltum* becomes a myth, and alternative ontogenies a mechanism of evolution. Environ Biol Fish 65:1–35

Bell WJ, Roth LM, Nalepa CA (2007) Cockroaches: ecology, behavior, and natural history. The Johns Hopkins University Press, Baltimore, MD

Blackburn DG, Evans HE (1986) Why there are no viviparous birds? Am Nat 128:165–190

Bonner JT (2006) Why size matters. Princeton University Press, Princeton, NJ

Bordereau C (1985) The role of pheromones in caste differentiation. In: Watson JAL, Okot-Kotber BM, Noirot C (eds) Caste differentiation in social insects. Pergamon Press, Oxford, pp 221–226

Brown JL, Vleck CM (1998) Prolactin and helping in birds: has natural selection strengthened helping behavior? Behav Ecol 9:541–545

Buchli H (1950) Recherche sur la fondation et le développement des nouvelles colonies chez le termite lucifuge (*Reticulitermes lucifugus* Ressi). Physiol Comp Oecol 2:145–160

Buchli HHR (1958) L'origine des castes et les potentialités ontogénétiques des termites européens du genre *Reticulitermes* Holmgren. Ann Sci Nat Zool (11)20:263–429

Buczkowski G, Wang C, Bennett G (2007) Immunomarking reveals food flow and feeding relationships in the eastern subterranean termite, *Reticulitermes flavipes* (Kollar). Environ Entomol 36:173–182

Burley NT, Johnson K (2002) The evolution of avian parental care. Phil Trans R Soc Lond B 357:241–250

Calow P, Townsend CR (1981) Resource utilization in growth. In: Townsend CR, Calow P (eds) Physiological ecology: an evolutionary approach to resource use. Sinauer, Sunderland, pp 220–244

Carpenter KJ, Chow L, Keeling PJ (2009) Morphology, phylogeny, and diversity of *Trichonympha* (Parabasalia: Hypermastigida) of the wood feeding cockroach *Cryptocercus punctulatus*. J Eukaryot Microbiol 56:305–313

Case TJ (1978) On the evolution and adaptive significance of postnatal growth rates in the terrestrial vertebrates. Q Rev Biol 53:243–282

Cleveland LR, Hall SR, Sanders EP, Collier J (1934) The wood feeding roach *Cryptocercus*, its protozoa, and the symbiosis between protozoa and roach. Mem Am Acad Arts Sci 17:185–342

Clutton-Brock TH (1991) The evolution of parental care. Princeton University Press, Princeton, NJ

Clutton-Brock TH, Gaynor D, Kansky R et al (1998) Costs of cooperative behavior in suricates (*Suricata suricatta*). Proc R Soc Lond B 265:185–190

Clutton-Brock TH, Godfray C (1991) Parental investment. In: Krebs JR, Davies NB (eds) Behavioural ecology: an evolutionary approach. Blackwell, Boston, MA, pp 234–262

Cockburn A (2004) Mating systems and sexual conflict. In: Koenig WD, Dickenson JL (eds) Ecology and evolution of cooperative breeding in birds. Cambridge University Press, Cambridge, pp 81–101

Craig JL, Jamieson IG (1990) Pukeko: different approaches and some different answers. In: Stacey PB, Koenig WD (eds) Cooperative breeding in birds: long term studies of ecology and behavior. Cambridge University Press, Cambridge, pp 385–412

Daly M, Wilson M (1978) Sex, evolution, and behavior. Duxbury Press, North Scituate, RI

Danks HV (1992) Long life-cycles in insects. Can Entomol 124:167–187

Derrickson EM (1992) Comparative reproductive strategies of altricial and precocial eutherian mammals. Funct Ecol 6:57–65

Eggleton P, Beccaloni G, Inward D (2007) Response to Lo et al. Biol Lett 3:564–565

Farmer CG (2000) Parental care: the key to understanding endothermy and other convergent features in birds and mammals. Am Nat 155:326–334

Fong DW, Kane TC, Culver DC (1995) Vestigialization and loss of nonfunctional characters. Annu Rev Ecol Syst 26:249–268

Gill FB (1995) Ornithology. WH Freeman, New York, NY

Gillespie JP, Kanost MR, Trenczek T (1997) Biological mediators of insect immunity. Annu Rev Entomol 42:611–643

Grassé P-P (1952) Rôle des flagellés symbiotiques chez les Blattes et les Termites. Tijdschr Entomol 95:70–80

Grassé PP, Noirot C (1945) La transmission des flagellés symbiotiques et les aliments des termites. Bull Biol France Belg 79:273–297

Grassi B, Sandias A (1897) The constitution and development of the society of termites: observations on their habits; with appendices on the parasitic protozoa of Termitidae, and on the Embiidae. Q J Microsc Soc 40:1–75

Guyer C, Slowinski JB (1993) Adaptive radiation and the topology of large phylogenies. Evolution 47:253–263

Hadley NF (1984) Cuticle: ecological significance. In: Bereiter-Hahn J, Matoltsy AG, Richards KS (eds) Biology of the integument I. Invertebrates. Springer-Verlag, Berlin, pp 685–702

Hajek AE, St. Leger RJ (1994) Interactions between fungal pathogens and insect hosts. Annu Rev Entomol 39:293–322

Hall BK (1998) Evolutionary developmental biology, 2nd edn. Chapman & Hall, London

Hall AR, Colegrave N (2008) Decay of unused characters by selection and drift. J Evol Biol 21:610–617
Hall BK, Miyake T (1995) How do embryos measure time? In: McNamara KJ (ed) Evolutionary change and heterochrony. Wiley, Chichester, pp 3–20
Hatle JD, Borst DW, Juliano SA (2004) Plasticity and canalization in the control of reproduction in the lubber grasshopper. Integr Comp Biol 43:635–645
Haverty MI (1977) The proportion of soldiers in termite colonies: a list and a bibliography. Sociobiology 2:199–216
Heinsohn R, Cockburn A (1994) Helping is costly to young birds in cooperatively breeding white-winged choughs. Proc R Soc Lond B 256:293–298
Heinsohn R, Cockburn A, Mulder RA (1990) Avian cooperative breeding: old hypotheses and new direction. Trends Ecol Evol 5:403–407
Heinsohn R, Legge S (1999) The cost of helping. Trends Ecol Evol 14:53–57
Hensley FR (1993) Ontogenetic loss of phenotypic plasticity of age at metamorphosis in tadpoles. Ecology 74:2405–2412
Hentschel BT (1999) Complex life cycles in a variable environment: predicting when the timing of metamorphosis shifts from resource dependent to developmentally fixed. Am Nat 154:549–558
Higashi M, Yamamura N, Abe T, Burns TP (1991) Why don't all termite species have a sterile worker caste? Proc R Soc Lond B 246:25–29
Hölldobler B, Wilson EO (2005) Eusociality: origin and consequences. Proc Natl Acad Sci U S A 102:13367–13371
Hölldobler B, Wilson EO (2008) The superorganism. The beauty, elegance, and strangeness of insect societies. WW Norton & Co, New York, NY
Hopkins TL, Kramer KJ (1992) Insect cuticle sclerotization. Annu Rev Entomol 37:273–302
Horiuchi S, Yamamura N, Abe T (2002) Soldier production strategy in lower termites: from young instars or old instars? J Theor Biol 218:195–205
Huber I (1976) Evolutionary trends in *Cryptocercus punctulatus* (Blattaria: Cryptocercidae). J N Y Entomol Soc 84:166–168
Hunt JH (2007) The evolution of social wasps. Oxford University Press, Oxford
Ikeda-Ohtsubo W, Brune A (2009) Cospeciation of termite gut flagellates and their bacterial endosymbionts: *Trichonympha* species and '*Candidatus* Endomicrobium trichonymphae'. Mol Ecol 18:332–342
Imms AD (1931) Recent advances in entomology. P Blakiston's Son & Co, Philadephia, PA
Inward D, Beccaloni G, Eggleton P (2007a) Death of an order: a comprehensive molecular phylogenetic study confirms that termites are eusocial cockroaches. Biol Lett 3:331–335
Inward DJG, Vogler AP, Eggleton P (2007b) A comprehensive phylogenetic analysis of termites (Isoptera) illuminates key aspects of their evolutionary biology. Mol Phylogenet Evol 44: 953–967
Jacob F (1977) Evolution and tinkering. Science 196:1161–1166
Jamieson IG (1989) Behavioral heterochrony and the evolution of birds' helping at the nest: an unselected consequence of communal breeding? Am Nat 133:394–406
Johnston ML, Wheeler DE (2005) The role of storage proteins in colony-founding in termites. Insectes Soc 54:383–387
Kennedy CH (1947) Child labor of the termite society versus adult labor of the ant society. Sci Mon 65:309–324
Kidder GW (1937) The intestinal protozoa of the wood-feeding roach *Panesthia*. Parasitology 29:163–219
Kikukawa S, Tobe SS (1986) Critical periods for juvenile hormone sensitivity during larval life of female *Diploptera punctata*. J Insect Physiol 32:1035–1042
Klass K-D, Nalepa CA, Lo N (2008) Wood-feeding cockroaches as models for termite evolution (Insecta: Dictyoptera): *Cryptocercus* vs. *Perisphaeria*. Mol Phylogenet Evol 46:809–817
Kölliker M (2005) Ontogeny in the family. Behav Genet 35:7–18

Koshikawa S, Matsumoto T, Miura T (2002) Morphometric changes during soldier differentiation of the damp-wood termite *Hodotermes japonica* (Isoptera: Termopsidae). Insectes Soc 49: 245–250

Kupfer A, Müller H, Antoniazzi MM et al (2006) Parental investment by skin feeding in a caecilian amphibian. Nature 440:926–929

LaFage JP, Nutting WL (1978) Nutrient dynamics of termites. In: Brian MV (ed) Production ecology of ants and termites. Cambridge University Press, Cambridge, pp 165–232

Langen TA (2000) Prolonged offspring dependence and cooperative breeding in birds. Behav Ecol 11:367–377

Ligon JD, Burt DB (2004) Evolutionary origins. In: Koenig WD, Dickinson JL (eds) Ecology and evolution of cooperative breeding in birds. Cambridge University Press, Cambridge, pp 5–34

Lo N, Bandi C, Watanabe H et al (2003) Evidence for cocladogenesis between diverse dictyopteran lineages and their intracellular endosymbionts. Mol Biol Evol 20:907–913

Lo N, Engel MS, Cameron S et al (2007) Save Isoptera. A comment on Inward et al. Biol Lett 3:562–563

Lo N, Tokuda G, Watanabe H et al (2000) Evidence from multiple gene sequences indicated that termites evolved from wood-feeding cockroaches. Curr Biol 10:801–804

Lüscher M (1953) The termite and the cell. Sci Am 188:74–78

Machida M, Kitade O, Miura T, Matsumoto T (2001) Nitrogen recycling through proctodeal trophallaxis in the Japanese damp-wood termite *Hodotermes japonica* (Isoptera: Termopsidae). Insectes Soc 48:52–56

Maekawa K (1997) Molecular phylogeny and ecology of the wood-feeding cockroaches. Master of Science Thesis, Department of Biology. University of Tokyo, Tokyo

Maekawa K, Lo N, Rose HA, Matsumoto T (2003) The evolution of soil-burrowing cockroaches (Blattaria: Blaberidae) from wood-burrowing ancestors following an invasion of the latter from Asia into Australia. Proc R Soc Lond B 270:1301–1307

Maekawa K, Matsumoto T, Nalepa CA (2008) Social biology of the wood-feeding cockroach genus *Salganea* (Dictyoptera: Blaberidae: Panesthiinae): did ovoviviparity prevent the evolution of eusociality in the lineage? Insectes Soc 55:107–114

Matsuda R (1979) Abnormal metamorphosis and arthropod evolution. In: Gupta AP (ed) Arthopod phylogeny. Van Nostrand Reinhold, New York, NY, pp 135–256

May RM, Rubenstein DI (1984) Reproductive strategies. In: Austin CR, Short RV (eds) Reproduction in mammals 4: reproductive fitness. Cambridge University Press, Cambridge, pp 1–23

McKinney ML, Gittleman JL (1995) Ontogeny and phylogeny: tinkering with covariation in life history, morphology, and behaviour. In: McNamara KJ (ed) Evolutionary change and heterochrony. Wiley, Chichester, pp 21–47

Metcalf NB, Monaghan P (2001) Compensation for a bad start: grow now, pay later? Trends Ecol Evol 16:254–260

Minelli A (2003) The development of animal form: ontogeny, morphology, and evolution. Cambridge University Press, Cambridge

Minelli A (2009) Perspectives in animal phylogeny and evolution. Oxford University Press, Oxford

Mitter C, Farrell B, Weigmann B (1988) The phylogenetic study of adaptive zones: has phytophagy promoted insect diversification? Am Nat 132:107–128

Moczek AP (2008) On the origins of novelty in development and evolution. BioEssays 30:432–447

Nalepa CA (1988) Cost of parental care in *Cryptocercus punctulatus* Scudder (Dictyoptera: Cryptocercidae). Behav Ecol Sociobiol 23:135–140

Nalepa CA (1990) Early development of nymphs and establishment of hindgut symbiosis in *Cryptocercus punctulatus* (Dictyoptera: Cryptocercidae). Ann Entomol Soc Am 83: 786–789

Nalepa CA (1991) Ancestral transfer of symbionts between cockroaches and termites: an unlikely scenario. Proc R Soc Lond B 246:185–189

Nalepa CA (1994) Nourishment and the evolution of termite eusociality. In: Hunt JH, Nalepa CA (eds) Nourishment and evolution in insect societies. Westview Press, Boulder, CO, pp 57–104

Nalepa CA (2010) Altricial development in subsocial cockroach ancestors: foundation for phenotypic plasticity in extant termites. Evol Dev 12:95–105

Nalepa CA, Bandi C (2000) Characterizing the ancestors: paedomorphosis and termite evolution. In: Abe T, Bignell DE, Higashi M (eds) Termites: Evolution, sociality, symbioses, ecology. Kluwer Academic Publishers, Dordrecht, pp 53–75

Nalepa CA, Bell WJ (1997) Postovulation parental investment and parental care in cockroaches. In: Choe JC, Crespi BJ (eds) Social behavior in insects and arachnids. Cambridge University Press, Cambridge, pp 26–51

Nalepa CA, Bignell D, Bandi C (2001) Detritivory, coprophagy, and the evolution of digestive mutualisms in Dictyoptera. Insectes Soc 48:194–201

Nalepa CA, Jones SC (1991) Evolution of monogamy in termites. Biol Rev 66:83–97

Nalepa CA, Maekawa K, Shimada K et al (2008) Altricial development in subsocial wood-feeding cockroaches. Zool Sci 25:1190–1198

Nalepa CA, Mullins DE (1992) Initial reproductive investment and parental body size in *Cryptocercus punctulatus* Scudder (Dictyoptera: Cryptocercidae). Physiol Entomol 17: 255–259

Nalepa CA, Mullins DE (2009) Egg size, hatching asynchrony, survivorship and development in young colonies of the subsocial cockroach *Cryptocercus punctulatus* (Dictyoptera: Cryptocercidae). Sociobiology 54:489–508

Nice MM (1962) Development of behavior in precocial birds. Trans Linnean Soc N Y 8:1–211

Noirot C (1982) La caste des ouvriers, élément majeur du succès évolutif des termites. Riv Biol 75:157–196

Noirot C (1985a) Pathways of caste development in the lower termites. In: Watson JAL, Okot-Kotber BM, Noirot C (eds) Caste differentiation in social insects. Pergamon Press, Oxford, pp 41–57

Noirot C (1985b) The caste system in higher termites. In: Watson JAL, Okot-Kotber BM, Noirot C (eds) Caste differentiation in social insects. Pergamon Press, Oxford, pp 75–86

Noirot C, Darlington JPEC (2000) Termite nests: architecture, regulation and defence. In: Abe T, Bignell DE, Higashi M (eds) Termites: Evolution, sociality, symbioses, ecology. Kluwer Academic Publishers, Dordrecht, pp 121–139

Noirot C, Pasteels JM (1987) Ontogenetic development and the evolution of the worker caste in termites. Experientia 43:851–952

Noirot C, Pasteels JM (1988) The worker caste is polyphyletic in termites. Sociobiology 14: 15–20

Nutting WL (1969) Flight and colony foundation. In: Krishna K, Weesner FM (eds) Biology of termites, vol 1. Academic Press, New York, NY, pp 233–282

Obata Y (1988) Behavioral and ecological studies on life history traits and familial relationship in the wood-feeding cockroaches *Panesthia angustipennis spadica* Shiraki, *Salganea esakii* and *S. taiwanensis* Roth (Blattaria: Blaberidae, Panesthiinae). Master of Science Thesis, Department of Biology, University of Tokyo, Tokyo

O'Connor RJ (1984) The growth and development of birds. Wiley, Chichester

Ohkuma M, Noda S, Hongoh Y et al (2009) Inheritance and diversification of symbiotic trichonymphid flagellates from a common ancestor of termites and the cockroach *Cryptocercus*. Proc R Soc Lond B 276:239–245

Olson VA, Liker A, Freckleton RP, Szekely T (2008) Parental conflict in birds: comparative analyses of offspring development, ecology and mating opportunities. Proc R Soc Lond B 275:301–307

O'Reilly JC, Fenolio D, Raina LC, Wilkinson M (1998) Altriciality and extended parental care in the West African Caecilian *Geotrypetes seraphini* (Gymnophiona: Caeciliidae). Am Zool 38:187A

Park Y-C, Choe JC (2003) Territorial behavior of the Korean wood-feeding cockroach, *Cryptocercus kyebangensis*. J Ethol 21:79–85

Prestwich GD (1984) Defense mechanisms of termites. Annu Rev Entomol 29:201–232

Price T, Langen T (1992) Evolution of correlated characters. Trends Ecol Evol 7:307–310

Raff RA (1996) The shape of life: genes, development, and the evolution of animal form. The University of Chicago Press, Chicago, IL

Raff RA (2008) Origins of the other metazoan body plans: the evolution of larval forms. Phil Trans R Soc Lond B 363:1473–1479

Raina A, Osbrink W, Park YI (2004) Nymphs of the Formosan subterranean termite (Isoptera: Rhinotermitidae): aspects of formation and transformation. Ann Entomol Soc Am 97:757–764

Reilly SM (1994) The ecological morphology of metamorphosis: heterochrony and the evolution of feeding mechanisms in salamanders. In: Wainright PC, Reilly SM (eds) Ecological morphology. Integrative organismal biology. The University of Chicago Press, Chicago, IL, pp 319–338

Ricklefs RE (1974) Energetics of reproduction in birds. In: Paynter RA (ed) Avian energetics 15. Nuttall Ornithol Club, Cambridge, MA, pp 152–297

Ricklefs RE (1983) Avian postnatal development. J Avian Biol 7:1–83

Roisin Y (1999) Philopatric reproduction, a prime mover in the evolution of termite sociality? Insectes Soc 46:297–305

Roisin Y (2000) Diversity and evolution of caste patterns. In: Abe T, Bignell DE, Higashi M (eds) Termites: Evolution, sociality, symbioses, ecology. Kluwer Academic Publishers, Dordrecht, pp 95–119

Roisin Y (2006) Life history, life types and caste evolution in termites. In: Kipyatkov VE (ed) Life cycles in social insects: Behaviour, ecology and evolution. St. Petersburg University Press, St. Petersburg, pp 85–95

Rollo CD (1986) A test of the principle of allocation using 2 sympatric species of cockroaches. Ecology 67:616–628

Rombough P (2006) Developmental costs and the partitioning of metabolic energy. In: Warburton SJ et al (eds) Comparative developmental physiology. Contributions, tools, and trends. Oxford University Press, Oxford, pp 99–123

Rosengaus RB, Maxmen AB, Coates LE, Traniello JFA (1998) Disease resistance: a benefit of sociality in the dampwood termite *Zootermopsis angusticollis* (Isoptera: Termopsidae). Behav Ecol Sociobiol 44:125–134

Roux EA, Korb J (2004) Evolution of eusociality and the soldier caste in termites: a validation of the intrinsic benefit hypothesis. J Evol Biol 17:869–875

Rugg D, Rose HA (1990) Nymphal development and adult longevity of the Australian wood-feeding cockroach *Panesthia cribrata* (Dictyoptera: Blaberidae). Ann Entomol Soc Am 83:766–775

Russell AF, Langmore NE, Cockburn A et al (2007) Reduced egg investment can conceal helper effects in cooperatively breeding birds. Science 317:941–944

Sands WA (1965) Alate development and colony foundation in five species of *Trinervitermes* (Isoptera, Nasutitermitinae) in Nigeria, West Africa. Insectes Soc 12:117–130

Scholtz OI, Macleod N, Eggleton P (2008) Termite defence strategies: a reassessment of Prestwich's classification and an examination of the evolution of defence morphology using extended eigenshape anayses of head morphology. Zool J Linn Soc 153:631–650

Scriber JM, Slansky F Jr (1981) The nutritional ecology of immature insects. Annu Rev Entomol 26:183–211

Seamans L, Woodruff LC (1939) Some factors influencing the number of molts of the German roach. J Kans Entomol Soc 12:73–76

Seelinger G, Seelinger U (1983) On the social organization, alarm and fighting in the primitive cockroach *Cryptocercus punctulatus*. Z Tierpsychol 61:315–333

Shaw G, Renfree MB (2006) Parturition and perfect prematurity: birth in marsupials. Aust J Zool 54:139–149

Shimada K, Maekawa K (2008) Correlation between social structure and nymphal wood-digestion ability in the xylophagous cockroaches *Salganea esakii* and *Panesthia angustipennis*. Sociobiology 52:417–427

Smith KK (2001) Early development of the neural plate, neural crest and facial region of marsupials. J Anat 199:121–131
Spicer JI, Burggren WW (2003) Development of physiological regulatory systems: altering the timing of crucial events. Zoology 106:91–99
Staal GB (1986) Anti-juvenile hormone agents. Annu Rev Entomol 31:391–429
Starck JM (1993) Evolution of avian ontogenies. Curr Ornithol 10:275–366
Sugumaran M (2002) Comparative biochemistry of eumelanogenesis and the protective roles of phenoloxidase and melanin in insects. Pigment Cell Res 15:2–9
Taborsky M (1984) Broodcare helpers in the cichlid fish *Lamprologus richardi*: their costs and benefits. Anim Behav 32:1236–1252
Tallamy DW (2005) Egg dumping in insects. Annu Rev Entomol 50:347–370
Tanaka K (1981) Regulation of body size during larval development in the German cockroach *Blattella germanica*. J Insect Physiol 27:587–592
Temrin H, Tullberg BS (1995) A phylogenetic analysis of the evolution of avian mating systems in relation to altricial and precocial young. Behav Ecol 6:296–307
Thomas G, Székely T (2005) Evolutionary pathways in shorebird breeding systems: sexual conflict, parental care, and chick development. Evolution 59:2222–2230
Thompson GJ, Kitade O, Lo N, Crozier RH (2000) Phylogenetic evidence for a single, ancestral origin of a 'true' worker caste in termites. J Evol Biol 13:869–881
Thorne BL, Breisch NL, Muscedere ML (2003) Evolution of eusociality and the soldier caste in termites: Influence of intraspecific competition and accelerated inheritance. Proc Natl Acad Sci U S A 100:12808–12813
Traniello JFA, Rosengaus RB, Savoie K (2002) The development of immunity in a social insect: evidence for the group facilitation of disease resistance. Proc Natl Acad Sci U S A 99: 6838–6842
True JR, Haag ES (2001) Developmental system drift and flexibility in evolutionary trajectories. Evol Dev 3:109–119
Walser ES (1977) Maternal behaviour in mammals. Symp Zool Soc Lond 41:313–331
Ware JL, Litman J, Klass K-D, Spearman LA (2008) Relationships among major lineages of Dictyoptera: the effect of outgroup selection on dictyopteran tree topology. Syst Entomol 33:429–450
Wcislo WT (1989) Behavioral environments and evolutionary change. Annu Rev Ecol Syst 20:137–169
Wesolowski T (1994) On the origin of parental care and the early evolution of male and female parental roles in birds. Am Nat 143:39–58
West MJ, Alexander RD (1963) Sub-social behavior in a burrowing cricket *Anurogryllus muticus* (De Geer). Ohio J Sci 63:19–24
West-Eberhard MJ (1987) The epigenetical origins of insect sociality. In: Eder J, Rembold H (eds) Chemistry and biology of social insects. Verlag J. Peperny, München, pp 369–372
West-Eberhard MJ (2005) Phenotypic accommodation: adaptive innovation due to developmental plasticity. J Exp Zool 304B:610–618
Wheeler WM (1928) The social insects. Their origin and evolution. Harcourt, Brace and Company, New York, NY
Williams GC (1966) Adaptation and natural selection. Princeton University Press, Princeton, NJ
Wolf JB, Brodie ED III (1998) The coadaptation of parental and offspring characters. Evolution 52:299–308
Woodrow RJ, Grace JK, Nelson LJ, Haverty MI (2000) Modification of cuticular hydrocarbons of *Cryptotermes brevis* (Isoptera: Kalotermitidae) in response to temperature and relative humidity. Environ Entomol 29:1100–1107
Yanagawa A, Shimizu S (2007) Resistance of the termite, *Coptotermes formosanus* to *Metarhizium anisopliae* due to grooming. BioControl 52:75–85
Zeveloff SI, Boyce MS (1982) Why human neonates are so altricial. Am Nat 120:537–542

Chapter 5
Eusocial Evolution in Termites and Hymenoptera

Kenneth J. Howard and Barbara L. Thorne

Abstract The evolution of eusociality remains an intriguing mystery. Why do individuals forego their own reproduction to help others produce offspring? Complex colonies have evolved in ants, bees, and wasps (Hymenoptera) and in termites (Isoptera). We discuss theories applied to explain eusocial traits in the Isoptera and the Hymenoptera in order to learn more about how and why eusociality evolved from subsocial ancestors. Striking overlap occurs in the genes controlling caste and the shift of parental care to offspring via heterochrony, suggesting ecological pressures lead to co-option of similar developmental mechanisms. There is also overlap in factors that predispose groups to eusociality, including extended parental care, defense of and the inheritance of a valuable nest, and enhanced direct and indirect benefits from cooperative behavior. However, differences in developmental paths and ecological traits are also informative, such as the evolution of the soldier form and adult workers. Multiple selective processes may favor eusociality over alternatives, but the degree to which each process was involved in eusocial evolution may have varied in each taxon. Reduced emphasis on haplodiploidy to explain eusocial evolution in the Hymenoptera and the similarities in developmental control allow for greater overlap of theories explaining eusocial evolution in the Hymenoptera and Isoptera.

5.1 Introduction

The evolution of eusociality in insects remains one of the great mysteries in biology. The central reason for excitement among social biologists relates to the evolution of reproductive division of labor: Why have some individuals evolved to restrict or forego their own opportunities to reproduce in order to help others? Eusociality has evolved independently in a diversity of taxa, including Hymenoptera (ants, bees, and wasps), Isoptera (termites), Thysanoptera (thrips), Coleoptera (beetles), crustaceans

K.J. Howard (✉)
Department of Entomology, University of Maryland, College Park, MD 20742, USA
e-mail: howark2@sage.edu

D.E. Bignell et al. (eds.), *Biology of Termites: A Modern Synthesis*,

DOI 10.1007/978-90-481-3977-4_5,

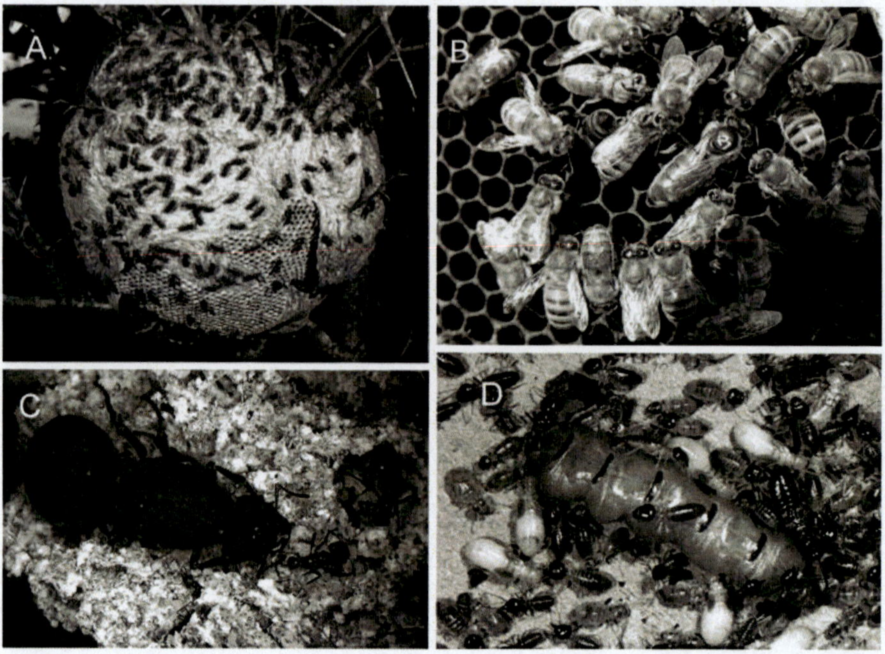

Fig. 5.1 Complex eusocial conlonies have evolved independently in the Hymenoptera: (**a**) wasps, e.g. *Brachygastra mellifica* (photo copyright R. L. Jeanne); (**b**) bees, e.g. *Apis mellifera* (photo copyright D. R. Tarpy); and (**c**) ants, e.g. *Atta cephalotes* (photo copyright M. Poulsen); and the Isoptera; (**d**) termites, e.g. *Nasutitermes corniger* (photo copyright B.L. Thorne)

(shrimp), and mammals (naked mole rats) (Sherman et al. 1991; Duffy 1996; Choe and Crespi 1997). That eusociality occurs in diverse taxa, yet is a phylogenetically rare event, suggests that uncommon suites of factors are responsible (Hansell 1987; Crespi 1996; Thorne 1997; Hunt 1999; Wilson and Hölldobler 2005a). Large colonies with complex castes and behaviors are even rarer, having arisen only in the Hymenoptera and Isoptera (Fig. 5.1).

Attention has focused on the Hymenoptera due to their haplodiploid genetic system and ease of study. Termites have received less attention, despite the fact that termites are the only diploid group that form highly complex colonies and have an extraordinary diversity of castes (Choe and Crespi 1997; Thorne 1997). Recent advances indicate great convergence in mechanisms underlying the development of worker forms (see below). In addition, similar ecological pressures on social colonies lead to striking similarities among taxa, and suggest both ancestral and modern eusocial forms have faced similar challenges. Here we compare and contrast theories for eusociality applied to termites and to Hymenoptera, allowing for differing phylogeny and life history (Crespi 1996; Hunt 1999), to reveal insights and new directions for investigating eusociality.

While numerous studies use Tinbergen's four questions to categorize behavioral research (Alcock and Sherman 1994), the questions are not well-matched to the two questions that drive most research into eusocial evolution: what factors promote the evolution of eusociality in some taxa and not others, and how can selection favor helper forms that sacrifice much, if not all, of their own reproduction? In this review, we first discuss theories focused on predisposing traits that include both proximate mechanisms and ecological pressures (as formulated by Gadagkar 2001), and then discuss selective processes that permit altruistic helpers.

A distinction is made between the initial appearance of eusociality and later maintenance within populations. Few species have both eusocial and subsocial forms. This is true of termites and ants, which have no modern groups that shift between subsocial and eusocial forms such as we see in the primitively eusocial bees (Wilson 1971; Thorne 1997; Schwarz et al. 2006), nor are many intermediates present in the fossil record due to rapid diversification following origins (Nalepa and Bandi 2000; Thorne et al. 2000; Hölldobler and Wilson 1990; Crozier et al. 1997). Most eusocial species, including extant termites, are unlikely to revert to solitary or subsocial lifestyles even when factors are present that would favor dissolution of groups at eusocial origins (Thorne and Traniello 2003; Wilson and Hölldobler 2005a). At origin, eusocial traits would be rare and many supporting traits absent, so initial conditions may differ from current ones (Charnov 1978; Crespi 1996; Thorne 1997). That said, many eusocial species, including primitive termites, have helpers with potential to reproduce, though they usually do not.

We begin by outlining the evolution of eusociality from subsocial ancestors and indicating distinctions between Hymenoptera and Isoptera. We then describe the most common forms of altruistic helpers in each taxon. We outline potentially predisposing traits, followed by selective processes that may favor helping behavior. Finally, we discuss insights from the comparisons.

5.2 Evolution from Subsocial Ancestors

5.2.1 Origins of Eusociality

The majority of, if not all, eusocial species are now thought to have arisen from parent-offspring groupings, referred to as subsocial (West-Eberhard 1987; Noirot 1985b; Alexander et al. 1991; Hunt 1994; Thorne 1997; Helanterä and Bargum 2007). Eusocial termites are thought to have evolved in the upper Jurassic or lower Cretaceous from a subsocial ancestor nesting and feeding in wood (Thorne et al. 2000). Whether these subsocial ancestors resembled modern wood roaches in the genus *Cryptocercus* has been the subject of debate (reviewed by Nalepa and Bandi 2000). Phylogenetic analyses indicate that *Cryptocercus* is the sister taxon to the termites (Lo et al. 2000; Inward et al. 2007a). *Cryptocercus* forms subsocial groupings that exhibit extensive parental care, have totipotent juvenile helpers, live in the wood they feed on, and also house in their guts assemblages of symbiotic microbiota similar to those of several termite families (Cleveland et al. 1934; Nalepa 1994; Thorne 1997; Shellman-Reeve 1997; Nalepa and Bandi 2000). However, as

Cryptocercus roaches have not evolved eusociality, we must be cautious in assuming termite ancestors had similar social traits (Myles 1988; Thorne 1997).

Termite species are generally grouped in three categories: one-piece nesting in which the termites nest in and feed on a single piece of wood, intermediate-type nesting in which the termites build subterranean galleries to new pieces of wood, and separate-piece nesting in which the termites always forage or feed away from the nest (Abe 1987). In the monotypic family Mastotermitidae, *Mastotermes darwiniensis* diverges at the base of the termite phylogenetic tree (Inward et al. 2007b). While *M. darwiniensis* has distinct plesiomorphic features, it also nests in the soil and has other characteristics typical of more derived separate-piece nesting termites (Thorne and Carpenter 1992; Lenz 1994; Thompson et al. 2000; Goodisman and Crozier 2002). Although the genera *Zootermopsis* and *Archotermopsis* within Termopsidae diverge from other termite families immediately after Mastotermitidae (Inward et al. 2007b; but see Legendre et al. 2008), termopsids are recognized by many to have colony sizes, social organization, nesting biology, and caste polyphenism most similar to putative ancestral termites (Imms 1919; Noirot 1985b; Grassé 1986; Thorne and Carpenter 1992; Nalepa 1994; Thorne 1997; Shellman-Reeve 1997; Inward et al. 2007b).

Eusociality is thought to have evolved independently at least seven times in the Hymenoptera (Wilson 1971; Gadagkar 2001; Wilson and Hölldobler 2005b; Wilson 2008). Bee and wasp species show relatively clear gradations from solitary to eusocial lives, including multiple origins and reversions in the halictid bees (Wilson 1971; Schwarz et al. 2006). Bees and wasps likely arose in the Lower Cretaceous period, though eusociality may have arisen much later in the vespid wasps (Upper Cretaceous) and halictine and allodapine bees (Tertiary) (Grimaldi and Engel 2005; Schwarz et al. 2006; Poinar and Danforth 2006). While ants lack extant intermediate forms, fossils of the extinct vespoid wasp subfamily Sphecomyrminae give clues to ancestral forms and link ants to the rest of the wasps (Wilson 1987; Hölldobler and Wilson 1990). Fossil and molecular evidence suggest that ants diverged from the Sphecomyrminae approximately 105–135 million years ago, followed by rapid diversification associated with the rise of angiosperm plants in the mid-Cretaceous period (Crozier et al. 1997; Wilson and Hölldobler 2005b; Brady et al. 2006).

5.2.2 Differences Between Hymenoptera and Isoptera Affecting Eusocial Evolution

One clear difference between Isoptera and Hymenoptera is the development path from juvenile to adult. Hymenoptera are holometabolous, transitioning from a helpless, vermiform larva into a morphologically dissimilar adult form via a pupal stage (Fig. 5.2). Termites are hemimetabolous, meaning individuals proceed through several stadiums during successive instars, each with the basic adult body plan, and becoming semi-independent relatively early in development. Juvenile termites are thus better equipped to perform tasks in a social colony (Kennedy 1947; Noirot and Pasteels 1987; Alexander et al. 1991). Aside from a few limited instances in which

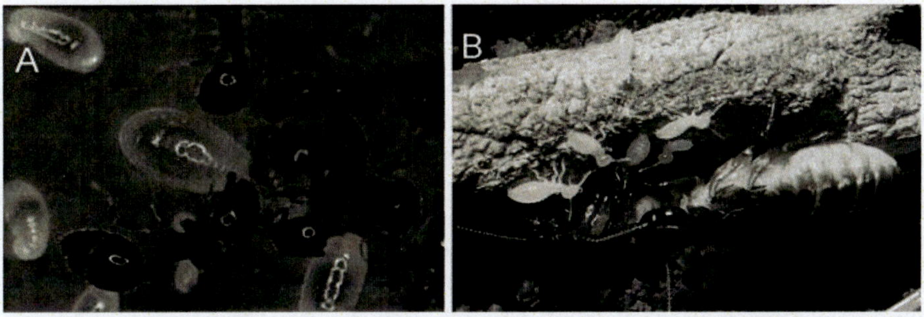

Fig. 5.2 Juvenile forms of Hymenoptera and Isoptera depend heavily on associations with their parents. (**a**) A worker (*Temnothorax longispinosus*) is exchanging liquids with a larva. Hymenopteran offspring are relatively helpless until the adult stage (photo copyright K. J. Howard). (**b**) Early instar termite larvae, such as these *Zootermopsis nevadensis*, are mobile, but must obtain gut symbionts from older individuals to digest the cellulose in wood efficiently (photo copyright B.L. Thorne)

larvae produce nutrients or silk (see Wilson 1975), most tasks in hymenopteran colonies are performed by adult workers.

The genetic system also differs. All hymenopterans are haplodiploid; all termites are diplodiploid. While termites do not have clear genetic differences between males and females as in the Hymenoptera, relatedness asymmetries based on other processes are potential predisposing factors (see below). Hamilton (1964) suggested that haplodiploidy may be the reason hymenopteran workers are all female, while most termites have both male and female workers, with less dramatic genetic differences between them. Hymenopteran female helpers may have contributed to eusocial evolution through maternal effects (Linksvayer and Wade 2005), but the pre-condition of female-only parental care and stingers in solitary hymenopteran ancestors are more likely to underlie the evolution of female-only helpers (Craig 1982; Andersson 1984; Pamilo 1991; Queller and Strassmann 1998; Wilson 2008, but see Sen and Gadagkar 2005).

Ecological pressures may also have differed, making universal ecological explanations of eusociality difficult (Queller and Strassmann 1998). Ancestral termites are thought to have nested in or near their food, making defense of this resource paramount and specialized foragers less critical. Soldiers evolved very early during the evolution of eusocial termites. In contrast, a key role for helpers in the Hymenoptera was likely foraging outside the nest rather than nest defense. Furthermore, most hymenopteran workers have stings, which may reduce the need for soldiers (Oster and Wilson 1978). Despite these differences, there still may be quite a bit of overlap in ecological pressures faced by ancestral forms of both groups.

Bees and wasps are distinct from both ants and termites for two additional reasons: their adults are winged throughout their lives, giving them greater mobility, and their offspring are raised in individual cells, which may constrain developmental plasticity.

5.3 Forms of Helpers

The function of helpers and their reproductive capacity varies widely among social insects. Generally, as colonies evolve to become larger and more complex at maturity, helpers lose reproductive options earlier in development (Bourke 1999, Fig. 5.3). This may stem from declining chances for direct reproduction, enhanced abilities to complete tasks while requiring less energy to develop, or less need to retain dispersal options as resources become more stable.

In all termites, there is a bifurcation during development into two pathways: the sexual (imaginal or nymphal) line identified by developing wing buds, and the apterous line which generally leads to neotenics, workers and soldiers. The developmental stage at which the two lines diverge is later and more plastic in the Termopsidae, Kalotermitidae, and some Rhinotermitidae, and earlier and more rigid in Mastotermitidae, Hodotermitidae, some Rhinotermitidae and Termitidae (Noirot 1985a, b; Noirot and Pasteels 1987, 1988). Caste determination generally involves a flexible response to hormonal changes within developing juveniles in response to queen and king signals, nutrition, and physical confrontations (reviewed by Roisin 2000), though there are exceptions in which inheritance or substances in the eggs play a role (e.g. Hayashi et al. 2007). Juvenile hormone (JH) generally impedes alate

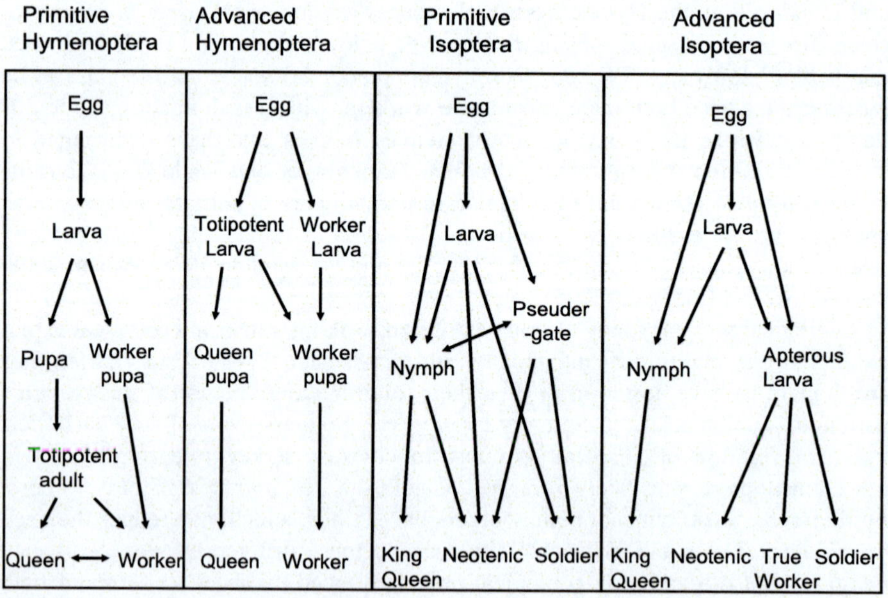

Fig. 5.3 Developmental plasticity within different hymenopteran and isopteran levels of eusociality. In ancestral-like forms, individuals retain totipotency later into development. In derived forms, reproductive and non-reproductive individuals are differentiated earlier in development and workers have fewer reproductive capabilities

or neotenic development, though the specific roles of different hormones in caste determination are unclear (Roisin 2000; Hartfelder and Emlen 2005). Hexamerin proteins favor juvenile stasis by sequestering JH and preventing soldier development (Zhou et al. 2006), and vitellogenin proteins may play similar roles (Scharf et al. 2005).

Hymenopteran juveniles exhibit similar developmental plasticity to that of termites, and a similar shift to earlier developmental bifurcations in caste as eusociality elaborates (Wilson 1971; Hartfelder and Emlen 2005). In primitively eusocial species, both behavioral and physiological interactions among adults and feeding levels of the larvae lead to individuals filling worker roles, with only moderate morphological distinctions between queens and workers (Wilson 1971; Michener 1990; O'Donnell 1998; Sledge et al. 2001; Pereboom et al. 2003; Bloch and Hefetz 1999, Fig. 5.2). In more derived species, workers tend to have specialized forms distinct from queens, and larvae are biased towards queen or worker development via differential feeding and biting by care-giving workers (Wilson 1971; Brian 1973; Wheeler 1986; Hartfelder and Emlen 2005), though genetic biases occur rarely (e.g. Kerr 1950; Fraser et al. 2000; Julian et al. 2002). With few exceptions, switch mechanisms in both primitive and advanced eusocial taxa lead to changes in JH and/or vitellogenin titers within the larvae or adult, shifting caste and egg productions (reviewed by Hartfelder and Emlen 2005).

Three important novelties occurred in the evolution and elaboration of termite eusociality: the origin of totipotent juvenile helpers, the evolution of the soldier caste, and the evolution of "true", non-reproductive workers (Roisin 1999; Higashi et al. 2000). These three forms also occur in hymenopteran social colonies, though not with the same frequencies or with such clear distinction. We review their major characteristics below.

5.3.1 Helpers Retaining Full Reproductive Potential

Though it remains unclear whether totipotent helpers are an ancestral or derived trait within termites (Inward et al. 2007b), the subsocial ancestor presumably started with totipotent offspring prior to the evolution of eusociality (Thorne 1997). All extant termite species have juvenile helpers, but all non-soldier helpers are totipotent in the one-piece nesting species (families Termopsidae, Kalotermitidae, and some Rhinotermitidae) (Thorne 1997; Shellman-Reeve 1997). Juvenile helpers in termites, which may include larvae, nymphs, and pseudergates, are often referred to as "workers" in the functional sense, though they are not considered "true" workers because they may develop later into alate forms (Noirot and Pasteels 1987; Roisin 2000; Thorne and Traniello 2003). There is some debate over how much work they perform (Korb 2007, but see Nalepa 1994; Shellman-Reeve 1997). Developmental plasticity of most helpers gives colonies tremendous flexibility in responses to environmental change and deterioration of the nest and food resources (Higashi et al. 1991; Abe 1991).

In primitive bees, wasps, and some ants, workers retain great reproductive potential (Wilson 1971; Michener 1975; West-Eberhard 1975; Ward 1983; Peeters and Ito 2001). For example, polistine wasps and halictid bees have workers capable of mating and reproducing when queens are removed (reviewed by O'Donnell 1998). In the ant subfamilies Myrmeciinae and Ponerinae, forms intermediate between queens and workers occur that may be inseminated (ergatogynes) and some workers have secondarily evolved the ability to be inseminated (gamergates; Peeters and Ito 2001). Workers with full reproductive potential are rare in ants (Wilson 1971).

5.3.2 Soldiers

The occurrence of soldiers, individuals specialized for nest defense or fighting (Fig. 5.4a), is monophyletic among extant termite groups (Hare 1937; Noirot 1985b; Noirot and Pasteels 1987). Only 1–10% of individuals are soldiers in most colonies (Haverty 1977), but at least one soldier tends to develop very early in colony ontogeny (Noirot 1969). Though soldiers in more derived lineages participate in defense against predators, the soldiers of more primitive families such as the Termopsidae and Kalotermitidae function more in interactions with conspecifics (Thorne et al. 2003), as the primary defense against predators appears to be the nest itself (Noirot and Darlington 2000). This may make sense in that soldiers arose prior to the ecological dominance of their most significant modern predator, social ants (Korb 2008). In the Termopsidae, individuals occur that have some soldier characteristics but are also reproductively capable (Heath 1928; Myles 1986, "soldier neotenics" from Thorne 1997). However, it does not appear that soldier

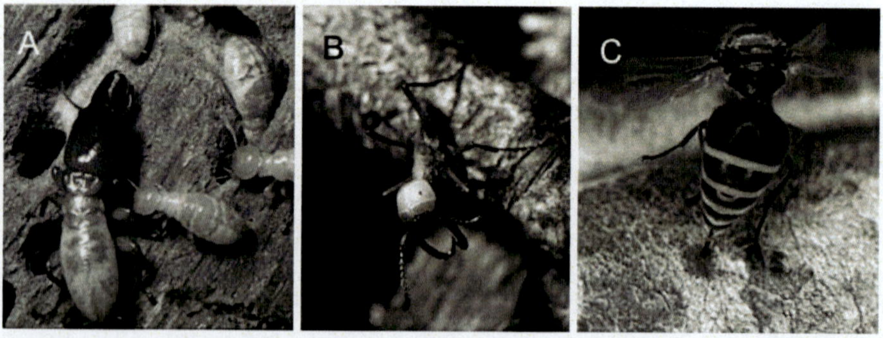

Fig. 5.4 **(a)** In termites, sterile soldiers with morphological specializations, such as the sclerotized head and large mandibles of this *Zootermopsis nevadensis* soldier, evolved prior to subfertile workers (photo copyright B.L. Thorne). **(b)** In ants, soldiers (*Eciton burchelli* major shown, photo copyright R. L. Jeanne) evolved specialized forms subsequent to subfertile workers. **(c)** Soldiers have not evolved in bees or wasps, though most hymenopteran workers have weapons in the form of a sting, such as this *Brachygastra mellifica* worker stinging a glove (photo copyright R. L. Jeanne)

neotenics function as normal soldiers (Roisin 1999). In *Zootermopsis nevadensis* soldier neotenics have greater survival during interactions between colonies, suggesting that soldiers first evolved so that colonies retain reproductives during conspecific challenges (Thorne et al. 2003). Also, in the termopsid *Archotermopsis wroughtoni*, all soldiers and some "worker-like" individuals have gonads that are as well-developed as in mature alates (Imms 1919), though it is not clear whether these forms can produce viable offspring (Thorne 1997). If so, *Archotermopsis* may represent a transitional form between subsocial ancestors and modern eusocial termites. Further evidence is needed on the reproductive capabilities of soldiers and the social organization of *Archotermopsis* colonies.

The evolution of soldiers in Hymenoptera differs markedly from that of the Isoptera (Fig. 5.5). While soldier castes occur in all extant termites except the more highly derived genera of Apicotermitinae (where they are assumed to have been secondarily lost), the majority of ants lack a distinct soldier caste (Wilson 1971). Larger-sized worker castes, referred to as majors, are called soldiers when specialized for defense of the colony (Fig. 5.4b). Thus, soldier castes in ants can have similar morphologies to the other workers, some level of allometry, or complete polymorphisms in which soldiers forms have greatly enlarged heads, mandibles, and defensive glands (Oster and Wilson 1978; Hölldobler and Wilson 1990). Soldiers mostly arise in species with early determination of queen or worker caste, which may facilitate caste diversity in the workers (Wheeler 1986, 1991; Fjerdingstad and Crozier 2006). Soldiers also tend to be absent from young colonies but appear later

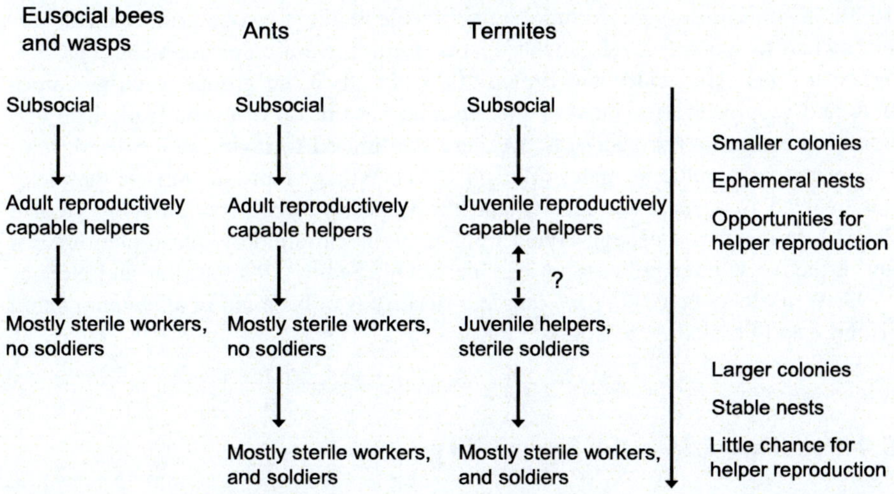

Fig. 5.5 The sequences in the evolution of helpers differed between taxa. The bees and wasps evolved mostly sterile workers, but never morphological soldiers. Ants evolved adult helpers, mostly sterile workers, and then soldiers as colony size increased. Termites evolved juvenile helpers and adult sterile soldiers, though which came first is not clear, then mostly sterile workers as nesting and feeding behavior changed

when colonies become larger (Wilson 1985; Tschinkel 1988). Distinct soldier castes are completely absent from the bees and wasps (Wilson 1971). In some honeybees, individuals that perform defensive tasks have distinct genetic profiles and may be called "soldiers", but it is not clear if they are specialized defenders (Breed et al. 2004). The lack of distinct soldiers in bees and wasps may relate to the defensive properties of strong stings (Oster and Wilson 1978), constrained development in cells, or limitations imposed by flight.

5.3.3 *Subfertile or Sterile Workers*

True workers that are subfertile or sterile may have evolved three times and were lost at least once in Isoptera (Myles 1988; Noirot and Pasteels 1988; Abe 1991; Inward et al. 2007b). The worker caste arises due to an early switch between the alate and apterous developmental lines, so that most of the helpers within the nest enter the apterous line in the egg or soon after (Fig. 5.3; Roisin 2000). True workers are never found in the one-piece nesting termites that must disperse when the wood resource deteriorates (Abe 1991; Shellman-Reeve 1997; Inward et al. 2007b). In most of the separate-piece nesters with large colonies, individuals in the nymphal pathway never act as workers, while individuals developing into workers and soldiers have effectively zero opportunities to reproduce (Shellman-Reeve 1997).

Workers in Hymenoptera have lost reproductive capability as colonies evolved to have larger worker populations and greater complexity (Ratnieks 1988; Bourke 1999). As in termites, the evolution of subfertile workers is associated with an earlier switch to worker developmental paths during juvenile development (Fig. 5.3; Wheeler 1991; Hartfelder and Emlen 2005). In advanced groups, such as epiponine and vespine wasps, most of the eusocial bees aside from the Halictidae and bumblebees, and almost all ants, workers are limited to laying unfertilized, haploid eggs that develop as males (Wilson 1971). Worker reproduction via thelytoky has evolved in several ant species and the Cape honeybee (reviewed by Heinze 2008b). In most species, egg-laying by workers is curtailed by queen pheromones and effective worker policing (Alexander 1974; Seeley 1995; Keller and Nonacs 1993; Wenseleers et al. 2004). Complete sterility is only found in a few ant genera (Bourke and Franks 1995; Bourke 1999).

5.4 Predispositions for Eusociality

The question underlying studies into predispositions is why eusociality arose in some taxa and not others. While the phylogenetic distance between termites and Hymenoptera means suites of predispositions could be unique to each, many causal factors are thought to be in common. Each of the following could have promoted or allowed eusociality to arise, and may also help maintain eusociality.

5.4.1 Extended Parental Care and Parent-Offspring Interaction

Hymenopteran ancestors progressively provisioned offspring, so that extended parental care occurred prior to eusocial colonies (Charnov 1978; Wilson 2008). To form eusocial colonies, offspring must remain at the nest and express parental care behaviors (e.g. nest-building, foraging) before ovipositing themselves (Gadagkar 2001). Recent genetic evidence suggests that shift occurred via altered timing of trait expression, or heterochrony, in bees and social wasps. Many of the same genes expressed in queens are also expressed in workers performing parental care (Amdam et al. 2006; Toth et al. 2007). Two hypotheses propose that physiological states of offspring eased the shift to eusociality. One, the ovarian groundplan hypothesis, suggests that ovarian cycles and parental care behaviors which are normally associated became decoupled, for example in groups of nest-sharing, primitively social wasps (West-Eberhard 1996). Some group members oviposited, while others performed parental care tasks, depending on their dominance status. This hypothesis does not explain why eusociality is always an elaboration of parent-offspring associations (Alexander et al. 1991).

The diapause groundplan hypothesis focuses on parent-offspring groups, suggesting a shift in behavior of a progressively provisioning solitary wasp with a bivoltine lifestyle (Hunt and Amdam 2005). A bivoltine lifestyle requires that some offspring are raised in preparation for diapause (prolonged development) and provisioned with more resources, similar to what is seen in gynes of polistine wasps that overwinter. Other individuals are raised in preparation for reproduction within the same season (quick development) with fewer stored resources, as is seen in polistine workers (Hunt 2007). It is argued that these developmental patterns in *Polistes* wasps are mirrored by patterns in the development of worker honeybees (Amdam et al. 2006). The diapause groundplan hypothesis by itself is not a complete explanation, as it still requires a shift in behavior by non-diapausing offspring to remain at the nest and perform tasks without ovipositing.

Similar transitional shifts have been proposed for termites. Pre-existing parental care may facilitate eusocial evolution in termites if it can be shifted to the juvenile stages, most likely also via heterochrony (Nalepa 1994). In termites, the responsibility for parental care is shifted to the older brood, which may become food donors rather than food recipients, leading to their developmental arrest in a still juvenile form due to food deficiency. Similar developmental plasticity and delays have been observed in non-eusocial cockroaches in response to social environment, food levels, and injuries, suggesting the plasticity was present in the termite ancestors (reviewed by Nalepa and Bandi 2000). This hypothesis is explained in more detail in Nalepa (1994). It is interesting that similar proteins involved in developmental arrest in termites are also involved in caste determination in bees and wasps (Hartfelder and Emlen 2005; Zhou et al. 2006). While this similarity across taxa is extraordinary, it could reflect the convenience of co-opting these developmental pathways for caste control rather than their importance as pre-requisites for eusociality (Wilson 2008).

Extended parental care also provides greater opportunities for adults to manipulate the development of offspring by limiting nutrition (Craig 1983; Hunt 1991; O'Donnell 1998), or to stress or destroy young with reproductive potential (Brian 1973; Hunt 1991; Vargo and Passera 1991; Jeanne 2009).

5.4.2 Incentives for Offspring to Remain at the Nest

Incentives promoting offspring philopatry (remaining at the point of origin, i.e. within the nest) may favor eusocial behavior by facilitating shifts in brood care from parents to offspring. In wasps, larvae in nests often exude nutritional droplets rich in amino acids which entices workers to remain longer on the natal nest (Roubaud 1916; Hunt 2007). Workers disperse when deprived of larval saliva (Kumano and Kasuya 2001). A delay in the maturation process could also favor philopatry (Gadagkar 1991b). If the offspring disperse, they must survive long enough to raise offspring to an independent stage, so the alternative of helping to raise siblings becomes more attractive (see the head-start hypothesis: Queller 1989).

Long development times and incentives for offspring to remain may have also predisposed termite juveniles to evolve helping behaviors (Thorne 1997; Shellman-Reeve 1997). Slow development in termites results from insufficient nutrients, specifically a lack of nitrogen and difficulty processing wood (LaFage and Nutting 1978; Collins 1983; Waller and LaFage 1987; Nalepa 1994; Shellman-Reeve 1997). Young termite colonies grow slowly because egg-production by reproductive females is limited while feeding offspring (Nalepa 1994). As development time increases, older brood overlap more with siblings, potentially facilitating a shift to alloparental care that supports greater egg production (Nalepa 1994).

An additional factor for termites in the families Mastotermitidae, Kalotermitidae, Termopsidae, Hodotermitidae, and Rhinotermitidae is that after every molt, juveniles must receive flagellated protists that improve the efficiency of cellulose digestion. The transfer is achieved by juveniles imbibing proctodeal fluids expelled from the hindguts of older individuals. In the subsocial roach *Cryptocercus*, which has similar gut symbionts, the protists encyst during molting, lessening juvenile dependence on social groups (Cleveland et al. 1934). Cleveland et al. (1934) first noted that required symbiont transfers would extend stays of juveniles and predispose termites to both parental care and social living (the "symbiont transfer" hypothesis; also Lin and Michener 1972; Nalepa 1994). However, termite symbionts appear to be killed by hormones associated with developmental stasis prior to molting, suggesting encysting symbionts were lost after eusociality arose rather than before (Nalepa 1994). Also, subsociality can arise or be maintained in Dictyoptera without extended juvenile persistence (Pellens et al. 2007). Nor does extensive interaction necessarily lead to reproductive division of labor, as evidenced by *Cryptocercus* roaches. Even if not involved in eusocial origins, the necessity of transferring endosymbionts may make reversion to solitary life difficult.

In all taxa, the difficulty of dispersing to found a new colony may promote philopatry (reviewed by Myles 1988). Unfortunately, few data exist for mortality and success rates of dispersing reproductives. Estimates of the survival of termite alates suggest extremely high mortality (more than 99%) prior to successful colony founding (Basalingappa 1970; Myles 1988). Estimates for Hymenoptera also suggest high mortality associated with dispersal and independent colony founding (e.g. Peeters and Ito 2001). With high mortality rates, selection may favor philopatry.

5.4.3 *Valuable, Protected Nests*

All social insects house their offspring in a protective nest, which can favor the evolution of social cooperation to construct and defend (Charnov 1978; Andersson 1984; Alexander et al. 1991; Crespi 1994; Wilson 2008); (see Fig. 5.6). A completed nest is valuable: it necessitates energy and mortality risk to construct or excavate, expand as the colony grows, maintain in the face of disturbances, and defend from predators and competitors. Individuals save time and energy by sharing construction costs (Waloff 1957; Andersson 1984). In addition, nests that produced offspring have demonstrated success, enhancing their inheritance value (Andersson 1984; Alexander et al. 1991). Many eusocial Hymenoptera and some termites store food within their nest, which also enhances its value (Myles 1988; Starr 1991; Breed et al. 2004). The nest also encourages relatives to remain in close proximity, facilitating the evolution of kin-selected (Hamilton 1964, 1978) and permitting parentally manipulated behaviors (Charnov 1978; Alexander et al. 1991).

The importance of the nest in one-piece type and intermediate type nesting termites is two-fold: the nest serves as the main protection against predators (Noirot and Darlington 2000) and as a primary food source (Abe 1987). But since many founding pairs may colonize a dead tree at the same time, the resource is extremely limited. Evidence suggests that in this competitive environment larger groups are more successful retaining space and food when meeting competitors (Thorne et al. 2003). In addition, the chance to retain this ready-made resource by inheritance could select for helpers to forestall dispersal (Myles 1988; Thorne 1997; Thorne et al. 2003).

5.4.4 *Weaponry*

One cost of living in a group is that the dense collection of developing offspring may attract predators, so characteristics that improve defense of the brood may facilitate social behavior (Starr 1990; Hansell 1996). Cooperative defense may be more effective than individual defense when some forms have weapons. For example, the sting may have predisposed Hymenoptera to eusociality by increasing effectiveness of group defense (Starr 1985, 1989; Kukuk et al. 1989; Seger 1991). It is striking that all cases of eusocial evolution in the Hymenoptera have occurred in the aculeates, or stinging wasps, despite a great diversity of species outside this group (Hunt 1999).

Termites have also evolved morphological weaponry in a soldier caste, one that varies tremendously among families and species and may include heavily sclerotized heads, enlarged mandibles, and nozzles to spray toxic or sticky liquids

Fig. 5.6 The nest and assembled offspring that make up the colony are a valuable resource. (**a**) A bumblebee nest (*Bombus impatiens*, photo copyright J. M. Jandt) requires energy and mortality risk to forage for materials, food is stored within the nest, and the developing brood represents invested resources. (**b**) In many cases, the nest structure itself requires significant resource investment to construct and also maintain as the colony grows (pictured: *Polistes instabilis*, photo copyright R. L. Jeanne). (**c**) The excavation of nests (*Lasius* sp. tumulus shown, photo copyright R. L. Jeanne) from a substrate requires energy and sometimes wear-and-tear on the workers. (**d**) Many termites nest inside their source of food, making nests valuable for habitation and for nutritional resources (*Reticulitermes flavipes* is shown, photo copyright B.L. Thorne)

(Noirot and Darlington 2000). It is unclear whether weapons evolved before or after the evolution of juvenile helper forms, as soldiers are monophyletic for all extant termite taxa (Thorne 1997; Roisin 1999; Inward et al. 2007b). The context favoring weaponry is different than that for Hymenoptera. Soldier weapons likely arose for advantage in intrafamilial battles over reproduction (Myles 1988) or due to intraspecific battles for resources (Thorne et al. 2003), while hymenopteran weapons initially evolved in solitary ancestors to subdue prey and repel larger predators.

5.4.5 Efficiency of Foragers

For helpers to be beneficial, they must be efficiently produced and provide meaningful benefits to the group (West-Eberhard 1975; Craig 1983). A primary benefit of helpers is that they bring food to nestmates and brood, so the type of food foraged may limit efficiency if it requires larger foragers. This may be one reason so many bees which feed on pollen that can be carried by smaller-sized helpers are eusocial, while so few of their sister taxon, predatory sphecid wasps, have reached the same level (Strohm and Liebig 2008). The one sphecid wasp that has evolved eusociality, *Microstigmus comes*, feeds on very small collembolan prey (Matthews 1991). It is not clear how this theory could apply to termites. The cloistered nature of primitive termites within their food may permit smaller offspring by limiting exposure to the elements and predators as compared to non-eusocial, non-wood-feeding roaches. On the other hand, competition with neighbors or size constraints related to processing wood may favor larger offspring.

5.4.6 Relatedness Asymmetries

The haplodiploid genetic system results in higher relatedness of the female workers to their sisters than to their brothers. This led Hamilton (1964) to propose that females could benefit more by raising their sisters than their own daughters. A difficulty arises when considering males, to which the female helpers are much less related than their own sons, giving less incentive to help raise their brothers. Trivers and Hare (1976) suggested that biasing toward raising more female reproductives returned the benefits to the female workers. However, eusociality was later discovered in several diploid taxa aside from termites (Choe and Crespi 1997; Wilson and Hölldobler 2005a). Thus the relatedness advantages of haplodiploidy are not so clear over multiple generations (reviewed by Queller and Strassmann 1998), and eusociality is highly clustered within the aculeate Hymenoptera (Hunt 1999). Therefore, current consensus is that haplodiploidy does not favor eusocial evolution via relatedness asymmetries (Gadagkar 1991a, 2001; Bourke and Franks 1995; Wilson and Hölldobler 2005a), though it could have promoted maternal care in ancestral forms (Linksvayer and Wade 2005).

Despite the fact that termites are diploid, two mechanisms have been suggested to increase relatedness among family members above that seen in most non-eusocial families. First, chromosomal translocations occur that could lead to inheritance asymmetries, similar to haplodiploids, that favor helpers raising siblings (Lacy 1980, 1984). This hypothesis is undermined by several lines of evidence (reviewed by Thorne 1997); thus chromosomal translocations were probably not important to eusocial evolution. Second, cycles of inbreeding and outbreeding could result in higher relatedness between helpers and the parents' offspring, favoring the evolution of sibling helpers (see Bartz 1979). While this scenario is plausible, studies in a variety of species have failed to find high enough levels of inbreeding to support the hypothesis (in Termopsidae: Shellman-Reeve 2001, in Rhinotermitidae: Husseneder et al. 1999; in Termitidae: Atkinson and Adams 1997; Thompson and Hebert 1998).

5.4.7 Mating Behavior and the Number of Reproductives

While relatedness asymmetries are not fundamental to eusocial evolution, high relatedness may still be important. Single mating and single reproductive pairs within colonies could maintain high-enough relatedness so that raising siblings provides significant indirect fitness. In Hymenoptera, monogyny is common among ancestral forms, and may be a critical predisposing factor (Hamilton 1964; Strassmann 2001; Boomsma 2007, 2009). A recent phylogenetic study of six hymenopteran eusocial taxa found that monandry is ubiquitous in eusocial ancestors (Hughes et al. 2008). All modern instances of polyandry and most instances of polygyny appear to be derived traits (Boomsma 2007; Hughes et al. 2008). Monogyny and monandry keep relatedness levels within family groups at or near 0.5, so that helpers are about equally related to their siblings and their own offspring. Unfortunately, data from solitary or subsocial species were not included, making it unclear whether these levels of monandry were exceptional relative to other Hymenoptera. Multiple inseminations may be uncommon in solitary Hymenoptera generally (Charnov 1978). Whether high relatedness is critical or not, there is a clear link between the evolution of multiple mating and multiple queens and the later elaboration of eusociality into large colonies with subfertile workers (Fjerdingstad and Crozier 2006; Hughes et al. 2008; Boomsma 2009).

Termites exhibit a similar pattern because founding pairs generally consist of a single king and queen in the primitive taxa (Boomsma 2007, but see Shellman-Reeve 1997). In this situation, helpers are equally related to siblings and offspring, rendering them indifferent as to which they raise (Hamilton 1972; Charnov 1978; Boomsma 2007). Additional reproductives in the form of neotenics are quite common in termites, but they are inbred, so that new alleles are not introduced when neotenics reproduce (Boomsma 2007). But new alleles may enter via colony fusions, if reproductives from the two original colonies interbreed. Fusions are known to be common in two one-piece nesting termites, *Zootermopsis nevadensis* (Termopsidae: Broughton 1995; Thorne et al. 2003; Howard et al, unpublished observations) and *Cryptotermes secundus* (Kalotermitidae; Korb and Schneider 2007), often resulting in neotenic forms from one colony mating with reproductives from the other. Fused colonies are also suggested in *Archotermopsis* (Imms 1919) and *Stolotermes* (Morgan 1959; Gay and Calaby 1970; Thorne and Lenz 2001). When colonies of *Z. nevadensis* are allowed to interact in the lab, replacement reproductives and alates emerge that originated in either original colony or from interbreeding between the colonies (Johns et al. 2009; Howard et al, unpublished observations). While some of these fusions may occur between kin or related groups, outbred founding pairs appear to be the norm, and neighboring colonies of *Z. nevadensis* are generally unrelated (Shellman-Reeve 1997, 2001). These fusions may have been typical of termite ancestors, which would have faced similar competition for space and food

(Thorne et al. 2003). The phylogenetically basal termite *Mastotermes darwiniensis* also has colonies headed by multiple reproductives possibly resulting from fusions followed by inbreeding, causing relatedness values greater than zero, but lower than 0.5 (Goodisman and Crozier 2002). Genetic evidence of colony fusions have been reported from more derived species of termites, but the incidence appears to be lower (Clément 1986; Bulmer et al. 2001; DeHeer and Vargo 2004). Therefore, the importance of high relatedness to the origin and elaboration of eusociality in termites remains unclear.

5.4.8 High Chromosome Numbers

Social Hymenoptera and termites tend to have higher chromosome numbers than non-social taxa (Sherman 1979). By having more chromosomes, the variance in relatedness between siblings is reduced, making it less likely that potential helpers could detect and favor more closely related sibs (Sherman 1979; Templeton 1979; Seger 1983). Eliminating potential for favoritism may make it more likely for cooperative care to be maintained in large family groups. The importance of this is uncertain, as there are some non-social groups with even higher chromosome numbers (Sherman 1979) and the ant genus *Myrmecia* has only one chromosome (Crosland and Crozier 1986). It is also not clear whether this trait appeared subsequent to the evolution of eusociality. This characteristic has seen renewed interest due to findings that high levels of recombination in social insect genomes promote caste diversity and disease resistance (reviewed by Wilfert et al. 2007).

5.4.9 Disease Resistance

Social groups may face increased incidence of disease, so taxa with pre-existing mechanisms of preventing disease may be more likely to evolve sociality (Rosengaus et al. 2000). For a more detailed explanation see Rosengaus et al. (Chapter 7 by Rosengaus et al., this volume).

5.5 Selective Processes Promoting Eusocial Helpers

The second common question about eusociality is how helper or worker forms with limited reproductive opportunities could be favored by selection. This sacrifice includes increasing mortality from foraging and nest defense, and reduced reproductive capabilities of worker forms. In some cases, the sacrifice clearly precludes any chance for reproduction, such as autothysis (e.g. Mill 1984) or workers lacking gonads (e.g. Fletcher and Ross 1985). These traits likely benefit the colony, but how can they be favored by selection if the individual expressing them has little or no direct fitness?

Recently, possible selective processes have been proposed as drivers of eusocial evolution: indirect fitness benefits within kin selection (e.g. Foster et al. 2006; Boomsma 2007; Hughes et al. 2008), group selection (e.g. Wilson and Wilson 2007), and to a lesser extent parental manipulation (e.g. Alonso and Schuck-Paim 2002). Multiple selective processes may operate at any given time (e.g. "fitness-centered hypotheses" below), but each may vary in effect according the developmental, ecological and phylogenetic context (Michener and Brothers 1974). Therefore, it may be less important to discuss which process is key or driving than to attempt to measure the relative contributions of each to social evolution in different situations.

Models traditionally applied to one process can include others (e.g. Lehmann et al. 2007), but there is a danger in making models so broad that they lose explanatory power (Wilson 2008). Here we present theoretical ideas organized by the selective process emphasized, while recognizing that there can be broad overlap in the parameters they include. We direct the reader to detailed treatments of these topics (Reeve 1991; Queller and Strassmann 1998; Alonso and Schuck-Paim 2002, Bourke 2005; Wilson and Wilson 2007).

5.5.1 Theories Focused on Kin Selection and Indirect Fitness

A helper may have direct fitness in the form of its own offspring, or indirect fitness gained by raising relatives likely to share helper alleles ("inclusive fitness" from Hamilton 1964). Excepting gene linkages or epistatic interactions, we expect average relatedness to be proportional to the likelihood that eusocial traits are inherited by offspring. Formal models to describe the evolution of cooperation generally include a mixture of direct and indirect fitness benefits (Gardner and Foster 2008), though most studies purposing to investigate kin selection have focused on the indirect fitness benefits of helping. Theories have taken two forms: relatedness-centered and fitness-centered hypotheses (from Queller and Strassmann 1998).

Relatedness-centered hypotheses focus on the indirect benefits possible through heightened relatedness. If relatedness among cooperating members of the family groups is higher between siblings than between parents and offspring, then selection could favor the evolution of helping behavior with little or no added benefits from group living. Ways in which relatedness may be heightened were discussed above, though each has weaknesses. The modern consensus is that relatedness asymmetries have not significantly contributed to eusocial evolution (see above). Individuals could heighten relatedness to offspring by preferentially helping more closely related nest mates when multiple lineages are present (Hamilton 1964; Ratnieks and Reeve 1991). Despite many investigations, there have been no clear examples of nepotism in any social insect (reviewed by Atkinson et al. 2008).

Fitness-centered hypotheses focus more on processes that raise indirect fitness even at normal levels of relatedness. In most cases, theories focus on the fitness of the helpers via social versus non-social alternatives, rather than on the fitness of recipients of the help (e.g. queens/kings). We focus on several fitness-centered hypotheses below.

5.5.1.1 Intrinsic Benefits to the Group

Group cooperation may provide such significant benefits that the resulting increases to indirect fitness within eusocial groups will be sufficient to compensate for lower relatedness of helpers to the young being raised (Hamilton 1964; Bourke and Franks 1995; Foster et al. 2006). In most cases, group cooperation provides significant benefits in competition with other family groups and solitary organisms. One difficulty is in quantifying how great intrinsic benefits need to be, such that below that level we could conclude the offspring do not actually have higher fitness by helping than by attempting independent reproduction. For this we need more accurate measures of success for dispersal and independent reproduction.

The evolution of the soldier morph in termites may represent an example of intrinsic benefits to the group (Noirot 1990; Roisin 1999). In primitive termites, all functional workers are juvenile forms, so they retain potential for direct fitness. Immatures that develop into soldiers cannot reproduce and can only gain fitness indirectly. If sterile soldiers provide enough benefits to colony productivity, the inclusive fitness of the soldiers can increase despite lacking opportunity for reproduction. Hymenopteran workers, which often have effectively zero opportunities for reproduction, may provide similar benefits, enough to offset losses due to the evolution of reduced reproductive capabilities.

5.5.1.2 Transactional Skew

In social groups, a primary or dominant reproductive may allow helper or subordinate individuals limited reproduction, giving them incentive to help with brood care. The difference in reproduction between dominants and subordinates is termed reproductive skew. Transactional models predict equilibrium levels of skew at which helping becomes worthwhile for subordinates. One individual is assumed to control allocation of reproduction and allocates just enough to subordinates to retain their help (Buston and Zink 2009). In the eusocial Hymenoptera, skew models are usually applied to groups of females initiating new colonies. One female dominates cofoundresses and becomes the primary egg-layer, while subordinate females perform tasks in the nest, forage or contribute resources. The concessions needed to encourage helpers to stay become smaller if the benefits of helping increase, within-group relatedness increases and/or solitary founding is more difficult (Vehrencamp 1983; Reeve 1991; Reeve and Ratnieks 1993). One difficulty is measuring concessions during founding stages, often done by quantifying egg-laying rates. Most offspring act primarily as workers during the founding stage, aside from rare exceptions (Yanega 1988; Reeve et al. 1998). Therefore, it is unclear whether most measures of concessions accurately estimate direct fitness. Also, skew models may not explain parent-offspring groups, due to inherent advantages of the parents in biasing skew.

Some measures of skew have found support for transactions occurring in foundress associations of wasps, based on relatedness similarity and colony productivity (Reeve et al. 2000; Reeve and Keller 2001). On the other hand, Nonacs et al. (2006) found that skew models correctly predicted when co-founding should be favored in several populations of the wasp genera *Polistes* and *Liostenogaster*, but that unrelated and distantly related females exhibited high cooperation despite the dominant individual obtaining the vast majority of reproduction. Transactional skew appears to be rare in ant foundress associations (Heinze 2008a). A recent variant on transactional skew models is the tug-of-war model, in which dominant individuals have incomplete control and sharing results from competition for reproduction within the group at levels dependent on the selfishness of others, group productivity, and relatedness (Reeve and Hölldobler 2007; Buston and Zink 2009). These models await empirical testing.

No attempts have been made to test skew models in termites. There may be opportunities for reproductive transactions to promote cooperation in founding groups and neotenic development, particularly if the within-colony relatedness is lowered by colony fusions (see below).

5.5.1.3 Assured Fitness Returns

If adults have high mortality while raising offspring, then solitary individuals will often fail to successfully reproduce. The helpers may gain more immediate advantage by raising close relatives than by dispersing and raising their own offspring (Queller 1989). The helpers therefore can gain at least some fitness as opposed to zero fitness if they fail when on their own. This assurance is enough to provide significant benefits for helping behavior (Queller 1994). The helper individual must have high relatedness to the reproductive individuals to gain inclusive fitness by helping raise relatives. Evidence supporting the importance of assured fitness returns has been found in hover wasps (Field et al. 2000) and facultatively social sweat bees (Smith et al. 2003, 2007), but not in independent-founding vespid wasps (Nonacs et al. 2006).

For the most part, termites do not face the prospect of complete failure if the king or queen dies before offspring mature, due to greater independence and developmental plasticity of the offspring. However, young colonies face heavy competition from neighbors for space and resources, so success depends on how quickly their colony can grow to a large size (e.g. Thorne et al. 2003). Thus a head start would be beneficial to termites due to competition with fast-growing neighbors rather than dependent offspring.

5.5.2 Theories Focused on Group Selection and Mutualisms

5.5.2.1 Group Selection or Mutualism Between Individuals

Traditional theories of group selection that assumed individual fitness could be subsumed for the benefit of the group (e.g. Wynne-Edwards 1962) have long been discarded. A more formal treatment in the form of "levels-of-selection" theories is reviewed by Okasha (2006). The different levels refer to between-group and within-group selection. Selection on cooperating members of a group can favor some individuals sacrificing to benefit others, as long as their direct fitness is higher because of this sacrifice than if they attempt leaving the group. Recently the pre-eminent biologist E.O. Wilson came out in support of group selection as of greater importance than indirect benefits in kin selection (Wilson and Hölldobler 2005a; Wilson 2005; Wilson and Wilson 2007). As long as all individuals in the group have non-zero direct fitness, selfishness generally outperforms altruism within groups, but altruism outperforms selfishness among groups (Wilson and Wilson 2007).

Group selection theory has generated much discussion. Authors have argued that kin selection and group selection are describing the same phenomenon if groups are made up of relatives (e.g. Bourke and Franks 1995; Lehmann et al. 2007). Others suggest that group selection has less bearing on eusocial evolution because it cannot produce sterile castes (e.g. Crozier 2008), though this may not apply at eusocial origins. Group selection models include relatedness as a factor that creates variation between groups, though kinship is considered to have a supporting, non-pivotal role (Wilson and Wilson 2007). This means that the differences boil down to the relative importance of direct versus indirect fitness benefits to eusocial evolution. Group selection may become more important as social colonies evolve greater complexity and size (Korb and Heinze 2008) or less important as workers lose reproductive options (Bourke 1999).

Termites are a test case for group selection theory, because primitive termite families have helpers that are totipotent and yet are part of relatively large, successful colonies. Therefore, selection could favor either development into a dispersing alate, development into a neotenic reproductive, or remaining as a juvenile worker. Juveniles can also develop into sterile soldiers, but indirect fitness or parental manipulation is needed to favor that. If there is variation in the propensity for juveniles to develop into different forms, we could observe selection among groups based on cooperative versus selfish traits. Colonies can have variable levels of relatedness, thus potentially allowing careful observations to tease apart the importance of direct versus indirect benefits to helping.

5.5.2.2 Future Reproductive and Inheritance Opportunities

The opportunity to inherit the reproductive role and the resources within a colony may be a powerful incentive for remaining near the natal nest in eusocial and cooperative breeding animals (Andersson 1984; Myles 1988; Alexander et al. 1991; Thorne 1997; Kokko and Johnstone 1999; Queller et al. 2000). Inheritance can be modeled as transactional skew in which opportunities to inherit the nest provide incentive for mutualistic behaviors to evolve despite no reproductive opportunities being ceded by dominants while they are still alive (e.g. Ragsdale 1999). Nest philopatry is favored by low probabilities of success in dispersal and initiating reproduction (Myles 1988). Indirect fitness is not necessary if inheritance opportunities are common, though it may help when inheritance is less common or helping less productive (Kokko and Johnstone 1999). Philopatry may also increase the odds that kin are assisted (Myles 1988).

Inheritance opportunities may favor pleometrosis in independent-founding social Hymenoptera, because presumably each female joining to found the new colony purposes to become the primary reproductive (West-Eberhard 1978). After intense early aggression, one queen emerges as the primary reproductive, while the remaining females take on worker roles (Pardi 1948; Röseler 1991; Gadagkar 2001). The dominance hierarchy determines the order in which subordinates move into a reproductive role following the death of the dominant individual. Nest inheritance may have been involved in the origins of eusocial wasps if offspring, like co-foundresses, had opportunities to inherit colonies. It is not clear how often inheritance occurs. In co-foundress situations in eusocial wasps, from 0 to 10% of subordinates move into a dominant role, leading Nonacs et al. (2006) to conclude this is unlikely to be high enough to favor philopatry. However, the argument assumes that solitary-founding success is higher and some individuals are not subfertile (see below), for which current evidence is unclear. In the primitive ant genera *Nothomyrmecia*, *Amblyopone*, and *Myrmecia*, daughters frequently inherit colonies after parents die (Peeters and Ito 2001; Sanetra and Crozier 2002; Thorne and Traniello 2003). Primitively eusocial ants, bees, wasps, and termites tend to have small mature colony size (Imms 1919; Wilson 1971; Bourke 1999), which would make nest inheritance more feasible (Bourke 1999; Thorne and Traniello 2003). In some Hymenoptera, workers have lost the ability to produce females, but it is possible for male production to make up for this loss (Lin and Michener 1972).

In termites, a combination of nest inheritance, high mortality of dispersing alates, and inclusive fitness benefits may facilitate the evolution of delayed maturation and reproduction (Myles 1988; Thorne 1997; Shellman-Reeve 1997; Roisin 1999). Large numbers of juveniles develop into neotenics following the death of kings or queens. In the Mastotermitidae, Termopsidae, and Rhinotermitidae, multiple neotenics can co-exist (reviewed in Goodisman and Crozier 2002), while in other families, the neotenics have battles until only a single pair remains (Myles 1988; Noirot 1989). This means that a large proportion of the colony attempts to become a replacement reproductive. The development of neotenics primarily occurs after the death of the founding king and queen, due to their pheromones previously inhibiting neotenic formation (Light 1943; Lüscher 1952). However founding kings and queens often have the potential to live a long time (Thorne et al. 2002), so inheritance opportunities would be rare. Thorne et al. (2003) proposed a potential factor that could favor juvenile philopatry: accelerated inheritance due to frequent queen and king mortality early in natural colony ontogeny. In two different one-piece nesting species, *Zootermopsis nevadensis* (Termopsidae) and *Cryptotermes secundus* (Kalotermitidae), evidence indicates that colonies frequently interact with neighbors as they expand (Thorne et al. 2003; Korb and Schneider 2007). If the colonies are similar in size, they merge and one or both pairs of kings and queens are killed. In *Z. nevadensis*, juveniles from both colonies have opportunities to develop into multiple replacement reproductives following the merger (Johns et al. 2009). Life history traits and competition were likely similar in prototermites, so these colony interactions could have played a role in the origin of termite eusociality (Thorne et al. 2003). A separate selective process must be involved in the evolution of sterile soldiers or mostly sterile workers.

5.5.3 *Theories Focused on Coercion and Parental Manipulation*

5.5.3.1 Manipulation that Favors Parental Fitness at Cost to Helper Offspring

Parents can manipulate their offspring to benefit the parent's fitness at a cost to the inclusive fitness of those offspring (Alexander 1974; Charnov 1978). This can occur by coercing offspring behavior or by passing on alleles that actually benefit the parents, but are expressed in the offspring workers, such as conditionally expressed sterility. Offspring have difficulty escaping this treatment because the parents are often larger and more powerful. Therefore, parents can always withdraw parental care from unresponsive or inappropriately behaving offspring (Alexander 1974). Parents may also tap into pre-existing developmental responses to environmental stimuli to bias offspring towards subfertile forms, such as vibrating or attacking larvae to trigger a stress response (Brian 1973; Jeanne 2009). Selection could quickly eliminate parents that produce noncompliant offspring early in colony development, when competition from other colonies, young and mature, is likely to be extremely high. Multiple-mating, multiple founding pairs, or the founding of colonies with the help of workers from the natal colony may reduce the chances that non-compliance alleles are exposed to selection if compliant alleles are also present and expressed in some workers. Non-compliant alleles may also be expressed late in colony development, when colonies are more stable, leading to greater worker insubordination in older, larger colonies. Developmental plasticity can facilitate manipulation by only expressing costly subfertile traits in offspring intended to help, rather than in all offspring (Charnov 1978). The manipulation may operate as a supplement to kin selection (Michener and Brothers 1974; West-Eberhard 1975) or may be the primary force (Alexander 1974). The idea is intriguing because it could explain the widespread use of coercive treatment in social insects (Ratnieks and Wenseleers 2008), facilitating the spread of eusociality through populations more easily than via indirect fitness benefits (Michod 1982; Crespi and Ragsdale 2000; Wade 2001). Coercion is thus a mechanism to create compliant workers (Alexander 1974; Charnov 1978). Such manipulation is generally thought to contribute to worker sterility (Bourke and Franks 1995; Ratnieks et al. 2006; Foster et al. 2006).

Various authors have questioned parental manipulation as an explanation for eusocial evolution. First, it may require parent-offspring relatedness, which makes parental manipulation dependent on or a form of kin selection (e.g. West-Eberhard 1975; Craig 1979; Crozier 2008). However, parental manipulation may cause offspring to develop and behave against their interests as measured by inclusive fitness (e.g. West-Eberhard 1987; Richards et al. 2005). In addition, the family member within which alleles are expressed differs: kin selection is focused on alleles expressed by the helpers, while parental manipulation is focused on alleles expressed in parents, though some may occur in helpers (Charnov 1978; Linksvayer and Wade 2005). While some manipulative behaviors are costly enough to require indirect fitness benefits to be maintained (Lehmann and Keller 2006), this is not always the case. Feeding offspring less food is a common form of caste control (Wheeler 1986; Hunt 1991, see subfertility below) and is less costly to parents than the alternative.

Second, cheaters are expected to arise that evade parental controls (e.g. Keller and Nonacs 1993; Crozier 2008). Two factors could prevent this: the presence of effective coercive behaviors that are difficult to evade, possibly the same that control cheating within kin conflict, and strong selection favoring parents better able to produce compliant helpers (Alonso and Schuck-Paim 2002). Effective controls need only occur in a subset of the workers (Bourke 2007). Parental manipulation may be more important in the maintenance of eusociality, as parents evolve to produce workers with fewer capabilities to escape parental controls and reproduce on their own (Ratnieks and Wenseleers 2008).

Little is known about the extent of manipulative behaviors in termites. In most primitive termites, king and queen pheromones suppress reproductive development (Springhetti 1969; Lüscher 1974) and juvenile workers molt into replacement reproductives upon removal or death of the king or queen (Light and Weesner 1951; Lüscher 1964; Nagin 1972; Lenz et al. 1985; Thorne and Noirot 1982; Noirot and Thorne 1988; Myles 1988). However, these pheromones may help offspring respond to maximize inclusive fitness, rather than coercing them into remaining juveniles (Shellman-Reeve 1997). Neotenics are often cannibalized when kings and queens are present (Lüscher 1952; Miyata et al. 2004), suggesting a form of policing occurs to control neotenic development. Even if not key, manipulation may have tipped the balance towards remaining to help, as long as there is little difference in relatedness between siblings and offspring (Myles 1988; Thorne 1997).

5.5.3.2 Subfertility

Offspring with limited capability for dispersal and reproduction may find advantage in remaining at the nest to raise relatives (West-Eberhard 1975; Craig 1983). The limitations could be imposed by environmental fluctuations during development or manipulations in the nutrients available to offspring (Hunt 1991). Evidence comes from the primitively eusocial wasp *Ropalidia marginata*, in which around half of females isolated in the lab were unable to found nests on their own (Gadagkar et al. 1988; Gadagkar 1991a, but see Nonacs and Reeve 1995) and from work on *Polistes* species (Hunt 2007). However, in some studies workers have not been found to be subfertile (Röseler 1985; Field and Foster 1999). The widespread observation that feeding rates influence caste differentiation suggests that subfertility may have been involved in the evolution of more specialized worker forms (Wilson 1971; Michener 1974; Wheeler 1986; Reuter and Keller 2001). Subfertility is self-evident when workers lack reproductive organs, trapping them once they have passed caste developmental switches (West-Eberhard 1975; Queller 1996). Separating the roles of manipulation and kin selection in the evolution of helping traits can be difficult. For example, parent-offspring conflict models suggest offspring should always prefer reproductive to worker development based on inclusive fitness (Reuter and Keller 2001). Therefore, development to a subfertile form is manipulation. However, indirect fitness may favor helping behaviors in subfertile forms and therefore reduce the likelihood of cheaters arising.

In termites, subfertility may prevent development into dispersing reproductives, as individuals with fewer resources remain as juveniles until the next swarming season (Nalepa 1994; Korb and Schmidinger 2004). It is not clear how other colony members influence resource levels in nestmates. Food sharing is common, but this may lead to juveniles having fewer food stores if they are frequent food donors (Nalepa 1994). It is not clear if sharing is enforced.

5.5.3.3 Within Group Competition for Reproductive Opportunities

Coercion and kin selection could favor eusocial evolution in termites if it arises via competition for reproductive opportunities. Conflict among late instar nymphs developing into alate forms could lead to damage of some individuals, thereby preventing their reaching the alate stage (Zimmerman 1983; Myles 1988; Roisin 1994). Kalotermitid, termopsid, and some rhinotermitid nymphs sustain wingbud damage due to attacks from other colony members in the lab (Roisin 1994). Nymphs are unable to molt directly into alates, though additional molts circumvent this constraint (Springhetti 1969; Sewell and Watson 1981; Roisin 1988; Roisin and Pasteels 1991). Damaged individuals remain as juveniles longer and help raise offspring. Eventually, undamaged individuals begin to help for kin-selected benefits. Several issues need to be addressed: mutilated individuals can still become neotenics, termite juveniles help prior to wingbud development, and the wingbud scars in the Termopsidae may actually be self-abscission resulting from lab rearing conditions (Thorne 1997). More evidence is needed to determine whether competition favors eusocial helpers (Thorne 1997), and whether this is a form of coercion (Shellman-Reeve 1997). Similar theories relating competitive interactions to reproductive options have been proposed for eusocial bees and wasps (see Michener 1985; West-Eberhard 1987).

5.6 Synthesis

5.6.1 Similarities Suggesting Common Principles

Similarities between hymenopteran and isopteran eusocial evolution suggest there are general principles. In every eusocial species offspring exhibit parental care towards siblings, suggesting extended parental care is a necessary predisposing trait. Helping behaviors in offspring generally involve the expression of parental care behaviors by individuals that have not reproduced, indicating that heterochrony is critical. In addition, both taxa shift to earlier caste determination in more derived species with large, complex colonies. Developmental pathways have recruited similar mechanisms for caste determination, including roles for juvenile hormone, vitellogenin and hexamerin proteins, and control by nutrition and social environment.

Colonies in both taxa defend valuable nests. Though primitively eusocial Hymenoptera may have experienced stronger selection due to foraging needs (Queller and Strassmann 1998), defense of the nest and the brood inside are clearly important. Also, the nest is a food resource in its own right (basal, wood-feeding termites and derived soil-feeders) or a food store. As a result, the chance to inherit the nest may favor helpers staying in both taxa, as may be the case generally for cooperatively breeding animals including insects, fish, birds, and mammals (Kokko and Johnstone 1999; Pen and Weissing 2000; Queller et al. 2000). Low rates of dispersal success and assured fitness may favor philopatry in both taxa, though the importance of competition in creating these conditions may differ.

Relatedness-centered hypotheses have met with mixed success in both taxa. Monogamy and monandry appear to be standard in ancestral forms of both hymenopterans and termites, though applying this to explain termite eusociality may be difficult (see above). Theories that rely solely on heightened relatedness have been poorly supported. Indirect fitness, selection among groups, and selection among parents all play a role, and potentially a simultaneous role, in eusocial evolution.

5.6.2 Distinctions Suggesting Unique Properties Within Each Group

Differences in developmental and genetic systems between hymenopterans and termites, once emphasized in all discussions of eusociality, have now diminished with the discovery of similar mechanisms of creating helpers that have yet to reproduce and the downfall of the haplodiploid hypothesis. However, haplodiploidy improves the spread of maternal effects alleles (Linksvayer and Wade 2005), so parental care may be less likely to evolve in diplo-diploids. The independence of termite juveniles makes progressive provisioning less important as a precondition, but other mechanisms may have increased parent-offspring contact. There are clear differences in the evolution of specialized soldiers, which in termites evolved early on, while hymenopteran societies either lack them (bees and wasps) or evolved them only in derived groups (ants). This may be due to different preconditions, such as the hymenopteran stings, or due to higher competition for space in the termites (Queller and Strassmann 1998). However, many extant species in both taxa have soldiers and subfertile workers, suggesting that ecological pressures converge as colonies become larger and more complex. Though subfertile forms are common in both taxa, complete sterility appears to be less common in the Hymenoptera than the Isoptera.

While the above distinctions suggest there may be minor differences in the factors and processes involved, the lessening importance placed on haplodiploidy and the striking similarities in developmental pathways allow for much greater overlap of theories to explain eusocial evolution in the Hymenoptera and Isoptera. In particular, we should consider whether transactional skew, policing and coercion, and/or nest inheritance contribute to the evolution of termite helpers which forego reproductive opportunities. Termites represent a fascinating opportunity for comparative studies into why helpers forego reproductive opportunities, given multiple shifts between totipotent and subfertile helpers within the clade. Future work on the evolution of eusociality will benefit by incorporating evidence from both Hymenoptera and Isoptera when discussing the origins and elaboration of eusocial traits.

References

Abe T (1987) Evolution of life types in termites. In: Kawano S, Connell JH, Hidaka T (eds) Evolution and coadaptation in biotic communities. University of Tokyo Press, Tokyo, pp 125–148

Abe T (1991) Ecological factors associated with the evolution of worker and soldier castes in termites. Ann Entomol 9:101–107

Alcock J, Sherman P (1994) The utility of the proximate-ultimate dichotomy in ethology. Ethology 96:58–62

Alexander RD (1974) The evolution of social behavior. Annu Rev Ecol Syst 5:325–383

Alexander RD, Noonan KM, Crespi BJ (1991) The evolution of eusociality. In: Sherman PW, Jarvis JUM, Alexander RD (eds) The biology of the Naked Mole-rat. Princeton University Press, Princeton, NJ, pp 3–44

Alonso WJ, Schuck-Paim C (2002) Sex-ratio conflicts, kin selection, and the evolution of altruism. Proc Natl Acad Sci U S A 99:6843–6847

Amdam GV, Csondes A, Fondrk MK, Page RE (2006) Complex social behaviour derived from maternal reproductive traits. Nature 439:76–78

Andersson M (1984) The evolution of eusociality. Annu Rev Ecol Syst 15:165–189

Atkinson L, Adams ES (1997) The origins and relatedness of multiple reproductives in colonies of the termite *Nasutitermes corniger*. Proc R Soc Lond B 264:1131–1136

Atkinson L, Teschendorf G, Adams ES (2008) Lack of evidence for nepotism by workers tending queens of the polygynous termite *Nasutitermes corniger*. Behav Ecol Sociobiol 62:805–812

Bartz SH (1979) Evolution of eusociality in termites. Proc Natl Acad Sci U S A 76:5764–5768

Basalingappa S (1970) Environmental hazards to reproductives of *Odontotermes assmuthi* Holmgren. Indian Zool 1:45–50

Bloch G, Hefetz A (1999) Regulation of reproduction by dominant workers in bumblebee (*Bombus terrestris*) queenright colonies. Behav Ecol Sociobiol 45:125–135

Boomsma JJ (2007) Kin selection versus sexual selection: why the ends do not meet. Curr Biol 17:673–683

Boomsma JJ (2009) Lifetime monogamy and the evolution of eusociality. Phil Trans R Soc Lond B 364:1–17

Bourke AFG (1999) Colony size, social complexity and reproductive conflict in social insects. J Evol Biol 12:245–257

Bourke AFG (2005) Genetics, relatedness and social behaviour in insect societies. In: Fellowes MDE, Holloway GJ, Rolff J (eds) Insect evolutionary ecology. Proceedings of the Royal Entomological Society, 22nd Symposium. CABI Publishing, Cambridge, Massachusetts, pp 1–30

Bourke AFG (2007) Social evolution: community policing in insects. Curr Biol 17:519–520

Bourke AFG, Franks NR (1995) Social evolution in ants. Princeton University Press, Princeton, NJ
Brady SG, Schultz TR, Fisher BL, Ward PS (2006) Evaluating alternative hypotheses for the early evolution and diversification of ants. Proc Natl Acad Sci U S A 103:18172–18177
Breed MD, Guzmán-Novoa E, Hunt GJ (2004) Defensive behavior of honey bees: organization, genetics, and comparisons with other social insects. Annu Rev Entomol 49:271–298
Brian MV (1973) Caste control through worker attack in the ant *Myrmica*. Insectes Soc 20:87–102
Broughton RE (1995) Mitochondrial DNA variation within and among species of termites in the genus *Zootermopsis* (Isoptera: Termopsidae). Ann Entomol Soc Am 88:120–128
Bulmer MS, Adams ES, Traniello JFA (2001) Variation in colony structure in the subterranean termite *Reticulitermes flavipes*. Behav Ecol Sociobiol 49:236–243
Buston PM, Zink AG (2009) Reproductive skew and the evolution of conflict resolution: a synthesis of transactional and tug-of-war models. Behav Ecol 20:672–684
Charnov EL (1978) Evolution of eusocial behavior: offspring choice or parental parasitism? J Theor Biol 75:451
Choe JC, Crespi BJ (1997) The evolution of social behavior in insects and arachnids. Cambridge University Press, Cambridge, MA
Clément JL (1986) Open and closed societies in *Reticulitermes* termites (Isoptera, Rhinotermitidae): geographic and seasonal variations. Sociobiology 11:311–323
Cleveland LR, Hall SR, Sanders EP, Collier J (1934) The wood-feeding roach *Cryptocercus*, its protozoa, and the symbiosis between protozoa and roach. Mem Am Acad Arts Sci 17:185–342
Collins NM (1983) The utilization of nitrogen resources by termites (Isoptera). In: Lee JA, McNeill S, Rorison IH (eds) Nitrogen as an ecological factor. Blackwell Scientific Publications, Oxford, pp 381–412
Craig R (1979) Parental manipulation, kin selection, and the evolution of altruism. Evolution 33:319–334
Craig R (1982) Evolution of male workers in the Hymenoptera. J Theor Biol 94:95–105
Craig R (1983) Subfertility and the evolution of eusociality by kin selection. J Theor Biol 100: 379–397
Crespi BJ (1994) Three conditions for the evolution of eusociality: are they sufficient? Insectes Soc 41:395–400
Crespi BJ (1996) Comparative analysis of the origins and losses of eusociality: causal mosaics and historical uniqueness. In: Martins E (ed) Phylogenies and the comparative method in animal behavior. Oxford University Press, New York, pp 253–287
Crespi BJ, Ragsdale JE (2000) A skew model for the evolution of sociality via manipulation: why it is better to be feared than loved. Proc Biol Sci 267:821–828
Crosland MWJ, Crozier RH (1986) *Myrmecia pilosula*, an ant with only one pair of chromosomes. Science 231:1278
Crozier RH (2008) Advanced eusociality, kin selection and male haploidy. Aust J Entomol 47:2–8
Crozier RH, Jermiin LS, Chiotis M (1997) Molecular evidence for a Jurassic origin of ants. Naturwissenschaften 84:22–23
DeHeer CJ, Vargo EL (2004) Colony genetic organization and colony fusion in the termite *Reticulitermes flavipes* as revealed by foraging patterns over time and space. Mol Ecol 13:431–441
Duffy JE (1996) Eusociality in a coral-reef shrimp. Nature 381:512–514
Field J, Foster W (1999) Helping behaviour in facultatively eusocial hover wasps: an experimental test of the subfertility hypothesis. Anim Behav 57:633–636
Field J, Shreeves G, Sumner S, Casiraghi M (2000) Insurance-based advantage to helpers in a tropical hover wasp. Nature 404:869–871
Fjerdingstad EJ, Crozier RH (2006) The evolution of worker caste diversity in social insects. Am Nat 167:390–400
Fletcher DJC, Ross KG (1985) Regulation of reproduction in eusocial Hymenoptera. Annu Rev Entomol 30:319–343

Foster KR, Wenseleers T, Ratnieks FLW (2006) Kin selection is the key to altruism. Trends Ecol Evol 21:57–60
Fraser VS, Kaufmann B, Oldroyd BP, Crozier RH (2000) Genetic influence on caste in the ant *Camponotus consobrinus*. Behav Ecol Sociobiol 47:188–194
Gadagkar R (1991a) On testing the role of genetic asymmetries created by haplodiploidy in the evolution of eusociality in the Hymenoptera. J Genet 70:1–31
Gadagkar R (1991b) Demographic predisposition to the evolution of eusociality: a hierarchy of models. Proc Natl Acad Sci U S A 88:10993–10997
Gadagkar R (2001) The social biology of *Ropalidia marginata*: toward understanding the evolution of eusociality. Harvard University Press, Cambridge, MA
Gadagkar R, Vinutha C, Shanubhogue A, Gore AP (1988) Pre-imaginal biasing of caste in a primitively eusocial insect. Proc R Soc Lond B 233:175–189
Gardner A, Foster KR (2008) The evolution and ecology of cooperation – history and concepts. In: Korb J, Heinze J (eds) Ecology of social evolution. Springer, Berlin, pp 1–36
Gay FJ, Calaby JH (1970) Termites of the Australian region. In: Krishna K, Weesner FM (eds) Biology of termites, vol 2. Academic Press, New York, pp 393–448
Goodisman MAD, Crozier RH (2002) Population and colony genetic structure of the primitive termite *Mastotermes darwiniensis*. Evolution 56:70–83
Grassé PP (1986) Termitologia, vol 3. Masson, Paris
Grimaldi DA, Engel MS (2005) Evolution of the insects. Cambridge University Press, Cambridge
Hamilton WD (1964) The genetical evolution of social behavior I, II. J Theor Biol 7:1–52
Hamilton WD (1972) Altruism and related phenomena, mainly in social insects. Annu Rev Ecol Syst 3:193–232
Hamilton WD (1978) Evolution and diversity under bark. In: Mound LA, Waloff N (eds) Diversity of insect faunas. Symposium of the Royal Entomological Society of London, vol 9. Halsted, New York, NY, pp 154–175
Hansell MH (1987) Nest building as a facilitating and limiting factor in the evolution of eusociality in the Hymenoptera. In: Harvey PH, Partridge L (eds) Oxford surveys in evolutionary biology, vol 4. Oxford University Press, Oxford, pp 155–181
Hansell MH (1996) Wasps make nests: nests make conditions. In: Turillazzi S, West-Eberhard MJ (eds) Natural history and evolution of paper-wasps. Oxford University Press, Oxford, pp 272–289
Hare L (1937) Termite phylogeny as evidenced by soldier mandible development. Ann Entomol Soc Am 30:459–486
Hartfelder K, Emlen DJ (2005) Endocrine control of insect polyphenism. In: Gilbert LI, Iatrou K, Gill S (eds) Comprehensive molecular insect science, vol 3. Elsevier, Oxford, pp 651–703
Haverty MI (1977) The proportion of soldiers in termite colonies: a list and a bibliography (Isoptera). Sociobiology 2:199–216
Hayashi Y, Lo N, Miyata H et al (2007) Sex-linked genetic influence on caste determination in a termite. Science 318:985–987
Heath H (1928) Fertile termite soldiers. Biol Bull 54:324–326
Heinze J (2008a) Social plasticity: ecology, genetics, and the structure of ant societies. In: Korb J, Heinze J (eds) Ecology of social evolution. Springer, Berlin, pp 129–150
Heinze J (2008b) The demise of the standard ant. Myrmecol News 11:9–20
Helanterä H, Bargum K (2007) Pedigree relatedness, not greenbeard genes, explains eusociality. Oikos 116:217
Higashi M, Yamamura N, Abe T, Burns TP (1991) Why don't all termite species have a sterile worker caste? Proc R Soc Lond B 246:25–29
Higashi M, Yamamura N, Abe T (2000) Theories on the sociality of termites. In: Abe T, Bignell DE, Higashi M (eds) Termites: evolution, sociality, symbioses, ecology. Kluwer Academic Publishers, Dordrecht, pp 169–188
Hölldobler B, Wilson EO (1990) The Ants. Belknap Press, Cambridge, MA

Hughes WOH, Oldroyd BP, Beekman M, Ratnieks FLW (2008) Ancestral monogamy shows kin selection is key to the evolution of eusociality. Science 320:1213–1216
Hunt JH (1991) Nourishment and the evolution of the social Vespidae. In: Ross KG, Matthews RW (eds) The social biology of wasps. Cornell University Press, Ithaca, NY, pp 426–450
Hunt JH (1994) Nourishment and social evolution in wasps sensu lato. In: Hunt JH, Nalepa CA (eds) Nourishment and evolution in insect societies. Westview Press, Boulder, CO, pp 211–244
Hunt JH (1999) Trait mapping and salience in the evolution of eusocial vespid wasps. Evolution 53:225–237
Hunt JH (2007) The evolution of social wasps. Oxford University Press, Oxford
Hunt JH, Amdam GV (2005) Bivoltinism as an antecedent to eusociality in the paper wasp genus *Polistes*. Science 308:264–267
Husseneder C, Brandl R, Epplen C et al (1999) Within-colony relatedness in a termite species: genetic roads to eusociality? Behaviour 136:1045–1063
Imms AD (1919) On the structure and biology of *Archotermopsis*, together with descriptions of new species of intestinal protozoa, and general observations on the Isoptera. Phil Trans R Soc Lond 209:75–180
Inward DJG, Beccaloni G, Eggleton P (2007a) Death of an order: a comprehensive molecular phylogenetic study confirms that termites are eusocial cockroaches. Biol Lett 3:331
Inward DJG, Vogler AP, Eggleton P (2007b) A comprehensive phylogenetic analysis of termites (Isoptera) illuminates key aspects of their evolutionary biology. Mol Phylogenet Evol 44: 953–967
Jeanne RL (2009) Vibrational signals in social wasps: a role in caste determination? In: Gadau J, Fewell JH (eds) Organization of insect societies: from genome to sociocomplexity. Harvard University Press, Cambridge, MA, 243–265
Johns PM, Howard KJ, Breisch NL et al (2009) Nonrelatives inherit colony resources in a primitive termite. Proc Natl Acad Sci U S A 106:17452–17456
Julian GE, Fewell JH, Gadau J et al (2002) Genetic determination of the queen caste in an ant hybrid zone. Proc Natl Acad Sci U S A 99:8157–8160
Keller L, Nonacs P (1993) The role of queen pheromones in social insects: queen control or queen signal? Anim Behav 45:787–794
Kennedy JS (1947) Child labor of the termite society versus adult labor of the ant society. Sci Mon 65:309–324
Kerr WE (1950) Evolution of the mechanism of caste determination in the genus *Melipona*. Evolution 4:7–13
Kokko H, Johnstone RA (1999) Social queuing in animal societies: a dynamic model of reproductive skew. Proc R Soc Lond B 266:571–578
Korb J (2007) Workers of a drywood termite do not work. Front Zool 4:7
Korb J (2008) The ecology of social evolution in termites. In: Korb J, Heinze J (eds) Ecology of social evolution. Springer, Berlin, pp 151–174
Korb J, Heinze J (2008) The ecology of social life: a synthesis. In: Korb J, Heinze J (eds) Ecology of social evolution. Springer, Berlin, pp 245–260
Korb J, Schmidinger S (2004) Help or disperse? Cooperation in termites influenced by food conditions. Behav Ecol Sociobiol 56:89–95
Korb J, Schneider K (2007) Does kin structure explain the occurrence of workers in a lower termite? Evol Ecol 21:817–828
Kukuk PF, Eickwort GC, Raveret-Richter M et al (1989) Importance of the sting in the evolution of sociality in the Hymenoptera. Ann Entomol Soc Am 82:1–5
Kumano N, Kasuya E (2001) Why do workers of the primitively eusocial wasp *Polistes chinensis antennalis* remain at their natal nest? Anim Behav 61:655–660
LaFage JP, Nutting WL (1978) Nutrient dynamics of termites. In: Brian MV (ed) Production ecology of ants and termites. Cambridge University Press, Cambridge, pp 165–244
Lacy RC (1980) The evolution of eusociality in termites: a haplodiploid analogy? Am Nat 116:449–451

Lacy RC (1984) The evolution of termite sociality: reply to Leinaas. Am Nat 123:876–879
Legendre F, Whiting MF, Bordereau C et al (2008) The phylogeny of termites (Dictyoptera: Isoptera) based on mitochondrial and nuclear markers: implications for the evolution of the worker and pseudergate castes, and foraging behaviors. Mol Phylogenet Evol 48:615–627
Lehmann L, Keller L (2006) The evolution of cooperation and altruism-a general framework and a classification of models. J Evol Biol 19:1365–1376
Lehmann L, Keller L, West S, Roze D (2007) Group selection and kin selection: two concepts but one process. Proc Natl Acad Sci U S A 104:6736–6739
Lenz M (1994) Food resources, colony growth and caste development in wood-feeding termites. In: Hunt JH, Nalepa CA (eds) Nourishment and evolution in insect societies. Westview Press, Boulder, CO, pp 159–209
Lenz M, Barrett RA, Williams ER (1985) Reproductive strategies in *Cryptotermes*: neotenic production in indigenous and 'tramp'species in Australia (Isoptera: Kalotermitidae). In: Watson J, Okot-Kotber B, Noirot C (eds) Caste differentiation in social insects. Pergamon Press, Oxford, pp 147–163
Light SF (1943) The determination of caste of social insects. Q Rev Biol 18:46–63
Light SF, Weesner FM (1951) Further studies on the production of supplementary reproductives in *Zootermopsis* (Isoptera). J Exp Zool 117:397–414
Lin N, Michener CD (1972) Evolution of sociality in insects. Q Rev Biol 47:131
Linksvayer TA, Wade MJ (2005) The evolutionary origin and elaboration of sociality in the aculeate Hymenoptera: maternal effects, sib-social effects, and heterochrony. Q Rev Biol 80:317–336
Lo N, Tokuda G, Watanabe H et al (2000) Evidence from multiple gene sequences indicates that termites evolved from wood-feeding cockroaches. Curr Biol 10:801–804
Lüscher M (1952) Die Produktion und Elimination von Ersatzgeschlechtstieren bei der Termite *Kalotermes flavicollis* (Fabr.). Z Vgl Physiol 34:123–141
Lüscher M (1964) Die spezifische Wirkung männlicher und weiblicher Ersatzgeschlechtstiere auf die Entstehung von Ersatzgeschlechtstieren bei der Termite *Kalotermes flavicollis* (Fabr.). Insectes Soc 11:79–90
Lüscher M (1974) Kasten und Kastendifferenzierung bei niederen Termiten. In: Schmidt GH (ed) Sozialpolymorphismus bei Insekten. Wissenschaftliche Verlagsgesellschaft, Stuttgart, pp 694–739
Matthews RW (1991) Evolution of social behavior in sphecid wasps. In: Ross KG, Matthews RW (eds) The social biology of wasps. Cornell University Press, Ithaca, NY, pp 570–602
Michener CD (1974) The social behavior of the bees. Belknap Press, Cambridge, MA
Michener CD (1975) A taxonomic study of African allodapine bees. Bull Am Mus Nat Hist 155:67–240
Michener CD (1985) From solitary to eusocial: need there be a series of intervening species? In: Hölldobler B, Lindauer M (eds) Experimental behavioral ecology and sociobiology. Gustav Fischer, Stuttgart, pp 293–305
Michener CD (1990) Reproduction and castes in social halictine bees. In: Engels W (ed) Social insects: an evolutionary approach to castes and reproduction. Springer, Heidelberg, pp 77–121
Michener CD, Brothers DJ (1974) Were workers of eusocial Hymenoptera initially altruistic or oppressed? Proc Natl Acad Sci U S A 71:671–674
Michod RE (1982) The theory of kin selection. Annu Rev Ecol Syst 13:23–55
Mill AE (1984) Exploding termites – an unusual defensive behaviour. Entomol Mon Mag 120: 179–183
Miyata H, Furuichi H, Kitade O (2004) Patterns of neotenic differentiation in a subterranean termite, *Reticulitermes speratus* (Isoptera: Rhinotermitidae). Entomol Sci 7:309–314
Morgan FD (1959) The ecology and external morphology of *Stolotermes ruficeps* Brauer (Isoptera: Hodotermitidae). Trans R Soc N Z 86:155–195
Myles TG (1986) Reproductive soldiers in the Termopsidae (Isoptera). Pan-Pac Entomol 62: 293–299

Myles TG (1988) Resource inheritance in social evolution from termites to man. In: Slobodchikoff CN (ed) The ecology of social behavior. Academic Press. New York, NY, pp 379–423
Nagin R (1972) Caste determination in *Neotermes jouteli* (Banks). Insectes Soc 19:39–61
Nalepa CA (1994) Nourishment and the origin of termite eusociality. In: Hunt JH, Nalepa CA (eds) Nourishment and evolution in insect societies. Westview Press, Boulder, CO, pp 57–104
Nalepa CA, Bandi C (2000) Characterizing the ancestors: paedomorphosis and termite evolution. In: Abe T, Bignell DE, Higashi M (eds) Termites: evolution, sociality, symbioses, ecology. Kluwer Academic Publishers, Dordrecht, pp 53–75
Noirot C (1969) Formation of castes in the higher termites. In: Krishna K, Weesner FM (eds) Biology of termites (vol 1). Academic Press, New York, NY, pp 311–350
Noirot C (1985a) Differentiation of reproductives in higher termites. In: Watson JAL, Okot-Kotber BM, Noirot C (eds) Caste differentiation in social insects. Oxford Pergamon, Oxford, pp 177–186
Noirot C (1985b) Pathways of caste development in the lower termites. In: Watson JAL, Okot-Kotber BM, Noirot C (eds) Caste Differentiation in Social Insects. Oxford Pergamon, Oxford, 41–58
Noirot C (1989) Social structure in termite societies. Ethol Ecol Evol 1:1–17
Noirot C (1990) Sexual castes and reproductive strategies in termites. In: Engels W (ed) Social insects: an evolutionary approach to castes and reproduction. Springer, Berlin, pp 5–35
Noirot C, Darlington JPEC (2000) Termite nests: architecture, regulation and defence. In: Abe T, Bignell DE, Higashi M (eds) Termites: evolution, sociality, symbioses, ecology. Kluwer Academic Publishers, Dordrecht, pp 121–140
Noirot C, Pasteels JM (1987) Ontogenetic development and evolution of the worker caste in termites. Cell Mol Life Sci 43:851–860
Noirot C, Pasteels JM (1988) The worker caste is polyphyletic in termites. Sociobiology 14:15–20
Noirot C, Thorne BL (1988) Ergatoid reproductives in *Nasutitermes columbicus* (Isoptera, Termitidae). J Morphol 195:83–93
Nonacs P, Liebert AE, Starks PT (2006) Transactional skew and assured fitness return models fail to predict patterns of cooperation in wasps. Am Nat 167:467–480
Nonacs P, Reeve HK (1995) The ecology of cooperation in wasps: causes and consequences of alternative reproductive decisions. Ecology 76:953–967
O'Donnell S (1998) Reproductive caste determination in eusocial wasps (Hymenoptera: Vespidae). Annu Rev Entomol 43:323–346
Okasha S (2006) Evolution and the levels of selection. Oxford University Press, Oxford
Oster GF, Wilson EO (1978) Caste and ecology in the social insects. Princeton University Press, Princeton, NJ
Pamilo P (1991) Evolution of the sterile caste. J Theor Biol 149:75–95
Pardi L (1948) Dominance order in *Polistes* wasps. Physiol Zool 21:1–13
Peeters C, Ito F (2001) Colony dispersal and the evolution of queen morphology in social Hymenoptera. Annu Rev Entomol 46:601–630
Pellens R, D'Haese CA, Bellés X et al (2007) The evolutionary transition from subsocial to eusocial behaviour in Dictyoptera: phylogenetic evidence for modification of the "shift-in-dependent-care" hypothesis with a new subsocial cockroach. Mol Phylogenet Evol 43: 616–626
Pen I, Weissing FJ (2000) Towards a unified theory of cooperative breeding: the role of ecology and life history re-examined. Proc R Soc Lond B 267:2411–2418
Pereboom JJM, Velthuis HHW, Duchateau MJ (2003) The organisation of larval feeding in bumblebees (Hymenoptera, Apidae) and its significance to caste differentiation. Insectes Soc 50:127–133
Poinar GO, Danforth BN (2006) A fossil bee from Early Cretaceous Burmese amber. Science 314:614
Queller DC (1989) The evolution of eusociality: reproductive head starts of workers. Proc Natl Acad Sci U S A 86:3224–3226

Queller DC (1994) Extended parental care and the origin of eusociality. Proc Roy Soc Lond B 256:105–111

Queller DC (1996) The origin and maintenance of eusociality: the advantage of extended parental care. In: Turillazzi S, West-Eberhard MJ (eds) Natural history and evolution of paper wasps (vol 12). Oxford University Press, Oxford, pp 218–234

Queller DC, Strassmann JE (1998) Kin selection and social insects. BioScience 48:165–175

Queller DC, Zacchi F, Cervo R et al (2000) Unrelated helpers in a social insect. Nature 405: 784–787

Ragsdale JE (1999) Reproductive skew theory extended: the effect of resource inheritance on social organization. Evol Ecol Res 1:859–874

Ratnieks FLW (1988) Reproductive harmony via mutual policing by workers in eusocial Hymenoptera. Am Nat 132:217–236

Ratnieks FLW, Foster KR, Wenseleers T (2006) Conflict resolution in insect societies. Ann Rev Entomol 51:581–608

Ratnieks FLW, Reeve HK (1991) The evolution of queen-rearing nepotism in social Hymenoptera: effects of discrimination costs in swarming species. J Evol Biol 4:93–115

Ratnieks FLW, Wenseleers T (2008) Altruism in insect societies and beyond: voluntary or enforced? Trends Ecol Evol 23:45–52

Reeve HK (1991) *Polistes*. In: Ross KG, Matthews RW (eds) The social biology of wasps. Cornell University Press, Ithaca, NY, pp 99–148

Reeve HK, Hölldobler B (2007) The emergence of a superorganism through intergroup competition. Proc Natl Acad Sci U S A 104:9736

Reeve HK, Keller L (2001) Tests of reproductive-skew models in social insects. Annu Rev Entomol 46:347–385

Reeve HK, Peters JM, Nonacs P, Starks PT (1998) Dispersal of first workers in social wasps: causes and implications of an alternative reproductive strategy. Proc Natl Acad Sci U S A 95:13737–13742

Reeve HK, Ratnieks F (1993) Queen-queen conflict in polygynous societies: mutual tolerance and reproductive skew. In: Keller L (ed) Queen number and sociality in insects. Oxford University Press, Oxford, pp 45–85

Reeve HK, Starks PT, Peters JM, Nonacs P (2000) Genetic support for the evolutionary theory of reproductive transactions in social wasps. Proc R Soc Lond B 267:75–79

Reuter M, Keller L (2001) Sex ratio conflict and worker production in eusocial Hymenoptera. Am Nat 158:166–177

Richards MH, French D, Paxton RJ (2005) It's good to be queen: classically eusocial colony structure and low worker fitness in an obligately social sweat bee. Mol Ecol 14:4123–4133

Roisin Y (1988) Morphology, development and evolutionary significance of the working stages in the caste system of *Prorhinotermes* (Insecta, Isoptera). Zoomorphology 107:339–347

Roisin Y (1994) Intragroup conflicts and the evolution of sterile castes in termites. Am Nat 143:751–765

Roisin Y (1999) Philopatric reproduction, a prime mover in the evolution of termite sociality? Insectes Soc 46:297–305

Roisin Y (2000) Diversity and evolution of caste patterns. In: Abe T, Bignell DE, Higashi M (eds) Termites: evolution, sociality, symbioses, ecology. Kluwer Academic Publishers, Dordrecht, pp 95–119

Roisin Y, Pasteels JM (1991) Polymorphism in the giant cocoa termite, *Neotermes papua* (Desneux). Insectes Soc 38:263–272

Röseler PF (1985) Endocrine basis of dominance and reproduction in polistine paper wasps. In: Hölldobler B, Lindauer M (eds) Experimental behavioral ecology and sociobiology. Sinauer, Sunderland, MA, pp 259–272

Röseler PF (1991) Reproductive competition during colony establishment. In: Ross KG, Matthews RW (eds) The social biology of wasps. Cornell University Press, Ithaca, NY, pp 309–335

Rosengaus RB, Lefebvre ML, Carlock DM, Traniello JFA (2000) Socially transmitted disease in adult reproductive pairs of the dampwood termite *Zootermopsis angusticollis*. Ethol Ecol Evol 12:419–433

Roubaud E (1916) Recherches biologiques sur les guëpes solitaires et sociales d'Afrique. La genèse de la vie sociale et l'évolution de l'instinct maternel chez les vespides. Ann Sc Nat Zool 1:1–160

Sanetra M, Crozier RH (2002) Daughters inherit colonies from mothers in the 'living-fossil' ant *Nothomyrmecia macrops*. Naturwissenschaften 89:71–74

Scharf ME, Ratliff CR, Wu-Scharf D et al (2005) Effects of juvenile hormone III on *Reticulitermes flavipes*: changes in hemolymph protein composition and gene expression. Insect Biochem Mol Biol 35:207–215

Schwarz MP, Richards MH, Danforth BN (2006) Changing paradigms in insect social evolution: insights from halictine and allodapine bees. Annu Rev Entomol 52:127–150

Seeley TD (1995) The wisdom of the hive: the social physiology of Honey Bee colonies. Harvard University Press, Cambridge, MA

Seger J (1983) Conditional relatedness, recombination, and the chromosome number of insects. In: Rhodin AGJ, Miyata K (eds) Advances in herpetology and evolutionary biology. Essays in honor of Ernst E. Mayr. Harvard University Press, Cambridge, MA, pp 596–612

Seger J (1991) Cooperation and conflict in social insects. In: Krebs JR, Davies NB (eds) Behavioural ecology: an evolutionary approach. Blackwell Scientific Publications, Oxford, pp 338–373

Sen R, Gadagkar R (2005) Males of the social wasp *Ropalidia marginata* can feed larvae, given an opportunity. Anim Behav 71:345–350

Sewell JJ, Watson JAL (1981) Developmental pathwavs in Australian species of *Kalotermes* Hagen (Isoptera). Sociobiology 6:243–324

Shellman-Reeve JS (1997) The spectrum of eusociality in termites. In: Choe J, Crespi B (eds) The evolution of social behavior in insects and arachnids. Cambridge University Press, Cambridge, UK, pp 52–93

Shellman-Reeve JS (2001) Genetic relatedness and partner preference in a monogamous, wood-dwelling termite. Anim Behav 61:869–876

Sherman PW (1979) Insect chromosome numbers and eusociality. Am Nat 113:925–935

Sherman PW, Jarvis JUM, Alexander RD (1991) Biology of the Naked Mole-Rat. Princeton University Press, Princeton, NJ

Sledge MF, Boscaro F, Turillazzi S (2001) Cuticular hydrocarbons and reproductive status in the social wasp *Polistes dominulus*. Behav Ecol Sociobiol 49:401–409

Smith AR, Wcislo WT, O'Donnell S (2003) Assured fitness returns favor sociality in a mass-provisioning sweat bee, *Megalopta genalis* (Hymenoptera: Halictidae). Behav Ecol Sociobiol 54:22–29

Smith AR, Wcislo WT, O'Donnell S (2007) Survival and productivity benefits to social nesting in the sweat bee *Megalopta genalis* (Hymenoptera: Halictidae). Behav Ecol Sociobiol 61: 1111–1120

Springhetti A (1969) Il controllo sociale della differenziazione degli alati In *Kalotermes flavicollis* (Isoptera). Ann Univ Ferrara (Sez 3) 3:73–96

Starr CK (1985) Enabling mechanisms in the origin of sociality in the Hymenoptera – the sting's the thing. Ann Entomol Soc Am 78:836–840

Starr CK (1989) In reply, is the sting the thing? Ann Entomol Soc Am 82:6–8

Starr CK (1990) Holding the fort: colony defense in some primitively social wasps. In: Evans DL, Schmidt JO (eds) Insect defenses: adaptive mechanisms and strategies of prey and predators. State University of New York Press, Albany, NY, pp 421–463

Starr CK (1991) The nest as the locus of social life. In: Ross KG, Matthews RW (eds) The social biology of wasps. Comstock Publishing Associates. Ithaca, New York, NY, pp 520–539

Strassmann J (2001) The rarity of multiple mating by females in the social Hymenoptera. Insectes Soc 48:1–13

Strohm E, Liebig J (2008) Why are so many bees but so few digger wasps social? The effect of provisioning mode and helper efficiency on the distribution of sociality among the Apoidea. In: Korb J, Heinze J (eds) Ecology of social evolution. Springer, Berlin, pp 109–128

Templeton AR (1979) Chromosome number, quantitative genetics and eusociality. Am Nat 113:937–941

Thompson GJ, Hebert PDN (1998) Population genetic structure of the neotropical termite *Nasutitermes nigriceps* (Isoptera: Termitidae). Heredity 80:48–55

Thompson GJ, Kitade O, Lo N et al (2000) Phylogenetic evidence for a single, ancestral origin of a 'true' worker caste in termites. J Evol Biol 13:869–881

Thorne BL (1997) Evolution of eusociality in termites. Annu Rev Ecol Syst 28:27–54

Thorne BL, Breisch NL, Haverty MI (2002) Longevity of kings and queens and first time of production of fertile progeny in dampwood termite (Isoptera; Termopsidae; *Zootermopsis*) colonies with different reproductive structures. J Anim Ecol 71:1030–1041

Thorne BL, Breisch NL, Muscedere ML (2003) Evolution of eusociality and the soldier caste in termites: influence of intraspecific competition and accelerated inheritance. Proc Natl Acad Sci U S A 100:12808–12813

Thorne BL, Carpenter JM (1992) Phylogeny of the Dictyoptera. Syst Entomol 17:253–268

Thorne BL, Grimaldi DA, Krishna K (2000) Early fossil history of the termites. In: Abe T, Bignell DE, Higashi M (eds) Termites: evolution, sociality, symbioses, ecology. Kluwer Academic Publishers, Dordrecht, pp 77–93

Thorne BL, Lenz M (2001) Population and colony structure of *Stolotermes inopinus* and *S. ruficeps* (Isoptera: Stolotermitinae) in New Zealand. N Z Entomol 24:63–70

Thorne BL, Noirot C (1982) Ergatoid reproductives in *Nasutitermes corniger* (Motschulsky) (Isoptera: Termitidae). Int J Insect Morphol Embryol 11:213–226

Thorne BL, Traniello JFA (2003) Comparative social biology of basal taxa of ants and termites. Annu Rev Entomol 48:283–306

Toth AL, Varala K, Newman TC et al (2007) Wasp gene expression supports an evolutionary link between maternal behavior and eusociality. Science 318:441

Trivers RL, Hare H (1976) Haploidploidy and the evolution of the social insect. Science 191: 249–263

Tschinkel WR (1988) Colony growth and the ontogeny of worker polymorphism in the fire ant, *Solenopsis invicta*. Behav Ecol Sociobiol 22:103–115

Vargo EL, Passera L (1991) Pheromonal and behavioral queen control over the production of gynes in the Argentine ant *Iridomyrmex humilis* (Mayr). Behav Ecol Sociobiol 28: 161–169

Vehrencamp SL (1983) A model for the evolution of despotic versus egalitarian societies. Anim Behav 31:667–682

Wade MJ (2001) Maternal effect genes and the evolution of sociality in haplo-diploid organisms. Evolution 55:453–458

Waller DA, LaFage JP (1987) Nutritional ecology of termites. In: Slansky F Jr, Rodriguez JG (eds) Nutritional ecology of insects, mites, spiders, and related invertebrates. Wiley, New York, NY, pp 487–532

Waloff N (1957) The effect of the number of queens of the ant *Lasius flavus* (Fab.) (Hym., Formicidae) on their survival and on the rate of development of the first brood. Insectes Soc 4:391–408

Ward PS (1983) Genetic relatedness and colony organization in a species complex of ponerine ants. Behav Ecol Sociobiol 12:285–299

Wenseleers T, Helanterä H, Hart A, Ratnieks FLW (2004) Worker reproduction and policing in insect societies: an ESS analysis. J Evol Biol 17:1035–1047

West-Eberhard MJ (1975) The evolution of social behavior by kin selection. Q Rev Biol 50:1–33

West-Eberhard MJ (1978) Polygyny and the evolution of social behavior in wasps. J Kansas Entomol Soc 51:832–856

West-Eberhard MJ (1987) The epigenetical origins of insect sociality. In: Eder J, Rembold H (eds) Chemistry and biology of social insects. Verlag J Peperny, München, pp 369–372
West-Eberhard MJ (1996) Wasp societies as microcosms for the study of development and evolution. In: Turillazzi S, West-Eberhard MJ (eds) Natural history and evolution of paper-wasps. Oxford University Press, New York, NY, pp 290–317
Wheeler DE (1986) Developmental and physiological determinants of caste in social Hymenoptera: evolutionary implications. Am Nat 128:13–34
Wheeler DE (1991) The developmental basis of worker caste polymorphism in ants. Am Nat 138:1218–1238
Wilfert L, Gadau J, Schmid-Hempel P (2007) Variation in genomic recombination rates among animal taxa and the case of social insects. Heredity 98:189–197
Wilson EO (1971) The insect societies. Harvard University Press, Cambridge, MA
Wilson EO (1975) Sociobiology: the new synthesis. Harvard University Press, Cambridge, MA
Wilson EO (1985) The sociogenesis of insect colonies. Science 228:1489–1495
Wilson EO (1987) The earliest known ants: an analysis of the Cretaceous species and an inference concerning their social organization. Paleobiology 13:44–53
Wilson EO (2005) Kin selection as the key to altruism: its rise and fall. Soc Res 72:1–8
Wilson EO (2008) One giant leap: how insects achieved altruism and colonial life. BioScience 58:17–25
Wilson EO, Hölldobler B (2005a) Eusociality: origin and consequences. Proc Natl Acad Sci U S A 102:13367–13371
Wilson EO, Hölldobler B (2005b) The rise of the ants: a phylogenetic and ecological explanation. Proc Natl Acad Sci U S A 102:7411–7414
Wilson DS, Wilson EO (2007) Rethinking the theoretical foundation of sociobiology. Q Rev Biol 82:327–348
Wynne-Edwards VC (1962) Animal dispersion in relation to social behaviour. Oliver and Boyd, Edinburgh
Yanega D (1988) Social plasticity and early-diapausing females in a primitively social bee. Proc Natl Acad Sci U S A 85:4374–4377
Zhou X, Oi FM, Scharf ME (2006) Social exploitation of hexamerin: RNAi reveals a major caste-regulatory factor in termites. Proc Natl Acad Sci U S A 103:4499–4504
Zimmerman RB (1983) Sibling manipulation and indirect fitness in termites. Behav Ecol Sociobiol 12:143–145

Chapter 6
Social Organisation and the Status of Workers in Termites

Yves Roisin and Judith Korb

Abstract Whereas the soldier caste is clearly ancestral to all extant termite lineages and constitutes the hallmark of termite eusociality, the origin and evolution of workers is much more difficult to comprehend. Here, we first review the status of working individuals throughout the Isoptera, insisting on the need for a consistent terminology. Pseudergates sensu stricto are defined ontogenetically as individuals stepping aside the straight egg-to-alate pathway, through regressive or stationary moults. They reveal an ancestral developmental flexibility, since they are common in small-colony wood-dwelling termites, but may also occur in large-colony Rhinotermitidae or Termitidae. They may participate in social tasks but never constitute a functional category by themselves. Pseudergates sensu lato, or false workers, are defined as a functional category of individuals performing work, while remaining developmentally flexible. This latter property distinguishes them from true workers, which result from an early and irreversible developmental bifurcation, and have lost the possibility to resume alate development. We emphasize three major kinds of social organisation, with respect to the worker caste. In the drywood-dwelling *Cryptotermes* (Kalotermitidae), all immatures (pseudergates sensu lato) participate in colony work in a cooperative rather than altruistic manner, with very little if any direct fitness loss. In *Prorhinotermes* (Rhinotermitidae) and *Glossotermes* (Serritermitidae), wing bud development is postponed until the last pre-alate nymph instar and the preceding instars form a large functional caste of pseudergates (sensu lato), performing partially altruistic colony work. In the Termitidae, tasks are done by a specialized, altruistic true worker caste. After reviewing the proximate developmental and regulatory mechanisms determining the fate of termite immatures, we discuss the ultimate causes of worker evolution. The potential role of kin-selected benefits of helping versus direct benefits of inheritance is examined in detail. Both

Y. Roisin (✉)
Evolutionary Biology and Ecology, Faculté des Sciences, Université Libre de Bruxelles, B–1050 Brussels, Belgium
e-mail: yroisin@ulb.ac.be

D.E. Bignell et al. (eds.), *Biology of Termites: A Modern Synthesis*,
DOI 10.1007/978-90-481-3977-4_6, © Springer Science+Business Media B.V. 2011

approaches still face some difficulties and it is probably a combination of both that explains the early social evolution of termites. We stress the need for new theoretical models as well as broader comparative data on life history and social behaviour.

6.1 Introduction

Unlike the eusocial Hymenoptera, which are all basically wasps, termites are social cockroaches. This convergent evolution of eusociality in termites and social Hymenoptera is reflected in several important differences (Chapter 5 by Howard and Thorne, this volume). One of these differences is that termites have two main altruistic castes, soldiers and workers, both of which are sometimes polymorphic while the social Hymenoptera, with few exceptions, have only workers. The presence of complex patterns of social polymorphism, including castes of reproductives, soldiers and workers (or "labourers"), has been known in termites for centuries (Smeathman 1781). However, as early as 1874, F. Müller (in Darwin 1874) pinpointed that soldiers were lacking in some species (e.g. *Anoplotermes* spp.) and workers in others (such as *Kalotermes*).

6.1.1 The Soldier Caste

The whole Isoptera clade is eusocial for the possession of a caste of soldiers. There is little doubt about the evolutionary origin, development and function of soldiers. Judging from phylogeny, soldiers evolved once in a basal lineage ancestral to all living termites. Only some derived soil-dwelling termitids have lost them secondarily. Soldiers are a terminal caste that is always sterile. They always develop from apterous or wing-budded instars through an intermediate, unsclerotised instar, the presoldier. This development seems proximately to be associated with a high juvenile hormone concentration (Lüscher 1969; Hrdý et al. 2006; Elliott and Stay 2008). The soldiers' role mainly consists in defending their colony against predators and/or inter- and intraspecific competitors (Haverty and Howard 1981; Thorne and Haverty 1991; Roux and Korb 2004), but they may also take part in exploration, recruitment (Traniello 1981; Kaib 1990; Rupf and Roisin 2008), and even egg care (Hanus et al. 2005). Whereas the soldier caste is well-defined and poses no terminological difficulty, the development, evolutionary origin, and function of the termite workers is more debated.

6.1.2 The "Worker" Caste

There is still considerable confusion over terminology applied to active immatures in termites. The terms "worker" and "pseudergate" are extensively used to characterize different concepts, but their definitions and limits may vary, implicitly or explicitly, according to the authors concerned. Definitions are based on a mixture of ontogenetical, anatomical, morphological, and behavioural criteria applicable to the individual. In addition, the notion of "caste" carries the idea of a collective function. This confusion is a special hindrance to the search for homologies and the reconstruction of evolutionary pathways. For instance, the way of including the character states "presence/absence of workers" and "presence/absence of pseudergates" in phylogenetic reconstructions was central to the exchange of views between Thompson et al. (2000a, 2004) and Grandcolas and D'Haese (2002, 2004) on the evolution of the working castes (see below).

6.1.2.1 Larvae and Nymphs

Termites exhibit a hemimetabolous development, typically comprising a series of instars characterised by growing wing rudiments (called wing tips, buds or pads) and ending in a winged imago (alate). A well-accepted peculiarity of isopteran terminology is that those immature instars, which are called "nymphs" in all other hemimetabolous insects, deserve this title in termites only if they display visible wing rudiments, otherwise they are called "larvae".

6.1.2.2 Pseudergates

The synonyms *false workers* (*faux-ouvriers*) and *pseudergates* were coined by Grassé and Noirot (1947) to designate individuals deviating from the straight egg-to-alate pathway through "regressive" or "stationary" moults in the drywood termite *Kalotermes flavicollis*. These terms specifically referred to the fate of the wing rudiments during the moult. This is an individual-based, ontogenetic definition. Although it has the merit of pinpointing an exceptional feature of termite developmental pathways, this definition of pseudergates sensu stricto is very impractical. Firstly, on morphological grounds pseudergates are often difficult to distinguish from regular larvae or nymphs. The presence or absence of tiny wing tips is often hard to establish objectively, and normal variations between individuals in wing bud development make it tricky to recognize nymphs with a slight degree of wing bud atrophy. Even more confusion is introduced when pseudergates resume wing development. Thus, short of following individual developments, it is often impossible to decide whether any given immature is a pseudergate or not (Korb and Hartfelder 2008). Secondly, these pseudergates sensu stricto do not constitute a functional category, but take part in the same activities as larvae (past the dependent earliest instars) or nymphs (except those approaching the imaginal moult). The interpretation of the pseudergate stage was somewhat altered by Lüscher (1952), who represented it as a central, obligatory stage in the development of *K. flavicollis*. This led to a broader and even more imprecise usage of the term for totipotent immature working stages which have the capability to develop into all other castes (i.e. alates, neotenic reproductives or soldiers) (= pseudergate sensu lato as stated by Korb and Hartfelder 2008).

6.1.2.3 True Workers

Detailed studies of caste systems in Termitidae (Noirot 1955; 1956) outlined sharp contrasts with the *Kalotermes* pattern. In particular, the development of the Termitidae systematically features an early and irreversible decision point, at the first moult, between the fundamental hemimetabolous nymphal–alate pathway and the apterous (sometimes called "neuter") line. This apterous line gives rise to all workers (also called "true workers" to distinguish them from the pseudergates sensu lato) and soldiers. Further studies of Rhinotermitidae (Buchli 1958; Renoux 1976; Roisin 1988a), Hodotermitidae (Clément 1953), and Mastotermitidae (Watson et al. 1977) revealed a forked Termitidae-like pattern. However, the Termopsidae and *Prorhinotermes* (Rhinotermitidae) share the *Kalotermes* pattern (Miller 1942; Roisin 1988b). It then appeared easy and very convenient to classify termite caste patterns in two groups, one characterised by a flexible development with pseudergates sensu lato, the other characterised by an early developmental fork separating the nymphal–alate line on the one hand and the apterous line (true worker and soldier) on the other (Fig. 6.1). Yet, problems arise as many species with a bifurcated development have pseudergates sensu stricto (e.g. Rhinotermitidae) and authors are often not explicit about the definition they apply.

The developmental caste patterns correspond well to distinct nesting and food acquisition strategies (called "life types" by Abe 1987). Species with pseudergates sensu lato, such as *Kalotermes flavicollis*, are wood-dwellers (Korb 2007a; Korb and Hartfelder 2008) that spend their colony life in a single piece of wood serving as both food source and shelter (= "one-piece type" of Abe (1990), "single-site nesters" of Shellman-Reeve 1997). The alternative life type is that of foraging termites (Korb 2007a; Korb and Hartfelder 2008) with workers that live in a nest more or less separated from their feeding sites. It comprises the "separate" and "intermediate" types of Abe (1990), or "central-site" and "multiple-site nesters" of Shellman-Reeve (1997). To get access to food, workers sooner or later have to forage outside the nest. This latter category includes *Mastotermes*, the Hodotermitidae, Termitidae, and several Rhinotermitidae. This distinction is important as selective forces fundamentally change depending on whether individuals do or do not leave their protective piece of wood.

6.1.2.4 Difficulties in the Interpretation of Termite Castes

It seemed intuitive that the simple, flexible pattern exhibited by the small *Kalotermes* colonies should be considered as ancestral, and the complex and more rigid pattern of the large Termitidae colonies as derived. However, this nice dualistic picture became blurred for the following reasons.

(1) Watson and Sewell (1981; see also Sewell and Watson 1981) noticed wing bud polymorphism in early instars of Australian *Kalotermes* species. They interpreted this as evidence for the ancestral nature of an early developmental bifurcation between the nymphal-alate line and an apterous line, and concluded that true workers preceded pseudergates sensu lato in the evolution.

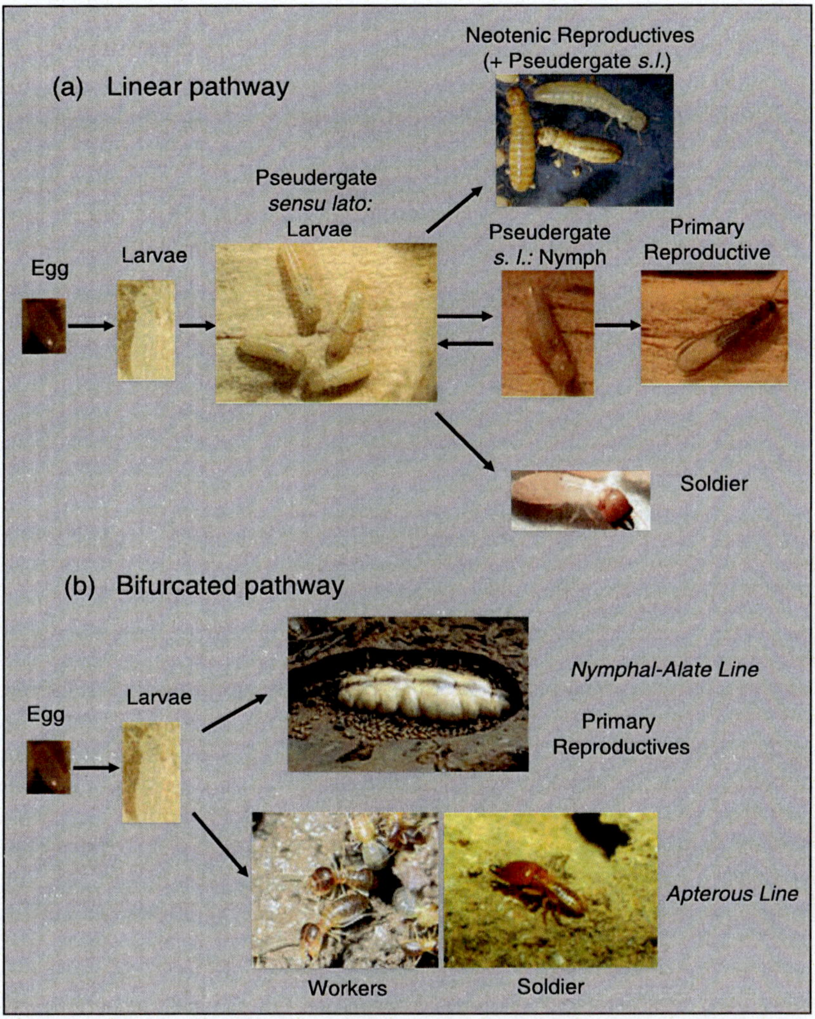

Fig. 6.1 Simplified developmental pathways of (**a**) wood-dwelling termites with a linear development and (**b**) foraging termites with a bifurcated development. In wood-dwellers, pseudergates sensu lato are totipotent to develop into primary reproductives, neotenic reproductives and soldiers. In foraging species, the developmental lines separate early on into a nymphal and an apterous line, the former leading to winged sexuals and the latter to workers and soldiers. In many species workers still have the potential to become reproductives

(2) The status of the apterous and nymphal lines in the closely related rhinotermitids *Reticulitermes* and *Coptotermes* is still subject to diverging interpretations (see below).
(3) Casual observations revealed the possibility of regressive moults by nymphs of some Termitidae, giving rise to worker-like individuals (Roisin and Pasteels 1987; Noirot and Thorne 1988; Roisin 1990a) which suggested possible transitions between the nymphal and the apterous pathways.

Using a taxonomical approach that aims at reflecting phylogenetic position (see Chapters 1 by Eggleton and Chapter 2 by Lo and Eggleton, this volume), we summarize here what is known about the development and the function of the working castes in termites. First, we detail the developmental patterns separately for different termite taxa. In the second part, we address the origin and evolution of the true worker caste. We start by showing that pseudergates occur in three different situations across the termite phylogeny, in which their abundance and ergonomic importance largely differs. In the next part, we outline what is known about the proximate mechanisms underlying termite development. This helps to shed light on the homology of different kinds of pseudergates and the evolutionary transition to a bifurcated development.

6.2 Mastotermitidae

The development schema of *Mastotermes darwiniensis* presents a bifurcation between the worker and the nymphal line (Watson et al. 1977). Workers become active late, from instar 7 onwards (instar 4 in incipient colonies). They can go through several instars (at least four), and give rise to presoldiers or ergatoid neotenics (i.e. worker derived reproductives). There are at least seven nymphal instars, recognizable by their wing rudiments, before the alate. According to Watson et al. (1977), future nymphs can already be distinguished from future workers at instar 2, by the development of their setae and pigmentation. It is hard to take a firm stand on this point, because the youngest nymphal instars are very difficult to find, and actual observations of transitions between instars are wanting. Experiments aiming at determining the ontogenetic potentialities of *Mastotermes* nymphs are also urgently needed: for instance, whether or not such nymphs can be induced to enter regressive or stationary moults remains unknown. However, the presence of a long series of successive nymphal instars parallel to a long series of worker instars is not subject to caution. The sex ratio is balanced in all worker instars and in soldiers, but varies somewhat among colonies for nymphs and alates (Parmentier 2006). Interestingly, only workers are able to develop into neotenic reproductives (Watson and Abbey 1985). Neotenics are very common in nature, where they contribute to the spread of colonies by underground galleries over extensive territories. Microsatellite markers revealed that the foraging area of a single colony can extend over 300 m (Goodisman and Crozier 2002).

6.3 Wood-Dwelling Termites

Wood-dwelling termites of the families Termopsidae (dampwood termites) and Kalotermitidae (drywood termites) have the most flexible development among termites (Fig. 6.1a). Immature individuals (late instar larvae; pseudergates sensu lato) that stay in the nest are totipotent to become (i) sterile soldiers, (ii) winged sexuals (alates) that leave the nest and found new colonies as primary reproductives or (iii) neotenic reproductives that either inherit the natal breeding position when the same sex reproductive of the colony dies (replacement reproductives, mainly in the Kalotermitidae) or that breed in addition to other reproductives (supplementary reproductives, mainly in the Termopsidae). Neotenic reproductives are characterized by the absence of wings and compound eyes and a less sclerotized cuticle than in alates. They may originate via a single moult from any instar after the third. At this neotenic moult, their gonads grow and they develop some imaginal characters while maintaining an otherwise larval appearance; some characters, such as wing pads, may even regress. This stands in contrast to the development of winged sexuals which is gradual and generally occurs via several nymphal instars with increasing wing pad development.

6.3.1 Kalotermitidae

In the drywood termite *Cryptotermes secundus* alates develop after passing through five nymphal instars with increasing wing pad development. In its natural habitat in northern Australia, the development is seasonally triggered and starts about 8 months before the annual nuptial flight in August (Korb and Katrantzis 2004). During this development there seem to be deadlines, as individuals which have not reached a certain nymphal instar by a certain date cease progressive development and will not become winged sexuals that same year (Korb and Katrantzis 2004). Instead, they moult regressively "back" into larval instars (pseudergates sensu stricto). These individuals can resume nymphal development the next year or can stay for another year (Korb, unpublished observations). When the wood block size is reduced below colony size-specific thresholds, nymphal development can be triggered at any time of the year (Korb and Katrantzis 2004). Thus more individuals pass the developmental deadlines in time resulting in more individuals leaving the colony as alates. In the penultimate and ultimate nymphal instars development can be temporally "arrested", so that no precocious nuptial flights occur. Strikingly, behavioural observations revealed that under such limited food conditions, those individuals which provided most proctodeal trophallaxis (anal feeding) to other late instar larvae or nymphs developed progressively along the nymphal line (Korb and Schmidinger 2004). This might be indicative for an honest signalling process: only well-fed, competent individuals can afford to feed others. If inhibitory pheromones are transmitted via proctodeal feeding, as suggested by Springhetti (1969) and Lüscher (1974), this process would be an evolutionary stable mechanism for regulating alate development. Only the most competent individuals leave the colony while the others stay to disperse later or to become neotenic replacement reproductives when the same-sex reproductive of the colony dies. The wood thresholds triggering alate development are well above limiting and wood still lasts for several years, so that less competent individuals can accumulate resources for the next nuptial flight. The data imply that alate development in *C. secundus* is a regulatory feedback process between colony size and wood availability: in large pieces of wood which outlast the longevity of the reproductives, individuals stay as long as the probability to inherit the colony as neotenic reproductive is larger than the probability to found a new colony (Korb and Schneider 2007). This leads to monogamous colonies with about 300–400 individuals (Korb and Schneider 2007). If the size of the piece of wood declines the optimal colony size declines, correspondingly leading to increased numbers of individuals leaving the colony, especially when wood sizes are reduced suddenly, as during thunderstorms or in many experiments (Korb and Lenz 2004). With an annual growth rate of about 100 individuals this means that in small pieces of wood the development of alates is expected each year as long as there are old enough individuals (note that individual development from egg to larval instars that can resume nymphal development lasts at least 2 years). The sex ratio in larval and nymphal instars is 1:1, while soldiers are highly female biased (Muller and Korb 2008; Roux et al. 2009).

Caste patterns in the kalotermitid genus *Neotermes* differ from those of *Cryptotermes* in some minor respects. There are only three nymphal instars, typically preceded by four apterous larval instars, the fourth of which is fully active (Myles and Chang 1984; Roisin and Pasteels 1991). Stationary and regressive moults do occur. When homogenous groups of wing-padded nymphs are constituted, some individuals complete their normal development and moult into alates, whereas others regress into pseudergates sensu stricto, which may resume alate development later on (Roisin 1990b). Meso- or metathoracic mutilations, whose occurrence is accidental in *Cryptotermes* (Korb 2005), may affect more than 40% of the large individuals in *Neotermes papua*, where they might constitute a means of caste regulation by impeding alate development (Roisin 1994).

6.3.2 Termopsidae

For Termopsidae, the development seems to be equally plastic (Heath 1927; Light 1944; Light and Illg 1945; Stuart 1979). Using morphometric measurements from preserved individuals of different sized *Zootermopsis nevadensis* colonies, Heath (1927) found a homogenous group of first instars with no indication of a separation into two developmental pathways. There are five larval and two nymphal instars with soldiers developing from the 5th instar upwards, depending on colony size. Nymphal instars only occurred in colonies with a minimum size of 450 individuals and there were indications of regressive moults. Such a delimitation of instars, from morphometrics alone, has limitations as the developmental trajectories of individuals are not followed (Korb and Katrantzis 2004). Yet further studies confirmed the linear developmental pathway of *Zootermopsis* and proved their high developmental flexibility. For instance, all individuals above the 3rd instar can develop into neotenic reproductives. The main difference between Kalotermitidae and Termopsidae is that in the former generally only one pair inherits the breeding position while there are often several neotenic reproductives in the latter (Light and Illg 1945). In *Zootermopsis nevadensis*, the extant to which supplementaries develop can be influenced by the composition of the group, the group's origin and the season (Light and Illg 1945). There are two wing-budded nymphal instars in *Zootermopsis*, but only one in *Hodotermopsis* (Miura et al. 2004). Large individuals, the size of nymphs but devoid of wing pads, are present. In *Zootermopsis* and *Archotermopsis*, the wing pads are sometimes cast off, a phenomenon called "abscission" (Heath 1927). One pecularity of the Termopsidae are the so-called reproductive soldiers (Myles 1986), more appropriately named soldier-like neotenics (Roisin 2000). They are neotenics with soldier-like traits. They seem to develop especially after intraspecific encounters when two colonies nest in the same log (Thorne et al. 2003).

6.4 Hodotermitidae

This family of harvester termites is phylogenetically close to the Termopsidae, and might be nested within them (Inward et al. 2007; Legendre et al. 2008). Hodotermitids have long been known to possess a true worker caste. The apterous (worker-soldier) line diverges from the nymphal line at the second moult. There is a sexual dimorphism among workers in *Hodotermes* (Luamba 1980), whereas all workers and soldiers are male in *Anacanthotermes* (Clément 1953; Mednikova 1977). There are at least 6 nymphal instars (Luamba 1980). Whether or not nymphs can undergo regressive moults to worker-like forms has never been investigated, but they can differentiate into neotenic reproductives (Luamba 1980; Mednikova 1977).

6.5 Rhinotermitidae, Serritermitidae and Termitidae

The precise branching pattern at the base of the [Rhinotermitidae + Serritermitidae + Termitidae] clade remains uncertain. There is however strong evidence (e.g. Inward et al. 2007; Legendre et al. 2008) that the following groups are monophyletic: (i) the Termitidae family; (ii) the Rhinotermitinae, comprising *Schedorhinotermes*, *Parrhinotermes*, and the neotropical *Rhinotermes*-group; (iii) *Reticulitermes* + [*Heterotermes* + *Coptotermes*]; (iv) *Serritermes* + *Glossotermes*, grouped in the Serritermitidae family (Cancello and DeSouza 2005). The remaining genera are *Stylotermes*, *Psammotermes*, *Prorhinotermes*, and *Termitogeton*. The caste system of *Stylotermes* is virtually unknown. The status of workers in the other genera is reviewed below.

6.5.1 Psammotermes

This genus makes deep subterranean nests in arid regions of Africa and Asia. Soldiers and workers of *Psammotermes hybostoma* are highly polymorphic: for instance, the head width of soldiers from a single colony varied from 1.07 to 3.35 mm, with two major modes near 1.6 and 2.9 mm. Worker head width presented a first mode near 1.15 mm, and several lower modes at larger values, with a few individuals reaching over 2 mm (Clément 1952). Whether these are true workers or not cannot be determined, as no data is available regarding the origin and development of nymphs.

6.5.2 *Prorhinotermes*

The development of *Prorhinotermes* species follows a linear pathway, comprising two inactive larval instars, three or more instars of active apterous individuals, one wing-padded nymphal instar and the alate (Miller 1942; Roisin 1988b). There is no developmental bifurcation towards true workers, hence individuals would classify as pseudergates sensu lato. The sex ratio in nymphs and soldiers is balanced. Older larvae and nymphs may give rise to presoldiers or to neotenics. Nymphs may also undergo regressive moults. Mutilations of the meso- or metathorax are commonly observed in wing-padded nymphs, and are a potential cause of regressive moults (Roisin 1988b).

Although the development of *Prorhinotermes* closely matches the Kalotermitidae model, their biology is different. They nest in a single-piece of wood like wood-dwellers, but in case of food shortage, soldiers and workers leave the nest to explore its surroundings. If they find a suitable piece of wood, they establish a trail along which the colony, or part of it, including neotenics, will move into the new site (Rupf and Roisin 2008).

6.5.3 *Termitogeton*

The development of *Termitogeton planus* resembles that of *Prorhinotermes* species. A minor difference is the emergence of tiny wing buds at the first moult, in all larvae. Those buds are conserved at the next (third) instar but regress afterwards to disappear at the fifth instar. However, long wing pads reappear at once at the moult to the single nymphal instar, which precedes the alate (Parmentier and Roisin 2003). There is no bifurcation towards a true worker caste. The sex ratio is balanced in all castes except soldiers, among which females are approximately twice as numerous as males (Parmentier and Roisin 2003).

The biology of *Termitogeton* is poorly known. In Indonesian Papua, *T. planus* was found in large logs like wood-dwelling termites. Attempts to census whole colonies resulted in the collection of apparently complete colonial units, with reproductives, comprising up to 103 individuals (Parmentier and Roisin 2003). However, actual inter-colonial limits were hard to assess. Neotenics do occur, and colony expansion through satellite, neotenic-headed units is possible.

6.5.4 *Glossotermes*

In *Glossotermes oculatus*, as in *Prorhinotermes* and *Termitogeton*, there is a single wing-padded nymphal instar, directly derived from apterous individuals. A remarkable feature of these individuals is that, except at the time of nymph production, they are almost exclusively male. Soldiers are all male as well. Like *Termitogeton*, *Glossotermes* seem to be wood-dwellers that live in a diffuse network of galleries in large logs, and neotenics are commonplace (Bourguignon et al. 2009).

6.5.5 Reticulitermes

Although *Reticulitermes* and *Coptotermes* are economically important genera, and are probably the most studied of all termites, the knowledge of their caste patterns is still marked by serious inconsistencies. The closely related genus *Heterotermes* has never been specifically investigated in this respect.

The presence of an early decision point in the development was demonstrated half a century ago by Buchli (1958) in *Reticulitermes* species from southwestern France. However, this detailed study also revealed considerable developmental flexibility. For both *R. grassei* (then known as *R. lucifugus*) and the French population of *R. flavipes* (then known as *R. lucifugus santonensis*), Buchli (1958) established a basal developmental schema comprising two undifferentiated larval instars, followed by a bifurcated development into either (i) six wing-budded nymphal instars and the alate, or (ii) a series of up to nine worker instars, from which presoldiers and soldiers originate. The last three nymphal instars can moult into wing-budded (brachypterous) nymphoid neotenic reproductives, whereas late worker instars can in a similar way moult into apterous ergatoid neotenics.

However, experiments with small artificial colonies revealed several possibilities of shift between the nymphal and apterous lines. Wing-budded nymphs of the first four instars placed into small laboratory groups often experience a reduction of their wing buds and shift towards worker aspect and behaviour, giving rise to pseudergates sensu stricto (*pseudo-ouvriers*) (Buchli 1958: pp. 284–285). Further experiments suggested that resource shortage is a determining factor inducing nymph regression to pseudergates (Buchli 1958: pp. 319–323). Pseudergates sensu stricto do occur in natural colonies, sometimes in "important quantities" (Buchli 1958: p. 358). They seem able to resume their nymphal development, but in the laboratory, those who did so invariably ended up as neotenics, never as alates.

The reciprocal shift, from the apterous line to the nymphal line, is not as convincingly documented. Workers can transform into ergatoid neotenics, through a process requiring two moults. The intermediate instar may display small wing buds, and is then called pseudonymph (*pseudonymphe*, Buchli 1958: p. 326 sqq.). Buchli (1958: p. 328) mentioned the observation of two such pseudonymphs of *R. grassei* that moulted into normal nymphs instead of moulting into neotenics. These individuals went through two additional nymphal instars before moulting to brachypterous neotenics. A similar observation was made on *R. flavipes* (Buchli 1958: p. 338).

Sex ratio data are surprisingly few for *Reticulitermes* workers. In a population of *R. flavipes* from Massachusetts, Zimet and Stuart (1982) found a slight excess of males (1.30:1) among workers, whereas in Texas, Dean and Gold (2004) reported balanced sex ratios for all castes, except soldiers which were slightly female-biased (1.27:1). Within the genus, soldier sex ratio varies from balanced to all-female (Matsuura 2006).

In conclusion, the caste patterns of *R. grassei* and *R. flavipes* appear as extremely flexible. Transitions between apterous and nymphal lines are possible during all but the latest instars. Individuals from both lines may develop into neotenic reproductives, either in a straightforward way (nymph → brachypterous, nymphoid neotenic; worker → apterous, ergatoid neotenic) or after an apparent shift to the other line (nymph → pseudergate → apterous neotenic; worker → pseudonymph → brachypterous neotenic). *Reticulitermes* thus basically displays a bifurcated development with true workers, although switches between lines are possible and pseudergates sensu stricto occur. Remarkably, it is in this genus (in the Japanese species *R. speratus*) that evidence for a genetic control of caste determination was provided for the first time in termites (Hayashi et al. 2007). An X-linked allelic polymorphism would regulate the decision between the early diverging apterous and nymphal lines. Nymph-derived reproductives (i.e. nymphoid neotenics or imagos) can only produce workers, but worker-derived reproductives (ergatoids) produce both nymphs and workers. The persistence of this system rests on an obligatory phase of worker reproduction.

6.5.6 *Coptotermes*

At first glance, *Coptotermes lacteus* nicely matches the basic *Reticulitermes* pattern: two undifferentiated larval instars are followed either by a long series of true worker instars or by six nymphal instars and the alate (Roisin and Lenz 1999). The sex ratio of workers is balanced. Soldiers derive only from female workers in *C. lacteus*, but this is not true for all *Coptotermes* species. In orphaned colonies, nymphoid neotenics develop, but workers seem unaffected and do not transform into ergatoids. However, colonies headed by neotenics produce almost exclusively male alates throughout the year, whereas colonies headed by primary reproductives produce alates seasonally and with a balanced sex ratio (Lenz and Runko 1993). Neotenic-headed colonies actually produce an excess of male larvae; at the second moult, all female larvae and an equivalent number of male larvae moult into workers, whereas the remaining male larvae proceed towards the nymphal line (Roisin and Lenz 2002). However, nymphs can revert towards pseudergate sensu stricto forms, as shown by the finding of worker-like individuals with a nymph-like brain and wing buds in several orphaned colonies.

The mechanism responsible for this sex ratio shift in neotenic-headed colonies is still unknown. Lo et al. (2009) suggested that a genetic polymorphism similar to that inferred in *R. speratus* could be involved, but in the case of *C. lacteus*, there is so far no evidence that ergatoids exist, and it appears even less likely that they would be involved in an obligatory alternation with nymph-derived reproductives since most natural colonies are headed by primary reproductives and produce alates of both sexes (Lenz and Runko 1993).

A different pattern of nymph development was proposed for *C. formosanus* by Raina et al. (2004), featuring moults from worker instars 1 to 5 into nymphs with small wing buds. Evidence for such moults was obtained from trap-collected foraging parties, kept in the laboratory after all nymphs were removed: some individuals with small wing buds then developed from workers (Fig. 3 of Raina et al. 2004). However, whether these laboratory obtained individuals are normal nymphs is not fully clear, as there is no information as to their subsequent development. The possibility that they represent a transitional instar towards ergatoid neotenics should not be discarded, as Raina et al. (2004) also reported the finding of one such neotenic. In any case, these results show that workers of *C. formosanus* do possess some potentialities for sexual development.

6.5.7 *Rhinotermitinae*

Schedorhinotermes and *Parrhinotermes* are known to possess almost exclusively female true workers and soldiers (Renoux 1976; Roisin 1988a), a pattern probably common to the whole subfamily (Roisin 2000). Female and the rare male workers may develop into ergatoid neotenics (Renoux 1976; Roisin 1988a). Likewise, nymphs can give rise to nymphoid neotenics. Whether nymphs can revert to worker-like forms is not known.

6.5.8 *Termitidae*

The developmental pathway of all Termitidae bifurcates at the first moult after the first, externally undifferentiated larval instar (Noirot 1969). A nymphal line of five instars with growing wing buds, followed by the alate, probably constitutes the common pattern for all species. The apterous line comprises one or two apterous larval instars followed by workers (Fig. 6.1b). The number of successive worker instars, their sex, their polymorphism, and the origin of soldiers are highly diversified (Roisin 2000).

Workers are often totally sterile, but in some species they are still able to differentiate into ergatoid neotenics. These ergatoids are usually less fertile than nymphoid neotenics (reproductives derived from nymphal instars), but in some species, they are the only known form of replacement reproductives (reviewed in Roisin 2000). Ergatoids may show some nymphal characters, such as small wing buds and rudimentary eyes, but there is no reported case of workers acquiring a normal nymphal morphology. In the opposite direction, the acquisition of worker traits by nymphs has been observed, though rarely, in orphaned colonies suggesting the occurrence of regressive moults, and thus of pseudergates sensu stricto (Roisin and Pasteels 1987; Roisin 1990a, Fig. 6.2).

Fig. 6.2 Normal third-instar nymph (*top*) and worker-like individual derived from a nymph after a regressive moult (= pseudergate sensu stricto) (*bottom*) in an orphaned colony of *Nasutitermes novarumhebridarum* (colony No. 465 as given in Roisin and Pasteels 1987). Note darker head capsule, rudimentary compound eye, reduced fat body, small wing buds and wood-filled digestive tube in pseudergate

6.6 Origin and Evolution of the Worker Caste

6.6.1 Evolutionary Transitions Between Pseudergates and Workers

Whether workers or pseudergates arose first in the evolution of termites has been a recurrent question since Watson and Sewell (1981) argued that the caste pattern of the Kalotermitidae represented a derived condition, a view challenged by Noirot and Pasteels (1987, 1988). Parsimony analyses focusing on the number of evolutionary events fail to provide a definite answer, because (1) some polytomies remain unresolved in the phylogeny of the Isoptera (Lo et al. 2004; Ohkuma et al. 2004), (2) some caste patterns still give rise to diverging interpretations (see above, *Reticulitermes* and *Coptotermes*), (3) the coding of character states may influence the conclusions (see Thompson et al. 2000a, 2004, vs. Grandcolas and D'Haese 2002, 2004). For instance, the presence of workers and the presence of pseudergates (sensu stricto) should be coded as two independent binary characters, rather than as different states of a single character: here we have shown that pseudergates sensu stricto may occur in all major lineages, with the possible exception of the Mastotermitidae. Pseudergates sensu stricto may even occur among specialized Rhinotermitidae and Termitidae, such as some *Coptotermes* and *Nasutitermes* species, alongside the true worker caste. Finally (4), the presence or absence of workers and pseudergates depends on the definition of these terms, which is not always consistent. In particular, pseudergates occur in three different situations, in which their status, abundance and ergonomic importance dramatically differ.

6.6.1.1 Facultative Pseudergates Sensu Stricto in Small-Colony Kalotermitidae

These are the pseudergates as originally defined by Grassé and Noirot (1947), resulting from facultative deviations of the egg-larva-nymph-alate developmental pathway. They commonly occur, for instance, in *Kalotermes* and *Cryptotermes* species and probably in many Termopsidae. In *Cryptotermes secundus*, these individuals result from variability in the timing of wing development induced by season or food availability (Korb and Katrantzis 2004), but they do not constitute a functional category distinct from wing-budded nymphs or late instar larvae. Altogether, pseudergates sensu stricto, larval and nymphal instars represent pseudergates sensu lato.

6.6.1.2 Functional Caste: Pseudergates *Sensu Lato* in Basal Rhinotermitidae

Prorhinotermes and *Termitogeton* among the Rhinotermitidae, but also *Glossotermes* (Serritermitidae) and *Hodotermopsis* (Termopsidae) are characterised by a late differentiation of wing buds, postponed until the last pre-alate instar. The preceding instars are apterous and resemble true workers as found, e.g. in *Reticulitermes*. These individuals carry out worker tasks: for instance in *Hodotermopsis*, they are the most frequent food donors in intra-colonial trophallaxis (Machida et al. 2001), and in *Prorhinotermes* they take part in the search for new food sources (Rupf and Roisin 2008).

Such individuals do not fit the definition of nymphs, because they do not possess wing buds; they do not fit the definition of true workers, because they remain on the linear pathway to the alate stage; considering that they develop through stationary moults, they do not fit the definition of pseudergates sensu stricto, because they do not result from an actual deviation from the normal egg-to-alate line of development. They are pseudergates sensu lato.

The reduction of the number of wing-budded instars to one is very unusual for hemimetabolous insects (Sehnal et al. 1996), but happened at least once in *Hodotermopsis*, once in *Glossotermes* and once or twice in the Rhinotermitidae (*Termitogeton* and *Prorhinotermes*). This suggests that apterism among immatures has been selected probably in connection with their social working activity, perhaps because wing pads may constitute a hindrance in the narrow nest chambers. No such reduction of the number of wing-budded nymphal instars occurred in termites with true workers, whose nymphs do not work, nor in small-colony Kalotermitidae, where altruistic, social behaviours are rudimentary (Korb 2007b, 2008a). It is interesting here to note that *Zootermopsis* has two nymphal instars (Heath 1927; Miura et al. 2004). This might suggest that the genus is intermediate between the Kalotermitidae and *Hodotermopsis* with regard to social activity, as indeed seems to be the case (see Section 6.6.1.4 below).

6.6.1.3 Occasional Pseudergates Sensu Stricto Amid True Workers

Worker-like individuals deriving from nymphs after a regressive moult (i.e. pseudergates sensu stricto) have now been reported from several genera possessing a true worker caste: *Reticulitermes*, *Coptotermes* (Rhinotermitidae), *Microcerotermes*, *Nasutitermes* (Termitidae). In *Reticulitermes*, such individuals were obtained under laboratory conditions but also occur in natural colonies (Buchli 1958). In the other genera, they were always found in colonies recovering after orphaning and reproductive replacement. Although they reveal interesting developmental potentialities, their actual function is probably circumstantial and they cannot be considered a distinct caste. Their origin could be best understood as a response to critical changes in colony condition triggered by orphaning: an arrest in egg production entails a dearth of workers, which in turn translates into a shortage of resources and impedes the completion of alate development by remaining nymphs. If these nymphs cannot develop into neotenics (because other replacement reproductives already developed, or because of insufficient resources), they have no other option than reverting towards worker-like forms and perhaps save some indirect reproductive success by contributing to the colony's recovery.

6.6.1.4 Origins of True Workers

There is little doubt that termite evolution proceeded from small wood-dwelling colonies towards larger colonies foraging outside the wood. Furthermore, true workers are consistently absent in colonies of the former category and present in the latter. Pseudergates sensu stricto are well known in the former category, but may still occur in termites with true workers. The most likely evolutionary scenario is therefore an ancestral developmental flexibility, revealed by the presence of facultative stationary or regressive moults in basal lineages with small colonies (e.g. *Cryptotermes*; Korb 2007a; Korb and Hartfelder 2008). From there on, larger colonies developed, requiring a more specialized working caste. This requirement could be fulfilled to an extent by an increased social activity of the immatures, which remained totipotent but postponed the development of their wing rudiments until the last two (e.g. *Zootermopsis*) or ultimate pre-alate instar (e.g., *Hodotermopsis*, *Prorhinotermes*), thereby constituting a functional caste of false workers, or pseudergates sensu lato. However, it was the invention of the true worker caste, permanently diverted from the alate pathway, that allowed termites to develop huge colonies and reach their present-day diversity and omnipresence (Noirot 1982). The onset of true workers corresponds to the fixation of an early divergence point between wing-budded (nymphs) and apterous individuals (workers), and occurred at least three times in termite evolution (Noirot and Pasteels 1987, 1988; Korb and Hartfelder 2008).

A question central to this evolutionary scenario is that of the homology between the transformations leading to the various kinds of pseudergates. Are the developmental mechanisms triggering the formation of a facultative pseudergate in *Cryptotermes* homologous to those maintaining apterism through successive working instars of *Prorhinotermes*, and to those splitting workers from wing-budded nymphs in *Nasutitermes* or in *Hodotermes*? From a developmental point of view they all are a deviation from the direct egg-nymph-alate development typical for hemimetabolous insects (Nijhout and Wheeler 1982; Nijhout 1994). Although it is too early to answer this question, endocrinological and gene-expression studies have allowed recent progress in our understanding of caste development and regulation. However such studies are so far limited to few termite species, mainly *Cryptotermes secundus* (pseudergates sensu stricto with five nymphal instars), *Hodotermopsis sjostedti* (pseudergates sensu lato with one nymphal instar) and *Reticulitermes flavipes* (true workers with a flexible development).

6.6.2 Proximate Developmental and Regulatory Mechanisms

The flexibility in development of termites is achieved through a unique combination of progressive, stationary and regressive moults, reflected, respectively, by an increase, a standstill or a decrease in morphometric size and wing development. Corresponding to the different caste fates of immature wood-dwelling termites, progressive moults of particular importance are (i) the transition from late instar larva to presoldier/soldier, (ii) the transition from late instar larva to nymph (i.e. the transition that can lead to the development of winged sexuals), and (iii) the transition from late instar larva/nymph to neotenic reproductive. Of these, the presoldier-soldier transition is at present the best understood. Since we concentrate our chapter on "workers", we refer to recent studies and reviews for soldiers (e.g. Miura et al. 1999; Miura 2001, 2005; Scharf et al. 2003, 2005; Koshikawa et al. 2005; Zhou et al. 2006, 2007; Cornette et al. 2008; Korb and Hartfelder 2008, see also Chapters 9 by Miura and Scharf and Chapter 10 by Matsuura, this volume). The transition from late instar larva to nymph has been less studied (Cornette et al. 2008; Korb et al. 2009a). In the dampwood termite *Hodotermopsis sjostedti* (Cornette et al. 2008) and in the drywood termite *Cryptotermes secundus* (Korb et al. 2009a) late instar larvae without wing buds are characterized by higher JH titres than nymphal instars, despite the fact that there is a single nymphal instar in the former while there are five nymphal instars in the latter. Although more data are clearly needed, this suggests that nymphal development is triggered by low JH titres and that larval and nymphal instars are homologous regardless of whether there is only one or several nymphal instars. The importance of low JH titres has been emphasized because high JH titres during nymphal stages prevent the development of termite immatures into sexuals (see Fig. 6.3). Further support for this hypothesis comes from recent evidence that the unique regressive moults of termites, characterizing pseudergates sensu stricto, are associated with elevated JH titres in *C. secundus* (Korb et al. 2009a). It is, however, important to add a caveat here as the data presented in the last study are mean values of hormone titres for developmental stages that can last several weeks. As recently shown for the subterranean termite *Reticulitermes flavipes* (Rhinotermitidae) and as known for insects in general (Nijhout 1999), rates of juvenile hormone synthesis can vary considerably during developmental stages (Elliott and Stay 2008).

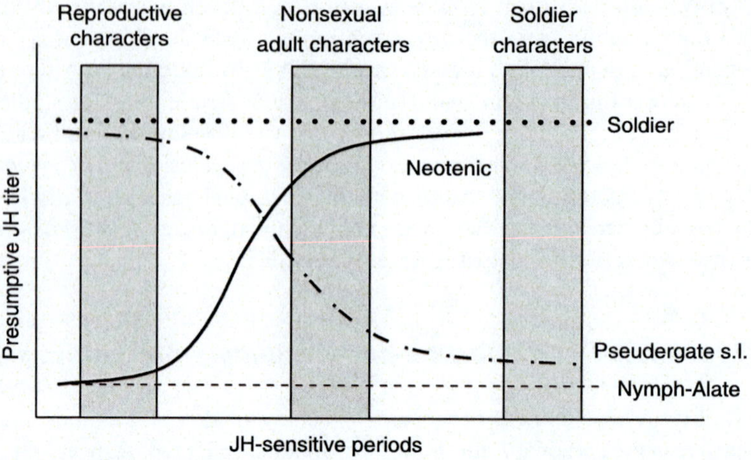

Fig. 6.3 Model for the control of caste polyphenism in lower termites (after Lüscher 1974; Nijhout and Wheeler 1982). The pseudergate sensu lato is a larval form that undergoes moults. During each moulting cycle, it has the option to develop into any of three final stages (soldier, neotenic reproductive, or winged sexual) or to remain a pseudergate sensu lato. These developmental decisions are controlled by JH titres during at least three JH-sensitive periods (*grey bars*). The JH titre *curves* shown are hypothetical. If JH titre is high during the appropriate sensitive period, reproductive characters and nonsexual adult characters (such as wings) are suppressed and soldier characters are induced. Pseudergates sensu lato induced to develop into neotenic reproductives have a shortened instar and moult before the soldier-sensitive period begins

The transition towards neotenic reproductives is even less well understood. Most endocrinological data concentrate on comparing neotenic reproductives with other castes including "worker" instars. Yet, these results give no consistent pattern: in the dampwood termite *H. sjostedti*, neotenics have lower JH titres than larvae and nymphs (Cornette et al. 2008), while in the drywood termite *C. secundus* they were more than 10 times higher than in either nymphs or primary reproductives (Korb et al. 2009a). This difference is interesting when keeping in mind that JH is a major regulator of fecundity in insects (Nijhout 1999), and that Termopsidae generally have several neotenics (per colony), with low fecundity, while Kalotermitidae have only one pair of neotenics which seems to be as fecund as primary reproductives (Lenz 1987; Korb and Schneider 2007). The high JH titres in *C. secundus* are also in accordance with JH synthesis rates in the subterranean termite *R. flavipes* in which neotenic rates far exceeded those of workers, soldiers and even presoldiers (Elliott and Stay 2008). The same study showed an increase in JH synthesis around the time some of the older instar larvae were expected to develop into neotenic reproductives or presoldiers. In the pharate stage (that is during the subsequent moulting phase after new cuticle deposition), corpora allata activity was considerably lower in presoldiers than in late instar larvae or neotenics. The authors interpreted this as similar to the terminal stage of development in cockroaches. But if JH titres in the haemolymph reflect corpora allata activity, the general model proposed for termite caste development must be modified (Fig. 6.3). Based on the drywood termite *Kalotermes flavicollis* it was proposed that there are three sensitive periods before a moult which determine caste fate (Lüscher 1953, 1974; Nijhout and Wheeler 1982, Fig. 6.3). During the first period low JH titres determine sexual characters and this leads to the development of either alates or neotenic reproductives. During the second period, adult development is triggered and a rise of JH titres leads to the development of neotenic reproductives, while constantly high JH titres trigger soldier development. Maybe with the exception of soldier development, this model still needs more empirical data for its support.

The molecular basis of development from late instar larvae to reproductives (especially into neotenic replacement reproductives since this requires only a single moult) is currently under investigation in *Cryptotermes* drywood termites. Using a highly sensitive suppression substractive hybridisation approach (representational difference analysis), studies on *C. secundus* identified five genes (*Neofem 1–5*, respectively, a member of the esterase-lipase family, a putative β-glycosidase, a cytochrome P450 gene, vitellogenin, and an unknown gene) that are highly overexpressed in female neotenics compared to pseudergates sensu lato and three of these genes are specifically overexpressed in female reproductives only (Weil et al. 2007). A similar expression analysis in a closely related species, *Cryptotermes cynocephalus*, revealed 13 genes that are highly overexpressed in queens (Weil et al. 2009). Strikingly, in most cases the caste-specific expression patterns were not conserved between species, despite the close phylogenetic relationship of the two *Cryptotermes* species (Thompson et al. 2000b). It is likely that these genes are associated with known species-specific differences in caste development modes (Weil et al. 2009). Only three genes (*Neofem 1–3*) showed a conserved and highly queen-specific expression pattern, suggesting that their products may play important roles in female reproductives, in particular in controlling caste determination and reproductive division of labour. This has been supported for *Neofem 2*. Silencing the gene with RNA interference resulted in behavioural changes in pseudergates sensu lato that were characteristic of queenless colonies (Korb et al. 2009b). A behavioural assay showed that the pseudergates sensu lato did not recognize their queen anymore, although the queen was present and did not change her behaviour. Thus, *Neofem 2* seems to be a gene necessary for the reproductive suppression of pseudergates sensu lato, at least in *C. secundus*. *Neofem 2*'s putative 532-amino acid gene product suggests that it is a member of the glycosyl hydrolase family 1 and its closest homologue plays a role in wood consumption in the kalotermitid *Neotermes koshunensis* (Tokuda et al. 2002). β-Glycosidases also can release volatiles from substrates composed of sugars and may play a role in egg pheromones in *Reticulitermes speratus* and cockroach species (Matsuura et al. 2009), and sex-specific signalling in the Maderian cockroach (Cornette et al. 2003). Altogether, this suggests that *Neofem 2* may have evolved from an ancestral role of wood digestion to one of queen-worker pheromonal communication in modern termites.

6.6.3 Ultimate Causes of Worker Caste Evolution

6.6.3.1 Properties of Wood as Habitat and Food Source

The wood-dwelling life type is associated with some idiosyncratic properties that must have been present when the termites evolved from a cockroach ancestor (Korb 2008a, 2009).

(1) They live in a bonanza-type food source. The nest of wood-dwelling termites constitutes a massive and long-lasting resource, allowing the feeding of many individuals for much longer than the time necessary to complete their development. Hence offspring can stay with their parents as there is no local resource competition which normally selects for offspring dispersal (Hamilton and May 1977). Only when the wood block becomes exploited does resource competition occur; in line with this, under such food conditions, pseudergates sensu lato of wood-dwelling termites develop into winged sexuals that leave the nest during the annual nuptial flight to found new colonies (see above; Lenz 1976, 1994; LaFage and Nutting 1978; Korb and Lenz 2004). This is possible because termites can sense food availability through vibrations generated during gnawing, and these act as reliable indicators of wood size (Evans et al. 2005).

(2) Living in a bonanza-type food source reduces the value of food provisioning for nestmates. Wood-dwelling termites sit inside their food. Thus, there is no necessity for costly foraging and food is easily accessible to all individuals. In addition, the young of termites become able to feed themselves at an early age, corresponding to the third or fourth instar. Termites are altricial compared to solitary cockroaches (Chapter 4 by Nalepa, this volume), but precocial compared to social Hymenoptera. Young can move around and do not rely on intensive brood care. Correspondingly, a lack of specialized brood care has been recorded for at least five wood-dwelling species: *Zootermopsis nevadensis* (Howse 1968), *Z. angusticollis* (Rosengaus and Traniello 1993), *Cryptotermes secundus*, *C. domesticus* and *C. cynocephalus* (Korb 2007b, 2008a; see also below). By contrast, all foraging termite species have true workers that care intensively for the brood (Korb 2007a).

(3) Wood is a resource that is hard to chew into and difficult to digest (Nalepa 1994). Wood hardness may explain why it takes a couple of moults until young termites become able to feed on wood by themselves. Additionally, wood hardness creates a challenge for imaginal or neotenic reproductives. Although reproductives are in principle able to feed directly on wood, they cannot replace their mandibles as they get worn, because they are unable to moult. By contrast, immatures are not faced with such a limitation, their mandible cutting edge being renewed at each moult. The existence of several successive chewing instars in wood-feeding termites, where workers of soil- and litter-feeders often stop moulting (Noirot 1955, 1982; Roisin 1996; 2000), suggests that mandible wear may impose a strong constraint on feeding on hard materials. Indeed,

drywood feeding termites have the hardest mandibles among all termites (Cribb et al. 2008).

(4) Individuals in all termites need to be infected with gut symbionts through proctodeal trophallaxis to exploit the common resource. This (re-)infestation after each moult, initially for wood-feeding but subsequently also for other trophic modes (see Chapter 15 by Ohkuma and Brune, this volume) selects for the parental care which also occurs in woodroaches, and which termites and woodroaches most likely inherited from their common ancestors (Cleveland et al. 1934).

(5) The nest presents a safe haven (sensu Kokko and Ekman 2002) where individuals are largely protected against predators and hostile environmental conditions and where they can inherit the natal breeding position (Myles 1988; Thorne et al. 2003; Korb 2007b, 2009; Korb and Schneider 2007). This contrasts with a risky dispersal option. As has been suggested for termites in general (Nutting 1969), for *C. secundus* it has been shown that the probability of successfully founding a nest is less than 1% (Korb and Schneider 2007). The chances of inheriting the colony as neotenic reproductive are of the same order in this species, and the number of individuals developing into winged sexuals that leave the nest versus those that stay can be explained by the relative probability of successfully founding a new colony versus inheriting the nest (Korb and Schneider 2007; Korb 2008a).

(6) Differences between dry- and dampwood conditions might exert somewhat different selective pressures on social evolution (Korb 2008b). As the names imply, drywood termites (Kalotermitidae) generally nest in sound wood, while the nest of most dampwood termites (Termopsidae) is at least partially decayed by fungus. In the former, the nest is a solid rather parasite- and pathogen-free environment, while the nests of dampwood termites are afflicted by high microbial loads (Rosengaus et al. 2003, see also Chapter 7 by Rosengaus et al., this volume). Although data are scarce so far, available evidence indicates that this might have consequences for brood care. In the drywood termites *Cryptotermes secundus*, *C. cynocephalus* and *C. domesticus*, there is no evidence of care for eggs and young, while similar data for the dampwood termite *Zootermopsis angusticollis* indicate that there is some brood care, especially allogrooming (Rosengaus and Traniello 1993). Further studies are clearly needed to elucidate whether this difference in grooming behaviour and pathogen load between damp- and drywood termites proves to be consistent. Based on the more basal phylogenetic position of the dampwood termites in most recent phylogenetical trees (Inward et al. 2007; Engel et al. 2009, but see notable exception: Legendre et al. 2008) and the fact that the termites' sister taxon, the woodroaches, inhabit damp wood, allogrooming to remove pathogens might have been the first component of brood care in the early evolution of termite sociality and was already present in the termites' ancestors.

In wood-dwelling termites, and probably in their common ancestor with the woodroaches, the protective, bonanza food type environment selected for offspring staying at the nest. These offspring were re-infested with gut symbionts after each moult. The importance of other parental care, and whether it was costly probably depended on the nesting environment. Humid, pathogen-rich rotten wood probably favoured allogrooming. It is reasonable to propose that the constant environment together with the parental care and the difficulty of processing wood as a food source selected for altricial young.

From there on, schemas of social evolution in termites follow two major lines of interpretation. The main point of divergence is the ultimate reason for immatures to stay in the nest instead of proceeding to the alate stage and leave. One approach focuses on the kin-selected benefits of helping, whereas the other emphasizes the direct benefits of inheritance. Both approaches face some difficulties and it is probably a combination of both that explains the early social evolution of termites.

6.6.3.2 Kin-Selected Benefits of Helping

Indirect benefits of helping were highlighted by several authors, including Nalepa (Chapter 4 by Nalepa, this volume) and Roisin (1994, 1999, 2006). Nalepa's approach emphasizes altriciality of termite neonates, which renders them dependent on help provided by other group members. By analogy with *Cryptocercus* woodroaches, Nalepa (1988) hypothesizes that caring for neonates entails a cost for older immatures, temporarily impeding their further development. Roisin's (1994, 2006) models outline that immatures have a choice between investing into their own growth and dispersal, equivalent to the release of one alate related to themselves by $r = 1$, or into the growth of younger siblings, thereby increasing the release of alates related to themselves by $r = 0.5$ (assuming monogamy).

The strength of this hypothesis lies in the high benefit/cost ratio it assumes. Because young termites become self-sufficient at an early age, the actual investment required to bring an embryo to this self-sufficient stage is low. Nutrients can be transferred through the reproductives into the eggs, or provided directly to young larvae. The price paid by the altruists is assumed to be a postponement of their own maturation and dispersal, but the actual cost of this postponement in terms of personal reproductive success may remain low, because the closed environment in which the termites live is relatively risk-free.

One weakness of this hypothesis is that because the costs allegedly incurred by the altruists are low, they are hard to demonstrate. For instance, in *Cryptotermes secundus*, the addition of extra brood to colonies had no detectable influence on the number of immatures staying at the nest as could have been expected if individuals stayed to provide help on demand, and no brood care could be observed during several 100 h of behavioural observations (Korb 2007b, 2008a and unpublished observations). Older immatures feed soldiers and reproductives, and they have reciprocal exchanges with other nestmate immatures (Korb 2007b). Yet, the feeding rates of soldiers (which make up < 2% of the colony members, that is, 2–5 individuals) and reproductives are low (for both castes together: less than once in 24 h; Korb, unpublished observations) and those with nestmate immatures are reciprocal.

Thus, in *C. secundus*, there is no evidence that brood care negatively influences alate development. On the contrary, under limited food conditions those individuals which provide most proctodeal feeding to nestmates are those which develop progressively along the nymphal line (Korb and Schmidinger 2004). However, the fact that *Cryptotermes* immatures mostly behave selfishly, providing only help that does not incur any obvious developmental costs, does not mean that this help is not important from the nestmate's viewpoint. For instance in *C. brevis*, Lenz (1987) showed that it is the presence of active immatures ("helpers") rather than the type of reproductives (imagos or neotenics) that accounts for colony growth rates. To quantify more precisely the help *Cryptotermes* immatures provide, an approach focused on the flow of nutrients through the various individuals composing the colonies is advisable. In any case, the difficulty to detect costly helping indicates that immatures do not incur large direct fitness costs, even if their presence offers benefits to relatives: thus they appear more cooperative than altruistic. More data on other wood-dwellers are clearly needed.

6.6.3.3 Direct Benefits of Inheritance

A major difference between woodroaches and all basal termite species is the occurrence of neotenic reproductives (Korb and Hartfelder 2008). Neotenic reproductives occur in all major termite lineages (review in Myles 1999), leaving no doubt that the evolution of this alternative reproductive morph occurred early at the base of the isopteran tree and neotenic reproduction can be considered a synapomorphy of termites. In case of death of the primary reproductives, the advantages of taking their place are evident: the successful replacement reproductives inherit the shelter and food bonanza provided by the nest. Considering that the period of availability of a piece of wood may commonly outlast the life span of termite imagos, being able to replace them is obviously highly profitable. Strikingly, this option to reproduce within the nest is present in all lower termites, including even species with a bifurcated development such as *Mastotermes darwiniensis* (Watson and Abbey 1985; Korb and Hartfelder 2008). These potential future direct benefits of reproducing within the natal nest present an important incentive for staying there.

Open questions are (1) whether the perspectives of future inheritance may induce immatures to postpone their growth and maturation as alates, and (2) if so, whether such a postponement could have provided the basis for the evolution of a helping, alloparental, worker caste There is little doubt that the answer to question (1) is yes. As outlined above, the nest environment is relatively risk-free, which keeps costs of postponing maturation low, whereas the potential benefits of inheriting are very high. In some species with a bifurcated development, a fraction of the immatures engage into the neotenic pathway, and so irreversibly forego the alate option, even while the primaries are still fully active. Inactive neotenics of this kind are found, e.g. in *Coptotermes* species (Lenz and Barrett 1982). The compensation for the loss of developmental flexibility is probably a higher chance of successful inheritance, should an opportunity arise. Question (2) is more difficult to answer. Because staying at the nest to inherit seems a much safer strategy than dispersal to found a new colony independently, several authors have suggested that neotenic perspectives were the main selective force inducing immatures to stay at home and postpone wing development (Myles 1988; Shellman-Reeve 1997; Thorne 1997; Thorne et al. 2003; Johns et al. 2009; Korb 2009). At the same time, stayers could gain indirect fitness benefits from reciprocal interactions with relatives, at little or no cost. Costly helping (as in the case of true workers) would have followed because it yielded an indirect fitness bonus to stayers.

The critical question that arises from the "direct benefit of inheritance" explanation is why those immatures do not proceed to the neotenic stage immediately and start to breed without delay. Assuming wood-dwellers live in a massive source of food, it seems that the founders as well as their neotenic offspring would so be able to raise a maximum number of offspring in a minimum amount of time. Local competition could arise, but should remain limited as long as the colony's expansion is not hindered by competing neighbouring colonies, or by the size of the food source: constraints that would also heavily bear upon the future benefits of nest inheritance. Theoretical modelling of such a situation is badly needed, to reveal the potential fitness effects of intra- and inter-colonial competition on individual developmental strategies. Another explanation for why immatures refrain from reproduction might be inbreeding depression (DeHeer and Vargo 2006); although experimental results are not unambiguous so far, inbred offspring of neotenic pairs may be less fit than outbred offspring of the founders (Shellman-Reeve 2001; Calleri et al. 2006). Even though some termite species, especially invasive Rhinotermitidae, overcame this difficulty most termites still routinely outbreed (Vargo and Husseneder 2009 and Chapter 12 by Vargo and Husseneder, this volume). Costs of inbreeding may compensate for, or at least reduce potential benefits of higher relatedness between neotenic parents and their inbred offspring ($r = 0.75$).

6.6.3.4 Why Be a False Worker Rather than a True One?

Discussions about caste evolution in termites have been mostly focused on the evolution of altruistic help. Ultimate reasons for being a false worker (pseudergate sensu lato) instead of a true one have seldom been addressed. Higashi et al. (1991) emphasized nest stability as a decisive factor for the evolution of sterile workers: nesting within an long-lasting but nevertheless exhaustible food source, as wood dwellers do (e.g. kalotermitids), favours the preservation of dispersal abilities, whereas exploiting renewable resources outside the nest (as foraging termitids do) favours sterile workers. An implicit condition in Higashi et al. (1991) model is that true workers make better workers than false ones.

The argumentation of Higashi et al. (1991) regarding the respective nest stability of wood dwellers and separate nest-builders seems reasonable, but actual data on nest longevity in nature are badly needed. The turnover of colonies in some nest-building termitids may actually be remarkably fast (Soki et al. 1996; Korb and Linsenmair 1999). As to worker efficiency, it is obvious that *Cryptotermes* immatures are poor workers (Korb 2007b), but a question remains: do they behave selfishly because they cannot be good workers, or do they perform poorly as workers because they favour selfish opportunities? Further insight into this question could come from future studies of intermediate situations such as *Prorhinotermes*, whose false workers constitute a real functional group involved, for instance, in exploration and gallery building (Rupf and Roisin 2008); or such as *Glossotermes*, whose false workers' heavily male-biased sex ratio leaves them with little hope of personal reproduction (Bourguignon et al. 2009).

Workers of foraging species leave the protective nest. This increases mortality risk and reduces worker longevity. According to life history theory, this ultimately selects for an early investment in "reproduction", i.e. gain of indirect fitness benefits through helping. Proximately, selection acts on improved resistance against hostile environmental conditions which goes hand in hand with increased sclerotization and thus reduced moulting abilities and reduced developmental flexibility.

6.7 Conclusion: What Is a Worker?

One purpose of this review was to emphasize that developmental versatility involving active individuals is broadly distributed throughout the Isoptera, making it very difficult to provide a consistent and accurate terminology. Individuals matching the original definition of pseudergates (sensu stricto: Grassé and Noirot 1947), that is, nymphs with regressed or undersized wing buds, are much more widespread than previously recognized (e.g. Legendre et al. 2008). Whether or not they constitute a caste, i.e. a functional category, is variable: in *Cryptotermes*, pseudergates sensu stricto are part of a pool of mostly selfish individuals, comprising also older larvae and nymphs; in *Nasutitermes*, they occur only sporadically and probably integrate with the true worker caste functionally. By contrast, true workers always constitute a well-defined functional group wherever they are present. Difficulties arise in genera such as *Prorhinotermes*, *Glossotermes* or *Hodotermopsis*, in which the development of wing buds is systematically postponed until the last pre-alate stage. The bulk of the actual working caste comprises such individuals, which fit neither the definition of pseudergates sensu stricto nor that of true workers, but can be labelled pseudergates sensu lato.

The term worker may be used indiscriminately to designate any kind of non-reproductive individual which feeds itself and can provide some assistance to other individuals. The level of altruism displayed by workers varies from near-zero in *Cryptotermes* to 100% in many Termitidae, where they are completely sterile. It may then be useful to distinguish false workers (= pseudergates sensu lato), which retain a full range of developmental options, from true workers which are definitively excluded from the nymph-alate pathway. Pseudergates sensu stricto correspond to the developmental definition of Grassé and Noirot (1947). The important point is to make clear whether definitions are based on morphology, development or function. Future studies should be explicit in the definition they use.

References

Abe T (1987) Evolution of life types in termites. In: Kawano S, Connell JH, Hidaka T (eds) Evolution and coadaptation in biotic communities. University of Tokyo Press, Tokyo, pp 125–148

Abe T (1990) Evolution of worker caste in termites. In: Veeresh GK, Mallik B, Viraktamath CA (eds) Social insects and the environment. Proceedings 11th International Congress IUSSI 1990, Bangalore, India. Oxford & IBH, New Delhi, pp 29–30

Bourguignon T, Šobotník J, Hanus R, Roisin Y (2009) Developmental pathways of *Glossotermes oculatus* (Isoptera, Serritermitidae): at the cross-roads of worker caste evolution in termites. Evol Dev 11:659–668

Buchli HHR (1958) L'origine des castes et les potentialités ontogénétiques des termites européens du genre *Reticulitermes* Holmgren. Ann Sci Nat Zool 20:263–429

Calleri DV, Reid DM, Rosengaus RB et al (2006) Inbreeding and disease resistance in a social insect: effects of heterozygosity on immunocompetence in the termite *Zootermopsis angusticollis*. Proc R Soc Lond B 273:2633–2640

Cancello EM, DeSouza O (2005) A new species of *Glossotermes* (Isoptera): reappraisal of the generic status with transfer from the Rhinotermitidae to the Serritermitidae. Sociobiology 45:31–51

Cleveland LR, Hall SR, Sanders EP, Collier J (1934) The wood-feeding roach *Cryptocercus punctulatus*, its protozoa, and the symbiosis between protozoa and roach. Mem Am Acad Arts Sci 17:185–342

Clément G (1952) Recherches sur le polymorphisme de *Psammotermes hybostoma* Desneux. Ann Sci Nat Zool (11) 14:95–116

Clément G (1953) Sur la différenciation d'ovocytes dans les testicules des "neutres" d'*Anacanthotermes ochraceus*. C R Acad Sci 236:1095–1096

Cornette R, Farine J-P, Abed-Viellard D et al (2003) Molecular characterization of a male-specific glycosyl hydrolase, Lma-p72, secreted on to the abdominal surface of the Madeira cockroach *Leucophaea maderae* (Blaberidae, Oxyhaloinae). Biochem J 372:535–541

Cornette R, Gotoh H, Koshikawa S, Miura T (2008) Juvenile hormone titers and caste differentiation in the damp-wood termite *Hodotermopsis sjostedti* (Isoptera, Termopsidae). J Insect Physiol 54:922–930

Cribb BW, Stewart A, Huang H et al (2008) Unique zinc mass in mandibles separates drywood termites from other groups of termites. Naturwissenschaften 95:433–441

Darwin C (1874) Recent researches on termites and stingless honey-bees. Am Nat 8:553–556

DeHeer CJ, Vargo EL (2006) An indirect test of inbreeding depression in the termites *Reticulitermes flavipes* and *Reticulitermes virginicus*. Behav Ecol Sociobiol 59:753–761

Dean SR, Gold RE (2004) Sex ratios and development of the reproductive system in castes of *Reticulitermes flavipes* (Kollar) (Isoptera: Rhinotermitidae). Ann Entomol Soc Am 97: 147–152

Elliott KL, Stay B (2008) Changes in juvenile hormone synthesis in the termite *Reticulitermes flavipes* during development of soldiers and neotenic reproductives from groups of isolated workers. J Insect Physiol 54:492–500

Engel MS, Grimaldi DA, Krishna K (2009) Termites (Isoptera): their phylogeny, classification, and rise to ecological dominance. Am Mus Novit 3650:1–27

Evans TA, Lai JCS, Toledano E et al (2005) Termites assess wood size by using vibration signals. Proc Natl Acad Sci U S A 102:3732–3737

Goodisman MAD, Crozier RH (2002) Population and colony genetic structure of the primitive termite *Mastotermes darwiniensis*. Evolution 56:70–83

Grandcolas P, D'Haese C (2002) The origin of a 'true' worker caste in termites: phylogenetic evidence is not decisive. J Evol Biol 15:885–888

Grandcolas P, D'Haese C (2004) The origin of a 'true' worker caste in termites: mapping the real world on the phylogenetic tree. J Evol Biol 17:461–463

Grassé P-P, Noirot C (1947) Le polymorphisme social du termite à cou jaune (*Calotermes flavicollis* F.). Les faux-ouvriers ou pseudergates et les mues régressives. C R Acad Sci 224:219–221

Hamilton WD, May RM (1977) Dispersal in stable habitats. Nature 269:578–581

Hanus R, Šobotník J, Cizek L (2005) Egg care by termite soldiers. Insectes Soc 52:357–359

Haverty MI, Howard RW (1981) Production of soldiers and maintenance of soldier proportions by laboratory experimental groups of *Reticulitermes flavipes* (Kollar) and *Reticulitermes virginicus* (Banks) (Isoptera: Rhinotermitidae). Insectes Soc 28:32–39

Hayashi Y, Lo N, Miyata H, Kitade O (2007) Sex-linked genetic influence on caste determination in a termite. Science 318:985–987

Heath H (1927) Caste formation in the termite genus *Termopsis*. J Morphol Physiol 43: 387–425

Higashi M, Yamamura N, Abe T, Burns TP (1991) Why don't all termite species have a sterile worker caste? Proc R Soc Lond B 246:25–29

Howse PE (1968) On the division of labour in the primitive termite *Zootermopsis nevadensis* (Hagen). Insectes Soc 15:45–50

Hrdý I, Kuldová J, Hanus R, Wimmer Z (2006) Juvenile hormone III, hydroprene and a juvenogen as soldier caste differentiation regulators in three *Reticulitermes* species: potential of juvenile hormone analogues in termite control. Pest Manag Sci 62:848–854

Inward DJG, Vogler AP, Eggleton P (2007) A comprehensive phylogenetic analysis of termites (Isoptera) illuminates key aspects of their evolutionary biology. Mol Phylogenet Evol 44: 953–967

Johns PM, Howard KJ, Breisch NL et al (2009) Nonrelatives inherit colony resources in a primitive termite. Proc Natl Acad Sci U S A 106:17452–17456

Kaib M (1990) Intra- and interspecific chemical signals in the termite *Schedorhinotermes*. Production sites, chemistry, and behaviour. In: Gribakin FG, Wiese K, Popov AV (eds) Sensory systems and communication in arthropods. Birkhäuser Verlag, Basel, pp 26–32

Kokko H, Ekman J (2002) Delayed dispersal as a route to breeding: territorial inheritance, safe havens, and ecological constraints. Am Nat 160:468–484

Korb J (2005) Regulation of sexual development in the basal termite *Cryptotermes secundus*: mutilation, pheromonal manipulation or honest signal? Naturwissenschaften 92:45–49

Korb J (2007a) Termites. Curr Biol 17:R995–R999

Korb J (2007b) Workers of a drywood termite do not work. Front Zool 4:7

Korb J (2008a) The ecology of social evolution in termites. In: Korb J, Heinze J (eds) Ecology of social evolution. Springer-Verlag, Berlin, Germany, pp 151–174

Korb J (2008b) Termites, hemimetabolous diploid white ants? Front Zool 5:e15

Korb J (2009) Termites: An alternative road to eusociality and the importance of group benefits in social insects. In: Gadau J, Fewell JH (eds) Organization of insect societies: from genome to socio-complexity. Harvard University Press, Cambridge, MA, pp 128–147

Korb J, Hartfelder K (2008) Life history and development – a framework for understanding developmental plasticity in lower termites. Biol Rev 83:295–313

Korb J, Hoffmann K, Hartfelder K (2009a) Endocrine signatures underlying plasticity in postembryonic development of a lower termite, *Cryptotermes secundus* (Kalotermitidae). Evol Dev 11:269–277

Korb J, Katrantzis S (2004) Influence of environmental conditions on the expression of the sexual dispersal phenotype in a lower termite: implications for the evolution of workers in termites. Evol Dev 6:342–352

Korb J, Lenz M (2004) Reproductive decision-making in the termite, *Cryptotermes secundus* (Kalotermitidae), under variable food conditions. Behav Ecol 15:390–395

Korb J, Linsenmair KE (1999) Reproductive success of *Macrotermes bellicosus* (Isoptera, Macrotermitinae) in two neighbouring habitats. Oecologia 118:183–191

Korb J, Schmidinger S (2004) Help or disperse? Cooperation in termites influenced by food conditions. Behav Ecol Sociobiol 56:89–95

Korb J, Schneider K (2007) Does kin structure explain the occurrence of workers in a lower termite? Evol Ecol 21:817–828

Korb J, Weil T, Hoffmann K et al (2009b) A gene necessary for reproductive suppression in termites. Science 324:758

Koshikawa S, Cornette R, Hojo M et al (2005) Screening of genes expressed in developing mandibles during soldier differentiation in the termite *Hodotermopsis sjostedti*. FEBS Lett 579:1365–1370

LaFage JP, Nutting WL (1978) Nutrient dynamics of termites. In: Brian MV (ed) Production ecology of ants and termites. Cambridge University Press, Cambridge, UK, pp 165–232

Legendre F, Whiting MF, Bordereau C et al (2008) The phylogeny of termites (Dictyoptera: Isoptera) based on mitochondrial and nuclear markers: implications for the evolution of the worker and pseudergate castes, and foraging behaviors. Mol Phylogenet Evol 48: 615–627

Lenz M (1976) The dependence of hormone effects in termite caste determination on external factors. In: Lüscher M (ed) Phase and caste determination in insects. Pergamon Press, Oxford, UK, pp 73–90

Lenz M (1987) Brood production by imaginal and neotenic pairs of *Cryptotermes brevis* (Walker): the significance of helpers (Isoptera: Kalotermitidae). Sociobiology 13:59–66

Lenz M (1994) Food resources, colony growth and caste development in wood-feeding termites. In: Hunt JH, Nalepa CA (eds) Nourishment and evolution in insect societies. Westview Press, Boulder, CO, pp 159–209

Lenz M, Barrett RA (1982) Neotenic formation in field colonies of *Coptotermes lacteus* (Froggatt) in Australia, with comments on the roles of neotenics in the genus *Coptotermes* (Isoptera: Rhinotermitidae). Sociobiology 7:47–59

Lenz M, Runko S (1993) Long-term impact of orphaning on field colonies of *Coptotermes lacteus* (Froggatt) (Isoptera: Rhinotermitidae). Insectes Soc 40:439–456

Light SF (1944) Experimental studies on ectohormonal control of the development of supplementary reproductives in the termite genus *Zootermopsis* (formerly *Termopsis*). Univ Calif Publ Zool 43:413–454

Light SF, Illg PI (1945) Rate and extent of development of neotenic reproductives in the termite genus *Zootermopsis*. Univ Calif Publ Zool 53:1–40

Lo N, Hayashi Y, Kitade O (2009) Should environmental caste determination be assumed for termites? Am Nat 173:848–853

Lo N, Kitade O, Miura T et al (2004) Molecular phylogeny of the Rhinotermitidae. Insectes Soc 51:365–371

Luamba JLN (1980) Recherches sur le polymorphisme et aperçu sur l'influence de l'analogue de l'hormone juvénile sur le développement d'un termite, *Hodotermes mossambicus* (Isoptera, Hodotermitidae). Biol Ecol Médit 7:169–171

Lüscher M (1952) Untersuchungen über das individuelle Wachstum bei der Termite *Kalotermes flavicollis* Fabr. (Ein Beitrag zum Kastenbildungsproblem). Biol Zentralbl 71:529–543

Lüscher M (1953) Kann die Determination durch eine monomolekulare Reaktion ausgelöst werden? Rev Suisse Zool 60:524–528

Lüscher M (1969) Die Bedeutung des Juvenilhormons für die Differenzierung der Soldaten bei der Termite *Kalotermes flavicollis*. Proceedings VI Congress IUSSI, Bern, pp 165–170

Lüscher M (1974) Kasten und Kastendifferenzierung bei niederen Termiten. In: Schmidt GH (ed) Sozialpolymorphismus by Insekten. Probleme der Kastenbildung im Tierreich. Wissenschaftliche Verlagsgesellschaft MBH, Stuttgart, pp 694–739

Machida M, Kitade O, Miura T, Matsumoto T (2001) Nitrogen recycling through proctodeal trophallaxis in the Japanese damp-wood termite *Hodotermopsis japonica* (Isoptera, Termopsidae). Insectes Soc 48:52–56

Matsuura K (2006) A novel hypothesis for the origin of the sexual division of labor in termites: which sex should be soldiers? Evol Ecol 20:565–574

Matsuura K, Yashiro T, Shimizu K et al (2009) Cuckoo fungus mimics termite eggs by producing the cellulose-digesting enzyme beta-glucosidase. Curr Biol 19:30–36

Mednikova TK (1977) Caste differentiation in the termite *Anacanthotermes ahngerianus* Jacobson (Isoptera, Hodotermitidae). In: Velthuis HHW, Wiebes JT (eds) Proc VIII Congress IUSSI Centre for Agricultural Publishing and Documentation. Wageningen Academic Publishers, Wageningen, pp 118–120

Miller EM (1942) The problem of castes and caste differentiation in *Prorhinotermes simplex* (Hagen). Bull Univ Miami 15:1–27

Miura T (2001) Morphogenesis and gene expression in the soldier-caste differentiation of termites. Insectes Soc 48:216–223

Miura T (2005) Developmental regulation of caste-specific characters in social-insect polyphenism. Evol Dev 7:122–129

Miura T, Kamikouchi A, Sawata M et al (1999) Soldier caste-specific gene expression in the mandibular glands of *Hodotermopsis japonica* (Isoptera: Termopsidae). Proc Natl Acad Sci U S A 96:13874–13879

Miura T, Koshikawa S, Machida M, Matsumoto T (2004) Comparative studies on alate wing formation in two related species of rotten-wood termites: *Hodotermopsis sjostedti* and *Zootermopsis nevadensis* (Isoptera, Termopsidae). Insectes Soc 51:247–252

Muller H, Korb J (2008) Male or female soldiers? An evaluation of several factors which may influence soldier sex ratio in lower termites. Insectes Soc 55:213–219

Myles TG (1986) Reproductive soldiers in the Termopsidae (Isoptera). Pan-Pac Entomol 62: 293–299

Myles TG (1988) Resource inheritance in social evolution from termites to man. In: Slobodchikoff C (ed) The ecology of social behavior. Academic Press, San Diego, CA, pp 379–423

Myles TG (1999) Review of secondary reproduction in termites (Insecta: Isoptera) with comments on its role in termite ecology and social evolution. Sociobiology 33:1–91

Myles TG, Chang F (1984) The caste system and caste mechanisms of *Neotermes connexus* (Isoptera: Kalotermitidae). Sociobiology 9:163–319

Nalepa CA (1988) Cost of parental care in the woodroach *Cryptocercus punctulatus* Scudder (Dictyoptera: Cryptocercidae). Behav Ecol Sociobiol 23:135–140

Nalepa CA (1994) Nourishment and the origin of termite eusociality. In: Hunt JH, Nalepa CA (eds) Nourishment & Evolution in Insect Societies. Westview Press, Boulder, CO, pp 57–104

Nijhout HF (1994) Insect Hormones. Princeton University Press, Princeton, NJ

Nijhout HF (1999) Hormonal control in larval development and evolution – insects. In: Hall BK, Wake MH (eds) The origin and evolution of larval forms. Academic Press, San Diego, CA, pp 217–254

Nijhout HF, Wheeler DE (1982) Juvenile hormone and the physiological basis of insect polyphenisms. Q Rev Biol 57:109–133

Noirot C (1955) Recherches sur le polymorphisme des termites supérieurs (Termitidae). Ann Sci Nat Zool (11) 17:399–595

Noirot C (1956) Les sexués de remplacement chez les termites supérieurs (Termitidae). Insectes Soc 3:145–158

Noirot C (1969) Formation of castes in the higher termites. In: Krishna K, Weesner FM (eds) Biology of termites, vol I. Academic Press, New York, NY, pp 311–350

Noirot C (1982) La caste des ouvriers, élément majeur du succès évolutif des termites. Riv Biol 75:157–195

Noirot C, Pasteels JM (1987) Ontogenetic development and the evolution of the worker caste in termites. Experientia 43:851–860

Noirot C, Pasteels JM (1988) The worker caste is polyphyletic in termites. Sociobiology 14:15–20

Noirot C, Thorne BL (1988) Ergatoid reproductives in *Nasutitermes columbicus* (Isoptera, Termitidae). J Morphol 195:83–93

Nutting WL (1969) Flight and colony foundation. In: Krishna K, Weesner FM (eds) Biology of termites, vol I. Academic Press, New York, NY, pp 233–282

Ohkuma M, Yuzawa H, Amornsak W et al (2004) Molecular phylogeny of Asian termites (Isoptera) of the families Termitidae and Rhinotermitidae based on mitochondrial COII sequences. Mol Phylogenet Evol 31:701–710

Parmentier D (2006) Developmental flexibility and evolution of the worker caste in termites. PhD dissertation, Université Libre de Bruxelles

Parmentier D, Roisin Y (2003) Caste morphology and development in *Termitogeton* nr. *planus* (Insecta, Isoptera, Rhinotermitidae). J Morphol 255:69–79

Raina A, Osbrink WLA, Park YI (2004) Nymphs of the Formosan subterranean termite (Isoptera: Rhinotermitidae): aspects of formation and transformation. Ann Entomol Soc Am 97:757–764

Renoux J (1976) Le polymorphisme de *Schedorhinotermes lamanianus* (Sjöstedt) (Isoptera-Rhinotermitidae). Essai d'interprétation. Insectes Soc 23:279–494

Roisin Y (1988a) The caste system of *Parrhinotermes browni* (Isoptera: Rhinotermitidae). Sociobiology 14:21–28

Roisin Y (1988b) Morphology, development and evolutionary significance of the working stages in the caste system of *Prorhinotermes* (Insecta, Isoptera). Zoomorphology 107:339–347

Roisin Y (1990a) Queen replacement in the termite *Microcerotermes papuanus*. Entomol Exp Appl 56:83–90

Roisin Y (1990b) Reversibility of regressive molts in the termite *Neotermes papua*. Naturwissenschaften 77:246–247

Roisin Y (1994) Intragroup conflicts and the evolution of sterile castes in termites. Am Nat 143:751–765

Roisin Y (1996) Castes in humivorous and litter-dwelling neotropical nasute termites (Isoptera, Termitidae). Insectes Soc 43:375–389

Roisin Y (1999) Philopatric reproduction, a prime mover in the evolution of insect sociality? Insectes Soc 46:297–305

Roisin Y (2000) Diversity and evolution of caste patterns. In: Abe T, Bignell DE, Higashi M (eds) Termites: evolution, sociality, symbioses, ecology. Kluwer Academic Publishers, Dordrecht, pp 95–119

Roisin Y (2006) Life history, life types and caste evolution in termites. In: Kipyatkov VE (ed) Life Cycles in Social Insects: Behaviour, Ecology and Evolution. St. Petersburg University Press, Saint Petersburg, pp 85–95

Roisin Y, Lenz M (1999) Caste developmental pathways in colonies of *Coptotermes lacteus* (Froggatt) headed by primary reproductives (Isoptera, Rhinotermitidae). Insectes Soc 46:273–280

Roisin Y, Lenz M (2002) Origin of male-biased sex allocation in orphaned colonies of the termite, *Coptotermes lacteus*. Behav Ecol Sociobiol 51:472–479

Roisin Y, Pasteels JM (1987) Caste developmental potentialities in the termite *Nasutitermes novarumhebridarum*. Entomol Exp Appl 44:277–287

Roisin Y, Pasteels JM (1991) Polymorphism in the giant cocoa termite, *Neotermes papua* (Desneux). Insectes Soc 38:263–272

Rosengaus RB, Moustakas JE, Calleri DV, Traniello JFA (2003) Nesting ecology and cuticular microbial loads in dampwood (*Zootermopsis angusticollis*) and drywood termites (*Incisitermes minor, I. schwarzi, Cryptotermes cavifrons*). J Insect Sci 3:31

Rosengaus RB, Traniello JFA (1993) Temporal polyethism in incipient colonies of the primitive termite *Zootermopsis angusticollis*: a single multiage caste. J Insect Behav 6:237–252

Roux EA, Korb J (2004) Evolution of eusociality and the soldier caste in termites: a validation of the intrinsic benefit hypothesis. J Evol Biol 17:869–875

Roux EA, Roux M, Korb J (2009) Selection on defensive traits in a sterile caste–caste evolution: a mechanism to overcome life-history trade-offs? Evol Dev 11:80–87

Rupf T, Roisin Y (2008) Coming out of the woods: do termites need a specialized worker caste to search for new food sources? Naturwissenschaften 95:811–819

Scharf ME, Wu-Scharf D, Pittendrigh BR, Bennett GW (2003) Caste- and development-associated gene expression in a lower termite. Genome Biol 4:R62

Scharf ME, Wu-Scharf D, Zhou X et al (2005) Gene expression profiles among immature and adult reproductive castes of the termite *Reticulitermes flavipes*. Insect Mol Biol 14:31–44

Sehnal F, Švácha P, Zrzavý J (1996) Evolution of insect metamorphosis. In: Gilbert LI, Tata JR, Atkinson BG (eds) Metamorphosis. Postembryonic reprogramming of gene expression in amphibian and insect cells. Academic Press, San Diego, CA, pp 3–58

Sewell JJ, Watson JAL (1981) Developmental pathways in Australian species of *Kalotermes* Hagen (Isoptera). Sociobiology 6:243–323

Shellman-Reeve JS (1997) The spectrum of eusociality in termites. In: Choe JC, Crespi BJ (eds) The evolution of social behavior in insects and arachnids. Cambridge University Press, Cambridge, UK, pp 52–93

Shellman-Reeve JS (2001) Genetic relatedness and partner preference in a monogamous, wood-dwelling termite. Anim Behav 61:869–876

Smeathman H (1781) Some account of the termites which are found in Africa and other hot climates. Phil Trans R Soc Lond 71:139–192

Soki K, Josens G, Loreau M (1996) Growth and demography of *Cubitermes speciosus* mounds (Isoptera: Termitidae). Insectes Soc 43:189–200

Springhetti A (1969) Influenza dei reali sulla differenziazione dei soldati di *Kalotermes flavicollis* Fabr. (Isoptera). Proceedings VI Congress IUSSI, Bern, pp 267–273

Stuart AM (1979) The determination and regulation of the neotenic reproductive caste in the lower termites (Isoptera): with special reference to the genus *Zootermopsis* (Hagen). Sociobiology 4:233–237

Thompson GJ, Kitade O, Lo N, Crozier RH (2000a) Phylogenetic evidence for a single, ancestral origin of a 'true' worker caste in termites. J Evol Biol 13:869–881

Thompson G, Kitade O, Lo N, Crozier R (2004) On the origin of termite workers: weighing up the phylogenetic evidence. J Evol Biol 17:217–220

Thompson GJ, Miller LR, Lenz M, Crozier RH (2000b) Phylogenetic analysis and trait evolution in Australian lineages of drywood termites (Isoptera, Kalotermitidae). Mol Phylogenet Evol 17:419–429

Thorne BL (1997) Evolution of eusociality in termites. Annu Rev Ecol Syst 28:27–54

Thorne BL, Breisch NL, Muscedere ML (2003) Evolution of eusociality and the soldier caste in termites: influence of intraspecific competition and accelerated inheritance. Proc Natl Acad Sci U S A 100:12808–12813

Thorne BL, Haverty MI (1991) A review of intracolony, intraspecific, and interspecific agonism in termites. Sociobiology 19:115–145

Tokuda G, Saito H, Watanabe H (2002) A digestive beta-glucosidase from the salivary glands of the termite, *Neotermes koshunensis* (Shiraki): distribution, characterization and isolation of its precursor cDNA by 5′- and 3′-RACE amplifications with degenerate primers. Insect Biochem Mol Biol 32:1681–1689

Traniello JFA (1981) Enemy deterrence in the recruitment strategy of a termite: soldier-organized foraging in *Nasutitermes costalis*. Proc Natl Acad Sci U S A 78:1976–1979

Vargo EL, Husseneder C (2009) Biology of subterranean termites: insights from molecular studies of *Reticulitermes* and *Coptotermes*. Annu Rev Entomol 54:379–403

Watson JAL, Abbey HM (1985) Development of neotenics in *Mastotermes darwiniensis* Froggatt: an alternative strategy. In: Watson JAL, Okot-Kotber BM, Noirot C (eds) Caste differentiation in social insects. Pergamon Press, Oxford, pp 107–124

Watson JAL, Metcalf EC, Sewell JJ (1977) A re-examination of the development of castes in *Mastotermes darwiniensis* Froggatt (Isoptera). Aust J Zool 25:25–42

Watson JAL, Sewell JJ (1981) The origin and evolution of caste systems in termites. Sociobiology 6:101–118

Weil T, Korb J, Rehli M (2009) Comparison of queen-specific gene expression in related lower termite species. Mol Biol Evol 26:1841–1850

Weil T, Rehli M, Korb J (2007) Molecular basis for the reproductive division of labour in a lower termite. BMC Genomics 8:28

Zhou X, Oi FM, Scharf ME (2006) Social exploitation of hexamerin: RNAi reveals a major caste-regulatory factor in termites. Proc Natl Acad Sci U S A 103:4499–4504

Zhou X, Tarver MR, Scharf ME (2007) Hexamerin-based regulation of juvenile hormone-dependent gene expression underlies phenotypic plasticity in a social insect. Development 134:601–610

Zimet M, Stuart AM (1982) Sexual dimorphism in the immature stages of the termite, *Reticulitermes flavipes* (Isoptera: Rhinotermitidae). Sociobiology 7:1–7

Chapter 7
Ecology, Behavior and Evolution of Disease Resistance in Termites

Rebeca B. Rosengaus, James F.A. Traniello, and Mark S. Bulmer

Abstract The nesting and feeding habits of termites create the risk of contact with microbial and invertebrate pathogens and parasites. Additionally, termite life history can result in cyclical decreases in nestmate genetic heterogeneity, increasing susceptibility to parasites, and sociality may elevate transmission rates of infection within colonies. Current research indicates that ecology and group living have selected for disease resistance in both basal and derived termite families, which have evolved diverse immune adaptations deployed sequentially or simultaneously at both individual and societal levels. These include inducible behavioral, biochemical, immunological and social mechanisms of infection control. Mortality from disease can be significant for reproductives, pseudergates and sterile castes, and influences colony fitness through impacts on colony size, demography, polymorphism, division of labor, communication, development, reproduction, colony foundation, and colony and population genetics. The hemimetabolic development, diplodiploid genetics, microbial symbioses and recalcitrant diets of termites present unique opportunities to model the effects of disease on immune function, including the adaptive design of immune molecules, life-history traits and social evolution. Comparisons can also be made between termite and hymenopteran immunocompetences, highlighting phylogenetic and ecological differences. We advocate a multidisciplinary approach to disease resistance in termites, focusing simultaneously on cellular and humoral immunity, antibiotic prophylaxis and social modes of infection control.

7.1 Introduction

Although social interaction can increase exposure to pathogen and parasites, various aspects of "herd immunity" and "herd effect" (John and Samuel 2000) function in the prevention and control of infection (Rosengaus et al. 1998b; Schmid-Hempel

R.B. Rosengaus (✉)
Department of Biology, Northeastern University, Boston, MA 02115, USA
e-mail: r.rosengaus@neu.edu

D.E. Bignell et al. (eds.), *Biology of Termites: A Modern Synthesis*,
DOI 10.1007/978-90-481-3977-4_7, © Springer Science+Business Media B.V. 2011

1998; Wilson and Reeson 1998; Rosengaus and Traniello 2001; Traniello et al. 2002; Wilson and Cotter 2008; Cremer et al. 2007; Ugelvig and Cremer 2007; Fefferman and Traniello 2009; Cremer and Sixt 2009). Termites, as social insects, pose particularly interesting problems for the study of the evolution of disease control: they frequently live in confined, densely populated homeostatic nest environments which, in principle, will promote microbial growth and facilitate the transmission of infection (Hendee 1933, 1934; Sands 1969; Blackwell and Rossi 1986; Cruse 1998; Rosengaus et al. 2003; Roose-Amsaleg et al. 2004; Hughes et al. 2008; Jackson and Hart 2009). Despite this, termites are extraordinarily diverse, abundant and have significant ecological impact (Bignell and Eggleton 2000).

Termites feed on an wide array of nutritionally poor cellulosic materials, from wood and/or soil to grass, leaf-litter, fungi, algae and lichens (Abe 1987; Bignell and Eggleton 2000), and have coevolved microbial communities to aid in their digestion (Brune et al. 2006; Ohkuma et al. 2009 and references therein). The ancestors of termites were dictyopterans, likely similar to present-day woodroaches (Inward et al. 2007; Lo et al. 2007), and are therefore hemimetabolic diploids with sterile worker and soldier castes composed of both males and females. Basal termite species show cycles of inbreeding and outbreeding that can alter the genetic structure of colonies during growth and reproduction (Thorne and Traniello 2003; Calleri et al. 2006a). Because immature stages have miniature adult morphology and are motile, termite larvae are capable of providing labor, in contrast to most social Hymenoptera (Rosengaus and Traniello 1993a; Crosland and Traniello 1997 and references within). Termite cuticular armament is thin and less sclerotized than in other social insects, rendering them more vulnerable to parasites that invade through the integument. Colony foundation is generally haplometrotic; their primary mating system is monogamous and the male ("king") not only inseminates the female periodically, but is also a full participant in parental care, at least during early colony establishment (Shellman-Reeve 1990, 1997; Rosengaus and Traniello 1991). We expect individual and social immune systems to vary adaptively as a function of these distinctive life-history traits (Dunn 1990; Rowley and Powell 2007) across the roughly 2,600 species (281 genera; Eggleton 2000) of termites.

The many parameters involved in termite epidemiology can be modeled computationally (Naug and Camazine 2002; Pie et al. 2004; Fefferman et al. 2007; Fefferman and Traniello 2009). In addition to immunological responsiveness, parameters can include colony size and demography, nest architecture and the spatial distribution of nestmates, worker density, activity level and rates of interaction, hygienic behavior such as self grooming and allogrooming, cannibalism, physical caste and temporal division of labor. In this chapter, we consider the evolution of termite disease resistance in light of their ecology, phylogenetic history and behavior. We focus on the contributions of the immune system of individual termites and social mechanisms of infection control from which colony-level immunocompetence emerges.

7.2 Phylogeny, Eusociality, and the Evolution of Disease Resistance in Termites

Termites originated in the Upper Jurassic (Thorne et al. 2000). The role pathogenic microorganisms may have played in termite social evolution can therefore only be inferred from our knowledge of extant species. Because all living termites are eusocial, comparisons associating immunocompetence with a gradation of social organizations within the Isoptera cannot be made. Also, the phylogeny of the group is in a state of flux (Inward et al. 2007; Lo et al. 2007). However, it is possible to contrast phylogenetically basal and derived termites and compare their respective immune systems with those of ancestral taxa.

Why, when and how did disease resistance evolve in termites? How has the coevolution of termite hosts and their pathogens impacted immune defenses and the virulence of pathogens and parasites? How can we interpret the immunocompetence of individuals and colonies in the Isoptera? When did the first methods of disease prevention and resistance emerge and under what social and environmental circumstances? How was social prophylaxis expressed? Unfortunately, important extant and basal genera are geographically restricted, and phylogenetic gaps and the absence of grades of sociality further challenge our understanding of how disease resistance evolved among transitional forms. However, given their similarity in nesting habits, the subsocial *Cryptocercus*-like prototermite probably lived under significant pathogenic constraints (Rosengaus et al. 2003). *Cryptocercus* is colonized by a diverse microbial community: on average 234 colony forming units (CFUs)/cm^2 of fungi and bacteria were isolated from cuticular washes, a microbial load comparable to that of the basal dampwood termite *Zootermopsis* (Rosengaus et al. 2003). Other roaches are known to harbor a high diversity and density of microbes on their exoskeleton (42 species of fungi and 85 species of bacteria, averaging 10^5 CFUs/mg; Le Guyader et al. 1989). We assume that prior to the inception of termite sociality, prototermite forms must have possessed biochemical and physiological adaptations to cope with disease agents, similar to those found in modern day roaches (Brossut 1983; Karp and Duwel-Eby 1991; Faulhaber and Karp 1992; Karp et al. 1994). As ancestral species evolved from subsociality and group living to eusociality, existing disease–resistance traits were altered and molded by sociality (Fig. 7.1). If the risks of pathogen transmission became more significant after the inception of eusociality, infection-control traits present in ancestral forms would have been further fine-tuned within the context of a more complex social life. We have little understanding of the nature of the events that influenced the evolution of physiological immunity, which traits were involved and how newly adaptive disease-resistance mechanisms evolved, but the mechanisms of innate cellular and humoral immunity characterizing solitary and presocial ancestors of derived termite taxa were likely highly conserved. Thus, we propose that the basal immunological attributes of these species served as preadaptations for individual and group strategies of infection control found in lower and higher termite species.

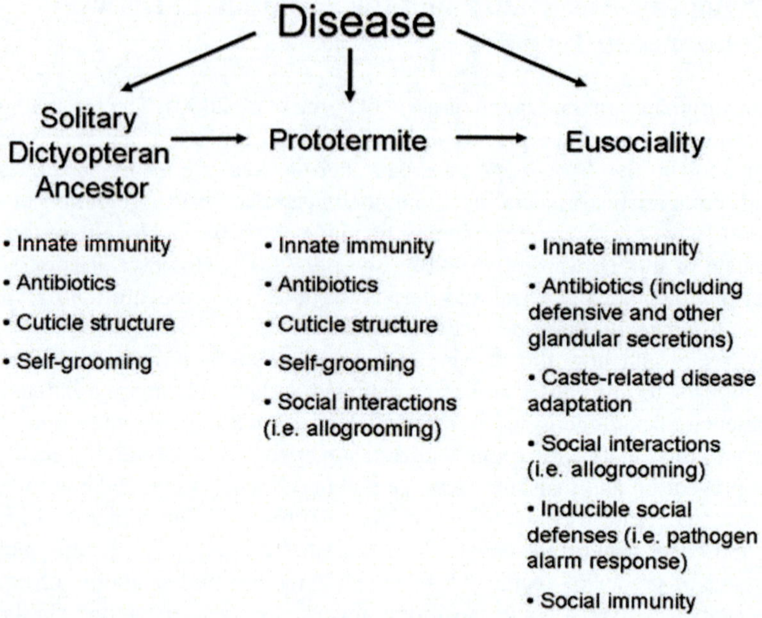

Fig. 7.1 Hypothesized anatomical, biochemical, physiological and behavioral adaptations to resist disease during the transition from solitary and presocial termite ancestors to eusocial termite species

A comprehensive understanding of the impact of pathogens and parasites on sociality can best be achieved through integrative research. A multidisciplinary approach, *socioecoimmunology*, considers the environmental dynamics of pathogens and parasites and how these influence disease-resistance in individuals living in groups, their genetic structure, parasite/host balances, and social organization (Fig. 7.2). Research in termite socioecoimmunology should: (1) identify and quantify microbes naturally associated with social termite hosts and the

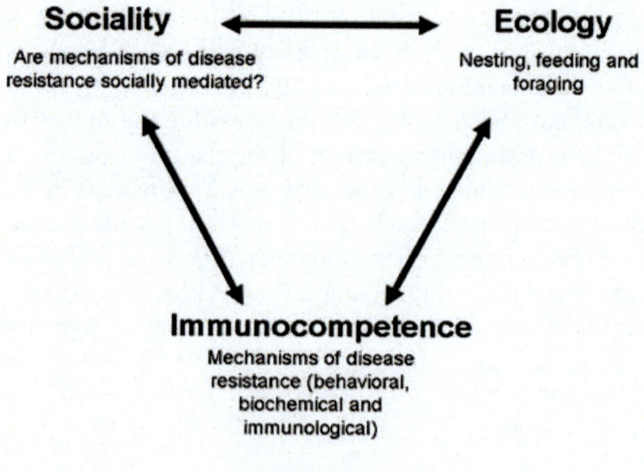

Fig. 7.2 The socioecoimmunological approach to the study of disease control

environments they exploit; (2) examine how microbes trigger immune responses in those hosts, and determine which responses are induced; (3) assess the balance of modes of immunocompetence (immunological, behavioral, biochemical) and their synergisms; (4) establish the extent to which individual immune responses are socially mediated.

7.3 Termite Microbial Ecology, Disease Risk and Immunocompetence

7.3.1 Nesting Ecology and Exposure to Disease Agents

Termite nests and their surrounding soils are laden with a variety of potential pathogens and parasites including bacteria, nematodes, viruses, protozoa and rickettsiae (reviewed by Schmid-Hempel 1998; Rosengaus et al. 2003; Roose-Amsaleg et al. 2004), and termites have a high incidence of interactions with fungi (Blackwell and Rossi 1986). Furthermore, termites may control microclimatic conditions within their nests, making it ideal not only for the development of offspring but also for the growth of microorganisms (Blackwell and Rossi 1986), many of which could be detrimental to colony fitness. Yet termites thrive in these environments. Quantifying the microorganisms associated with termites having different nesting, feeding and foraging ecologies is not an easy task. Cruse (1998) and Rosengaus et al. (2003) quantified cultivable nest and cuticular microbial loads of selected termite species (mostly bacteria and fungi). Results support the hypothesis that inoculum potential varies with termite nesting and feeding biology. Moreover, high variation in microbial loads was found between colonies of the same species (for example, *Reticulitermes flavipes*; Fig. 7.3a) suggesting that immune mechanisms may have

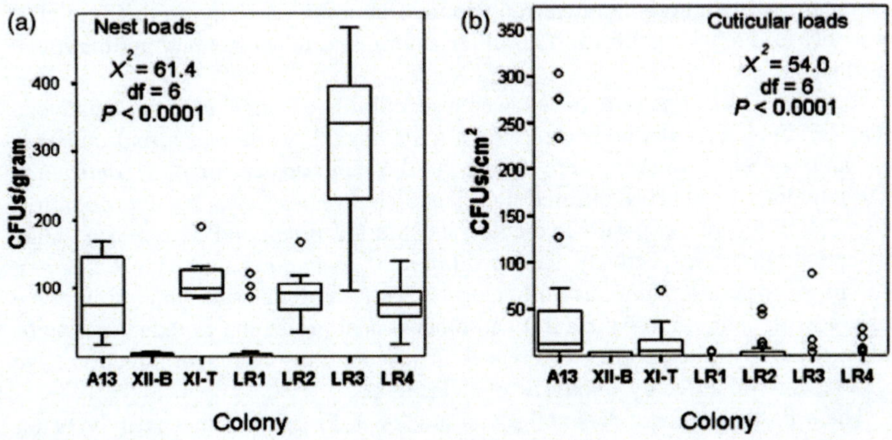

Fig. 7.3 Nest (**a**) and cuticle (**b**) microbial loads of *Reticulitermes flavipes* workers originating from seven colonies

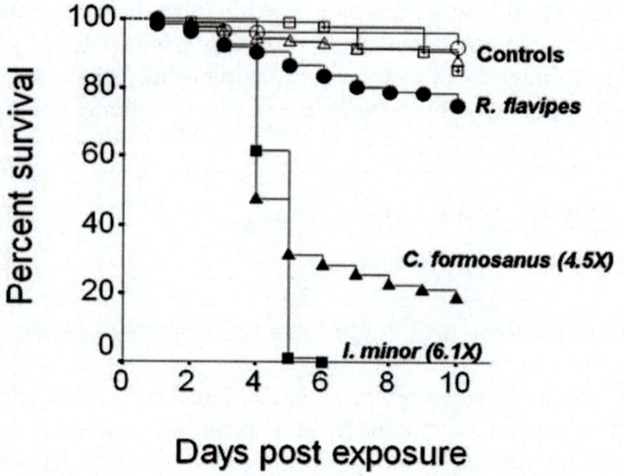

Fig. 7.4 Survival distributions of control (*open symbols*) and conidia-exposed (*filled symbols*) workers of three termite species (*circles* represent *Reticulitermes flavipes*, *triangles* indicate *Coptotermes formosanus* and *squares* denote *Incisitermes minor*). Hazard ratio of dead relative to conidia-exposed *R. flavipes* are in parentheses

to adapt to particular sets of pathogens (Rosengaus and Traniello 1993b). This may in part explain why in vivo experiments show that colony of origin is usually a significant and independent predictor of termite survival (Rosengaus et al. 1998b; Rosengaus and Traniello 2001; Calleri et al. 2006b, 2007; Wilson-Rich et al. 2007) and immune function (Rosengaus et al. 2007). Remarkably, higher nest loads in *R. flavipes* are not necessarily reflected in higher cuticular loads (Fig. 7.3b): CFUs isolated from cuticular washes of termites originating from the same seven colonies did not match the abundance of CFUs isolated from their nest material (see for example the pattern in colony LR3). This discrepancy between nest loads and cuticular loads suggests termites can suppress microbial activity on their cuticles, probably through a combination of allogrooming and integumentary antimicrobial chemistry.

R. flavipes also appears to have lower susceptibility to fungal infection relative to other termites of similar body size. When *R. flavipes, Coptotermes formosanus* and *Incisitermes minor* workers were exposed to the same suspension of *M. anisopliae* (10^7 conidia/mL), survival differed significantly between species, but not in controls (Fig. 7.4). Following exposure to the fungal conidia, *I. minor* and *C. formosanus* had 6.1 and 4.5 times, respectively, the hazard ratio of death compared with *R. flavipes.* It is likely that *R. flavipes* has particularly effective cellular and humoral immune responses in addition to behavioral and biochemical protection to reduce susceptibility. Similarly, interspecific differences in susceptibility to fungal infection have been recently reported by Chouvenc et al. (2009a).

The evolution of individual and social modes of disease resistance can be better understood through the application of molecular techniques. PCR amplification of mixed DNAs can be a powerful tool for estimating microbial diversity and the specificity and stability of associated communities (Hughes et al. 2001; Corby-Harris et al. 2007). Other useful culture-independent techniques include 454 amplicon pyrosequencing, a procedure that has helped in the identification of microbial communities in ants (Sen et al. 2009). Clone libraries provide "snap-shots" of the dynamic microbial assemblages present throughout the life of a termite colony. Although these techniques may have limitations (Polz et al. 1999), molecular assessment of microbial diversity and abundance should advance understanding of the impact that microorganisms have on termite biology. In one key study, Bulmer and Crozier (2004) linked the diversification of antifungal peptides of Australian *Nasutitermes* with nesting ecology and assumed differences in microbial loads.

7.3.2 Immunocompetence in Termites

In accordance with the broad definition of Owens and Wilson (1999), we define immunocompetence as the quantitative measure of the ability of an organism to resist infection in the absence of disease or parasitism, that ultimately minimizes the fitness costs of an infection. Thus, immunocompetence includes resistance to infection through biochemical and physiological mechanisms (including antibiotic prophylactic protection and innate constitutive and inducible immune responses), as well as individual and societal behavioral responses (Owens and Wilson 1999; Schmid-Hempel and Ebert 2003). We next discuss the nature and effectiveness of these mechanisms of disease resistance.

7.3.2.1 Antibiotic Prophylaxis

Termites produce antimicrobial compounds in rectal fluids and feces, defensive glandular secretions and body exudates (Rosengaus et al. 1998a, 2000b, 2004 and references therein), but such secretions are less well understood than those of social Hymenoptera. Despite thorough histological investigation, termites appear to lack a structure analogous to the metapleural gland of ants (Noirot 1969).

Terpenoids compounds from the cephalic glands of *Nasutitermes* (α- and β-pinene, limonene) as well as trinervitane diterpenes and the compounds of the sternal gland secretions of *Z. angusticollis* (mainly 4,6-dimethyldodecanal; see Chapter 11 by Bordereau and Pasteels, this volume) have potent antifungal and antibacterial properties (Rosengaus et al. 2000b, 2004; Zhao et al. 2004). Prior to recent and more accurate chemical identification of the *Z. angusticollis* sternal gland secretion, n-hexanoic acid was thought to be an important product. However, while n-hexanoic acid is present on the cuticle of pseudergates, it is apparently not secreted by the sternal gland (Pasteels and Bordereau 1998), although it has antifungal properties (Rosengaus et al. 2004). Interestingly, both terpenoids and n-hexanoic acid provide topical protection of the cuticle and fumigate the nest environment, inhibiting the germination of fungal conidia (Rosengaus et al. 2000b, 2004). Naphthalene, found in *Coptotermes* carton nests at concentrations of at 50.56–214.6 μg/kg, has also been considered a fumigant, although the source of this compound has not been determined (Chen et al. 1998a, 1998b). In *Anacanthotermes ahngerianus*, nest soil and chambers are impregnated with secretions that reduce the development of pathogens (Lutikova 1990).

Although waste material can accumulate quickly in nests of social insects and promote detection by predators and/or disease susceptibility (Jackson and Hart 2009), fecal material is widely used in nest construction by termites. *Z. angusticollis* and other termite species may have solved the predicament of feces accumulation by producing fecal waste with significant antifungal activity to construct its nest galleries (Rosengaus et al. 1998a). The fungistatic compounds of feces have yet to be completely identified, but recent research indicates that β-(1,3)-glucanases are present (Bulmer et al. 2009). These enzymes are known to breakdown components of fungal cell walls.

7.3.2.2 Immune Function in Termites

Although arthropods appear to lack an equivalent to vertebrate acquired immunity, they do produce immune molecules functionally comparable to mammalian immunoglobins (Watson et al. 2005; Dong et al. 2006; Pham et al. 2007; Rowley and Powell 2007 and references therein) that can protect individual for weeks rather than days (Haine et al. 2008b). More sophisticated immune responses, characterized by higher degrees of specificity and longer immunological memory, are expected in termites because they are long-lived as individuals and colonies and thus have a high probability of re-encountering the same pathogens. Colony longevity and disease encounter may be further extended when neotenic or supplementary reproductives are produced. Short-lived and ecologically ephemeral social insect species (for example those with annual life cycles) may have no need for complex immune responses because they may "escape" pathogens in time or by behavioral and/or biochemical mechanisms (Dunn 1990).

7.3.3 Immunity in Individuals

Although relatively little is known about immune function in individual termites (Rosengaus et al. 1999b, 2007; Lamberty et al. 2001; Thompson et al. 2003; Bulmer and Crozier 2004, 2006), we assume that termites exhibit the highly conserved recognition, signaling and effector molecules found in other insects, including *Drosophila*. Because different components of the immune response appear to be closely linked to life history (Zuk and Stoehr 2002), questions about immunity in relation to termite life history merit further research. Below we explore termite immune responses relative to those of other insects.

7.3.3.1 Cellular Immune Responses

Responses in the cellular compartment of the immune system include opsonization, phagocytosis, melanization, nodulation, encapsulation and coagulation

(Schmid-Hempel 2005). The significance of these processes in insect immunity has received relatively little attention given that their combined action may clear more than 99.5% of microbes within the first few hours of an infection (Haine et al. 2008a). Although preliminary evidence indicates that termites have cellular immunity, little empirical data exist to evaluate its importance.

7.3.3.2 Phagocytosis

Phagocytosis is the primary response to bacteria and other small foreign particles (Gillespie et al. 1997). In termites, no information exists on the speed of the process or how phagocytic activity differs as a function of gender, caste, fertility, reproductive form, instar and nutritional state. Pilot studies indicate that termite hemocytes are phagocytic: when *Z. angusticollis* nymphs were injected with bacteria-sized fluorescent microlatex beads and then bled 12 h post-injection, cell membrane extensions (filopodia) typically associated with phagocytosis were clearly identifiable and engulfed particles seen within (Fig. 7.5).

Phagocytes appear essential to priming the specific and long-lived immune response of insects (Dong et al. 2006; Pham et al. 2007). Impaired phagocytes were incapable of generating the hypervariable hemoglobin-like pattern recognition molecules needed to generate immunity with specificity and memory (reviewed by Rowley and Powell 2007). Termites may produce diverse isoforms through the same alternative splicing strategies used by fruit flies and mosquitoes, but this remains to be determined. Nevertheless, evidence shows that *Zootermopsis* can be immunologically primed (Rosengaus et al. 1999b, 2007; see Section 7.3.3.4).

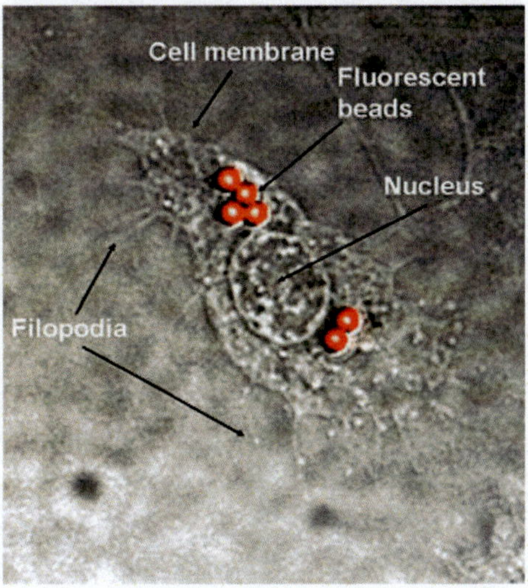

Fig. 7.5 Confocal microscope image of a *Zootermopsis angusticollis* hemocyte in the process of phagocytosis of fluorescent microlatex beads (2 μm in diameter) injected into the termite's hemocoel

7.3.3.3 Encapsulation and the Phenoloxidase Proteolytic Cascade

Encapsulation, which is mainly directed against multicellular parasites (Siva-Jothy et al. 2005), involves the participation of hemocytes and the activation of the phenoloxidase proteolytic cascade. Phenoloxidase (PO) is a multifunctional enzyme involved in the sclerotization of newly molted exoskeleton as well as immune defense and wound healing (Gillespie et al. 1997; Siva-Jothy et al. 2005). During encapsulation, hemocytes adhere to the surface of parasites. The layered cell mass becomes melanized after the activation of the PO cascade and results in the production of cytotoxic by-products like quinones. Ultimately, an encasing layer of melanin is formed, causing parasite death (Söderhall and Cerenius 1998).

PO activity and rates of encapsulation have been used as indirect measures of immune investment in a variety of species. Low levels of PO activity (and therefore encapsulation) are considered to represent reduced immunocompetence (Reeson et al. 1998), reflecting preferential investment of nutritional resources in other life history traits (Fellowes et al. 1998; Adamo et al. 2001; Siva-Jothy et al. 2001). Termites encapsulate an inert sterile nylon thread implanted experimentally in the abdomen to mimic a nematode infection (Fig. 7.6). In *Z. angusticollis*, the

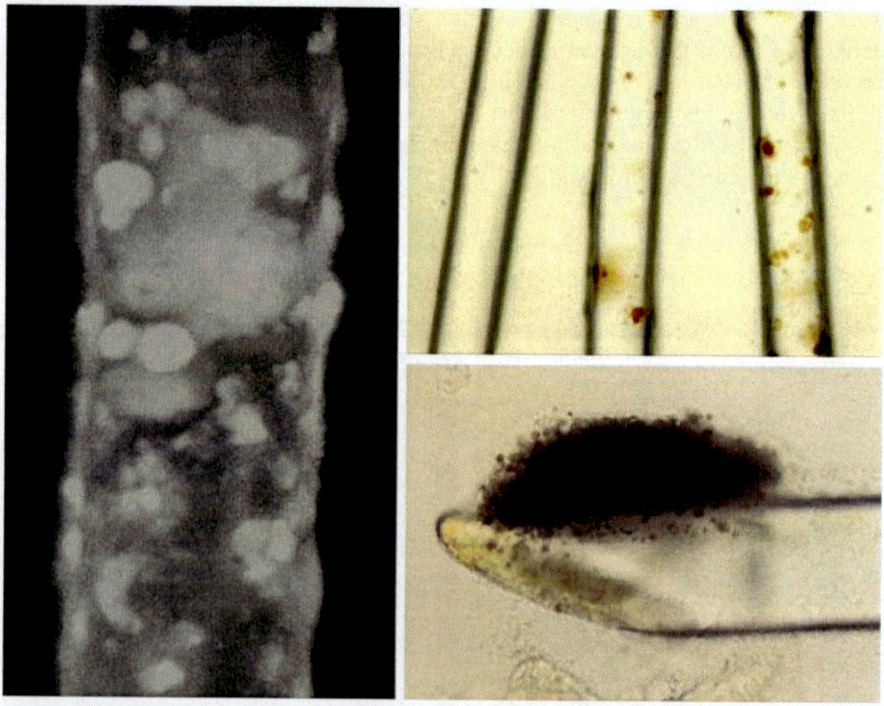

Fig. 7.6 Confocal scan (*left*) and photomicrograph (*upper right* and *lower right*) of a nylon implant showing melanin deposits. Note that the nylon implant to the *left* in the *upper right* image was a non-inserted control whereas the *middle* and *right* threads were implanted into the termite's hemocoel. Deposits result from melanization and provide a relative measure of encapsulation

implant will be surrounded with melanin deposits, which can be quantified using confocal microscopy (Fig. 7.6). Calleri et al. (2007) used this technique to test potential trade-offs between encapsulation and reproduction during colony foundation in *Zootermopsis*. PO activity differs significantly among termite species, colonies and castes (Reichheld, unpublished observations). This variation suggests a role for colony genetics and ecology in the evolution of termite immunocompetence.

7.3.3.4 Humoral Immune Defenses

Rosengaus et al. (1999b) studied the in vivo immune priming effect of exposure to either dead bacteria or sublethal dosages of a fungal pathogen in *Z. angusticollis*. Termites immunized by injection of killed *Pseudomonas aeruginosa* (Fig. 7.7) or exposure to non-lethal dosages of *M. anisopliae* conidia (Fig. 7.8) had significantly lower susceptibility to future challenges with active and lethal dosages of the same pathogens than controls. Such immunizations afforded a higher resistance to disease for at least 7 days (Rosengaus et al. 1999b). Electrophoretic evidence suggests that the in vivo protection results from the upregulation of constitutive proteins (62–85 kDa) and the production of novel proteins (Rosengaus et al. 2007). The novel proteins (28–48 kDa) have antifungal properties, as demonstrated after their excision and elution from native isoelectric gels and incubation with conidia of *M. anisopliae*. Germination rates of treated conidia were reduced by up to 28% relative to control conidia, indicating that the proteins were fungistatic, but not necessarily fungicidal.

Another basal termite, *Mastotermes darwiniensis*, responds to infection by *M. anisopliae* by upregulating the expression of a gene for transferrin. This 73 kDa protein sequesters circulating iron, potentially reducing its availability to invading microbes which require it for growth. Interestingly, transferrin has also been

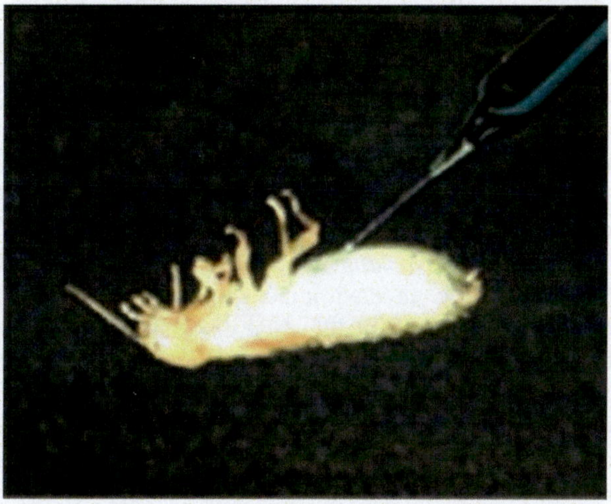

Fig. 7.7 Termites can be injected with killed *Pseudomonas aeruginosa* or active bacteria to immunize or challenge them

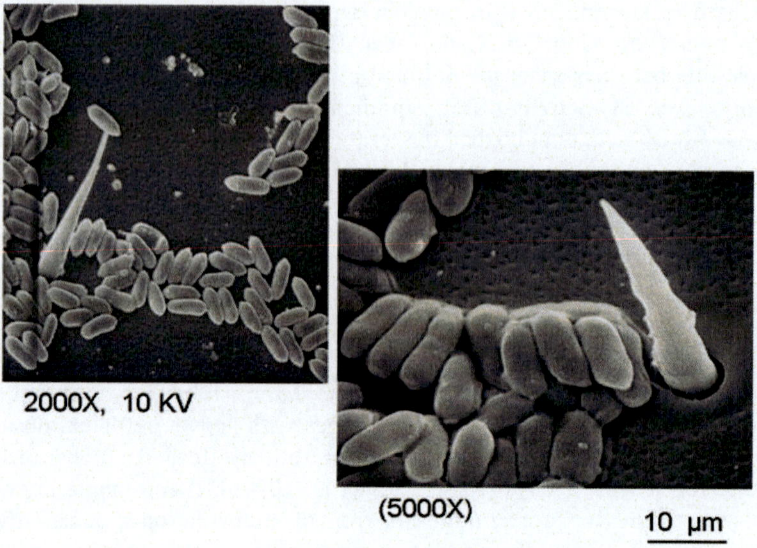

Fig. 7.8 Scanning electron micrograph of *Metarhizium anisopliae* conidia on the cuticle of *Zootermopsis angusticollis*

identified in the higher termite *Nasutitermes graveolus* (Bulmer unpublished observations), suggesting that the sequestration of iron is a ubiquitous immune defense strategy in termites. Further, the immune proteins termicin and spinigerin have been identified in *Pseudacanthotermes spiniger* (Lamberty et al. 2001) and variants of termicin have been found in *Nasutitermes* (Bulmer and Crozier 2004). Termicin, a small effector molecule (~4 kDa), is most active against filamentous fungi (Lamberty et al. 2001; Da Silva et al. 2003) and could be a means of prophylactic protection. In *N. corniger*, Gram-negative binding proteins (GNBPs) are bifunctional pattern recognition receptors with unusual antifungal β-(1,3)-glucanase activity (Bulmer et al. 2009). In combination with small peptides (less than 10 kDa), they synergistically arrest the development of fungi conidia, germ tubes and developing hyphae of *M. anisopliae* (Fig. 7.9). Bulmer et al. (2009) suggested that the enzymatic activity of termite GNBPs first disrupt the epispore and fungal cell walls, allowing smaller effector molecules to access and destroy fungal cells.

Remarkably, *Nasutitermes* GNBPs are incorporated into nest materials, potentially providing an environmental immune sensor/effector system that directly damages fungal cell walls while also releasing partially digested β-(1,3)-glucans that could in turn activate innate immune pathways, thus serving as immune-elicitors (Bulmer et al. 2009). Termites thus appear to have co-opted components of the internal immune system for external defense. The efficacy of these secreted immune proteins is likely enhanced by allogrooming. This contrasts with ants,

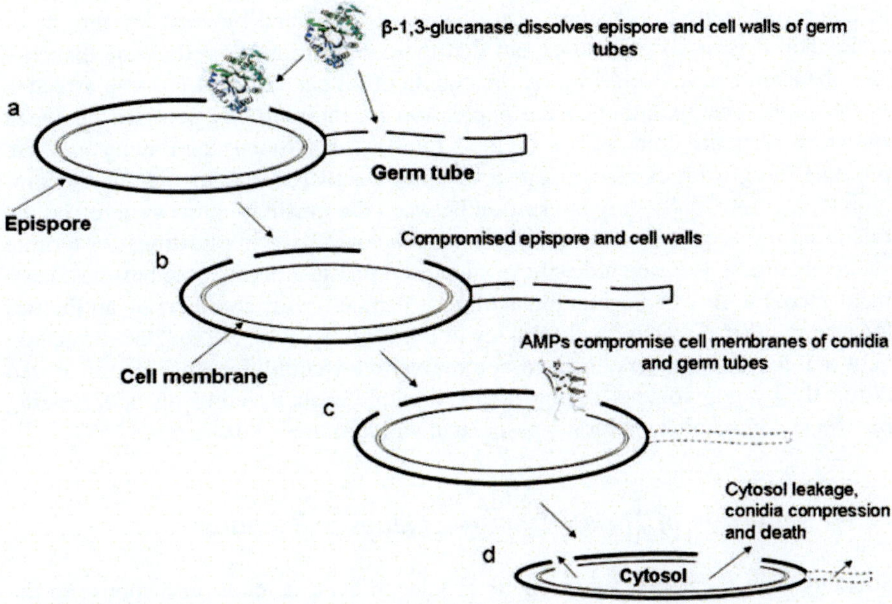

Fig. 7.9 Schematic representation of the hypothesized synergistic antifungal activity between β-1,3-glucanases and a small peptide, putatively termicin

which use antibiotic secretions from the specialized metapleural gland (Maschwitz 1974; Beattie et al. 1986; Fernández-Marín et al. 2006; Schlüns and Crozier 2009). However, mechanisms of antibiotic defense in ants and termites do appear to converge: both employ a secreted cocktail that can be disseminated among nestmates by grooming. These antibiotics may limit a pathogen's ability to quickly evolve counter responses. Bees also use secreted antibiotics for defense. the efficacy of which increase with the degree of sociality (Stow et al. 2007). Diversification or directional selection in these compounds could contribute to an increase in their antimicrobial potency.

Lysozymes are innate eukaryote immune proteins used by termites in a number of applications. In *Reticulitermes speratus* they serve as egg recognition pheromones and presumably protect them from bacterial pathogens (Matsuura et al. 2007, 2009b). The enzymes are secreted by salivary glands and spread by grooming. This suggests that salivary glands in termites can function in an analagous manner to the metapleural gland in ants. Importantly, the social dissemination of secretions from these glands increases after exposure to disease agents (Rosengaus et al. 1998b, 2000b; Fernández-Marín et al. 2006), accentuating the critical link between sociality and defense against pathogens.

The innate immune system of termites may also influence symbiotic microbial communities. Colony genetic diversity appears to influence cuticular microbial load in dampwood termites (Calleri et al. 2006a). If mediated by secreted antimicrobials, then it is reasonable to expect that the composition as well as the diversity and abundance of microbes lining the cuticle or alimentary tract are also affected. In fruit flies, manipulation of the expression of antimicrobial peptides has been shown to alter the composition of their intestinal microbial community and can promote the growth of rare endogenous microbes that are pathogenic when common (Ryu et al. 2008). Any interaction between the innate immune system and the microbial intestinal and cuticular communities could have implications for termite social evolution. Recent studies have suggested that in at least some termites, nestmate recognition can be manipulated by treatment with commercial antibiotics (Matsuura 2001; Kirchner and Minkley 2003; Minkley et al. 2006). Discrimination between colony members and non-colony members could therefore be influenced indirectly by components of the innate immune system acting on odor-bearing microbial communities naturally associated with termites (Adams et al. 2007).

7.3.4 Evolution of Innate Immune Genes in Termites

Sociality may accelerate selection for innate immune function in response to the evasive tactics deployed by pathogens. A simple test of this hypothesis would be to compare molecular evolution in immune genes among solitary and presocial species and their eusocial relatives, an analysis still under development (Stow et al. 2007; Viljakainen et al. 2009). In termites, comparative study of this type is limited by the lack of extant social intermediates (Thorne and Traniello 2003; Fefferman et al. 2007). Nevertheless, broad phylogenetic comparisons indicate that pathogen receptors and antimicrobial peptides evolve at higher rates in termites than fruit flies (Sackton et al. 2007). Pathogen recognition proteins (such as Gram-negative bacterial-binding proteins or GNBPs) and antimicrobial peptides (termicins) show clear evidence of adaptive evolution (positive selection) in a group of Australian *Nasutitermes* (Bulmer and Crozier 2004, 2006). Changes in equivalent proteins in *Drosophila* appear to have been constrained (i.e. have experienced purifying selection) presumably because their molecular targets are also conserved (Jiggins and Hurst 2003; Jiggins and Kim 2005, 2006). These targets include recognition epitopes such as lipopolysaccharides and β-glucans, pathogen cell wall components in which functionality constrains molecular variation (Janeway and Medzhitov 2002). Positive selection in termite GNBPs and termicins is therefore exceptional and supports the expectation that greater selection pressure acts on termites than solitary insect species.

Adaptive change in termite innate immune genes appears to have been driven by shifts in nesting and feeding ecology that exposed termites to new pathogenic microbes during the course of their adaptive radiation. For example, in Australian *Nasutitermes*, the highest rate of positively selected change in three immune genes occurred during the transition from feeding on dry grass stored in epigeic nest mounds to feeding on heavily decayed wood accessed from subterranean nests (Bulmer and Crozier 2006, Fig. 7.10).

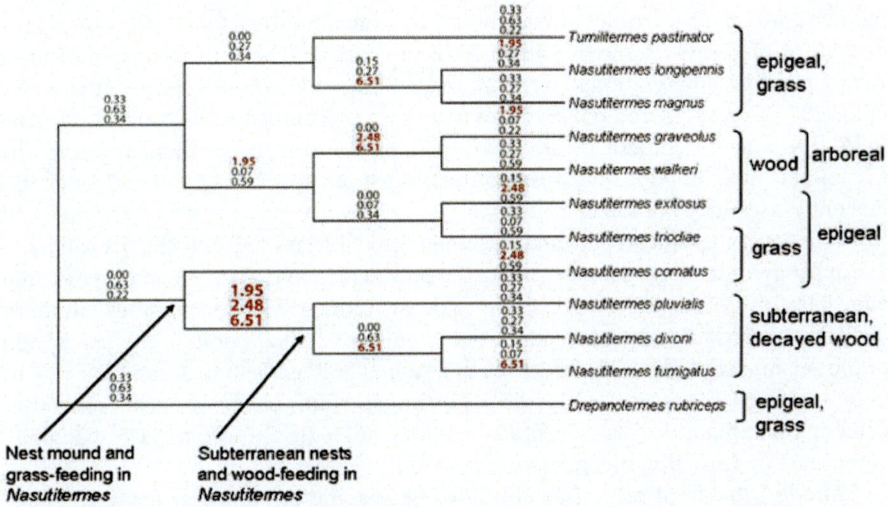

Fig. 7.10 Results from the PAML free-ratio codon selection model (Yang 1997). Values above different termite lineages represent estimates of the ratio of nonsynonymous to synonymous nucleotide substitution rates for three immune genes. *Top* and *middle* values correspond to estimates for GNBP1 and GNBP2, respectively and the *bottom* values for *Relish*. Ratios greater > 1 (*in bold*) suggest selective pressure from pathogens drove adaptive change in the proteins. The GNBPs are involved in pathogen recognition and GNBP2 is also a direct antifungal protein. *Relish* is a transcription activator of antimicrobial peptides. Adapted from Fig. 4 of Bulmer and Crozier (2006)

7.4 Social Behavior and Infection Control

Although social interactions such as allogrooming, proctodeal and stomodeal trophallaxis, coprophagy and cannibalism of diseased and/or injured nestmates increase the dissemination of pathogens among nestmates (Kramm et al. 1982; Rosengaus and Traniello 1997), these behaviors may actually also decrease infection risk (Rosengaus et al. 1998b; Traniello et al. 2002; Cremer et al. 2007; Yanagawa and Shimizu 2007; De Souza et al. 2008). Termites avoid contaminated areas or infected nestmates and may relocate a nest after encounter with pathogenic or parasitic microbes (Wang et al. 2002). Typically, infected cadavers are isolated and covered with soil, and moribund or dead individuals are quarantined by constructing walls around them with fecal pellets and/or soil (Logan et al. 1990). These behaviors, which diminish the probability of sporulation by fungi, may be triggered by rapid changes in the cuticular chemistry of infected and/or dead nestmates (Choe et al. 2009).

Termites also have the ability to recognize pathogenic conidia through antennal contact (Yanagawa et al. 2009), then initiate and maintain intense self-grooming and allogrooming behavior of contaminated body parts (Cruse 1998; Rosengaus et al. 1998b, 2000a; Yanagawa and Shimizu 2007; Wilson-Rich et al. 2007), which restricts the spread of infection (Yanagawa et al. 2009). Allogrooming may also include glandular secretion onto the mouthparts, and thus deactivate infective agents as well as physically removing them from the cuticle. The salivary origin of antifungal peptides spinigerin and termicin in *Pseudacanthotermes spiniger* (Lamberty et al. 2001), termicin and tGNBP-2 in *N. corniger* (Bulmer et al. 2009), lysozyme in *R. speratus* (Matsuura et al. 2007) and β-(1,3)-glucanase in *Z. angusticollis* (Schultheis et al. unpublished observations) illustrate how behavioral and biochemical antibioses may combine.

Termites communicate information about infection risk and engage in a pathogen alarm behavior to alert nestmates about the presence (and possibly location) of lethal concentrations of pathogens in the nest (Rosengaus et al. 1999a). Vigorous vibratory displays are produced by termites in the vicinity of fungal conidia; the perception of the seismic signal causes nestmates to abscond and tends to increase their rate of allogrooming (Rosengaus et al. 1999a). Pathogen alarm behavior occurs only after detecting lethal doses of *M. anisopliae* conidia and is distinct from "head-banging" stimulated by breaching the nest (Stuart 1969).

Termite behavioral responses appear to be specific to particular disease agents. For example, *Z. angusticollis* respond to nematodes by increasing the frequency and duration of self-grooming, and repeatedly opening and closing the mandibles, seemingly to attack infective juvenile nematodes invading the oral cavity. Most striking is the response of *Z. angusticollis* to the nematode *Steinernema carpocapsae*: termites raise the abdomen and orient its tip orthogonally (Wilson-Rich et al. 2007). This latter behavior has never been observed in the context of any other pathogen exposure and may represent an attempt by the termite to reduce the probability of infection via the anus.

7.4.1 Social Mediation of Immunocompetence

Sociality increases the complexity of disease dynamics. Although disease risk has been considered to increase with group size due to higher rates of transmission (Wilson and Cotter 2008), termites living in groups may in fact be less susceptible to infection than isolated individuals (Fig. 7.11; Rosengaus et al. 1998b, 2000a; Rosengaus and Traniello 2001; Traniello et al. 2002; Wilson-Rich et al. 2007; Calleri et al. 2010). In *Z. angusticollis*, Rosengaus et al. (1998b) found that isolated individuals exposed to *M. anisopliae* succumb to disease significantly faster (2–6 days) compared with similarly exposed nestmates maintained in groups of 10 or 25. Grouped, exposed termites had a 90% reduction in their hazard of death relative to termites maintained in isolation (Fig. 7.11; Rosengaus et al. 1998b).

A possible explanation for the reduced susceptibility of grouped termites is density dependent prophylaxis (DDP), a phenomenon known to occur in some facultative social locusts, lepidopteran larvae and beetle larvae during the gregarious phase of their life cycle. Density-dependent reduction in susceptibility is partly due to increased investment in innate immunity, mediated via the phenoloxidase enzymatic cascade (Wilson and Cotter 2008), which may prevent epizootics. Despite living in groups that vary in density throughout colony ontogeny, *Z. angusticollis*

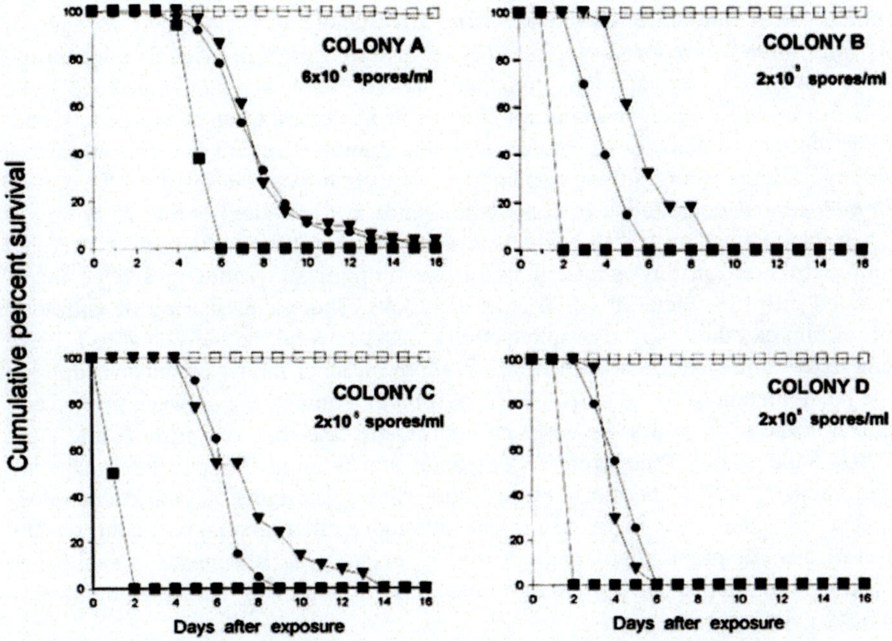

Fig. 7.11 Survival distributions of 25 (*solid triangles*), 10 (*solid circles*) or isolated (*solid squares*) *Z. angusticollis* nymphs after exposure to *M. anisopliae* conidia. Survivorship of control nymphs (*open squares*) exposed to a Tween 80 solution lacking fungal conidia is included for reference. Colony A exposed to 6×10^6 conidia/mL; colonies B, C and D exposed to 2×10^8 conidia/mL. With kind permission from Springer Science + Business Media (Rosengaus et al. 1998b, Fig. 1)

appears to lack DDP, for reasons that are unclear (Pie et al. 2005). Given their social nature, termites, as well as other social insects, should require a continuous heightened constitutive immunity investment, and since this form of immunity is costly (Sheldon and Verhulst 1996; Schmid-Hempel 2003), termites may have alternative but highly effective forms of prophylactic protection that can be deployed rapidly when needed (Pie et al. 2005). Less costly prophylactic adaptations might involve increases the frequency of allogrooming and pathogen alarm communication. The production and secretion of antibiotics could also be increased as a function of colony density.

Interactions among individuals in termite colonies could favor emergent properties that enhance disease resistance and promote survival (Traniello et al. 2002; Miramontes and DeSouza 2008; Wilson-Rich et al. 2009). "Collective immunocompetence" could arise from the additive effects of individual immunological responses and their social mediation and amplification. Indeed, grouped termites can cope with disease better than isolated individuals (Rosengaus et al. 1998b; Calleri et al. 2010), a finding that led to the discovery of the social enhancement of resistance (Traniello et al. 2002). Naïve nymphs of *Z. angusticollis* allowed to socially interact with immunized nestmates prior to a challenge with lethal dosages of *M. anisopliae* had reduced susceptibility relative to control individuals not exposed to fungal conidia. Social interactions may thus facilitate immunocompetence at the group-level because naïve termites benefit from the distribution of non-lethal inoculate dosages or constitutive and/or inducible immune proteins and related factors through allogrooming and/or trophallaxis. A second example involves the spread of endogenous constitutive immune compounds in the higher termite *N. corniger*. These compounds protect individuals against fungal infection; they can be excreted and distributed on the cuticle of nestmates through allogrooming and be incorporated into nest material (Bulmer et al. 2009). The social sharing of antibiotic protection can thus expand prophylactic defenses beyond the individual to achieve colony-level prevention of infection. Research on other model organisms supports the phenomenon of social transmission of immunocompetence between primed and naïve workers, as well as between primed parents and their offspring (Little et al. 2003; Sadd et al. 2005; Moret 2006; Sadd and Schmid-Hempel 2007; Ugelvig and Cremer 2007). The mechanism(s) underlining transgenerational immunological priming have yet to be described, although oral trophallaxis (Hamilton and Rosengaus, in preparation) and proctodeal exchanges (Rosengaus, unpublished observations) appear to play a significant role in achieving social immunization among colony members.

7.5 Termite Life History, Genetic Diversity and Disease Resistance

Termites are of interest to the study of life-history traits and disease-related adaptations. Most basal termite species are monogamous and undergo alternating generations of outbreeding and inbreeding during colony foundation and growth (Bartz 1979; Thorne 1997; Thorne et al. 1999; Thorne and Traniello 2003). Although these strategies could potentially result in a progressive reduction of genetic diversity within colonies (and thus increase disease susceptibility), recent data indicate that at least in *Reticulitermes speratus*, secondary queens are the result of parthenogenetic reproduction by primary queens (Matsuura et al. 2009a), which maintains levels of genetic diversity that may reduce disease susceptibility. Rosengaus and Traniello (1993b) addressed the fitness cost of inbreeding in basal termites by comparing the survivorship and growth of colonies headed by nestmate (assumed to be genetically inbred) and non-nestmate primary reproductives (assumed to be outbred) in laboratory colonies of *Z. angusticollis*. Fitness measures recorded throughout the first 4 years of colony establishment and growth included mortality of primaries, colony size and biomass. Results indicated no differences in colony size and biomass between surviving inbred and outbred pairs, and no negative effects of inbreeding on fitness were detected during colony foundation and the growth of incipient colonies. However, outbred pairs had significantly higher mortality during the first 40 days following pairing, apparently due to infection. It was inferred that exposure to disease agents during social contact with mates enhanced disease risks during the outbreeding phase of the colony life cycle, selecting for local adaptation to pathogens involving colony-specific immunological profiles. Rosengaus and Traniello (1993b) suggested that risks of infection should be considered as important agents of selection for inbreeding, genetic relatedness of colony members and the evolution of termite sociality. Studies by Fei and Henderson (2003) and Calleri et al. (2005) support this model. Termites provide significant opportunities to further evaluate the role of life history, reproductive plasticity, colony genetics and the infection environment of colonies to further our understanding of disease and sociality.

7.6 Disease and Colony Foundation

Infections carried by dispersing alates or acquired from a mate or the environment can have a significant impact on reproduction and colony success. The risk of infection between mates (either genetically related or unrelated) during social interactions associated with mate attraction, construction of the copularium and the initial stages of colony foundation may be additional challenges that help explain the high failure rates of newly established colonies (Rosengaus et al. 2000b and references therein). Reproductives may also encounter novel pathogens as they move through soil or under or within decayed wood.

The consequences of infection during colony establishment may be compounded by the low quality of the termite diet (Waller and La Fage 1987). Because nitrogen is critical for the synthesis of vitellogenin and oogenesis, as well as the production of immune proteins, the nitrogen-poor diet may compromise the ability of females to simultaneously allocate resources to reproductive and immune functions. Calleri et al. (2006b) demonstrated that the number of eggs produced by *Z. angusticollis* reproductives exposed to *M. anisopliae* conidia was significantly reduced, but that the impact of fungal exposure was transient, with all surviving colonies having similar fitness 300 days post-pairing. The short-term effect of immune induction (assumed to be at the humoral level) on reproductive output suggests that reproductives first cope with the immediate threat of disease, diverting energy and resources away from reproduction, and then subsequently reallocate resources back to reproduction. Calleri et al. (2007) detected significant trade-offs at the time when cellular immunity was activated. A nylon implant (as in Fig. 7.6) was inserted into male and female reproductives to induce encapsulation at three phases of colony foundation: shortly after pairing; during copulation and subsequent oocyte maturation, and during oviposition. Higher encapsulation rates were evident soon after pairing, apparently as a prophylactic response anticipating the need for heightened immunity in the event of being infected by a mate. During copulation and oocyte maturation, in contrast, mated females had significantly reduced encapsulation, supporting the hypothesis that mating compromises immunity (Rolff and Siva-Jothy 2002). Finally, the onset of oviposition was significantly delayed in mated females that had received an implant. These findings indicate that reproduction and cellular immunity are costly and that trade-offs result when the two processes are simultaneously engaged during colony foundation.

When larvae in the first brood are sufficiently mature to provide labor, there is a change in infection dynamics. Primary reproductives are soon emancipated from brood care and other tasks, as immatures tend, groom and pile eggs, maintain the nest and initiate foraging. The responsibility for maintaining an antiseptic nest therefore shifts from reproductives to the first brood. The reduced behavioral repertoire of primaries that now specialize on reproduction should decrease their risk of infection and lower task-related energetic demands. At this point, reproduction should be less compromised by the opposing demands of immunity and the need for colony labor, because the burden of allocating energy to task performance is taken over by helpers. Age-based pattern of disease susceptibility in *Z. angusticollis* (Rosengaus and Traniello 2001) may be explained by the insufficiency of their immune system (Rheins and Karp 1985). Young termites are particularly susceptible to disease: third and fourth instar larvae, respectively, have 3.6 and 2.0 times the hazard ratio of death of nymphs, the oldest and largest individuals (Rosengaus and Traniello 2001). Even fifth instar termites were still more susceptible to fungal infection than sixth instars and nymphs, the latter stages not differing in susceptibility. Even more remarkable, group composition, in addition to age, plays an important role in determining susceptibility to mycosis. Mixed-instar groups exhibited the least susceptibility relative to same-instar groups, even after controlling for density. In the same challenges, isolated termites had the highest susceptibility. Colony demography therefore influences the impact of fungal pathogens. Rosengaus and Traniello (2001) attributed the lower susceptibility in mixed-instar subcolonies to the benefits derived from the asynchronous development of termites hatching during successive bouts of oviposition. Younger individuals may benefit from engaging larger and older nestmates in allogrooming. Younger nestmates in mixed-age groups may also benefit from more abundant, potent and/or diverse biochemical protection provided by their older siblings (Rosengaus et al. 1998b, 2000a, 2004). Colony demography can reduce disease risk; this is a previously unrecognized adaptive effect of multiple overlapping generations. Rates of cannibalism increase when termites are exposed to fungal pathogens and cannibalism is preferentially directed toward younger instars (Rosengaus and Traniello 2001). This could be inferred as an attempt by older nestmates to cull the most susceptible individuals in the nest to reduce the reservoir of infection (Fefferman et al. 2007). Ingestion of conidia during cannibalism does not necessarily pose a major infection risk because conidia appear to be deactivated in the gut (Batra and Batra 1966, 1979; Rich 1969; Thomas 1987; Wood and Thomas 1989; Yanagawa and Shimizu 2007; Chouvenc et al. 2009b).

7.7 Conclusions

To control infection within microbially rich and densely populated nests, termites have evolved prophylactic and molecular inducible immune responses. Social behavior and group living also contribute significantly to disease resistance.

Defenses of ancestral forms appear to have been conserved in part, serving as preadaptations, whereas other mechanisms of resistance are secondarily acquired protective functions. In contrast to ants, in which the origin of the metapleural gland and its antiseptic secretions was key to nesting and hence to their adaptive radiation, infection control in termites does not appear to be linked to a single, preeminent evolutionary innovation. Microbial symbiosis, long noted for its importance in termite nutrition, may have yielded disease prevention and treatment traits that arose from coevolution. This hypothesis awaits critical evaluation.

Because of the great variation in nesting ecology and life history, termite immune function may have pathogen-specific elements and therefore be characterized by immunological memory. In support of this concept, life-type evolution appears to parallel the diversification of immune defenses. How termites cope with multiple immune challenges is poorly understood. A socioecoimmunological approach, spanning molecules, behavior, life history and ecology, can offer insight into the nature of immune function in the Isoptera. The significance of genetic heterogeneity, the cycles of inbreeding and outbreeding, and infection in termite colonies requires further analysis. Comparisons of the mechanisms of disease resistance in termites and other eusocial, presocial and solitary insects can advance our understanding of the significance of nest type, nutritional ecology, microbial load, life history and sociality to termite health.

Acknowledgments We thank Dr. Kenneth Grace for noting useful references and Brian Lejeune for help with formatting. This work was supported by an NSF CAREER Development grant (DEB-0447316) to RB Rosengaus and NSF IBN-9632134 and IBN-0116857 to JFA Traniello and JFA Traniello and RB Rosengaus, respectively.

References

Abe T (1987) Evolution of life types in termites. In: Kawano S, Connell JH, Hidaka T (eds) Evolution and coadaptation in biotic communities. University of Tokyo Press, Tokyo, pp 125–148

Adamo S, Jensen M, Younger M (2001) Changes in lifetime immunocompetence in male and female *Gryllus taxensis* (formerly *G. integer*): trade-offs between immunity and reproduction. Anim Behav 62:417–425

Adams ES, Atkinson L, Bulmer MS (2007) Relatedness, recognition errors, and colony fusion in the termite *Nasutitermes corniger*. Behav Ecol Sociobiol 61:1195–1201

Bartz SH (1979) Evolution of eusociality in termites. Proc Natl Acad Sci U S A 76:5764–5768

Batra LR, Batra SWT (1966) Fungus-growing termites of tropical India and associated fungi. J Kans Entomol Soc 39:725–738

Batra LR, Batra SWT (1979) Termite-fungus mutualism. In: Batra LR (ed) Insect-fungus symbiosis, nutrition, mutualisms, and commensalism. Wiley, New York, NY, pp 117–163

Beattie A, Turnbull C, Hough T, Knox B (1986) Antibiotic production: a possible function for the metapleural gland of ants (Hymenoptera: Formicidae). Ann Entomol Soc Am 79:448–450

Bignell DE, Eggleton P (2000) Termites in ecosystems. In: Abe T, Bignell DE, Higashi M (eds) Termites: evolution, sociality, symbiosis, ecology. Kluwer Academic Publishers, Dordrecht, pp 363–387

Blackwell M, Rossi W (1986) Biogeography of fungal ectoparasites of termites. Mycotaxon 25:581–601

Brossut R (1983) Allomonal secretions in cockroaches. J Chem Ecol 9:143–158
Brune A (2006) Symbiotic associations between termites and prokaryotes. In: Dworkin M, Falkow S, Rosenberg E et al (eds) The Prokaryotes: symbiotic associations, biotechnology, applied microbiology, vol 1. Springer, New York, NY, pp 439–474
Bulmer MS, Bachelet I, Raman R et al (2009) Targeting antimicrobial effector function in insect immunity as a pest control strategy. Proc Natl Acad Sci U S A 106:12652–12657
Bulmer MS, Crozier RH (2004) Duplication and diversifying selection among termite antifungal peptides. Mol Biol Evol 21:2256–2264
Bulmer MS, Crozier RH (2006) Variation in positive selection in termite GNBPs and Relish. Mol Biol Evol 23:317–326
Calleri DV, Reid E, Rosengaus RB et al (2006a) Inbreeding and disease resistance in a social insect: effects of genetic variation on immunocompetence in the termite *Zootermopsis angusticollis*. Proc R Soc Lond B 273:2633–2640
Calleri DV, Rosengaus RB, Traniello JFA (2005) Disease and colony foundation in the dampwood termite *Zootermopsis angusticollis*: the survival advantage of nestmate pairs. Naturwissenschaften 92:300–304
Calleri DV, Rosengaus RB, Traniello JFA (2006b) Disease and colony establishment in the dampwood termite *Zootermopsis angusticollis*: survival and fitness consequences of infection in primary reproductives. Insectes Soc 53:204–211
Calleri DV, Rosengaus RB, Traniello JFA (2007) Immunity and reproduction during colony foundation in the dampwood termite *Zootermopsis angusticollis*. Physiol Entomol 32: 136–142
Calleri DV, Rosengaus RB, Traniello JFA (2010) Disease resistance in the drywood termite *Incisitermes schwarzi* (Isoptera: Kalotermitidae): does nesting ecology affect immunocompetence? J Insect Sci 10:Article 44
Chen J, Henderson G, Grimm CC et al (1998a) Naphthalene in Formosan subterranean termite carton nests. J Agric Food Chem 46:2337–2339
Chen J, Henderson G, Grimms CC et al (1998b) Termites fumigate their nests with naphthalene. Nature 392:558–559
Choe D-H, Millar JG, Rust MK (2009) Chemical signals associated with life inhibit necrophoresis in Argentine ants. Proc Natl Acad Sci U S A 106:8251–8255
Chouvenc T, Su N-Y, Robert A (2009a) Susceptibility of seven termite species (Isoptera) to the pathogenic fungus *Metarhizium anisopliae*. Sociobiology 54:723–748
Chouvenc T, Su N-Y, Robert A (2009b) Inhibition of *Metarhizium anisopliae* in the alimentary tract of the eastern subterranean termite *Reticulitermes flavipes*. J Invertebr Pathol 101: 130–136
Corby-Harris V, Pontaroli AC, Shimkets LJ et al (2007) Geographical distribution and diversity of bacteria associated with natural populations of *Drosophila melanogaster*. Appl Environ Microbiol 73:3470–3479
Cremer S, Armitage S, Schmid-Hempel P (2007) Social immunity. Curr Biol 17:R693–R702
Cremer S, Sixt M (2009) Analogies in the evolution of individual and social immunity. Philos Trans R Soc Lond B 364:129–142
Crosland MWJ, Traniello JFA (1997) Behavioral plasticity in division of labor in the lower termite *Reticulitermes fukienensis*. Naturwissenschaften 84:208–211
Cruse A (1998) Termite defences against microbial pathogens. PhD Thesis, Macquarie University, Australia
Da Silva P, Jouvensal L, Lamberty M et al (2003) Solution structure of termicin, an antimicrobial peptide from the termite *Pseudacanthotermes spiniger*. Protein Sci 12:438–446
De Souza DJ, Van Vlaenderen J, Moret Y, Lenoir A (2008) Immune response affects ant trophallactic behaviour. J Insect Physiol 54:828–832
Dong Y, Taylor HE, Dimopoulos G (2006) AgDscam, a hypervariable immunoglobulin domain-containing receptor of the *Anopheles gambiae* innate immune system. PLoS Biol 4:e229
Dunn PE (1990) Humoral immunity in insects. BioScience 40:738–744

Eggleton P (2000) Global patterns of termite diversity. In: Abe T, Bignell DE, Higashi M (eds) Termites: evolution, sociality, symbioses, ecology. Kluwer Academic Publishers, Dordrecht, pp 25–54

Faulhaber LM, Karp RD (1992) A diphasic immune response against bacteria in the American cockroach. Immunology 75:378–381

Fefferman NH, Traniello JFA (2009) Social insects as models in epidemiology: establishing the foundation for an interdisciplinary approach to disease and sociality. In: Gadau J, Fewell J (eds) Organization of insect societies: from genome to sociocomplexity. Harvard University Press, Cambridge, MA, pp 545–571

Fefferman NH, Traniello JFA, Rosengaus RB, Calleri DV (2007) Disease prevention and resistance in social insects: modeling the survival consequences of immunity, hygienic behavior and colony organization. Behav Ecol Sociobiol 61:565–577

Fei HX, Henderson G (2003) Comparative study of incipient colony development in the Formosan subterranean termite, *Coptotermes formosanus* Shiraki (Isoptera, Rhinotermitidae). Insectes Soc 50:226–233

Fellowes MDE, Kraaijeveld AR, Godfray HCJ (1998) Trade-offs constraining the evolution of *Drosophila melanogaster* defence against attack by the parasitoid *Leptopilina boulardi*. Proc R Soc Lond B 265:1553–1558

Fernández-Marín H, Zimmerman JK, Rehner SA, Wcislo WT (2006) Active use of the metapleural glands by ants in controlling fungal infection. Proc R Soc Lond B 273:1689–1695

Gillespie JP, Kanost MR, Trenczek T (1997) Biological mediators of insect immunity. Annu Rev Entomol 42:611–643

Haine ER, Moret Y, Siva-Jothy M, Rolff J (2008a) Antimicrobial defense and persistent infection in insects. Science 322:1257–1259

Haine ER, Pollitt LC, Moret Y et al (2008b) Temporal patterns in immune responses to a range of microbial insults (*Tenebrio molitor*). J Insect Physiol 54:1090–1097

Hendee EC (1933) The association of the termites, *Kalotermes minor, Reticulitermes hesperus,* and *Zootermopsis angusticollis* with fungi. Univ Calif Publ Zool 39:111–134

Hendee EC (1934) The association of termites and fungi. In: Kofoid CA (ed) Termites and termite control. Berkeley University, Berkeley, CA, pp 105–116

Hughes DP, Pierce NE, Boomsma JJ (2008) Social insect symbionts: evolution in homeostatic fortresses. Trends Ecol Evol 22:672–677

Hughes JB, Hellmann JJ, Ricketts TH, Bohannan BJM (2001) Counting the uncountable: statistical approaches to estimating microbial diversity. Appl Environ Microbiol 67:4399–4406

Inward D, Beccaloni G, Eggleton P (2007) Death of an order: a comprehensive molecular phylogenetic study confirms that termites are eusocial cockroaches. Biol Lett 3:331–335

Jackson DE, Hart AG (2009) Does sanitation facilitate sociality? Anim Behav 77:e1–e5

Janeway CA, Medzhitov R (2002) Innate immune recognition. Annu Rev Immunol 20:197–216

Jiggins FM, Hurst GD (2003) The evolution of parasite recognition genes in the innate immune system: purifying selection on *Drosophila melanogaster* peptidoglycan recognition proteins. J Mol Evol 57:598–605

Jiggins FM, Kim KW (2005) The evolution of antifungal peptides in *Drosophila*. Genetics 171:1847–1859

Jiggins FM, Kim KW (2006) Contrasting evolutionary patterns in *Drosophila* immune receptors. J Mol Evol 63:769–780

John TJ, Samuel R (2000) Herd immunity and herd effect: new insights and definitions. Eur J Epidemiol 16:601–606

Karp RD, Duwel LE, Faulhaber LM, Harju MA (1994) Evolution of adaptive immunity: inducible responses in the American cockroach. Ann N Y Acad Sci 712(1):82–91

Karp RD, Duwel-Eby LE (1991) Adaptive immune responses in insects. In: Warr GW, Cohen N (eds) Phylogenesis of immune functions. CRC, Boca Raton, FL, pp 1–18

Kirchner WH, Minkley N (2003) Nestmate discrimination in the harvester termite *Hodotermes mossambicus*. Insectes Soc 50:222–225

Kramm KR, West DF, Rockenbach PG (1982) Termite pathogens: transfer of the entomopathogen *Metarhizium anisopliae* between *Reticulitermes* sp. termites. J Invertebr Pathol 40:1–6

Lamberty M, Zachary D, Lanot R et al (2001) Insect immunity – constitutive expression of a cysteine-rich antifungal and a linear antibacterial peptide in a termite insect. J Biol Chem 276:4085–4092

Le Guyader A, Rivault C, Chaperon J (1989) Microbial organisms carried by brown-banded cockroaches in relation to their spatial distribution in a hospital. Epidemiol Infect 102:485–492

Little TJ, O'Connor B, Colegrave N et al (2003) Maternal transfer of strain-specific immunity in an invertebrate. Curr Biol 13(6):489–492

Lo N, Engel MS, Cameron S et al (2007) Save Isoptera: A comment on Inward et al. Biol Lett 3:562–563

Logan JWM, Cowie RH, Wood TG (1990) Termite (Isoptera) control in agriculture and forestry by non-chemical methods: a review. Bull Entomol Res 80:309–330

Lutikova LI (1990) The influence of soil substrates from the *Anacanthotermes ahngerianus* Jac. termite-house on the development of entomopathogenic fungi. Mycol Phytopathol 24:520–528

Maschwitz U (1974) Vergleichende Untersuchungen zur Funktion der Ameisenmetathorakaldrüse. Oecologia 16:303–310

Matsuura K (2001) Nestmate recognition mediated by intestinal bacteria in a termite, *Reticulitermes speratus*. Oikos 92:20–26

Matsuura K, Tamura T, Kobayashi N et al (2007) The antibacterial protein lysozyme identified as the termite egg recognition pheromone. PLoS ONE 2:e813. 10.137/journal.pone.0000813

Matsuura K, Vargo EL, Kawatsu K et al (2009a) Queen succession through asexual reproduction in termites. Science 323:1687

Matsuura K, Yashiro T, Shimizu K et al (2009b) Cuckoo fungus mimics termite eggs by producing the cellulose-digesting enzyme β-glucosidase. Curr Biol 19:1–7

Minkley N, Fujita A, Brune A, Kirchner WH (2006) Nest specificity of the bacterial community in termite guts (*Hodotermes mossambicus*). Insectes Soc 53:339–344

Miramontes O, DeSouza O (2008) Individual basis for collective behavior in the termite, *Cornitermes cumulans*. J Insect Sci 8:1–11

Moret Y (2006) Trans-generational immune priming: specific enhancement of the antimicrobial immune response in the mealworm beetle, *Tenebrio molitor*. Proc Biol Sci 273:1399–1405

Naug D, Camazine S (2002) The role of colony organization on pathogen transmission in social insects. J Theor Biol 215:427–439

Noirot C (1969) Glands and secretions. In: Krishna K, Weesner FM (eds) Biology of termites, vol 1. Academic Press Inc, New York, NY, pp 89–123

Ohkuma M, Noda S, Hongoh Y et al (2009) Inheritance and diversification of symbiotic trichonymphid flagellates from a common ancestor of termite and the cockroach *Cryptocercus*. Proc R Soc Lond B 276:239–245

Owens IPF, Wilson K (1999) Immunocompetence: a neglected life history trait or conspicuous red herring? Trends Ecol Evol 14:170–172

Pasteels JM, Bordereau C (1998) Releaser pheromones in termites. In: Vander Meer RK, Breed MD, Espelie KE, Winston ML (eds) Pheromone communication in social insects. Westview Press, Boulder, CO, pp 193–215

Pham LN, Dionne MS, Shirasu-Hiza M, Schneider DS (2007) A specific primed immune response in *Drosophila* is dependent on phagocytes. PLoS Pathog 3(3):e26

Pie MR, Rosengaus RB, Calleri DV, Traniello JFA (2005) Density and disease resistance in group-living insects: do eusocial species exhibit density-dependent prophylaxis? Ethol Ecol Evol 17:41–50

Pie MR, Rosengaus RB, Traniello JFA (2004) Nest architecture, activity pattern, worker density and the dynamics of disease transmission in social insects. J Theor Biol 226(1):45–51

Polz MF, Harbison C, Cavanaugh CM (1999) Diversity and heterogeneity of epibiotic bacterial communities on the marine nematode *Eubostrichus dianae*. Appl Environ Microbiol 65: 4271–4275

Reeson AF, Wilson K, Gunn A et al (1998) Baculovirus resistance in the noctuid *Spodoptera exempta* is phenotypically plastic and responds to population density. Proc R Soc Lond B 265:1787–1791

Rheins LA, Karp RD (1985) Ontogeny of the invertebrate humoral immune response: studies on various developmental stages of the American cockroach (*Periplaneta americana*). Dev Comp Immunol 9:395–406

Rich S (1969) Quinones. In: Torgeson DC (ed) Fungicides: an advanced treatise, vol 2. Academic Press Inc, New York, NY, pp 647–648

Rolff J, Siva-Jothy MT (2002) Copulation corrupts immunity: a mechanism for a cost of mating in insects. Proc Natl Acad Sci U S A 99:9916–9918

Roose-Amsaleg C, Brygoo Y, Harry M (2004) Ascomycete diversity in soil-feeding termite nests and soils from a tropical rainforest. Environ Microbiol 6:462–469

Rosengaus RB, Cornelisse T, Guschanski K, Traniello JFA (2007) Inducible immune proteins in the dampwood termite *Zootermopsis angusticollis*. Naturwissenschaften 94:25–33

Rosengaus RB, Guldin MR, Traniello JFA (1998a) Inhibitory effect of termite fecal pellets on fungal spore germination. J Chem Ecol 24:1697–1706

Rosengaus RB, Lefebvre ML, Carlock DM, Traniello JFA (2000a) Socially transmitted disease in adult reproductive pairs of the dampwood termite *Zootermopsis angusticollis*. Ethol Ecol Evol 12:419–433

Rosengaus RB, Lefebvre ML, Jordan C, Traniello JFA (1999a) Pathogen alarm behavior in a termite: A new form of communication in social insects. Naturwissenschaften 86: 544–548

Rosengaus RB, Lefebvre ML, Traniello JFA (2000b) Inhibition of fungal spore germination by *Nasutitermes*: evidence for a possible antiseptic role of soldier defensive secretions. J Chem Ecol 26:21–39

Rosengaus RB, Maxmen AM, Coates LE, Traniello JFA (1998b) Disease resistance: a benefit of sociality in the dampwood termite *Zootermopsis angusticollis* (Isoptera: Termopsidae). Behav Ecol Sociobiol 44:125–134

Rosengaus RB, Moustakas JE, Calleri DV, Traniello JFA (2003) Nesting ecology and cuticular microbial loads in dampwood (*Zootermopsis angusticollis*) and drywood termites (*Incisitermes minor, I. schwarzi, Cryptotermes cavifrons*). J Insect Sci 3:31

Rosengaus RB, Traniello JFA (1991) Biparental care in incipient colonies of the dampwood termite *Zootermopsis angusticollis* Hagen (Isoptera:Termopsidae). J Insect Behav 4:633–647

Rosengaus RB, Traniello JFA (1993a) Temporal polyethism in incipient colonies of the primitive termite *Zootermopsis angusticollis*: a single multi-age caste. J Insect Behav 6:237–252

Rosengaus RB, Traniello JFA (1993b) Disease risk as a cost of outbreeding in the termite *Zootermopsis angusticollis*. Proc Natl Acad Sci U S A 90:6641–6645

Rosengaus RB, Traniello JFA (1997) Pathobiology and disease transmission in dampwood termites *Zootermopsis angusticollis* (Isoptera: Termopsidae) infected with the fungus *Metarhizium anisopliae* (Deuteromycotina:Hypomycetes). Sociobiology 30:185–195

Rosengaus RB, Traniello JFA (2001) Disease susceptibility and the adaptive nature of colony demography in the dampwood termite *Zootermopsis angusticollis*. Behav Ecol Sociobiol 50:546–556

Rosengaus RB, Traniello JFA, Chen T et al (1999b) Immunity in a social insect. Naturwissenschaften 86:588–591

Rosengaus RB, Traniello JFA, Lefebvre ML, Maxmen AB (2004) Fungistatic activity of the sternal gland secretion of the dampwood termite *Zootermopsis angusticollis*. Insectes Soc 51:1–6

Rowley AF, Powell A (2007) Invertebrate immune systems–specific, quasi-specific, or nonspecific? J Immunol 179:7209–7214

Ryu JH, Kim SH, Lee HY et al (2008) Innate immune homeostasis by the homeobox gene *Caudal* and commensal-gut mutualism in *Drosophila*. Science 319:777–782

Sackton TB, Lazzaro BP, Schlenke TA et al (2007) Dynamic evolution of the innate immune system in *Drosophila*. Nature Genetics 39:1461–1468

Sadd BM, Kleinlogel Y, Schmid-Hempel R, Schmid-Hempel P (2005) Trans-generational immune priming in a social insect. Biol Lett 1:386–388
Sadd BM, Schmid-Hempel P (2007) Facultative but persistent trans-generational immunity via the mother's eggs in bumblebees. Curr Biol 17:1046–1047
Sands WL (1969) The association of termites and fungi. In: Krishna K, and Weesner FM (eds) Biology of termites, vol 1. Academic Press, New York, NY, pp 495–524
Schlüns H, Crozier R (2009) Molecular and chemical immune defenses in ants (Hymenoptera: Formicidae). Myrmecol News 12:237–249
Schmid-Hempel P (1998) Parasites in social insects. Princeton University Press, Princeton, NJ
Schmid-Hempel P (2003) Variation in immune defence as a question of evolutionary ecology. Proc R Soc Lond B 270:357–366
Schmid-Hempel P (2005) Evolutionary ecology of insect immune defenses. Annu Rev Entomol 50:529–551
Schmid-Hempel P, Ebert D (2003) On the evolutionary ecology of specific immune defence. Trends Ecol Evol 18:27–32
Sen R, Ishak HD, Estrada D et al (2009) Generalized antifungal activity and 454-screening of *Pseudonocardia* and *Amycolatopsis* bacteria in nests of fungus-growing ants. Proc Natl Acad Sci U S A 106:17805–17810
Sheldon BC, Verhulst S (1996) Ecological immunology: costly parasite defenses and trade-offs in evolutionary ecology. Trends Ecol Evol 11:217–321
Shellman-Reeve JS (1990) Dynamics of biparental care in the dampwood termite, *Zootermopsis nevadensis* (Hagen): Response to nitrogen availability. Behav Ecol Sociobiol 26:389–397
Shellman-Reeve JS (1997) Advantages of biparental care in the wood-dwelling termite, *Zootermopsis nevadensis*. Anim Behav 54:163–170
Siva-Jothy MT, Moret Y, Rolff J (2005) Insect immunity: an evolutionary ecology perspective. Adv Insect Physiol 32:1–48
Siva-Jothy MT, Tsubaki Y, Hooper R, Plaistow SJ (2001) Immune function in the face of chronic and acute parasite burdens. Physiol Entomol 26:1–6
Stow A, Briscoe D, Gillings M et al (2007) Antimicrobial defences increase with sociality in bees. Biol Lett 3:422–424
Stuart AM (1969) Social behavior and communication. In: Krishna K, Weesner FM (eds) Biology of termites, vol 1. Academic, New York, NY, pp 193–232
Söderhall K, Cerenius L (1998) Role of prophenoloxidase activating system in invertebrate immunity. Curr Opin Immunol 10:23–28
Thomas RJ (1987) Factors affecting the distribution and activity of fungi in the nests of Macrotermitinae (Isoptera). Soil Biol Biochem 19:343–349
Thompson GJ, Crozier YC, Crozier RH (2003) Isolation and characterization of a termite transferrin gene up-regulated on infection. Insect Mol Biol 12:1–7
Thorne BL (1997) Evolution of eusociality in termites. Annu Rev Ecol Syst 28:27–54
Thorne BL, Grimaldi DA, Krishna K (2000) Early fossil history of the termites. In: Abe T, Bignell DE, Higashi M (eds) Termites: evolution, sociality, symbioses, ecology. Kluwer Academic Publishers, Dordrecht, pp 77–93
Thorne BL, Traniello JFA (2003) Comparative social biology of basal taxa of ants and termites. Annu Rev Entomol 48:283–306
Thorne BL, Traniello JFA, Adams ES, Bulmer M (1999) Reproductive dynamics and colony structure of subterranean termites of the genus *Reticulitermes* (Isoptera Rhinotermitidae): a review of the evidence from behavioral, ecological, and genetic studies. Ethol Ecol Evol 11: 149–169
Traniello JFA, Rosengaus RB, Savoie K (2002) The development of immunity in a social insect: evidence for the group facilitation of disease resistance. Proc Natl Acad Sci U S A 99: 6838–6842
Ugelvig LV, Cremer S (2007) Social prophylaxis: group interaction promotes collective immunity in ant colonies. Curr Biol 17:1967–1971

Viljakainen L, Evans JD, Hasselmann M et al (2009) Rapid evolution of immune proteins in social insects. Mol Biol Evol 26:1791–1801

Waller DA, LaFage JP (1987) Nutritional ecology of termites. In: Slansky F, Rodriguez JG (eds) Nutritional ecology of insects, mites and spiders. Wiley, New York, NY, pp 487–532

Wang C, Powell JE, O'Connor BM (2002) Mites and nematodes associated with three subterranean termite species (Isoptera: Rhinotermitidae). Fla Entomol 85:499–506

Watson FL, Püttmann-Holgado R, Thomas F et al (2005) Extensive diversity of Ig-superfamily proteins in the immune system of insects. Science 309:1874–1878

Wilson K, Cotter SC (2008) Density-dependent prophylaxis in insects. In: Ananthakrishnan TN, Whitman DW (eds) Insects and phenotypic plasticity: mechanisms and consequences. Science Publishers Inc, Plymouth, pp 137–176

Wilson K, Reeson AF (1998) Density-dependent prophylaxis: evidence from Lepidoptera-baculovirus interactions. Ecol Entomol 23:100–101

Wilson-Rich N, Spivak M, Fefferman NH, Starks PT (2009) Genetic, individual, and group facilitation of disease resistance in insect societies. Annu Rev Entomol 54:405–423

Wilson-Rich N, Stuart RJ, Rosengaus RB (2007) Susceptibility and behavioral responses of the dampwood termite *Zootermopsis angusticollis* to the entomopathogenic nematode *Steinernema carpocapsae*. J Invertebr Pathol 95:17–25

Wood TG, Thomas RJ (1989) The mutualistic association between Macrotermitinae and Termitomyces. In: Wilding N, Collins NM, Hammond PM, Webber JF (eds) Insect-fungus interactions. Academic Press, San Diego, CA, pp 69–92

Yanagawa A, Shimizu S (2007) Resistance of the termite, *Coptotermes formosanus* Shiraki to *Metarhizium anisopliae* due to grooming. BioControl 52:75–85

Yanagawa A, Yokohari F, Shimizu S (2009) The role of antennae in removing entomopathogenic fungi from the cuticle of the termite, *Coptotermes formosanus*. J Insect Sci 9:6

Yang Z (1997) PAML: a program package for phylogenetic analysis by maximum likelihood. CABIOS 13:555–556

Zhao C, Rickards RW, Trowell SC (2004) Antibiotics from Australian terrestrial invertebrates. Part 1: Antibacterial trinervitadienes from the termite *Nasutitermes triodiae*. Tetrahedron 60: 10753–10759

Zuk M, Stoehr AM (2002) Immune defense and host life history. Am Nat 160:S9–S22

Chapter 8
Comparative Biology of Fungus Cultivation in Termites and Ants

Tânia Nobre, Corinne Rouland-Lefèvre, and Duur K. Aanen

Abstract We review the two known mutualistic symbioses between basidiomycete fungi and social insects: the attine ants and macrotermitine termites, comparing their origin, history and patterns of co-evolution, and stability. It is argued that ants are "specialised farmers of unspecialised crops", whereas termites are "specialised farmers of specialised crops". Furthermore, despite differences in symmetry and symbiont transmission mode, in both relationships there is a moderate specificity between partners. The unresolved debate about the main role of the symbiotic fungus in the fungus-growing termites is summarised and contrasted with the role in the fungus in attine ants, which is little debated. We compare colony foundation and structure, and the modes of symbiotic interaction between the two groups of social insects, highlighting gaps in our understanding of both systems. Finally, we discuss how these symbioses can be evolutionarily stable and the mechanisms by which the ant and termite symbionts ensure monopolies of host care. We conclude by identifying some lines of future research within the fungus-growing termite symbiosis.

8.1 Introduction

Interactions between termites and fungi are found in multiple groups, ranging across both lower and higher termites. In some cases, fungi play a role in termite nutrition, but they can also influence termite survival and caste development. Fungal interactions with subterranean termites, for example, involve more than 50 species of wood-inhabiting fungi (Amburgey 1979) which can influence the palatability of wood, either by conditioning or by producing compounds that act as attractants or repellents. However, these interactions are usually ephemeral and not obligate and therefore not symbiotic.

T. Nobre (✉)
Laboratory of Genetics, Wageningen University, 6708 PB Wageningen, The Netherlands
e-mail: tania.mesquitanobre@wur.nl

D.E. Bignell et al. (eds.), *Biology of Termites: A Modern Synthesis*,
DOI 10.1007/978-90-481-3977-4_8, © Springer Science+Business Media B.V. 2011

The term symbiosis was first used in 1879 by Anton de Bary, to describe close, long-term associations between different organisms living together (Wilkinson 2001). Applying this definition, only a single lineage of termites, the fungus-growing termites (subfamily Macrotermitinae, family Termitidae) has developed a symbiosis with fungi (the basidiomycete genus *Termitomyces*). The fungus is used for food provisioning, and both partners are obligately dependent on each other.

Several other groups of insects have independently also evolved symbiotic relationships with fungi. However, behaviourally complex systems of insect agriculture have only evolved in three insect orders (Mueller et al. 2005): once in termites (Aanen et al. 2002), once in ants (Mueller et al. 2001) and seven times in ambrosia beetles (Farrell et al. 2009). All these groups have a certain level of sociality: most ambrosia beetles are subsocial, in which a single female cares for her brood, or communal, in which several reproductive females cooperate in brood care and gardening (Kirkendall et al. 1997), and one (*Austroplatypus incompertus*) is known to be eusocial (Kent and Simpson 1992). Fungus-growing termites and ants are both eusocial, and share many similarities, at least superficially.

8.2 Evolutionary History of Fungiculture

The fungus-growing termites are affiliated in a single subfamily, the Macrotermitinae, which has been divided into 12 genera and ca. 330 species (Eggleton 2000). However, the Asiatic genus *Hypotermes* has now been shown to be nested within the genus *Odontotermes*, leaving only 11 of the described genera monophyletic (Aanen et al. 2002). All species of fungus-growing termites live in obligate mutualistic symbiosis with basidiomycete fungi of the genus *Termitomyces*. A comparison between the number of described species in both partners – 40 in *Termitomyces* (Kirk et al. 2001) and 330 in fungus-growing termites – suggests that many of these symbionts are shared between termite species. However, *Termitomyces* taxonomy is largely based on the morphology of sexual fruiting bodies, which are rare and possibly absent in some *Termitomyces* species. Furthermore, recent molecular data have suggested that there may also be many morphologically indistinguishable sibling species of *Termitomyces* (Frøslev et al. 2003).

The fungus-growing termites originated in Africa (Aanen and Eggleton 2005) and occur in the Old-World tropics. Most of the diversity occurs in Africa, where all genera are present (except for *Hypotermes*), while only four genera are found in Asia. A recent study has shown that a single colonisation has occurred to Madagascar, of termites belonging to the genus *Microtermes* (Nobre et al. 2010). The habitat of fungus-growing termites is variable and ranges from savannah to rain forest. Although the fungus-growing termites are found abundantly in the African rain forests, which likely represents their ancestral habitat (Aanen and Eggleton 2005), their relative contribution to ecosystem decomposition is relatively low (ca. 1–2% of all carbon mineralization; Bignell and Eggleton 2000). This contrasts with their role in savannah ecosystems where up to 20% of all carbon mineralization is caused by fungus-growing termites (Wood and Sands 1978, see also Chapter 17 by Jones and Eggleton, this volume), and where species numbers are also higher.

In ants also, a single clade has evolved an obligate mutualistic symbiosis with basidiomycete fungi (Chapela et al. 1994; Schultz and Brady 2008). The attine ants (subfamily Myrmicinae, tribe Attini) comprise a monophyletic group of more than 230 species, exclusively New World and primarily neotropical (Schultz and Brady 2008). Similarly to the macrotermitine termites, all attine ants obligately depend on the cultivation of fungi for food. In contrast to the termite fungi, the cultivated fungi of the attine ants do not form a monophyletic group. The largest group of cultivated fungi belongs to the family Lepiotaceae (parasol mushrooms). However, phylogenetic studies have shown a close relationship between some extant symbionts and free-living relatives (Mueller et al. 1998; Vo et al. 2009). This indicates repeated recent acquisitions of these fungal symbionts. A completely independently domesticated group of fungi, belonging to coral fungi (family Pterulaceae), is found with some *Apterostigma* species (Munkacsi et al. 2004).

When comparing patterns of co-evolution between insects and fungi in both fungus-growing ants and termites, the following overall picture emerges (schematically illustrated in Fig. 8.1). In fungus-growing termites, the symbiosis is "symmetric", meaning that the macrotermitine termites and their *Termitomyces* fungi both constitute monophyletic groups, and that specificity occurs at higher taxonomic levels (mainly the genus level). In contrast, the symbiosis of fungus-growing ants is highly asymmetric, meaning that a single clade of ants has specialised on a diverse group of fungi, some of which have a close relationship to free-living relatives.

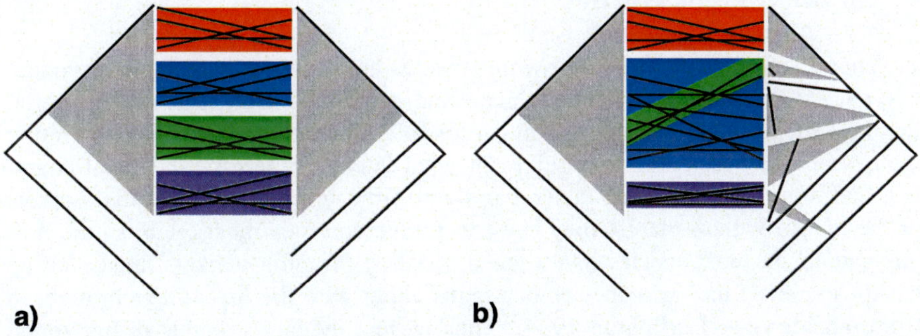

Fig. 8.1 Schematic, highly simplified representation of the general patterns of co-evolution between insects and fungi, highlighting patterns of symmetry and recent domestications and reversals to non-symbiotic lifestyles. *Black lines* indicate non-symbiotic species, whereas *grey* indicates symbiotic clades. (**a**) Fungus-growing termites – monophyly of both symbiotic partners and broad-scale patterns of coevolution; (**b**) fungus-growing ants – the ants constitute a monophyletic group but the fungal symbionts are polyphyletic. Phylogenetic reconstructions indicate that multiple recent domestications of fungal symbionts have occurred as well as multiple recent "escapes" from a symbiotic to a non-symbiotic lifestyle. In both symbioses, co-evolution occurs between specific clades of insects associated with specific clades of symbiotic fungi, but within these groups host switching occurs. (Based on Chapela et al. 1994; Mueller et al. 1998; Schultz and Brady, 2008; Mikheyev et al. 2010; Aanen et al. 2002, 2007; Rouland-Lefèvre and Bignell 2001)

This implies that some of the ant fungi must have remained largely unspecialised. Overall, in both mutualisms the insects have specialised on their symbiotic (agricultural) lifestyle, and no reversals to non-symbiotic lifestyles have occurred. Yet, in the fungus-ant mutualism, the fungi have remained unspecialised to some degree, while in fungus-growing termites the fungi have completely specialised to the symbiotic lifestyle. The following summary is therefore apt: "ants are specialised farmers of unspecialised crops", whereas termites are "specialised farmers of specialised crops" (Aanen et al. 2002).

It has long been believed that the fungi of one specialised group of fungus-growing ants, the leaf cutter ants (belonging to the so-called "higher attines") were exclusively clonally and vertically transmitted, leading to strict co-speciation with their host (Chapela et al. 1994). However, the seminal work of Mikheyev and Mueller has shown that this view can not be maintained (Mikheyev et al. 2006, 2010). Their work shows that the fungal symbionts of leaf-cutter ants can and occasionally do reproduce sexually and disperse horizontally, which is also reflected in incongruence between phylogenies of ants and fungi (Mikheyev et al. 2006).

Despite the differences in symmetry and the normal mode of transmission mode between the two fungus-growing mutualisms in both cases there is moderate specificity: monophyletic groups of insects (i.e. discrete groups of genera, are associated with monophyletic groups of fungi). However, within these groups there is frequent host switching.

8.3 Colony Foundation and Establishment of the Fungus Garden

In general, colonies of fungus-growing termites are founded by a single reproductive pair. When a male and female alate find each other after their nuptial flight, they lose their wings and, upon finding a suitable substratum, they begin excavating their nest. They then seal themselves in what later will become the royal chamber, and after some time the queen can then start laying eggs. When the workers of the first brood are mature they leave to forage and then must establish the fungus garden. In most species investigated, the first foraging workers must pick up basidiospores of the right species of fungus along with the first forage brought in from outside (e.g. Darlington 1994). Thus current evidence is thus of horizontal transmission in a majority of cases, implying that the reproduction of the two partners is decoupled. One of the main consequences of this is that in each generation new combinations of termite and fungal lineages occur (Korb and Aanen 2003). Such transmission appears to represent the ancestral state in fungus-growing termites (Aanen et al. 2002), but more work is needed as the overall number of species documented is small. In contrast, the between-nest fungal symbiont transmission mode in ants is vertical, clonal and uniparental (Mueller et al. 1998) – a typical ant colony is founded by a single mated female who takes the fungal inoculum from her maternal colony in her infrabuccal pocket and starts the garden of her new colony with it.

In the fungus-growing termites, two cases of vertical transmission are known as an exception to the general pattern of horizontal transmission. In both cases, vegetative spores (conidia) of *Termitomyces* are transported in the gut of one of the reproductives from a parental colony and used to inoculate the fungus comb of the newly founded colonies. This vertical transmission mode is uniparental: it occurs via the female alate in the *Microtermes* species complex, and via the male alate in *Macrotermes bellicosus* (Grassé and Noirot 1955; Johnson 1981; Johnson et al. 1981; Aanen et al. 2002, reviewed in Korb and Aanen, 2003). Under strict vertical transmission of fungal symbionts, host-switching would not occur, so that we would expect that the fungal symbionts of the termite clades with vertical transmission would form monophyletic groups. Surprisingly, however, this is not the case and occasional events of horizontal transmission must be inferred both for the genus *Microtermes* (Aanen et al. 2002, 2007; Nobre et al. 2010) and for *M. bellicosus* (Aanen et al. 2002). Furthermore, a study focussing on *M. bellicosus* has shown a largely clonal population structure with occasional rare events of recombination (Nobre et al. in preparation).

Also throughout the attine ant–fungus symbiosis, ubiquitous cultivar sharing and switching has been shown (Mueller et al. 1998; Bot et al. 2001; Green et al. 2002; Mikheyev et al. 2007, 2010). Furthermore, a large group of ants (sometimes indicated as "lower attines") have been shown to recruit free-living strains (Mueller et al. 1998; Vo et al. 2009). Recently, Poulsen et al. (2009) showed that new *Acromyrmex* queens provided with a novel fungus fragment can accept and nurture the alien fungal symbiont, suggesting that the colony founding stage may provide an efficient but transient window for horizontal transmission.

A similar transient window for horizontal transmission can be hypothesized for the otherwise vertical transmitted *Termitomyces* strains (cultivars of *Microtermes* and *Macrotermes bellicosus*). Aanen et al. (2009) have shown that single-strain monocultures of *Termitomyces* within single nests are maintained through positive frequency-dependent selection (thus preventing subsequent colonization by other strains). If we consider that at the initial stages of colony formation, the fungus still has a low biomass, it would be possible for a new, horizontally transmitted symbiont strain to establish itself. In this particular group of fungus-growing termites, future experimental studies should focus on the details of symbiont transmission and the frequency of horizontal transmission. For all the vertically transmitted strains, the presence of a certain level of horizontal transmission, associated with sexual reproduction, benefits the symbiosis at a longer term, as there seems to be a critical level of horizontal transfers below which natural selection is unable to purge deleterious mutations, leading to an expected loss of fitness over time (O'Fallon and Hansen 2008).

Once the fungus-comb is established and the colonies begin to grow, both fungus-growing termites and fungus-growing ants show division of labour among castes (Fig. 8.2a, b). Although not all termites show marked age-related polyethism, the general pattern is different workers carrying out complementary tasks to manage the fungus garden, which is added to at the top and is consumed at the middle and bottom (Rouland-Lefèvre and Bignell 2001). For some fungus-growing termites, such

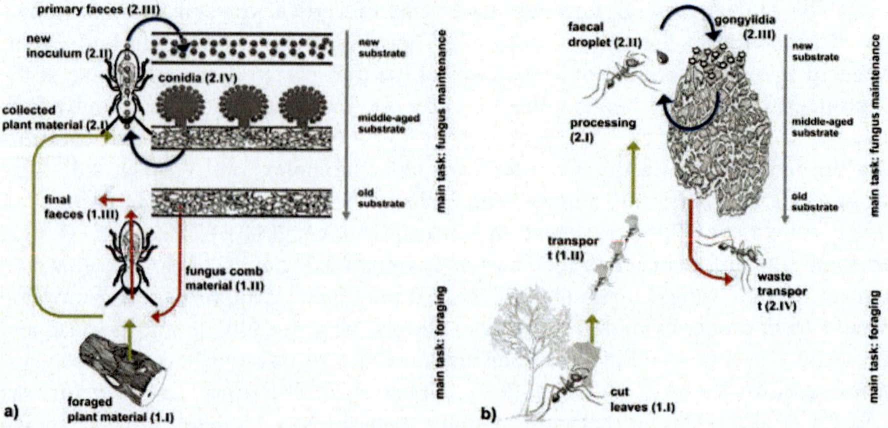

Fig. 8.2 Schematic representation of food-processing pathways in: (**a**) Fungus-growing termites (*Macrotermes*). (*1.I*) plant-material is brought into the nest by older workers that feed on aged comb-material (*1.II*) and produce final faeces (*1.III*). Collected plant material (*2.I*) is inoculated with new spores (*2.II*) directly in the gut of young workers before being deposited on the fungus comb as so-called primary faeces (*2.III*). These gut-resistant asexual spores (*2.IV*; conidia) are produced in the fungus nodules consumed by young workers. (**b**) Fungus-growing ants (*Atta,* leaf cutter ant). (*1.I*) leaves are cut from the vicinity and brought into the nest along trails (*1.II*). A group of specialized workers process the leaves (*2.I*), in part through the application of faecal droplets (*2.II*) that contain enzymes produced by the fungus and passing through the ant gut, and the end result, a mulch, is incorporated in the top section of the fungus garden and is used as growth substrate. Once all nutrients have been used, the waste is transported to special dump chambers (*2.IV*). In *Atta*, the fungus produces special nutritional bodies (*2.III*; gongylidia) which are fed to the larvae by specialized nurse workers and are also the main diet of the adult ants

as the species within the *Macrotermes* and *Odontotermes* genera, there is a continuous turnover of comb material, with primary faeces (including fungal inoculum) being deposited on top and the older comb material at the bottom being consumed. At the other extreme of the comb turn-over dynamics gradient, species belonging to the genera *Pseudacanthotermes* consume the entire comb before building a new one in an empty chamber (Rouland-Lefèvre 2000). The food material collected by fungus-growing ants, once deposited in the fungus garden by the specialized workers and depleted of nutrients by the fungi, is transported as waste to special chambers (where dead ants and dead fungus are also placed). In the derived leafcutter ants, the specialization of workers is remarkable, with a range of differently sized castes, each specialized for a different task (Wilson 1980; Currie 2001b; Hart et al. 2002a).

8.4 Role of Fungal Symbiont

In fungus-growing termites, the fungal symbiont is thought to be the main decomposer of lignocellulose, the predominant component of (mostly) dead plant material collected. However, the exact function of the symbiont is still unclear and seems to differ between genera and species (Rouland-Lefèvre et al. 1991; Bignell 2000; Rouland-Lefèvre and Bignell 2001; Hyodo et al. 2003; Ohkuma 2003). The major constituents of lignocellulose are cellulose (28–50%), hemicelluloses (20–30%) and lignin (18–30%), with the cellulose and hemicellulose polymers tightly bound to the lignin (Breznak and Brune 1994). Therefore, lignin degradation enhances the digestibility of the cellulose, and it has been suggested that this is the main role of *Termitomyces* in some species (such as *Macrotermes gilvus*; Hyodo et al. 2000). For other species, however, it seems that the fungus serves mainly as food source (Rouland-Lefèvre et al. 1991; Hyodo et al. 2000, 2003). Overall, the proposed functions fall into four categories, which are not mutually exclusive (Bignell 2000):

(1) *Termitomyces* is an additional protein-rich food source (mainly the fungal nodules);
(2) *Termitomyces* has a role in lignin degradation, which facilitates the access to cellulose;
(3) *Termitomyces* decreases the C/N ratio of foraged products by metabolising carbohydrates;
(4) *Termitomyces* provisions cellulases and xylanases to work synergistically and/or complementarily with endogenous termites enzymes (Martin and Martin 1978; Rouland-Lefèvre et al. 1991).

Some degree of mutualistic diversification seems to have occurred and whether the different roles of the symbiont are related to differences in the foraged plant materials used for comb construction or on termite taxonomy cannot be derived from the available data. However, a correlation between *Termitomyces* metabolism and the symbiosis co-evolutionary history is suggested by Rouland-Lefèvre et al. (2006). They have classified *Termitomyces* into two groups according to their metabolism: (1) relatively generalist fungi, cultivated by several species of termites and varying the production of enzymes depending on the substrate, e.g. the symbiont shared by *Microtermes*, *Ancistrotermes* and *Synacanthotermes* species (Aanen et al. 2002) and (2) relatively specialist fungi, found in association with single termite species and having a restricted variety of substrates inducing enzyme production e.g. the symbiont of *M. natalensis*, only found in association with its specific host (De Fine Licht et al. 2006; Aanen et al. 2007, 2009).

In contrast to the ongoing discussion about roles of the fungus in fungus-growing termites, in the fungus-growing ants the role of the fungus has been less controversial. The symbiont is the only source of nutrients for the larvae of all species (Weber 1972; Wetterer 1994). Furthermore, the available evidence suggests that the fungus is also an essential source of food for the adult ants, especially as a nitrogen source (Mattson 1980), but less as an energy source (< 10% of their energy requirements come from the symbiont (Bass and Cherrett 1995)). Some work has been done on the digestive interactions between the ants and their symbiotic fungi in the leaf-cutter genera *Atta* and *Acromyrmex*. Here there is a distinct division of labour between ants and fungi, as the enzymatic activities are complementary (D'Ettorre et al. 2002; Richard et al. 2005). However, it should be noted that most of this work has been done on the so-called higher attines. With the development of new molecular and genomic techniques, new insights into the role of the fungal symbiont are starting to emerge. It has recently been shown that the role of the fungus also differs among genera of fungus-growing ants (De Fine Licht and Boomsma 2010) and recent studies have started to explore the precise respective roles and capacities of the insect and fungal partners in the attine ants (Schiött et al. 2008; De Fine Licht et al. 2010)

Whatever the exact role of *Termitomyces*, both partners in the fungus-growing termite mutualism are reciprocally dependent on each other. Not only do the termites depend on the fungus for long-term survival but also the fungi are obligatorily dependent on the nursing termites (Korb and Aanen 2003). Advantages for the fungi are (1) access to plant material with an increased surface area; (2) an optimal microclimate for their growth within the termite hive, and (3) the selective inhibition of other fungi and the prevention of microbial infections, creating a competitor free niche. Furthermore, the very specific mechanism of fungal inoculation of forage by fungus-growing termites (Leuthold et al. 1989; Aanen 2006; for details, see below), in which asexual conidiospores are added at high density, is very unlike fungal growth in non-symbiotic settings, which is predominantly by mycelial growth. It is easy to imagine that this cultivation mode has led to dependence on cultivation by termites.

Martin and Martin (1978) suggested that *Termitomyces* provides the termites with a subset of the enzymes required for cellulose degradation (Rouland-Lefèvre et al. 2006 and references therein). According to this hypothesis, fungal enzymes derived from consumed nodules (clusters of conidia-bearing hyphae) act synergistically with termite derived enzymes in such a way that the degradation of cellulose is accomplished in the worker guts by the combined action of fungal and termite enzymes. On the other hand, Veivers et al. (1991), Slaytor (1992) and Bignell et al. (1994) suggested that cellulolytic enzymes of termite origin provided sufficient cellulose degradation to meet all energy requirements. The issue remains unresolved, but new insights into the role of the symbionts may be possible from genomic investigations. This has already begun in non-fungus growing termites with the metagenomic sequencing of the (intestinal) symbiont community of *Reticulitermes speratus* (Todaka et al. 2007), *Reticulitermes flavipes* (Scharf et al. 2003, 2005; Zhou et al. 2007) and a *Nasutitermes* species (Warnecke et al. 2007). In these studies, a large number of cellulose-degrading genes have been found, and their putative origins have been identified. In fungus growers, a bacterial gut microbiota is also present (Yara et al. 1989; Paul et al. 1993; Anklin-Mühlemann et al. 1995; Hongoh et al. 2006; Mackenzie et al. 2007), but has not been fully defined. Since considerable evidence for co-evolution between gut bacteria and their termite hosts has now been found in other termite groups (e.g. Hongoh et al. 2006; Shinzato et al. 2007, see Chapter 14 by Bignell and Chapter 15 by Ohkuma and Brune, this volume) a symbiont role for the bacteria in fungus-growers is likely. Studies of the digestive process in fungus-growers have shown that forage passes through the termite intestine twice (Rouland-Lefèvre 2000, Fig. 8.2a) and it is assumed that the substrate is not (or hardly) affected during the first gut passage, apart from being diced and inoculated with fungal spores. The hypothesis that during the first gut passage gut-derived enzymes (either from bacterial or termite origin) condition the substrate for fungal growth, by an initial breakdown of lignocellulose should be tested. Furthermore, the general assumption that during the second passage fungal-derived enzymes complementary to termite-derived enzymes would enable termites to utilise the cellulose present in the plant material, does not take into account the possible contribution of the gut bacteria.

8.5 Fungus Garden Protection

There is a wide range of organisms that can be considered as parasites of fungus gardens and their entrance must be prevented. Not only other fungi, but also mites and nematodes are common invaders of fungus gardens as they feed on the fungus and (potentially) contaminate it with alien spores (Mueller and Gerardo 2002). Task partitioning within the colony is likely to help in protecting the fungus garden, as foragers are not allowed in fungus-chambers creating a separation mechanism from the outside environment. To further prevent the entrance of parasites in the nest, allogrooming has proved to be effective in removing parasites from the insect cuticule and seems to be directed at individuals that were indeed exposed to parasites (Rosengaus et al. 1998; Yanagawa et al. 2008; Walker and Hughes 2009). In the termites, the gut bacteria directly interact with substrate microorganisms passing through the gut, before incorporation into the garden (Fig. 8.2a). This potentially differentially affects the survival of non-mutualistic fungi or bacteria, but remains to be studied. In the ants, such gut passage does not occur, but considerable effort is spent on cleaning substrate surfaces which may result in a partial sterilization of the garden (Weber 1972). Task partitioning may also help defend the fungus from nest and garden robbers and in monitoring gardens for diseases. Garden usurpation and colony joining in attine ants seems to emerge as critical behavioural adaptations to garden loss (Adams et al. 2000a, b) but to our knowledge, no equivalent behaviour has been documented in the fungus-growing termites.

From the termite nest, several fungi other than *Termitomyces* can be isolated (Thomas 1987b, c, d; Moriya et al. 2005; Shinzato et al. 2005; Guedegbe et al. 2009a, b; Visser et al. 2009). Apart from the common unspecialized soil fungi likely to get inside the nests via foragers or on the collected food, a number of *Xylaria* species (ascomycete) have been frequently reported from fungus-growing termite nests (Rogers et al. 2005; Ju and Hsieh 2007; Visser et al. 2009). *Xylaria* seems to grow in the nests only when these are, for any reason, dead or decaying, and it does not seem to be a common inhabitant from the nests' surrounding soil, but it immediately sprouts from fresh comb of *Macrotermes bellicosus* when this is excised from a mound (Thomas 1987b). This has led to the hypothesis of a parasitic association between *Xylaria* and termite combs (Rouland-Lefèvre 2000), as it is being suppressed in the fungus garden (Thomas 1987d). However, recent data suggest that *Xylaria* is not likely to be a parasite of *Termitomyces*, as in vitro experiments have shown that *Xylaria* does not significantly interfere with the growth of the fungal symbiont (Visser et al. in preparation). Furthermore, no mycoparasitic members of the Xylariales are known. Rather, *Xylaria* is a competitor of *Termitomyces* for substrate, and these two fungi have specialised on the same niche, as phylogenetic studies indicate some specificity for the fungus-growing termite niche (Guedegbe et al. 2009a; Visser et al. 2009).

For the fungus-growing ants, however, a specialised mycoparasitic symbiont has been identified, the ascomycete fungus *Escovopsis*. This fungus is prevalent in fungus-growing ant nests, but usually suppressed, and highly specialized on the ant fungus garden, showing some coevolution with its host. *Escovopsis*, directly targets and consumes the ant's mutualistic fungus (Currie et al. 1999b; Reynolds and Currie 2004). *Escovopsis* has a major impact on the ant's garden: even in the presence of the ants it can overgrow the garden, killing the colony, or after establishment of the infection it can maintain itself, with a substantial decrease in the growth rate of the colony (Currie et al. 1999a; Currie 2001a, c).

The ants have evolved several behavioural responses to an infection, such as changing colony-level task allocation (Hart et al. 2002b) or fungus grooming and weeding (Currie 2001b). Furthermore, the ants harbour antibiotic-producing actinobacteria on their cuticules, which aid in the defence against *Escovopsis* (Currie et al. 1999b). It has been suggested that there has been a long history of tight co-evolution between these actinobacteria (genus *Pseudonocardia*) and *Escovopsis* (Currie et al. 1999a, b; Mueller et al. 2005), leading to a moderate degree of host-specificity. However, it has more recently been shown that multiple acquisitions of the bacteria by the fungus-growing ants have occurred (Cafaro et al. 2005; Kost et al. 2007; Mueller et al. 2008) and it seems that the ant-associated actinobacteria are more diverse, both within single colonies and across colonies, than previously thought (Sen et al. 2009). Therefore, the current paradigm of the role of actinobacteria as antibiotics producers against *Escovopsis* for the benefit of ants in a co-evolutionary arms race with *Escovopsis* may need some modification (Sen et al. 2009; Boomsma and Aanen 2009).

Recently, two more symbiotic partners have been described in the fungus-growing ant symbiotic community: a black yeast (Ascomycota, *Philaphora*), which exploits the ant-actinomycete mutualism (Little and Currie 2007, 2008) and nitrogen-fixing bacteria (*Klebsiella* and *Pantoea*) that mediate the acquisition of atmospheric nitrogen (Pinto-Tomas et al. 2009).

In fungus-growing termites fewer additional symbionts or specialised parasites are known. One possibility is that this is due to insufficient investigation. Alternatively, if future research confirms this as a real difference between the two social-insect fungus mutualisms, it may be explained by differences in their biology. First, an important difference is that in fungus-growing termites the substrate passes through the gut before it is being deposited on the fungus garden. This allows a higher control over the fungus garden, either by the termites themselves, or by their gut symbionts. Second, the majority of the fungus-growing termites have horizontal symbiont transmission, associated with sexual reproduction. The resulting genetic variability of the fungal symbionts may confer stronger resilience and long-term protection of these fungi to parasites.

8.6 Evolutionary Stability

The evolution of mutualisms – reciprocally beneficial interactions between different species, or "reciprocal exploitations that nonetheless provide net benefits to each partner" is difficult to explain (Herre et al. 1999). Theory predicts that in the short term, cheating individuals, which profit but do not pay the cost of reciprocating the help of their partner, will be selected because they have an advantage in competition with reciprocating competitors (Fletcher and Doebeli 2009).

For fungus-growing social insects, the problem has two explicit components (Aanen 2006). First, how do the partners reproduce? In fungus-growing ants, the default symbiont transmission mode is vertical and uniparental. This means that the interests between insects and fungi are aligned, which stabilises the interaction (Frank 1996). In contrast, in fungus-growing termites, symbiont transmission is horizontal (barring the two known derived lineages that evolved vertical transmission; see above). This means that there is a fundamental conflict of interest between the mutualistic partners. Although they are obligatorily interdependent within each single colony, they are in conflict over the amount of resources to be allocated when reproducing sexually and independently: the termites have no interest in their fungal symbiont producing costly mushrooms outside the nest that they cannot eat (Korb and Aanen 2003), whereas the resident fungal symbiont has no interest in the termites producing winged dispersing reproductives instead of workers that would provide them with more resources (Aanen 2006).

Second, fungi are modular organisms that consist of colonies of cells each having the potential to reproduce independently. Genetic variation among symbionts within a termite colony thus creates competition over reproduction between genetically distinct lineages of symbiont and such variation can arise both by mutation and horizontal acquisition. Assuming that there is a trade-off between competition within the group of fungal symbionts and the success of the group against other groups (Frank 1996; Taylor and Frank 1996), genetic variation of this type implies a conflict between two levels of selection. As a consequence, it will be in the interest of the termite hosts to minimize symbiont genetic variation when competitive interactions between symbiont strains would reduce group productivity (Frank 1996; Korb and Aanen 2003; Aanen 2006). Alternatively, they could have evolved mechanisms to prevent genetically different symbionts from expressing their competitive traits.

Recent studies have shown that individual colonies of fungus-growing termites are associated with a single strain of *Termitomyces* only, despite horizontal transmission (Aanen et al. 2002; Katoh et al. 2002; De Fine Licht et al. 2006; Aanen et al. 2007, 2009). The mechanism that leads to monoculture has also been identified (Aanen 2006; Aanen et al. 2009). The key to this mechanism lies in the way termites cultivate their fungus. In all genera studied so far, the termite workers consume the nitrogen-rich fungal nodules that grow on the combs (Fig. 8.2a). These nodules are unripe mushrooms that are harvested long before they reach sexual maturity (Bathellier 1927; Heim 1977; De Fine Licht et al. 2005). Possibly as an adaptation to this early harvesting by termites, nodules contain asexual spores that survive gut passage and serve as inocula for newly constructed comb substrate (Leuthold et al. 1989). The continuous seeding with asexual spores allows rapid growth of a new mycelium and of new nodules, which are then consumed again. This mode of propagation is universal for the entire clade, although the details differ between genera (Thomas 1987a; Leuthold et al. 1989)

This propagation mode has two consequences. First, it is associated with genetic bottlenecks, as in each propagation cycle only a subset of the genetic diversity of the fungus will form nodules that are inoculated (Aanen 2006). Second, it has been shown that genotypes that are common in a garden become even more common at the expense of rarer genotypes (Aanen et al. 2009). This happens not because common genotypes are better direct competitors, but because they have a higher chance of having an identical genotype as neighbour. Every time this happens, genetically identical mycelia merge, which enhances the efficiency by which they produce asexual spores, which are eaten by the termites and deposited in new garden material of the colony. This process of positive reinforcement makes every colony end up with a life-time commitment to a single fungal symbiont in spite of the population at large having many fungal genotypes.

In contrast to the horizontally transmitted fungi of fungus-growing termites, in fungus-growing ants the expected diversity in a colony is low from the start because of clonal vertical symbiont transmission (at least in monogynous species). However, it has been shown that horizontal transmission does occur infrequently; this suggests that active mechanisms must be in place to prevent mixed cultures. It has indeed been shown that fungi have maintained antagonistic traits against genetically different clones (Bot et al. 2001; Poulsen and Boomsma 2005). Furthermore, the ants deposit allelopathic compounds, derived from the resident fungus, on new fragments of their fungus garden, which prevents other genotypes from colonizing the garden (Poulsen and Boomsma 2005). These two mechanisms maintain monoculture in fungus-growing ants. In agreement with this idea, a recent study has found that horizontal transmission is most easily achieved at the start of a colony, when the biomass of the fungus is still low (Poulsen et al. 2009).

Although the principles are similar, the mechanisms by which the ant and termite symbionts ensure monopolies of host care are different. In both systems, positive reinforcement of the monopoly of the resident (majority) symbiont is driven by the circulation of fecal material of the farming host (Aanen 2006; Aanen et al. 2009; Poulsen et al. 2009). This enables the fungus-garden symbionts to successfully protect their residency against any secondary colonization by an alternative minority symbiont. The ant symbionts deposit allelopathic compounds that cause extensive cell death upon contact with unrelated mycelia and these same compounds survive passage through the ant guts, so that intruding symbionts can also be eliminated by interference competition at a distance when they are "fertilized" with ant faecal droplets. The termite symbionts do not need aggressive mycelial incompatibility because the biased propagation of asexual spores suspended in the termite faeces continues to favour the same symbiont clone that achieved residential monopoly by positive frequency dependent selection shortly after colony foundation. The insect and fungal partners in both systems share a short-term interest in keeping the resident fungus free of competing strains, but the ensuing symbiont monopoly severely constrains the acquisition of potentially more productive strains later in colony life.

8.7 Concluding Remarks

The Old World fungus-growing termites and the New World fungus-growing ants symbioses have been differentially studied through time: the fungus-growing ant symbiosis has been studied in more detail, and we now know that it involves a network of insect farmers and at least five microbial symbionts, including antibiotic-producing bacteria. It is essential to identify all the symbionts involved, to characterize their roles in the symbioses and their impacts, and to explore their respective interactions with the host and between themselves. The recent technological advances, particularly new genomic and metagenomic tools, are expected to make possible the study of complete symbiotic communities associated with a given host species. The present momentum of research is thus ideal to further explore the fungus-growing termite symbiosis and to identify putative additional symbionts.

Research focusing on the diversity and function of gut-microbiota of termites will possibly reveal microorganisms that help stabilize the symbiosis by aiding in disease suppression. It will also elucidate the nutritional aspect of the gut-microbiota. Research on termite digestive genomics should also be directed towards solving the controversy about the role of *Termitomyces* across the symbiosis. Although all symbiotic partners profit from the synergy of combined abilities to exploit the food source, they also compete for its components. Knowing the details of partner complementarity and redundancy in substrate degradation will lead to a better understanding the functional co-evolution and stability of nutritional mutualisms.

References

Aanen DK (2006) As you reap, so shall you sow: coupling of harvesting and inoculating stabilizes the mutualism between termites and fungi. Biol Lett 2:209–212

Aanen DK, de Fine Licht HH, Debets AJM et al (2009) High symbiont relatedness stabilizes mutualistic cooperation in fungus-growing termites. Science 326:1103–1106

Aanen DK, Eggleton P (2005) Fungus-growing termites originated in African rain forest. Curr Biol 15:851–855

Aanen DK, Eggleton P, Rouland-Lefèvre C et al (2002) The evolution of fungus-growing termites and their mutualistic fungal symbionts. Proc Natl Acad Sci U S A 99:14887–14892

Aanen DK, Ros V, de Fine Licht H et al (2007) Patterns of interaction specificity of fungus-growing termites and *Termitomyces* symbionts in South Africa. BMC Evol Biol 7:115

Adams RMM, Mueller UG, Holloway AK et al (2000a) Garden sharing and garden stealing in fungus-growing ants. Naturwissenschaften 87:491–493

Adams RMM, Mueller UG, Schultz TR, Norden B (2000b) Agro-predation: usurpation of attine fungus gardens by *Megalomyrmex* ants. Naturwissenschaften 87:549–554

Amburgey TL (1979) Review and checklist of the literature on interactions between wood-inhabiting fungi and subterranean termites:1960–1978. Sociobiology 4:279–296

Anklin-Mühlemann R, Bignell DE, Veivers PC et al (1995) Morphological, microbiological and biochemical studies of the gut flora in the fungus-growing termite *Macrotermes subhyalinus*. J Insect Physiol 41:929–940

Bass M, Cherrett JM (1995) Fungal hyphae as a source of nutrients for the leaf-cutting ant *Atta sexdens*. Physiol Entomol 20:1–6

Bathellier J (1927) Contribution à l' etude systématique et biologique de termites de l'Indo-Chine. *Faune Colonies Francaises* 1:125–365

Bignell DE (2000) Introduction to symbiosis. In: Abe T, Bignell DE, Higashi M (eds) Termites: evolution, sociality, symbioses, ecology. Kluwer Academic Publishers, Dordrecht, pp 189–208

Bignell DE, Eggleton P (2000) Termites in ecosystems. In: Abe T, Bignell DE, Higashi M (eds) Termites: evolution, sociality, symbioses, ecology. Kluwer Academic Publishers, Dordrecht, pp 363–387

Bignell DE, Slaytor M, Veivers PC et al (1994) Functions of symbiotic fungus gardens in higher termites of the genus *Macrotermes*: evidence against the acquired enzyme hypothesis. Acta Microbiol Hung 41:391–401

Boomsma JJ, Aanen DK (2009) Rethinking crop-disease management in fungus-growing ants. Proc Natl Acad Sci U S A 106:17611–17612

Bot ANM, Rehner SA, Boomsma JJ (2001) Partial incompatibility between ants and symbiotic fungi in two sympatric species of *Acromyrmex* leaf-cutting ants. Evolution 55:1980–1991

Breznak JA, Brune A (1994) Role of microorganisms in the digestion of lignocellulose by termites. Annu Rev Entomol 39:453–487

Cafaro M, Matias J, Currie CR, Cameron R (2005) Phylogenetic analysis of mutualistic filamentous bacteria associated with fungus-growing ants. Can J Microbiol 51:441–446

Chapela IH, Rehner SA, Schultz TR, Mueller UG (1994) Evolutionary history of the symbiosis between fungus-growing ants and their fungi. Science 266:1691–1694

Currie CR (2001a) A community of ants, fungi and bacteria: a multilateral approach to studying symbiosis. Annu Rev Microbiol 55:357–380

Currie CR (2001b) Weeding and grooming of pathogens in agriculture by ants. Proc R Soc Lond B 268:1033–1039

Currie RC (2001c) Prevalence and impact of a virulent parasite on a tripartite mutualism. Oecologia 128:99–106

Currie CR, Mueller UG, Malloch D (1999a) The agricultural pathology of ant fungus gardens. Proc Natl Acad Sci U S A 96:7998–8002

Currie CR, Scott JA, Summerbell RC, Malloch D (1999b) Fungus-growing ants use antibiotic-producing bacteria to control garden parasites. Nature 398:701–704

Darlington J (1994) Nutrition and evolution in fungus-growing termites. In: Hunt, JH, Nalepa, CA (eds) Nourishment and evolution in insect societies. Westview Press, Boulder, CO, pp 105–130

D'Ettorre PD, Mora P, Dibangou V, Rouland C, Errard C (2002) The role of the symbiotic fungus in the digestive metabolism of two species of fungus-growing ants. J Comp Physiol B 172: 169–176

De Fine Licht HH, Andersen A, Aanen DK (2005) *Termitomyces* sp. associated with the termite *Macrotermes natalensis* has a heterothallic mating system and multinucleate cells. Mycol Res 109:314–318

De Fine Licht HH, Boomsma JJ (2010) Forage selection in fungus-growing ants. Ecol Entomol 35:259–269

De Fine Licht HH, Boomsma JJ, Aanen DK (2006) Presumptive horizontal symbiont transmission in the fungus-growing termite *Macrotermes natalensis*. Mol Ecol 15:3131–3138

De Fine Licht HH, Schiøtt M, Mueller UG, Boomsma JJ (2010) Evolutionary transitions in enzyme activity of ant fungus gardens. Evolution 64:2055–2067

Eggleton P (2000) Global patterns of termite diversity. In: Abe T, Bignell DE, Higashi M (eds) Termites: evolution, sociality, symbioses, ecology. Kluwer Academic Publishers, Dordrecht, pp 25–51

Farrell BD, Sequeira AS, O'Meara BC et al (2001) The evolution of agriculture in beetles (Curculionidae: Scolytinae and Platypodinae). Evolution 55:2011–2027

Fletcher JA, Doebeli M (2009) A simple and general explanation for the evolution of altruism. Proc R Soc Lond B 276:13–19

Frank SA (1996) Host-symbiont conflict over the mixing of symbiotic lineages. Proc R Soc Lond B 263:339–344

Frøslev TG, Aanen DK, Læssøe T, Rosendahl S (2003) Phylogenetic relationships of *Termitomyces* and related taxa. Mycol Res 107:1277–1286

Grassé P-P, Noirot C (1955) La fondation de nouvelles sociétés par *Bellicositermes natalensis* Hav. Insect Soc 2:213–220

Green AM, Mueller UG, Adams RMM (2002) Extensive exchange of fungal cultivars between sympatric species of fungus-growing ants. Mol Ecol 11:191–195

Guedegbe HJ, Miambi E, Pando A (2009a) Molecular diversity and host specificity of termite-associated *Xylaria*. Mycologia 101:686–691

Guedegbe HJ, Miambi E, Pando A (2009b) Occurrence of fungi in combs of fungus-growing termites (Isoptera: Termitidae, Macrotermitinae). Mycol Res 113(Pt 10):1039–1045

Hart A, Anderson C, Ratnieks F (2002a) Task partitioning in leafcutting ants. Acta Ethologica 5:1–11

Hart A, Bot A, Brown M (2002b) A colony-level response to disease control in a leaf-cutting ant. Naturwissenschaften 89:275–277

Heim R (1977) Termites et champignons. Société nouvelle des Èditions Boubée, Paris

Herre EA, Knowlton N, Mueller UG, Rehner SA (1999) The evolution of mutualisms: exploring the paths between conflict and cooperation. Trends Ecol Evol 14:49–53

Hongoh Y, Ekpornprasit L, Inoue T et al (2006) Intracolony variation of bacterial gut microbiota among castes and ages in the fungus-growing termite *Macrotermes gilvus*. Mol Ecol 15: 505–516

Hyodo F, Inoue T, Azuma JI et al (2000) Role of the mutualistic fungus in lignin degradation in the fungus-growing termite *Macrotermes gilvus* (Isoptera; Macrotermitinae). Soil Biol Biochem 32:653–658

Hyodo F, Tayasu I, Inoue T et al (2003) Differential role of symbiotic fungi in lignin degradation and food provision for fungus-growing termites (Macrotermitinae: Isoptera). Funct Ecol 17:186–193

Johnson R (1981) Colony development and establishment of the fungus comb in *Microtermes* sp. nr *usambaricus* (Sjöstedt) (Isoptera: Macrotermitinae) from Nigeria. Insect Soc 28:3–12

Johnson RA, Thomas RJ, Wood TG, Swift MJ (1981) The inoculation of the fungus comb in newly founded colonies of some species of the Macrotermitinae (Isoptera) from Nigeria. J Nat Hist 15:751–756

Ju Y-M, Hsieh H-M (2007) *Xylaria* species associated with nests of *Odontotermes formosanus* in Taiwan. Mycologia 99:936–957

Katoh H, Miura T, Maekawa K et al (2002) Genetic variation of symbiotic fungi cultivated by the macrotermitine termite *Odontotermes formosanus* (Isoptera: Termitidae) in the Ryukyu Archipelago. Mol Ecol 11:1565–1572

Kent DS, Simpson JA (1992) Eusociality in the beetle *Austroplatypus incompertus* (Coleoptera: Curculionidae). Naturwissenschaften 79:86–87

Kirk PM, Cannon PF, David JC, Stalpers JA (2001) Ainsworth & Bigby's dictionary of the Fungi. CAB International, Wallingford

Kirkendall LR, Kent DS, Raffa KF (1997) Interactions among males, females and offspring in bark and ambrosia beetles: the significance of living in tunnels for the evolution of social behavior. In: Choe JC, Crespi BJ (eds) The evolution of social behavior in insects and arachnids. Cambridge University Press, Cambridge, pp 181–215

Korb J, Aanen DK (2003) The evolution of uniparental transmission of fungal symbionts in fungus-growing termites (Macrotermitinae). Behav Ecol Sociobiol 53:65–71

Kost C, Lakatos T, Böttcher I et al (2007) Non-specific association between filamentous bacteria and fungus-growing ants. Naturwissenschaften 94:821–828

Leuthold RH, Badertscher S, Imboden H (1989) The inoculation of newly formed fungus comb with *Termitomyces* in *Macrotermes* colonies (Isoptera, Macrotermitinae). Insect Soc 36: 328–338

Little AEF, Currie CR (2007) Symbiotic complexity: discovery of a fifth symbiont in the attine ant microbe symbiosis. Biol Lett 3:501–504

Little AEF, Currie CR (2008) Black yeast symbionts compromise the efficiency of antibiotic defenses in fungus-growing ants. Ecology 89:1216–1222

Mackenzie LM, Muigai AT, Osir EO et al (2007) Bacterial diversity in the intestinal tract of the fungus-cultivating termite *Macrotermes michaelseni* (Sjöstedt). Afr J Biotechnol 6: 658–667

Martin MM, Martin JS (1978) Cellulose digestion in the midgut of the fungus-growing termite *Macrotermes natalensis*: the role of acquired digestive enzymes. Science 199: 1453–1455

Mattson WJ (1980) Herbivory in relation to plant nitrogen content. Annu Rev Ecol Syst 11: 119–161

Mikheyev AS, Mueller UG, Abbot P (2006) Cryptic sex and many-to-one coevolution in the fungus-growing ant symbiosis. Proc Natl Acad Sci U S A 103:10702–10706

Mikheyev AS, Mueller UG, Abbot P (2010) Evolution of the symbiosis between attine ants and their lepiotaceous cultivars. Am Nat 175:E126–E133

Mikheyev AS, Mueller UG, Boomsma JJ (2007) Population genetic signatures of diffuse co-evolution between leaf-cutting ants and their cultivar fungi. Mol Ecol 16:209–216

Moriya S, Inoue T, Ohkuma M (2005) Fungal community analysis of fungus gardens in termite nests. Microbes Environ 20:243–252

Mueller UG, Dash D, Rabeling C, Rodrigues A (2008) Coevolution between attine ants and actinomycete bacteria: a reevaluation. Evolution 62:2894–2912

Mueller UG, Gerardo N (2002) Fungus-farming insects: multiple origins and diverse evolutionary histories. Proc Natl Acad Sci U S A 99:15247–15249

Mueller UG, Gerardo NM, Aanen DK (2005) The evolution of agriculture in insects. Annu Rev Ecol Evol Syst 36:563–595

Mueller UG, Rehner SA, Schultz TR (1998) The evolution of agriculture in ants. Science 281:2034–2038

Mueller UG, Schultz TR, Cameron RC (2001) The origin of the attine ant-fungus mutualism. Q Rev Biol 76:169–197

Munkacsi AB, Pan JJ, Villesen P (2004) Convergent coevolution in the domestication of coral mushrooms by fungus-growing ants. Proc R Soc Lond B 271:1777–1782

Nobre T, Eggleton P, Aanen DK (2010) Vertical transmission as the key to the colonization of Madagascar by fungus-growing termites? Proc R Soc Lond B 277:359–365

O'Fallon B, Hansen T (2008) Population structure, levels of selection, and the evolution of intracellular symbionts. Evolution 62:361–373

Ohkuma M (2003) Termite symbiotic systems: efficient bio-recycling of lignocellulose. Appl Microbiol Biotechnol 61:1–9

Paul J, Saxena S, Varma A (1993) Ultrastructural studies of the termite (*Odontotermes obesus*) gut microflora and its cellulolytic properties. World J Microbiol Biotechnol 9:108–112

Pinto-Tomas AA, Anderson MA, Suen G et al (2009) Symbiotic nitrogen fixation in the fungus gardens of leaf-cutter ants. Science 326:1120–1123

Poulsen M, Boomsma JJ (2005) Mutualistic fungi control crop diversity in fungus-growing ants. Science 307:741–744

Poulsen M, Fernández-Marín H, Currie CR, Boomsma JJ (2009) Ephemeral windows of opportunity for horizontal transmission of fungal symbionts in leaf-cutting ants. Evolution 63:2235–2247

Reynolds HT, Currie CR (2004) Pathogenicity of *Escovopsis weberi*: the parasite of the attine ant-microbe symbiosis directly consumes the ant-cultivated fungus. Mycologia 96:955–959

Richard F-J, Mora P, Errard C, Rouland C (2005) Digestive capacities of leaf-cutting ants and the contribution of their fungal cultivar to the degradation of plant material. J Comp Physiol B 175:297–303

Rogers JD, Ju Y-M, Lehmann J (2005) Some *Xylaria* species on termite nests. Mycologia 97: 914–923

Rosengaus RB, Maxmen AB, Coates LE, Traniello JFA (1998) Disease resistance: a benefit of sociality in the dampwood termite *Zootermopsis angusticollis* (Isoptera: Termopsidae). Behav Ecol Sociobiol 44:125–134

Rouland-Lefèvre C (2000) Symbiosis with fungi. In: Abe T, Bignell DE, Higashi M (eds) Termites: evolution, sociality, symbioses, ecology. Kluwer Academic Publishers, Dordrecht, pp 289–306

Rouland-Lefèvre C, Bignell DE (2001) Cultivation of symbiotic fungi by termites of the subfamily Macrotermitinae. In: Seckbach J (ed) Symbiosis. Kluwer Academic Publishers, Dordrecht, pp 731–756

Rouland-Lefèvre C, Inoue T, Johjima T (2006) *Termitomyces*/termite interactions. In: König, H, Varma, A (eds) Intestinal microorganisms of termites and other invertebrates. Springer, Berlin, pp 335–350

Rouland-Lefèvre C, Lenoir F, Lepage M (1991) The role of the symbiotic fungus in the digestive metabolism of several species of fungus-growing termites. Comp Biochem Physiol 99A: 657–663

Scharf M, Wu-Scharf D, Pittendrigh B, Bennett G (2003) Caste- and development-associated gene expression in a lower termite. Genome Biol 4:R62

Scharf ME, Wu-Scharf D, Zhou X et al (2005) Gene expression profiles among immature and adult reproductive castes of the termite *Reticulitermes flavipes*. Insect Mol Biol 14:31–44

Schiött M, De Fine Licht H, Lange L, Boomsma J (2008) Towards a molecular understanding of symbiont function: identification of a fungal gene for the degradation of xylan in the fungus gardens of leaf-cutting ants. BMC Microbiol 8:40

Schultz TR, Brady SG (2008) Major evolutionary transitions in ant agriculture. Proc Natl Acad Sci U S A 105:5435–5440

Sen R, Ishak HD, Estrada D et al (2009) Generalized antifungal activity and 454-screening of *Pseudonocardia* and *Amycolatopsis* bacteria in nests of fungus-growing ants. Proc Natl Acad Sci U S A 106:17805–17810

Shinzato N, Muramatsu M, Matsui T, Watanabe Y (2007) Phylogenetic analysis of the gut bacterial microflora of the fungus-growing termite *Odontotermes formosanus*. Biosci Biotechnol Biochem 71:906–915

Shinzato N, Muramatsu M, Watanabe Y, Matsui T (2005) Termite-regulated fungal monoculture in fungus combs of a macrotermitine termite *Odontotermes formosanus*. Zool Sci 22: 917–922

Slaytor M (1992) Cellulose digestion in termites and cockroaches: what role do symbionts play? Comp Biochem Physiol 103B:775–784

Taylor PD, Frank SA (1996) How to make a kin selection model. J Theor Biol 180:27–37

Thomas RJ (1987a) Distribution of *Termitomyces* and other fungi in the nests and major workers of several Nigerian Macrotermitinae. Soil Biol Biochem 19:335–341

Thomas RJ (1987b) Distribution of *Termitomyces* Heim and other fungi in the nests and major workers of *Macrotermes bellicosus* (Smeathman) in Nigeria. Soil Biol Biochem 19:29–333

Thomas RJ (1987c) Distribution of *Termitomyces* Heim and other fungi in the nests and major workers of several Nigerian Macrotermitinae. Soil Biol Biochem 19:335–341

Thomas RJ (1987d) Factors affecting the distribution and activity of fungi in the nests of Macrotermitinae (Isoptera). Soil Biol Biochem 19:343–349

Todaka N, Moriya S, Saita K et al (2007) Environmental cDNA analysis of the genes involved in lignocellulose digestion in the symbiotic protist community of *Reticulitermes speratus*. FEMS Microbiol Ecol 59:592–599

Veivers PC, Mühlemann R, Slaytor M et al (1991) Digestion, diet and polyethism in two fungus-growing termites: *Macrotermes subhyalinus* Rambur and *M. michaelseni* Sjøstedt. J Insect Physiol 37:675–682

Visser AA, Ros V, Beer ZWD et al (2009) Levels of specificity of *Xylaria* species associated with fungus-growing termites: a phylogenetic approach. Mol Ecol 18:553–567

Vo TL, Mueller UG, Mikheyev AS (2009) Free-living fungal symbionts (Lepiotaceae) of fungus-growing ants (Attini: Formicidae). Mycologia 101:206–210

Walker TN, Hughes WOH (2009) Adaptive social immunity in leaf-cutting ants. Biol Lett 5: 446–448

Warnecke F, Luginbuhl P, Ivanova N et al (2007) Metagenomic and functional analysis of hindgut microbiota of a wood-feeding higher termite. Nature 450:560–565

Weber NA (1972) Gardening ants: the attines. American Philosophical Society, Philadelphia, PA

Wetterer JK (1994) Nourishment and evolution in fungus-growing ants and their fungi. In: Hunt, JH, Nalepa, CA (eds) Nourishment and evolution in insect societies. Westview Press, Boulder, CO, pp 309–328

Wilkinson DM (2001) At cross purposes. Nature 412:485–485

Wilson EO (1980) Caste and division of labor in leaf-cutter ants (Hymenoptera: Formicidae: Atta): I. The overall pattern in *A. sexdens*. Behav Ecol Sociobiol 7:143–156

Wood TG, Sands WA (1978) The role of termites in ecosystems. In: Brian MV (ed) Production ecology of ants and termites. Cambridge University Press, Cambridge, UK, pp 245–292

Yanagawa A, Yokohari F, Shimizu S (2008) Defense mechanism of the termite, *Coptotermes formosanus* Shiraki, to entomopathogenic fungi. J Invertebr Pathol 97:165–170

Yara K, Jahana K, Hayashi H (1989) In situ morphology of the gut microbiota of the fungus-growing termite *Odontotermes formosanus* (Termitidae: Macrotermitinae). Sociobiology 15:2247–2260

Zhou X, Smith JA, Oi FM (2007) Correlation of cellulase gene expression and cellulolytic activity throughout the gut of the termite *Reticulitermes flavipes*. Gene 395:29–39

Chapter 9
Molecular Basis Underlying Caste Differentiation in Termites

Toru Miura and Michael E. Scharf

Abstract Termite caste differentiation is a multifaceted process that is under the control of a range of intrinsic and extrinsic factors, and it has challenged researchers for decades. Advances in molecular, genomic, and integrative or "systems" biology in the past decade have greatly facilitated efforts to begin to understand this process. Using molecular tools, it is now possible to investigate caste differentiation through hypothesis-driven mechanistic studies at the sub- and super-organismal levels. This chapter provides examples of approaches to study the molecular bases of caste differentiation; it describes molecular biology approaches for gene and protein discovery and characterization, relevant genes and proteins that have been identified, and a contextual foundation on which genome sequencing can now be considered. Once whole genome sequences of termites are available it will be possible to conduct highly detailed comparative, integrative, functional and translational genomics studies that define: (1) the complex milieu of intrinsic and extrinsic factors that interact to drive caste differentiation (including genetic and environmental factors), (2) how genes used in solitary life have been co-opted for social functions, and (3) social evolution in termites and their ancestors.

9.1 Introduction

In the last decade, social insect research has been transformed by molecular biology and genomics. Although fewer studies have focused on termites than social hymenopterans, we can now review advances in our understanding of termite castes and caste differentiation as the field has moved from "observational" behavioral ecology research (organism level) to basic mechanistic studies of gene function (molecular level). Further, the search for caste-regulatory genes has unexpectedly led to the identification of many physiological genes linked to termite sociality (e.g. nutrition and endocrinology).

T. Miura (✉)
Laboratory of Ecological Genetics, Graduate School of Environmental Science, Hokkaido University, Sapporo 060-0810, Japan
e-mail: miu@ess.hokudai.ac.jp

D.E. Bignell et al. (eds.), *Biology of Termites: A Modern Synthesis*,
DOI 10.1007/978-90-481-3977-4_9,

Genomics research also provides an excellent platform for understanding the innovations used by termites to build upon pre-existing social systems acquired from their solitary wood roach ancestors. Although the molecular bases of caste differentiation and sociality are still not fully understood, a roadmap for future genomics and integrative functional genomics research can be provided.

9.2 A Historical View of Classic Work on Caste Determination and Differentiation

9.2.1 Castes in Termites

Historically, termites and ants have often been considered analogous because they share some characteristics in morphology and social organization, despite being phylogenetically distant (Wilson 1971). However, an important difference is seen in their postembryonic development: termites are hemimetabolous, while ants are holometabolous. Furthermore, although alate termites show essentially "normal" hemimetabolous insect development, juveniles and other castes (such as workers, nymphs, and pseudergates) can have unpredictable and unique molting patterns, such as stationary/regressive, reproductive, and soldier molts. Supplemental or replacement reproductives also sometimes emerge, but with less predictability. Supplemental reproductives (also referred to as "neotenics") can function as queens or kings, but often lack imaginal characteristics like wings or well-developed compound eyes (Weesner 1969; Wilson 1971; Thorne 1996). Soldier differentiation is the most specific developmental process seen in termites. Some ants also produce caste phenotypes specialized in defense but they are categorized as major workers in most cases, and are essentially larger versions of normal workers (Hölldobler and Wilson 1990). Thus, the worker/pseudergate, soldier and neotenic pathways are unique to termites and are a large component of termite sociality.

9.2.2 Division of Labor and Caste Developmental Patterns

Noirot (1989) proposed that termite social organization is deeply related to the developmental regulation of caste differentiation. In termite colonies, tasks are allocated between colony members. These include nursing, nest construction, sanitation, foraging and defense. Task allocation, or division of labor, is accomplished through the production of morphological castes and/or behavior-based age polyethism (Stuart 1969; Wilson 1975; Traniello and Leuthold 2000) and is strongly associated with caste developmental pathways in which specific phenotypes differentiate according to a combination of sex, developmental stage (e.g. Miura et al. 1998), environmental, social and other factors (e.g. Zhou et al. 2007b; Scharf et al. 2007). Because of their hemimetabolous development, termites pass through successive larval instars with progressive growth and little morphological change.

Immature termites possess similar body patterns to adults, especially with respect to legs and mouthparts (Noirot 1989, 1991; Roisin 2000; Miura 2005). Therefore, termite immatures are quite mobile and can accomplish tasks as readily as mature individuals.

There are diverse patterns of caste differentiation among the seven termite families (Miller 1969; Noirot 1969a; Noirot and Pasteels 1987; Roisin 2000). The pathways of the seven extant termite families can be separated into linear and forked (or bifurcated) pathways (Noirot and Pasteels 1987; Roisin 2000) and the underlying physiological mechanisms that control development should therefore also differ among termite lineages. Most of our pre-genomic knowledge of the physiology of caste determination is, however, limited to lower termites (eg, Lüscher 1960; Miller 1969; Watson and Sewell 1985; Nijhout 1994).

In termite taxa in which individual reproductive development is repressed (e.g. worker, nymph, soldier), immatures can behave as helpers with future reproductive potential, especially in lower termites (Myles and Nutting 1988; Shellman-Reeve 1997; Roisin 2000; Korb and Hartfelder 2008). Thus, although there are still some arguments against it (e.g. Thompson et al. 2000), developmental repression can be seen as a driving force in termite sociality and the maintenance of caste homeostasis (see Section 9.7.2 below).

9.2.3 Classical Work on Caste Differentiation

Early work on termite caste differentiation concentrated on observing changes in morphological characters, and to a degree, behavioral ecology (reviewed in Miller 1969; Noirot 1969a, 1985; Watson and Sewell 1985; Roisin 2000). The most extensive physiological work was that of Martin Lüscher, using *Kalotermes flavicollis*, where the developmental trajectories of individually-marked termites were followed to define caste developmental pathways (Lüscher 1953, 1961). In addition, he demonstrated that the implantation of corpora allata from reproductives could induce soldier development (Lüscher 1958).

Based on these studies, Noirot (1989, 1991) hypothesized that differential gene expression may occur in response to various extrinsic cues, like social interactions and physical environmental conditions (e.g. temperature, day length, etc.). Although there is evidence suggesting genetic caste determination in some termites (e.g. Goodisman and Crozier 2003; Hayashi et al. 2007), these findings may not apply under field conditions and should be cautiously interpreted (Vargo and Husseneder 2009). Also, in some termite lineages, for example in Termitidae, neuter castes like soldiers arise only from one sex (Noirot 1969a). This indicates some involvement of genetic factors in caste determination, however genetic factors are thought to be affected by environmental factors and much evidence suggests that social environments, together with physical environments, are the more broadly important factors underlying termite caste determination.

Termites and termite societies are thought to have evolved from subsocial ancestral woodroaches (Lo et al. 2000; Nalepa and Bandi 2000). Pre-existing genes responsible for developmental regulation, metabolism and communication in solitary insects would have been recruited to new functions in sociality (West-Eberhard 1987, 1996). Therefore, molecular and genetic data should provide us with deeper insights of the evolutionary and developmental/physiological mechanisms of caste differentiation. The concept of co-opting solitary genes for social functions (West-Eberhard 1987, 2003) rationalises molecular investigations of termite caste differentiation.

9.3 Screening of Genes Responsible for Caste Differentiation: Gene Discovery and Genomics

9.3.1 Molecular and Genomic Approaches to Study Termite Caste Differentiation

Genomics can be defined as the study of the sequence and structure of multiple genes simultaneously. High throughput "shotgun style" genomics can be conducted at the genome level (Adams et al. 2000; Holt et al. 2002; Honeybee Genome Sequencing Consortium 2006; Nene et al. 2007; Cameron and Whiting 2007), at the mRNA/transcriptome level (Whitfield et al. 2002; Sabater-Muñoz et al. 2006; Todaka et al. 2007; Warnecke et al. 2007; Tartar et al. 2009), or specific gene families can be targeted with more directed approaches (Snyder et al. 1996; Byrne et al. 1999; Tokuda et al. 2002; Liénard et al. 2006).

Almost 20 years ago, termite scientists began to hypothesize that there must be differential gene expression among castes, or during the course of caste differentiation (e.g. Noirot 1991), but no investigations were reported until the late 1990s (Miura et al. 1999). Several examples of the identification of termite genes are described below.

9.3.2 Differential Display

The first approach used for comparing gene expression between different castes was mRNA differential display (Miura et al. 1999). A gene specifically expressed in soldiers was identified and termed *SOL1* (see details below). Differential display is an easy and convenient technique that uses PCR amplification with arbitrary primers to compare cDNAs which are reverse-transcribed from mRNAs (Liang and Pardee 1992; Liang et al. 1993 Fig. 9.1A). However, before improved protocols became available (e.g. Zegzouti et al. 1997), a major problem with differential display was that false positives were frequently encountered. Another issue is that confirmation of differential expression profiles requires secondary verification; for example by northern blotting or real-time quantitative PCR (e.g. Miura et al. 1999; Koshikawa et al. 2005; Hojo et al. 2005a, b, 2007; Cornette et al. 2006).

In comparisons between soldier and worker heads in the dampwood termite *Hodotermopsis sjostedti* (previously *H. japonica*), differentially expressed transcripts were estimated at nearly 1% of total transcript abundance (Miura et al.

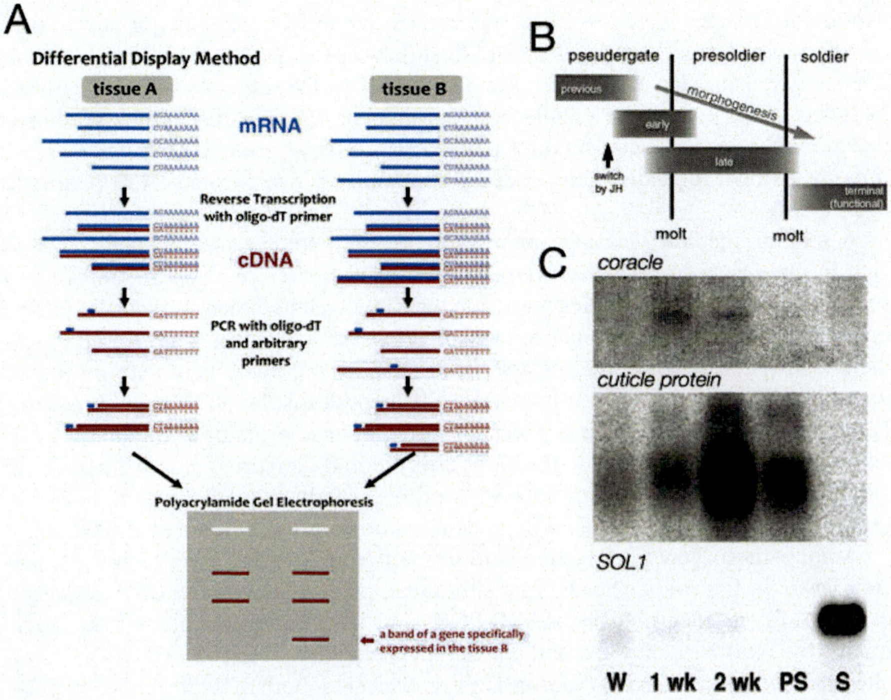

Fig. 9.1 Brief schema of differential display and identified genes specific to the soldier differentiation. (**a**) Differential display compares transcripts from different tissues by isolating mRNA, followed by reverse transcription, PCR with arbitrary primers and polyacrylamide gel electrophoresis. (**b**) In the identification of genes that are expressed during soldier caste differentiation, some genes are pseudergate- or soldier-specific, while others are up-regulated during the differentiation process. (**c**) Under such working hypothesis, genes expressed during and after the soldier differentiation were identified. Expression patterns detected by northern hybridization are shown

1999). Alternatively, in the case of a nasute termite *Nasutitermes takasagoensis*, the percentage of differentially expressed head transcripts between soldiers and workers was about 7% (Hojo et al. 2005a). This discrepancy probably results from the nasute soldier's specialized frontal glands that produce defensive substances. Therefore, nasute soldiers apparently require a diversity of "soldier-specific" genes for producing defensive secretions (see details below).

Although the above-mentioned differences in gene expression were determined from comparisons of workers and terminally-differentiated soldiers, in subsequent studies, gene expression profiles during the course of soldier differentiation were also examined in *H. sjostedti* (Miura 2004, Koshikawa et al. 2005; Cornette et al. 2006, Fig. 9.1b, c). In studies of termite soldier differentiation, the artificial induction of soldier differentiation (i.e. induction of the presoldier molt) by the application of juvenile hormone analogue (JHA) (Ogino et al. 1993) is utilized to induce and synchronize development. The presoldier differentiation process usually takes about 14 days after JHA application until the presoldier molt. Many gene expression changes were identified between days 7 and 14, when major physiological and morphogenetic changes occur (Koshikawa et al. 2003, 2005; Cornette et al. 2006, 2007; Ishikawa et al. 2008; see also Fig. 9.4). Genes that were up-regulated included genes coding for cuticle proteins, nucleic acid-binding proteins, ribosomal proteins, and an actin-binding protein (Koshikawa et al. 2005); whereas, an important down-regulated gene coded for a family 6 cytochrome P450 (Cornette et al. 2006).

A similar method was also applied to identify soldier-specific gene expression in the nasute termite *Nasutitermes takasagoensis* (Hojo et al. 2005a). In this study, gene expression profiles were compared between soldier and worker heads, because the frontal projections from which defensive substances are released in soldiers (Prestwich 1984) contain glandular tissues, where many genes are presumed to be specifically expressed. Fluorescent differential display (FDD) in *N. takasagoensis* identified 21 candidate genes encoding diverse proteins with homology to eukaryotic initiation factor 4A-I, a 3β-hydroxysteroid dehydrogenase/isomerase, an alkaline phosphatase homolog, a GTP-binding protein, ribosomal protein S13, a putative family 4 cytochrome P450, and an unknown protein (Hojo et al. 2005a).

Among these genes, one candidate termed *Nts19-1* shows very high expression levels in the soldier heads. In a subsequent study, the whole cDNA sequence of *Nts19-1* was identified by RACE-PCR, and its expression pattern was determined by Northern blotting and in situ hybridization (Hojo et al. 2007). Two alternatively-spliced *Nts19-1* isoforms were identified, both of which encoded putative homologues of the geranylgeranyl diphosphate (GGPP) synthase gene. GGPP synthase plays a central role in the mevalonate/terpenoid biosynthesis pathway (Bellés et al. 2005). In situ hybridization verified that this gene is expressed in frontal gland tissue. Considering that the soldiers discharge diterpenes (Prestwich 1984), this gene is thought to be involved in the synthesis of defensive secretions. Another hypothesis, however, is that soldier secretions serve caste regulatory or primer pheromone functions (Tarver et al. 2009; also see Section 9.7.1 below). In Rhinotermitidae and Termitidae, many species produce terpenoids. Thus, future analyses comparing terpene biosynthetic genes across species should reveal the evolutionary process behind such elaborate defensive and/or caste differentiation strategies.

9.3.3 Polyphenic Library and Arrays

Macro-arrays of cDNA were used with *R. flavipes* to search for genes involved in caste differentiation and caste regulation/homeostasis. The *Reticulitermes flavipes* arrays contained 18,000 positions and were printed from a polyphenic cDNA library representing all castes and phenotypes except primary reproductives (Wu-Scharf et al. 2003). Subsequently, when the arrays were probed with labeled mRNAs from various castes and developmental stages, this facilitated the direct comparison of gene expression among the castes and developmental stages. Using this approach, numerous genes were identified as differentially expressed between

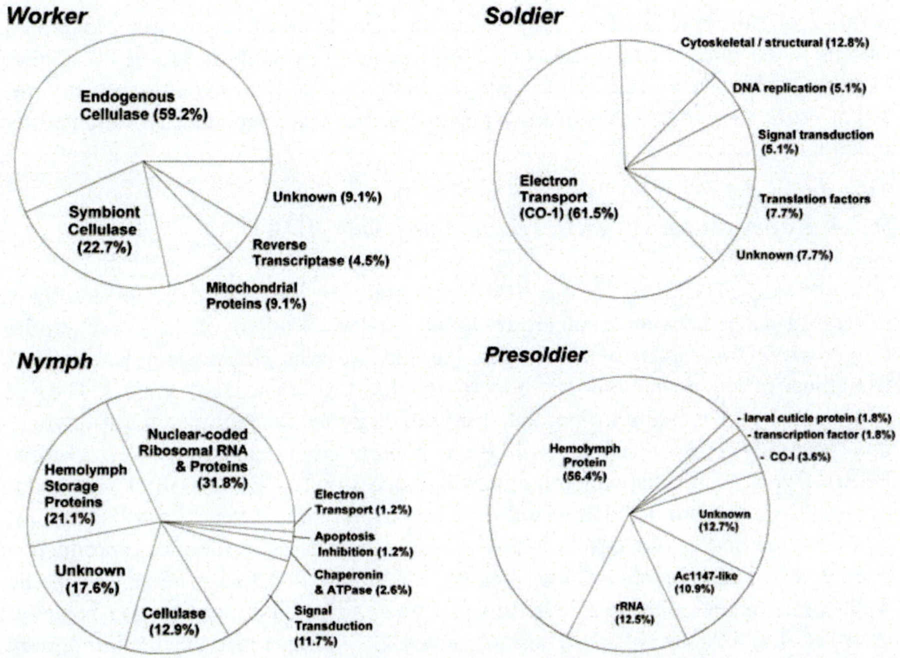

Fig. 9.2 Comparison of *R. flavipes* EST sequences from directed array screens showing sequence compositions for differentially expressed genes among castes and developmental stages. Over 400 individual ESTs are represented. Shown are categories of EST sequences identified from workers (Scharf et al. 2003b), soldiers (Scharf et al. 2003b), nymphs (Scharf et al. 2005b), and presoldiers (unpublished)

workers, soldiers (Scharf et al. 2003b), nymphs (Scharf et al. 2005b) and presoldiers (Scharf et al. unpublished observations; Zhou et al. 2006c) (Fig. 9.2). Over 50 genes were identified from this approach, several of which proved to be relevant to both caste regulation and nutrition; for example hexamerins (see below) and cellulases (Wheeler et al. 2007; Zhou et al. 2007a, 2008).

Four of the differentially expressed genes encoded hemolymph proteins (two vitellogenins and two hexamerins) with potential roles in hormone or primer pheromone signaling (see below). The hexamerin genes, named *hexamerin (Hex) 1* and *2*, were identified from nymph arrays (predominantly *Hex 1*) and presoldier arrays (predominantly *Hex 2*). Significantly, the two hexamerin genes are part of the first apparent caste regulatory mechanisms to be identified from a termite (see Section 9.4.3 below). In addition to hemolymph proteins, several other potentially important categories of genes were identified from nymph and worker array screens, including cellulases (both endogenous and symbiotic), mitochondrial/electronic transport proteins, ribosomal RNAs, possible pseudo-rRNAs, riboproteins, and signal transduction factors (Scharf et al. 2003b, 2005b). Presoldier arrays also led to the identification of hexamerin genes, as well as a larval cuticle protein and a member of the COP-9 signalsome complex involved in protein degradation (Zhou et al. 2007b). From soldier arrays, distinct categories of genes were identified that included many genes associated with increased muscular mass of the soldier head (mitochondrial electron transport components and cytoskeletal/structural proteins). With this approach, all array-positive genes were secondarily validated by quantitative real-time PCR.

9.3.4 Representation Difference Analysis (RDA)

Weil et al. (2007) used the RDA method to identify five differentially expressed genes between neotenic reproductives and workers of the lower termite *Cryptotermes secundus*. With the RDA method, worker cDNAs were "subtracted" from those of female neotenic reproductives before cloning and sequencing. The five genes identified as being up-regulated in female neotenics included the following: one carboxylesterase, one β-glucosidase, one vitellogenin, one family 4 cytochrome P450 ("Cyp4"), and an apparent transferrin (Accession No. EF029058). Using quantitative PCR, the authors also examined expression of these genes in body regions, and in whole bodies of workers relative to a range of male and female reproductive phenotypes. Later functional analyses by RNAi revealed that a neotenic-specific β-glucosidase gene plays a role in apparent neotenic regulation of worker behavior (Korb et al. 2009; see below). These *C. secundus* findings represented an important new direction for termite genomic caste differentiation research.

9.3.5 Identification of Cytochrome P450 Genes

Cytochrome P450 (P450) is a ubiquitous heme-containing oxidative enzyme found in all cellular organisms. In insects, P450s play diverse roles in the oxidation of endogenous substrates such as hormones and pheromones, as well as xenobiotic substrates that include plant secondary compounds and insecticides (Feyereisen 2005). The P450 gene family is extremely diverse and well-represented in insects for example, there are 83 P450s in *Drosophila* (Tijet et al. 2001), 111 in *Anopheles* (Ranson et al. 2002), and 46 in the honeybee (Claudianos et al. 2006). Several insect P450s have also been linked to juvenile hormone (JH) biosynthesis or degradation (Andersen et al. 1997; Sutherland et al. 2000; Helvig et al. 2004; Feyereisen 2005). Thus, because of the roles of JH in termite caste differentiation and related terpenes in defense and possibly caste-regulation, P450 would be expected to play some role in termite caste differentiation and caste regulation/homeostasis.

The P450 family #4 (*Cyp4*) is numerically the most highly represented insect P450 subfamily (Feyereisen 2005). Due to two highly conserved sequence motifs, family 4 can be easily PCR-amplified using degenerate primers targeted to the *i-helix* and *heme-binding* motifs (Snyder et al. 1996). This degenerate PCR approach allows gene fragments of about ~450 nucleotide bases to be amplified from cDNA templates (potentially larger if using genomic DNA templates because of intronic sequences). The first effort in termites to use this approach for P450 gene discovery was in *Mastotermes darwiniensis*; it resulted in identification and description of the first termite P450 (Cyp4C8) (Falckh et al. 1997). In apparently related but unpublished work, four additional *M. darwiniensis* P450s (3 = Cyp4; 1 = Cyp6) and five from *Coptotermes acinaciformis* (Cyp4) were sequenced and deposited in Genbank (www.ncbi.nlm.nih.gov). To date, none of these 9 Australian termite P450s have been linked to any physiological processes.

The same degenerate PCR approach was employed nearly a decade later in *R. flavipes* (Zhou et al. 2006b). By combining degenerate PCR with high-throughput sequencing of 96 clones, this study identified 7 new termite CYP4 genes and 1 processed pseudogene. These genes were from the families CYP4U and 4C. Interestingly, allelic variation was observed for 4 of the genes, and introns were identified for one processed gene and for the pseudogene. Using quantitative real-time PCR, the tissue distribution of expression was determined for the seven functional CYP4s, as well as temporal expression changes in workers released from colony influences, and during JH-induced soldier caste differentiation. One P450 had head-specific expression (*CYP4C46*), suggesting it may be implicated in corpora allata or antennal function. Several fat body-related CYP4 genes were differentially expressed in workers after JH treatment (*CYP4U3*, *4C46*, *4C47* and *4C48*), and a different complement changed expression in workers in association with removal from colony influences (*CYP4U3*, *4C45* and *4C46*). This latter result suggested that primer pheromones and/or other colony influences can directly impact P450 expression. However, no specific roles of any of the *R. flavipes* P450s have yet been determined in caste differentiation.

By screening gene expression over the course of soldier differentiation in *Hodotermopsis sjostedti* by differential display, a family-6 P450 categorized as *CYP6AM1* was identified (Cornette et al. 2006). Northern blotting analysis showed that the transcript of this gene was repressed in response to artificial JHA application. Further experimental analyses also indicated that *CYP6AM1* is specifically expressed in pseudergates/workers and soldiers and not in other castes. The high expression level was shown to be restricted in fat bodies. These patterns of *CYP6AM1* expression support the hypothesis that it plays a role in basic metabolism, which can be related to caste regulation or developmental suppression.

9.3.6 *Cytochrome c Oxidase Subunit III:* Reticulitermes santonensis *and* flavipes

Liénard et al. (2006) took a candidate gene approach and investigated differential expression of the mitochondrial gene cytochrome oxidase III (Cox III) in *R. santonensis*, which is the European synonym of *R. flavipes*. The authors chose Cox III because of its affiliation with neuronal development in other insects. Results from quantitative real-time PCR revealed differential Cox III gene expression among workers, larvae, nymphs and terminally-developed soldiers, with lowest levels in the soldiers. However, further studies in *R. flavipes* investigating Cox III expression in workers during JH III-induced presoldier differentiation, as well as in semiochemical and live soldier treatments, did not reveal any significant expression changes (Tarver et al. in preparation). Finally, another mitochondrial gene, cytochrome oxidase I, was compared across workers, soldiers, presoldiers and nymphs in *R. flavipes* (Scharf et al. 2003b; also see Fig. 9.2). In contrast to Cox III results, only Cox I was found to have significantly elevated expression in soldiers.

9.3.7 Shotgun Library Sequencing

The termite gut is highly interesting because of the ability of many termites to digest lignocellulose (Scharf and Tartar 2008) and the potential for flow of hormones, pheromones and semiochemicals to nestmates through trophallaxis (Wilson 1971). A small number of studies have focused on symbiont transcriptome/metagenome sequencing from termite guts (Todaka et al. 2007; Warnecke et al. 2007; Hongo et al. 2008), but none has reported caste-regulatory genes. However, a recent sequencing study that focused equally on the termite gut and symbiont transcriptomes offered clarification of lignocellulose digestion, hormonal processing, and potentially caste regulation (Scharf and Tartar 2008; Tartar et al. 2009). This work identified previously unknown, host-derived genes that apparently contribute to lignin degradation, the products providing protection from toxicity, as well as a non-overlapping complement of host and symbiont glycohydrolase genes that participate in α-carbohydrate, cellulose, hemicellulose, and chitin metabolism. These results revive previous evidence suggesting termite-derived lignin degradation and complementary host-symbiont collaboration in lignocellulose digestion (reviewed by Slaytor 2000).

With respect to hormonal processing and caste regulation, the *R. flavipes* gut library sequencing also revealed a number of endogenous (termite-derived) genes that encode enzymes that participate in JH biosynthesis and degradation (Tartar et al. 2009; Tarver et al. 2010). JH biosynthetic genes included several enzymes of the mevalonic acid pathway, including HMG Co-A reductase, two mevalonate kinases, and two farnesoic acid methyltransferases. Two cytochrome P450 genes (*CYP15A1, 15F1*) were also identified with homology to P450s that play roles in the final step in JH biosynthesis: *methyl farnesoate epoxidation* (Feyereisen 2005). Finally other endogenous gut genes were identified with strong homology to insect genes that participate in JH degradation. These JH degradation genes include P450s, JH esterases and epoxide hydrolases. As postulated by Henderson (1998), many of these genes could play important caste regulatory roles by modulating JH/semiochemical/terpene biosynthesis and availability (Tarver et al. 2010). Similar gut expression of mevalonic pathway and P450 genes has been identified in association with pheromone production in bark beetles (Keeling et al. 2004, 2006), and it is highly interesting that such genes are also associated with the termite gut.

9.4 Investigations of Gene Functions in Termites: Functional Genomics

Functional genomics is an approach that seeks to assign function to genes (and corresponding proteins) identified through genomics/gene discovery and proteomics research, as well as to identify gene-gene interactions. An overview of functional genomics research and procedures executed in termites is provided below.

9.4.1 Gene Expression Changes in Response to Experimental Treatments

9.4.1.1 Experimental Treatments

Once genes have been identified and sequenced, functional genomics research can proceed by characterizing gene expression changes in response to experimental treatments. Experimental approaches to gene expression profiling include quantitative real-time PCR, semi-quantitative PCR, Northern blotting, oligonucleotide macro-arrays, microarrays, and most recently, pyrosequencing. All of these approaches, except microarrays and pyrosequencing, have been employed in studying termite gene expression. While microarrays have not yet been developed for termites, they have addressed sociogenomics questions in other social insects (Whitfield et al. 2003; Grozinger et al. 2003; Goodisman et al. 2005).

When quantifying gene expression, the identification and use of reference genes that do not change expression in the experimental treatments is critical (Pfaffl et al. 2004; Scharf et al. 2008). Examples of such reference genes in termite functional genomics research include *β-actin*, *NADH-dh*, and *HSP-70* in *R. flavipes* (Zhou et al. 2006b; Scharf et al. 2008), *actin* in *R. santonensis/flavipes* (Liénard et al. 2006) and *18S rRNA* in *H. sjostedti* (Koshikawa et al. 2005; Cornette et al. 2006), *N. takasagoensis* (Hojo et al. 2005a, 2007) and *C. secundus* (Weil et al. 2007).

By tracking gene expression in association with treatments that induce caste differentiation, several studies have now identified genes linked to caste differentiation. For example, gene expression changes in *R. flavipes* in response to JH treatments have been noted for the hemolymph protein-coding genes *hexamerin 1* and *2* and *vitellogenin 1* and *2* (Scharf et al. 2005a). A P450 gene identified from *H. sjostedti* (Cyp6AM1) is expressed in fat body, upregulated in soldiers, and down-regulated in workers during juvenoid (pyriproxyfen)-induced presoldier differentiation (Cornette et al. 2006). Cyp6AM1 expression is also down-regulated or absent in worker midgut and salivary gland, presoldiers, nymphs, alates and neotenic reproductives (Cornette et al. 2006). Also, the responses of several *R. flavipes* Cyp4 P450 genes to colony release and JH-treatment were investigated by quantitative real-time PCR (Zhou et al. 2006b). Differential responses among these 7 Cyp4 genes were identified in colony-release controls, as well as JH treatments.

Results of independent studies on termite P450 genes reveal solid links between P450 genes and JH-dependent worker-to-presoldier differentiation. Hemolymph proteins are also linked to JH action and termite caste differentiation. In *R. flavipes* other downstream genes have been linked pleiotropically to major effect hemolymph proteins.

9.4.1.2 Pleiotropic Effects

In very broad genetic terms, pleiotropy occurs when one genetic locus has an influence over another. Thus, pleiotropy can be the influence of one gene on the expression of another gene or gene network, or when that gene's product has multiple functions. *R. flavipes* hexamerins provide a seminal example of termite genes that exert pleiotropic effects on other genes: in this case by binding JH and apparently sequestering it (Zhou et al. 2006a, c), hexamerin proteins can limit JH impacts on downstream networks of JH-responsive genes (Zhou et al. 2007b). This conclusion was made possible by a 3-tiered approach that combined RNA interference (Section 9.4.3), JH bioassays, and gene expression profiling. By comparing gene expression profiles between hexamerin-silenced and JH treatments, Zhou et al. (2007b) observed a high degree of correlation. While this phenomenon has been documented only once to date, it could represent a potentially important theme in termite caste differentiation and caste regulation.

9.4.2 Controlling for Colony and Environmental Effects

9.4.2.1 Colony Release Effects

Because of social and physiological coordination among nestmates, additional experimental design considerations are necessary in termites beyond those normally used in functional genomics studies on non-social insects. It is necessary to control for colony-release effects on (i) hormone titers and (ii) gene expression (e.g., Zhou et al. 2006b; 2007b). Regarding hormone titers, Okot-Kotber et al. (1993) observed that, upon removing *R. flavipes* workers from colonies, JH and ecdysteroid titers rose significantly within 1 day and peaked at 7–14 days. A similar 1-day effect on JH biosynthesis was noted for isolated groups of *R. flavipes* workers (Elliott and Stay 2008). Changes in JH-responsive gene expression have been noted after colony release. Three examples of "colony release" responses from *R. flavipes* are with hexamerin genes, Cyp4 cytochrome P450s, and a number of JH-responsive genes that act downstream of the hexamerins as part of an apparent pleiotropic network (Scharf et al. 2005a; Zhou et al. 2006b, 2007b). In addition to helping to elucidate the roles of genes in caste differentiation (see below), these findings illustrate the importance of having proper controls in functional genomics research on isolated groups of termites, particularly when attempting to characterize gene expression profiles or define mechanisms through RNA interference studies. In both kinds of studies, relevant controls include: (i) colony individuals (to reveal baseline gene expression levels at "time-0" or later), and (ii) untreated/sham controls that move through time with experimental treatments (Zhou et al. 2007b; Scharf et al. 2008).

9.4.2.2 Environmental Effects

The environmental conditions under which functional genomic assays are conducted also require careful consideration. In particular, temperature and seasonality are known to have significant influences on caste composition (e.g. Haverty and Howard 1981; Howard and Haverty 1981). In *R. flavipes*, JH-dependent presoldier differentiation from workers is profoundly impacted by temperature; no responses occur at 22°C, while significant presoldier/soldier differentiation occurs at 27°C (Scharf et al. 2007). Similar seasonal and temperature-dependent effects have been noted with *C. formosanus* soldier proportions (Haverty 1977), soldier differentiation (Waller and LaFage 1988, Henderson 1998, Fei and Henderson 2002) and JH titers (Liu et al. 2005a, b). Thus, since caste differentiation seems to be a seasonal, temperature-dependent, "ecological-developmental" (Gilbert 2001) phenomenon, building an understanding of the molecular basis of caste differentiation will require study of candidate genes under controlled environmental conditions.

9.4.3 RNA Interference

9.4.3.1 RNAi in Termites

Targeted gene silencing by RNA interference (RNAi) is an extremely powerful functional genomics research approach for studying the molecular basis of termite caste differentiation. RNAi is broadly defined as the process by which foreign double-stranded (ds) RNAs trigger the destruction of corresponding mRNAs in cells (Fig. 9.3). This method of gene silencing, first demonstrated in nematodes (Fire et al. 1998; Timmons and Fire 1998), led to a Nobel Prize in 2006 because of its broad utility in animals (Hopkin 2006). As noted in other arthropods (e.g. Amdam et al. 2003; Ramakrishnan et al. 2005; Ciudad et al. 2006) and now demonstrated in *R. flavipes* (Zhou et al. 2006a) and *C. secundus* (Korb et al. 2009), injection of dsRNA into the hemocoel can effectively silence gene expression. However, RNAi via feeding is also achievable in termites (Zhou et al. 2008). As a functional genomics tool, RNAi via feeding represents a significant advance because it is highly amenable for use in a high-throughput format. As discussed below, important outcomes of RNAi research in termites include target gene identification/selection, bioassays that effectively identify phenotypic impacts, and the application of proper controls in assays that enable identification of sub-organismal impacts by RNAi on gene and protein expression.

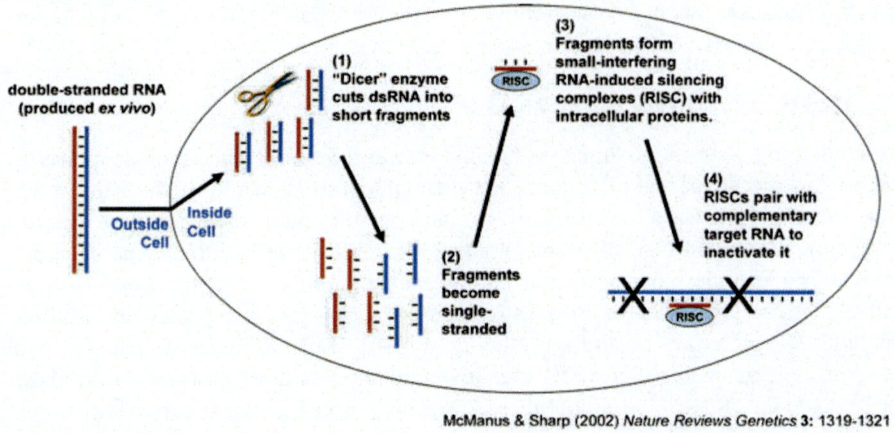

Fig. 9.3 Model for the proposed mechanism for RNA interference. Modeled after McManus and Sharp (2002)

9.4.3.2 RNAi Target Genes and Silencing Options

Four genes have now been silenced by RNAi in *R. flavipes*: *Hex-1* and *-2* (Zhou et al. 2006a, c, 2007b, 2008), *Cell-1* (Zhou et al. 2006a, 2008), and *Deviate* (Schwinghammer et al. 2011). For RNAi to be most effective, genes with high levels of constitutive expression are the best targets. Genes with narrow windows of expression, which are expressed in non-abundant tissues, and/or which are expressed at low levels are more difficult to silence. For these reasons, the temporal and tissue expression profiles for potential target genes need to be precisely determined if they are to be effectively targeted by RNAi. The hexamerin genes are abundantly expressed in fat body and readily inducible by JH (Zhou et al. 2006c), and thus, are highly susceptible RNAi targets. Likewise, the *Cell-1* gene is highly expressed in the *R. flavipes* salivary gland and its protein product is secreted in large quantities into the foregut (Zhou et al. 2007a). The *Deviate* gene is a putative odorant binding protein and a member of the *takeout*-homologous gene family of insects (related to NtSP1; see Section 9.4.4). *Deviate* also has ubiquitous tissue expression and has been effectively silenced through double stranded (ds) RNA injection (Schwinghammer et al. 2011).

However, RNAi is not a completely gene-specific phenomenon and caution is warranted when designing RNAi studies and interpreting results. For example, in initial studies on hexamerin RNAi in *R. flavipes*, small-interfering RNAs (siRNAs) designed around an internal region of the *Hex-2* gene also caused substantial *Hex-1* silencing because of ~50% nucleotide sequence overlap between the two genes (Zhou et al. 2006c). Thus, careful validation of possible sequence homologies before designing dsRNAs and conducting RNAi studies is important. Finally, in RNAi studies there is a choice between using long dsRNA fragments, or ~15–25 base-pair siRNA fragments obtained by ribonuclease digestion of long dsRNA fragments. Studies on hexamerins have suggested that digested pools of siRNAs offer much more complete, but less-specific gene silencing (Zhou et al. 2006a); whereas, long dsRNA fragments seem to offer more gene-specific silencing, but with less potency (Zhou et al. 2008).

9.4.3.3 Phenotypic Impacts of RNAi

Documenting phenotypic impacts of RNAi requires well-designed bioassays. RNAi bioassays should permit maximum survival of controls, while at the same time enable straightforward observation of impacts on caste differentiation and/or behavior. Effective RNAi bioassays control for colony-release effects and successfully reproduce the environmental conditions required for optimal gene expression/function. Examples of bioassays used in *R. flavipes* gene silencing studies include JH bioassays, cellulose feeding assays, trail-following bioassays, and worker aggression bioassays. JH and juvenoid bioassays as described by Scharf et al. (2003a, 2005a) and Cornette et al. (2006) are effective because they force developmental synchronization of workers in small assay groups. JH bioassays revealed increased JH-dependent presoldier differentiation in hexamerin siRNA-injected termite workers, suggesting that the hexamerins function as part of a

JH-binding status quo caste-regulatory mechanism (Zhou et al. 2006a). Controls in these experiments included non-injected and siRNA-injected workers held without JH, and water-injected workers held with JH.

A second gene targeted by RNAi encodes the endogenous cellulase *Cell-1*. Feeding assays were conducted by feeding worker termites filter papers treated with *Cell-1* dsRNA; due to digestive inhibition, significant mortality resulted (Zhou et al. 2008). Additionally, RNAi studies were conducted to investigate the role of the *Deviate* gene (encoding a putative odorant binding protein from the *takeout* gene family) in trail-following behavior. Following dsRNA injection, termites were subjected to trail-following assays once per day for a period of several days. Relative to water-injected controls, *Deviate*-silenced termites deviated significantly from trails, implicating the *Deviate* gene in trail-following behavior (Schwinghammer et al. 2011). Finally, behavioral assays in which *C. secundus* worker "butting" behaviors were monitored in the presence or absence of β-glucosidase gene silencing in neotenic queens revealed a novel role for β-glucosidase in neotenic suppression of worker aggression (Korb et al. 2009).

9.4.3.4 Sub-organismal Impacts of RNAi

Characterizing sub-organismal RNAi impacts on gene and protein expression are also necessary components of RNAi studies. Taking such steps is critical to demonstrate that phenotypic impacts observed in bioassays are correlated with gene silencing. As noted above, having proper controls and reference genes with stable expression are essential pre-requisites for quantifying silencing effects. Effects at the mRNA level can be determined by quantitative real-time PCR, semi-quantitative PCR, or Northern blotting. In *R. flavipes*, gene silencing lasted for a maximum of 2–4 days after injection-based RNAi, relative to water-injected controls (Zhou et al. 2006a; Schwinghammer et al. 2011). Because of potential colony-release effects on gene expression, water-injected controls (that move through time with treatments) are preferable to colony workers sampled from the colony on the day of injection or initiation of feeding assays (Scharf et al. 2008; Zhou et al. 2008).

Verifying gene silencing effects at the protein level can also be informative, provided that the protein is identifiable on polyacrylamide gels, a selective antiserum is available, and/or a suitable enzyme assay is available. An important point is that gene silencing effects at the protein and mRNA levels may not be correlated. Specifically, while gene silencing has been observed to last for 2–4 days at the mRNA level, protein attenuation can last for at least twice as long (Zhou et al. 2006a). The mechanisms for this discrepancy are not clear at the present time; however hexamerin silencing impacts at the protein level are well-correlated with phenotypic effects on presoldier differentiation (Zhou et al. 2006a, 2008).

9.4.4 Protein Studies: Proteomics

Proteomic approaches are useful for gene identification (by deducing gene sequence from protein sequence), and vice-versa for functional genomics (to link unknown proteins to known gene sequences). One example of going from protein to gene is the *Ntsp1* protein/gene from the higher termite *Nasutitermes takasagoensis* (Hojo et al. 2005b). The *Ntsp1* protein was initially found to be highly expressed in soldier heads by SDS-PAGE, and was then transferred directly from PAGE gels to PVDF membrane for N-terminal peptide sequencing. The N-terminal sequence revealed 28 of the first 30 amino acids of the processed protein. Subsequently, the full-length sequence of the gene was determined by RACE-PCR, and from the sequence, the protein was determined to be a member of the *takeout*-homologous family of ligand-binding proteins. Subsequently, by in situ hybridization, the *Ntsp1* gene was determined to be highly expressed in epithelial cells of the frontal gland reservoir (Hojo et al. 2005b). These findings have clear biological significance, as *takeout* proteins bind terpenoids, such as those produced and stored in the soldier head reservoir of nasute termites (Prestwich et al. 1981). Another important terpenoid linked to termite caste differentiation is the morphogenetic hormone JH (see Section 9.5 below).

Proteomic approaches have also been applied in *R. flavipes* in an effort to link genes to proteins observed by SDS-PAGE. In particular, N-terminal sequencing of hemolymph proteins proved very effective for identification of the Hex-1 and -2 proteins (Zhou et al. 2006a). This identification has not only made it possible to track hexamerin protein expression/induction in response to experimental JH treatments (Scharf et al. 2005a), but also to document sub-organismal effects and identify significant influences by JH and temperature on hexamerin expression (Zhou et al. 2006a; Scharf et al. 2007). Other proteomic efforts in *R. flavipes* focused on internal peptide sequencing of hemolymph proteins by mass spectrometry yielded only limited results because of a lack of corresponding gene sequences in available sequence databases (Scharf et al. 2005a).

However, once a termite genome sequence becomes available, high-throughput proteomic approaches should be able to provide a wealth of functional genomic data. For example, once the full-length hexamerin cDNA sequences became available (Zhou et al. 2006a), some unknown peptide sequences reported by Scharf et al. (2005a) were readily identifiable (Scharf, unpublished observations). For the time being, options for conducting high-throughput proteomics in termites are limited.

9.5 Hormonal Regulation of Caste Differentiation

In the following section, we give an overview of known and suspected roles for JH and ecdysteroids in termite caste differentiation. Genes and genomics can be studied in an effort to elucidate the mechanisms of action for JH, ecdysteroids, and insulin signaling, as well as possible primer pheromones (see also Section 9.7.1).

9.5.1 Overview of JH Literature in the Last Five Decades

JH is a morphogenetic hormone secreted by the corpus allatum, a paired endocrine gland that is closely associated with the insect brain (Wigglesworth 1935). The dominant JH homolog in insects is JH III (Baker 1990; Nijhout 1994; Gilbert et al. 2000). JH III has been identified from a number of lower termites and thus seems to be the JH of Isoptera (Meyer et al. 1976; Lanzrein et al. 1985; Greenberg and Tobe 1985, Park and Raina 2004; Mao et al. 2005; Yagi et al. 2005; Cornette et al. 2008). Several other JH forms have been identified in higher insects, such as Lepidoptera, where at least four homologs are synthesized (JH 0, I, II and III; Gilbert et al. 2000).

9.5.2 Multiple Roles for Hormones, Especially JH (Soldier Differentiation vs. Vitellogenesis)

9.5.2.1 JH in Soldier Caste Differentiation

When its titers are elevated, JH induces presoldier differentiation from lower termite workers (Park and Raina 2005, Mao et al. 2005; Elliott and Stay 2007, Cornette et al. 2008). The presoldier stage lasts a single instar and is directly followed by soldier emergence after approximately 2 weeks (Thorne 1996; Lainé and Wright 2003). Because it is a developmental end point, the differentiation of the soldier caste in response to JH is in direct contrast to the status quo role of JH in maintaining immature features in all other non-social insects (Henderson 1998, Gilbert et al. 2000). In a manner consistent with "normal" hemimetabolous insect development, in nymphs developing along the imaginal route, elevated JH titers may also suppress molting to supplementary reproductives (Henderson 1998; Brent et al. 2005; Elliott and Stay 2008). Interestingly, some evidence also suggests that the soldier caste can serve as a JH "sink" that reduces JH titers in termite colonies to the point that reproductive caste differentiation can occur (Henderson 1998; Mao et al. 2005). Therefore, JH and the soldier caste are together considered as central in termite caste differentiation and caste homeostasis.

It is also well established that ectopic application of natural JH homologs and synthetic juvenoids to lower termite workers induces presoldier differentiation (Howard and Haverty 1979; Okot-Kotber 1980; Jones 1984; Hrdý 1985; Haverty et al. 1989; Ogino et al. 1993; Scharf et al. 2003a, 2005a; Cornette et al. 2006; Tsuchiya et al. 2008). Consistent with JH III being the JH of Isoptera, commercially available JH III readily induces differentiation of morphologically normal presoldiers from workers in both *C. formosanus* and *R. flavipes* (Park and Raina 2003, Scharf et al. 2003a, 2005a). However, this effect is not achievable by JH III treatment of *H. sjostedti* workers, possibly because of rapid JH metabolism (Cornette et al. 2006, 2008). Similar weak induction activity of JH III was also noted in *R. lucifugus*, *R. santonensis* and *R. virginicus* (Hrdý et al. 2006), suggesting differences in physiological regulation among species. Exposure to synthetic juvenoids typically results in morphological abnormalities in *Reticulitermes* presoldiers and soldiers (Jones 1984, Lelis and Everaerts 1993, Scharf et al. 2003a). Additionally, JH II was shown to induce soldier-reproductive intercastes in *R. tibialis* and *R. flavipes*, while JH I was found to have lethal effects (Scharf et al. 2003a).

9.5.2.2 JH and Ecdysteroid Roles in Vitellogenesis

Vitellogenin is the dominant egg yolk protein not only of insects, but all animals (Sappington and Raikhel 1998). To date, four independent studies have investigated correlations between JH and vitellogenesis in termites. Brent et al. (2005) investigated JH and ecdysteroid levels in *Zootermopsis angusticollis* immature female nymphs and de-alate females that were paired with neotenic males. Their findings suggested that JH plays a dual role in reproduction, by (i) inhibiting development in immature reproductives and (ii) by inducing ovarian activity in mature female reproductives. Brent et al. (2005) also found that JH and ecdysteroid levels were inversely proportional in female reproductives after pairing at day 4 (JH low, ecdysteroids high), but then had reversed by day 30 in association with ovary activation and vitellogenesis (JH high, ecdysteriods low). These findings suggest that ecdysteriods may help modulate JH biosynthesis by the corpus allatum (Brent et al. 2005). A subsequent study in *R. flavipes* by Elliott and Stay (2007) investigated in vitro rates of JH biosynthesis by corpora allata of apterous and brachypterous neotenic females. This study found that rates of JH biosynthesis correlated with the size of the corpus allatum, regardless of reproductive phenotype or status. Also, JH synthesis was high in early vitellogenic (non-physogastric) reproductives, while JH synthesis was variable in more mature physogastric female reproductives. These findings led the authors to conclude that mature and early vitellogenic oocytes may interact to regulate JH biosynthesis (Elliott and Stay 2007). A second study by Elliott and Stay (2008) in *R. flavipes* documented higher levels of JH biosynthesis in pre-eclosion and pre-vitellogenic apterous neotenics than non-differentiating colony workers, supporting findings of Brent et al. (2005) and earlier hypotheses of Henderson (1998) that JH suppresses adult development in termite nymphs. Finally, a recent study by Cornette et al. (2008) investigated JH titers among castes and developmental stages of *Hodotermopsis sjostedti*, including immature and adult reproductives, revealing results similar to earlier studies noted above. Together, these studies establish a firm link between JH and vitellogenesis in termites.

9.5.3 Recent Analytical Identification of JH: Quantification of JH Titer and Related Gene Expression

On the basis of classical work involving transplantation of corpora allata or applications of JH and synthetic JH analogues, Nijhout and Wheeler (1982) proposed a model for termite caste differentiation, in which (i) a high JH titer at sensitive periods would induce soldier differentiation, and (ii) alate production would result from continuously low JH titer. Additionally, an initially high JH titer followed by a lower titer would induce a stationary pseudergate molt, and the differentiation of neotenics. Until recently, however, only limited information was available on actual JH titers in termites and transitions during development. Quantifications of actual JH titers were carried out using radiochemical assays, GS-MS and LC-MS (Greenberg and Tobe 1985; Park and Raina 2004; Liu et al. 2005a,b; Mao et al. 2005; Cornette et al. 2008; Gotoh et al. 2008). The general trends emerging from all of these investigations are low JH levels in workers, pseudergates and soldiers, and high JH levels in presoldiers and queens. In particular, several reports have confirmed that JH rises during the soldier differentiation process (Park and Raina 2004; Liu et al. 2005a; Mao et al. 2005; Elliott and Stay 2008; Cornette et al. 2008). However, seasonal changes of JH titer have also been noted (e.g. Cornette et al. 2008). Cornette et al. (in preparation) also identified genes, by suppressive subtractive hybridization, that responded to JHA application in *Hodotermopsis sjostedti*. These responsive genes coded hexamerins, lysosomal protein, chymotrypsin, endoglucanase, and several types of esterases. The identification of hexamerins in this species is in agreement with JH-responsive hexamerins identified in *R. flavipes* (see Section 9.7.2 below), but physiological mechanisms acting downstream of JH remain to be uncovered.

9.5.4 Insulin Signaling in Termite Caste Differentiation

Besides JH, ecdysone and related endocrine factors, insulin signaling has recently received focus in relation to caste development in social insects, and more broadly, to insect polyphenism (Emlen et al. 2006). In general, insulin and insulin-like growth factor (IGF) are known to govern growth, metabolism, nutrition, reproduction and longevity. A more general link between insulin and JH action is now well-established in insects (Tatar et al. 2003), which suggests a logical connection between JH and insulin signaling in termite caste differentiation. Additionally, many studies in insects have now connected insulin signaling to overall body and organ size (Edgar 2006). Insulin signaling is deeply related to nutrition, so this physiological pathway unsurprisingly also affects feeding behaviors in some animals (Cheung et al. 2004; Williams et al. 2004). In honeybees, recent studies show the link between insulin signaling and social regulation; for example, queen-worker differentiation and worker behavioral polyethism (Bloch et al. 2002; Wheeler et al. 2006; Hunt et al. 2007a; Ament et al. 2008). In termites, experimental results have shown that differential expression of some genes in the insulin-signaling pathway occurs during soldier caste differentiation, as induced by the ectopic application of a JH analogue (Hattori et al. in preparation).

9.6 Morphogenesis in Caste Differentiation

9.6.1 Evo-Devo and Termites (Modularity and Heterochrony)

The field of evolutionary developmental ("evo-devo") biology focuses on the phenotypic evolution of animal body plans. In this respect, termites are intriguing subjects. Within termite developmental pathways, the morphologies of body parts are selectively modified and specific body patterns are generated which are suitable for caste-specific tasks (Miura 2005). The most specialized termite castes are alates and soldiers. Alates have well-developed wings, compound eyes, and reproductive organs (Weesner 1969; Katoh et al. 2007; Maekawa et al. 2008). In addition, alates (but not workers) of most lower termites possess a specialized adhesive pad at the tip of each leg called the arolium, which apparently plays a role in locomotion and behavior (Crosland et al. 2005). In the case of soldiers, defensive characters like enlarged mandibles, frontal projections and secretory glands are characteristic (Weesner 1969; Deligne et al. 1981; Miura and Matsumoto 2000; Koshikawa et al. 2003; Watanabe and Maekawa 2008).

Another area of evo-devo biology focuses on the developmental mechanisms that produce morphological variations between species (Carroll et al. 2001). In social insects, such mechanisms can direct phenotypic variations between castes, although all individuals share similar or even identical genetic backgrounds within a species. In this sense, the developmental regulation of caste differentiation in termites is a good model of evo-devo mechanisms that produce phenotypic diversity (Nijhout 1999, 2003; West-Eberhard 2003; Miura 2005; Toth and Robinson 2007; Yang 2007; Korb and Hartfelder 2008).

Heterochrony, or the evolutionary change of developmental timing, is also a key event in the evolution of termite caste developmental pathways. For example, the relative timing of caste differentiation and/or organ development has been diversified among termite lineages (Miura et al. in preparation). In addition, allometric growth, where different parts of an organism grow at different rates, is also a fundamental focus of evo-devo biology (Stern and Emlen 1999; Emlen and Nijout 2000). Some insects show extremely exaggerated morphology like horns or enlarged appendages; these morphological characters are thought to have evolved through either natural or sexual selection. Exaggerated characters are also seen in the case of social insect castes, for example, the enlarged mandibles or frontal projections in termite soldiers. In termites, these characters are classically assumed to have evolved through kin selection. In all these cases, preexisting developmental mechanisms together with highly versatile "tool kit" genes are co-opted away from their normal/ubiquitous functions in animal development to achieve new functions viz. the specialized morphological characteristics of social insect castes.

9.6.2 Histological and Morphological Changes During Soldier Morphogenesis

Termite soldier differentiation provides an excellent model experimental system. This is mainly for two reasons: (i) morphology of termite soldiers is distinct from other castes, and (ii) soldier differentiation is artificially inducible by the application of JH analogues and/or JH (e.g., Cornette et al. 2006, 2008; Scharf et al. 2003a). Under natural conditions, the timing of caste differentiation is asynchronous and controlled by endogenous JH titers in response to numerous extrinsic factors such as social interactions, seasonality and the physical environment (e.g. Scharf et al.

2007; Mao and Henderson 2007; Dong et al. 2009). Therefore, without artificially inducing soldier differentiation, it is virtually impossible to obtain an objective time schedule of development for comparisons among individuals or between castes. In addition, it is also not yet possible to induce alates artificially. Although experimental evidence suggests a genetic basis to alate/worker differentiation in one termite species (Hayashi et al. 2007), there is also evidence that alates develop in response to seasonal cues (Roisin 2000). However, these cues and others that trigger L2–L3 nymphal differentiation (e.g. Lainé and Wright 2003) remain mostly undefined.

For the evaluation of morphological differences among species, or allometric growth within a species, morphometric analysis is one of the most powerful tools. For example, Koshikawa et al. (2002) measured various body parts in developing soldiers of *Hodotermopsis sjostedti* and analyzed the data by means of principal component analysis, finding that anterior body parts are extremely exaggerated as a general tendency. However, among anterior body parts, only distal portions of mandibles were elongated. Subsequent morphological and histological investigations revealed that epithelial cell proliferation and morphogenesis were concentrated in the distal part of the developing mandibles just prior to the presoldier molt (Koshikawa et al. 2003). These two studies documented, at a cellular level, the profound morphological changes taking place in mandibles of lower termites during presoldier morphogenesis.

Interestingly, similar morphogenetic processes are also seen in nasute soldiers of the subfamily Nasutitermitinae (Termitidae). In this case, morphogenesis in accordance with soldier differentiation is more drastic because of the greater specialization of nasute soldiers. Miura and Matsumoto (2000) first described the "soldier-nasus disc", that is the primordial structure giving rise to the soldier nasus. This structure is highly analogous to imaginal discs of holometabolous insects that give rise to adult appendages like wings and antennae (Minelli 2003). Nasus discs and other imaginal disc structures of insects share similar characteristics, such as proliferation of epithelial tissues and expansion of compressed epidermis at molting events. As in the case of horned beetles, which also express dramatic anterior polyphenism, typical appendage patterning genes seem to be conserved (Moczek and Nagy 2005; Moczek et al. 2006). Consistent with this, similar morphogenetic genes are up-regulated in the developing head of both mandibulate and nasute termite soldiers (Koshikawa et al. in preparation; Toga et al. in preparation).

In addition to morphological specialization of soldier termites, correlated behavioral changes are also seen, for example aggressive behavior (Stuart 1969). In this connection, Ishikawa et al. (2008) recently reported that the suboesophageal ganglion (SOG) of *Hodotermopsis sjostedti* is larger in soldiers than in pesudergates. Further, the mandibular motor neurons in the SOG are largely developed during the process of soldier differentiation (Ishikawa et al. 2008). These changes are common in all termite families that have been examined, suggesting a neurophysiological adaptation that underlies a caste-associated behavior pattern, and another potential model system for genomic investigation.

9.6.3 Factors Responsible for Morphogenetic Changes

During caste differentiation, there must be some regulatory mechanisms controlling cell growth and morphogenesis in tissues that become exaggerated or reduced

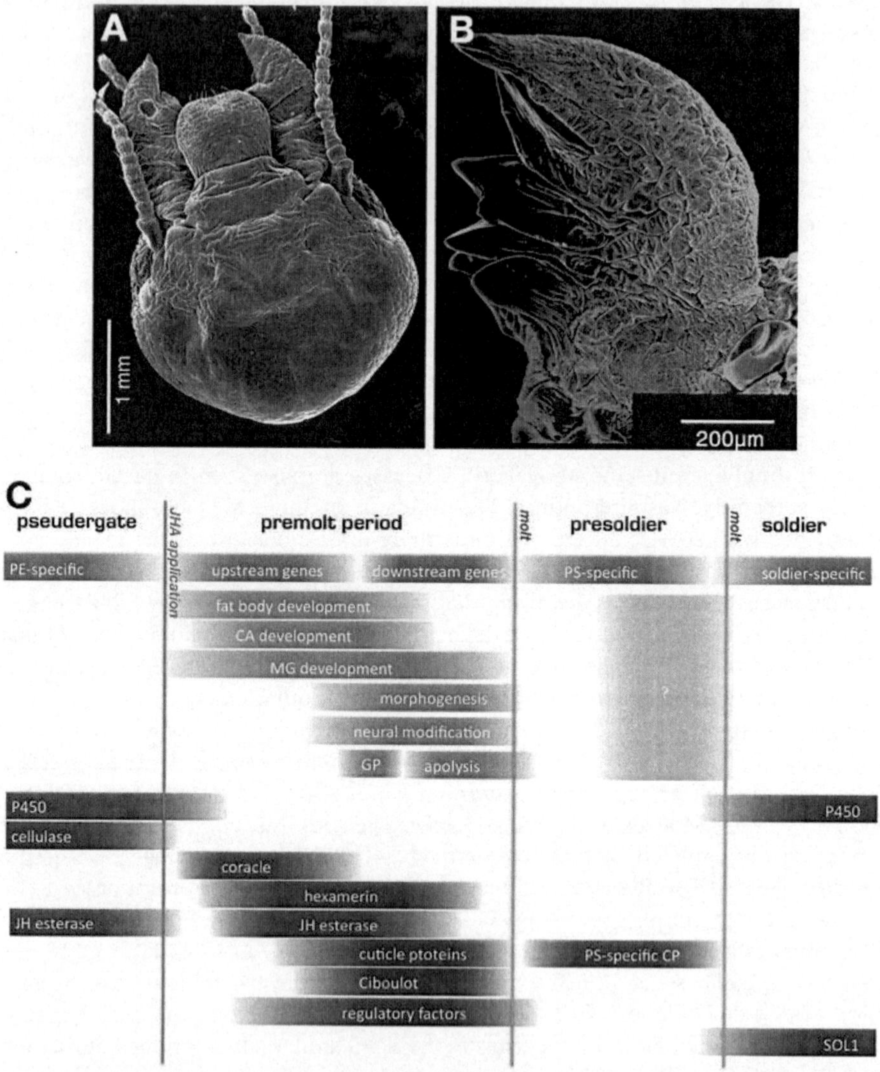

Fig. 9.4 Soldier morphogenesis (**a**, **b**) and summarized gene expression profiles during the course of soldier differentiation (**c**) in *Hodotermopsis sjostedti*. Prior to the molt into presoldier, dynamic morphogenetic process with the folding structures of cuticles and epidermis is seen in the anterior part of the head (**a**), especially mandibles (**b**). So far, during the soldier differentiation in *Hodotermopsis sjostedti*, which can be induced by the application of pyriproxyfen, many physiological and histological events have been observed together with the gene expression profiles (**c**)

in terminal caste phenotypes, such as soldier mandibles and alate wings (Miura 2001). In soldier differentiation in *H. sjostedti*, genes that are both up- and down-regulated in response to JH treatment have been identified (Koshikawa et al. 2005; Cornette et al. 2006). These genes are thought to be components of downstream pathways in the morphogenetic process, such as cuticle proteins and actin-binding protein (Koshikawa et al. 2005; Fig. 9.4). In future studies, the JH-inducible presoldier model system should allow a precise correlation of tissue changes with gene and protein expression prior to the presoldier molt. Such studies will be important for understanding the molecular basis of caste differentiation.

In ants, the differentiation of reproductive (winged) and sterile (wingless) caste phenotypes also represents a model system in which to search for convergently evolved caste regulatory genes and for investigating differential gene expression. Using this system, Abouheif and Wray (2002) examined the expression of genes linked to wing formation in model insects (e.g. *Drosophila melanogaster*), relative to imaginal wing discs of developing queens and workers from several ant species. This revealed that many conserved genes were expressed in workers across species, but interestingly, the repressed genes were variable, or "evolutionarily labile" between species. In workers of *Diacamma* and *Pheidole* wing discs are present during the larval stages, but an apoptotic process eliminates the wing tissues at the time of pupation (Sameshima et al. 2004; Gotoh et al. 2005). These examples from ants suggest that, in termites, both similarities and differences in the expression of morphogenetic factors or "toolkit" genes are likely to be identified. Termite imaginal differentiation, therefore, represents a promising model system for specifying the developmental mechanism producing reproductive phenotypes.

9.7 Social Regulation of Caste Ratios

9.7.1 Primer and Releaser Pheromones: An Overview

Primer pheromones are chemical messengers passed among individuals, especially in social insect colonies, which trigger physiological responses in recipients (Wilson and Bossert 1963). Primer pheromones are distinct from "releaser" pheromones, which elicit rapid behavioral responses in recipients (Vander Meer and Alonso 1998). Two examples of releaser pheromones in termites are the trail pheromone (*Z*,*Z*,*E*)-3,6,8-dodecatrien-1-ol (Matsumura et al. 1968) and the phagostimulatory pheromone hydroquinone (Reinhard et al. 2002). From honeybees, three examples of socially-relevant primer pheromones are worker behavioral maturation inhibitory pheromone (Leoncini et al. 2004), brood pheromone (Le Conte et al. 2006), and queen mandibular pheromone (Grozinger et al. 2007).

Although inhibitory effects by colony members were noted (Lüscher 1974, 1976), some evidence indicates termite primer pheromones. As one example, JH exhibits primer pheromone-like qualities (Lüscher 1960, 1976; Noirot 1969b, 1985; 1990; Henderson 1998). The ability of exogenously applied JH to pass through the termite cuticle and elicit developmental, molecular and physiological effects supports the idea that JH may be a primer pheromone (in rhinotermitids; Park and Raina 2003, Scharf et al. 2003a, 2005a).

No direct evidence yet exists to show that JH is transferred among nestmates to exert primer pheromone-like effects. However, considerable evidence suggests that non-JH primer pheromones exist that influence JH production by individual colony members. First, JH titers rise in workers upon removal from colonies, resulting in presoldier differentiation (Okot-Kotber et al. 1993; Park and Raina 2005, Mao et al. 2005; Elliott et al. 2009); however, JH titers and presoldier differentiation are both reduced when workers are held with soldiers (Park and Raina 2005; Mao et al. 2005; Tsuchiya et al. 2008). Second, in *R. flavipes* ecdysteroid titers rise in groups of isolated workers held away from the colony (Okot-Kotber et al. 1993). Third, soldier head extracts have been shown to inhibit presoldier differentiation (Lefeuve and Bordereau 1984; Okot-Kotber et al. 1991; Korb et al. 2003). Contrary to this, crude extracts of *R. flavipes* soldier heads containing mostly γ-cadinene and its aldehyde γ-cadinenal, as well as many other previously identified soldier head-derived terpenes (originally reported by Bagnères et al. 1990; Nelson et al. 2001, and Quintana et al. 2003) all enhance JH-dependent presoldier differentiation by *R. flavipes* workers (Tarver et al. 2009). Thus, although the mechanism is not defined, *Reticulitermes* soldiers are a source of terpenes that so far meet all the expected criteria of caste-regulatory primer pheromones (Tarver et al. 2009).

As reviewed by Tittiger (2004), a mixture of chemical ecology and functional genomics is a good way to define the function of any potential termite primer pheromones. Understanding how termite semiochemicals extrinsically impact developmental and/or caste-regulatory gene expression will be critical for distinguishing primer and releaser pheromones in termites, as well as ultimately determining if JH is a primer pheromone, endogenous hormone, or both.

9.7.2 Termite Hexamerins as a Caste Regulatory Mechanism

9.7.2.1 Experimental Summary

Understanding of the role of termite hexamerins in caste regulation has increased over the years. Prior to precise identification, two hemolymph-soluble hexamerin proteins were found to increase in abundance during JH-induced soldier caste differentiation (Scharf et al. 2005a). From nymph and presoldier array screens, expressed sequence tags representing the two hexamerins were identified and subsequently, full length cDNA sequences were obtained (Scharf et al. 2005b; Zhou et al. 2006c; Fig. 9.2). The protein sequences deduced from the full length cDNAs revealed that the Hex-1 protein has a unique hydrophobic tail with a putative JH-binding prenylation motif (Zhou et al. 2006c) that may confer covalent JH-binding (Wheeler and Nijhout 2003). The Hex-2 protein has a unique hydrophilic region of unknown function, as well as hexamerin-receptor-like qualities (Zhou et al. 2006c).

Dual silencing of both hexamerins by RNAi causes significant increases in JH-dependent presoldier differentiation, and even significant presoldier differentiation in the absence of ectopic JH (Zhou et al. 2006a). These findings provided unprecedented evidence that the hexamerins serve a "status-quo" presoldier-inhibitory function in workers, rather than an inductive function driving caste differentiation. Additionally, using a combination of RNAi and gene expression profiling, the effects of hexamerin silencing on downstream gene expression were found to correlate significantly with JH-dependent changes in gene expression (Zhou et al. 2007b). This body of evidence supports the hypothesis that hexamerins modulate JH availability and limit its induction of worker differentiation to soldier caste phenotypes. In a socio-evolutionary context (Myles and Nutting 1988), such a mechanism is highly advantageous because it sustains high degrees of colony fitness through maintenance of a large work force. This is the first evidence of a molecular caste-regulatory mechanism in termites.

In a follow-up study, Scharf et al. (2007) tested the hypothesis that the hexamerins are part of an environmentally- and/or nutritionally-responsive switching mechanism (Wheeler 1986) that regulates caste polyphenism/homeostasis. Significant impacts were found for both environment (ambient temperature) and nutrition (feeding) on hexamerin protein levels in older members of the colony work force, which are considered the most sociologically relevant members of the worker caste (Crosland et al. 1997). In older workers, hexamerin expression was significantly impacted by several factors, suggesting that hexamerin abundance is under the control of a complex array of intrinsic and extrinsic factors. Impacts on hexamerin levels in younger members of the work force are not the same (Scharf, unpublished observations). This explains previous observations in *C. formosanus* suggesting a correlation between soldier caste proportions, JH titers, temperature and food quality (Fei and Henderson 2002; Liu et al. 2005b). In a socio-biological context, findings on the *R. flavipes* hexamerins support the idea that hexamerin proteins serve interconnected proximate functions in hormonal and nutritional signaling (Fig. 9.5). Despite this understanding, studies on hexamerins from a greater number of termite species are needed to test these hypotheses.

9.7.2.2 Hexamerins and Co-option of a Juvenile Ground Plan

The idea of groundplan co-option is not new in the field of insect socio-evolutionary biology (West-Eberhard 1987, 1996). This concept is best summarized by the statement "solitary life histories provide building blocks from which complex social phenotypes can be assembled" (Amdam 2006). One example of a solitary mechanism co-opted for social functions is the honeybee vitellogenin protein. Honeybee vitellogenin has been implicated in a wide range of behavioral and physiological processes in non-reproductive worker bees (Amdam et al. 2004; Nelson et al. 2007). Another co-option example comes from *Polistes* wasps, which have co-opted a solitary developmental/diapause switch during their social evolution (Hunt and Amdam 2005). Interestingly, hexamerin storage proteins have been implicated as part of the developmental/diapause switch in *Polistes* (Hunt et al. 2003, 2007b).

Hexamerins of solitary insects are traditionally considered as storage proteins that sequester nutrients during periods of active larval feeding, and which release

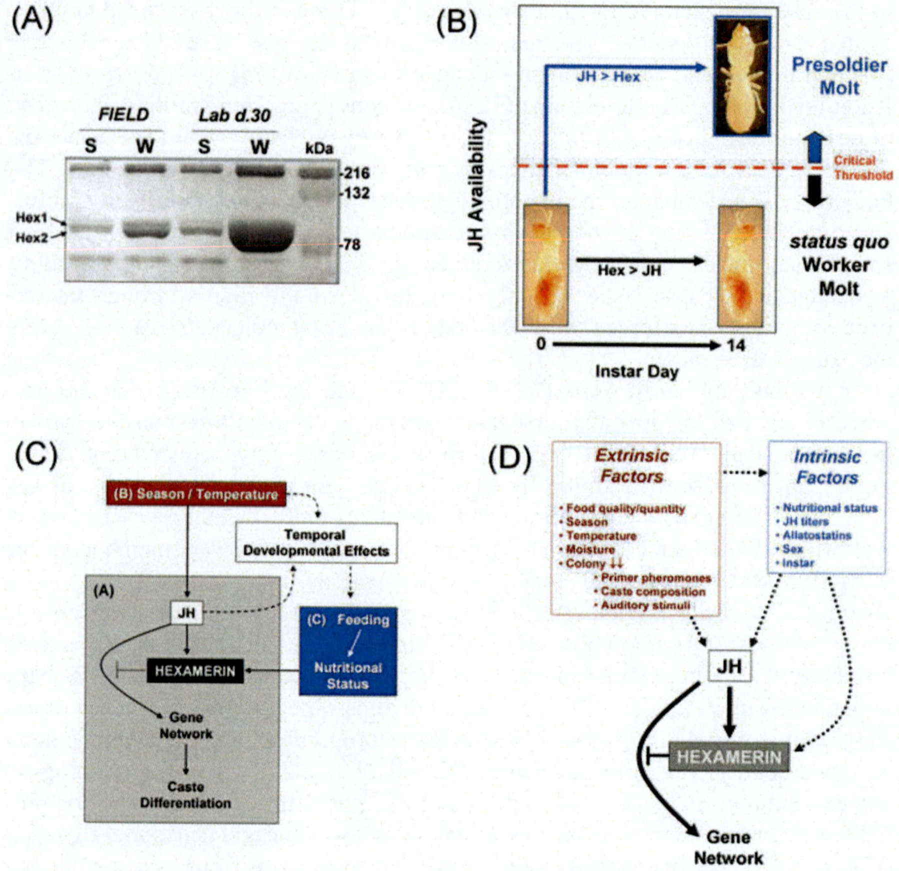

Fig. 9.5 Experimental evidence (**a**) and various models (**b**, **c**, **d**) proposed in regard to the role of termite hexamerins in caste regulation/caste homeostasis. (**a**) SDS-PAGE of *R. flavipes* hemolymph proteins with Coomassie staining, showing high level hexamerin induction in worker termites (W) relative to soldiers (S) after 30 days away from the parent field colony. (**b**) Inter-molt status quo model suggesting how presoldier morphogenesis might ensue when JH titers exceed the binding/sequestration capacity of the hexamerin mechanism (*top*) or status quo worker-to-worker molts may occur when the hexamerins can successfully sequester available JH (*bottom*) and keep it below a critical threshold (*dashed red line*). (**c**) Theoretical model depicting hexamerin influence on JH-dependent gene expression, as well as potential intrinsic and extrinsic factors that impact the degree of pleiotropic influence by hexamerins over JH-responsive gene networks (taken from Zhou et al. 2007b). (**d**) Refined model summarizing results of experiments that investigated the influence of environment (season/temperature) and nutrition/feeding on hexamerin expression and JH-dependent gene networks (taken from Scharf et al. 2007)

nutrients during stressful periods like metamorphosis and diapause (Burmester and Scheller 1999). In an ultimate sense, hexamerins have been exploited or co-opted during social evolution to inhibit the final worker molt to terminally developed soldier phenotypes. In a proximate sense, the inter-connected roles of termite hexamerins in environmental, hormonal, and nutritional signaling together accomplish caste homeostasis. Thus, this body of evidence lends support to the hypothesis initially proposed by Nalepa (1994) that co-option of a nutritional-juvenile ground plan promotes worker caste maintenance and underlies termite sociality. Although evidence also supports the existence of roles by termite hexamerins in reproduction (Scharf et al. 2005b) and incipient colony founding (Johnston and Wheeler 2007), the juvenile ground plan hypothesis seems ripe for future testing using molecular research approaches.

9.7.2.3 Cockroach Hexamerins as a Model

It was recently shown that in holometabolous insects like *Manduca sexta*, JH functions in suppressing adult tissue differentiation by inhibiting intrinsic signaling independent of nutritional state (Truman et al. 2006). Conversely, findings to date for *R. flavipes* hexamerins support that in termites, and possibly other hemimetabolous insects, the hexamerins are a bridging mechanism that brings together nutritional and JH signaling. In this context, seeking to understand how hexamerins, nutrition and development are inter-connected in other hemimetabolous insects would likely be a worthwhile endeavor. Cockroaches would seem to be an excellent starting point in such investigations (Nalepa 1994, Holbrook and Schal 2004). Investigating cockroach hexamerins through development and under different temperature and nutritional regimes would, in addition, likely reveal important new information.

9.7.3 Caste-Specific Signals Secreted from Exocrine Glands

There are many types of exocrine glands on the termite body surface that secrete various semiochemicals (Noirot 1969b, Henderson 1998). The sternal gland, the source of trail pheromones, is probably the best known. In addition, frontal glands are described in the head capsule and show caste specificity in many termite species. These glands are specialized for defensive functions; they usually discharge sticky repellent substances that consist of a complex of chemical compounds (Prestwich 1984). Tergal glands in alates that secrete sex pheromones have also been relatively well studied (Ampion and Quennedey 1981, Pasteels and Bordereau 1998). Considering these well-defined functions it seems likely that caste-specific or soldier-specific gene expression might be involved.

Miura et al. (1999) identified a gene, termed *SOL1*, that is specifically expressed in the mandibular glands of soldiers in *Hodotermopsis sjostedti*. These are exocrine glands showing some caste polyphenism; for example, they can be more developed in reproductives than workers (Noirot 1969b). The identification of the *SOL1* gene and its expression site showed that mandibular glands are largely developed in the course of soldier differentiation, and contain large amounts of proteinaceous substances inside the reservoir (Miura et al. 1999; Miura in preparation; Fig. 9.6). *SOL1* encodes a protein categorized in the lipocalin family. This protein family, characterized by tertiary structures consisting of beta barrels made of repeating beta sheets, has diverse and important biological functions. In particular, most lipocalin protein

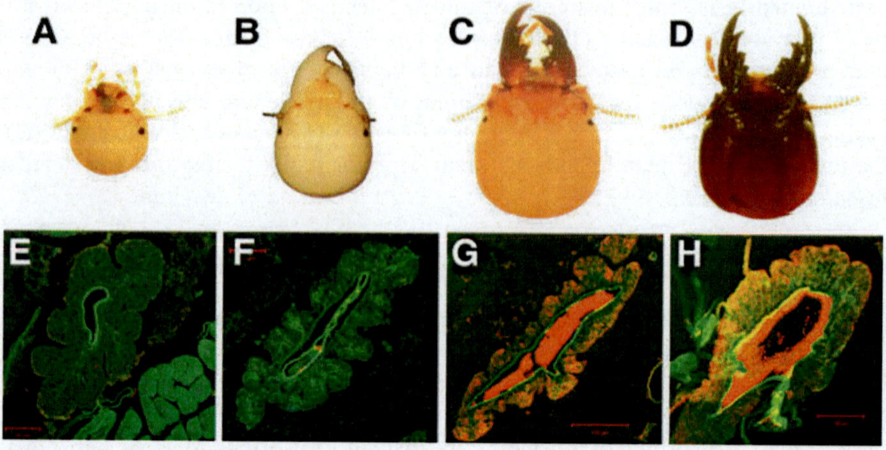

Fig. 9.6 Development of mandibular glands during soldier differentiation in *Hodotermopsis sjostedti*, and SOL1 expression in the mandibular glands. **a–d** Outer head morphology. **e–f** Mandibular glands analyzed by immunohistochemistry with the anti-SOL1 andibody (*red*: SOL1 protein; *green*: actin showing cell membrane). **a**, **e**: pseudergate, **b**, **f**: presoldier, **c**, **g**: young soldier, **d**, **h**: mature soldier

functions are related to hormone signaling; they possess binding pockets that can carry hormone ligands and deliver them to their targets.

Subsequent genomic analyses revealed that the *SOL1* gene is highly duplicated in the termite genome. The different gene copies are apparently used in different exocrine glands, such as salivary and mandibular glands. Interestingly, proteomic analysis revealed that the protein products encoded by the different *SOL1* gene copies expressed in different glands are differentially glycosylated through post-translational modification, and as a result they display a variety of molecular weights among their different source tissues (Miura, in preparation). Immunological investigations such as western blotting and immunohistochemistry strongly suggested that the SOL1 protein is transferred among colony members and is most strongly secreted in association with defensive behaviors. Lipocalin proteins have signaling functions; thus the functions of SOL1 proteins most likely relate to defense (Miura in preparation).

9.8 Sociogenomics in Termites

9.8.1 Sociogenomics as Defined by Robinson

"Sociogenomics" is a term proposed by Robinson (1999), defined as the integrative study of the molecular genetics of social behavior, especially in social insects. The conception of sociogenomics coincided with the honeybee genome project (Honeybee Genome Sequencing Consortium 2006). As already discussed, genome-wide tools like whole-genome microarrays have been developed and tested in the honeybee (Whitfield et al. 2003; Grozinger et al. 2003). The quest to understand behavior based on molecular and genetic information is not restricted to social insects, but extends to other animals like mammals and birds (reviewed in Robinson et al. 2005). It has been shown that some molecular mechanisms are shared by animal groups that are phylogenetically distant. For example, the gene *foraging* (*for*) which is related to feeding behavior, is shared by *Caenorhabditis elegans* (roundworm), *Drosophila melanogaster* (fruitfly) and *Apis mellifera* (honeybee). With this perspective in mind, we now consider some common non-behavioral genes identified across termite species, as well as other possibilities for large-scale sociogenomics (behavioral and physiological) in Isoptera.

9.8.2 Similar Genes Identified Across Termite Species

Caste developmental pathways, a reproductive division of labor, and sterile altruistic castes were apparently present in the common ancestor of all extant termites (Thorne and Traniello 2003). Therefore, common physiological, developmental and molecular mechanisms that underlie social regulation and eusocial polyphenism should be shared by all termite species. Thus, identification of common genes responsible for termite sociality, and their functional characterization will be important for inferring the evolutionary trajectories of termite sociality. In this respect, several homologous genes have been identified from different termites to date (e.g. coding hexamerins, vitellogenins, P450s, larval cuticle proteins, lignocellulases). In an evolutionary sense, these genes can be considered as potentially common genes with broad contributions to termite sociality and eusocial polyphenism.

9.8.2.1 Hexamerin Genes

As noted above (see Section 9.7.2 above), hexamerin genes act as part of a nutritionally and environmentally responsive switching mechanism that regulates caste polyphenism. To date hexamerins from *R. flavipes* have been extensively characterized, and more recently, they have been identified in *H. sjostedti* (Cornette et al. in preparation). Because of their important role in caste-regulation and possibly social evolution, there is good rationale for investigating hexamerin genes and proteins throughout the Isoptera (Scharf et al. 2007).

9.8.2.2 Vitellogenin Genes

Three termite vitellogenin genes have been sequenced to date; two from *R. flavipes* (Scharf et al. 2003b, 2005a, b) and one from *C. secundus* (Weil et al. 2007). Of these three termite vitellogenins, the *C. secundus* vitellogenin (EF029054) and *R. flavipes* vitellogenin-1 (BQ788169) are highly similar and probably orthologous. Both *R. flavipes* vitellogenins are inducible from 500- to >10,000-fold in workers in response to JH treatment (Scharf et al. 2005b; Tarver et al. in preparation), and both genes are also significantly upregulated in presoldiers and brachypterous female nymphs (Scharf et al. 2005a). Given the important socio-regulatory role identified for honeybee vitellogenin (Amdam et al. 2003, 2004; Guidugli et al.

2005; Seehuus et al. 2006; Nelson et al. 2007), vitellogenin upregulation in non-reproductive termite phenotypes suggests broader socio-biological and evolutionary roles for vitellogenin beyond egg provisioning.

9.8.2.3 P450 Genes

Insect P450s are an incredibly diverse and well-represented gene family (Feyereisen 2005). As noted throughout this chapter, the differential expression of P450 genes among castes and in response to experimental treatments has been independently identified in multiple termite species. The *C. secundus* Cyp4 P450 (EF029057; Weil et al. 2007) is significantly similar to seven Cyp4s and their allelic variants identified in *R. flavipes* (DQ279461–DQ279462; Zhou et al. 2006b), but not at a level where they could be considered orthologous. In addition, Cyp6AM1 was identified in *Hodotermopsis sjostedti* (Cornette et al. 2006), and new P450 genes similar to Cyp6AM1 were found in the *R. flavipes* gut (Tartar et al. 2009).

Additionally, the honeybee genome contains the fewest number of P450 and other detoxification-related genes of any insect genome sequenced to date (Claudianos et al. 2006). Identification of all P450s, as well as related detoxification genes in termites will test the hypothesis that a deficit of detoxification enzymes is linked to insect social evolution.

9.8.2.4 Larval Cuticle Protein (LCP) Genes

LCPs are major components of arthropod cuticle (Willis 1999, Takeda et al. 2001; Cornman et al. 2008). They occur in differentiating tissues that are sensitive to JH and ecdysone (i.e. imaginal discs), and they are involved in processes related to chitin binding and cuticle hardening (Willis 1999). LCPs (4 total) were first identified during juvenoid-induced presoldier differentiation in *H. sjostedti* (Koshikawa et al. 2005). Three of these genes peaked in expression during the second week of presoldier differentiation, while a fourth peaked in presoldiers. Another LCP from *R. flavipes* (Accession no. DN792534) was identified from presoldier arrays (Fig. 9.2). The *R. flavipes* LCP is most similar to the HsjCP1 LCP of *H. sjostedti*; however, all of the termite LCPs show similar temporal expression profiles with peaks in expression during the second week of presoldier differentiation, and/or in presoldiers. Because they are part of a large gene family, and because of important morphogenetic changes in termite cuticle during caste differentiation (see Section 9.6 above), LCP genes may prove to be important genes for future study across Isoptera.

9.8.2.5 Lignocellulase Genes

The topic of cellulose and lignocellulose digestion receives detailed attention in Chapter 3 by Lo et al. and Chapter 16 by Brune and Ohkuma, this volume. Cooperative living on a nutritionally poor food source like wood is thought to be one of the critical factors favoring social evolution in termite ancestors (Nalepa 1994; Thorne and Traniello 2003). Additionally, not only are symbiotic digestive genes important in termite nutrition, but also endogenous termite genes (e.g., Nakashima et al. 2002; Scharf and Tartar 2008; Fujita et al. 2008). Thus, because of the potential importance of nutrition in caste differentiation and caste homeostasis (Nalepa 1994; Scharf et al. 2007), lignocellulase genes likely played important roles in termite social evolution and maintenance of present-day termite sociality. To date, many lignocellulase genes have been identified across Isoptera that could be considered for comparative investigation in caste regulatory processes.

9.8.3 Genome Sizes in Termites

Genome size is an important consideration in whole genome sequencing. The genome sizes of some termites, cockroaches and related insects have been estimated (Gregory 2002; Koshikawa et al. 2008). Normally, eukaryotic genome size is shown by C-value, the haploid DNA content, expressed as picogram (pg) or base pairs, where $1\ \text{pg} = 0.987 \times 10^9$ base pairs (Dolzel et al. 2003; Greilhuber et al. 2005). The known C-values of termites range from 0.58 to 1.90 pg, a range which also includes the wood roach *Cryptocercus,* but is smaller than in non-social relatives (Table 9.1), suggesting that reduction of genome size is related to social evolution (Koshikawa et al. 2008).

Of the existing estimates, *Zootermopsis nevadensis* has the smallest genome (0.58 pg); therefore, this species is a good candidate for full sequencing. However, because of their economic importance, *Reticulitermes* spp. and *Coptotermes* are also good candidates.

9.8.4 The Need for a Genome Sequence and Other Robust Genomic Tools and Resources

Kin selection theory based on the concepts of inclusive fitness and increased relatedness adequately explains the maintenance of eusocial behavior in haplodiploid insects (social hymenopterans, thrips) and in clonal forms such as aphids (Hamilton 1964). However, in termites both males and females are diploid; and no single hypothesis adequately accounts for the origin and maintenance of their eusocial status (Thorne 1997, see also Chapter 5 by Howard and Thorne, this volume). Furthermore, whereas the Hymenoptera includes taxa that exhibit a continuum of sociality ranging from solitary to highly eusocial, all extant termites are highly eusocial. As a result, comparisons among termite taxa exhibiting a range of eusocial behaviors are not possible and there is thus a significant missing link between solitary woodroach-like ancestors and eusocial termites. As a consequence, genomic approaches probably represent one of the few remaining options for gaining insights into the evolution of eusociality in termites. Robust genomic tools and resources, such as ESTs, and genome sequence of multiple species, once available, will lead to a more comprehensive understanding of termite eusociality through comparative studies among termite taxa and other eusocial insects. Genomic information may also contribute to pest management and biofuel development.

Table 9.1 Summary of the known C-value (genome size) data of termites, cockroaches and mantids

Order	Family	Species name	C-value (pg)[a]	references
Isoptera	Termitidae	*Nasutitermes takasagoensis*	1.68	Koshikawa et al. (2008)
		Odontotermes formosanus	1.48	Koshikawa et al. (2008)
		Pericapritermes nitobei	1.90	Koshikawa et al. (2008)
	Rhinotermitidae	*Coptotermes formosanus*	1.01/0.86	Gregory (2002)/ Koshikawa et al. (2008)
		Reticulitermes flavipes	1.14	Gregory (2002)
		Reticulitermes speratus kyushuensis	1.07	Koshikawa et al. (2008)
		Reticulitermes speratus speratus	1.01	Koshikawa et al. (2008)
		Reticulitermes yaeyamanus	1.00	Koshikawa et al. (2008)
	Kalotermitidae	*Glyptotermes fuscus*	1.56	Koshikawa et al. (2008)
		Glyptotermes nakajimai	0.89	Koshikawa et al. (2008)
		Cryptotermes secundus	1.27	Koshikawa et al. (in preparation)
		Neotermes koshunensis	1.87	Koshikawa et al. (2008)
	Termopsidae	*Hodotermopsis sjostedti*	1.27	Koshikawa et al. (2008)
		Zootermopsis nevadensis	0.58	Koshikawa et al. (2008)
	Hodotermitidae	*Hodotermes mossambicus*	1.00	Koshikawa et al. (2008)
	Mastotermitidae	*Mastotermes darwiniensis*	1.30	Koshikawa et al. (2008)
Blattaria	Cryptocercidae	*Cryptocercus kyebangensis*	1.16	Koshikawa et al. (2008)
		Cryptocercus punctulatus	1.32	Koshikawa et al. (2008)
	Blaberidae	*Blaberus fuscus*	3.40	Bier and Müller (1969)
		Blaptica dubia	4.54	Koshikawa et al. (2008)
		Gromphadorhina portentosa	3.80	Gregory (2002)
		Nauphoeta cinerea	5.15	Koshikawa et al. (2008)
	Blattellidae	*Blattella germanica*	2.00	Bier and Müller (1969)
		Blattella nipponica	2.13	Koshikawa et al. (2008)
		Parcoblatta sp.	3.80	Gregory (2002)
	Blattidae	*Blatta orientalis*	3.03	Gregory (2007)
		Periplaneta americana	2.72	Gregory (2007)
Mantodea	Hymenopodidae	*Acromantis japonica*	4.53	Koshikawa et al. (2008)
	Mantidae	*Mantis religiosa*	3.15	Gregory (2002)
		Statilia maculata	3.05	Koshikawa et al. (2008)
		Tenodera aridifolia	2.92	Koshikawa et al. (2008)

[a]Average values between male and female are cited when measured for each sex.

9.9 Conclusions and Perspectives

Termite genomics embraces molecular developmental mechanisms, social communication via chemical substances, control of the nervous system and physiological regulation. In this chapter, we have summarized studies of termite caste differentiation that span the range of classical observational work to recent molecular and molecular-genetic studies, discussing both the mechanisms conserved across the termite lineages, and those that are divergent.

Future studies can use more integrative approaches that include bioinformatics and systems biology to give a fuller picture of termite sociality. In addition, more detailed work on termite developmental regulation will also be required to understand the specific morphogenetic processes which accompany caste differentiation and provide deeper insights into evo-devo phenomena such as phenotypic plasticity, modularity, and heterochrony. A model species should be selected for whole genome sequencing with *Reticulitermes*, *Coptotermes*, *Hodotermopsis* and *Zootermopsis* as the main candidates. New data should clarify the cockroach-to-termite social evolution, as well as many other issues in termite sociobiology.

References

Abouheif E, Wray G (2002) Evolution of gene network underlying wing polyphenism in ants. Science 297:249–252

Adams MD et al (2000) The genome sequence of *Drosophila melanogaster*. Science 287: 2185–2195

Amdam GV (2006) The making of a social insect – the regulatory architecture of social design. Proceedings of XV International IUSSI Congress, 5–8

Amdam GV, Norberg K, Fondrk MK, Page RE Jr (2004) Reproductive ground plan may mediate colony-level selection effects on individual foraging behavior in honey bees. Proc Natl Acad Sci U S A 101:11350–11355

Amdam GV, Norberg K, Hagen A, Omholt SW (2003) Social exploitation of vitellogenin. Proc Natl Acad Sci U S A 100:1799–1802

Ament SA, Corona M, Pollock HS, Robinson GE (2008) Insulin signaling is involved in the regulation of worker division of labor in honey bee colonies. Proc Natl Acad Sci U S A 105:4226–4231

Ampion M, Quennedey A (1981) The abdominal epidermal glands of termites and their phylogenetic significance. In: Howse PE, Clément JL (eds) Biosystematics of social insects. Academic Press, London, pp 249–261

Andersen JF, Walding JK, Evans PH et al (1997) Substrate specificity for the epoxidation terpenoids and active site topology of house fly cytochrome P450 6A1. Chem Res Toxicol 10:156–164

Bagnères AG, Clément JL, Blum MS et al (1990) Cuticular hydrocarbons and defensive compounds of *Reticulitermes flavipes* (Kollar) and *R. santonensis* (Feytaud): polymorphism and chemotaxonomy. J Chem Ecol 16:3213–3244

Baker FC (1990) Techniques for identification and quantification of juvenile hormones and related compounds in arthropods. In: Gupta AP (ed) Morphogenetic hormones of arthropods. Rutgers University Press, New Brunswick, NJ and London, pp 389–453

Bellés X, Martín D, Piulachs MD (2005) The mevalonate pathway and the synthesis of juvenile hormone in insects. Annu Rev Entomol 50:181–199

Bier K, Müller W (1969) DNA-Messungen bei Insekten und eine Hypothese über retardierte Evolution und besonderen DNA-Reichtum in Tierreich. Biol Zbl 88: 425–449

Bloch G, Wheeler DE, Robinson GE (2002) Endocrine influences on the organization of insect societies. In: Brent CS, Schal C, Pfaff DW et al (eds) Hormones, brain and behavior. Academic Press, San Diego, CA, pp 195–235

Brent CS, Schol C, Vargo EL (2005) Endocrine changes in maturing primary queens of *Zootermopsis angusticollis*. J Insect Physiol 51:1200–1209

Burmester T, Scheller K (1999) Ligands and receptors: common theme in insect storage protein transport. Naturwissenschaften 86:468–474
Byrne KA, Lehnert SA, Johnson SE, Moore SS (1999) Isolation of a cDNA encoding a putative cellulase in the red claw crayfish *Cherax quadricarinatus*. Gene 239:317–324
Cameron SL, Whiting MF (2007) Mitochondrial genomic comparisons of the subterranean termites from the genus *Reticulitermes* (Insecta: Isoptera: Rhinotermitidae). Genome 50:188–202
Carroll SB, Grenier JK, Weatherbee SD (2001) From DNA to diversity: molecular genetics and the evolution of animal design. Blackwell Publishing, Malden, MA
Cheung BHH, Arellano-Carbajal F, Rybicki I, de Bono M (2004) Soluble guanylate cyclases act in neurons exposed to the body fluid to promote *C. elegans* aggregation behavior. Curr Biol 14:1105–1111
Ciudad L, Piulachs MD, Bellés X (2006) Systemic RNAi of the cockroach vitellogenin receptor results in a phenotype similar to that of the *Drosophila* yolkless mutant. FEBS J 273:325–335
Claudianos C, Ranson H, Johnson RM et al (2006) A deficit of detoxification enzymes: pesticide sensitivity and environmental response in the honeybee. Insect Mol Biol 15:615–636
Cornette R, Gotoh H, Koshikawa S, Miura T (2008) Juvenile hormone titers and caste differentiation in the damp-wood termite *Hodotermopsis sjostedti* (Isoptera, Termopsidae). J Insect Physiol 54:922–930
Cornette R, Koshikawa S, Hojo M et al (2006) Caste-specific cytochrome P450 in the damp-wood termite *Hodotermopsis sjostedti* (Isoptera, Termopsidae). Insect Mol Biol 15:235–244
Cornette R, Matsumoto T, Miura T (2007) Histological analysis on fat body development and molting events during soldier differentiation in the damp-wood termite *Hodotermopsis sjostedti* (Isoptera, Termopsidae). Zool Sci 24:1066–1074
Cornman RS, Togawa T, Dunn WA et al (2008) Annotation and analysis of a large cuticular protein family with the R&R Consensus in *Anopheles gambiae*. BMC Genomics 9:22
Crosland MWJ, Lok CM, Wong TC et al (1997) Division of labour in a lower termite: the majority of tasks are performed by older workers. Anim Behav 54:999–1012
Crosland MWJ, Su N-Y, Scheffrahn RH (2005) Arolia in termites (Isoptera): functional significance and evolutionary loss. Insectes Soc 52:63–66
Deligne J, Quennedey A, Blum MS (1981) The enemies and defense mechanisms of termites. In: Hermann HR (ed) Social insects, vol 2. Academic Press, New York, NY, pp 1–76
Dolzel J, Bartos J, Voglmayr H, Greihuber J (2003) Nuclear DNA conetent and genome size of trout and human. Cytometry A 51:127–128
Dong SL, Mao L, Henderson GR (2009) Physical contact between soldier and worker is essential in soldier self-regulation of *Coptotermes formosanus* (Isoptera, Rhinotermitidae). Insectes Soc 56:28–34
Edgar BA (2006) How flies get their size: genetics meets physiology. Nat Rev Genet 7:907–916
Elliott KL, Chan KK, Teesch L et al (2009) Identification of Phe-Gly-Leu-amide type allatostatin-7 in *Reticulitermes flavipes*: its localization in tissues and relation to juvenile hormone synthesis. Peptides 30:495–506
Elliott KL, Stay B (2007) Juvenile hormone synthesis as related to egg development in neotenic reproductives of the termite *Reticulitermes flavipes*, with observations on urates in the fat body. Gen Comp Endocrinol 152:102–110
Elliott KL, Stay B (2008) Changes in juvenile hormone synthesis in the termite *Reticulitermes flavipes* during development of soldiers and neotenic reproductives from groups of isolated workers. J Insect Physiol 54:492–500
Emlen DJ, Nijout HF (2000) The development and evolution of exaggerated morphologies in insects. Annu Rev Entomol 45:661–708
Emlen DJ, Szafran Q, Corley LW, Dworkin I (2006) Insulin signaling and limb-patterning: candidate pathways for the origin and evolutionary diversification of beetle horns. Heredity 97:179–191
Falckh PH, Balcombe W, Haritos VS, Ahokas JT (1997) Isolation and identification of a cytochrome P450 sequence in an Australian termite, *Mastotermes darwiniensis*. Biochem Biophys Res Commun 241:579–583

Fei H, Henderson G (2002) Formosan subterranean termite (Isoptera: Rhinotermitidae) wood consumption and worker survival as affected by temperature and soldier proportion. Environ Entomol 31:509–514

Feyereisen R (2005) Insect cytochrome P450. In: Gilbert LI, Iatrou K, Gill SS (eds) Comprehensive Molecular Insect Science, vol 4. Elsevier, Oxford, pp 1–77

Fire A, Xu S, Montgomery MK et al (1998) Potent and specific genetic interference by double-stranded RNA in *Caenorhabditis elegans*. Nature 391:744–745

Fujita A, Miura T, Matsumoto T (2008) Differences in cellulose digestive systems among castes in two termite lineages. Physiol Entomol 33:73–82

Gilbert SF (2001) Ecological developmental biology: developmental biology meets the real world. Dev Biol 233:1–12

Gilbert LI, Granger NA, Roe RM (2000) The juvenile hormones: historical facts and speculations on future research directions. Insect Biochem Mol Biol 30:617–644

Goodisman MAD, Crozier RH (2003) Association between caste and genotype in the termite *Mastotermes darwiniensis* Froggatt. Aust J Entomol 42:1–5

Goodisman MAD, Isoe J, Wheeler DE, Wells MA (2005) Evolution of insect metamorphosis: a microarray-based study of larval and adult gene expression in the ant *Camponotus festinatus*. Evolution 59:858–870

Gotoh H, Cornette R, Koshikawa S, Miura T (2008) Effects of precocenes on the corpora allata and the JH titer in the damp-wood termite *Hodotermopsis sjostedti* (Isoptera: Termopsidae). Sociobiology 52:345–356

Gotoh A, Sameshima S, Tsuji K et al (2005) Apoptotic wing degeneration and formation of an altruism-regulating glandular appendage (gemma) in the ponerine ant *Diacamma* sp. from Japan (Hymenoptera, Formicidae, Ponerinae). Dev Genes Evol 215:69–77

Greenberg S, Tobe SS (1985) Adaptation of a radiochemical assay for juvenile hormone biosynthesis to study caste differentiation in a primitive termite. J Insect Physiol 31: 347–352

Gregory TR (2002) The C-value enigma. Ph.D. Thesis, Department of Zoology, University of Guelph, Guelph, Ontario, Canada. 894 pp. http://www.genomesize.com/rgregory/thesis/

Gregory TR (2007) Animal Genome Size Database [online]. Available from http://www.genomesize.com

Greilhuber J, Doležel J, Lysák MA, Bennett MD (2005) The origin, evolution and proposed stabilization of the terms 'genome size' and 'C-value' to describe nuclear DNA contents. Ann Bot 95:255–260

Grozinger CM, Fan Y, Hoover SE, Winston ML (2007) Genome-wide analysis reveals differences in brain gene expression patterns associated with caste and reproductive status in honey bees (Apis mellifera). Mol Ecol 16:4837–4848

Grozinger CM, Sharabash NM, Whitfield CW, Robinson GE (2003) Pheromone-mediated gene expression in the honey bee brain. Proc Natl Acad Sci U S A 100:14519–14525

Guidugli KR, Nascimento AM, Amdam GV et al (2005) Vitellogenin regulates hormonal dynamics in the worker caste of a eusocial insect. FEBS Lett 579:4961–4965

Hamilton WD (1964) The genetical evolution of social behaviour, I, II. J Theor Biol 7:1–52

Haverty MI (1977) The proportion of soldiers in termite colonies: a list and bibliography. Sociobiology 2:199–216

Haverty MI, Howard RW (1981) Production of soldiers and maintenance of soldier proportions by laboratory experimental groups of *Reticulitermes flavipes* (Kollar) and *Reticulitermes virginicus* (Banks) (Isoptera: Rhinotetrmitidae). Insectes Soc 28:32–39

Haverty MI, Su N-Y, Tamashiro M, Tamamoto R (1989) Concentration-dependent presoldier induction and feeding deterrency: potential of two insect growth regulators for remedial control of the Formosan subterranean termite (Isoptera: Rhinotermitidae). J Econ Entomol 82:1370–1374

Hayashi Y, Lo N, Miyata H, Kitade O (2007) Sex-linked genetic influence on caste determination in a termite. Science 318:985–987

Helvig C, Koener JF, Unnithan GC, Feyereisen R (2004) CYP15A1, the cytochrome P450 that catalyzes epoxidation of methyl farnesoate to juvenile hormone III in cockroach corpora allata. Proc Natl Acad Sci U S A 101:4024–4029

Henderson G (1998) Primer pheromones and possible soldier caste influence on the evolution of sociality in lower termites. In: Vander Meer RK, Breed MD, Winston ML, Espelie KE (eds) Pheromone communication in social insects. Westview Press, Boulder, CO, pp 314–330

Hojo M, Koshikawa S, Cornette R et al (2005a) Identification of soldier-specific genes in the nasute termite *Nasutitermes takasagoensis* (Isoptera: Termitidae). Entomol Sci 8:379–387

Hojo M, Matsumoto T, Miura T (2007) Cloning and expression of a geranylgeranyl diphosphate synthase gene: insights into the synthesis of termite defense secretion. Insect Mol Biol 16: 121–131

Hojo M, Morioka M, Matsumoto T, Miura T (2005b) Identification of soldier caste-specific protein in the frontal gland of nasute termite *Nasutitermes takasagoensis* (Isoptera: Termitidae). Insect Biochem Mol 35:347–354

Holbrook GL, Schal C (2004) Maternal investment affects offspring phenotypic plasticity in viviparous cockroach. Proc Natl Acad Sci U S A 101:5595–5597

Hölldobler B, Wilson EO (1990) The Ants. Belknap Press, Harvard, MA

Holt RA, Subramianian M, Halpern A et al (2002) The genome sequence of the malaria mosquito *Anopheles gambiae*. Science 298:129–149

Honeybee Genome Sequencing Consortium (2006) Insights into social insects from the genome of the honeybee *Apis mellifera*. Nature 443:931–949

Hongo Y, Sharma VK, Prakash T et al (2008) Genome of an endosymbiont coupling N_2 fixation to cellulolysis within protist cells in termite gut. Science 322:1108–1109

Hopkin M (2006) RNAi scoops medial Nobel. Nat News. doi:10.1038/news061002-2

Howard RW, Haverty MI (1979) Termites and juvenile hormone analogues: a review of methodology and observed effects. Sociobiology 4:269–278

Howard RW, Haverty MI (1981) Seasonal variation in caste proportions of field colonies of *Reticulitermes flavipes* (Kollar). Environ Entomol 10:546–549

Hrdý I (1985) The role of juvenile hormone and juvenoids in soldier formation in Rhinotermitidae. In: Watson JAL, Okot-Kotber BM, Noirot C (eds) Caste differentiation in social insects. Pergamon Press, Oxford, pp 245–249

Hrdý I, Kuldová J, Hanus R, Wimmer Z (2006) Juvenile hormone III, hydroprene and a juvenogen as soldier caste differentiation regulators in three *Reticulitermes* species: poteintial of juvenile hormone analogues in termite control. Pest Manag Sci 62:848–854

Hunt JH, Amdam GV (2005) Bivoltinism as an antecedent to eusociality in the paper wasp genus *Polistes*. Science 308:264–267

Hunt GJ, Amdam GV, Schlipalius D et al (2007a) Behavioral genomics of honey bee foraging and nest defense. Naturwissenschaften 94:247–267

Hunt JH, Buck NA, Wheeler DE (2003) Storage proteins in vespid wasps: characterization, developmental pattern, and occurrence in adults. J Insect Physiol 49:785–794

Hunt JH, Kensinger BJ, Kossuth JA et al (2007b) A diapause pathway underlies the gyne phenotype in *Polistes* wasps, revealing an evolutionary route to caste-containing insect societies. Proc Natl Acad Sci U S A 104:14020–14025

Ishikawa Y, Aonuma H, Miura T (2008) Soldier-specific modification of the mandibular motor neurons in termites. PLoS ONE 3:e2617

Johnston ML, Wheeler DE (2007) The role of storage proteins in colony-founding in termites. Insectes Soc 54:383–387

Jones SC (1984) Evaluation of two insect growth regulators for the bait-block method of subterranean termite (Isoptera: Rhinotermitidae) control. J Econ Entomol 77:1086–1091

Katoh H, Matsumoto T, Miura T (2007) Alate differentiation and compound-eye development in the dry-wood termite *Neotermes koshunensis* (Isoptera, Kalotermitidae). Insectes Soc 54:11–19

Keeling CI, Bearfield JC, Young S et al (2006) Effect of juvenile hormone on gene expression in the pheromone-producing midgut of the pine engraver beetle, *Ips pini*. Insect Mol Biol 15:207–216

Keeling CI, Blomquist GJ, Tittiger C (2004) Coordinated gene expression for pheromone biosynthesis in the pine engraver beetle, *Ips pini* (Coleoptera: Scolytidae). Naturwissenschaften 91:324–328

Korb J, Hartfelder K (2008) Life history and development – a framework for understanding developmental plasticity in lower termites. Biol Rev 83:295–313

Korb J, Roux EA, Lenz M (2003) Proximate factors influencing soldier development in the basal termite *Cryptotermes secundus* (Hill). Insectes Soc 50:299–303

Korb J, Weil T, Hoffman K et al (2009) A gene necessary for reproductive suppression in termites. Science 324:758–759

Koshikawa S, Cornette R, Hojo M et al (2005) Screening of genes expressed in developing mandibles during soldier differentiation in the termite *Hodotermopsis sjostedti*. FEBS Lett 579:1365–1370

Koshikawa S, Matsumoto T, Miura T (2002) Morphometric changes during soldier differentiation of the damp-wood termite *Hodotermopsis japonica* (Isoptera: Termopsidae). Insectes Soc 49:245–250

Koshikawa S, Matsumoto T, Miura T (2003) Mandibular morphogenesis during soldier differentiation in the damp-wood termite *Hodotermopsis sjoestedti* (Isoptera: Termopsidae). Naturwissenschaften 90:180–184

Koshikawa S, Miyazaki S, Cornette R et al (2008) Genome size of termites (Insecta, Dictyoptera, Isoptera) and wood roaches (Insecta, Dictyoptera, Cryptocercidae). Naturwissenschaften 95:859–867

Lainé LV, Wright DJ (2003) The life cycle of *Reticulitermes* spp. (Isoptera: Rhinotermitidae): what do we know? Bull Entomol Res 93:267–278

Lanzrein B, Gentinetta V, Fehr R (1985) Titers of juvenile hormone and ecdysteroids in reproductives and eggs of *Macrotermes michaelseni*: relation to caste determination? In: Watson JAL, Okot-Kotber BM, Noirot C (eds) Caste differentiation in social insects. Pergamon Press, Oxford, pp 307–327

Le Conte Y, Bécard JM, Costagliola G et al (2006) Larval salivary glands are a source of primer and releaser pheromone in honey bee (*Apis mellifera* L.). Naturwissenschaften 93:237–241

Lefeuve P, Bordereau C (1984) Soldier formation regulated by a primer pheromone from the soldier frontal gland in a higher termite, *Nasutitermes lujae*. Proc Natl Acad Sci U S A 81:7665–7668

Lelis AT, Everaerts C (1993) Effects of juvenile hormone analogues upon soldier differentiation in the termite *Reticulitermes santonensis* (Rhinotermitidae, Heterotermitinae). J Morphol 217:239–261

Leoncini I, Le Conte Y, Costagliola G et al (2004) Regulation of behavioral maturation by a primer pheromone produced by adult worker honey bee. Proc Natl Acad Sci U S A 101:17559–17564

Liang P, Averboukh L, Pardee AB (1993) Distribution and cloning of eukaryotic mRNAs by means of differential display: refinements and optimization. Nucleic Acids Res 21:3269–3275

Liang P, Pardee AB (1992) Differential display of eukaryotic messenger RNA by means of the polymerase chain reaction. Science 257:967–971

Liénard MA, Lassance JM, Paulmier I et al (2006) Differential expression of cytochrome c oxidase subunit III gene in castes of the termite *Reticulitermes santonensis*. J Insect Physiol 52:551–557

Liu Y, Henderson G, Mao L, Laine RA (2005a) Seasonal variation of juvenile hormone titers of the Formosan suberranean termite, *Coptotermes formosanus* (Rhinotermitidae). Environ Entomol 34:557–562

Liu Y, Henderson G, Mao L, Laine RA (2005b) Effects of temperature and nutrition on juvenile homone titers of *Coptotermes formosanus* (Isoptera: Rhinotermitidae). Ann Entomol Soc Am 98:732–737

Lo N, Tokuda G, Watanabe H et al (2000) Evidence from multiple gene sequences indicates that termites evolved from wood-feeding cockroaches. Curr Biol 10:801–804

Lüscher M (1953) The termite and the cell. Sci Am 188:74–78

Lüscher M (1958) Von der Gruppe zum "Staat" bei Insekten. In: Lehmann FE (ed) Gestaltungensozialen Lebens bei Tier und Mensch. Francke Verlag, Bern, pp 48–65

Lüscher M (1960) Hormonal control of caste differentiation in termites. Ann N Y Acad Sci 89: 549–556

Lüscher M (1961) Social control of polymorphism in termites. Symp R Entomol Soc Lond 1:57–67

Lüscher M (1974) Kasten und Kastendifferenzierung bei niederen Termiten. In: Schmidt GH (ed) Sozialpolymorphisms bei Insekten. Wiss Verl Ges, Stuttgart, pp 695–739

Lüscher M (1976) Evidence for an endocrine control of caste determination in higher termites. In: Lüscher M (ed) Phase and caste determination in insects – endocrine aspects. Pergamon Press, Oxford, pp 91–103

Maekawa K, Mizuno S, Koshikawa S, Miura T (2008) Compound eye development during caste differentiation of the termite *Reticulitermes speratus* (Isoptera: Rhinotermitidae). Zool Sci 25:699–705

Mao L, Henderson G (2007) A case for a free-running circannual rhythm in soldier developmetal time of Formosan subterranean termites. Insectes Soc 54:388–392

Mao L, Henderson G, Liu Y, Laine RA (2005) Formosan subterranean termite (Isoptera: Rhinotermitidae) soldiers regulate juvenile homone levels and caste differentiation in workers. Ann Entomol Soc Am 98:340–345

Matsumura F, Coppel HC, Tai A (1968) Isolation and identification of termite trail-following pheromone. Nature 219:963–964

McManus MT, Sharp PA (2002) Gene silencing in mammals by small interfering RNAs. Nat Rev Genet 3:737–747

Meyer D, Lanzrein B, Lüscher M, Nakanishi K (1976) Isolation and identification of a juvenile hormone in termites. Experientia 32:773–780

Miller EM (1969) Caste differentiation in the lower termites. In: Krishna K, Weesner FM (eds) Biology of termites, vol I. Academic Press, New York, NY, pp 283–310

Minelli A (2003) The development of animal form – ontogeny, morphology, and evolution. Cambridge University Press, Cambridge

Miura T (2001) Morphogenesis and gene expression in the soldier-caste differentiation of termites. Insectes Soc 48:216–223

Miura T (2004) Proximate mechanisms and evolution of caste polyphenism in social insects: from society to genes. Ecol Res 19:141–148

Miura T (2005) Developmental regulation of caste-specific characters in social-insect polyphenism. Evol Dev 7:122–129

Miura T, Kamikouchi A, Sawata M et al (1999) Soldier caste-specific gene expression in the mandibular glands of *Hodotermopsis japonica* (Isoptera: Termopsidae). Proc Natl Acad Sci U S A 96:13874–13879

Miura T, Matsumoto T (2000) Soldier morphogenesis in a nasute termite: discovery of a disk-like structure forming a soldier nasus. Proc R Soc Lond B 267:1185–1189

Miura T, Roisin Y, Matsumoto T (1998) Developmental pathways and polyethism of neuter castes in the processional nasute termite *Hospitalitermes medioflavus* (Isoptera: Termitidae). Zool Sci 15:843–848

Moczek AP, Nagy LM (2005) Diverse developmental mechanisms contribute to different levels of diversity in horned beetles. Evol Dev 7:175–185

Moczek AP, Rose D, Sewell W, Kesselring BR (2006) Conservation, innovation, and the evolution of horned beetle diversity. Dev Genes Evol 216:655–665

Myles TG, Nutting WL (1988) Termite eusocial evolution: a re-examination of Bartz's hypothesis and assumptions. Quart Rev Biol 63:1–23

Nakashima KI, Watanabe H, Saitoh H (2002) Dual cellulose digesting system of the wood feeding termite, *Coptotermes formosanus*. Insect Biochem Mol Biol 32:777–784

Nalepa CA (1994) Nourishment and the evolution of termite eusociality. In: Hunt, JH and Nalepa, CA (eds) Nourishment and evolution in insect societies. Westview Press, Boulder, CO, pp 57–104

Nalepa CA, Bandi C (2000) Characterizing the ancestors: paedomorphosis and termite evolution. In: Abe T, Bignell DE, Higashi M (eds) Termites: evolution, sociality, symbioses, ecology. Kluwer Academic Publishers, Dordrecht, pp 53–75

Nelson LJ, Cool LG, Forschler BT, Haverty MI (2001) Correspondence of soldier difence secretion mixtures with cuticular hydrocarbon phenotypes for chemotaxonomy of the termite genus *Reticulitermes* in North America. J Chem Ecol 27:1449–1479

Nelson CM, Ihle KE, Fondrk MK et al (2007) The gene vitellogenin has multiple coordinating effects on social organization. PLoS Biol 5(3):e62

Nene V, Wortman JR, Lawson D et al (2007) Genome sequence of *Aedes aegypti*, a major arbovirus vector. Science 316:1718–1723

Nijhout HF (1994) Insect hormones. Princeton University Press, Princeton, NJ

Nijhout HF (1999) Control mechanisms of polyphenic development in insects. Bioscience 49: 181–192

Nijhout HF (2003) Development and evolution of adaptive polyphenism. Evol Dev 5:9–18

Nijhout HF, Wheeler DE (1982) Juvenile hormone and the physiological basis of insect polyphenism. Quart Rev Biol 57:109–133

Noirot C (1969a) Formation of castes in the higher termites. In: Krishna K, Weesner FM (eds) Biology of termites, vol I. Academic Press, New York, NY, pp 311–350

Noirot C (1969b) Glands and secretions. In: Krishna K, Weesner FM (eds) Biology of termites, vol I. Academic Press, New York, NY, pp 89–123

Noirot C (1985) The caste system in higher termites. In: Watson JAL, Okot-Kotber BM, Noirot C (eds) Caste differentiation in social insects. Pergamon Press, Oxford, pp 75–86

Noirot C (1989) Social structure in termite societies. Ethol Ecol Evol 1:1–17

Noirot C (1990) Sexual castes and reproductive strategies in termites. In: Engels W (ed) Social insects: an evolutionary approach to castes and reproduction. Springer, Berlin, pp 5–35

Noirot C (1991) Caste differentiation in Isoptera: basic features, role of pheromones. Ethol Ecol Evol Special Issue 1:3–7

Noirot C, Pasteels JM (1987) Ontogenetic development and evolution of the worker caste in termites. Experientia 43:851–860

Ogino K, Hirono Y, Matsumoto T, Ishikawa H (1993) Juvenile hormone analogue, S-31183, causes a high level induction of presoldier differentiation in the Japanese damp-wood termite. Zool Sci 10:361–366

Okot-Kotber BM (1980) The influence of juvenile hormone analogue on soldier differentiation in the higher termite, *Macrotermes michaelseni*. Physiol Entomol 5:407–416

Okot-Kotber M, Prestwich GD, Strambi A, Strambi C (1993) Changes in morphogenetic hormone titers in isolated workers of the termite *Reticulitermes flavipes* (Kollar). Gen Comp Endocrinol 90:290–295

Okot-Kotber M, Ujvary I, Mollaaghababa R et al (1991) Physiological influence of fenoxycarb pro-insecticides and soldier head extracts of various termite species on soldier differentiation in *Reticulitermes flavipes*. Sociobiology 19:77–89

Park YI, Raina AK (2003) Factors regulating caste differentiation in the Formosan subterranean termite with emphasis on soldier formation. Sociobiology 41:1–12

Park YI, Raina AK (2004) Juvenile hormone III titers and regulation of soldier caste in *Coptotermes formosanus* (Isoptera: Rhinotermitidae). J Insect Physiol 50: 561–566

Park YI, Raina AK (2005) Regulation of juvenile hormone titers by soldiers in the Formosan subterranean termite, *Coptotermes formosanus*. J Insect Physiol 51:385–391

Pasteels JM, Bordereau C (1998) Releaser pheromones in termites. In: Van der Meer RK, Breed MD, Winston ML, Espelie KE (eds) Pheromone communication in social insects. Westview Press, Boulder, CO, pp 193–215

Pfaffl MW, Tichopad A, Progmet C, Neuvians TP (2004) Determination of stable housekeeping genes, differentially regulated target genes and sample integrity: bestkeeper – excel-based tool using pair-wise correlations. Biotechnol Lett 26:509–515

Prestwich GD (1984) Defense mechanisms of termites. Annu Rev Entomol 29:201–232

Prestwich GD, Jones RW, Collins MS (1981) Terpene biosynthesis by nasute termite soldiers (Isoptera: Nasutitermitinae). Insect Biochem 11:331–336

Quintana A, Reinhard J, Faure R et al (2003) Interspecific variation in terpenoid composition of defensive secretions of European *Reticulitermes* termites. J Chem Ecol 29:639–652

Ramakrishnan VG, Aljamali MN, Sauer JR, Essenberg RC (2005) Application of RNA interference in tick salivary gland research. J Biomol Tech 16:297–305

Ranson H, Nikou D, Hutchinson M et al (2002) Molecular analysis of multiple cytochrome P450 genes from the malaria vector, *Anopheles gambiae*. Insect Mol Biol 11:409–418

Reinhard J, Lacey MJ, Ibarra F et al (2002) Hydroquinone: a general phagostimulating pheromone in termites. J Chem Ecol 28:1–14

Robinson GE (1999) Integrative animal behaviour and sociogenomics. Trends Ecol Evol 14: 202–205

Robinson GE, Grozinger CM, Whitfield CW (2005) Sociogenomics: social life in molecular terms. Nat Rev Genet 6:257–270

Roisin Y (2000) Diversity and evolution of caste patterns. In: Abe T, Bignell DE, Higashi M (eds) Termites: evolution, sociality, symbioses, ecology. Kluwer Academic Publishers, Dordrecht, pp 95–119

Sabater-Muñoz B, Legeai F, Rispe C et al (2006) Large-scale gene discovery in the pea aphid *Acyrthosiphon pisum* (Hemiptera). Genome Biol 7:R21

Sameshima S, Miura T, Matsumoto T (2004) Wing disc development during caste differentiation in the ant *Pheidole megacephala* (Hymenoptera: Formicidae). Evol Dev 6:336–341

Sappington TW, Raikhel AS (1998) Molecular characteristics of insect vitellogenin and vitellogenin receptors. Insect Biochem Mol Biol 28:277–300

Scharf ME, Buckspan CE, Grzymala TL, Zhou X (2007) Regulation of polyphenic caste differentiation in the termite *Reticulitermes flavipes* by interaction of intrinsic and extrinsic factors. J Exp Biol 210:4390–4398

Scharf ME, Ratliff CR, Hoteling JT et al (2003a) Caste differentiation responses of two sympatric *Reticulitermes* termite species to juvenile hormone homologs and synthetic juvenoids in two laboratory assays. Insectes Soc 50:346–354

Scharf ME, Ratliff CR, Wu-Scharf D (2005a) Effects of juvenile hormone III on *Reticulitermes flavipes*: changes in hemolymph protein composition and gene expression. Insect Biochem Mol Biol 35:207–215

Scharf ME, Tartar A (2008) Termite digestomes as sources of novel lignocellulases. Biofuels Bioprod Bioref 2:540–552

Scharf ME, Wu-Scharf D, Pittendrigh BR, Bennett GW (2003b) Caste- and development-associated gene expression in a lower termite. Genome Biol 4(10):R62

Scharf ME, Wu-Scharf D, Zhou X et al (2005b) Gene expression profiles among immature and adult reproductive castes of the termite *Reticulitermes flavipes*. Insect Mol Biol 14:31–44

Scharf ME, Zhou X, Schwinghammer MA (2008) Application of RNA interference in functional genomics studies of a social insect. Methods Mol Biol 442:205–229

Schwinghammer MA, Zhou X, Kambhampati S et al (2011) A novel gene from the takeout family involved in termite trail-following behavior. Gene: In press

Seehuus S-C, Norberg K, Gimsa U et al (2006) Reproductive protein protects sterile honey bee workers from oxidative stress. Proc Natl Acad Sci U S A 103:962–967

Shellman-Reeve JS (1997) The spectrum of eusociality in termites. In: Choe JC, Crespi BJ (eds) The evolution of social behavior in insects and arachnids. Cambridge University Press, Cambridge, pp 52–93

Slaytor M (2000) Energy metabolism in the termite and its gut microbiota. In: Abe T, Bignell DE, Higashi M (eds) Termites: evolution, sociality, symbioses, ecology. Kluwer Academic Publishers, Dordrecht, pp 307–332

Snyder MJ, Scott JA, Andersen JF, Feyereisen R (1996) Sampling P450 diversity by cloning polymerase chain reaction products obtained with degenerate primers. Meth Enzymol 272: 304–312

Stern DL, Emlen DJ (1999) The developmental basis for allometry in insects. Development 126:1091–1101

Stuart AM (1969) Social behavior and communication. In: Krishna K, Weesner FM (eds) Biology of termites, vol I. Academic Press, New York, NY, pp 233–282

Sutherland TD, Unnithian GC, Feyereisen R (2000) Terpenoid-hydroxylase (*CYP4C7*) messenger RNA levels in the corpora allata: a marker for ovarian control of juvenile hormone synthesis in *Diploptera punctata*. J Insect Physiol 46:1219–1227

Takeda M, Mita K, Quan GX et al (2001) Mass isolation of cuticle protein cDNAs from wing discs of *Bombyx mori* and their characterizations. Insect Biochem Mol Biol 31:1019–1028

Tartar A, Wheeler MM, Zhou X et al (2009) Parallel metatranscriptome analyses of host and symbiont gene expression in the gut of the termite *Reticulitermes flavipes*. Biotechnol Biofuels 2:25

Tarver MR, Schmelz EA, Scharf ME (2009) Effects of soldier-derived terpenes on soldier caste differentiation in the termite *Reticulitermes flavipes*. J Chem Ecol 35:256–264

Tarver MR, Zhou X, Scharf ME (2010) Socio-environmental and semiochemical impacts on phenotypic plasticity and gene expression in a social insect, *R. flavipes*. BMC Mol Biol 11:28

Tatar M, Bartke A, Antebi A (2003) The endocrine regulation of aging by insulin-like signals. Science 229:1346–4351

Thompson GJ, Kitade O, Lo N, Crozier RH (2000) Phylogenetic evidence for a single, ancestral origin of a 'true' worker caste in termites. J Evol Biol 13:869–881

Thorne BL (1996) Termite terminology. Sociobiology 28:253–263

Thorne BL, Traniello JFA (2003) Comparative social biology of basal taxa of ants and termites. Annu Rev Entomol 48:283–306

Tijet N, Helvig C, Feyereisen R (2001) The cytochrome P450 gene superfamily in *Drosophila melanogaster*: annotation, intron-exon organization and phylogeny. Gene 262:189–198

Timmons L, Fire A (1998) Specific interference by ingested dsRNA. Nature 395:854

Tittiger C (2004) Functional genomics and insect chemical ecology. J Chem Ecol 30:2335–2358

Todaka N, Moriya S, Saita K et al (2007) Environmental cDNA analysis of the genes involved in lignocellulose digestion in the symbiotic protist community of *Reticulitermes speratus*. FEMS Microbiol Ecol 59:592–599

Tokuda G, Saito H, Watanabe H (2002) A digestive beta-glucosidase from the salivary glands of the termite, *Neotermes koshunensis* (Shiraki): distribution, characterization and isolation of its precursor cDNA by 5/- and 3/-RACE amplifications with degenerate primers. Insect Biochem Mol Biol 32:1681–1689

Toth AL, Robinson GE (2007) Evo-devo and the evolution of social behavior. Trends Genet 23:334–341

Traniello JFA, Leuthold RH (2000) Behavior and ecology of foraging in termites. In: Abe T, Bignell DE, Higashi M (eds) Termites: evolution, sociality, symbioses, ecology. Kluwer Academic Publishers, Dordrecht, pp 141–168

Truman JW, Hiruma K, Allee JP (2006) Juvenile hormone is required to couple imaginal disc formation with nutrition in insects. Science 312:1385–1388

Tsuchiya M, Watanabe D, Maekawa K (2008) Effect on mandibular length of juvenile hormones and regulation of soldier differentiation in the termite *Reticulitermes speratus* (Isoptera: Rhinotermitidae). Appl Entomol Zool 43:307–314

Vander Meer RK, Alonso LE (1998) Pheromone directed behavior in ants. In: Vander Meer RK, Breed MD, Winston ML, Espelie KE (eds) Pheromone communication in social insects. Westview Press, Boulder, CO, pp 159–192

Vargo EL, Husseneder C (2009) Biology of subterranean termites: insights from molecular studies of *Reticulitermes* and *Coptotermes*. Annu Rev Entomol 54:379–403

Waller DA, LaFage JP (1988) Environmental influence on soldier differentiation in *Coptotermes formosanus* Shiraki (Rhinotermitidae). Insectes Soc 35:144–152

Warnecke F et al (2007) Metagenome and functional analysis of hindgut microbiota of a wood-feeding higher termite. Nature 450:487–488

Watanabe D, Maekawa K (2008) Frontal-pore formation during soldier differentiation induced by juvenile hormone III in the termite *Reticulitermes speratus* (Isoptera: Rhinotermitidae). Sociobiology 52:437–447

Watson JAL, Sewell JJ (1985) Caste development in *Mastotermes* and *Kalotermes*. In: Watson JAL, Okot-Kotber BM, Noirot C (eds) Caste differentiation in social insects. Pergamon Press, Oxford, pp 27–40
Weesner FM (1969) External anatomy. In: Krishna K, Weesner FM (eds) Biology of termites, vol I. Academic Press, New York, NY, pp 19–47
Weil T, Rehli M, Korb J (2007) Molecular basis for the reproductive division of labour in a lower termite. BMC Genomics 8:198
West-Eberhard MJ (1987) Flexible strategy and social evolution. In: Ito Y, Brown JL, Kikkawa J (eds) Animal societies: theories and facts. Japan Scientific Societies Press, Tokyo, pp 35–51
West-Eberhard MJ (1996) Wasp societies as microcosms for the study of developmental and evolution. In: Turillazzi S, West-Eberhard MJ (eds) Natural history and evolution of paper wasps. Oxford University Press, New York, NY, pp 35–51
West-Eberhard MJ (2003) Developmental plasticity and evolution. Oxford University Press, Oxford
Wheeler DE (1986) Developmental and physiological determinants of caste in social Hymenoptera: evolutionary implications. Am Nat 128:13–34
Wheeler DE, Buck N, Evans JD (2006) Expression of insulin pathway genes during the period of caste determination in the honey bee, *Apis mellifera*. Insect Mol Biol 15:597–602
Wheeler DE, Nijhout HF (2003) A perspective for understanding the modes of juvenile hormone action as a lipid signaling system. Bioessays 25:994–1001
Wheeler MM, Zhou X, Scharf ME, Oi FM (2007) Molecular and biochemical markers for monitoring dynamic shifts of cellulolytic protozoa in *Reticulitermes flavipes*. Insect Biochem Mol Biol 37:1366–1374
Whitfield CW, Band MR, Bonaldo MF (2002) Annotated expressed sequence tags and cDNA microarrays for studies of brain and behavior in the honey bee. Genome Res 12: 555–566
Whitfield CW, Cziko AM, Robinson GE (2003) Gene expression profiles in the brain predict behavior in individual honey bees. Science 302:296–299
Wigglesworth VB (1935) Functions of the corpus allatum of insects. Nature 136:338–339
Williams G, Cai XJ, Elliot JC, Harrold JA (2004) Anabolic neuropeptides. Physiol Behav 81: 211–222
Willis J (1999) Cuticular proteins in insects and crustaceans. Am Zool 39:600–609
Wilson EO (1971) The insect societies. Harvard University Press, Cambridge, MA
Wilson EO (1975) Sociobiology. Belknap press, Cambridge, MA
Wilson EO, Bossert WH (1963) Chemical communication among insects. Recent Prog Horm Res 19:673–716
Wu-Scharf D, Scharf ME, Pittendrigh BR, Bennett GW (2003) Expressed sequence tags from a polyphenic *Reticulitermes flavipes* cDNA library. Sociobiology 41:479–490
Yagi KJ, Kwok R, Chan KK (2005) Phe-Gly-Leu-amide allatostatin in the termite *Reticulitermes flavipes*: content in brain and corpus allatum and effect on juvenile hormone synthesis. J Insect Physiol 51:357–365
Yang AS (2007) Thinking outside the embryo: the superorganism as a model for evo-devo studies. Biol Theor 2:398–408
Zegzouti H, Marty C, Jones B et al (1997) Improved screening of cDNAs generated by mRNA differential display enables the selection of true positives and the isolation of weakly expressed messages. Plant Mol Biol Rep 15:236–245
Zhou X, Oi FM, Scharf ME (2006a) Social exploitation of hexamerin: RNAi reveals a major caste-regulatory factor in termites. Proc Natl Acad Sci U S A 103:4499–4504
Zhou X, Smith JA, Oi FM et al (2007a) Correlation of cellulase gene expression and cellulolytic activity throughout the gut of the termite *Reticulitermes flavipes*. Gene 395:29–39
Zhou X, Song C, Grzymala TL et al (2006b) Juvenile hormone and colony conditions differentially influence cytochrome P450 gene expression in the termite *Reticulitermes flavipes*. Insect Mol Biol 15:749–761

Zhou X, Tarver MR, Bennett GW (2006c) Two hexamerin genes from the termite *Reticulitermes flavipes*: sequence, expression, and proposed functions in caste regulation. Gene 376:47–58

Zhou X, Tarver MR, Scharf ME (2007b) Hexamerin-based regulation of juvenile hormone-dependent gene expression underlies phenotypic plasticity in a social insect. Development 134:601–610

Zhou X, Wheeler MM, Oi FM, Scharf ME (2008) RNA interference in the termite *Reticulitermes flavipes* through ingestion of double-stranded RNA. Insect Biochem Mol Biol 38:805–815

Chapter 10
Sexual and Asexual Reproduction in Termites

Kenji Matsuura

Abstract The evolution and maintenance of sexual reproduction is believed to involve important tradeoffs. Queens of social insects face a dilemma over the costs and benefits of sexual and asexual reproduction. Asexual reproduction by a queen doubles her contribution to the gene pool. However, overuse of asexual reproduction reduces the offspring's genetic diversity and thus the colony's ability to adapt to environmental stress. Recent investigations on breeding systems using molecular markers revealed that queens of some termites can solve this tradeoff by their conditional use of sexual and asexual reproduction, where queens produce additional (and/or subsequent) queens by parthenogenesis, but use sexual reproduction to produce workers. The asexual queen succession (AQS) system enables the primary queen to maintain her full genetic contribution to the next generation, while avoiding any loss in genetic diversity from inbreeding. In other words, this system gives, in effect, genetically eternal lives to the primary queens. In this chapter, I discuss how eusociality, with its attendant caste structure and unique life histories, can generate novel reproductive and genetic systems with mixed modes of reproduction. This provides important insights into the advantages and disadvantages of sexual reproduction.

10.1 Introduction

Thelytoky is a type of parthenogenesis in which females are produced from unfertilized eggs. Among the 12,500 ant and 2,400 described termite species, thelytoky has been reported in only 14 ant species and seven termite species (Table 10.1). The percentages of species with thelytokous capability are 0.01 and 0.29% in ants

K. Matsuura (✉)
Laboratory of Insect Ecology, Graduate School of Environmental Science, Okayama University, Okayama 700-8530, Japan
e-mail: kenjijpn@cc.okayama-u.ac.jp

D.E. Bignell et al. (eds.), *Biology of Termites: A Modern Synthesis*,
DOI 10.1007/978-90-481-3977-4_10,

Table 10.1 Thelytokous parthenogenesis in termites and social Hymenoptera

Order/family	Species	Frequency	Modes of ploidy restoration	References
Isoptera				
Termopsidae	*Zootermopsis angusticollis*	Facultative	Unknown	Light (1944)
	Z. nevadensis	Facultative	Unknown	Light (1944)
Kalotermitidae	*Kalotermes flavicollis*	Facultative	Unknown	Grassé (1949)
	Bifiditermes beesoni	Facultative	Unknown	Chhotani (1962) and Afzal and Salihah (1985)
Rhinotermitidae	*Reticulitermes virginicus*	Facultative, AQS	Terminal fusion	Howard et al. (1981) and Matsuura et al. (unpublished)
	R. speratus	Facultative, AQS	Terminal fusion	Matsuura and Nishida (2001) and Matsuura et al. 2004, 2009)
Termitidae	*Velocitermes* sp.	Facultative	Unknown	Stansly and Korman (1993)
Hymenoptera:				
Formicidae	*Platythyrea punctata*	Obligate/ facultative	Central fusion	Heinze and Hölldobler (1995) and Hartmann et al. (2003, 2005)
	Cerapachys biroi	Obligate	Central fusion	Tsuji and Yamauchi (1995)
	Messor capitatus	Facultative	Unknown	Grasso et al. (2000)
	Pristomyrmex pungens	Obligate	Central fusion	Tsuji (1988)
	Cataglyphis cursor	AQS	Central fusion	Cagniant (1979) and Pearcy et al. (2004)
	Cataglyphis piliscapus	Facultative	Unknown	Lenoir et al. (1988)
	Cataglyphis sabulosa	Facultative	Unknown	Timmermans et al. (2008)
	Wasmannia auropunctata	AQS	Central fusion	Fournier et al. (2005)
	Vollenhovia emeryi	AQS	Central fusion	Ohkawara et al. (2006)
	Mycocepurus smithii	Obligate	Unknown	Rabeling et al. (2009)
	Pyramica membranifera	Unknown	Unknown	Ito et al. (2010)
	Monomorium triviale	Obligate	Unknown	Tsuji (personal Communication)
	Anoplolepis gracilipes	Facultative and AQS	Central fusion and apomixis	Gruber et al. (2010)
	Paratrechina longicornis	AQS	Unknown	Pearcy et al. (2010)
Apidae	*Apis mellifera capensis*	Obligate	Central fusion	Tucker (1958) and Verma and Ruttner (1983)

AQS: Asexual Queen Succession

and termites, respectively. These values are much lower than 2%, which is the percentage of all insect species with thelytokous capabilities (Bell 1982). By taking advantage of asexual reproduction, queens of eusocial insects should be able to realize a twofold advantage over sexual reproduction by allowing the transmission of twice the number of genes to offspring. This leads to the very interesting question of why thelytoky is so rare among eusocial insects.

One possible explanation is that only a limited number of studies have carefully examined the possibility of thelytokous parthenogenesis in eusocial insects. Thelytoky may be more difficult to detect in eusocial insects than in other insects because its presence is sometimes concealed or suppressed by social structure. Future studies may reveal thelytoky in many more eusocial species. Possibly, the rarity of thelytoky in eusocial insects is attributable to the importance of genetic diversity among colony members. While asexual reproduction by a queen would increase within-colony relatedness, the resulting reduction in genetic diversity within colonies would lower homeostasis in colonies (Oldroyd and Fewell 2007). Once workers lose reproductive totipotency, the evolution of the breeding system is directed towards higher intracolonial genetic diversity rather than higher relatedness (Hughes et al. 2008). For example, colonies low in genetic diversity are more afflicted by disease than genetically diverse colonies in bumblebees (Baer and Schmid-Hempel 2001), honeybees (Tarpy 2003; Seeley and Tarpy 2007) and leaf-cutting ants (Hughes and Boomsma 2004). In honeybees, high-diversity colonies maintained more uniform temperatures in their brood nests than did the uniform ones due to a system of genetically based task specialization (Jones et al. 2004). Thus, reduction in genetic diversity would render the colonies less resilient to environmental perturbation.

Perhaps the best solution to the dilemma over the costs and benefits of sexual and asexual reproduction is to use both modes of reproduction conditionally and therefore to experience the advantages of both. Indeed, recent studies have uncovered unusual modes of reproduction both in ants (Pearcy et al. 2004; Fournier et al. 2005; Ohkawara et al. 2006) and termites (Matsuura et al. 2009), in which advantage is taken of the social caste system to use sex for somatic growth, but parthenogenesis for germ line production.

The capability of parthenogenesis in Isoptera was first reported by Light (1944). However, the reproductive biology of termite parthenogenesis has not been examined in detail beyond its notional documentation until recently. This is largely because parthenogenetic reproduction has been regarded as an unusual case with little adaptive significance in nature. Even after the adaptive significance of termite parthenogenesis was recognized, researchers believed that the function of parthenogenesis in termites was no more than "the best of a bad job", that is, females used parthenogenesis only when they failed to mate with males (Matsuura and Nishida 2001; Matsuura et al. 2002). Most recently, however, it was revealed that parthenogenesis in some termite species plays a much greater role than has been previously understood. In the Japanese subterranean termite *Reticulitermes speratus* (Matsuura et al. 2009) and the North American subterranean termite *R. virginicus*

(Matsuura and Vargo unpublished observations), queens exclusively use parthenogenesis to produce secondary (neotenic) queens. On the other hand, queens produce workers and alates by outcrossing with the primary king. This model of asexual queen succession (AQS) or conditional use of sexual and asexual reproduction can be studied to understand the advantages and disadvantages of thelytoky in termites.

10.2 Facultative Parthenogenesis in Maleless Colony Foundation

Facultatively parthenogenetic species, in which females reproduce parthenogenetically if they fail to mate with males, are widespread in animals (Roth and Willis 1956; Nur 1971; Kurup and Prabhoo 1977; Kramer and Templeton 2001). The importance of studying facultative parthenogenesis has been recognized as a means to investigate the evolution and maintenance of sex because it is possible to directly compare individuals with the capacity to reproduce both sexually and asexually (e.g. Corley and Moore 1999; Matsuura and Nishida 2001; Corley et al. 2001; Ball 2002). Facultative parthenogenesis enables females to produce offspring by themselves, whereas females would only die if mating failed and they could not reproduce parthenogenetically (Cuellar 1977). Therefore, facultative parthenogenesis is likely to be advantageous in certain situations, even if ultimately it may be inferior to sexual reproduction in terms of long-term fitness.

In *R. speratus*, females that fail to pair with males found colonies cooperatively with partner females, or even alone and reproduce by parthenogenesis (Matsuura and Nishida 2001; Matsuura et al. 2002). After swarming, an alate breaks off its wings as soon as it alights, and the resultant dealated reproductive runs on the ground until it encounters a potential partner with which to form a tandem running pair. The pair then locates a suitable nest site, with the male following the female. Homosexual tandems often occur when two consexual dealates encounter one another (Matsuura et al. 2002). Excavation into nest wood, i.e., initiation of colony foundation, is significantly more delayed in single female than female–female (FF) and female–male (FM) pairs. Actually, FF pairs initiate colony foundation earlier than FM pairs.

Observation of spermatheca of female alates confirmed the absence of insemination in natal nests before swarming, indicating that females in maleless colonies have no way to produce offspring by sexual reproduction (Fig. 10.1). Females of FF pairs begin to lay eggs at the same time as those of FM pairs. The parthenogenetic eggs are significantly larger than sexual eggs (Matsuura and Kobayashi 2007). This might be an adaptive adjustment of egg size by the queens founding the colony without kings' contributions to brood care, because it is generally known that larger offspring do better under adverse environments than small offspring (Mousseau and Fox 1998). While the two types of eggs have similar hatching rates, parthenogenetic eggs have longer hatching periods (36.36±0.16 days) than sexual eggs (34.95±0.12).

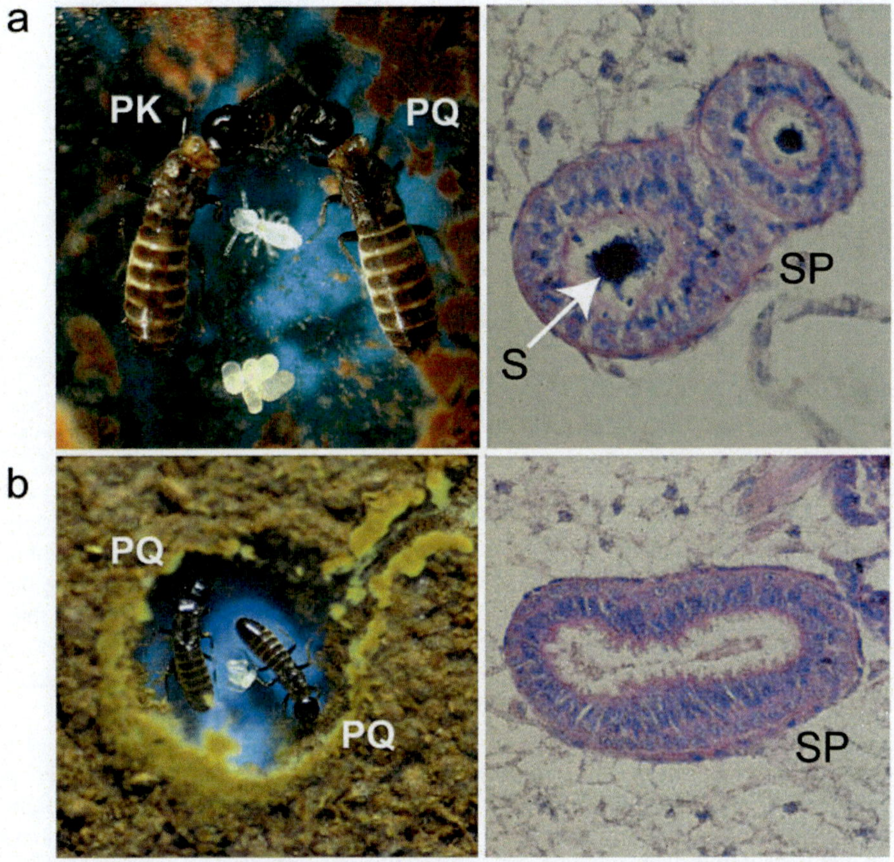

Fig. 10.1 Colony foundation by a monogamous pair (**a**) and a female–female pair (**b**). Sperms (S) were clearly observed in the spermatheca (SP) of the female of the monogamous pair, while no sperm was observed in the spermatheca of the female of the female–female pair

The survival rate of single females was significantly lower than that of FF- and FM-pair females, although there was no difference between FF and FM pairs. This result demonstrates that cooperation, even female–female, promotes female survivorship. The number of progeny per female was significantly lower in FF pairs than in FM pairs, possibly because females of FF pairs must share reproductive output. These results lead us to the conclusion that a normal monogamous pair is the best unit for colony foundation. Nevertheless, females alone can establish colonies by parthenogenesis, and even female–female cooperation promotes colony-foundation success if pairing with males is not possible.

In FF incipient colonies, females spent a lot of time allogrooming, i.e. grooming other individuals (Matsuura et al. 2002, Fig. 10.2a). It has been reported that the saliva of termites has fungistatic properties (Thomas 1987; Rosengaus et al. 1998; Matsuura et al. 2000) and anti-bacterial enzymes (Matsuura et al. 2007). This cooperative hygienic behavior should therefore be very important for termite

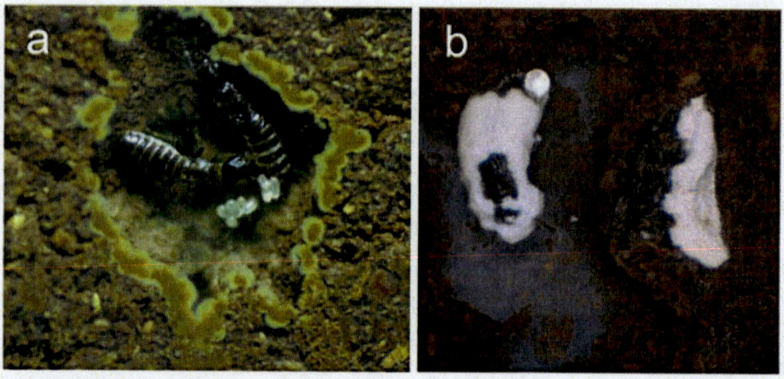

Fig. 10.2 Female–female cooperation for survivorship. (**a**) Allogrooming of a female–female pair. (**b**) Dead females of a female–female killed by a fungal pathogen. If a female dies, the other female fails to survive without a partner

foundresses, because, unlike the social hymenopterans, they are able to clean only their antennae by self-grooming. Termites typically establish colonies in rotten wood with abundant bacteria and fungi, including natural pathogens. Therefore the absence of grooming by partners may render single females unsanitary and thus increase their risk of infection. Without the help of partners, single females are burdened with all the labor associated with initial colonization. This heavy workload of single females may also explain their high mortality.

In FF colonies, the second female rarely survives if the other female dies (Fig. 10.2b). Therefore, the two females behave altruistically with each other in early foundation stage. Initially, smaller females gain significantly more weight than initially larger females resulting from altruistic behavior of the larger females, including asymmetric trophallaxis and labor allocation. On the other hand, female–female cooperation involves reproductive conflict because two females must share reproductive output. The developmental and ecological constraints of asexual reproduction would offset the twofold fitness advantage of asexual reproduction; female–female colony foundation can thus be considered a “best of a bad job” strategy for termites.

10.3 Mechanism of Termite Parthenogenesis

Research on chromosome numbers of Isoptera showed that higher termites (Termitidae) are karyotypically uniform ($2n=42$), while lower termites are more variable, with diploid numbers ranging from 28 to 56 (Bergamaschi et al. 2007). *Reticulitermes* termites have a fixed number of chromosomes ($2n=42$; Fig. 10.3a, b). Karyotypic chromosome observations showed that parthenogenetic offspring are diploid, with $2n=42$ chromosomes in *R. speratus* (Fig. 10.3c). Although termite sex determination mechanisms are not completely understood, males commonly appear heterogametic in termites (Roisin 2001). Interchange

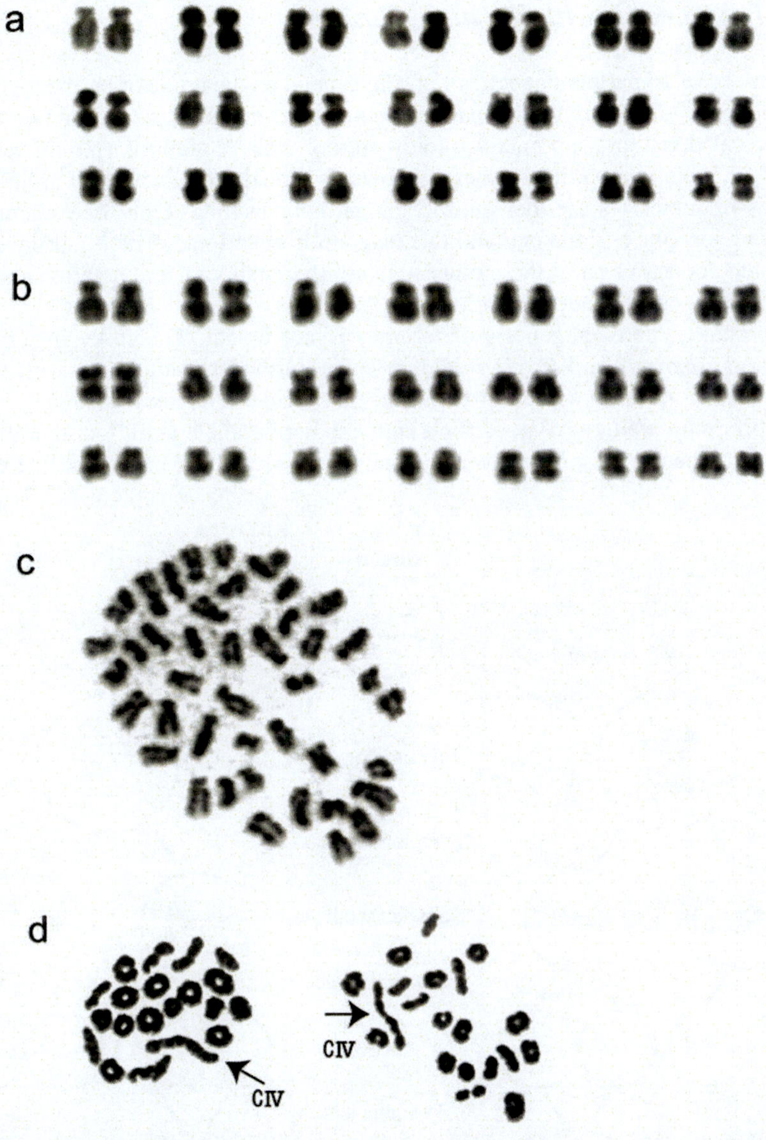

Fig. 10.3 Giemsa-stained karyotype of *Reticulitermes speratus*. (**a**) Male; $2n = 42$. (**b**) Female; $2n = 42$. (**c**) Parthenogenetically produced offspring (all female); $2n = 42$. (**d**) Chromosome multivalent chain (CIV) is observed in male meiosis

multiples (chains or rings of chromosome) in the male meiosis is common in termites (Syren and Luykx 1977). Chromosome multivalents have been observed in male meiosis in *R. speratus*, generating a multiple-X, multiple-Y system (Fig. 10.3d; Matsuura 2002). Because of the XY sex determination system, termite parthenogenesis produces only female progeny.

10.3.1 Automixis with Terminal Fusion

The genotypes of parthenogenetic offspring depend on the mode of parthenogenesis (Templeton 1982). Thelytokous parthenogenesis can be categorized into two major cytological divisions, "apomixis (ploidy stasis)" and "automixis (ploidy restoration)." In apomictic parthenogenesis, known as clonal reproduction in aphids, the features of meiosis are either entirely or partially lacking. Only one maturation division takes place in the egg and this division is equational. The offspring retain the genetic constitution of the mother (excluding mutations), and heterozygosity is maintained in subsequent generations.

Importantly, parthenogenesis of termites is not clonal. Thelytoky in termites, such as *R. speratus* and *R. virginicus*, is accomplished by automixis with terminal fusion, in which two haploid pronuclei that divide at meiosis fuse (Fig. 10.4). Thus, offspring are homozygous for a single maternal allele at all loci that did not crossover, whereas offspring have the same genotype as their mother at loci where

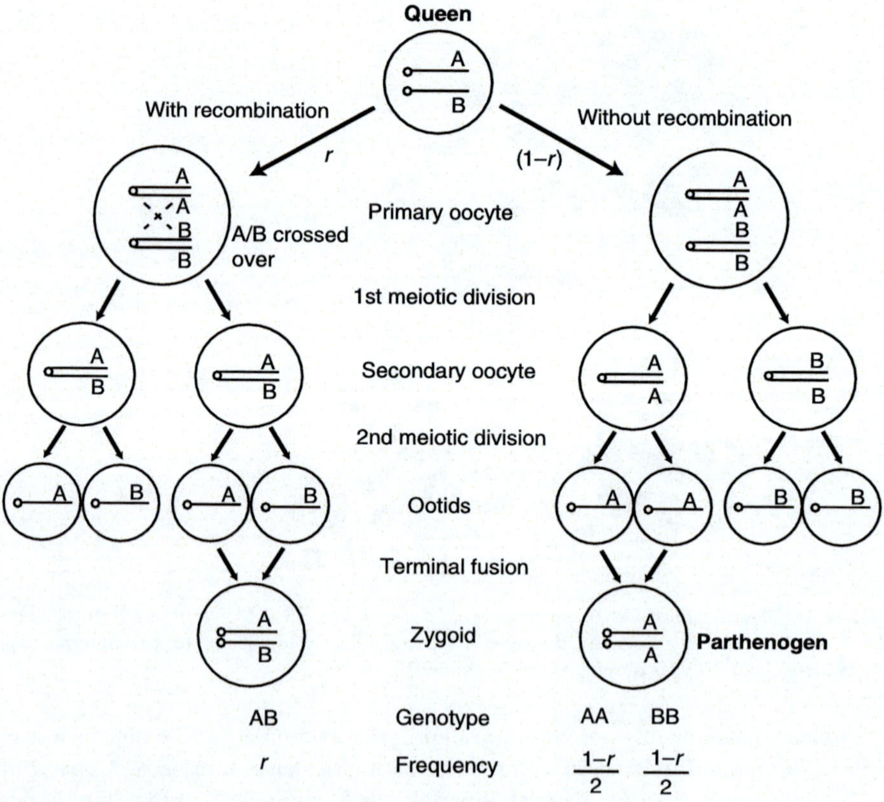

Fig. 10.4 Scheme of ploidy restoration by automixis with terminal fusion. This type of parthenogenesis results in rapid reduction of heterozygosity. *r*, recombination rate

crossover occurred. This causes a rapid reduction of heterozygosity (Matsuura et al. 2004).

10.3.2 Developmental Constraints

It is known that parthenogens are generally less viable than sexually produced offspring in various insect species (Lamb and Willey 1987; Corley and Moore 1999). There is no significant difference in hatching rate between sexually produced eggs and parthenogenetically produced eggs (Matsuura and Kobayashi 2007). However, comparison of the number of first-brood offspring between 50 and 400 days showed a significant reduction in number of individuals in FF colonies, but no difference in FM colonies. In *Reticulitermes*, most of the first-brood eggs are deposited within 50 days after pairing (e.g. Weesner 1956). All of the first-brood offspring develop into the neuter pathway. Therefore, this result suggests a lower survival rate of parthenogenetically produced workers than sexually produced workers. The survival rate of parthenogens at 400 days was 68.5%, compared with sexually produced offspring which survived at 100%. The higher mortality of parthenogenetically produced workers than sexually produced workers is most likely to be explained by the near-total homozygosity due to terminal fusion. This suggests that producing workers by parthenogenesis imposes much higher costs on the colonies.

10.4 Asexual Queen Succession (AQS)

In most termite species, one king and one queen usually found colonies. In termites, especially in lower termites, it has long been believed that the inbreeding cycles of generations of neotenic reproductives (offspring of the colony) propagate the colony after the death of the primary king and queen (Bartz 1979; Thorne et al. 1999). Evidence for inbreeding depression in termites is mounting, such as higher mortality in inbred incipient colonies and lower lifetime fecundity in inbred colonies (DeHeer and Vargo 2006; Fei and Henderson 2003). Like most subterranean termites, *Reticulitermes* species have cryptic nesting habits with transient, hidden royal chambers underground or deep inside wood, making it difficult to reliably collect reproductives. Therefore, the breeding system of subterranean termites has been primarily estimated by genotyping workers or culturing laboratory colonies rather than making censuses of field colonies.

10.4.1 Composition of Reproductives in Field Colonies

Reticulitermes speratus is the most common termite in Japan. To date, we have collected more than 1,000 nests in the field to obtain reproductives from a sufficient number of natural colonies. We successfully found the royal chambers, where reproductives and young broods were protected, of 47 colonies (updated data after Matsuura et al. 2009). In nearly all cases, primary kings were continuously present

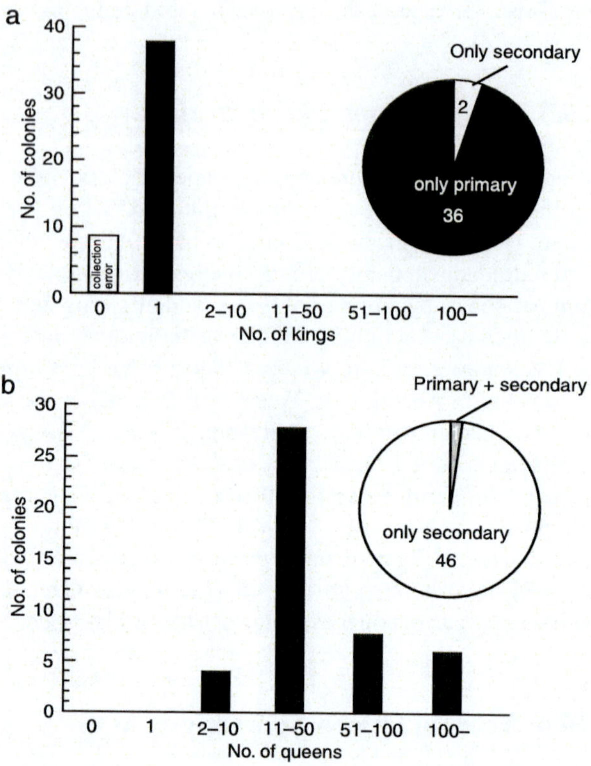

Fig. 10.5 Number of reproductives in field colonies of *Reticulitermes speratus*. (**a**) Number of kings in a colony. The *pie chart* shows that primary kings were present in most colonies. (**b**) Number of queens in a colony. Primary queens had been replaced by secondary queens in nearly all field colonies. Data were updated after Matsuura et al. (2009)

(Fig. 10.5a), but primary queens had been replaced by an average of 64.4±16.1 secondary queens (Fig. 10.5b). The largest colony had 676 secondary queens but had only a single primary king (Matsuura et al. unpublished data; Fig. 10.6). This is a new record of the largest harem in nature since it exceeds the largest harem size

Fig. 10.6 The biggest harem in nature, where a single primary king (*arrow*) and 676 secondary queens reproduce

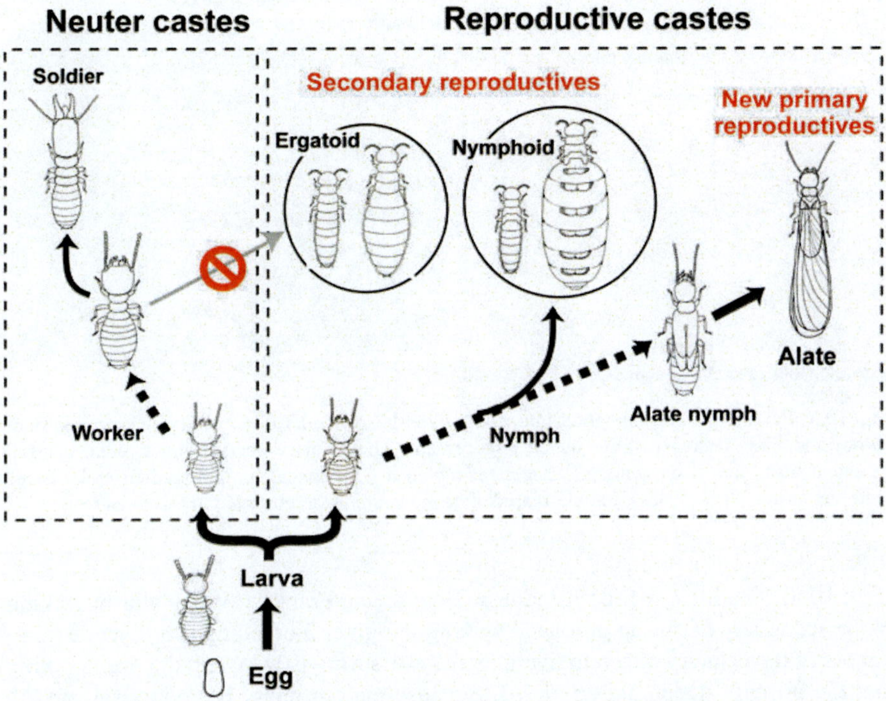

Fig. 10.7 Differentiation pathways of primary and secondary reproductives in the genus *Reticulitermes*. In field colonies of *R. speratus*, secondary queens always differentiate from nymphs but never from workers ($n = 3{,}029$)

(up to 250 females) in elephant seals (Modig 1996). These results indicate that primary kings live much longer than primary queens; replacement of the primary king is rare, whereas replacement of the primary queen is the rule at a certain point in colony development. Rare occurrence of secondary kings in natural colonies shows that colonies terminate soon after the death of primary kings. In addition, secondary reproductives always differentiate from nymphs but never from workers in natural colonies of *R. speratus* ($n = 3{,}029$) (Fig. 10.7).

10.4.2 The Paradox of the King-Daughter Inbreeding Hypothesis

Sexual reproduction can lead to important conflicts between sexes and within genomes. In monogamous termites, conflicts between the primary king and queen can arise over parental investment and genetic contribution to offspring. Our finding that primary queens are replaced much earlier than primary kings in *R. speratus* leads to a paradox if the secondary queens are the daughters of the primary king. King-daughter inbreeding should result in uneven genetic contribution to the secondary offspring (offspring of secondary queens) by the primary king and queen

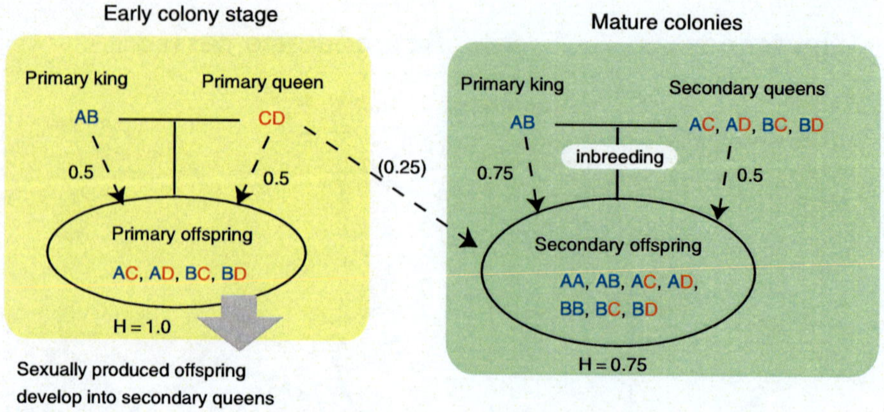

Fig. 10.8 Paradox of king-daughter inbreeding hypothesis. Under this system, heterozygosity of colony members decreases after queen replacement. The primary queen cannot retain her full genetic contribution to descendants after she is replaced. This sexual conflict predicts male-biased alate sex ratios. Nevertheless, female-biased alate sex ratios were observed in this termite

(Fig. 10.8). The life-for-life relatedness (Bourke and Franks 1995) of a primary king to the secondary offspring produced by king-daughter inbreeding is 5/8, while relatedness of the primary queen to the secondary offspring is 1/4 when the primary king and the primary queen are unrelated. Because male primary reproductives are 2.5 times more related to offspring than is the primary queen under this system, colonies are expected to bias alate (new primary reproductives that disperse) production in favor of males. Contrary to this prediction, the actual alate sex ratio is slightly but significantly female-biased in this species (numerical ratio of male = $0.43 \pm 0.02_{\mathrm{SE}}$). Because of the larger size of females relative to males, the biomass sex ratio was even more biased toward females (investment ratio of male: 0.415 ± 0.02). The inconsistency between king-daughter inbreeding and sex investment ratio in alate production suggests that there is a different breeding system in which the king and queen have more equal genetic contributions to offspring.

10.4.3 Parthenogenesis for Secondary Queens, but Sex for Workers and Alates

While examining genotypes within nests of *R. speratus*, an extraordinary mode of reproduction was uncovered (Matsuura et al. 2009). Secondary queens are almost exclusively produced parthenogenetically by the founding primary queens, whereas workers and alates were produced by sexual reproduction (Fig. 10.9). By using parthenogenesis to produce secondary queens, primary queens are able to retain the transmission rate of their genes to descendants while maintaining genetic diversity in the workers and new primary reproductives even after the primary queen is replaced (Fig. 10.10). The relatedness of the primary queens to workers (r = 0.49, $\mathrm{SE}_{\mathrm{jackknife}}$ = 0.04) and to alate nymphs (r = 0.58,

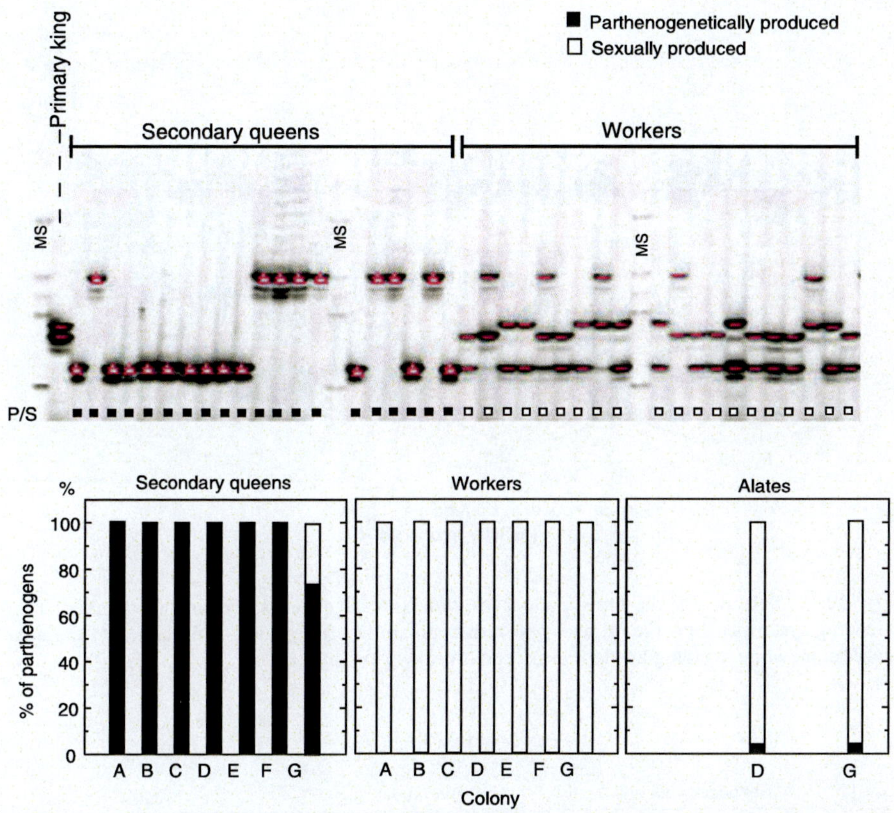

Fig. 10.9 Proportion of parthenogenetically produced offspring (*black*) and sexually produced offspring (*white*) in secondary queens, workers and alates. Secondary queens have only maternal (primary queens') alleles, while workers and alates have both maternal and paternal alleles indicating conditional use of sexual and asexual reproduction. MS, molecular standard. Electrophoresis photo by E. L. Vargo

$SE_{jackknife} = 0.079$) is not significantly different from 0.5, the value expected between a female and her sexual offspring. This is twice the expected genetic contribution queens would make to colony members under king-daughter inbreeding ($r_{\text{primary queen to king-daughter offspring}} = 0.25$).

Parthenogenetic production of secondary queens allows *R. speratus* to undergo queen succession without inbreeding. Heterozygosity of workers in colonies headed by secondary queens was as high ($H_o = 0.733$) as that expected for offspring produced by outcrossing of the primary king and the primary queen ($H_e = 0.736$) (Fig. 10.11). Likewise, there was no significant reduction of heterozygosity in nymphs produced in colonies with secondary queens. Further evidence of the lack of inbreeding in *R. speratus* colonies is provided by the low inbreeding coefficient of workers, which did not differ significantly from zero ($F_{IT} = 0.014$, $SE_{jackknife} = 0.048$, over all loci). The lack of consanguineous mating in this breeding system

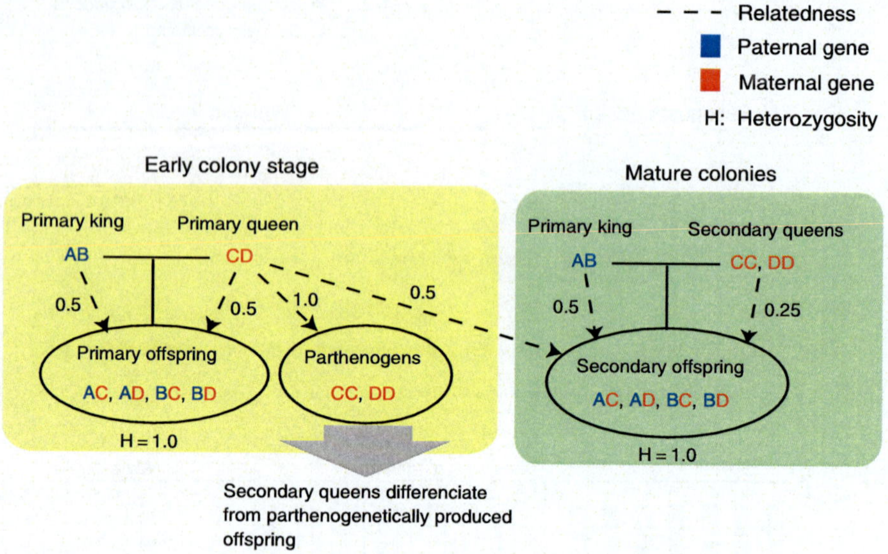

Fig. 10.10 Scheme for the breeding system with asexual queen succession in termites. This breeding system enables the primary queen to maintain her full genetic contribution to the next generation, while avoiding any loss in genetic diversity from inbreeding

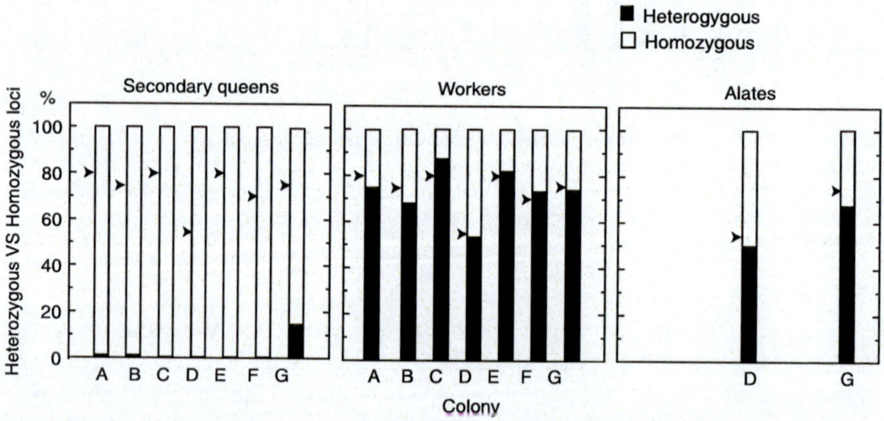

Fig. 10.11 Proportion of heterozygous loci (*black*) and homozygous loci (*white*) in secondary queens, workers and alates. The amount of heterozygosity expected for offspring produced by outcrossing of the primary king and the primary queen are indicated by *arrowheads*

may also benefit primary kings. The offspring produced by outcrossing between the king and parthenogenetic queens may have greater fitness than those produced by king-daughter inbreeding.

The production of secondary queens through conditional parthenogenesis effectively extends the reproductive life of the primary queen, greatly expanding her

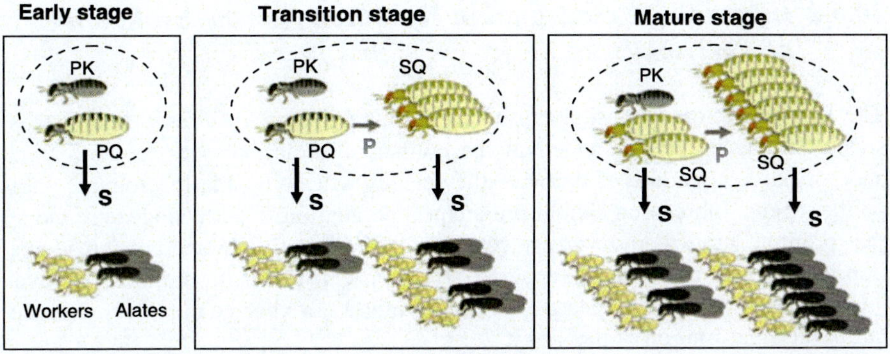

Fig. 10.12 Schematic diagram of asexual queen succession. PK, primary king; PQ, primary queen; SQ, secondary queen; P, parthenogenesis; S, sexual reproduction

reproductive capacity (Fig. 10.12). This process of queen succession allows the colony to boost its size and possibly its growth rate without suffering any loss in genetic diversity or diminishing the transmission rate of the queen's genes to her grand offspring, feats that would not be possible if secondary queens were produced by normal sexual reproduction.

10.5 Parthenogenesis and Recessive Deleterious Genes

10.5.1 Purging Selection

A faster rate of accumulation of deleterious mutations is a major cost of asexual reproduction (Kondrashov 1993). In haplodiploid organisms, deleterious alleles are directly exposed to selection each generation in the haploid males, and there is no masking effect of dominance. Therefore, purging selection will cause a more rapid decrease in the frequency of deleterious alleles at haplodiploid loci than at comparable loci in diploid organisms (Goldstein 1994).

In termites, paradoxically, asexual queen succession can function to purge deleterious mutations. Parthenogenetic offspring are homozygous for a single maternal allele at most of loci due to terminal fusion under a low recombination rate (see Fig. 10.4). Therefore, deleterious recessive genes are exposed to selection in homozygous parthenogens. Parthenogens carrying homozygous recessive deleterious alleles should not be able to survive or develop into functional secondary queens. The obligate occurrence of parthenogenesis in the normal life cycle of this species can eliminate recessive deleterious genes in every generation, much like the genetic purging that haploid males of Hymenoptera undergo, eliminating the transmission of deleterious recessive alleles to the sexual offspring.

10.5.2 *Inbreeding Preadaptation Hypothesis for the Evolution of Thelytoky*

The near-total homozygosity caused by terminal fusion requires that all recessive lethal genes be eliminated in the course of evolution prior to the appearance of parthenogenesis. Otherwise the species will be unable to produce viable parthenogens. Inbreeding should pre-adapt a population to parthenogenetic modes that promote rapid homozygosity (Cuellar 1977), but this inbreeding preadaptation hypothesis has not been supported by empirical studies. In termites, however, genetic purging through repeated inbreeding might have been a necessary preadaptation for the evolution of thelytoky.

10.6 Genetic Basis of AQS

10.6.1 *Selfish Genetic Elements Involved in AQS*

During meiosis, chromosomes copy themselves, line up with their matching partners, and then split up. At meiosis I, pairs of homologous chromosome, each made up of two sister chromatids, separate into two cells. Because of the ploidy restoration by terminal fusion, parthenogenesis of termites causes strong genetic conflict at the first meiotic division. Genes on the chromosomes present in secondary oocytes can be transmitted to parthenogens, while the chromosomes in first polar bodies cannot (Fig. 10.13). A primary queen of the genotype AB produces either AA or BB secondary queens. If any selfish gene elements influence the first meiotic division, e.g. AA c BB, it results in a biased frequency of genotypes in alates, which are produced by outcrossing between the primary king and secondary queens. Data from some *R. speratus* colonies suggested the existence of such a strong genetic conflict (All genotyping data were presented in the supplementary online material of Matsuura et al. 2009). In the colony G, for instance, the primary queen had a heterozygous genotype (170/152) at the microsatellite locus Rf6-1. This colony had 11 parthenogenetically-produced secondary queens. Under the assumption of random meiotic division, equal frequencies of 170/170 and 152/152 are expected among parthenogens. However, 10 secondary queens were 170/170, while only a single secondary queen was 152/152 ($P < 0.05$, two-tailed binominal test). Consequently, alates produced by outcrossing of these secondary queens with the primary king (167/185) showed a significantly biased genotypic frequency. Among 20 alates genotyped, 19 carried the maternal allele 170, whereas only a single alate had the maternal allele 152 ($P < 0.0001$, two-tailed binominal test). This clearly shows that struggle at the first meiotic division (maternal allele vs. maternal allele) is amplified in termites with AQS. In *R. virginicus*, another termite species with AQS (Matsuura and Vargo unpublished data), all secondary queens in each colony showed a single

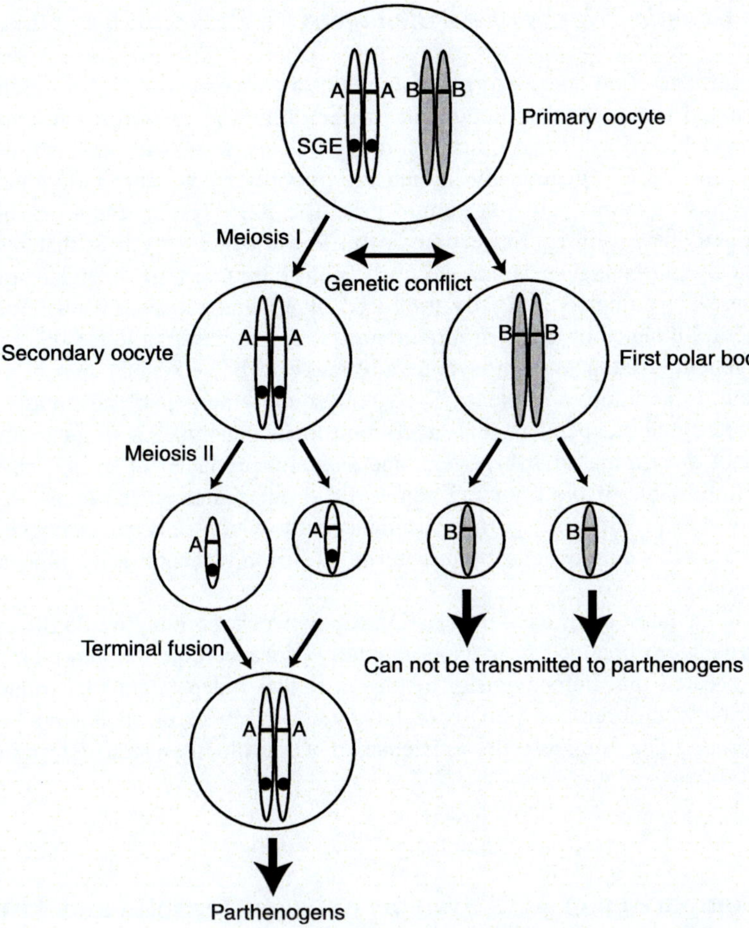

Fig. 10.13 Scheme showing genetic conflict and presumable selfish genetic element (SGE) involved in termite parthenogenesis. This driver gene causes bias in parthenogens' genotypes AA c BB

homozygous genotype at some microsatellite loci suggesting the existence of very strong meiotic driver genes.

Recently, researchers have discovered a few mechanisms by which one gene can thwart a rival during meiosis (Pennisi 2003). Montchamp-Moreau and her group revealed a mechanism behind meiotic drive, which seems to be widespread among insects (Cazemajor et al. 2000). In sexual reproduction, an X-linked selfish driver gene causes female-biased progeny. In termites with AQS, however, a meiotic driver gene can function as the selfish element to increase its own transmission rate to the next generation through biasing the chromosome pairs in secondary queens (Fig. 10.13). In other words, AQS is really a fight between maternal chromosomes.

10.6.2 Genetic Priority of Parthenogens To Be Secondary Queens

Genetic influences on queen-worker differentiation are essential to the conditional use of sexual and asexual reproduction. In the termite *R. speratus*, parthenogens are strongly biased to develop into secondary queens, suggesting that differentiation into this caste is genetically influenced, possibly by whether individuals are heterozygous or homozygous at certain loci. The AQS system can work only if parthenogens have priority to become secondary queens. Why is it that nymphs produced by parthenogenesis can exclusively differentiate into secondary queens when there are numerous sexually produced nymphs at the same time? Without any genetic influence on caste differentiation, it seems impossible for this to happen. A genetic system of homozygous advantage to be secondary queens makes it possible. In addition, a stable AQS system requires the genetic advantage to be determined by an independent (unlinked) multilocus genotype. A single locus system cannot discriminate parthenogens and sexually produced offspring, and thus AQS is impossible, if the founding pair had the same allele at the locus. A multilocus system provides a rigorous mechanism by which only parthenogens can develop into secondary queens, because terminal fusion yields progeny of near-total homozygosity.

Indeed, the latest analysis of the relationship between reproductive dominance of secondary queens obtained from experimentally orphaned colonies and their genotypes suggested that homozygosity at (at least) two independent loci influenced the priority to differentiate into secondary queens (Matsuura et al. unpublished observations). This suggests the existence of a multilocus queen determination system.

10.7 Comparison of AQS Systems Between Termites and Ants

Termites have often been compared with ants. However, termites are basically social cockroaches, whereas ants evolved from wasps. Their different ancestries provided both groups with different life history pre-adaptations for social evolution. These phylogenetically divergent insects differ in fundamental ways. Unlike the haplodiploid Hymenoptera, both sexes of termites are diploid. Termites are hemimetabolous, whereas ants are holometabolous. Nevertheless, the structural elements of social organization in ants and termites are highly convergent, suggesting common selection factors in their social evolution. One of the most amazing convergences can be seen in the fact that conditional use of sexual and asexual reproduction evolved not only in ants but also in termites.

Asexual royal succession in *R. speratus* is in some ways analogous to the conditional use of sex found in the ants *Cataglyphis cursor* (Pearcy et al. 2004), *Wasmannia auropunctata* (Fournier et al. 2005), and *Vollenhovia emeryi* (Ohkawara et al. 2006). In these ant and termite species, queens do not require sperm from mates to produce diploid (female) offspring. Nevertheless, they retain sexual reproduction for production of workers and thus the genetic diversity in the worker force is maintained. By using alternative modes of reproduction for the queen and worker castes, genetic diversity in the worker population can be maintained.

As discussed above, reduced genetic diversity in the worker force may be detrimental for the colony because it leads to reduced defense against parasites, less efficient division of labor, and a decreased range of environmental conditions that a colony can tolerate. These costs are akin to those thought to lead to the instability of parthenogenetic reproduction in nonsocial organisms. Thus, sexual reproduction might have important benefits for colony function through increased defense against parasites, more efficient division of labor, and an increased range of environmental conditions that a colony can tolerate. Queens of these species take advantage of the social caste system to use sex for producing workers, which amounts to somatic growth, but parthenogenesis, which does not involve the evolutionary cost of sex, is used for germ line production.

If workers retain reproductive potential (totipotency) and have the chance to reproduce, queens may increase direct fitness by producing workers by parthenogenesis to some extent. In *C. cursor*, a small proportion (2.5%) of workers is produced by parthenogenesis, and unmated workers have the chance of reproduction in colonies that have lost the queen. Under complete worker sterility, however, queens gain no fitness benefit by using parthenogenesis for worker production. In *W. auropunctata*, which exhibits complete worker sterility, all workers are produced by sexual reproduction. In the termite *R. speratus*, workers are always produced by sexual reproduction. All of the 3,027 secondary queens collected from field colonies were nymphoid, i.e., neotenic reproductives with wing buds differentiated from nymphs. This indicates that workers have no chance to develop into secondary queens in nature (Fig. 10.7), and suggests that worker sterility relaxes the sexual conflict over worker production, and thus favors the use of sexual reproduction for worker production.

Another commonality is that the parthenogenetic queens produced in both systems are cared for their entire lives by workers. In ants, the conditional use of parthenogenesis for new queen production primarily occurs in dependent-founding species, most likely because the presence of workers compensates for the negative effects of parthenogenesis. Automictic parthenogenesis with central fusion gradually increases homozygosity over time due to crossing-over. By increasing the levels of homozygosity, parthenogenesis should result in reduced queen survival and fitness. In line with this prediction, queens of *Cataglyphis sabulosa*, in which colony reproduction proceeds through flight dispersal and independent colony founding, produce new queens by sexual reproduction despite their ability of thelytokous parthenogenesis (Timmermans et al. 2008).

In *R. speratus*, only the non dispersing neotenic (secondary) queens that develop within established colonies are produced asexually, whereas alates (adult primary queens) are produced sexually. Parthenogenesis with terminal fusion results in near complete homozygosity, which should reduce the viability and fitness of the secondary queens. However, the consequences for secondary queens in this termite species may be minimal, because secondary queens stay in the natal nest, protected and cared for by the existing worker force, unlike independent colony founding by the primary king and queen. Contrary to neotenic queens, alates of termites disperse and found colonies independently, and are thus subjected to a number of environmental contingencies in which genetic diversity is likely to be advantageous.

10.8 Clues to Find New AQS Species

AQS is, in effect, a "cryptic parthenogenesis". Termites with AQS avoid inbreeding even after queen replacement and thus maintain colony genetic structure for almost the entire life of the colony. Therefore, it is impossible to identify AQS by genotyping workers. Analyses of colony breeding structure based only on worker genotyping could wrongly estimate a colony with many secondary queens and a primary king to be a simple family, if the species has AQS. This is the reason why the vast majority of *R. virginicus* colonies have been estimated to be simple families, although their colony sizes are so huge that a single primary queen cannot explain the reproductive capacity (Vargo and Husseneder 2009). If such an inconsistency was found in other termite species, it may also suggest AQS.

The frequent difficulties in finding royal chambers in termite nests have concealed AQS. However, a reasonable clue to the existence of AQS is alate sex ratio. AQS system balances the genetic contribution to the next generation between the primary king and the primary queen for as long as the primary king is alive. However, inbreeding is inevitable after the primary king's death, and should result in uneven genetic contribution to the offspring by the primary king and queen. The effect seems to be marginal because secondary kings cannot continue the colony so long (Fig. 10.5), but even a small bias in genetic contribution to next generation can cause an unequal alate sex ratio. The AQS species, *R. speratus* and *R. virginicus*, have significantly female-biased alate sex ratios, whereas a non AQS species *R. flavipes* shows an equal alate sex ratio. Under this hypothesis, it can be predicted that an Italian subterranean termite *R. lucifugus* and an East African termite *Schedorhinotermes lamanianus* should be AQS species because it has been reported that these species have significantly female-biased alate sex ratios (reviewed by Jones et al. 1988).

Acknowledgments The study on asexual queen succession was conducted by collaboration with Edward L. Vargo at North Carolina State University and Kazuki Tsuji at University of Ryukyus. I am grateful to L. Keller. K. Ross, N. E. Pierce, D. Haig, P. E. Labadie, D. J. C. Kronauer, K. Shimizu, T. Yashiro, K. Kawatsu and H. Nakano for helpful discussion.

References

Afzal M, Salihah Z (1985) Sex ratio, occurrence of parthenogenesis, ovarian development and oviposition behaviour of the primary reproductives of *Bifiditermes beesoni* (Gardner) (Isoptera, Kalotermitidae). Z Angew Entomol 100:132–146

Baer B, Schmid-Hempel P (2001) Unexpected consequences of polyandry for parasitism and fitness in the bumblebee, *Bombus terrestris*. Evolution 55:1639–1643

Ball SL (2002) Population variation and ecological correlates of tychoparthenogenesis in the mayfly *Stenonema femoratum*. Biol J Linn Soc 75:101–123

Bartz SH (1979) Evolution of eusociality in termites. Proc Natl Acad Sci U S A 76:5764–5768

Bell G (1982) The masterpiece of nature: the evolution and genetics of sexuality. University of California Press, San Francisco, CA

Bergamaschi S, Dawes-Gromadzki TZ, Scali V, et al (2007) Karyology, mitochondrial DNA and the phylogeny of Australian termites. Chromosome Res 15:735–753

Bourke AFG, Franks NR (1995) Social evolution in ants. Princeton University Press, Princeton, NJ

Cagniant H (1979) Thelytoky and arrhenotoky in the ant *Cataglyphis cursor* Fonsc (Hymenoptera: Formicidae): biological cycle in laboratory rearing of colonies with a queen and colonies without a queen. Insect Soc 26:51–60

Cazemajor M, Joly D, Montchamp-Moreau C (2000) Sex-ratio meiotic drive in *Drosophila simulans* is related to equational nondisjunction of the Y chromosome. Genetics 154:229–236

Chhotani OB (1962) Further observations on biology and parthenogenesis in the termite *Kalotermes beesoni* (Kalotermitidae). In: Termites in the humid tropics: proceedings of the New Delhi symposium. Paris, UNESCO, pp 73–75.

Corley LS, Blankenship JR, Moore AJ (2001) Genetic variation and asexual reproduction in the facultatively parthenogenetic cockroach *Nauphoeta cinerea*: implication for the evolution of sex. J Evol Biol 14:68–74

Corley LS, Moore AJ (1999) Fitness of alternative modes of reproduction: developmental constraints and the evolutionary maintenance of sex. Proc R Soc Lond B Biol 266: 471–476

Cuellar O (1977) Animal parthenogenesis. Science 197:837–843

DeHeer CJ, Vargo EL (2006) An indirect test of inbreeding depression in the termites *Reticulitermes flavipes* and *Reticulitermes virginicus*. Behav Ecol Sociobiol 59:753–761

Fei HX, Henderson G (2003) Comparative study of incipient colony development in the Formosan subterranean termite, *Coptotermes formosanus* Shiraki (Isoptera, Rhinotermitidae). Insect Soc 50:226–233

Fournier D, Estoup A, Orivel J, et al (2005) Clonal reproduction by males and females in the little fire ant. Nature 435:1230–1234

Goldstein DB (1994) Deleterious mutations and the evolution of male haploidy. Am Nat 144: 176–183

Grasso DA, Wenseleers T, Mori A, et al (2000) Thelytokous worker reproduction and lack of *Wolbachia* infection in the harvesting ant *Messor capitatus*. Ethol Ecol Evol 12:309–314

Grassé PP (1949) Ordre des Isoptères ou termites. In: Grassé P-P (ed) Traité de zoologie, vol 9. Masson, Paris, pp 408–544

Gruber M, Hoffmann B, Ritchie P, Lester P (2010) Crazy ant sex: genetic caste determination, clonality and inbreeding in a population of invasive yellow crazy ants. Proceedings of the 16th Congress of IUSSI: 93

Hartmann A, Wantia J, Torres JA, Heinze J (2003) Worker policing without genetic conflicts in a clonal ant. Proc Natl Acad Sci U S A 100:12836–12840

Hartmann A, Wantia J, Heinze J (2005) Facultative sexual reproduction in the parthenogenetic ant *Platythyrea punctata*. Insectes Soc 52:155–162

Heinze J, Hölldobler B (1995) Thelytokous parthenogenesis and dominance hierarchies in the ponerine ant *Platythyrea punctata*. Naturwissenschaften 82:40–41

Howard RW, Mallette EJ, Haverty MI, Smythe RV (1981) Laboratory evaluation of within-species, between-species, and parthenogenetic reproduction in *Reticulitermes flavipes* and *Reticulitermes virginicus*. Psyche 88:75–87

Hughes WOH, Boomsma JJ (2004) Genetic diversity and disease resistance in leaf-cutting ant societies. Evolution 58:1251–1260

Hughes WOH, Oldroyd BP, Beekman M, Ratnieks FLW (2008) Ancestral monogamy shows kin selection is key to the evolution of eusociality. Science 320:1213–1216

Ito F, Touyama Y, Gotoh A, Kitahiro S, Billen J (2010) Thelytokous parthenogenesis by queens in the dacetine ant *Pyramica membranifera* (Hymenoptera: Formicidae). Naturwissenschaften 97:725–728

Jones SC, LaFage JP, Howard RW (1988) Isopteran sex-ratios: phylogenetic trends. Sociobiology 14:89–156

Jones JC, Myerscough MR, Graham S, Oldroyd BP (2004) Honey bee nest thermoregulation: diversity promotes stability. Science 305:402–404

Kondrashov AS (1993) Classification of hypotheses on the advantage of amphimixis. J Hered 84:372–387

Kramer MG, Templeton AR (2001) Life-history changes that accompany the transition from sexual to parthenogenetic reproduction in *Drosophila mercatorum*. Evolution 55:748–761

Kurup NC, Prabhoo NR (1977) Facultative parthenogenesis in *Cyphoderus javanus* (Collembola: Insecta). Curr Sci India 46:168–168

Lamb RY, Willey RB (1987) Cytological mechanism of thelytokous parthenogenesis in insects. Genome 29:367–369

Lenoir A, Querard L, Pondicq N, Berton F (1988) Reproduction and dispersal in the ant *Cataglyphis cursor* (Hymenoptera, Formicidae). Psyche 95:21–44

Light SF (1944) Parthenogenesis in termites of the genus *Zootermopsis*. Univ Calif Publ Zool 43:405–412

Matsuura K (2002) A test of the haplodiploid analogy hypothesis in the termite *Reticulitermes speratus* (Isoptera: Rhinotermitidae). Ann Entomol Soc Am 95:646–649

Matsuura K, Fujimoto M, Goka K (2004) Sexual and asexual colony foundation and the mechanism of facultative parthenogenesis in the termite *Reticulitermes speratus* (Isoptera, Rhinotermitidae). Insect Soc 51:325–332

Matsuura K, Fujimoto M, Goka K, Nishida T (2002) Cooperative colony foundation by termite female pairs: altruism for survivorship in incipient colonies. Anim Behav 64: 167–173

Matsuura K, Kobayashi N (2007) Size, hatching rate, and hatching period of sexually and asexually produced eggs in the facultatively parthenogenetic termite *Reticulitermes speratus* (Isoptera: Rhinotermitidae). Appl Entomol Zool 42:241–246

Matsuura K, Nishida T (2001) Comparison of colony foundation success between sexual pairs and female asexual units in the termite *Reticulitermes speratus* (Isoptera: Rhinotermitidae). Popul Ecol 43:119–124

Matsuura K, Tamura T, Kobayashi N, et al (2007) The antibacterial protein lysozyme identified as the termite egg recognition pheromone. PLos ONE 2:e813. doi:10.1371/journal.pone.0000813

Matsuura K, Tanaka C, Nishida T (2000) Symbiosis of a termite and a sclerotium-forming fungus: sclerotia mimic termite eggs. Ecol Res 15:405–414

Matsuura K, Vargo EL, Kawatsu K, et al (2009) Queen succession through asexual reproduction in termites. Science 323:1687–1687

Modig AO (1996) Effects of body size and harem size on male reproductive behaviour in the southern elephant seal. Anim Behav 51:1295–1306

Mousseau TA, Fox CW (1998) The adaptive significance of maternal effects. Trends Ecol Evol 13:403–407

Nur U (1971) Parthenogenesis in Coccids (Homoptera). Am Zool 11:301–308

Ohkawara K, Nakayama M, Satoh A, et al (2006) Clonal reproduction and genetic caste differences in a queen-polymorphic ant, *Vollenhovia emeryi*. Biol Lett 2:359–363

Oldroyd BP, Fewell JH (2007) Genetic diversity promotes homeostasis in insect colonies. Trends Ecol Evol 22:408–413

Pearcy M, Aron S, Doums C, Keller L (2004) Conditional use of sex and parthenogenesis for worker and queen production in ants. Science 306:1780–1783

Pearcy M, Goodisman M, Keller L (2010) Sib-mating without inbreeding in the crazy ant. Proceedings of the 16th Congress of IUSSI: 97

Pennisi E (2003) Bickering genes shape evolution. Science 301:1837–1839

Rabeling C, Lino-Neto J, Cappellari SC, et al (2009) Thelytokous parthenogenesis in the fungus-gardening ant *Mycocepurus smithii* (Hymenoptera: Formicidae). PLoS ONE 4:e6781. doi:10.1371/journal.pone.0006781

Roisin Y (2001) Caste sex ratios, sex linkage, and reproductive strategies in termites. Insect Soc 48:224–230

Rosengaus RB, Guldin MR, Traniello JFA (1998) Inhibitory effect of termite fecal pellets on fungal spore germination. J Chem Ecol 24:1697–1706

Roth LM, Willis ER (1956) Parthenogenesis in cockroaches. Ann Entomol Soc Am 49: 31–37

Seeley TD, Tarpy DR (2007) Queen promiscuity lowers disease within honeybee colonies. Proc R Soc Lond B Biol 274:67–72

Stansly PA, Korman AK (1993) Parthenogenetic development in *Velocitermes* spp (Isoptera, Nasutitermitidae). Sociobiology 23:13–24

Syren RM, Luykx P (1977) Permanent segmental interchange complex in the termite *Incisitermes schwarzi*. Nature 266:167168

Tarpy DR (2003) Genetic diversity within honeybee colonies prevents severe infections and promotes colony growth. Proc R Soc Lond B Biol 270:99–103

Templeton A (1982) The prophecies of parthenogenesis. In: Dingle H, Hegmann JP (eds) Evolution and genetics of life histories. Springer-Verlag, Berlin, pp 75–102

Thomas RJ (1987) Factors affecting the distribution and activity of fungi in the nest of Macrotermitinae (Isoptera). Soil Biol Biochem 19:343–349

Thorne BL, Traniello JFA, Adams ES, Bulmer M (1999) Reproductive dynamics and colony structure of subterranean termites of the genus *Reticulitermes* (Isoptera Rhinotermitidae): a review of the evidence from behavioral, ecological, and genetic studies. Ethol Ecol Evol 11:149–169

Timmermans I, Hefetz A, Fournier D, Aron S (2008) Population genetic structure, worker reproduction and thelytokous parthenogenesis in the desert ant *Cataglyphis sabulosa*. Heredity 101:490–498

Tsuji K (1988) Obligate parthenogenesis and reproductive division of labor in the Japanese queenless ant *Pristomyrmex pungens*: comparison of intranidal and extranidal workers. Behav Ecol Sociobiol 23:247–255

Tsuji K, Yamauchi K (1995) Production of females by parthenogenesis in the ant *Cerapachys biroi*. Insect Soc 42:333–336

Tucker KW (1958) Automictic parthenogenesis in the honeybee. Genetics 43:299–316

Vargo EL, Husseneder C (2009) Biology of subterranean termites: insights from molecular studies of *Reticulitermes* and *Coptotermes*. Annu Rev Entomol 54:379–403

Verma S, Ruttner F (1983) Cytological analysis of the thelytokous parthenogenesis in the cape honeybee (*Apis mellifera capensis* Escholts). Apidologie 14:41–57

Weesner FM (1956) The biology of colony foundation in *Reticulitermes hesperus* Banks. Univ Calif Publ Zool 61:253–314

Chapter 11
Pheromones and Chemical Ecology of Dispersal and Foraging in Termites

Christian Bordereau and Jacques M. Pasteels

Abstract Pheromones play a crucial role in the ecology of dispersal and foraging in termites. Sex-pairing pheromones possess a double role of long-range attraction to unite sexual partners and a short-range or contact attraction to maintain the pair during the tandem behaviour. Sex-pairing pheromones most often comprise a single compound capable of eliciting both behavioural effects. They appear very conservative in their evolution, and their role in the reproductive isolation of sympatric species greatly varies according to species. Species-specific sex-pairing pheromones consist of different major compounds or of a common major compound with species-specific minor components. Foraging is a collective behaviour mainly regulated by trail-following pheromones secreted from only one glandular source, the sternal gland. Trail-following pheromones may be also used by "one-piece" termites to colonise additional food sources. Although trail-following pheromones of termites have a double role of orientation and recruitment, they appear most often composed of only one compound. An alternative hypothesis is given to the postulated existence of a volatile ephemeral compound of recruitment and a long-lasting compound of orientation. Trail-following pheromones appear highly conserved in their chemical evolution (only 8 different pheromones for 60 species), even if a clear separation is observed between basal termites and more derived termites. The major ecological event of the external foraging was not related to a chemical evolutionary step of the trail-following pheromones. Pheromonal parsimony (the utilisation of a same molecule for multiple functions) is common in termites in the behavioural context of dispersal and foraging. The same molecule is used in many species as a sex-pairing pheromone and a trail-following pheromone, depending upon the pheromone concentration and the caste involved.

C. Bordereau (✉)
Développement et Communication Chimique chez les Insectes, Université de Bourgogne, F-21000 Dijon, France
e-mail: christian.bordereau@u-bourgogne.fr

D.E. Bignell et al. (eds.), *Biology of Termites: A Modern Synthesis*,
DOI 10.1007/978-90-481-3977-4_11, © Springer Science+Business Media B.V. 2011

11.1 Introduction

Semiochemicals involved in biotic interactions are of prime importance in the evolution and biodiversity of termites where most sexual and social activities are chemically mediated. Our knowledge of chemical ecology of dispersal and foraging has advanced since Pasteels and Bordereau's review (1998) due to the application of solid phase microextraction (SPME) which allows selective extraction of substances secreted at the surface of pheromone glands. We are now able to present the main characteristics of the pheromones involved, and their evolutionary trends.

Dispersal and foraging behaviours appear unrelated. However, in termites, they are regulated mainly by secretions of the sternal gland, a small ectodermal abdominal gland present in every species and every caste (Noirot 1969). The sternal gland is involved, alone or combined with other sexual glands, in pairing and tandem behaviour of swarming imagoes. It is the only glandular source of trail communication in foragers. Furthermore, trail communication is involved during postflight behaviour of alates and, in several species, sex-pairing pheromones are identical to trail-following pheromones. According to Traniello and Leuthold (2000), the ancestral function of the sternal gland was in sex attraction and tandem behaviour of alates, but in neuter castes of derived species the gland became specialized for other social functions, particularly foraging.

11.2 Dispersal

11.2.1 Dispersal Sequence

Termites mainly reproduce by annual dispersal of alates which found new colonies from bisexual pairs. Dispersal is typically a seasonal phenomenon, with a major exodus, preceded and/or followed by minor flights (Nutting 1969; Grassé 1984; Nalepa et al. 2001). Releasing of alates is dependent upon environmental and internal factors. After a dispersal flight rarely exceeding a few hundred metres in the absence of wind, alates land on ground or on trees, probably attracted by olfactory environmental stimuli, especially from decaying wood (Nutting 1969).

Pairing of partners might result from fortuitous encounters in a few species with limited dispersal or forming dense swarms by hypsotaxis or phototaxis (Morgan 1959; Wilkinson 1962; Jones et al. 1981; Costa-Leonardo and Barsotti 1998; Raina et al. 2003), but is generally mediated by long-range olfactory attraction (Nutting 1969; Pasteels and Bordereau 1998). Visual cues might be also involved in some species (Mitchell 2007). Olfactory signals were first postulated by Fuller (1915), but sex pheromones were only demonstrated in the 1970s (Wall 1971; Pasteels 1972). There are relatively few detailed observations on pairing behaviour in termites (Fuller 1915; Grassé and Noirot 1951a; Williams 1959; Buchli 1960; Noirot and Bodot 1964; Leuthold 1975; Leuthold and Bruinsma 1977; Mitchell 2007) and undoubtedly they do not reflect the whole behavioural diversity of the phenomenon. Nothing is known in particular on the sexual behaviour of the most basal living termite, *Mastotermes darwiniensis*, whose imagoes possess several sternal and tergal glands in both sexes. Pairing is initiated by both sexes or by males in a few species (Clément 1956; Kalshoven 1960; Pasteels 1972; Leuthold 1975; Stuart 1975; Shellman-Reeve 1999), but the female calling behaviour is by far the most frequent (Fig. 11.1). Most of the observers report short distances of attraction between sexual partners in termites, from a few centimetres to a few tens of centimetres (Grassé 1984). In *Trinervitermes bettonianus*, the range of action of tergal gland secretion appears higher (10 cm) than that of the sternal gland (<4 cm) (Leuthold 1975). In natural conditions, calling females of *Macrotermes carbonarius* were observed to attract males from 20 cm downwind, but 1 cm only upwind, and males of *Syntermes praecellens* orient towards females calling at 40 cm (unpublished). The maximum distance of attraction reported until now is 2.5–3 m by males of *Hodotermes mossambicus* (Leuthold and Bruinsma 1977). As a crucial factor for the ecological success of species, the range of action of sex-pairing pheromones would deserve to be studied in detail under natural conditions. Mate choice was often considered absent in termites (Grassé 1984; Nalepa and Jones 1991). However, a few studies showed correlations between anatomical characteristics and mate choice (Shellman-Reeve 1999; Matsuura and Nishida 2001; Kitade et al. 2004; Husseneder and Simms 2008). Likewise, mate-seeking alates of *Zootermopsis nevadensis* avoid forming partnerships with nestmates (Shellman-Reeve 2001), and swarm-dispersal mating systems in *Coptotermes lacteus* can involve a notable degree of non-random mating (Thompson et al. 2007). In many species, males frequently change partners during the nuptial promenade (Williams 1959; Grassé 1984), which might reflect a mate choice strategy with potential relationships with levels of sex-pairing pheromones.

During the nuptial promenade in tandem that follows pairing, short-range or contact chemical and tactile stimuli play a crucial role (Stuart 1969; Grassé 1984). The basic pattern is a leader seeking a suitable nesting site, followed by a partner maintaining contact by antennating the posterior pleural membranes and licking with its mouthparts the last tergites of the leader. Tactile stimuli of the follower elicit moving of the leader, and the chemical stimuli released by the leader induce attraction and excitement of the follower. Depending on species, the leader lays a trail during the tandem behaviour or stops as soon as the follower loses contact and reinstates a calling posture (Nutting 1969; Bordereau et al. 1991; Robert et al. 2004).

The selection of a suitable nesting site is dependent upon environmental factors (Nutting 1969; Shellman-Reeve 1994; Traniello and Leuthold 2000; Lima et al. 2006). Mating only occurs after the pair are settled in their initial cell or copularium, several hours or several days after dispersal according to species. This behaviour is totally different from other insects, especially from cockroaches, where copulation immediately follows the pairing (Bell et al. 2007). For this reason, we distinguish *sex-pairing pheromones* inducing meeting of sexual partners, from *sex-mating pheromones* eliciting copulation. Sex-mating pheromones have not yet been demonstrated, but they are strongly suggested by the ritual behaviour of antennation and palpation preceding mating (Grassé 1984; Wood and Johnson 1986). Furthermore, mating is repeated regularly during the life of the pair, even in species

Fig. 11.1 (continued)

with physogastric queens (Hinze and Leuthold 1997) where sex-pairing glands are dedifferentiated (Noirot 1969; Quennedey and Leuthold 1978).

11.2.2 Sex-Pairing Pheromone Glands

The glandular origin of sex-pairing pheromones could be postulated from the simple calling behaviour (Fig. 11.1), but bioassays are the only clear means of conclusively demonstrating their origin and their behavioural effects (Ayasse et al. 2001). For sex-pairing pheromones in termites, the general basic pattern is a long-range olfactory attraction eliciting the recruitment and the location of partner, and a short-range attraction and/or contact chemoreception eliciting excitement with antennation and palpation. Sex attraction bioassays used until now essentially tested short range olfactory or contact attraction (Clément 1982; Clément et al. 1989; Bordereau et al. 1991; 2002; Peppuy et al. 2004).

11.2.2.1 Tergal Glands

The evolutionary trends of tergal glands which are not present in all species of termites remain obscure (Ampion and Quennedey 1981). However, being present only in alates and in newly differentiated neotenics, and undergoing a dedifferentiation in functional reproductives, tergal glands appear as typically sexual glands involved in pairing behaviour. Their role in long-range attraction has been shown in *Trinervitermes bettonianus* (Leuthold 1975), *Macrotermes annandalei* and *Macrotermes barneyi* (Peppuy et al. 2004). Tergal glands are also considered as crucial in maintaining tandem behaviour by short-range attraction or contact chemoreception (Barth 1955; Bordereau et al. 2002), although several observations challenge this precise role. They are unexposed during the tandem behaviour, and might be inactive unless their secretions spread over the tergites. Females of *T. bettonianus* with varnished tergal glands are still able to elicit tandem behaviour (Leuthold 1975). In *Coptotermes formosanus*, where females possess 2 tergal glands, there is tandem behaviour although tergal gland extracts do not cause

Fig. 11.1 Post-flight behaviour: calling postures of females and tandem behaviour. Sex-pairing pheromones are released from the sternal gland in *Odontotermes maesodensis* (**a**) and *Pseudacanthotermes spiniger* (**h**), from tergal glands in *Syntermes grandis* (**c**) and in a small undetermined Nasutitermitinae from Brazil possessing enlarged and evaginable glands (**e**), from both sternal and tergal glands in *Nasutitermes lujae* (**d**), from both tergal and posterior sternal glands in *Macrotermes annandalei* (**b**), from sternal and possibly abdominal pleural glands in *Cubitermes fungifaber* (**f**). In some Apicotermitinae from Brazil (undetermined species) (**g**), females lay tergal secretions on the substrate between calling postures. During tandem behaviour, males of *P. spiniger* lick with their mouthparts the ano-genital areas of females lacking tergal glands (**i**), whereas males of *M. annandalei* lick the last tergites of females possessing tergal glands (**j**). Photographs (**a, b**) and (**j**) by Alexis Peppuy, photographs **c–i** by Christian Bordereau. Figure 11.1b © Journal of Morphology a behavioural response from male alates (Raina et al. 2003; Bland et al. 2004; Park et al. 2004). Finally, tandem behaviour is also present in species without tergal glands, as in *Pseudacanthotermes spiniger* where males often lick the distal region of the abdomen of females (Bordereau et al. 1991). Therefore, other semiochemicals than those released by tergal glands, such as dermal gland secretions or ano-genital secretions could be involved in tandem behaviour. This hypothesis initially proposed by Buchli (1960) for *Reticulitermes lucifugus* deserves further investigations.

Females of small undetermined Apicotermitinae from Brazil and Vietnam intermittently touching the substrate with their last abdominal tergite between calling postures also show an original and unknown role of tergal glands in sexual behaviour of termites (unpublished observations).

11.2.2.2 Sternal Gland

In contrast to tergal glands, the sternal gland is present in alates of all species. It is generally hypertrophied when tergal glands are lacking (Ampion and Quennedey 1981), but the females of *Pseudacanthotermes militaris* have no tergal glands and a sternal gland of normal size (Bordereau et al. 1993), and the females of *Trinervitermes bettonianus* possess both tergal glands and an enlarged sternal gland (Leuthold and Lüscher 1974; Leuthold 1975; Quennedey and Leuthold 1978; McDowell and Oloo 1984). In *Zootermopsis nevadensis* (Bordereau et al. 2010), *Hodotermes mossambicus* (Leuthold and Bruinsma 1977), *Reticulitermes santonensis* (Laduguie et al. 1994), *Pseudacanthotermes spiniger* (Bordereau et al. 1991), *Ancistrotermes pakistanicus* (Robert et al. 2004), species without tergal glands, sternal gland secretions elicit attraction at a distance and strong excitement in short-range and contact bioassays. In *T. bettonianus*, the range of action of the sternal gland would be lower than that of tergal glands, and surprisingly, given its ventral location, it would be more efficient in maintaining the cohesion of the pair during tandem behaviour (Leuthold and Lüscher 1974; Leuthold 1975). In another species with tergal glands, *Cornitermes bequaerti*, no sexual role could be attributed to the sternal gland (Bordereau et al. 2002).

Secretions of the sternal gland play another important role during postflight behaviour in females laying a trail between successive calling periods, or during the nuptial promenade, undoubtedly increasing the rate of success of pairing or the reunion of partners in case of tandem disconnection.

11.2.2.3 Posterior Sternal Glands

In *Macrotermes annandalei* and *M. barneyi*, the posterior sternal glands release long-range and short-range attractants (Peppuy et al. 2004), and their structure is similar to that of tergal glands (Quennedey et al. 2004). Nothing is known about the role of posterior sternal glands present in both sexes of *Mastotermes darwiniensis*, in the males of the Termopsidae *Porotermes* and *Stolotermes* (Ampion and Quennedey 1981), and the Rhinotermitidae *Prorhinotermes simplex* (Šobotník et al. 2005).

11.2.2.4 Pleural Glands

The calling behaviour of females of *Cubitermes fungifaber* suggests the involvement of abdominal pleural glands (Ampion and Quennedey 1981), but bioassays could only demonstrate the action of the sternal gland (unpublished observations).

11.2.3 Chemical Nature of Sex-Pairing Pheromones

To date, sex-pairing pheromones have been completely identified in 10 species, and the major component of the sexual glands is known in other 10 species (Table 11.1, Fig. 11.2). They belong to 3 chemical classes, C13–C14 aldehydes, C12 alcohols and C20 hydrocarbons. Their molecular weight varies from 180 to 272. Female and male sex-pairing pheromones of *Zootermopsis nevadensis* and *Zootermopsis angusticollis* are identical ((*E*)-2,6,10-trimethylundeca-5,9-dienal, and 4,6-dimethyldodecanal respectively). The female sex-pairing pheromone of *Hodotermopsis sjostedti* is similar to that of *Zootermopsis*, whereas the male sex-pairing pheromone is postulated to be the C13 homologue of the male pheromone of *Zootermopsis*. The C12 alcohols (*Z*)-dodec-3-en-1-ol (dodecenol), (*Z,Z*)-dodeca-3,6-dien-1-ol (dodecadienol), and (3*Z*,6*Z*,8*E*)-dodeca-3,6,8-trien-1-ol (dodecatrienol) are sex-pairing pheromones of Rhinotermitidae, Termitidae (Macrotermitinae and Syntermitinae), and dodecatrienol is the major component of sex-pairing pheromone in many species. Neocembrene is the major component of sex-pairing pheromone in Nasutitermitinae. In *Trinervitermes bettonianus*, tergal secretions which are responsible for the long-range attraction (Leuthold 1975) remain unidentified. Trinervitatriene ((11*E*)-trinervita-1(14),2,11-triene) is secreted in abundance in tergal glands of female alates of *Nasutitermes ephratae* (Buděšínský et al. 2005). Preliminary bioassays showed that this compound is active on males, but secretions of the sternal gland containing a mixture of trinervitatriene, neocembrene and dodecatrienol also are involved in sex attraction (unpublished).

The same molecule used as sex-pairing pheromone may have different glandular sources. For example, dodecatrienol is secreted by the hypertrophied sternal gland of females of *Pseudacanthotermes spiniger* (Bordereau et al. 1991), and by tergal glands of *Cornitermes bequaerti* (Bordereau et al. 2002), and neocembrene is secreted by the sternal gland of *T. bettonianus* (McDowell and Oloo 1984) and by tergal glands of *Nasutitermes corniger* and *Nasutitermes voeltzkowi* (unpublished observations).

Sex-pairing pheromones could not be identified in several species or required further investigations. In *Prorhinotermes simplex*, dodecatrienol has a clear sex-attraction activity. GC-EAD analyses highlighted a strong physiological response on antennal receptors precisely at its retention time, but dodecatrienol could not be detected in sternal secretion, probably due to inadequate quantities (Hanus et al. 2009). The different chemical nature of sex-pairing pheromones of *Reticulitermes flavipes* (*n*-tetradecyl propionate) (Clément et al. 1989) and *Reticulitermes santonensis* (dodecatrienol) (Laduguie et al. 1994) is problematic, as molecular analyses

Table 11.1 Glandular origin and chemical nature of sex-pairing pheromones

Species	Sex	Glandular origin	Chemical nature	References
TERMOPSIDAE				
Zootermopsis nevadensis	Female	Sternal gland	(*E*)-2,6,10-trimethyl-5,9-undecadienal	Bordereau et al. (2010)
	Male	Sternal gland	4,6-dimethyldodecanal	Bordereau et al. (2010)
Zootermopsis angusticollis	Female	Sternal gland	(*E*)-2,6,10-trimethyl-5,9-undecadienal	Bordereau et al. (2010)
	Male	Sternal gland	4,6-Dimethyldodecanal	Bordereau et al. (2010)
Hodotermopsis sjostedti	Female	Sternal gland	(*E*)-2,6,10-trimethyl-5,9-undecadienal	Unpublished
	Male	Sternal gland	4,6-Dimethylundecanal[a]	Unpublished
HODOTERMITIDAE				
Hodotermes mossambicus	Male	Sternal gland	Unknown	Leuthold and Bruinsma (1977)
KALOTERMITIDAE				
Kalotermes flavicollis	Female	Sternal gland	Unknown	Wall (1971)
RHINOTERMITIDAE				
Prorhinotermes simplex	Female	Tergal gland	(*Z*,*Z*,*E*)-dodeca-3,6,8-trien-1-ol	Hanus et al. (2009)
Reticulitermes flavipes	Female	Sternal gland	*n*-Tetradecyl propionate	Clément et al. (1989)
Reticulitermes santonensis	Female	Sternal gland	(*Z*,*Z*,*E*)-dodeca-3,6,8-trien-1-ol	Laduguie et al. (1994)
TERMITIDAE				
Macrotermitinae				
Ancistrotermes pakistanicus	Female	Sternal gland	(*Z*,*Z*)-dodeca-3,6-dien-1-ol	Robert et al. (2004)
Macrotermes annandalei	Female	Tergal glands + posterior SG	Unknown	Peppuy et al. (2004)
Macrotermes barneyi	Female	Tergal glands + posterior SG	Unknown	Peppuy et al. (2004)
Pseudacanthotermes spiniger	Female	Sternal gland	(*Z*,*Z*,*E*)-dodeca-3,6,8-trien-1-ol	Bordereau et al. (1991)
Pseudacanthotermes militaris	Female	Sternal gland	(*Z*,*Z*,*E*)-dodeca-3,6,8-trien-1-ol	Bordereau et al. (1993)
Syntermitinae				
Cornitermes bequaerti	Female	Tergal glands	(*Z*,*Z*,*E*)-dodeca-3,6,8-trien-1-ol	Bordereau et al. (2002)
Cornitermes cumulans	Female	Tergal glands	(*Z*,*Z*,*E*)-dodeca-3,6,8-trien-1-ol + (*E*)-nerolidol	Unpublished

Table 11.1 (continued)

Species	Sex	Glandular origin	Chemical nature	References
Cornitermes snyderi	Female	Tergal glands	(*Z*,*Z*,*E*)-dodeca-3,6,8-trien-1-ol[a] + (*Z*)-dodec-3-en-1-ol + (*E*)-nerolidol	Unpublished
Armitermes euhamignathus	Female	Tergal glands	(*Z*,*Z*)-dodeca-3,6-dien-1-ol[a]	Unpublished
Procornitermes sp.	Female	Tergal glands	(*Z*,*Z*,*E*)-dodeca-3,6,8-trien-1-ol[a]	Unpublished
Embiratermes festivellus[a]	Female	Tergal glands	(*Z*,*Z*,*E*)-dodeca-3,6,8-trien-1-ol[a] + (*Z*)-dodec-3-en-1-ol + unidentified sesquiterpene	Unpublished
Syntermes grandis	Female	Tergal glands	(*Z*,*Z*,*E*)-dodeca-3,6,8-trien-1-ol	Unpublished
Syntermes praecellens	Female	Tergal glands	(*Z*)-dodec-3-en-1-ol	Unpublished
Nasutitermitinae				
Nasutitermes corniger	Female	Tergal glands	Neocembrene	Unpublished
			Traces neocembrene+trinervitatriene isomers	
Nasutitermes ephratae	Female	Tergal glands	Trinervitatriene	Buděšínský et al. (2005)
			Traces trinervitatriene+neocembrene isomers	
Nasutitermes voeltzkowi	Female	Tergal glands	Neocembrene[a]	Unpublished
Trinervitermes bettonianus	Female	Tergal glands	Unknown	
		Sternal gland	Neocembrene[a]	McDowell and Oloo (1984)

[a]This molecule was not completely identified.

definitely proved that *R. santonensis* is the same species as *R. flavipes* introduced into Europe at the end of nineteenth century from USA (Austin et al. 2005). Since the evolution of sex-pairing pheromones during this recent geographic separation remains highly improbable, the explanation might be the identification of two components isolated through two different bioassays (olfactometer for *R. flavipes*, short-range attraction bioassay for *R. santonensis*). However, the presence of *n*-tetradecyl propionate in female alates of *R. flavipes* from northern America needs to be confirmed by using the SPME. In *Coptotermes formosanus,* trilinolein, identified as a sex-specific component of tergal glands, does not elicit sex attraction

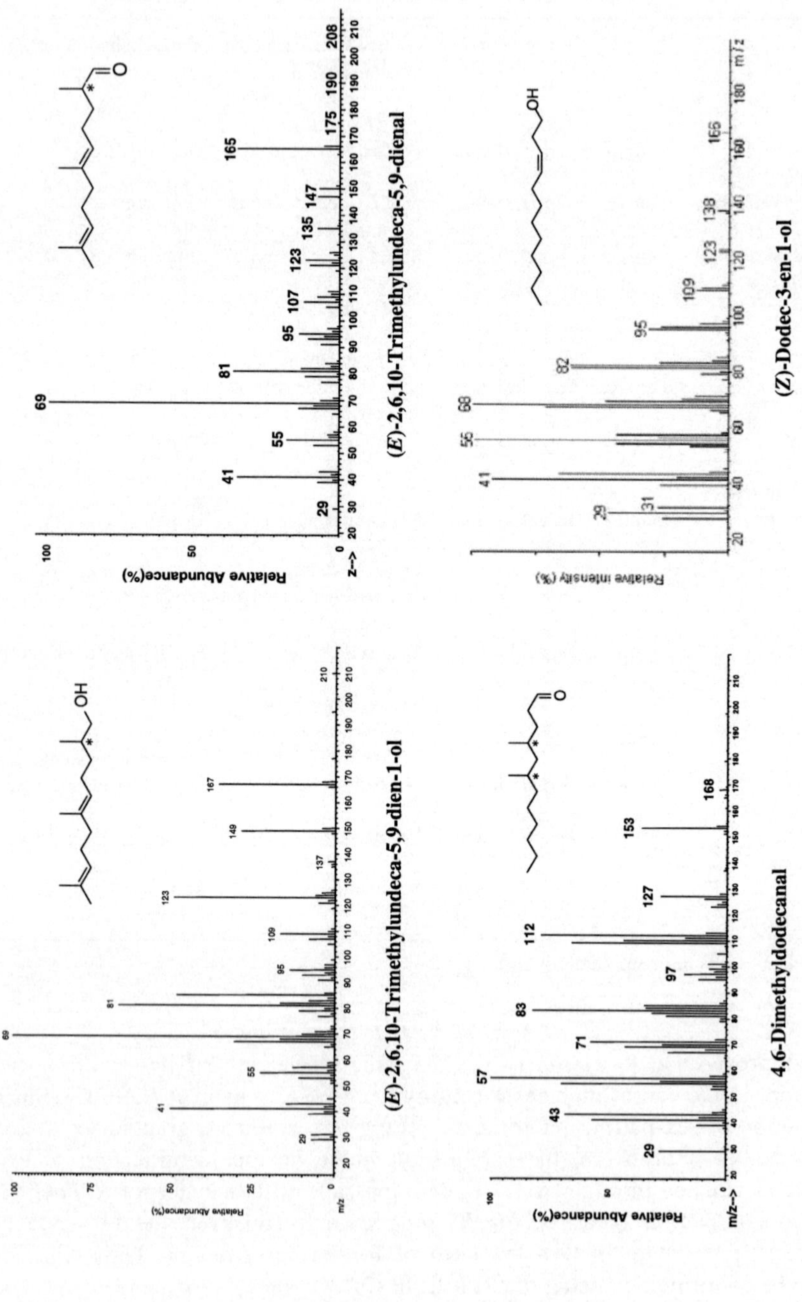

Fig. 11.2 (continued)

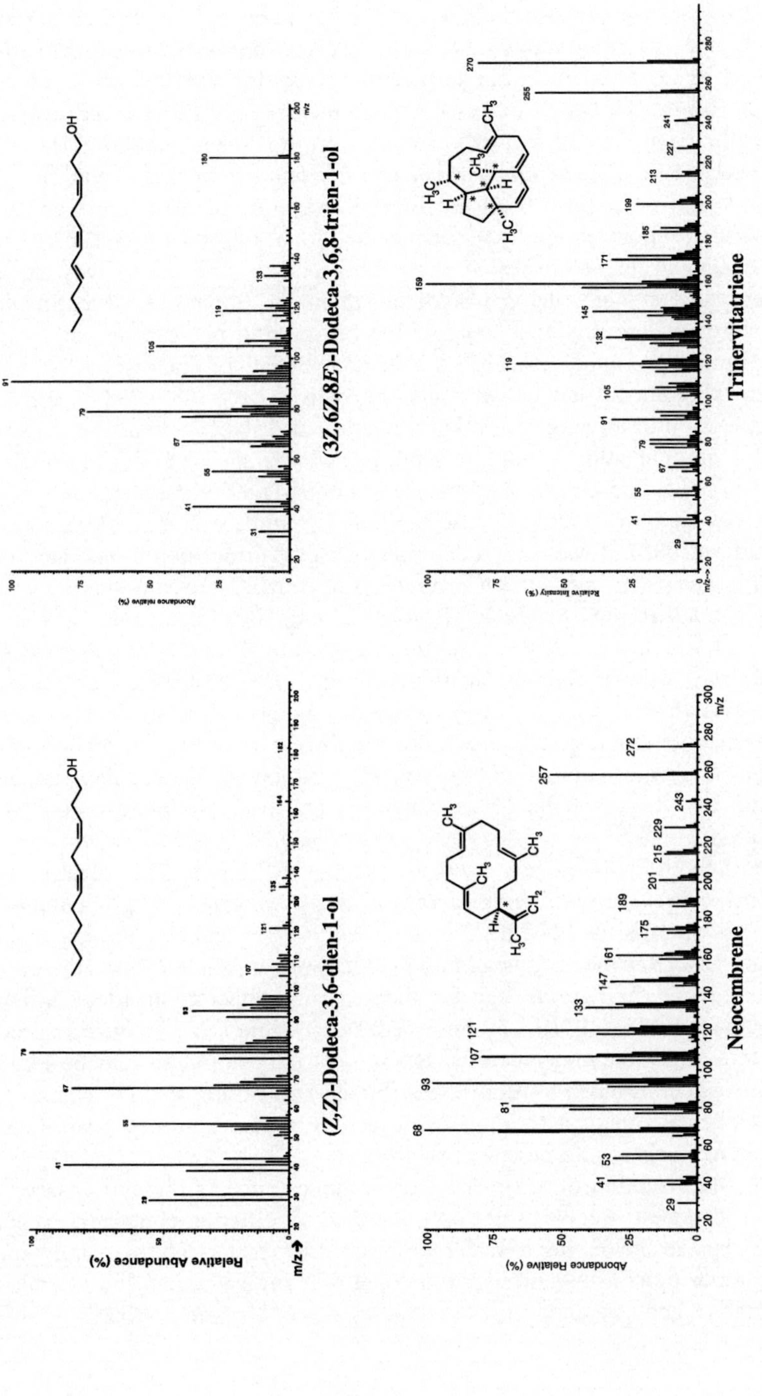

Fig. 11.2 EI mass spectra of sex-pairing pheromones and trail-following pheromones

behaviour (Raina et al. 2003; Bland et al. 2004), and might be a simple nuptial gift transferred from females to males during courtship behaviour (Park et al. 2004). In several species (*Mastotermes darwiniensis, Kalotermes flavicollis, Hodotermes mossambicus, Macrotermes annandalei, Macrotermes barneyi*), no specific components could be extracted from the surface of sex-pairing pheromone glands after SPME. In *H. mossambicus*, sex-pairing pheromone might be secreted only during the calling posture, as non-calling males are no longer attractive (Leuthold and Bruinsma 1977). In *M. annandalei* and *M. barneyi*, sexual secretion might be released entirely at once from the special reservoirs of their tergal and poststernal glands (Peppuy et al. 2004; Quennedey et al. 2004). However, for several species, chemical characteristics of sex-pairing pheromones themselves, such as high polarity or bonding with ligands, might explain the difficulties of identification.

It is worth underlining that most of the sex-pairing pheromones of termites identified until now appear to be a single component, in contrast to the multicomponent pheromonal blends generally found in insects. Remarkably, the same molecule, e.g. dodecatrienol, can elicit attraction at distance and also excitement, antennation and palpation at contact. Both behaviours appear to depend upon the pheromone concentration: low levels of pheromone trigger recruitment and orientation, whereas high levels elicit intense excitement (Bordereau et al. 1993; Pasteels and Bordereau 1998). However, multicomponent sex-pairing pheromones have been identified in sympatric species of *Cornitermes* in Brazil. Dodecatrienol is the only compound identified in *C. bequaerti* (Bordereau et al. 2002), but it is secreted along with (*E*)-nerolidol in *Cornitermes cumulans*, and with (*E*)-nerolidol and dodecenol in *Cornitermes snyderi* (unpublished). In *N. ephratae*, besides the major compound trinervitatriene, isomers of trinervitatriene and neocembrene are also present on the surface of tergal glands, but 4,000 times less abundant. In the sympatric *N. corniger*, the major compound of sex-pairing pheromone is neocembrene, associated with traces of isomers of neocembrene and trinervitatriene (unpublished observations).

Sex-pairing pheromones are often secreted at 10–100 ng/individual, e.g. in *Z. nevadensis*, *Pseudacanthotermes spiniger*, and *C. bequaerti*. High amounts are found in Nasutitermitinae, 1–2 μg/female in *N. corniger* and *N. ephratae* (unpublished), and 12 μg/female in *T. bettonianus* (McDowell and Oloo 1984). However, in *Reticulitermes* and *Coptotermes* species, the sex-pairing pheromone (dodecatrienol) is secreted at a low level (10–250 pg/female) (Laduguie et al. 1994; Bland et al. 2007). Interestingly, the molecule used as a sex-pairing pheromone may be secreted by both sexes, but in much lower amounts by the non-calling sex. For example, in *P. spiniger*, female alates assuming sex-attraction secrete 10 times more dodecatrienol than male alates (Bordereau et al. 1991).

Finally, the evolution of sex-pairing pheromones appears highly conserved in termites. The same molecule is not only secreted by different congeneric species, but also by species belonging to different families. Dodecatrienol especially enters into the composition of sex-pairing pheromones of many species of Rhinotermitidae and Termitidae, and neocembrene in several species of Nasutitermitinae.

11.2.4 Trail-Following Pheromones Involved in Tandem Behaviour

In numerous species, females deposit a trail-following pheromone during tandem behaviour. Most often, especially in species without tergal glands, the sex-pairing pheromone and the trail-following pheromone are identical, as in *Reticulitermes santonensis* (Laduguie et al. 1994), *Pseudacanthotermes spiniger* (Bordereau et al. 1991), and perhaps in *Trinervitermes bettonianus* (McDowell and Oloo 1984). However, when sex-pairing pheromone originates from tergal glands, the alate trail-following pheromone is generally different. For example, in *Nasutitermes corniger*, *N. ephratae* and *N. voeltzkowi*, the sternal gland of female alates secretes dodecatrienol, in contrast to tergal glands which secrete neocembrene or trinervitatriene (unpublished).

11.2.5 Sex-Pairing Pheromones and Reproductive Isolation

Insect sex pheromones are generally considered as species-specific and to be a key factor in their reproductive isolation. In the sympatric American *Zootermopsis nevadensis* and *Z. angusticollis* (Bordereau et al. 2010), as well as in the sympatric African fungus-growing *Pseudacanthotermes spiniger* and *P. militaris* (Bordereau et al. 1993), sex-pairing pheromones are identical. In bioassays of short-range attraction, males of *P. militaris* are preferentially attracted by extracts of sternal glands of *P. spiniger* which contain more dodecatrienol than their own glandular extracts (unpublished). In these cases, but also in the sympatric *Macrotermes subhyalinus* and *M. michaelseni* in Kenya (Darlington 1986), and in several species of *Microtermes* in Nigeria (Wood 1981), species are isolated by dispersal flights occurring at different times of the day or season.

In other species, sex-pairing pheromones are composed of both a common major compound and species-specific minor compounds. In the sympatric South America termites *Cornitermes bequaerti* and *C. cumulans*, sex-pairing pheromones are composed of dodecatrienol in both species, but the sex-pairing pheromone of *C. cumulans* also contains small amounts of (*E*)-nerolidol which, although sexually inactive, determines the species-specificity of the responses in bioassays (unpublished). A similar situation might explain results obtained by Clément (1982) on *Reticulitermes*.

Finally, sympatric species may secrete species-specific pheromones, such as *Nasutitermes corniger* secreting neocembrene (unpublished) and *N. ephratae* secreting trinervitatriene (Buděšínský et al. 2005). Sex-pairing pheromones of the southeastern Asiatic fungus-growing termites *M. annandalei*, and *M. barneyi* also are strictly species-specific, but their chemical nature remains unknown (Peppuy et al. 2004).

In all termites, reproductive isolation is undoubtedly reinforced by species-specific recognition factors. Under natural conditions, interspecific meeting of alates elicits either avoidance or, more often, strong agonism. Under experimental conditions, the simple application of a male antenna of *C. cumulans* on any instead of every part of the body of a calling female of *C. bequaerti* immediately triggers the aggressiveness of the female, whereas the application of a homospecific antenna initiates the tandem behaviour (unpublished). As in other social insects, species-specific recognition and nestmate discrimination in termites probably involve cuticular hydrocarbons (Clément and Bagnères 1998; Howard and Blomquist 2005), but this remains controversial and the phenomenon might rather result from a complex set of genetic, behavioural and environmental stimuli, or from odours released by symbionts (Thorne and Haverty 1991; Kirchner and Minkley 2003; Matsuura 2003; Florane et al. 2004; Dronnet et al. 2006; Adams et al. 2007; Wei et al. 2007). Recently, Dapporto et al. (2008) showed by the Maldi-Tof-MS technique that cuticular polar substances may be involved in nestmate recognition in the paper wasp *Polistes dominulus*. With the same technique, new cuticular hydrocarbons of high molecular weight could be detected for the first time in a termite (Cvačka et al. 2006). The complex story of nestmate recognition in termites needs further investigations.

A very unusual situation is observed in the Neotropical termite *Constrictotermes cyphergaster* and its inquiline species *Inquilinitermes microcerus*, whose alates do not show any aggressive behaviour when mixed under experimental conditions. Brief tandem behaviours of heterospecific pairs could be observed. The presence of abundant quantities of crotonic acid on the integument of both species might explain this exceptional and fascinating cohabitation (unpublished).

11.3 Foraging

In termites, foraging is always a collective behaviour, based upon the integration of diverse strategies of communication, mainly of chemical nature. The behavioural ecology of foraging has been excellently reviewed by Traniello and Robson (1995) and Traniello and Leuthold (2000). Pasteels and Bordereau (1998) described the state of knowledge on the chemical nature of trail-following pheromones before SPME applications. Emphasis is given here to the properties and the nature of foraging trails and trail-following pheromones.

11.3.1 Trails and Trail-Following Pheromones

The involvement of odorant stimuli in the organisation of trails has been postulated since the earliest studies on termites (Dudley and Beaumont 1889; Bugnion 1909; Andrews 1911). Grassé and Noirot (1950, 1951b) attributed the origin of the main olfactory stimulus of orientation to the "pavé" of trails, made of faecal and earth pellets. However, Lüscher and Müller (1960), and Stuart (1961) identified the sternal gland as the source of trail-following pheromones of termites. Stuart (1961, 1963, 1969), and Heidecker and Leuthold (1984) concluded that hindgut fluid, frequently deposited on trails, did not contain orientation factors.

11.3.1.1 Trails and Ecological Life Types

Termites of "one-piece" life type termites, nesting in their food, do not require orientation systems a priori. Secretions of their sternal gland are considered to function in the recruitment of nestmates to sources of disturbance within a nest but not sources of food (Stuart 1969; Traniello and Leuthold 2000). In fact, these termites undoubtedly also use trail-following pheromones to colonise new food sources, as they are able to build external galleries ("faecal fortresses", B. Thorne, personal communication) which may reach up to 20 cm long in *Zootermopsis nevadensis*, and *Hodotermopsis sjostedti* (unpublished) (Fig. 11.3). Under experimental conditions of trail-following "open-field" bioassays, pseudergates of "one-piece" nest termites are perfectly able to follow artificial trails. The existence of external foraging by pseudergates of *Prorhinotermes* (Rupf and Roisin 2008) demonstrated that this behaviour could evolve before the appearance of a true worker caste, and that the communication by pheromone trail was already present in termites without true workers.

The case of the most primitive extant termite *Mastotermes darwiniensis* is especially interesting. This termite of "intermediate" life type colonises different sources of food through subterranean connecting galleries in the field. However, under experimental conditions, its workers explore a new environment individually or in tandem, never in Indian files, and are not able to follow artificial trails if they are not guided in tunnels. This could represent a specific stage in the evolution of the recruitment strategies in termites, between the tandem running and mass recruitment (Traniello and Leuthold 2000; Sillam-Dussès et al. 2007).

In "separate" life type termites, nests are connected to food sources through a network of subterranean, open air or covered trails, all of which are built with the involvement of trail-following pheromones. In many species, trails are also impregnated with faecal and earth pellets ("pavé") (Grassé and Noirot 1950; Grassé and Noirot 1951b; Kalshoven 1958; Sugio 1995).

Fig. 11.3 Fifteen-centimetre long gallery built by pseudergates of *Hodotermopsis sjostedti* on the surface of a piece of wood after laying a trail. "One-piece" life type termites are able to use their sternal gland secretion as a trail-following pheromone to colonise additional close pieces of wood

11.3.1.2 Exploratory Trails and Foraging Trails

The establishment of foraging trails always begins with an exploratory phase of searching for food, prolonged by the recruitment of nestmates to exploit the food source. Exploratory trails appear less attractive than recruitment and foraging trails (MacFarlane 1983; Omo Malaka 1987; Runcie 1987; Reinhard et al. 1997a; Reinhard and Kaib 2001). In *Trinervitermes bettonianus*, a trail deposited by exploring workers is 5 times less attractive than a trail of food-finders (Oloo and Leuthold 1979). Whether the functional differences emanate from trail-following pheromone quantity or quality remains questionable. They would be essentially quantitative in many species (Tschinkel and Close 1973; Stuart 1981; Traniello and Busher 1985; Reinhard and Kaib 2001), but in *Reticulitermes flavipes*, different pheromones might be involved (Runcie 1987), and workers of *Macrotermes subhyalinus* may be able to quantitatively and qualitatively vary their trails depending on ecological context (Affolter and Leuthold 2000).

11.3.1.3 Trails and Polyethism

In termites, foraging and trail building can be initiated by workers only (Hewitt et al. 1969; Tschinkel and Close 1973; Leuthold and Lüscher 1974; Oloo and Leuthold 1979; Badertscher et al. 1983; Heidecker and Leuthold 1984; Lys and Leuthold 1991; Reinhard et al. 1997a), but columns of workers are most often preceded and protected by soldiers (Jander and Daumer 1974; Rickli and Leuthold 1987; Kaib 1990; Marcel et al. 1991; Miura and Matsumoto 1995, 1998; Casarin et al. 2008). In this latter case, soldier-organised foraging appears to be an adaptation to deter predation during the early phases of foraging. However, in *Nasutitermes corniger* (formerly *N. costalis*), only soldiers initiate searching and recruit workers after discovering the food (Traniello 1981; Traniello and Busher 1985). This behaviour is apparently exceptional in termites and cannot be generalised to all species of *Nasutitermes*. In *N. lujae*, soldiers also are the first individuals to reach a new food source, but they are always immediately followed by workers, and nestmates are recruited by workers. Moreover, under experimental conditions, workers are as efficient as mixed workers and soldiers in discovering new food sources (Marcel et al. 1991).

Caste-specific pheromone trails were postulated by Lys and Leuthold (1991) and Gessner and Leuthold (2001) in *Macrotermes bellicosus* where small workers preferentially follow trails established by small workers. In *Trinervitermes trinervoides* (Tschinkel and Close 1973), *Pseudacanthotermes spiniger* and *N. lujae* (Laduguie 1993), no caste-specific trails could be demonstrated. Differences in sensory physiology and pheromone production may underlie patterns of caste polyethism. In *N. corniger*, polyethism does not appear to depend on caste-specific properties of soldier and worker trail-following pheromones, but rather on quantitative differences in pheromone production between castes (Traniello and Busher 1985). The qualitative differences between castes postulated by Arab et al. (2006) for the trail pheromone in *Nasutitermes corniger* and *N. ephratae* need to be confirmed. The search for food is generally assumed by the oldest individuals, which possess the most active sternal glands (Pasteels 1965; Badertscher et al. 1983; Marcel et al. 1991). The threshold of sensitivity to trail-following pheromones varies with castes; in *N. lujae*, soldiers are 10 times more sensitive to trail-following pheromone than workers (Laduguie 1993).

11.3.1.4 Trail Persistence and Trail-Following Pheromone Longevity

Trail longevity is a parameter critical to the function of a trail as it may be assumed to be matched to the foraging ecology of individual species. In ants, short-lived trails can rapidly modulate recruitment to ephemeral food sources, whereas long-lived trails are more suited to persistent or recurrent food sources (Jackson et al. 2006). In termites, where food sources are most often abundant and persistent, trails remain active for a few hours to a few days, in the absence of reinforcement (Table 11.2). This longevity is compatible with trails renewed at every foraging expedition (Rickli and Leuthold 1987; Sugio 1995; Hoare and Jones 1998) or used every night (Oloo and Leuthold 1979; Jones and Gathorne-Hardy 1995). It is more difficult to explain the persistence of trails after 15 days of non-utilisation, as in *Hospitalitermes hospitalis* (Jones and Gathorne-Hardy 1995), and during several months or several years as in *N. corniger* (Traniello 1982) and *R. flavipes* (Runcie 1987). Trail-following pheromones composed of dodecatrienol and/or neocembrene cannot remain active for such long periods. They might be "dissolved" in cuticular lipids which would greatly diminish their evaporation, as in controlled-release lures. The fact that extracts of sternal glands of *Macrotermes annandalei* remain active longer than the synthetic pheromone (Peppuy et al. 2001a) partly supports this hypothesis. An alternative explanation might be that the pheromone is applied to the ground surface as the free molecule, and partly as a conjugate which would be persistent, slowly breaking down spontaneously and releasing the pheromone (M. Lacey, personal communication). Another possibility would be that the pheromone is bound to components present in faecal pellets of the "pavé". The intestinal fluid itself is not active in eliciting trail-following (Stuart 1961; 1963; Heidecker and Leuthold 1984), but in *Constrictotermes cyphergaster*, trails made of faecal pellets contaminated with trail-pheromone may conserve their activity for at least 3 months (Souto and Kitayama 2000). Finally, the presence in old trails of non-pheromonal components eliciting trail-following cannot be excluded.

11.3.1.5 Trail Polarity

How termites are able to distinguish between homeward and outward direction remains mostly unknown. The simplest hypothesis of a pheromone gradient was postulated for *Reticulitermes hesperus*, where workers were able to orient along a gradient of trail-following pheromone by longitudinal klinotaxis (Grace et al. 1988), but it was discounted by Saran et al. (2007). In *Hodotermes mossambicus*, workers would use, in darkness, gradients of trail-following pheromone established around the foraging holes and food sources (Heidecker and Leuthold 1984). In *T. trinervoides* (Tschinkel and Close 1973), *Hospitalitermes* spp. (Jander and Daumer 1974) or *Macrotermes michaelseni* (MacFarlane 1983), no trail polarity could be demonstrated. Nevertheless, the network of trails itself can be highly polarised by a decrease of the trail pheromone concentration from trunk trails to terminal trails (Leuthold 1975). In several species, the direction of the orientation

Table 11.2 Trail longevity

Species	Chemical nature of trail	Nutritive regime	Trail longevity	References
MASTOTERMITIDAE				
Mastotermes darwiniensis	(*E*)-2,6,10-trimethyl-5,9-undecadien-1-ol	Xylophagous	72 h	Sillam-Dussès (2004)
TERMOPSIDAE				
Porotermes adamsoni	(*E*)-2,6,10-trimethyl-5,9-undecadien-1-ol	Xylophagous	120 h	Sillam-Dussès (2004)
Zootermopsis nevadensis	4,6-dimethyldodecanal	Xylophagous	1–3 h	Lüscher (1961)
HODOTERMITIDAE				
Anacanthotermes ahngerianus	Unknown	Grass feeder	10–12 h	Shatov (1977)
Anacanthotermes turkestanicus	Unknown	Grass feeder	<1 h	Khamraev et al. (2008)
Hodotermes mossambicus	Unknown	Grass feeder	1 h 30	Hewitt et al. (1969)
KALOTERMITIDAE				
Kalotermes flavicollis	(*Z*)-dodec-en-1-ol	Xylophagous	3–4 h	Shatov (1977)
RHINOTERMITIDAE				
Reticulitermes flavipes	(*Z,Z,E*)-dodeca-3,6,8-trien-1-ol	Xylophagous	24 h to 1 year	Runcie (1987)
Reticulitermes hesperus	(*Z,Z,E*)-dodeca-3,6,8-trien-1-ol[a]	Xylophagous	24 h	Saran et al. (2007)
Reticulitermes lucifugus	(*Z,Z,E*)-dodeca-3,6,8-trien-1-ol	Xylophagous	6–7 h	Shatov (1977)
Reticulitermes speratus	(*Z,Z,E*)-dodeca-3,6,8-trien-1-ol	Xylophagous	6–7 h	Shatov (1977)
TERMITIDAE				
Macrotermitinae				
Macrotermes annandalei	(*Z*)-dodec-en-1-ol	Xylophagous + litter	72 h	Peppuy et al. (2001a)
Macrotermes michaelseni	Unknown	Xylophagous + litter	1–9 h	MacFarlane (1983)
Termitinae				
Amitermes rhizophagus	Unknown	Xylophagous	6–7 h	Shatov (1977)
Nasutitermitinae				
Nasutitermes corniger	(*Z,Z,E*)-dodeca-3,6,8-trien-1-ol + neocembrene	Xylophagous	Up to 10 years	Traniello (1982)
Trinervitermes trinervoides	(*Z,Z,E*)-dodeca-3,6,8-trien-1-ol + neocembrene	Grass feeder	6–15–25 min	Hewitt et al. (1969)
			7 h 30	Tschinkel and Close (1973)
Constrictotermes cyphergaster	Dodecatrienol + neocembrene	Lichens	1 h (3 months)	Souto and Kitayama (2000)
Hospitalitermes hospitalis	Unknown	Lichens	15 days	Jones and Gathorne-Hardy (1995)

[a]This molecule was not completely identified.

may be regulated by means other than semiochemicals (Jander and Daumer 1974; Leuthold et al. 1976; Rickli and Leuthold 1987, 1988).

11.3.1.6 Territoriality and Trail-Following Pheromones

As stated by Traniello and Robson (1995), "we are still far from fully integrating behavioural and chemical ecology in the analysis of termite trail and territorial communication", and if a territory is an area advertised by scent marking and used exclusively by a colony, the examples are rare, if not absent, in termites. Colonies of *Nasutitermes corniger* and *N. nigriceps* (Levings and Adams 1984; Adams and Levings 1987), *N. princeps* and *Microcerotermes biroi* (Leponce et al. 1997) defend exclusive territories intra- and interspecifically by agonistic interactions, but nothing is known of possible scent marking of their foraging areas. Most often, colonies protect and defend their network of subterranean passageways and surface galleries which can be considered as extensions of their nest (Darlington 1982; Jmashly and Leuthold 1999a, 1999b). However, the defence of permanent foraging territories in the open air is probably very rare. In contrast, workers from different colonies, or different species frequently forage close together in overlapping areas, as do the sympatric *Macrotermes annandalei* and *M. barneyi* (Peppuy 1999). Trail-following pheromones of sympatric species can be species-specific (see above), but no colony-specificity has yet been shown (Hewitt et al. 1969; Oloo 1981; Oloo and McDowell 1982; Omo Malaka 1987; Grace et al. 1995). Components other than the trail-following pheromones might be involved in the species-specificity of trails. Recent results support Grassé and Noirot's hypothesis (1951b) about a possible role of "pavé". Workers of the sympatric *M. annandalei* and *M. barneyi* are not able to recognise their respective trails made of only sternal gland secretions, but workers of *M. annandalei* recognise their own trails if their pheromone trail is mixed with contents of their digestive tract (unpublished). *Coptotermes formosanus* and *R. flavipes* cannot recognize fresh trails, but have a species-specific response to long-established trails mixed with salivary secretions and faeces (Cornelius and Bland 2001). Nevertheless, extracts from "pavé" did not elicit agonism between *N. corniger* and *N. nigriceps* (Traniello and Leuthold 2000). The role of the "pavé" in trail ecology needs further research.

11.3.1.7 Semiochemicals on Foraging Sites

Termites probably detect food over a distance, even in subterranean species, by the volatiles emanating from it. Interestingly, workers of *Schedorhinotermes lamanianus* and *Reticulitermes santonensis* release phagostimulating pheromones from labial glands (Reinhard and Kaib 1995; Reinhard et al. 1997b). However, whether hydroquinone is a general phagostimulating pheromone (Reinhard et al. 2002) needs confirmation by using appropriate bioassays. At least in *Coptotermes formosanus*, hydroquinone does not work as a phagostimulant (Raina et al. 2005).

11.3.2 Trail-Following Pheromone Glands

In termites, trail-following pheromones emanate from only one glandular source, the sternal gland (Noirot 1969). It is an essential difference from trail-following pheromones of ants, secreted by multiple glands, which may largely explain the behavioural and chemical specificities of the foraging ecology of termites. The sternal gland is a small abdominal gland present in all castes and located on the sternites 3–5 in the Mastotermitidae, on the 4th sternite in the Termopsidae and Hodotermitidae, and on the 5th sternite in the Kalotermitidae, Rhinotermitidae, Serritermitidae and Termitidae (Noirot 1969, 1995). *Mastotermes darwiniensis* is the only extant termite to possess 3 sternal glands; sternal gland 4 is 4–5 times more active than gland 3, and the activity of sternal gland 5 is very low or absent (Sillam-Dussès et al. 2007).

The anatomical evolution of the sternal gland has been recently reviewed (Quennedey et al. 2008). In *M. darwiniensis,* glands are located on the middle of sternites, whereas in all other termites the sternal gland is situated in front of sternites, and covered by the posterior part of the preceding sternite, which implies an active exposure of the gland during trail-laying. The intersternal fold also may provide a temporary external storage of pheromone. The most striking case is observed in *Hodotermes mossambicus* where the sternal gland is a true cushion evaginable and retractable (Quennedey et al. 2008), and probably related to the special trail marking of this termite (Hewitt et al. 1969; Heidecker and Leuthold 1984).

All sternal glands possess class 1 and 2 cells. Class 2 cells are especially abundant and active in the Rhinotermitidae and Termitidae where trail-following pheromones are the most efficient. Class 3 cells are present only in the Termopsinae, Serritermitidae and Rhinotermitidae (Noirot and Quennedey 1974; Quennedey et al. 2008).

11.3.3 Chemical Nature of Trail-Following Pheromones

Thanks to the utilisation of SPME (Figs. 11.4 and 11.5), trail-following pheromones are now known in 60 species. Their major characteristics are their low chemical diversity, involving only 8 different molecules, and their similarity with sex-pairing pheromones (Table 11.3, Fig. 11.2). Like sex-pairing pheromones, they belong to three chemical classes (alcohols, aldehydes, hydrocarbons) and their molecular weight also varies from 180 to 272, which is surprising given their distinctive behavioural effects. Alcohols may have 14 carbon atoms, such as (*E*)-2,6,10-trimethylundeca-5,9-dien-1-ol (*Mastotermes darwiniensis* and basal termites *Porotermes adamsoni* and *Stolotermes victoriensis*), 13 carbon atoms, such as 4,6-dimethylundecan-1-ol (*Hodotermopsis sjostedti* – this identification needs to be confirmed by synthesis) or 12 carbon atoms, such as (*Z*)-dodec-3-en-1-ol, (*Z,Z*)-dodeca-3,6-dien-1-ol, and (*Z,Z,E*)-dodeca-3,6,8-trien-1-ol, in many species. Aldehydes have 12 carbon atoms, such as 4,6-dimethyldodecanal (*Zootermopsis angusticollis* and *Zootermopsis nevadensis*), or 18 carbon atoms (*Hodotermes mossambicus*). In *H. sjostedti,* traces of 4,6-dimethyldodecanal are present in addition to the potential major component 4,6-dimethylundecan-1-ol. The similarity

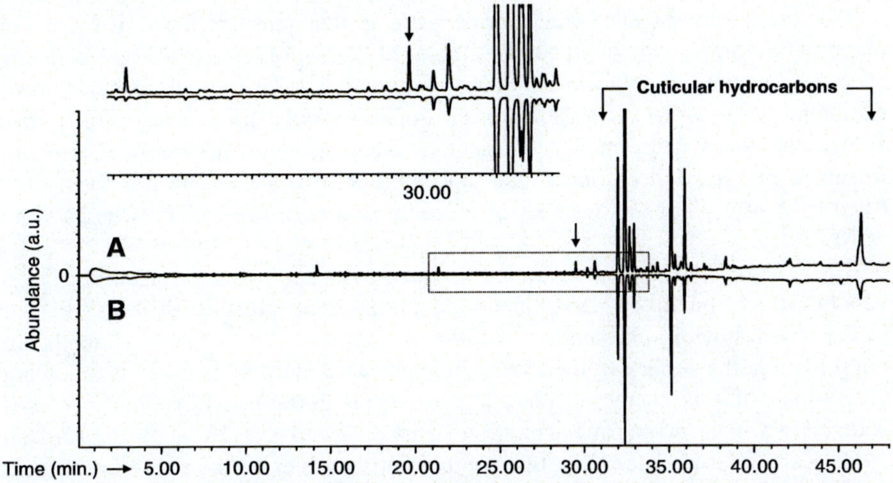

Fig. 11.4 Comparison of the GC profiles of SPME extracts of (*A*) the surface of sternal gland, and (*B*) the extra-glandular integument of pseudergates of *Porotermes adamsoni*. The majority of peaks are common to both surfaces and correspond to cuticular hydrocarbons from C22 to C30, whereas the peak marked by an *arrow* is specific to the glandular surface and corresponds to the trail-following pheromone ((*E*)-2,6,10-trimethylundeca-5,9-dien-1-ol). Detail of the window in the *upper part* of the figure. © Journal of Chemical Ecology

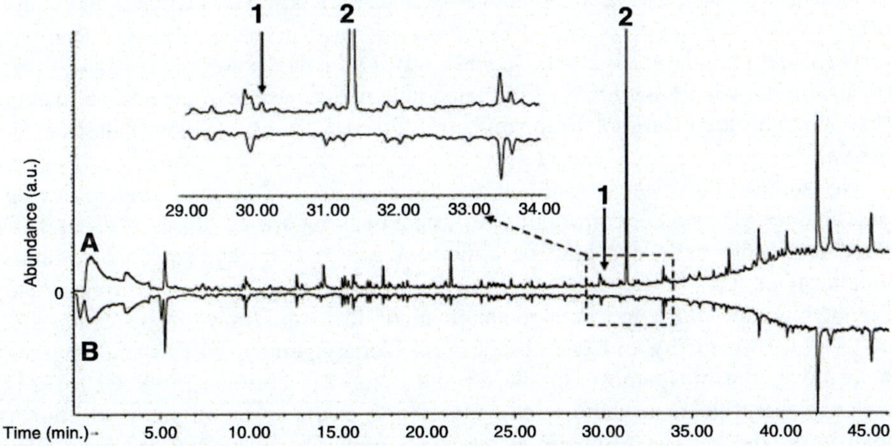

Fig. 11.5 GC profiles of SPME extracts of (*A*) the surface of sternal gland, and (*B*) the extra-glandular integument of workers of *Nasutitermes exitiosus* showing two compounds specific to the sternal gland secretion, dodecatrienol (*1*) and neocembrene (*2*). Only the association of these two compounds with their relative proportions elicits an optimum trail-following. Detail of the window in the upper part of the figure

between trail-following pheromones of *M. darwiniensis, P. adamsoni* and *S. victoriensis* (the C14 alcohol (*E*)-2,6,10-trimethylundeca-5,9-dien-1-ol) and the female sex-pairing pheromone of the Termopsinae *Z. nevadensis* and *H. sjostedti* (the C14 aldehyde (*E*)-2,6,10-trimethylundeca-5,9-dienal) must be underlined. The C20 hydrocarbon neocembrene is present in derived species, but also in *Prorhinotermes canalifrons* and *P. simplex* (Sillam-Dussès et al. 2005).

The second major chemical characteristic is that most of the trail-following pheromones are single compounds (Table 11.3, Fig. 11.4). However, in some species, especially in Nasutitermitinae, pheromones have two components, neocembrene with either dodecadienol or dodecatrienol (Table 11.3, Fig. 11.5), and minor components may be present in other species (Tokoro et al. 1992a). Recently, besides dodecatrienol and neocembrene, trinervitatriene has been identified in sternal glands of workers of *Nasutitermes corniger* (Sillam-Dussès et al. 2010).

The use of SPME has corrected some previous pheromone identifications. Klochkov and Zhuzhikov (1990) identified a mixture of saturated alcohols C9–C12 as the trail-following pheromone of *Kalotermes flavicollis*, but these alcohols are not present at the surface of their sternal gland, and dodecenol is the trail-following pheromone of this species (Sillam-Dussès et al. 2009b). *n*-Hexanoic acid was claimed to be the trail-following pheromone of *Z. nevadensis* (Hummel and Karlson 1968; Karlson et al. 1968), but this compound is present on the entire integument of pseudergates, and probably not a component of the trail-following pheromone, unless it spreads over the body from the sternal gland. 4,6-Dimethyldodecanal, specific to the sternal gland surface, is the trail-following pheromone of *Zootermopsis* (Bordereau et al. 2010). Neocembrene was identified as the only compound of the trail-following pheromone of *Nasutitermes exitiosus, N. graveolus, N. walkeri* (Moore 1966; Birch et al. 1972), and *Trinervitermes bettonianus* (McDowell and Oloo 1984). In fact, it is secreted both with dodecatrienol in *N. exitiosus* and many other species of *Nasutitermes* and *Trinervitermes* and probably also in *T. bettonianus* (Sillam-Dussès et al. 2010). Interestingly, GC-EAD experiments showed that dodecatrienol was also probably associated with neocembrene in the trail-following pheromone of the Prorhinotermitinae *Prorhinotermes simplex* (Sillam-Dussès et al. 2009a)

No studies have been done on the biosynthesis of termite trail-following pheromones. However, considering the chemical nature of these pheromones, three metabolic pathways may be hypothesized (M.J. Lacey, personal communication) (i) The mevalonate synthesis to yield geranylgeranyl pyrophosphate (geranylgeranyl-OPP) and then metabolism of the latter to yield neocembrene, or (*E*)-2,6,10-trimethylundeca-5,9-dien-1-ol. Geranylgeranyl-OPP is undoubtedly involved in the biosynthesis of the termite defence secretions (Prestwich 1981) and it is therefore reasonable to conclude that it is the common precursor, rather than farnesyl OPP. (ii) Aliphatic alcohols are probably biosynthesised from fatty acids (Tokoro et al. 1990, 1992b; Francke and Schulz 1999). Three cycles of beta oxidation of the fatty acid ligands, oleate and linoleate, following by reduction will yield the appropriate dodecenol and dodecadienol. Also, the action of a stereospecific omega-4 dehydrogenase on linoleate, followed by 3 cycles of beta oxidation then reduction, will yield dodecatrienol. (iii) Since saturated cuticular hydrocarbons are formed by fatty acid biosynthesis with combinations of acetate and propanoate, as is the insect pheromone 4,8-dimethyldecanal (Kim et al. 2005), analogous biosyntheses can be proposed for 4,6-dimethylundecan-1-ol and 4,6-dimethyldodecanal.

Table 11.3 Chemical nature of trail-following pheromones

Species	Origin	Ecology	Chemical nature	Quantity (pg/ind)	Activity threshold (ng/cm)	References
MASTOTERMITIDAE						
Mastotermes darwiniensis	Australian	2/W	(*E*)-2,6,10-trimethyl-5,9-undecadien-1-ol	20	1	Sillam-Dussès et al. (2007)
TERMOPSIDAE						
Porotermitinae						
Porotermes adamsoni	Australian	1/W	(*E*)-2,6,10-trimethyl-5,9-undecadien-1-ol	700	10^{-2}	Sillam-Dussès et al. (2007)
Stolotermitinae						
Stolotermes victoriensis	Australian	1/W	(*E*)-2,6,10-trimethyl-5,9-undecadien-1-ol	4	–	Sillam-Dussès et al. (2007)
Termopsinae						
Hodotermopsis sjostedti	Indomalayan	1/W	4,6-Dimethylundecan-1-ol[a] + traces 4,6-dimethyldodecanal	–	–	Unpublished
Zootermopsis nevadensis	Nearctic	1/W	*n*-Hexanoic acid	–		Hummel and Karlson (1968)
			4,6-Dimethyldodecanal	100–500	10^{-1}	Bordereau et al. (2010)
Zootermopsis angusticollis	Nearctic	1/W	4,6-Dimethyldodecanal	100–500	–	Bordereau et al. (2010)
HODOTERMITIDAE						
Hodotermes mossambicus	Aethiopian	3/G	C18 aldehyde	1,000		Unpublished
KALOTERMITIDAE						
Cryptotermes brevis	Cosmopolite	1/W	(Z)-dodec-3-en-1-ol	400–600	–	Sillam-Dussès et al. (2009b)
Cryptotermes darlingtonae	Neotropical	1/W	(Z)-dodec-3-en-1-ol	400–600	10^{-2}	Sillam-Dussès et al. (2009b)
Cryptotermes pallidus	Malagasy	1/W	(Z)-dodec-3-en-1-ol	400–600	10^{-1}	Sillam-Dussès et al. (2009b)
Incisitermes tabogae	Neotropical	1/W	(Z)-dodec-3-en-1-ol	–	–	Sillam-Dussès et al. (2009b)
Kalotermes flavicollis	Palearctic	1/W	Nonanol + decanol + undecanol + dodecanol	–	–	Klochkov and Zhuzhikov (1990)
			(Z)-dodec-3-en-1-ol	400–600	10^{-2}	Sillam-Dussès et al. (2009b)

Table 11.3 (continued)

Species	Origin	Ecology	Chemical nature	Quantity (pg/ind)	Activity threshold (ng/cm)	References
Neotermes holmgreni	Neotropical	1/W	(*Z*)-dodec-3-en-1-ol	–	–	Sillam-Dussès et al. (2009b)
Postelectrotermes howa	Malagasy	1/W	(*Z*)-dodec-3-en-1-ol	400–600	10^{-1}	Sillam-Dussès et al. (2009b)
Procryptotermes falcifer	Malagasy	1/W	(*Z*)-dodec-3-en-1-ol	400–600	10^{-2}	Sillam-Dussès et al. (2009b)
Procryptotermes leewardensis	Neotropical	1/W	(*Z*)-dodec-3-en-1-ol	–	–	Sillam-Dussès et al. (2009b)
RHINOTERMITIDAE						
Prorhinotermitinae						
Prorhinotermes canalifrons	Malagasy	2/W	Neocembrene	250	10^{-1}	Sillam-Dussès et al. (2005)
Prorhinotermes simplex	Neotropical	2/W	Neocembrene	–	10^{-1}	Sillam-Dussès et al. (2005)
			Neocembrene + (*Z,Z,E*)-dodeca-3,6,8-trien-1-ol			Sillam-Dussès et al. (2009a)
Heterotermitinae						
Heterotermes tenuis	Neotropical	2/W	(*Z,Z,E*)-dodeca-3,6,8-trien-1-ol	–	–	Sillam-Dussès et al. (2006)
Reticulitermes flavipes	Nearctic	2/W	(*Z,Z,E*)-dodeca-3,6,8-trien-1-ol	–	10^{-5}	Tai et al. (1969)
Reticulitermes lucifugus grassei	Palearctic	2/W	(*Z,Z,E*)-dodeca-3,6,8-trien-1-ol	1–10	10^{-3}	Wobst et al. (1999)
Reticulitermes santonensis	Palearctic	2/W	(*Z,Z,E*)-dodeca-3,6,8-trien-1-ol	1–10	10^{-4}	Laduguie et al. (1994)
Reticulitermes speratus	Palearctic	2/W	(*Z,Z,E*)-dodeca-3,6,8-trien-1-ol	3	10^{-4}	Tokoro et al. (1991)
Reticulitermes virginicus	Nearctic	2/W	(*Z,Z,E*)-dodeca-3,6,8-trien-1-ol	–	$<10^{-5}$	Matsumura et al. (1968)
Coptotermitinae						
Coptotermes formosanus	Indomalayan	3/W	(*Z,Z,E*)-dodeca-3,6,8-trien-1-ol	50	10^{-5}	Tokoro et al. (1989)
Coptotermes gestroi	Indomalayan	3/W	(*Z,Z,E*)-dodeca-3,6,8-trien-1-ol	–	10^{-4}	Sillam-Dussès et al. (2006)
Rhinotermitinae						
Rhinotermes marginalis	Neotropical	2/W	(*Z,Z,E*)-dodeca-3,6,8-trien-1-ol	–	10^{-2}	Sillam-Dussès et al. (2006)
Schedorhinotermes lamanianus	Aethiopian	3/W	(*Z,Z,E*)-dodeca-3,6,8-trien-1-ol	–	10^{-2}	Sillam-Dussès et al. (2006)
TERMITIDAE						
Macrotermitinae						
Ancistrotermes pakistanicus	Aethiopian	3/W+L	(*Z,Z*)-dodeca-3,6-dien-1-ol	1	10^{-4}	Robert et al. (2004)

Table 11.3 (continued)

Species	Origin	Ecology	Chemical nature	Quantity (pg/ind)	Activity threshold (ng/cm)	References
Macrotermes annandalei	Indomalayan	3/W+L	(*Z*)-dodec-3-en-1-ol	1,000	10^{-3}	Peppuy et al. (2001a)
Macrotermes barneyi	Indomalayan	3/W+L	(*Z*)-dodec-3-en-1-ol	1,400	10^{-2}	Peppuy et al. (2001a)
Macrotermes bellicosus	Aethiopian	3/W+L	(*Z*)-dodec-3-en-1-ol	750	10^{-3}	Peppuy et al. (2001b)
Macrotermes subhyalinus	Aethiopian	3/W+L	(*Z*)-dodec-3-en-1-ol	–	–	Peppuy et al. (2001b)
Odontotermes formosanus	Indomalayan	3/W+L	(*Z*,*Z*)-dodeca-3,6-dien-1-ol	–	–	Deng et al. (2002)
Odontotermes hainanensis	Aethiopian	3/W+L	(*Z*)-dodec-3-en-1-ol	100	10^{-2}	Peppuy et al. (2001b)
Odontotermes latericius	Aethiopian	3/W+L	Dodecenol + dodecadienol	–	–	Unpublished
Odontotermes maesodensis	Aethiopian	3/W+L	(*Z*)-dodec-3-en-1-ol	200	10^{-3}	Peppuy et al. (2001b)
Pseudacanthotermes militaris	Aethiopian	3/W+L	(*Z*,*Z*,*E*)-dodeca-3,6,8-trien-1-ol	20–30	10^{-4}	Bordereau et al. (1993)
Pseudacanthotermes spiniger	Aethiopian	3/W+L	(*Z*,*Z*,*E*)-dodeca-3,6,8-trien-1-ol	10	10^{-4}	Bordereau et al. (1993)
Apicotermitinae						
Apicotermes sp.	Aethiopian	3/S	(*Z*,*Z*,*E*)-dodeca-3,6,8-trien-1-ol	10	–	Unpublished
Termitinae						
Cubitermes sp.	Aethiopian	3/S	(*Z*,*Z*,*E*)-dodeca-3,6,8-trien-1-ol	<1	10^{-2}	Sillam-Dussès et al. (2006)
Drepanotermes perniger	Australian	3/G	(*Z*,*Z*,*E*)-dodeca-3,6,8-trien-1-ol	–	10^{-4}	Sillam-Dussès et al. (2006)
Termes hispaniolae	Neotropical	3/W	(*Z*,*Z*,*E*)-dodeca-3,6,8-trien-1-ol	<1	10^{-3}	Sillam-Dussès et al. (2006)
Amitermes evuncifer	Aethiopian	3/W+L	Dodecatrienol + neocembrene	–	–	Kotoklo et al. (2010)
Syntermitinae						
Armitermes euhamignathus	Neotropical	3/G	Dodecadienol+ neocembrene	–	–	Unpublished
Cornitermes bequaerti	Neotropical	3/G	(*Z*,*Z*,*E*)-dodeca-3,6,8-trien-1-ol	100	–	Sillam-Dussès et al. (2006)
Cornitermes cumulans	Neotropical	3/G	(*Z*,*Z*,*E*)-dodeca-3,6,8-trien-1-ol	100	10^{-3}	Sillam-Dussès et al. (2006)
Cornitermes snyderi	Neotropical	3/G	(*Z*,*Z*,*E*)-dodeca-3,6,8-trien-1-ol	–	10^{-3}	Sillam-Dussès et al. (2006)
Embiratermes festivellus	Neotropical	3/G	(*Z*,*Z*,*E*)-dodeca-3,6,8-trien-1-ol	–	–	Unpublished
Syntermes grandis	Neotropical	3/G+W	(*Z*,*Z*,*E*)-dodeca-3,6,8-trien-1-ol	300	10^{-4}	Sillam-Dussès et al. (2006)
Nasutitermitinae						
Constrictotermes cyphergaster	Neotropical	3/Lichens	Dodecatrienol + neocembrene	–	–	Sillam-Dussès et al. (2010)

Table 11.3 (continued)

Species	Origin	Ecology	Chemical nature	Quantity (pg/ind)	Activity threshold (ng/cm)	References
Nasutitermes corniger	Neotropical	3/W	Dodecatrienol + neocembrene + trinervitatriene	R:1/50–100	–	Sillam-Dussès et al. (2010)
Nasutitermes coxipoensis	Neotropical	3/W	Dodecatrienol + neocembrene	–	–	Unpublished
Nasutitermes diabolus	Aethiopian	3/W	Dodecatrienol + neocembrene	–	–	Sillam-Dussès et al. (2010)
Nasutitermes ephratae	Neotropical	3/W	Dodecatrienol + neocembrene	R:1/100	–	Sillam-Dussès et al. (2010)
Nasutitermes exitiosus	Australian	3/W	Neocembrene	–	10^{-2}	Moore (1966) and Birch et al. (1972)
			Dodecatrienol + neocembrene	R:1/100	–	Sillam-Dussès et al. (2010)
Nasutitermes graveolus	Australian	3/W	Neocembrene	–	–	Moore (1966) and Birch et al. (1972)
Nasutitermes guayanae	Neotropical	3/W	Dodecatrienol + neocembrene	R:1/1,000	–	Sillam-Dussès et al. (2010)
Nasutitermes kemneri	Neotropical	3/W	Dodecatrienol + neocembrene	–	–	Sillam-Dussès et al. (2010)
Nasutitermes lujae	Aethiopian	3/W	Dodecatrienol + neocembrene	–	–	Sillam-Dussès et al. (2010)
Nasutitermes voeltzkowi	Malagasy	3/W	Dodecatrienol + neocembrene	R:1/50	–	Sillam-Dussès et al. (2010)
Nasutitermes walkeri	Australian	3/W	Neocembrene	–	–	Moore (1966) and Birch et al. (1972)
Trinervitermes bettonianus	Aethiopian	3/G	Neocembrene	–	–	McDowell and Oloo (1984)
Trinervitermes geminatus	Aethiopian	3/G	Dodecatrienol + neocembrene	R:1/200	–	Sillam-Dussès et al. (2010)
Trinervitermes trinervoides	Aethiopian	3/G	Dodecatrienol + neocembrene	–	–	Sillam-Dussès et al. (2010)

[a]This molecule was not completely identified.

11.3.4 Biological Activity of Trail-Following Pheromones

Apart from *Mastotermes darwiniensis*, where stereoisomers (*E*) and (*Z*) of the trail-following pheromone are equally active in eliciting trail-following, although only the (*E*)-isomer is secreted, the stereoisomerism of trail-following pheromones is a major factor in their activity. In *Porotermes adamsoni*, which secretes the same trail-following pheromone as *M. darwiniensis*, the (*E*)-isomer is clearly more active than the (*Z*)-isomer (Sillam-Dussès et al. 2007). In *Macrotermes annandalei*, only the (*Z*)-isomer is active (Peppuy et al. 2001a). The stereoisomerism is also highly important for the widespread dodecatrienol: the natural (*Z,Z,E*)-isomer is 10,000 times more active than the (*Z,Z,Z*)-isomer (Tai et al. 1969, 1971; Matsumura et al. 1972). For neocembrene, the natural (*E,E,E*)-isomer in *Nasutitermes exitiosus* (Kodama et al. 1975) is 1,000 times more active than the (*E,Z,E*)-isomer and the (*E,E,Z*)-isomer is totally inactive (Kato et al. 1980). In contrast, the chirality seems to be insignificant in determining trail-laying activity of neocembrene in *N. exitiosus* (Kato et al. 1980). The chirality of trail-following pheromones composed of (*E*)-2,6,10-trimethylundeca-5,9-dien-1-ol, 4,6-dimethyldodecanal and neocembrene is still unknown.

The trail-following pheromones are active within a given range of concentrations (Table 11.3). In basal termites, the activity threshold ranges between 10^{-1} and 10^{-2} ng of pheromone/cm of trail, except in *M. darwiniensis* with a threshold at 1 ng/cm. In more derived termites, the threshold is between 10^{-2} and 10^{-6} ng/cm, the highest activity being observed for dodecatrienol. In *Reticulitermes hesperus*, dodecatrienol elicits trail-following from 10^{-7} to 10^{-8} ng/cm, but is as yet unidentified in the insect (Saran et al. 2007). For the majority of the species, optimal concentrations are between 10^{-2} and 1 ng of pheromone/cm of trail. Above 1 ng/cm, trail-following is greatly reduced, or even possibly totally suppressed, as if antennal sensory receptors were saturated. These highly concentrated pheromone trails never elicit repulsive behaviour. In species secreting bi-component pheromones, trail-following is only optimal with a mixture of components corresponding to relative proportions of the natural pheromone. For example, in *Nasutitermes corniger*, dodecatrienol alone does not elicit any trail-following, neocembrene alone elicits a moderate trail-following at 10^{-1} ng/cm, whereas a mixture of dodecatrienol at 10^{-3} ng/cm and neocembrene at 10^{-1} ng/cm is optimum (Sillam-Dussès et al. 2010).

The amounts of pheromone estimated per individual generally vary from less than 1 pg/individual, as in some Termitinae, to 1,400 pg/individual in *M. annandalei*. Amounts appear related both to the size of termites, and to the nature of the pheromone (Table 11.3). Dodecatrienol is generally secreted in lower amounts than dodecenol, probably due to its higher biological activity. Again, *M. darwiniensis* is an exception, its workers are of large size and secrete very low levels of trail-following pheromone. Amounts of pheromone and their optimal active concentrations clearly indicate that termites cannot continuously lay trail-following pheromone for long distances, without a permanent biosynthesis or intermittent trail-laying. For example, a worker of the subterranean termite *Reticulitermes speratus* would not be able to lay more than 30 cm of trail, and a worker of

Pseudacanthotermes spiniger more than 1.5 m. It might be one of the physiological constraints that explains the absence of individual foraging in termites. It is worth noting that for some species the estimated amounts of the pheromone cannot explain the biological activity of sternal glands extracts (Laduguie et al. 1994; Sillam-Dussès et al. 2005). This strongly suggests additional synergistic compounds (Pasteels and Bordereau 1998).

11.3.5 Chemical Nature and Behavioural Effects of Trail-Following Pheromones

From behavioural and ecological studies, it has often been suggested that termite trail-following pheromones play a role in recruitment in addition to their well known role in orientation (Oloo and Leuthold 1979; Kaib et al. 1982; Traniello and Busher 1985; Runcie 1987; Grace et al. 1988; Reinhard et al. 1997a; Affolter and Leuthold 2000). This was highlighted in *Nasutitermes corniger* by appropriate recruitment bioassays (Traniello 1982; Hall and Traniello 1985). To account for the respective behavioural effects of recruitment and orientation, pheromonal blends were postulated (Runcie 1987; Grace et al. 1988; Affolter and Leuthold 2000). In *N. corniger*, where neocembrene was ineffective for recruiting termites but had an orientation effect, Traniello (1982) and Hall and Traniello (1985) claimed the existence of both a volatile ephemeral component of recruitment and a long lasting orientation cue. To date, trail-following pheromones appear to be composed of only 1 or 2 compounds of similar volatility, but sparingly volatile components cannot be excluded. Bioassays performed by Traniello (1982), Hall and Traniello (1985), and Runcie (1987) used ethanolic extracts of sternal glands that might contain more polar and less volatile substances than the usual extracts made with non-polar solvents. Also, the standard SPME fibers generally used for the recent identifications of termite pheromones might be inappropriate for extracting highly polar or non-volatile substances. The non-detection of the polar sex-pairing pheromones of *Macrotermes annandalei* and *M. barneyi* after SPME-GC (Peppuy et al. 2004) supports this hypothesis. Nevertheless, artificial trails made of synthetic dodecatrienol (Laduguie 1993; Laduguie et al. 1994), dodecenol (Peppuy et al. 2001a) and neocembrene (Sillam-Dussès et al. 2005) were shown to have not only an orientation behavioural effect but also a clear recruitment effect, and performed as well as trails made with sternal gland extracts. Moreover, Peppuy (1999) used dodecenol to attract a column of foragers of *M. annandalei* in the field. For these reasons, an alternative hypothesis to that of a bi-component pheromone might be that, when pheromone extracts or solutions are initially deposited on the ground surface, vaporised molecules emanating from this application would form a short-lived aerial trail that would trigger recruitment. The portion of the pheromone adsorbed on the ground surface or that occluded by cuticular lipids would not contribute readily to the effluvium and therefore would provide a longer-lasting orientation effect.

11.3.6 Species-Specificity of Trail-Following Pheromones

The ecological success of an insect largely depends on the species-specificity of its semiochemicals, and species-specificity may be a key factor in pheromone evolution

(Symonds and Elgar 2008). Species-specificity of pheromones is generally achieved by combining a unique blend of simple components (Moore 1974; Wyatt 2003; Jackson et al. 2007). In termites, which most often forage under the protection of galleries, the ecological interest of species-specificity of trail-following pheromones was brought into question (Traniello and Leuthold 2000). From the chemical data known at present, the following situations can be envisaged.

In some sympatric species, species-specificity of trail-following pheromones could not be established. In both *Pseudacanthotermes spiniger* and *P. militaris,* the trail-following pheromone is made of dodecatrienol, but workers of *P. militaris* secrete 2 to 3 times more dodecatrienol than workers of *P. spiniger.* In trail-following bioassays, workers of *P. spiniger* preferentially follow trails made with sternal gland extracts of *P. militaris,* but responses can be reversed by increasing the concentration of *P. spiniger* trails. In this case, trail-following is clearly dependent upon the pheromone quantity (unpublished). Similar results were obtained in *Trinervitermes* (Oloo and McDowell 1982), and in *Reticulitermes santonensis* and *R. grassei* (Wobst et al. 1999).

In other species, species-specific components are clearly secreted with a common major compound (Howard et al. 1976; Kaib et al. 1982; Peppuy et al. 2001b). These species-specific signals are more pronounced between species from different genera than between congeneric species, and may even offset quantitative differences of the common signal. Workers of *Odontotermes hainanensis* prefer their trails to those of *M. annandalei* although they contain 10 times less dodecenol, the common major component of the trail-following pheromone (Peppuy et al. 2001b). Nevertheless, if the quantitative differences of the common signal are artificially modified, the responses of workers may then be ascribed to the quantity of this signal. Until now, no species-specific component of the trail-following pheromone could be identified. In species secreting two components, the relative proportions of their components might be the source of species-specificity of their trails (Sillam-Dussès et al. 2010).

Finally, the species-specificity of trail-following pheromones may arise from the different components in some sympatric termites. For example, *Ancistrotermes pakistanicus* secreting dodecadienol is sympatric with *M. annandalei* and *M. barneyi* secreting dodecenol (Peppuy et al. 2001a, 2001b; Robert et al. 2004). In the South American sympatric species *Nasutitermes corniger* and *N. erhardti*, trail-following pheromones are mutually repellent and could be involved in delimiting territories (unpublished).

11.3.7 Evolution of Trail-Following Pheromones and Phylogeny

Trail-following pheromones of termites appear to be highly conserved in their evolution, in great contrast to what is observed for allomones. Similar or closely related pheromones are present in species from different geographic origins, different ecological life types, various feeding groups, and unrelated taxa (Table 11.3). This is particularly illustrated by the trail-following pheromone of the "one piece" drywood Kalotermitidae, which is identical to that of several species of "separate" Macrotermitinae, which forage outside the nest. Major events in the ecological and developmental evolution of termites represented by the appearance of external foraging and the differentiation of a true worker caste are not related to the evolution of pheromones. Trail-following pheromones of basal termites are nevertheless different from those of more derived termites. Whereas in the former they are composed of branched C13, C14 alcohols or C14, C18 aldehydes, in the latter, from the Kalotermitidae, they are composed of unbranched C12 alcohols or a C20 diterpene. There is an interesting correlation with the location of the sternal gland situated on the 4th sternite in basal termites, but on the 5th sternite in Kalotermitidae, Rhinotermitidae, and Termitidae.

Remarkably, the most basal extant termite *Mastotermes darwiniensis* and the subsocial cockroach *Cryptocercus*, known to constitute the sister-group of the Isoptera (Eggleton 2001; Inward et al. 2007b; Legendre et al. 2008) secrete 2 isomers of trimethylundecadienol (Le Quéré et al. 1991; Grandcolas et al. 2001; Sillam-Dussès et al. 2007). However, convergence is not excluded since these isomers might arise from different metabolic pathways. Relationships between basal families remain disputed (Kambhampati and Eggleton 2000; Eggleton 2001; Inward et al. 2007b; Legendre et al. 2008). According to Inward et al. (2007b), Hodotermitidae would represent a monophyletic group nested within the paraphyletic group of Termopsidae, and a single family Termopsidae-Hodotermitidae is supported by recent molecular phylogenetic analyses (Legendre et al. 2008). Although the present taxon sampling for trail-following pheromones is still too scanty, it is interesting to underline the different trail-following pheromones in the Termopsidae and Hodotermitidae, and, within the Termopsidae, in the Porotermitinae and Stolotermitinae on one hand, and the Termopsinae on the other. That *Porotermes adamsoni* and *Stolotermes victoriensis* have the same trail-following pheromones as *M. darwiniensis* agrees with a basal position of the Termopsidae.

The homogenous nature of trail-following pheromones of Kalotermitidae agrees with the generally accepted monophyly of this family, and their chemical structure (C12 alcohol) agrees with its position as sister group of the Rhinotermitidae + Termitidae (Inward et al. 2007b). However, Legendre et al. (2008) report this family as the second diverging lineage within the termites after *Mastotermes*.

Trail-following pheromones of the Rhinotermitidae and Termitidae are all composed of dodecatrienol or neocembrene, supporting the close relationships of these 2 families. The paraphyly of the Rhinotermitidae is generally accepted, but relationships between major clades remain disputed. The trail-following pheromone of *Prorhinotermes* (neocembrene + dodecatrienol), in contrast to the other Rhinotermitidae (dodecatrienol), agrees with the isolation of Prorhinotermitinae among the Rhinotermitidae (Quennedey and Deligne 1975; Sillam-Dussès et al. 2005).

Among Termitidae, the trail-following pheromones of the fungus-growing Macrotermitinae, all composed of C12 alcohols probably biosynthesised from fatty acids, agree with the monophyly of the subfamily. It would be interesting to study the trail-following pheromone of *Sphaerotermes,* the Macrotermitinae that does not cultivate *Termitomyces* fungus and which has been recently isolated in its own subfamily, Sphaerotermitinae (Engel and Krishna 2004).

The separation between trail-following pheromones of Syntermitinae (formerly mandibulate Nasutitermitinae) composed of only dodecatrienol (*Armitermes euhamignathus* excepted), and trail-following pheromones of Nasutitermitinae (full nasutes) composed of both dodecatrienol and neocembrene agrees with the diphyly of the previous subfamily Nasutitermitinae (Bitsch and Noirot 2002; Inward et al. 2007b) and agrees with the proposition of the subfamily Syntermitinae by Engel and Krishna (2004). This needs to be confirmed with a more complete taxon sampling.

11.4 Pheromonal Parsimony

The pheromonal parsimony which consists in the utilisation of the same pheromone in different social contexts characterises the remarkable behavioural plasticity of social insects (Blum and Brand 1972; Pasteels 1976; Blum 1996). The phenomenon is well illustrated by the pheromone of the queen honeybee which possesses both social and sexual functions in workers and drones. Pheromonal parsimony is also well developed in termites. In many species, the same molecule is acting as a sex-pairing pheromone and a trail-following pheromone (Pasteels and Bordereau 1998). The best example is that of dodecatrienol, possessing this dual role in *Pseudacanthotermes spiniger*, *P. militaris* (Bordereau et al. 1993) and *Reticulitermes santonensis* (Laduguie et al. 1994). When secreted at lower concentrations, dodecatrienol functions as a trail-following pheromone for foragers and for male alates during tandem running. When released at higher levels by females, it attracts males at long distance and elicits a typical sexual excitement behaviour when males contact it (Bordereau et al. 1991, 1993). In *Cornitermes bequaerti*, the same compound also possesses both functions, but it is released from the sternal gland for trail-following behaviour, and from female tergal glands for sex-attraction (Bordereau et al. 2002). Dodecadienol also possesses the double function of trail-following pheromone and sex-pairing pheromone in *Ancistrotermes pakistanicus* (Robert et al. 2004). Dodecenol may also have two functions, but in different species. It is the trail-following pheromone of many species of Kalotermitidae and Macrotermitinae, and the major component of the tergal gland secretions in *Syntermes praecellens* (unpublished, Table 11.1). Finally, neocembrene functions as a sex-pairing pheromone in several species of Nasutitermitinae such as *Trinervitermes bettonianus* (McDowell and Oloo 1984) or *Nasutitermes corniger* (unpublished observations) and, associated with dodecatrienol, as a trail-following pheromone in many Nasutitermitinae.

The example of *Zootermopsis nevadensis* and *Z. angusticollis* secreting male and female sex-specific pheromones is a little different. In these termites, extracts of sternal glands of male alates, but not those of female alates, elicit trail-following behaviour in pseudergates. 4,6-Dimethyldodecanal is the major component specific to the sternal gland of male alates. Although not yet tested on females alates, it probably represents the major component of the male sex-pairing pheromone.

This substance is also the major specific component of the sternal gland of pseudergates and nymphs of both sexes, where it elicits trail-following. The female sex-pairing pheromone, ((*E*)-2,6,10-trimethylundeca-5,9-dienal) does not elicit any trail-following behaviour (Bordereau et al. 2010). The situation seems still more complex in *Hodotermopsis sjostedti* (unpublished). As in *Z. nevadensis*, extracts of male sternal glands of *H. sjostedti* elicit trail-following in pseudergates. Both castes possess two common components specific to the sternal gland, the postulated 4,6-dimethylundecan-1-ol, and traces of 4,6-dimethyldodecanal, which would represent the major components of the trail-following pheromone. However, male alates also secrete a male-specific component, presumed to be 4,6-dimethylundecanal, which probably is used as a sex-pairing pheromone. The female sex-pairing pheromone identical to that of *Z. nevadensis* does not possess any trail-following behavioural effect. The apparent absence of trail-following pheromone in female alates of *Zootermopsis* and *Hodotermopsis* remains difficult to understand, but it can be compared to the situation observed in *Hodotermes mossambicus*. In this harvester termite, pairing is assumed by the sternal gland of males which does not elicit trail following (Leuthold and Bruinsma 1977). Despite all these examples of pheromonal parsimony, the phenomenon cannot be generalised to all termites. In many species, trail-following pheromones and sex-pairing pheromones are clearly distinct (Leuthold and Bruinsma 1977; MacFarlane 1983; Peppuy et al. 2001a, 2004).

11.5 Conclusions

In the past decade, the use of SPME has permitted significant advances in the field of sex-pairing pheromones and trail-following pheromones of termites. Much remains to be done in the general chemical ecology of dipersal, as the determinism remains obscure for many behaviours. Is there really a pheromone eliciting the exodus of alates? A sex-mating pheromone? A mate choice pheromone? Until now, research on pheromones of termites was focused essentially on the emitters, but the physiological state of the receivers and the social context are also key factors in these biological phenomena. They might explain in particular the absence of behavioural responses occurring sometimes during sex attraction or trail-following bioassays despite the presence of stimuli known to be active. These studies require a more detailed knowledge of behavioural ecology and the development of appropriate bioassays.

For trail-following pheromones, crucial taxa (Hodotermitidae, Serritermitidae, Apicotermitinae, *Sphaerotermes*) and soil feeders with subterranean nests remain to be studied for a better understanding of their evolution. It will also be crucial to investigate potential additional pheromonal compounds which are essential in the evolution of the pheromones in insects and which might explain the various behavioural properties and the species-specificity of trail-following pheromones in some species of termites. For these studies, the EAD/GC-MS should be a fruitful technological approach. A chemical study of trails themselves also should be informative in understanding different aspects of their properties and the many obscure behaviours of foragers.

The comparison of the different strategies of communication developed by insect societies is very useful in understanding the keys to their respective ecological success. This is particularly true for trail communication, which has been developed by both termites and ants. Termite workers and soldiers are generally blind insects which collectively forage for food that is abundant, most often under cover and protected by soldiers. They are parsimonious insects as much for their glandular equipment (only one gland secretes trail-following pheromones) as for the chemical nature of their pheromones. Regarding all these aspects, they are quite distant from ants, and this probably explains in great part the respective specificities of their strategies of chemical communication (see review by Morgan 2009).

Finally, according to recent molecular phylogenetic studies (Inward et al. 2007a, 2007b; Ware et al. 2008), termites are nested within Blattodea, and are considered eusocial cockroaches (Inward et al. 2007a). The comparison of the origin, the nature and the role of pheromones of subsocial cockroaches (Bell et al. 2007) and basal termites could be also a fruitful perspective for studying the evolution of chemically based social behaviours.

Acknowledgments Unpublished results cited in this review are given with the permission of our collaborators, the biologists Alain Robert, David Sillam-Dussès (France), Eliana M. Cancello (Brazil), Anani Kotoklo (Togo), and the chemists Etienne Sémon (France), Michael J. Lacey (Australia), Jean-Claude Braekman and Jean Ghostin (Belgium), Irena Valterová (Czech Republic), all of whom we sincerely thank. We also thank M. J. Lacey, C.A. Nalepa, E. Cancello, R. Rosengaus, D. Sillam-Dussès for reviewing and editing the manuscript, A. Robert for preparing the plate of postflight behaviour, and E. Sémon for comments and preparing the plate of mass spectra and chemical structures.

References

Adams ES, Atkinson L, Bulmer MS (2007) Relatedness, recognition errors, and colony fusion in the termite *Nasutitermes corniger*. Behav Ecol Sociobiol 61:1195–1201

Adams ES, Levings SC (1987) Territory size and population limits in mangrove termites. J Anim Ecol 56:1069–1081

Affolter J, Leuthold RH (2000) Quantitative and qualitative aspects of trail pheromones in *Macrotermes subhyalinus* (Isoptera, Termitidae). Insectes Soc 47:256–262

Ampion M, Quennedey A (1981) The abdominal epidermal glands of termites and their phylogenetic significance. In: Howse PE, Clément JL (eds) Biosystematics of social insects. Academic Press, London, pp 249–261

Andrews EA (1911) Observations on termites in Jamaica. J Anim Behav 1:193–228

Arab A, Issa S, Alfonzo D, Jaffe K (2006) Caste, colony, and species specificity of the trail pheromone in two sympatric Nasutitermitinae. Sociobiology 47:345–351

Austin JW, Szalanski AL, Scheffrahn RH et al (2005) Genetic evidence for the synonymy of two *Reticulitermes* species: *Reticulitermes flavipes* and *Reticulitermes santonensis*. Ann Entomol Soc Am 98:395–401

Ayasse M, Paxton RJ, Tengö J (2001) Mating behavior and chemical communication in the order Hymenoptera. Annu Rev Entomol 46:31–78

Badertscher S, Gerber C, Leuthold RH (1983) Polyethism in food supply and processing in termite colonies of *Macrotermes subhyalinus* (Isoptera). Behav Ecol Sociobiol 12:115–119

Barth R (1955) Ueber der tergitalen Drusenfelder von '*Syntermes dirus*' (Isoptera). Rev Bras Biol 15:257–263

Bell WJ, Roth LM, Nalepa CA (2007) Cockroaches: ecology, behavior, and natural history. The Johns Hopkins University Press, Baltimore, MD

Birch AJ, Brown WV, Corrie JET, Moore BP (1972) Neocembrene-A, a termite trail pheromone. J Chem Soc Perkin Trans 1:2653–2658

Bitsch C, Noirot C (2002) Gut characters and phylogeny of the higher termites (Isoptera: Termitidae). A cladistic analysis. Ann Soc Entomol Fr 38:201–210

Bland JM, Park YI, Raina AK et al (2004) Trilinolein identified as a sex-specific component of tergal glands in alates of *Coptotermes formosanus*. J Chem Ecol 30:835–849

Bland JM, Raina AK, Carpita A, Dickens JC (2007) Comparative analysis of the sex/trail pheromone, 3,6,8-dodecatrien-1-ol, from three species of subterranean termites (Isoptera: Rhinotermitidae). Sociobiology 50:535–551

Blum MS (1996) Semiochemical parsimony in the arthropoda. Annu Rev Entomol 41: 353–374

Blum MS, Brand JM (1972) Social insect pheromones: their chemistry and function. Am Zoologist 12:553–576

Bordereau C, Cancello EM, Sémon E et al (2002) Sex pheromone identified after solid phase microextraction from tergal glands of female alates in *Cornitermes bequaerti* (Isoptera, Nasutitermitinae). Insectes Soc 49:209–215

Bordereau C, Lacey MJ, Sémon E et al (2010) Sex pheromones and trail-following pheromone in the basal termites *Zootermopsis nevadensis* (Hagen) and *Z. angusticollis* (Hagen) (Isoptera, Termopsidae, Termopsinae). Biol J Linn Soc 100:519–530

Bordereau C, Robert A, Bonnard O, Le Quéré JL (1991) (3*Z*,6*Z*,8*E*)-3,6,8-Dodecatrien-1-ol: sex pheromone in a higher fungus-growing termite, *Pseudacanthotermes spiniger* (Isoptera, Macrotermitinae). J Chem Ecol 17:2177–2191

Bordereau C, Robert A, Laduguie N et al (1993) Détection du (*Z*,*Z*,*E*)-3,6,8-dodecatrien-1-ol par les ouvriers et les essaimants de deux espèces de termites champignonnistes: *Pseudacanthotermes spiniger* et *P. militaris* (Termitidae, Macrotermitinae). Actes Coll Insect Soc 8:145–149

Buchli H (1960) Les tropismes lors de la pariade des imagos de *Reticulitermes lucifugus* R. Vie Milieu 11:308–315

Buděšínský M, Valterová I, Sémon E et al (2005) NMR structure determination of (11*E*)-trinervita-1(14),2,11-triene, a new diterpene from sexual glands of termites. Tetrahedron 61:10699–10704

Bugnion E (1909) Le termite noir de Ceylan, *Eutermes monoceros* (Koenig). Ann Soc Entomol Fr 78:271–281

Casarin FE, Costa-Leonardo AM, Arab A (2008) Soldiers initiate foraging activities in the subterranean termite, *Heterotermes tenuis*. J Insect Sci 8:2

Clément G (1956) Observations sur l'essaimage d'*Anacanthotermes ochraceus* Burm. (Isoptera). Bull Soc Entomol Fr 61:98–103

Clément JL (1982) Phéromones d'attraction sexuelle des termites européens du genre *Reticulitermes* (Rhinotermitinae). Mécanismes comportementaux et isolements spécifiques. Biol Behav 7:55–68

Clément JL, Bagnères AG (1998) Nestmate recognition in termites. In: Vander Meer RK, Breed MD, Espelie KE, Winston ML (eds) Pheromone communication in social insects. Westview Press, Boulder, CO, pp 223–255

Clément JL, Lloyd H, Nagnan P, Blum MS (1989) *n*-Tetradecyl propionate: identification as a sex pheromone of the eastern subterranean termite *Reticulitermes flavipes*. Sociobiology 15:19–24

Cornelius ML, Bland JM (2001) Trail-following behavior of *Coptotermes formosanus* and *Reticulitermes flavipes* (Isoptera: Rhinotermitidae): is there a species-specific response? Environ Entomol 30:457–465

Costa-Leonardo AM, Barsotti RC (1998) Swarming and incipient colonies of *Coptotermes havilandi* (Isoptera, Rhinotermitidae). Sociobiology 31:131–142

Cvačka J, Jiroš P, Šobotník J et al (2006) Analysis of insect cuticular hydrocarbons using matrix-assisted laser desorption ionization mass spectrometry. J Chem Ecol 32:409–434

Dapporto L, Lambardi D, Turillazzi S (2008) Not only cuticular lipids: first evidence of differences between foundresses and their daughters in polar substances in the paper wasp *Polistes dominulus*. J Insect Physiol 54:89–95

Darlington JPEC (1982) The underground passages and storage pits used in foraging by a nest of the termite *Macrotermes michaelseni* in Kajiado, Kenya. J Zool 198:237–247

Darlington JPEC (1986) Seasonality in mature nests of the termite *Macrotermes michaelseni* in Kenya. Insectes Soc 33:168–189

Deng XJ, Zhang JM, Hu JF et al (2002) Biological activity of a synthetic trail-pheromone analogue of the black-winged subterranean termite, *Odontotermes formosanus* Shiraki. Acta Entomol Sinica 45:739–742

Dronnet S, Lohou C, Christides JP, Bagnères AG (2006) Cuticular hydrocarbon composition reflects genetic relationship among colonies of the introduced termite *Reticulitermes santonensis* Feytaud. J Chem Ecol 32:1027–1042

Dudley PH, Beaumont J (1889) Observations on the termites, or white ants, of the isthmus of Panama. Trans N Y Acad Sci 8:85–114

Eggleton P (2001) Termites and trees: a review of recent advances in termite phylogenetics. Insectes Soc 48:187–193

Engel MS, Krishna K (2004) Family-group names for termites (Isoptera). Am Mus Novitates 3432:1–9

Florane CB, Bland JM, Husseneder C, Raina AK (2004) Diet-mediated inter-colonial aggression in the formosan subterranean termite *Coptotermes formosanus*. J Chem Ecol 30: 2559–2574

Francke W, Schulz S (1999) Pheromones. In: Barton D, Nakanishi K, Meth-Cohn O (eds) Comprehensive natural products chemistry, vol 8: Miscellaneous natural products including marine natural products, pheromones, plant hormones, and aspects of ecology. Elsevier, Amsterdam, pp 197–261

Fuller C (1915) Observations on some South African termites. Ann Natal Mus 3:329–505

Gessner S, Leuthold RH (2001) Caste-specificity of pheromone trails in the termite *Macrotermes bellicosus*. Insectes Soc 48:238–244

Grace JK, Wood DL, Frankie GW (1988) Trail-following behavior of *Reticulitermes hesperus* Banks (Isoptera: Rhinotermitidae). J Chem Ecol 14:653–667

Grace JK, Wood DL, Kubo I, Kim M (1995) Behavioural and chemical investigation of trail pheromone from the termite *Reticulitermes hesperus* Banks (Isopt., Rhinotermitidae). J Appl Entomol 119:501–505

Grandcolas P, Park YC, Choe JC et al (2001) What does *Cryptocercus kyebangensis*, n.sp. (Dictyoptera: Blattaria: Polyphagidae) from Korea reveal about *Cryptocercus* evolution? A study in morphology, molecular phylogeny, and chemistry of tergal glands. Proc Acad Nat Sci Philadelphia 151:61–79

Grassé PP (1984) Termitologia. Tome II. Fondation des sociétés-construction, Masson, Paris

Grassé PP, Noirot C (1950) Documents sur la biologie de l'*Odontotermes magdalenae* n.sp. Ann Sc Nat Zool 12:117–141

Grassé PP, Noirot C (1951a) Nouvelles recherches sur la biologie de divers termites champignonnistes (Macrotermitinae). Ann Sci Nat Zool 13:291–342

Grassé PP, Noirot C (1951b) Orientation et routes chez les termites. Le « balisage » des pistes. Ann Psychol 50:273–280

Hall P, Traniello JFA (1985) Behavioral bioassays of termite trail pheromones. Recruitment and orientation effects of cembrene A in *Nasutitermes costalis* (Isoptera: Termitidae) and discussion of factors affecting termite response in experimental contexts. J Chem Ecol 11:1503–1513

Hanus R, Luxová A, Šobotník J et al (2009) Sexual communication in the termite *Prorhinotermes simplex* (Isoptera, Rhinotermitidae) mediated by a pheromone from female tergal glands. Insectes Soc 56:111–118

Heidecker JL, Leuthold RH (1984) The organization of collective foraging in the harvester termite *Hodotermes mossambicus* (Isoptera). Behav Ecol Sociobiol 14:195–202

Hewitt PH, Nel JJC, Conradie S (1969) The rôle of chemicals in communication in the harvester termites *Hodotermes mossambicus* (Hagen) and *Trinervitermes trinervoides* (Sjöstedt). Insectes Soc 16:79–86

Hinze B, Leuthold RH (1997) Copulation of a physogastric queen in *Macrotermes bellicosus* (Isoptera: Termitidae). Sociobiology 30:243–246

Hoare A, Jones DT (1998) Notes on the foraging behaviour and taxonomy of the southeast Asian termite *Longipeditermes longipes* (Termitidae: Nasutitermitinae). J Nat Hist 32:1357–1366

Howard RW, Blomquist GJ (2005) Ecological, behavioral, and biochemical aspects of insect hydrocarbons. Annu Rev Entomol 50:371–393

Howard R, Matsumura F, Coppel HC (1976) Trail-following pheromones of the Rhinotermitidae: approaches to their authentication and specificity. J Chem Ecol 2:147–166

Hummel H, Karlson P (1968) Hexansäure als Bestandteil des Spurpheromons der Termite *Zootermopsis nevadensis* Hagen. Z Physiol Chem 349:725–727

Husseneder C, Simms DM (2008) Size and heterozygosity influence partner selection in the Formosan subterranean termite. Behav Ecol 19:764–773

Inward D, Beccaloni G, Eggleton P (2007a) Death of an order: a comprehensive molecular phylogenetic study confirms that termites are eusocial cockroaches. Biol Lett 3:331–335

Inward DJG, Vogler AP, Eggleton P (2007b) A comprehensive phylogenetic analysis of termites (Isoptera) illuminates key aspects of their evolutionary biology. Mol Phylogenet Evol 44: 953–967

Jackson DE, Martin SJ, Holcombe M, Ratnieks FLW (2006) Longevity and detection of persistent foraging trails in Pharaoh's ants, *Monomorium pharaonis* (L.). Anim Behav 71:351–359

Jackson DE, Martin SJ, Ratnieks FLW, Holcombe M (2007) Spatial and temporal variation in pheromone composition of ant foraging trails. Behav Ecol 18:444–450

Jander R, Daumer K (1974) Guide-line and gravity orientation of blind termites foraging in the open (Termitidae: *Macrotermes*, *Hospitalitermes*). Insectes Soc 21:45–69

Jmhasly P, Leuthold RH (1999a) The system of underground passages in *Macrotermes subhyalinus* and comparison of laboratory bioassays to field evidence on intraspecific encounters in *M. subhyalinus* and *M. bellicosus* (Isoptera, Termitidae). Insectes Soc 46:332–340

Jmhasly P, Leuthold RH (1999b) Foraging territories of *Macrotermes bellicosus* and mutual territory dynamics between *M. bellicosus* and *M. subhyalinus* (Isoptera: Termitidae). Sociobiology 34:23–33

Jones DT, Gathorne-Hardy FJ (1995) Foraging activity of the processional termite *Hospitalitermes hospitalis* (Termitidae: Nasutitermitinae) in the rain forest of Brunei, north-west Borneo. Insectes Soc 42:359–369

Jones SC, La Fage JP, Wright VL (1981) Studies of dispersal, colony caste and sexual composition, and incipient colony development of *Pterotermes occidentis* (Walker) (Isoptera: Kalotermitidae). Sociobiology 6:221–242

Kaib M (1990) Intra- and interspecific chemical signals in the termite *Schedorhinotermes*-Production sites, chemistry, and behaviour. In: Gribahin FG, Wiese K, Popov AV (eds) Sensory systems and communication in arthropods. Advances in life science. Birkhäuser, Basel, pp 26–31

Kaib M, Bruinsma O, Leuthold RH (1982) Trail-following in termites: evidence for a multicomponent system. J Chem Ecol 8:1193–1205

Kalshoven LGE (1958) Observations on the black termites, *Hospitalitermes* spp., of Java and Sumatra. Insectes Soc 5:9–30

Kalshoven LGE (1960) Biological notes on the *Cryptotermes* species of Indonesia. Acta Tropica 17:263–272

Kambhampati S, Eggleton P (2000) Taxonomy and phylogeny of termites. In: Abe T, Bignell DE, Higashi M (eds) Termites: evolution, sociality, symbioses, ecology. Kluwer Academic Publishers, Dordrecht, pp 1–23

Karlson P, Lüscher M, Hummel H (1968) Extraktion und biologische Auswertung des Spurpheromons der Termite *Zootermopsis nevadensis*. J Insect Physiol 14:1763–1771

Kato T, Suzuki M, Kobayashi T, Moore BP (1980) Synthesis and pheromone activities of optically active neocembrenes and their geometrical isomers, (*E,Z,E*)- and (*E,E,Z*)-neocembrenes. J Org Chem 45:1126–1130

Khamraev AS, Kuchkarova LS, Ahmerov RN et al (2008) Trail-following activity in extracts of sternal glands from *Anacanthotermes turkestanicus* (Isoptera: Hodotermitidae). Sociobiology 51:685–696

Kim J, Matsuyama S, Suzuki T (2005) 4,8-Dimethyldecanal, the aggregation pheromone of *Tribolium castaneum*, is biosynthesized through the fatty acid pathway. J Chem Ecol 31:1381–1400

Kirchner WH, Minkley N (2003) Nestmate discrimination in the harvester termite *Hodotermes mossambicus*. Insectes Soc 50:222–225

Kitade O, Hayashi Y, Kikushi Y, Kawarasaki S (2004) Distribution and composition of colony founding associations of a subterranean termite *Reticulitermes kanmonensis*. Entomol Sci 7:1–8

Klochkov SG, Zhuzhikov D (1990) Termite trail pheromones: specificity and biosynthesis. In: Gribakin FG, Wiese K, Popov AV (eds) Sensory systems and communication in arthropods. Advances in life science. Birkhäuser, Basel, pp 41–43

Kodama M, Matsuki Y, Itô S (1975) Synthesis of macrocyclic terpenoids by intramolecular cyclization I. (±)-Cembrene-a, a termite trail pheromone, and (±)-nephthenol. Tetrahedron Lett 16:3065–3068

Kotoklo EA, Sillam-Dussès D, Ketoh G et al (2010) Identification of the trail-following pheromone of the pest termite *Amitermes evuncifer* (Isoptera: Termitidae). Sociobiology 55:1–10

Laduguie N (1993) Phéromones de piste et phéromones sexuelles chez les termites. Thèse Université Bourgogne, Dijon

Laduguie N, Robert A, Bonnard O et al (1994) Isolation and identification of (3*Z*,6*Z*,8*E*)-3,6,8-dodecatrien-1-ol in *Reticulitermes santonensis* Feytaud (Isoptera, Rhinotermitidae): roles in worker trail-following and in alate sex-attraction behaviour. J Insect Physiol 40:781–787

Le Quéré JL, Brossut R, Nalepa CA, Bonnard O (1991) Isolation and identification of 4,6,8-trimethyl-7,9-undecadien-5-ol, a female specific compound in tergal gland secretion of *Cryptocercus punctulatus* Scudder (Dictyoptera. Cryptocercidae). J Chem Ecol 17: 811–821

Legendre F, Whiting MF, Bordereau C et al (2008) The phylogeny of termites (Dictyoptera: Isoptera) based on mitochondrial and nuclear markers: implications for the evolution of the worker and pseudergate castes, and foraging behaviors. Mol Phylogenet Evol 48:615–627

Leponce M, Roisin Y, Pasteels JM (1997) Structure and dynamics of the arboreal termite community in New Guinean coconut plantations. Biotropica 29:193–203

Leuthold RH (1975) Orientation mediated by pheromones in social insects. In: Noirot C, Howse PE, Le Masne G (eds) Pheromone and defensive secretions in social insects. Imprimerie Université de Dijon, Dijon, pp 197–211

Leuthold RH, Bruinsma O (1977) Pairing behavior in *Hodotermes mossambicus* (Isoptera). Psyche 84:109–119

Leuthold RH, Bruinsma O, van Huis A (1976) Optical and pheromonal orientation and memory for homing distance in the harvester termite *Hodotermes mossambicus* (Hagen). Behav Ecol Sociobiology 1:127–139

Leuthold RH, Lüscher M (1974) An unusual caste polymorphism of the sternal gland and its trail pheromone production in the termite *Trinervitermes bettonianus*. Insectes Soc 21:335–341

Levings SC, Adams ES (1984) Intra- and interspecific territoriality in *Nasutitermes* (Isoptera: Termitidae) in a Panamanian mangrove forest. J Anim Ecol 53:705–714

Lima MM, Gonçalves TT, DeSouza O, Reis R Jr (2006) Nesting site selection by *Coptotermes gestroi* (Insecta: Isoptera). Sociobiology 48:681–688

Lüscher M (1961) Demonstration of a trail pheromone in termites. Symp Gen Biol Ital 11:191–192

Lüscher M, Müller B (1960) Ein spurbildendes Sekret bei Termiten. Naturwissenschaften 21:503
Lys JA, Leuthold RH (1991) Task-specific distribution of the two worker castes in extranidal activities in *Macrotermes bellicosus* (Smeathman). Observation of behaviour during food acquisition. Insectes Soc 38:161–170
MacFarlane J (1983) Observations on trail pheromone, trail-laying and longevity of natural trails in the termite *Macrotermes michaelseni*. Insect Sci Appl 4:309–318
Marcel N, Lefeuve P, Bordereau C (1991) Recherche de nourriture et recrutement chez le termite Africain *Nasutitermes lujae* (Wasmann). Actes Coll Insect Soc 7:145–151
Matsumura F, Coppel HC, Tai A (1968) Isolation and identification of termite trail-following pheromone. Nature 219:963–964
Matsumura F, Jewett DM, Coppel HC (1972) Interspecific response of termite to synthetic trail-following substances. J Econ Entomol 65:600–602
Matsuura K (2003) Symbionts affecting termite behavior. In: Bourtzis K, Miller TA (eds) Insect symbiosis. CRC Press, Boca Raton, FL, pp 131–143
Matsuura K, Nishida T (2001) Comparison of colony foundation success between sexual pairs and female asexual units in the termite *Reticulitermes speratus* (Isoptera: Rhinotermitidae). Popul Ecol 43:119–124
McDowell PG, Oloo GW (1984) Isolation, identification, and biological activity of trail-following pheromone of termite *Trinervitermes bettonianus* (Sjöstedt) (Termitidae: Nasutitermitinae). J Chem Ecol 10:835–851
Mitchell JD (2007) Swarming and pairing in the fungus-growing termite, *Macrotermes natalensis* (Haviland) (Isoptera: Macrotermitinae). Afr Entomol 15:153–160
Miura T, Matsumoto T (1995) Worker polymorphism and division of labor in the foraging behavior of the black marching termite *Hospitalitermes medioflavus*, on Borneo Island. Naturwissenschaften 82:564–567
Miura T, Matsumoto T (1998) Foraging organization of the open-air processional lichen-feeding termite *Hospitalitermes* (Isoptera, Termitidae) in Borneo. Insectes Soc 45:17–32
Moore BP (1966) Isolation of the scent-trail pheromone of an Australian termite. Nature 211: 746–747
Moore BP (1974) Pheromones in termite societies. In: Birch MC (ed) Pheromones. American Elsevier, New York, NY, pp 250–266
Morgan FD (1959) The ecology and external morphology of *Stolotermes ruficeps* Brauer (Isoptera: Hodotermitidae). Trans R Soc N Zeal 86:155–195
Morgan ED (2009) Trail pheromones of ants. Physiol Entomol 34:1–17
Nalepa CA, Jones SC (1991) Evolution of monogamy in termites. Biol Rev 66:83–97
Nalepa CA, Miller LR, Lenz M (2001) Flight characteristics *of Mastotermes darwiniensis* (Isoptera, Mastotermitidae). Insectes Soc 48:114–148
Noirot C (1969) Glands and secretions. In: Krishna K, Weesner FM (eds) Biology of termites, vol 1. Academic Press, New York, NY, pp 89–123
Noirot C (1995) The sternal glands of termites: segmental pattern, phylogenetic implications. Insectes Soc 42:321–323
Noirot C, Bodot P (1964) L'essaimage d'*Allognathotermes hypogeus* Silvestri. C R Acad Sci Paris 258:3357–3359
Noirot C, Quennedey A (1974) Fine structure of insect epidermal glands. Annu Rev Entomol 19:61–80
Nutting WL (1969) Flight and colony foundation. In: Krishna K, Weesner FM (eds) Biology of termites, vol I. Academic Press, New York, NY, pp 233–282
Oloo GW (1981) Specificity of termite trails: analysis of natural trails of *Trinervitermes*, *Macrotermes*, and *Odontotermes* from sympatric populations. Entomol Exp Appl 29:162–168
Oloo GW, Leuthold RH (1979) The influence of food on trail-laying and recruitment behaviour in *Trinervitermes bettonianus* (Termitidae: Nasutitermitinae). Entomol Exp Appl 26:267–278
Oloo GW, McDowell PG (1982) Interspecific trail-following and evidence of similarity of trails of *Trinervitermes* species from different habitats. Insect Sci Appl 3:157–161

Omo Malaka SL (1987) Establishment of foraging trail and qualification of trail pheromone by *Amitermes evuncifer* Silvestri (Isoptera; Termitidae; Termitinae). Insect Sci Appl 8:343–350

Park YI, Bland JM, Raina AK (2004) Factors affecting post-flight behavior in primary reproductives of the Formosan subterranean termite, *Coptotermes formosanus* (Isoptera: Rhinotermitidae). J Insect Physiol 50:539–546

Pasteels JM (1965) Polyéthisme chez les ouvriers de *Nasutitermes lujae* (Termitidae Isoptères). Biol Gabonica 1:191–205

Pasteels JM (1972) Sex-specific pheromones in a termite. Experientia 28:105–106

Pasteels JM (1976) Evolutionary aspects in chemical ecology and chemical communication. Proceedings of the XV International Congress of Entomology, Washington, DC, pp 281–293

Pasteels JM, Bordereau C (1998) Releaser pheromones in termites. In: Vander Meer RK, Breed MD, Espelie KE, Winston ML (eds) Pheromone communication in social insects. Westview Press, Boulder, CO, pp 193–215

Peppuy A (1999) Termites du Nord Vietnam, communication chimique et isolement spécifique. Thèse Université Paris VI

Peppuy A, Robert A, Bordereau C (2004) Species-specific sex pheromones secreted from new sexual glands in two sympatric fungus-growing termites from northern Vietnam, *Macrotermes annandalei* and *M. barneyi*. Insectes Soc 51:91–98

Peppuy A, Robert A, Sémon E et al (2001b) Species specificity of trail pheromones of fungus-growing termites from northern Vietnam. Insectes Soc 48:245–250

Peppuy A, Robert A, Sémon E et al (2001a) (Z)-Dodec-3-en-1-ol, a novel termite trail pheromone identified after solid phase microextraction from *Macrotermes annandalei*. J Insect Physiol 47:445–453

Prestwich GD (1981) Terpene biosynthesis by nasute termite soldiers (Isoptera: Nasutitermitinae). Insect Biochem 11:331–336

Quennedey A, Deligne J (1975) L'arme frontale des soldats de termites. I. Rhinotermitidae. Insectes Soc 22:243–267

Quennedey A, Leuthold RH (1978) Fine structure and pheromonal properties of the polymorphic sternal gland in *Trinervitermes bettonianus* (Isoptera, Termitidae). Insectes Soc 25:153–162

Quennedey A, Peppuy A, Courrent A et al (2004) Ultrastructure of posterior sternal glands of *Macrotermes annandalei* (Silvestri): new members of the sexual glandular set found in termites (Insecta). J Morphol 262:683–691

Quennedey A, Sillam-Dussès D, Robert A, Bordereau C (2008) The fine structural organization of sternal glands of pseudergates and workers in termites (Isoptera): a comparative survey. Arthropod Struct Dev 37:168–185

Raina AK, Bland JM, Dickens JC et al (2003) Premating behavior of dealates of the Formosan subterranean termite and evidence for the presence of a contact sex pheromone. J Insect Behav 16:233–245

Raina AK, Bland JM, Osbrink W (2005) Hydroquinone is not a phagostimulant for the Formosan subterranean termite. J Chem Ecol 31:509–517

Reinhard J, Hertel H, Kaib M (1997a) Systematic search for food in the subterranean termite *Reticulitermes santonensis* de Feytaud (Isoptera, Rhinotermitidae). Insectes Soc 44:147–158

Reinhard J, Hertel H, Kaib M (1997b) Feeding stimulating signal in labial gland secretion of the subterranean termite *Reticulitermes santonensis*. J Chem Ecol 23:2371–2381

Reinhard J, Kaib M (1995) Interaction of pheromones during food exploitation by the termite *Schedorhinotermes lamanianus*. Physiol Entomol 20:266–272

Reinhard J, Kaib M (2001) Trail communication during foraging and recruitment in the subterranean termite *Reticulitermes santonensis* de Feytaud (Isoptera, Rhinotermitidae). J Insect Behav 14:157–171

Reinhard J, Lacey MJ, Ibarra F et al (2002) Hydroquinone: a general phagostimulating pheromone in termites. J Chem Ecol 28:1–14

Rickli M, Leuthold RH (1987) Spacial organisation during exploration and foraging in the harvester termite, *Trinervitermes geminatus*. Rev Suisse Zool 94:545–551

Rickli M, Leuthold RH (1988) Homing in harvester termites: evidence of magnetic orientation. Ethology 77:209–216

Robert A, Peppuy A, Sémon E et al (2004) A new C12 alcohol identified as a sex pheromone and a trail-following pheromone in termites: the diene (Z,Z)-dodeca-3,6-dien-1-ol. Naturwissenschaften 91:34–39

Runcie CD (1987) Behavioral evidence for multicomponent trail pheromone in the termite, *Reticulitermes flavipes* (Kollar) (Isoptera: Rhinotermitidae). J Chem Ecol 13: 1967–1978

Rupf T, Roisin Y (2008) Coming out of the woods: do termites need a specialized worker caste to search for new food sources? Naturwissenschaften 95:811–819

Saran RK, Millar JG, Rust MK (2007) Role of (3Z,6Z,8*E*)-dodecatrienol-1-ol in trail following, feeding, and mating behavior of *Reticulitermes hesperus*. J Chem Ecol 33:369–389

Shatov KS (1977) Chemical signals in the termite *Anacanthotermes ahngerianus* Jacobson (Isoptera, Hodotermitidae). Proceedings of the VIII Congress of IUSSI, Wageningen, The Netherlands, pp 50–51

Shellman-Reeve JS (1994) Limited nutrients in a dampwood termite: nest preference, competition and cooperative nest defence. J Anim Ecol 63:921–932

Shellman-Reeve JS (1999) Courting strategies and conflicts in a monogamous, biparental termite. Proc R Soc Lond B 266:137–144

Shellman-Reeve JS (2001) Genetic relatedness and partner preference in a monogamous, wood-dwelling termite. Anim Behav 61:869–876

Sillam-Dussès D (2004) Evolution des phéromones de piste chez les termites et leurs relations avec les phéromones sexuelles. Thèse Université Bourgogne, Dijon

Sillam-Dussès D, Kalinová B, Jiroš P et al (2009a) Identification by GC-EAD of the two-component trail-following pheromone of *Prorhinotermes simplex* (Isoptera, Rhinotermitidae, Prorhinotermitinae). J Insect Physiol 55:751–757

Sillam-Dussès D, Robert A, Sémon E et al (2006) Trail-following pheromones and phylogeny in termites. Proceedings of the XV Congress of IUSSI, Washington, DC, 100–101

Sillam-Dussès D, Sémon E, Lacey MJ et al (2007) Trail-following pheromones in basal termites, with special reference to *Mastotermes darwiniensis*. J Chem Ecol 33:1960–1977

Sillam-Dussès D, Sémon E, Moreau C et al (2005) Neocembrene A, a major component of the trail-following pheromone in the genus *Prorhinotermes* (Insecta, Isoptera, Rhinotermitidae). Chemoecology 15:1–6

Sillam-Dussès D, Sémon E, Robert A et al (2010) Identification of multi-component trail pheromones in the most evolutionarily derived termites, the Nasutitermitinae (Termitidae). Biol J Linn Soc 99:20–27

Sillam-Dussès D, Sémon E, Robert A, Bordereau C (2009b) (Z)-Dodec-3-en-1-ol, a common major component of the trail-following pheromone in the termites Kalotermitidae. Chemoecology 19:103–108

Šobotník J, Weyda F, Hanus R (2005) Ultrastructural study of tergal and posterior sternal glands in *Prorhinotermes simplex* (Isoptera: Rhinotermitidae). Eur J Entomol 102:81–88

Souto L, Kitayama K (2000) *Constrictotermes cyphergaster* (Isoptera: Termitidae: Nasutitermitinae) maintain foraging trails for a longer period by means of fecal droplets. Sociobiology 35:367–372

Stuart AM (1961) Mechanism of trail-laying in two species of termites. Nature 189:419

Stuart AM (1963) Origin of the trail in the termites *Nasutitermes corniger* (Motschulsky) and *Zootermopsis nevadensis* (Hagen), Isoptera. Physiol Zool 36:69–84

Stuart AM (1969) Social behavior and communication. In: Krishna K, Weesner FM (eds) Biology of termites, vol 1. Academic Press, New York, NY, pp 193–232

Stuart AM (1975) Some aspects of pheromone involvement in the post-flight behaviour of the termite *Zootermopsis angusticollis* (Hagen) and *Reticulitermes flavipes* (Kollar). In: Noirot C, Howse PE, Le Masne G (eds) Pheromone and defensive secretions in social insects. Imprimerie Université, Dijon, pp 219–223

Stuart AM (1981) The rôle of pheromones in the initiation of foraging, recruitment and defence by the soldiers of a tropical termite, *Nasutitermes corniger* (Motschulsky). Chem Senses 6: 409–420

Sugio K (1995) Trunk trail foraging of the fungus-growing termite *Macrotermes carbonarius* (Hagen) in Southeastern Thailand. Tropics 4:211–222

Symonds MRE, Elgar MA (2008) The evolution of pheromone diversity. Trends Ecol Evol 23: 220–228

Tai A, Matsumura F, Coppel HC (1969) Chemical identification of the trail-following pheromone for a southern subterranean termite. J Org Chem 34:2180–2182

Tai A, Matsumura F, Coppel HC (1971) Synthetic analogues of the termite trail-following pheromones, structure and biological activity. J Insect Physiol 17:181–188

Thompson GJ, Lenz M, Crozier RH, Crespi BJ (2007) Molecular-genetic analyses of dispersal and breeding behaviour in the Australian termite *Coptotermes lacteus*: evidence for non-random mating in a swarm-dispersal mating system. Aust J Zool 55:219–227

Thorne BL, Haverty MI (1991) A review of intracolony, intraspecific, and interspecific agonism in termites. Sociobiology 19:115–145

Tokoro M, Takahashi M, Tsunoda K et al (1991) Isolation and identification of the trail pheromone of the subterranean termite *Reticulitermes speratus* (Kolbe) (Isoptera: Rhinotermitidae). Wood Res 78:1–14

Tokoro M, Takahashi M, Tsunoda K, Yamaoka R (1989) Isolation and primary structure of trail pheromone of the termite, *Coptotermes formosanus* Shiraki (Isoptera: Rhinotermitidae). Wood Res 76:29–38

Tokoro M, Takahashi M, Yamaoka R (1992a) (*Z,E,E*)-Dodecatrien-1-ol: a minor component of trail pheromone of termite *Coptotermes formosanus* Shiraki. J Chem Ecol 20:199–215

Tokoro M, Takahashi M, Yamaoka R (1992b) Identification of trail pheromone precursors from subterranean termite, *Coptotermes formosanus* Shiraki (Isoptera: Rhinotermitidae). J Chem Ecol 18:517–526

Tokoro M, Yamaoka R, Hayashiya K et al (1990) Evidence for trail-pheromone precursor in termite *Reticulitermes speratus* (Kolbe) (Rhinotermitidae: Isoptera). J Chem Ecol 16:2549–2557

Traniello JFA (1981) Enemy-deterrence in the recruitment strategy of a termite: soldier-organized foraging in *Nasutitermes costalis*. Proc Natl Acad Sci USA 78:1976–1979

Traniello JFA (1982) Recruitment and orientation components in a termite trail pheromone. Naturwissenschaften 69:343–344

Traniello JFA, Busher C (1985) Chemical regulation of polyethism during foraging in the neotropical termite *Nasutitermes costalis*. J Chem Ecol 11:319–332

Traniello JFA, Leuthold RH (2000) Behavior and ecology of foraging in termites. In: Abe T, Bignell DE, Higashi M (eds) Termites: evolution, sociality, symbioses, ecology. Kluwer Academic Publishers, Dordrecht, pp 141–168

Traniello JFA, Robson S (1995) Trail and territorial pheromones in the social insects. In: Bell WJ, Cardé R (eds) Chemical ecology of insects, vol II. Chapman and Hall, New York, NY, pp 241–285

Tschinkel WR, Close PG (1973) The trail pheromone of the termite, *Trinervitermes trinervoides*. J Insect Physiol 19:707–721

Wall M (1971) Zur Geschlechtsbiologie der Termite *Kalotermes flavicollis* (Fabr.) (Isoptera). Acta Tropica 28:17–60

Ware JL, Litman J, Klass KD, Spearman LA (2008) Relationships among the major lineages of Dictyoptera: the effect of outgroup selection on dictyopteran tree topology. Syst Entomol 33:429–450

Wei J, Mo J, Pan C, Deng T, Cheng M, Chen C (2007) The intestinal microbes inducing the agonistic behavior of inter-colonial individuals in *Coptotermes formosanus* (Isoptera: Rhinotermitidae). Sociobiology 50:1–12

Wilkinson W (1962) Dispersal of alates and establishment of new colonies in *Cryptotermes havilandi* (Sjöstedt) (Isoptera, Kalotermitidae). Bull Entomol Res 53:265–286

Williams RMC (1959) Flight and colony foundation in two *Cubitermes* species (Isoptera: Termitidae). Insectes Soc 6:203–218

Wobst B, Farine JP, Giniès C et al (1999) (*Z*,*Z*,*E*)-3,6,8-Dodecatrien-1-ol, a major component of trail-following pheromone in two sympatric termite species *Reticulitermes lucifugus grassei* and *R. santonensis*. J Chem Ecol 25:1305–1318

Wood TG (1981) Reproductive isolating mechanisms among species of *Microtermes* (Isoptera, Termitidae) in the southern Guinea Savanna near Mokwa, Nigeria. In: Howse PE, Clément JL (eds) Biosystematics of social insects. Academic Press, London and New York, NY, pp 303–326

Wood TG, Johnson RA (1986) The biology, physiology, and ecology of termites. In: Vinson SB (ed) Control of social insects. Praeger, New York, NY, pp 1–68

Wyatt TD (2003) Pheromones and animal behaviour. Communication by smell and taste. Cambridge University Press, Cambridge

Chapter 12
Genetic Structure of Termite Colonies and Populations

Edward L. Vargo and Claudia Husseneder

Abstract Increasingly, researchers are using molecular markers to investigate the genetic structure of termite colonies and populations. The studies are providing insights into the natural history and evolution of Isoptera in unprecedented detail. At the level of the colony, genetic studies reveal the breeding structure and degrees of inbreeding. In many species colonies are often headed by monogamous pairs of reproductives, although there is considerable variation in breeding structure within and between species in the proportions of colonies with multiple functional reproductives, usually containing inbreeding neotenics. Recent studies have identified negative consequences of inbreeding with important consequences for colony breeding structure. Genetic evidence does not support budding as a common mode of reproduction in termites. In most cases studied to date, alates appear to disperse far enough to promote extensive gene flow among populations within about 10 km, while populations at 50–100 km often show moderate to strong genetic differentiation. There has also been considerable progress in phylogeographic studies, relating differentiation among populations and speciation of termites to geological events. The few studies to date of invasive termite species suggest that some successful invaders (e.g. *Reticulitermes flavipes*) may undergo changes in breeding structure in the introduced range toward larger, unicolonial societies, whereas other introduced populations (e.g. *Coptotermes formosanus*) do not exhibit unicolonial characteristics. The powerful approach to termite colony and population genetics afforded by molecular markers will address a wide range of issues of fundamental importance to termite biology and evolution. With continued advancement in the tools for characterizing genetic variation, we anticipate rapid progress in termite colony and population genetics.

E.L. Vargo (✉)
Department of Entomology, North Carolina State University, Raleigh, NC 27695, USA
e-mail: ed_vargo@ncsu.edu

D.E. Bignell et al. (eds.), *Biology of Termites: A Modern Synthesis*,
DOI 10.1007/978-90-481-3977-4_12,

12.1 Introduction

The aim of population genetics is to quantify genetic variation in natural populations and to understand the processes responsible for this variation. Since termites are social insects, the basic unit of social insect populations is the colony rather than the individual (Thorne et al. 1999; Lepage and Darlington 2000) and the genetic structure of their populations is tightly linked to colony genetic structure. Yet colonies are composed of genetically heterogeneous individuals. Thus social insect populations are hierarchically structured, with individuals forming colonies and colonies in turn composing populations. Therefore, levels of inbreeding and modes of dispersal are tightly linked to colony breeding structure, and both of these attributes have major impacts on population genetic structure (Ross and Keller 1995; Ross 2001). A comprehensive understanding of the role of factors shaping social insect population structure, both contemporary factors such as migration, dispersal and breeding structure and historical processes such as selection and genetic drift, requires careful quantification of genetic variation within and among colonies (Pamilo et al. 1997; Ross 2001).

There is now a large arsenal of molecular tools for measuring genetic variation at all levels from individuals to kin groups, from populations to species and higher taxonomic levels. These techniques together with advances in population genetic models and data analyses provide powerful methods for inferring colony genetic structure and breeding systems, determining levels of inbreeding, estimating dispersal distances, characterizing the degree of genetic differentiation at various spatial scales, and reconstructing historical patterns of distribution leading to speciation.

There is growing interest in the genetic structure of termite populations. Inbreeding is an integral part of the colony life cycle in many species and has been the focal point in discussions of the evolution of eusociality in the group (Hamilton 1972; Syren and Luykx 1977; Hamilton 1978). While it has long been assumed that termites often undergo extensive inbreeding, data from genetic studies in a number of species indicate that inbreeding levels are variable and often low (Vargo and Husseneder 2009), calling into question the role of inbreeding in termite social evolution. Moreover, inbreeding depression in termites may be more prevalent than has generally been appreciated, affecting foundation success (Rosengaus and Traniello 1993; Fei and Henderson 2003; Calleri et al. 2005; DeHeer and Vargo 2006), growth and long-term fecundity of termite colonies (Fei and Henderson 2003; Calleri et al. 2007), and worker size (Husseneder et al. 2005, 2007). Finally, a recent study found that secondary reproductives in *Reticulitermes speratus* are produced parthenogenetically (Matsuura et al. 2009), effectively reducing inbreeding to negligible levels even after queen replacement from within the colony. Thus, the degree to which termites undergo inbreeding and the potential negative consequences of inbreeding remain open questions.

Genetic markers are especially important in termites. Since nests containing reproductives are often difficult to find or access, particularly in species with cryptic nesting habits, the genotypes of workers can be used to infer the breeding structure of colonies, that is, the number of reproductives and the degree of relatedness among them. These same genetic tools can also be used to probe the structure of populations. In this chapter, we review the molecular tools that are used for studies of termite colony and population genetic structure, then discuss dispersal and colony founding, colony genetic structure and population genetic structure before briefly considering phylogeography and the genetic structure of invasive populations.

12.2 Genetic Tools

There is a growing battery of molecular techniques for discerning genetic structure of individuals, colonies and populations. Among other attributes, markers differ in their mode of inheritance (biparental or maternal), their variability, whether they are dominant (unable to distinguish heterozygotes from homozygotes) or codominant (heterozygotes can be distinguished from homozygotes), whether they are single locus or multilocus, and their repeatability (Sunnucks 2000). The best molecular marker to use for a given study will depend on the question being asked (e.g. parentage and kinship vs. population structure), the level of variability exhibited by the study population(s) at the class of markers being considered, and the existence of adequate statistical methods to estimate the desired population genetic parameters robustly. We briefly review the molecular techniques that have been used in studies of termite colony and population genetics, noting some of the advantages and limitations of each. For more complete discussions of molecular markers used in population studies see Loxdale and Lushai (1998), Sunnucks (2000), Zhang and Hewitt (2003), Avise (2004), and Behura (2006).

12.2.1 Allozyme Markers

Allozymes were used in the earliest studies of termite colony and population genetic structure. These co-dominant, biparentally inherited markers can provide information on parentage and population genetic structure and are relatively easy and inexpensive to run since the same protocols can be used on a wide range of taxa. Early application of allozymes provided some of the first information on the variation in colony breeding structure within populations (Clément 1981, 1984; Reilly 1987; Luykx 1993; Thompson and Hebert 1998a, 1998b; Bulmer et al. 2001; Clément et al. 2001); however, the relatively low levels of variability limited their power to differentiate colonies and to show a more detailed picture of colony genetic structure (DeHeer et al. 2005). In addition, some introduced populations of termites showed very limited variability at allozyme loci, greatly restricting their utility in studies of colony and population genetic structure in non native habitats (Strong and Grace 1993; Wang and Grace 2000). For these reasons researchers are now using primarily DNA markers.

12.2.2 Mitochondrial Genes

Mitochondrial genes have been used in many studies of colony and population genetic structure. A number of different regions have been used, most frequently the cytochrome oxidase I or II genes (COI and COII) (Kutnik et al. 2004; Park et al. 2006; Vargo and Carlson 2006; Lefebvre et al. 2008), but also the COIII gene (Park et al. 2006), 16S ribosomal (16S rDNA) gene (e.g., Austin et al. 2004a, 2004b; Kutnik et al. 2004), the NADH dehygrogenase-1 (ND1) gene (Kutnik et al. 2004; Uva et al. 2004b), and the AT-rich region (also called mitochondrial control region) (Atkinson and Adams 1997a, 1997b; Thompson and Hebert 1998b; Bulmer et al. 2001; Vargo 2003b). Most commonly, sequence data are employed, but other less direct methods of genetic variation have been used such as restriction fragment length polymorphism (e.g. Thompson and Hebert 1998b) and double-strand conformation polymorphism (e.g. Atkinson and Adams 1997a, 1997b; Bulmer et al. 2001). Mitochondrial DNA has been used to investigate colony breeding structure (Atkinson and Adams 1997b, Jenkins et al. 1999; Bulmer et al. 2001; DeHeer and Vargo 2008), population genetic structure (Thompson and Hebert 1998b; Vargo 2003b; Vargo and Carlson 2006; Fang et al. 2008; Dupont et al. 2009), phylogeography (Kutnik et al. 2004; Uva et al. 2004b; Luchetti et al. 2005, 2007; Lefebvre et al. 2008), and the sources of introduced populations (Jenkins et al. 2001, 2002, 2007; Austin et al. 2006b). Studies combining mtDNA markers with co-dominant nuclear markers, such as allozymes (Bulmer et al. 2001) or microsatellites (Vargo 2003b; Vargo and Carlson 2006; DeHeer and Vargo 2008), can be particularly informative about the breeding structure of colonies.

12.2.3 Nuclear Gene Sequences

The multicopy nuclear internal transcribed spacer (ITS) region of ribosomal DNA has been used in phylogeographic studies of European *Reticulitermes* spp. (Kutnik et al. 2004; Lefebvre et al. 2008) and *Coptotermes gestroi* (Jenkins et al. 2007), and for studies of species relationships and origins of introduced populations (Jenkins et al. 2001). However, these regions tend to be conserved within a species and therefore do not provide much information for studies of colony and population genetic structure.

12.2.4 Multilocus DNA Fingerprinting

Multilocus DNA fingerprinting offers a powerful method for differentiating colonies and determining the relative degree of relatedness among colony members (Kaib et al. 1996; Husseneder et al. 1998, 1999, 2003b; Husseneder and Grace 2001a, 2001b) and for determining the relatedness between males and females forming tandem pairs during colony foundation (Shellman-Reeve 2001). This technique involves Southern blotting of hypervariable DNA regions containing tandem repeats, often called minisatellites, subjected to restriction digest of genomic DNA to produce a complex individual-specific genetic fingerprint (Jeffreys et al. 1985). Important advantages of this technique are that it does not require any species-specific marker development and it is highly variable. The major disadvantage is that the bands produced tend to be dominant, preventing the assignment of heterozygosity or homozygosity at a specific locus thereby limiting accurate determinations of levels of relatedness or allele frequencies required for population genetic analysis.

12.2.5 Microsatellite Genotyping

Microsatellite genotyping also detects areas of tandem repetitive DNA, but in contrast to multilocus fingerprinting, microsatellite genotyping amplifies and analyzes individual loci. Microsatellites have become the markers of choice for many population genetic studies (Sunnucks 2000; Zhang and Hewitt 2003). This is because they are highly variable, co-dominant, highly repeatable, and provide information about discrete genetic loci. Thus, microsatellites can be useful in describing population structure, for determining the breeding structure of colonies, levels of inbreeding, parentage of individuals, and for assigning individuals to colonies (Husseneder et al. 2003b). The major disadvantage of microsatellites is that they need to be developed for each study species, or at least closely related species, although the methods for microsatellite development are becoming faster, easier, and relatively inexpensive (see e.g. Booth et al. 2008). Microsatellites have been developed for 14 termite

Table 12.1 Polymorphic microsatellite loci identified for termites

Family/species	No. loci	References
Mastotermitidae *Mastotermes darwiniensis*	10	Goodisman et al. (2001)
Termopsidae *Zootermopsis nevadensis* [a]	12	Aldrich and Kambhampati (2004)
Kalotermitidae *Incisitermes minor*	10	Indrayani et al. (2006)
	15	Booth et al. (2008)
Cryptotermes secundus	6	Fuchs et al. (2003)
Rhinotermitidae *Reticulitermes flavipes*	9	Vargo (2000)
	11[b]	Dronnet et al. (2004)
R. speratus	7	Hayashi et al. (2002)
R. hesperus	3	Copren (2007)
Coptotermes formosanus	12	Vargo and Henderson (2000)
C. lacteus	6	Thompson et al. (2000)
Termitidae		
Amitermes meridionalis	10	Schmidt et al. (2007)
Cubitermes subarquatus	7	Harry et al. (2001)
Labiotermes labralis	6	Harry et al. (2007)
Macrotermes michaelseni	4	Kaib et al. (2000)
Nasutitermes corniger	8	Atkinson et al. (2007)

[a] 10 loci were polymorphic in *Z. angusticollis*.
[b] Reported as *R. santonensis*.

species belonging to five families (Table 12.1. In addition, primers developed for one species often successfully amplify loci in other congeneric species (Aldrich and Kambhampati 2004; Matsuura et al. 2004; DeHeer et al. 2005; Vargo and Carlson 2006; Vargo et al. 2006b; Copren 2007) and sometimes in species in other genera (Goodisman et al. 2001; Vargo et al. 2006a).

12.3 Colony Genetic Structure

In any termite population, colony genetic structure will be a function of (1) the relatedness and (2) number of primary reproductives founding colonies, (3) the frequency with which colonies produce secondary reproductives to supplement or replace the primary reproductives, (4) the number and relatedness of the secondary reproductives, (5) the frequency with which colonies of a species fuse with other related or unrelated colonies or adopt foreign reproductives later in colony life. In the following section, we focus on colony foundation, the factors determining the degree of relatedness of primary kings and queens and the predominance of single pair founders over pleometrosis (multiple founding pairs). In the subsequent section, we discuss the numbers of secondary reproductives within colonies, their degree of relatedness and levels of inbreeding. We then conclude the section on colony genetic structure with a discussion of the occurrence of fused colonies.

12.3.1 Colony Founding

12.3.1.1 Relatedness of Colony Founders

The relatedness of colony founders, a key factor in determining the level of inbreeding and nestmate relatedness within colonies, is determined by one or more of the following mechanisms: (1) dispersal (total distance and dispersal differences between sexes), (2) sex-biased alate production at the colony level (colonies preferably investing in one sex or the other or releasing male or female alates at different times), and (3) kin discrimination during mate choice. Here we consider the possible role of each of these mechanisms in turn.

(1) Dispersal distances influence the likelihood that relatives will encounter each other during mate pair formation thus determining the degree of inbreeding within newly founded colonies. If male and female alates disperse only a short distance from the nest, then they are more likely to encounter each other during tandem pair formation. Only a few studies have actually directly measured the dispersal capabilities of alates, most of them in subterranean termites. Studies using fluorescent dyes (Messenger and Mullins 2005) and genetic markers (Simms and Husseneder 2009), showed that alates of *C. formosanus* in New Orleans, LA, an area with extremely high population densities, are capable of dispersing 500 m and more resulting in a high proportion of outbred colonies at the foundation stage (Husseneder et al. 2003a; Vargo et al. 2006a; Aluko and Husseneder 2007). As discussed below, the low degree of relatedness among

colony founders heading mature colonies suggest that in many species, dispersal distances are far enough to promote outbreeding. Sex-biased dispersal has been proposed as mechanism for inbreeding avoidance in termites. However, the only evidence so far that males and females differ in dispersal distances stems from the observation of male philopatry and female dispersal in the dampwood termite *Zootermopsis nevadensis* (Shellman-Reeve 1996). Husseneder et al. (2006) found that males and females in swarm aggregations of *C. formosanus* were genetically differentiated, but these authors concluded that this was more likely due to males originating from inbred colonies, as discussed below, rather than to differences in dispersal distances.

(2) While the sex ratio of alates at the population level is more or less even in most termite species, individual colonies do not necessarily invest equally in both sexes at any given time (Jones et al. 1988). This increases the likelihood that alates from different colonies will outcross. Whether colonies produce predominantly males or females might depend on the colony breeding structure. As mentioned above, male and female alates of *C. formosanus* were found to be genetically differentiated because males had significantly higher inbreeding levels than females and therefore must have come primarily from inbred colonies (Husseneder et al. 2006). This finding is consistent with the observation that colonies of Australian *Coptotermes* spp. that were headed by inbreeding neotenics produced males almost exclusively (Lenz and Runko 1993; Roisin and Lenz 2002). The opposite was found in *R. virginicus*, where female alates were produced primarily in inbred colonies as suggested by their lower levels of heterozygosity compared to males (DeHeer and Vargo 2006). One possible reason for biasing the sex of alates according to the degree of inbreeding in the colony could be sexual selection favoring heterozygosity in one sex. This hypothesis is supported by a study by Husseneder and Simms (2008) who found that genetic diversity confers a mating advantage in females of *C. formosanus*, which in turn would explain why outbred colonies preferably invest in female alates (see above). To date, only a few studies have investigated differences between sexes in levels of inbreeding. If such sex-biased heterozygosity were found to be widespread in termite species, it could lead to intriguing testable hypotheses concerning sex-biased investment by colonies and possible sexual selection operating in this group.

(3) While dispersal and sex-biased alate production are indirect mechanisms possibly influencing the likelihood that colony founders meet in swarms, the third potential mechanism determining relatedness among colony founders is active kin discrimination during mate choice. The role of kin recognition in partner selection in termites is not widely studied despite the obvious consequences of this behavior for colony genetic structure and kin selection in general. The only evidence for kin discrimination during colony foundation comes from a study by Shellman-Reeve (2001), who observed that alates of *Z. nevadensis* avoid forming partnerships with close relatives. In contrast, pair formation in the subterranean termites, *C. formosanus, R. flavipes, R. virginicus and R. kanmonenis*, appear to be random with respect to kinship (Kitade et al. 2004; DeHeer

and Vargo 2006; Husseneder and Simms 2008). Data on more species, especially those that undergo short-range dispersal flights, are needed to determine if kin discrimination is a common mechanism regulating inbreeding levels in colonies.

12.3.1.2 Number and Relatedness of Colony Founders

In the majority of termite species, colonies are founded by a single pair of male and female reproductives following a nuptial flight (Nutting 1969; Vargo and Husseneder 2009). Genetic studies show that the degree of relatedness of the founders in monogamous colonies (simple families) varies among species and populations within the same species. Simple family colonies of *N. corniger* (Atkinson and Adams 1997b), *Schedorhinotermes lamanianus* (Husseneder et al. 1999), *R. flavipes*, *R. virginicus* and *R. hesperus* are generally headed by a pair of unrelated colony founders, whereas the reproductive pairs of *R. hageni* and *R. malletei* colonies are often related (Copren 2004; Vargo and Carlson 2006; Vargo et al. 2006b; Parman and Vargo 2008). Simple family colonies in a French population of *R. grassei* have closely related reproductives (DeHeer et al. 2005) while colonies of a Portuguese population have largely unrelated reproductives (Nobre et al. 2008). Similar variation within a species was found in *C. formosanus* populations with mostly simple family colonies headed by siblings in a Japanese population to almost exclusively unrelated partners in a New Orleans population (Vargo et al. 2003, 2006a). In the Australian species *C. lacteus* the reproductive pairs were related, but at a low level (Thompson et al. 2007), suggesting a slight preference for mating with distant relatives. As mentioned above, there is little evidence of mate discrimination during tandem pair formation in termites. Thus the most likely mechanism determining the degree of relatedness of founders is the probability of related males and females encountering each other during mating flights as dictated by the distance they disperse from the nest. However, it should be noted that such genetic studies rarely involve genotyping of the reproductive pair, which are usually difficult to collect. Rather, they largely infer the genotypes of the reproductive pair from worker genotypes. Thus it is possible that in some colonies the monogamous pair heading the colony could be neotenics, in which case estimates of relatedness between primaries would be inflated.

Primary polygamy or pleometrosis, i.e., colony foundation by multiple same sex reproductives, is not as common in the Isoptera as in the Hymenoptera (Thorne 1984). It has been reported predominantly in the Termitidae (Nutting 1969; Thorne 1982; Atkinson and Adams 1997b) including about 35 species in 15 genera (Thorne 1985). The presence of unrelated primary queens has been confirmed by genetic studies in colonies of *Macrotermes michaelseni* (Hacker et al. 2005), *Zootermopsis* (Broughton 1995), and *Nasutitermes* (Atkinson and Adams 1997b). Possible mutualistic advantages of pleometrosis include the faster growth of an incipient colony due to the cumulative reproductive output and higher survival probability of the colony (Thorne 1984) and the risk minimization should one of the reproductives die (Thorne 1982; Darlington 1985).

12.3.2 Colony Breeding Structure

In some cases, nests can be censused to determine the numbers and types of reproductives (e.g. primary or neotenic). This method has been used primarily for termitids in which the nest and royal cell are often accessible (reviewed in Roisin 1993). However, in many species, nests are difficult to find and therefore direct census is not practical. Even when reproductives are found, the degree of relatedness among them cannot be determined without genetic markers. Therefore, in many situations the only way to determine colony breeding structure is to infer it from genotypes of foragers, and (when present) reproductives. Ideally, individuals are genotyped at numerous genetic loci using codominant markers such as microsatellites (e.g. Goodisman and Crozier 2002; Husseneder et al. 2003b; Vargo 2003b; DeHeer and Vargo 2004; Thompson et al. 2007) or allozymes (e.g. Thompson and Hebert 1998a; Bulmer et al. 2001). Microsatellites are preferred because they are generally highly polymorphic revealing great detail about colony genetic structure and parentage (Vargo and Husseneder 2009). Mitochondrial DNA can also be useful in detecting the presence of multiple matrilines within colonies (Atkinson and Adams 1997b; Jenkins et al. 1999; Bulmer et al. 2001). These genotypes are then subjected to pedigree analysis to determine whether colonies are simple families, extended families or mixed families (Bulmer et al. 2001; Vargo 2003b; DeHeer and Vargo 2004). *Simple families* are those with genotypes consistent with the progeny of a monogamous pair of reproductives. *Extended families* are headed by neotenics descended from the founding pair, as evidenced by having no more than four alleles at a locus but having genotypes inconsistent with those produced by a single monogamous pair (e.g. five or more genotypes or three classes of homozygotes) or having genotype frequencies deviating from those expected in simple families. *Mixed families* are the progeny of multiple unrelated reproductives, as evidenced by having more than four alleles at a locus and/or more than a single mtDNA haplotype present. More detailed information about the breeding structure can be inferred by estimating the coefficients of inbreeding and relatedness and comparing these values to those generated by computer models for different possible breeding systems (Thorne et al. 1999; Bulmer et al. 2001).

Colony breeding structure has been characterized by molecular markers and/or censuses of nests in populations of 24 species and subspecies representing five families (Table 12.2). On a broad phylogenetic scale, results of these studies so far largely reinforce the general conclusions from natural history studies about colony breeding structures and how they vary across clades. In particular, species of lower termites are more likely to form extended family colonies, a finding consistent with the far greater prevalence of neotenics in this group (Myles 1999). However, genetic studies of colony breeding structures also provide a far more detailed picture than was previously available, revealing greater complexity and variation in breeding structure than present in earlier views.

Half of the studied species are subterranean termites (*Reticulitermes* and *Coptotermes*), and these have recently been reviewed in detail elsewhere (Vargo and Husseneder 2009). There is considerable variation in colony breeding structure

Table 12.2 Colony family structure in termites as determined by molecular markers except where noted. For species in which multiple populations have been investigated, *n* indicates the number of populations studied. All populations are from native ranges unless otherwise noted

Family/species	Simple families	Extended families	Mixed families	Overall F_{IT}	References
Mastotermitidae					
Mastotermes darwiniensis ($n = 3$)	26%	47%	27%	–0.02 to 0.19	Goodisman and Crozier (2002)
Termopsidae					
Zootermopsis angusticollis ($n = 3$)	16–69%	36–79%	0–8%	0.15–0.16	Booth et al. (unpublished data)
Z. nevadensis	33%	67%	0%	0.17	Booth et al. (unpublished data)
Z. nevadensis nevadensis	0%	100%	0%	0.24	Aldrich and Kambhampati (2007)
Z. nevadensis nuttingi	45%	22%	33%	0.11	Aldrich and Kambhampati (2007)
Kalotermitidae					
Incisitermes schwarzi	75%	25%	0%	0.04	Luykx (1993)
Rhinotermitidae					
Reticulitermes flavipes ($n = 10$)	0–85%	15–74%	0–26%	0.03–0.44	Bulmer et al. (2001), Vargo (2003a, 2003b, unpublished data), DeHeer and Vargo (2004), Vargo and Carlson (2006), DeHeer and Kamble (2008), Parman and Vargo (2008)
Reticulitermes flavipes ($n = 2$)[a]	0%	100%	0%	0.17–0.39	Dronnet et al. (2005)
R. malletei	54%	46%	0%	0.19	Vargo and Carlson (2006)
R. hageni ($n = 3$)	87–95%	5–13%	0%	0.14–0.36	Vargo and Carlson (2006), Vargo et al. (2006b), Parman and Vargo (2008)
R. virginicus ($n = 2$)	75–100%	0–25%	0%	–0.04 to 0.04	Vargo (2003a), Vargo et al. (2006b), Parman and Vargo (2008)
R. hesperus	73%	27%	0%	0.08	Copren (2004)
R. grassei ($n = 7$)	0–82%	18–100%	0–13%	–0.02 to 0.31	DeHeer et al. (2005), Nobre et al. (2008), DeHeer et al. (unpblished data)
R. speratus	26.7%	26.7%	46.6%	NR	Hayashi et al. (2005)
R. urbis ($n = 18$)	0%	100%	0%	NR	Leniaud et al. (2010)
Coptotermes lacteus ($n = 6$)[b]	100%	0%	0%	NR	Thompson et al. (2007)

Table 12.2 (continued)

Family/species	Simple families	Extended families	Mixed families	Overall F_{IT}	References
C. formosanus	0%	100%	0%	0.18	Husseneder et al. (2008)
C. formosanus ($n = 6$)[a]	37–100%	0–63%	0%	0.16–0.46	Vargo et al. (2003, 2006a), Husseneder et al. (2005, 2007)
Termitidae					
Labiotermes labralis ($n = 3$)	100%	0%	0%	0.03–0.42	Dupont et al. (2009)
Macrotermes michaleseni [c]	50–95%	0%	12–24%	NR	Darlington (1985), Brandl et al. (2004), Hacker et al. (2005)
Nasutitermes nigriceps ($n = 3$)	95%	5%	0%	NR	Thompson and Hebert (1998a, 1998b)
N. corniger [c] ($n = 2$)	67–80%	4–16%	4–29%	NR	Thorne (1984), Atkinson and Adams (1997b), Atkinson et al. (2008)
N. corniger (reported as *N. costalis*)[d]	0%	?	?	NR	Roisin and Pasteels (1986)
N. corniger (reported as *N. polygynus*)[d]	7%	?	?	NR	Roisin and Pasteels (1986)
N. princeps [d]	40%	?	?	NR	Roisin and Pasteels (1985)

NR, not reported.
[a]Introduced populations.
[b]Based on small sample sizes (Range = 4–9 colonies).
[c]Based on nest censuses and genetic markers.
[d]Based on nest censuses only.

across species resulting in variation in the level of inbreeding among individuals (Table 12.2), with F_{IT} (measure of homozygosity of individuals relative to their population) ranging from zero in some species, or some populations of some species, up to 0.46 in the case of an introduced population of *C. formosanus* (Vargo et al. 2003). Higher levels of inbreeding occur in species in which males and females in founding pairs are related and/or in which extended family colonies are common. The different populations that have been studied varied in their compositions from exclusively simple families in *C. lacteus* (Thompson et al. 2007) to exclusively extended families in a native population of *C. formosanus* (Husseneder et al. 2008). Genetically complex mixed family colonies occurred in nearly half of the species (47%), including both higher and lower termites, and where they occurred their frequency varied from 2% in one population of *R. flavipes* (DeHeer and Vargo 2008) to 47% in *R. speratus* (Hayashi et al. 2005). However, populations of most species consisted primarily of simple family colonies.

12.3.2.1 Simple Families

There is considerable variation among species, and even among populations within a species, in the proportion of colonies forming simple families. For example, in *R. flavipes* simple family colonies have been found to compose 85% of all colonies in a population in South Carolina, USA (Vargo et al. 2006b), while a population in Nebraska, USA was found to have no simple families (DeHeer and Kamble 2008). The reasons for this variation, especially among species, are not well understood. However, as discussed below (Section 12.3.2.4), geographic variation in inbreeding depression may play a role in determining colony breeding structure in some species.

12.3.2.2 Extended Families

The frequency of colonies headed by secondary reproductives (neotenics in lower termites and primary reproductives in some termitids) will have a strong effect on the degree of inbreeding within a population. But the magnitude of this effect will be determined by the numbers of secondary reproductives within colonies and the number of generations over which these secondary reproductives inbreed. In general, the fewer the number of neotenics and the greater the number of generations of inbreeding, the more inbred colony members will be (i.e., the higher the F_{IT} value) (Thorne et al. 1999). As mentioned previously, extended family colonies are more common in lower termites than in termitids (Table 12.2), although very few species of higher termites have been closely examined using genetic markers. The lower frequency of extended families in the Termitidae might be expected based on the more limited occurrence of neotenics in this group, but extended families can be formed through the recruitment of primary reproductives from within the colony (Atkinson and Adams 1997b). Thus, the formation of extended families is not limited to species in which neotenics are produced.

Whereas both the number of secondary reproductives and the number of generations of inbreeding will decrease heterozygosity in individuals relative to their population, resulting in higher F_{IT} values, examination of the F_{IC} value (the coefficient of inbreeding in individuals relative to their colony) can help in determining whether the numbers of neotenics are few (<6 or so) or many (>10). This is because F_{IC} is highly sensitive to the numbers of reproductives, with strongly negative values (<–0.2) indicative of low numbers of reproductives (Thorne et al. 1999). There is considerable variation in the numbers of neotenics within extended families, often within a single genus or even within a species. Inferences from genetic studies of *Reticulitermes* spp. suggest colonies vary from a few neotenics on average (<6) in some populations of *R. flavipes*, *R. hageni* and *R. virginicus* (Vargo 2003a, 2003b; DeHeer and Vargo 2004; Vargo and Carlson 2006; Vargo et al. 2006b; Parman and Vargo 2008) to many neotenics (>10) in one native population of *R. flavipes* (Bulmer et al. 2001), introduced populations of *R. flavipes* (Dronnet et al. 2005), and a native population of *R. grassei* (DeHeer et al. 2005). Considerable intraspecific variation in inbreeding levels among extended family colonies has also been seen among introduced populations of *C. formosanus* (Vargo et al. 2003, 2006a; Husseneder et al.

2005, 2007, 2008) and native populations of *Mastotermes darwiniensis* (Goodisman and Crozier 2002).

The presence of multiple reproductives within extended family colonies allows for the possibility of having genetic substructuring within colonies through two potential processes. First, the presence of spatially separated groups of reproductives with little or no interbreeding between them coupled with limited movement of workers from their natal nest can lead to genetic differentiation within the colony. In an expansive colony of *C. formosanus*, Husseneder et al. (2005) found evidence of substructure among foragers feeding on monitoring stations located 25–100 m apart. Similar results were obtained by Husseneder et al. (1998) in a colony of the African subterranean termite *S. lamanianus*. In a very large introduced colony of *R. flavipes* in France, Dronnet et al. (2005) also found that foragers occurring further apart were genetically more differentiated than those occurring closer together.

Another process that can lead to differentiation among foraging groups is kin-biased foraging. In *S. lamanianus*, Kaib et al. (1996) found that groups of foragers collected away from the nest were more closely related than workers taken from the nest center, suggesting that workers sorted themselves into kin groups while foraging. It is not known whether genetic substructuring, either through kin-biased sorting or as a result of spatially separated reproductive centers, is widespread, but is certainly worthy of further investigation.

In the inferences of colony breeding structure and in the simulated breeding systems, reproduction is assumed to be sexual. However, a recent study of field-collected reproductives in the Japanese subterranean termite, *R. speratus*, has shown that secondary females in these colonies are produced parthenogenetically, while workers are produced sexually (Matsuura et al. 2009). The effect of having parthenogenetic secondary queens reproducing is to reduce levels of inbreeding among workers over those expected in colonies headed exclusively by sexually produced secondary queens. So far, parthenogenetically produced secondaries have only been found in a single species, but if they turn out to be more common, then the models estimating the numbers of secondary reproductives based on levels of inbreeding will have to be adjusted somewhat.

12.3.2.3 Mixed Family Colonies

Results of studies to date indicate that while mixed family colonies are infrequent or nonexistent in most taxa examined, they can comprise a substantial proportion of the colonies in some species, accounting for up to one-third of all colonies in *Zootermopsis nevadensis nuttingi* (Aldrich and Kambhampati 2007) and *Nasutitermes corniger* (Thorne 1984; Atkinson et al. 2008). There are several possible mechanisms leading to mixed families: (1) colony fusion, (2) pleometrotic associations, and (3) adoption of unrelated reproductives by established colonies. In fact, there is evidence that mixed family colonies can arise through all three of these mechanisms, sometimes in a single species, such as *N. corniger* (Atkinson and Adams 1997b). However, in subterranean termites where mixed family colonies have been shown to occur in only three species so far – *R. flavipes, R. grassei* and

R. speratus – colony fusion appears to be the only mechanism involved (DeHeer and Vargo 2004, 2008; Hayashi et al. 2005). Pleometrotic associations of multiple same sex reproductives have been found only in some termitids (Atkinson and Adams 1997b; Hacker et al. 2005).

It is worth noting that the long term stability of mixed family colonies, especially in subterranean termites, is unclear. Results of field (DeHeer and Vargo 2008) and lab (Fisher et al. 2004) studies indicate that the presence of multiple unrelated groups of reproductives in fused colonies of *R. flavipes* is generally rather short-lived; over time reproduction in fused colonies usually but not always is restricted to individuals from only one of the original source colonies, suggesting that many such colonies eventually revert to close family units. The mixed family colonies found in *Z. nevadensis* also may be a temporary phenomenon arising from colony fusion and the execution of the reproductives of one of the interacting colonies (Broughton 1995; Thorne et al. 2003).

The factors underlying colony fusion are not clear. Matsuura and Nishida (2001) proposed that colonies with numerous nymphs preparing to molt into alates would be more likely to accept individuals from foreign colonies, but this hypothesis has not been rigorously tested in the field. In a study of mixed family colonies of *R. flavipes* in North and South Carolina, DeHeer and Vargo (2008) showed that individuals originating from different families had identical or nearly identical mtDNA haplotypes, but were unrelated at nuclear microsatellite loci. These results suggest some maternally inherited factor underlying colony fusion, but the nature of this factor is not known. Because mixed family colonies have low genetic relatedness, and therefore lower inclusive fitness of colony members, the factors influencing colony fusion, including similarities in mtDNA haplotype, are certainly worthy of further study.

The presence of multiple unrelated reproductives within termite colonies decreases within-colony relatedness, thus lowering the inclusive fitness payoffs to workers. This raises the question of whether there is any nepotism in such colonies with workers preferring to tend kings and/or queens to which they are most closely related. The interactions among queens and workers in polygynous colonies of *N. corniger* have been studied in detail. About one-third of the colonies of this species in Panama contain multiple primary queens and sometimes multiple kings (Thorne 1984; Atkinson and Adams 1997b). Co-occurring queens in these colonies can be related or unrelated (Atkinson and Adams 1997b). Atkinson et al. (2008) investigated whether workers were more likely to tend closely related queens than unrelated queens in polygynous colonies. These authors found no evidence that workers preferentially tended queens to which they were closely related. These results support those on various ant species also showing a lack of nepotism in queen tending by workers (Keller 1997), but clearly additional studies are needed on other polygamous termites to determine whether nepotism can help explain the co-existence of multiple unrelated reproductives in some species.

12.3.2.4 Geographic Variation in Colony Breeding Structure

It is apparent from the results of several studies that there can be tremendous variation in colony breeding structure among populations of a species. This raises the question of the factors underlying this variation. There are only a few studies along these lines, but so far it seems that local ecological conditions can play a major role in breeding structure variation. As mentioned before, colonies of *Macrotermes michaelseni* are often founded by multiple males and females, and multiple primary reproductives persist in about 25% of established colonies but rarely do multiple males persist (Darlington 1985; Brandl et al. 2001, 2004; Kaib et al. 2001). The queens in such polygynous colonies are unrelated to each other (Hacker et al. 2005). Brandl et al. (2001, 2004) reported variation in the frequency of polygynous colonies as a function of both year and geographic location suggesting that local ecological conditions strongly influence the relative survival of pleometrotic associations.

Studies of numerous populations over large geographic areas of two subterranean termite species, *R. flavipes* along the east coast of the US and of *R. grassei* in Europe, show strong clinal variation with northern populations showing higher proportions of extended family colonies and greater levels of inbreeding (Fig. 12.1). These results suggest that colony breeding structure is responsive to local environmental conditions that vary along latitudinal gradients. If so, then this would open opportunities for determining the ecological factors shaping colony breeding structure, especially factors that select against inbreeding.

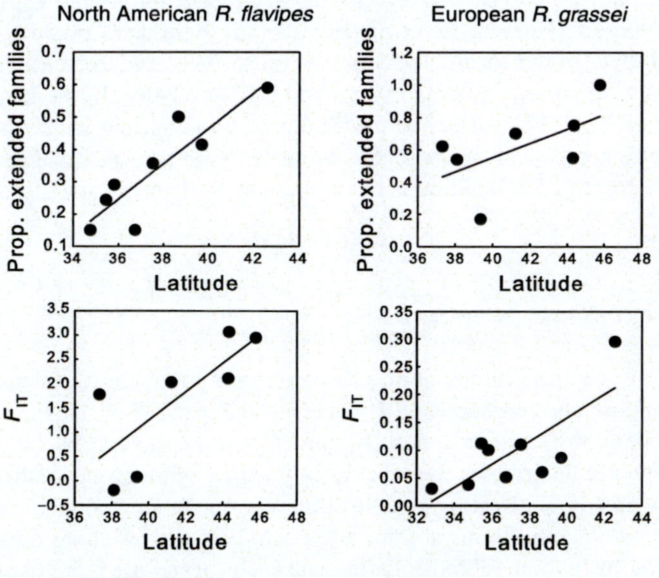

Fig. 12.1 Geographic inbreeding in two species of *Reticulitermes*

12.4 Population Genetic Structure

As mentioned above, dispersal and colony breeding structure, together with drift and selection, can significantly affect population genetics, that is the associations of genes within and among populations above the level of the colony. In this section we examine studies of the genetic structure both within (small scale) and among (large scale) populations.

12.4.1 Small-Scale Structure and the Issue of Budding

Short-range dispersal can potentially lead to significant genetic structure at the population level in termites. This could happen through short mating flights and/or budding. Both processes have been assumed to be common in some termite groups, especially subterranean termites (Shellman-Reeve 1997; Myles 1999; Thorne et al. 1999). If either short-range mating flights or budding were frequent, populations should show high population viscosity leading to greater levels of population structure at smaller spatial scales with colonies located near each other being more closely related than those further apart. There have been a number of fine scale genetic studies on a variety of subterranean species (Husseneder et al. 1998, 2005; Bulmer et al. 2001; Vargo 2003b; DeHeer and Vargo 2004; DeHeer et al. 2005; Vargo and Carlson 2006; Vargo et al. 2006a, 2006b) and one study on the mastotermitid *Mastotermes darwiniensis* (Goodisman and Crozier 2002), but in only two species – *R. malletei* (originally reported as *R. hageni*; Vargo and Carlson 2006) and *S. lamanianus* (Husseneder et al. 1998) – has significant population viscosity been reported. In the former species, it is most likely due to short-term mating flights rather than colony budding (Vargo and Carlson 2006). In the latter species, Husseneder et al. (1998) suggested it was due to budding, but short distance dispersal of alates could not be ruled out. Obviously, more studies are needed, but the evidence to date suggests budding is not a frequent mode of colony reproduction in termites.

12.4.2 Large-Scale Structure

The degree of genetic differentiation among termite populations is determined by both the contemporary processes of dispersal and gene flow and the historical forces of genetic drift and selection. Against this more general backdrop, colony breeding structure determines the level of inbreeding within a population and can influence the processes affecting genetic differentiation among populations. Species with short range mating flights are more likely to pair with relatives during colony founding, leading to more inbred colonies, and such species are more likely to experience restricted gene flow among populations. Such conditions are expected to enhance the role of genetic drift in population divergence. While significant population genetic structure across large spatial scales has been reported in several

Table 12.3 Levels of genetic differentiation among termite populations. Spatial scale refers to the distances separating populations. *n* is the number of populations studied

Family/species	F_{ST}	Spatial scale	References
Mastotermitidae			
Mastotermes darwiniensis ($n = 4$)	0.42–0.62	2–400 km	Goodisman and Crozier (2002)
Rhinotermitidae			
Coptotermes lacteus ($n = 6$)	0.01	3–11 km	Thompson et al. (2007)
($n = 2$)	0.04	150 km	
Reticulitermes flavipes ($n = 8$)	0–0.17	1–450 km	Bulmer et al. (2001), DeHeer and Vargo (2004), Vargo and Carlson (2006), Vargo (unpublished data)
R. grassei ($n = 3$)	0.0–0.14	30–120 km	DeHeer et al. (2005)
R. lucifugus corsicus ($n = 3$)	0.26–0.54	150–300 km[a]	Lefebvre et al. (2008)
Termitidae			
Labiotermes labralis ($n = 3$)	0.16–0.25	150–300 km	Dupont et al. (2009)
Macrotermes michaelseni ($n = 7$)	0.017	0–50 km	Brandl et al. (2005)
($n = 14$)	0.02–0.21	50–500 km	
Nasutitermes nigriceps	0.03–0.11	100–200 km	Thompson and Hebert (1998a)

[a]Populations separated by water.

termites, the degree to which this divergence is related to levels of inbreeding and/or dispersal has not been well characterized.

Results regarding differentiation among populations are summarized in Table 12.3. To date, we have data for only eight species representing three families. For the most part, these studies show low levels of differentiation at small spatial scales (<30–50 km) and moderate to strong differentiation at larger spatial scales (>100 km), suggesting relatively unfettered gene flow over distances of tens of kilometers in many species. However, it is often difficult to compare across studies since different sampling schemes were used and they were not conducted at comparable spatial scales. One exception to this rule was the South American soil-feeding termite *Labiotermes labralis* in which Dupont et al. (2009) found strong differentiation ($F_{ST} = 0.16 - 0.25$) among populations located 30–120 km apart, which the authors attributed to habitat fragmentation due to human disturbance. It is clear that additional studies are needed comparing levels of population genetic structure with studies of dispersal and colony breeding regime to gain a better understanding of the role of genetic drift and selection in the differentiation of termite populations, which may eventually lead to speciation.

12.5 Phylogeography

The study of historical processes and their potential effects on the contemporary distribution of species is known as phylogeography (Avise 2000). As the link between microevolutionary change leading to genetic differentiation among populations and macroevolutionary change resulting in speciation and higher level divergence, phylogeography has great potential to shed light on the evolutionary relationships among subterranean termite taxa, the processes leading to speciation, and the factors determining current distributions. These questions have taken on new relevance since termites appear to have undergone accelerated speciation compared to sister-group cockroach lineages (Davis et al. 2009), suggesting that eusociality has promoted radiation. The few studies that have been conducted to date show how geological history and vicariance events can explain speciation and species distributions. Studies of *Reticulitermes* spp. in Europe suggest that there were four refugia during the last glacial maximum, some 20,000–18,000 years ago: the southern Iberian Peninsula, southern Italy, the southern Balkan Peninsula, and the Middle East (Clément 1981; Clément et al. 2001; Marini and Mantovani 2002; Kutnik et al. 2004; Luchetti et al. 2004, 2005, 2007; Austin et al. 2006a). The current distributions of species and subspecies throughout southern Europe fit reasonably well with the northward expansion and radiation of populations from these refugia. Within the subspecies, *R. lucifugus corsicus*, Lefebvre et al. (2008) found a pattern of genetic differentiation consistent with both vicariance events and human-mediated dispersal for populations located in Corsica, in Tuscany and in Provence. A study of Japanese and Korean populations of *R. speratus* (Park et al. 2006) found that mtDNA haplotypes show a pattern consistent with the colonization of southern Japan by an ancestor of *R. speratus* from China and the subsequent migration into northern Japan and more recently into Korea from the southern Japanese islands. This pattern is consistent with geographic history of the region. Szalanski et al. (2004) conducted a phylogeographic study of *Heterotermes* spp. in the Caribbean. They identified six haplotype groupings that corresponded reasonably well with their geographic distribution and species status, but these authors also found evidence of an undescribed species as well as anthropogenic origins of some populations. Similar analyses of other groups of termites should help shed light on the historical events leading to speciation and current species ranges.

12.6 Population Genetics of Invasive Species

Although a number of termite species have been introduced and become established in new locations, especially species of drywood termites (Kalotermitidae) and subterranean termites (Rhinotermitidae), only a few species can be considered truly invasive in the sense that they have significant ecological impact in their introduced ranges. The most destructive of the invasive termites is the Formosan subterranean termite, *Coptotermes formosanus*, a native of mainland China. This species, considered one of the 100 worst invasive species in the world (Global Invasive Species Database 2010), has become established in South Africa, Sri Lanka, Japan and the US, where it is present in 11 states and actively spreading (Su 2003). In addition to *C. formosanus*, *C. gestroi* has also been introduced to several places around the world (Jenkins et al. 2007). *Reticulitermes flavipes* is well established in Europe, especially France (Austin et al. 2005), and in South America (Su et al. 2006). What little work has been done on invasive termites has mostly focused on *C. formosanus* and *R. flavipes* (Vargo and Husseneder 2009).

Population genetic approaches to invasive social insects have three main thrusts: (1) identification of potential source populations; (2) determining the genetic variability of introduced populations; (3) determining colony genetic structure. There has been limited work on determining the provenance of introduced populations, all involving mtDNA markers. DNA sequence data have clearly established that populations of *R. santonensis* in France and South America are introduced populations of *R. flavipes* (Austin et al. 2005; Su et al. 2006), a native of eastern and central US, but the precise locations of populations within the US serving as the sources of introduced populations have not been established. Efforts to identify the source populations and routes of introduction of invasive populations of *C. formosanus* (Austin et al. 2006b) and *C. gestroi* (Jenkins et al. 2007) have suffered from low variation in the genes used, making conclusions tentative at best (Muirhead et al. 2008). More thorough studies using much larger sample sizes and highly variable markers, such as microsatellites, are needed to identify likely source populations and routes of introduction of invasive subterranean termites (Cornuet et al. 1999; Pritchard et al. 2000).

Introduction events can have substantial effects on genetic diversity of introduced populations. Typically, introduced populations experience a genetic bottleneck (Dlugosch and Parker 2008) but in some cases, repeated introductions from multiple source populations can result in greater genetic diversity (Kolbe et al. 2004; Lavergne and Molofsky 2007). Most introduced populations of social insects (other than ants) show reduced genetic variation (Holway et al. 2002), but genetically diverse introduced populations of the paper wasp, *Polistes dominulus*, in the US (Johnson and Starks 2004) constitute a conspicuous exception. Studies of introduced populations of invasive termites show strong reductions in genetic diversity at nuclear microsatellite loci; *R. flavipes* populations in France possess 42% fewer alleles than native populations in the US (Vargo 2003b; Dronnet et al. 2005; Vargo et al. 2006b) and several *C. formosanus* populations have on average 26% fewer alleles than native populations in China (Vargo et al. 2003, 2006a; Husseneder et al. 2005, 2008).

The consequences of reduced genetic variation in introduced termite populations is not known but in several ants lower genetic diversity in introduced populations is often associated with reduced levels of intercolonial aggression, leading to large unicolonial populations that become ecologically dominant within the introduced ranges (Tsutsui et al. 2003). There is some evidence that introduced colonies of *R. flavipes* show signs of unicoloniality. Populations of this species in France form large, highly polygamous extended family and mixed family colonies (Dronnet et al. 2005; Perdereau, Bagnères, Dupont and Dedeine, unpublished data), whereas those in the native range tend to be primarily simple families with localized foraging areas. Thus, there are some intriguing parallels between these introduced populations of *R. flavipes* and those of some invasive ants. It is of interest to know whether other introduced populations of *R. flavipes*, such as those in Chile (Su et al. 2006), also form expansive, highly polygamous colonies.

In contrast to the situation in *R. flavipes*, invasive populations of *C. formosanus* do not form large, unicolonial societies. Studies across six introduced populations all show that colonies are genetically distinct family groups, either simple families or secondarily polygamous (extended) families derived from simple families (Vargo et al. 2003, 2006a; Husseneder et al. 2005, 2007, 2008). In contrast, the dozen or so colonies characterized to date from the native range were all polygamous (Husseneder et al. 2008). Thus, it appears that colonies in introduced populations of *C. formosanus* do not exhibit higher levels of polygamy than those in native populations. Interestingly, this may be different for other *Coptotermes* species. Lenz and Barrett (1982) reported that colonies of three species of mound building *Coptotermes*, *C. lacteus*, *C. acinaciformis* and *C. frenchi*, contain many neotenics in New Zealand, where they have been introduced, in contrast to forming almost exclusively simple families in their native Australia.

The study of invasion biology of termites is in its infancy and much more research is needed to understand the mechanisms responsible for the invasion success of certain species. Approaches that combine behavioral and ecological studies with genetic analyses of native and introduced populations should be especially fruitful.

References

Aldrich BT, Kambhampati S (2004) Microsatellite markers for two species of dampwood termites in the genus *Zootermopsis* (Isoptera: Termopsidae). Mol Ecol Notes 4:719–721

Aldrich BT, Kambhampati S (2007) Population structure and colony composition of two *Zootermopsis nevadensis* subspecies. Heredity 99:443–451

Aluko GA, Husseneder C (2007) Colony dynamics of the Formosan subterranean termite in a frequently disturbed urban landscape. J Econ Entomol 100:1037–1046

Atkinson L, Adams ES (1997a) Double-strand conformation polymorphism (DSCP) analysis of the mitochondrial control region generates highly variable markers for population studies in a social insect. Insect Mol Biol 6:369–376

Atkinson L, Adams ES (1997b) The origins and relatedness of multiple reproductives in colonies of the termite *Nasutitermes corniger*. Proc R Soc Lond B 264:1131–1136

Atkinson L, Adams ES, Crozier RH (2007) Microsatellite markers for the polygamous termite *Nasutitermes corniger* (Isoptera: Termitidae). Mol Ecol Notes 7:299–301

Atkinson L, Teschendorf G, Adams ES (2008) Lack of evidence for nepotism by workers tending queens of the polygynous termite *Nasutitermes corniger*. Behav Ecol Sociobiol 62:805–812

Austin JW, Szalanski AL, Ghayourfar R et al (2006a) Phylogeny and genetic variation of *Reticulitermes* (Isoptera: Rhinotermitidae) from the Eastern Mediterranean and Middle East. Sociobiology 47:1–19

Austin JW, Szalanski AL, Gold RE, Foster BT (2004a) Genetic variation and geographic distribution of the subterranean termite genus *Reticulitermes* in Texas. Southwest Entomol 29:1–11

Austin JW, Szalanski AL, Kard BM (2004b) Distribution and genetic variation of *Reticulitermes* (Isoptera: Rhinotermitidae) in Oklahoma. Fla Entomol 87:152–158

Austin JW, Szalanski AL, Scheffrahn RH et al (2005) Genetic evidence for the synonymy of two *Reticulitermes* species: *Reticulitermes flavipes* and *Reticulitermes santonensis*. Ann Entomol Soc Am 98:395–401

Austin JW, Szalanski AL, Scheffrahn RH et al (2006b) Genetic evidence for two introductions of the Formosan subterranean termite, *Coptotermes formosanus* (Isoptera: Rhinotermitidae), to the United States. Fla Entomol 89:183–193

Avise JC (2000) Phylogeography: the history and formation of species. Harvard University Press, Cambridge, MA

Avise JC (2004) Molecular markers, natural history, and evolution, 2nd edn. Sinauer Associates Inc, Sunderland, MA

Behura SK (2006) Molecular marker systems in insects: current trends and future avenues. Mol Ecol 15:3087–3115

Booth W, Lewis VR, Taylor RL et al (2008) Identification and characterization of 15 polymorphic microsatellite loci in the western dry-wood termite, *Incisitermes minor* (Hagen). Mol Ecol Resour 8:1102–1104

Brandl R, Hacker M, Bagine RKN, Kaib M (2001) Geographic variation of polygyny in the termite *Macrotermes michaelseni* (Sjöstedt). Insectes Soc 48:134–137

Brandl R, Hacker M, Bagine RKN, Kaib M (2004) Yearly variation in polygyny in the termite *Macrotermes michaelseni* (Sjöstedt). Insectes Soc 51:294–298

Brandl R, Hacker M, Epplen J, Kaib M (2005) High gene flow between populations of *Macrotermes michaelseni* (Isoptera, Termitidae). Insectes Soc 52:344–349

Broughton RE (1995) Mitochondrial DNA variation within and among species of termites in the genus *Zootermopsis* (Isoptera: Termopsidae). Ann Entomol Soc Am 88:120–128

Bulmer MS, Adams ES, Traniello JFA (2001) Variation in colony structure in the subterranean termite *Reticulitermes flavipes*. Behav Ecol Sociobiol 49:236–243

Calleri DV, Reid EM, Rosengaus RB et al (2007) Inbreeding and disease resistance in a social insect: effects of heterozygosity on immunocompetence in the termite *Zootermopsis angusticollis*. Proc R Soc Lond B 273:2633–2640

Calleri DV, Rosengaus RB, Traniello JFA (2005) Disease and colony foundation in the dampwood termite *Zootermopsis angusticollis*: The survival advantage of nestmate pairs. Naturwissenschaften 92:300–304

Clément J-L (1981) Enzymatic polymorphism in the European populations of various *Reticulitermes* species (Isoptera). In: Howse PE, Clément J-L (eds) Biosystematics of social insects. Academic Press, London and New York, NY, pp 49–61

Clément J-L (1984) Diagnostic alleles and systematics in termite species of the genus *Reticulitermes* in Europe. Experientia 40:283–285

Clément J-L, Bagnères A-G, Uva P et al (2001) Biosystematics of *Reticulitermes* termites in Europe: morphological, chemical and molecular data. Insectes Soc 48:202–215

Copren KA (2004) The genetic and social structure of the western subterranean termite, *Reticulitermes hesperus* (Isoptera: Rhinotermitidae). Ph.D. dissertation, University of California, Davis, CA

Copren KA (2007) Characterization of microsatellite loci in the western subterranean termite *Reticulitermes hesperus* and cross-amplification in closely related cryptic species. J Insect Sci 7:17

Cornuet J-M, Piry S, Luikart G et al (1999) New methods employing multilocus genotypes to select or exclude populations as origins of individuals. Genetics 153:1989–2000

Darlington JPEC (1985) Multiple primary reproductives in the termite *Macrotermes michaleseni* (Sjöstedt). In: Watson JAL, Okot-Kotber BM, Noirot C (eds) Caste differentiation in social insects. Pergamon Press, Oxford, pp 187–200

Davis RB, Baldauf SL, Mayhew PJ (2009) Eusociality and the success of the termites: insights from a supertree of dictyopteran families. J Evol Biol 22:1750–1761

DeHeer CJ, Kamble ST (2008) Colony genetic organization, fusion and Inbreeding in *Reticulitermes flavipes* from the Midwestern U.S. Sociobiology 51:307–325

DeHeer CJ, Kutnik M, Vargo EL, Bagnères A-G (2005) The breeding system and population structure of the termite *Reticulitermes grassei* in Southern France. Heredity 95: 408–415

DeHeer CJ, Vargo EL (2004) Colony genetic organization and colony fusion in the termite *Reticulitermes flavipes* as revealed by foraging patterns over time and space. Mol Ecol 13:431–441

DeHeer CJ, Vargo EL (2006) An indirect test of inbreeding depression in the termites *Reticulitermes flavipes* and *Reticulitermes virginicus*. Behav Ecol Sociobiol 59:753–761

DeHeer CJ, Vargo EL (2008) Strong mitochondrial DNA similarity but low relatedness at microsatellite loci among families within fused colonies of the termite *Reticulitermes flavipes*. Insectes Soc 55:190–199

Dlugosch KM, Parker IM (2008) Founding events in species invasions: genetic variation, adaptive evolution, and the role of multiple introductions. Mol Ecol 17:431–449

Dronnet S, Bagnères A-G, Juba TR, Vargo EL (2004) Polymorphic microsatellite loci in the European subterranean termite, *Reticulitermes santonensis* Feytaud. Mol Ecol Notes 4:127–129

Dronnet S, Chapuisat M, Vargo EL et al (2005) Genetic analysis of the breeding system of an invasive subterranean termite, *Reticulitermes santonensis*, in urban and natural habitats. Mol Ecol 14:1311–1320

Dupont L, Roy V, Bakkali A, Harry M (2009) Genetic variability of the soil-feeding termite *Labiotermes labralis* (Termitidae, Nasutitermitinae) in the Amazonian primary forest and remnant patches. Insect Conserv Divers 2:53–61

Fang R, Huang L, Zhong JH (2008) Surprising low levels of genetic diversity of Formosan subterranean termites in south China as revealed by the COII gene (Isoptera: Rhinotermitidae). Sociobiology 51:1–20

Fei HX, Henderson G (2003) Comparative study of incipient colony development in the Formosan subterranean termite, *Coptotermes formosanus* Shiraki (Isoptera, Rhinotermitidae). Insectes Soc 50:226–233

Fisher ML, Gold RE, Vargo EL, Cognato AI (2004) Behavioral and genetic analysis of colony fusion in *Reticulitermes flavipes* (Isoptera: Rhinotermitidae). Sociobiology 44:565–576

Fuchs A, Heinze J, Reber-Funk C, Korb J (2003) Isolation and characterization of six microsatellite loci in the drywood termite *Cryptotermes secundus* (Kalotermitidae). Mol Ecol Notes 3: 355–357

Global Invasive Species Database (2010) *Coptotermes formosanus* (insect). Available from: http://www.issg.org/database/species/ecology.asp?si=61&fr=1&sts=&lang=EN [Accessed 1 Jan 2010]

Goodisman MAD, Crozier RH (2002) Population and colony genetic structure of the primitive termite *Mastotermes darwiniensis*. Evolution 56:70–83

Goodisman MAD, Evans TA, Ewen JG, Crozier RH (2001) Microsatellite markers in the primitive termite *Mastotermes darwiniensis*. Mol Ecol Notes 1:250–251

Hacker M, Kaib M, Bagine RKM et al (2005) Unrelated queens coexist in colonies of the termite *Macrotermes michaelseni*. Mol Ecol 14:1527–1532

Hamilton WD (1972) Altruism and related phenomena, mainly in social insects. Annu Rev Ecol Syst 3:193–232

Hamilton WD (1978) Evolution and diversity under bark. In: Mound LA, Waloff N (eds) Diversity of insect faunas. Blackwell Scientific Publications, London, pp 154–175

Harry M, Roose CL, Garnier-Sillam E, Solignac M (2001) Microsatellite markers in soil-feeding termites (*Cubitermes subarquatus*, Isoptera, Termitidae, Termitinae). Mol Ecol Notes 1: 226–228

Harry M, Roy V, Mercier A et al (2007) Isolation and characterization of microsatellite markers in *Labiotermes labralis* (Isoptera, Termitidae, Nasutitermitinae). Mol Ecol Notes 7:121–123

Hayashi Y, Kitade O, Gonda M et al (2005) Diverse colony genetic structures in the Japanese subterranean termite *Reticulitermes speratus* (Isoptera: Rhinotermitidae). Sociobiology 46: 175–184

Hayashi Y, Kitade O, Kojima J-I (2002) Microsatellite loci in the Japanese subterranean termite, *Reticulitermes speratus*. Mol Ecol Notes 2:518–520

Holway DA, Lach L, Suarez AV et al (2002) The causes and consequences of ant invasions. Annu Rev Ecol Syst 33:181–233

Husseneder C, Brandl R, Epplen C et al (1998) Variation between and within colonies in the termite: morphology, genomic DNA, and behaviour. Mol Ecol 7:983–990

Husseneder C, Brandl R, Epplen C et al (1999) Within-colony relatedness in a termite species: genetic roads to eusociality? Behaviour 136:1045–1063

Husseneder C, Grace JK (2001a) Evaluation of DNA fingerprinting, aggression tests and morphometry as tools for colony delineation of the Formosan subterranean termite. J Insect Behav 14:173–186

Husseneder C, Grace JK (2001b) Similarity is relative: hierarchy of genetic similarities in the Formosan subterranean termite (Isoptera: Rhinotermitidae) in Hawaii. Environ Entomol 30:262–266

Husseneder C, Grace JK, Messenger MT et al (2003a) Describing the spatial and social organization of Formosan subterranean termite colonies in Armstrong Park, New Orleans. Sociobiology 41:61–65

Husseneder C, Messenger MT, Su N-Y et al (2005) Colony social organization and population genetic structure of an introduced population of Formosan subterranean termite from New Orleans, Louisiana. J Econ Entomol 98:1421–1434

Husseneder C, Powell JE, Grace JK et al (2008) Worker size in the Formosan subterranean termite in relation to colony breeding structure as inferred from molecular markers. Environ Entomol 37:400–408

Husseneder C, Simms DM (2008) Size and heterozygosity influence partner selection in the Formosan subterranean termite. Behav Ecol 19:764–773

Husseneder C, Simms DM, Riegel C (2007) Evaluation of treatment success and patterns of reinfestation of the Formosan subterranean termite (Isoptera: Rhinotermitidae). J Econ Entomol 100:1370–1380

Husseneder C, Simms DM, Ring DR (2006) Genetic diversity and genotypic differentiation between the sexes in swarm aggregations decrease inbreeding in the Formosan subterranean termite. Insectes Soc 53:212–219

Husseneder C, Vargo EL, Grace JK (2003b) Molecular genetic methods: new approaches to termite biology. In: Goodell B, Schultz TP, Nicholas DD (eds) Wood deterioration and preservation: advances in our changing world. American Chemical Society Symposium Series 845. American Chemical Society, Washington, DC, pp 358–370

Indrayani Y, Matsumura KEI, Yoshimura T et al (2006) Development of microsatellite markers for the drywood termite *Incisitermes minor* (Hagen). Mol Ecol Notes 6:1249–1251

Jeffreys AJ, Wilson V, Thein SL (1985) Inidividual-specific "fingerprints" of human DNA. Nature 316:76–79

Jenkins TM, Basten CJ, Kresovich S, Forschler BT (1999) Mitochondrial gene sequence questions *Reticulitermes* sp. social structure (Isoptera: Rhinotermitidae). Sociobiology 34: 161–172

Jenkins TM, Dean RE, Forschler BT (2002) DNA technology, interstate commerce, and the likely origin of Formosan subterranean termite (Isoptera: Rhinotermitidae) infestation in Atlanta, Georgia. J Econ Entomol 95:381–389

Jenkins TM, Dean RE, Verkerk R, Forschler BT (2001) Phylogenetic analyses of two mitochondrial genes and one nuclear intron region illuminate European subterranean termite (Isoptera: Rhinotermitidae) gene flow, taxonomy, and introduction dynamics. Mol Phylogenet Evol 20:286–293

Jenkins TM, Jones SC, Lee C-Y et al (2007) Phylogeography illuminates maternal origins of exotic *Coptotermes gestroi* (Isoptera: Rhinotermitidae). Mol Phylogenet Evol 42:612–621

Johnson RN, Starks PT (2004) A surprising level of genetic diversity in an invasive wasp: *Polistes dominulus* in the northeastern United States. Ann Entomol Soc Am 97:732–737

Jones SC, La Fage JP, Howard RW (1988) Isopteran sex ratios: phylogenetic trends. Sociobiology 14:89–118

Kaib M, Hacker M, Brandl R (2001) Egg-laying in monogynous and polygynous colonies of the termite *Macrotermes michaelseni* (Isoptera, Macrotermitidae). Insectes Soc 48:231–237

Kaib M, Hacker M, Over I et al (2000) Microsatellite loci in *Macrotermes michaelseni* (Isoptera: Termitidae). Mol Ecol 9:502–504

Kaib M, Husseneder C, Epplen C et al (1996) Kin-biased foraging in a termite. Proc R Soc Lond B 263:1527–1532

Keller L (1997) Indiscriminate altruism: unduly nice parents and siblings. Trends Ecol Evol 12: 99–103

Kitade O, Hayashi Y, Kikuchi Y, Kawarasaki S (2004) Distribution and composition of colony founding associations of a subterranean termite, *Reticulitermes kanmonensis*. Entomol Sci 7:1–8

Kolbe JJ, Glor RE, Schettino LR et al (2004) Genetic variation increases during biological invasion by a Cuban lizard. Nature 431:177–181

Kutnik M, Uva P, Brinkworth L, Bagnères A-G (2004) Phylogeography of two European *Reticulitermes* (Isoptera) species: the Iberian refugium. Mol Ecol 13:3099–3113

Lavergne S, Molofsky J (2007) Increased genetic variation and evolutionary potential drive the success of an invasive species. Proc Natl Acad Sci USA 104:3883–3888

Lefebvre T, Châline N, Limousin D et al (2008) From speciation to introgressive hybridization: the phylogeographic structure of an island subspecies of termite, *Reticulitermes lucifugus corsicus*. BMC Evol Biol 8:38

Leniaud L, Dedeine F, Pichon A et al (2010) Geographical distribution, genetic diversity and social organization of a new European termite, *Reticulitermes urbis* (Isoptera: Rhinotermitidae). Biol Invasions 12:1389–1402

Lenz M, Barrett RA (1982) Neotenic formation in field colonies of *Coptotermes lacteus* (Froggatt) in Australia, with comments on the roles of neotenics in the genus *Coptotermes* (Isoptera, Rhinotermitidae). Sociobiology 7:47–59

Lenz M, Runko S (1993) Long-term impact of orphaning on field colonies of *Coptotermes lacteus* (Froggatt) (Isoptera: Rhinotermitidae). Insectes Soc 40:439–456

Lepage M, Darlington JPEC (2000) Population dynamics of termites. In: Abe T, Bignell DE, Higashi M (eds) Termites: evolution, sociality, symbioses, ecology. Kluwer Academic Publishers, Dordrecht, The Netherlands, pp 333–361

Loxdale HD, Lushai G (1998) Molecular markers in entomology. Bull Entomol Res 88: 577–600

Luchetti A, Marini M, Mantovani B (2005) Mitochondrial evolutionary rate and speciation in termites: data on European *Reticulitermes* taxa (Isoptera, Rhinotermitidae). Insectes Soc 52:218–221

Luchetti A, Marini M, Mantovani B (2007) Filling the European gap: biosystematics of the eusocial system *Reticulitermes* (Isoptera, Rhinotermitidae) in the Balkanic Peninsula and Aegean area. Mol Phylogenet Evol 45:377–383

Luchetti A, Trenta M, Mantovani B, Marini M (2004) Taxonomy and phylogeny of north Mediterranean *Reticulitermes* termites (Isoptera, Rhinotermitidae): a new insight. Insectes Soc 51:117–122

Luykx P (1993) Allozyme markers and formal Mendelian genetics of the termite *Incisitermes schwarzi* (Isoptera: Kalotermitidae). Sociobiology 21:185–192

Marini M, Mantovani B (2002) Molecular relationships among European samples of *Reticulitermes* (Isoptera, Rhinotermitidae). Mol Phylogenet Evol 22:454–459

Matsuura K, Fujimoto M, Goka K (2004) Sexual and asexual colony foundation and the mechanism of facultative parthenogenesis in the termite *Reticulitermes speratus* (Isoptera, Rhinotermitidae). Insectes Soc 51:325–332

Matsuura K, Nishida T (2001) Colony fusion in a termite: what makes the society "open"? Insectes Soc 48:378–383

Matsuura K, Vargo EL, Kawatsu K et al (2009) Queen succession through asexual reproduction in termites. Science 323:1687

Messenger MT, Mullins AJ (2005) New flight distance recorded for *Coptotermes formosanus* (Isoptera: Rhinotermitidae). Fla Entomol 88:99–100

Muirhead JR, Gray DK, Kelly DW et al (2008) Identifying the source of species invasions: sampling intensity vs. genetic diversity. Mol Ecol 17:1020–1035

Myles TG (1999) Review of secondary reproduction in termites (Insecta: Isoptera) with comments on its role in termite ecology and social evolution. Sociobiology 33:1–87
Nobre T, Nunes L, Bignell D (2008) Colony interactions in *Reticulitermes grassei* population assessed by molecular genetic methods. Insectes Soc 55:66–73
Nutting WL (1969) Flight and colony foundation. In: Krishna K, Weesner FM (eds) Biology of termites, vol 1. Academic Press, New York, NY, pp 233–282
Pamilo P, Gertsch P, Thorén P, Seppä P (1997) Molecular population genetics of social insects. Annu Rev Ecol Syst 28:1–25
Park YC, Kitade O, Schwarz M et al (2006) Intraspecific molecular phylogeny, genetic variation and phylogeography of *Reticulitermes speratus* (Isoptera: Rhinotermitidae). Mol Cells 21: 89–103
Parman V, Vargo EL (2008) Population density, species abundance, and breeding structure of subterranean termite colonies in and around infested houses in central North Carolina. J Econ Entomol 101:1349–1359
Pritchard JK, Stephens M, Donnelly P (2000) Inference of population structure using multilocus genotype data. Genetics 155:945–959
Reilly LM (1987) Measurements of inbreeding and average relatedness in a termite population. Am Nat 130:339–349
Roisin Y (1993) Selective pressures on pleometrosis and secondary polygyny: a comparison of termites and ants. In: Keller L (ed) Queen number and sociality in insects. Oxford Science Publications, Oxford, pp 402–422
Roisin Y, Lenz M (2002) Origin of male-biased sex allocation in orphaned colonies of the termite, *Coptotermes lacteus*. Behav Ecol Sociobiol 51:472–479
Roisin Y, Pasteels JM (1985) Imaginal polymorphism and polygyny in the Neo-Guinean termite *Nasutitermes princeps* (Desneux). Insectes Soc 32:140–157
Roisin Y, Pasteels JM (1986) Reproductive mechanisms in termites: polycalism and polygyny in *Nasutitermes polygynus* and *Nasutitermes costalis*. Insectes Soc 33:149–167
Rosengaus RB, Traniello JFA (1993) Disease risk as a cost of outbreeding in the termite *Zootermopsis angusticollis*. Proc Natl Acad Sci USA 90:6641–6645
Ross KG (2001) Molecular ecology of social behaviour: analyses of breeding systems and genetic structure. Mol Ecol 10:265–284
Ross KG, Keller L (1995) Ecology and evolution of social organization: insights from fire ants and other highly eusocial insects. Annu Rev Ecol Syst 26:631–656
Schmidt AM, Trindl A, Korb J (2007) Isolation and characterization of 10 microsatellite loci in the magnetic termite, *Amitermes meridionalis* (Isoptera: Termitidae). Mol Ecol Notes 7:1045–1047
Shellman-Reeve JS (1996) Operational sex ratios and lipid reserves in the dampwood termite *Zootermopsis nevadensis* (Hagen) (Isoptera: Termopsidae). J Kans Entomol Soc 69: 139–146
Shellman-Reeve JS (1997) The spectrum of eusociality in termites. In: Choe JC, Crespi BJ (eds) The evolution of social behavior in insects and arachnids. Cambridge University Press, Cambridge, pp 52–93
Shellman-Reeve JS (2001) Genetic relatedness and partner preference in a monogamous wood-dwelling termite. Anim Behav 61:869–876
Simms D, Husseneder C (2009) Assigning individual alates of the Formosan subterranean termite to their colonies of origin within the context of an area-wide management program. Sociobiology 53:631–650
Strong KL, Grace JK (1993) Low allozyme variation in Formosan subterranean termite (Isoptera: Rhinotermitidae) colonies in Hawaii. Pan Pac Entomol 69:51–56
Su N-Y (2003) Overview of the global distribution and control of the Formosan subterranean termite. Sociobiology 41:7–16
Su N-Y, Ye W, Ripa R et al (2006) Identification of Chilean *Reticulitermes* (Isoptera: Rhinotermitidae) inferred from three mitochondrial gene DNA sequences and soldier morphology. Ann Entomol Soc Am 99:352–363

Sunnucks P (2000) Efficient genetic markers for population biology. Trends Ecol Evol 15: 199–203
Syren RM, Luykx P (1977) Permanent segmental interchange complex in the termite *Incisitermes schwarzi*. Nature 266:167–168
Szalanski AL, Scheffrahn RH, Austin JW et al (2004) Molecular phylogeny and biogeography of *Heterotermes* (Isoptera: Rhinotermitidae) in the West Indies. Ann Entomol Soc Am 97: 556–566
Thompson GJ, Hebert PDN (1998a) Population genetic structure of the neotropical termite *Nasutitermes nigriceps* (Isoptera: Termitidae). Heredity 80:48–55
Thompson GJ, Hebert PDN (1998b) Probing termite social systems through allozyme and mtDNA analysis: a case study of *Nasutitermes nigriceps* and *Nasutitermes costalis* (Isoptera, Termitidae). Insectes Soc 45:289–299
Thompson GJ, Lenz M, Crozier RH (2000) Microsatellites in the subterranean, mound-building termite *Coptotermes lacteus* (Isoptera: Rhinotermitidae). Mol Ecol 9:1932–1934
Thompson GJ, Lenz M, Crozier RH, Crespi BJ (2007) Molecular-genetic analyses of dispersal and breeding behaviour in the Australian termite *Coptotermes lacteus:* evidence for non-random mating in a swarm-dispersal mating system. Aust J Zool 55:219–227
Thorne BL (1982) Polygyny in termites: multiple primary queens in colonies of *Nasutitermes corniger* (Motschuls) (Isoptera: Termitidae). Insectes Soc 29:102–117
Thorne BL (1984) Polygyny in the Neotropical termite *Nasutitermes corniger*: life history consequences of queen mutualism. Behav Ecol Sociobiol 14:117–136
Thorne BL (1985) Termite polygyny: the ecological dynamics of queen mutualism. In: Hölldobler B, Lindauer M (eds) Experimental behavioral ecology and sociobiology. G. Fischer, Stuttgart, New York, NY, pp 325–342
Thorne BL, Breisch NL, Muscedere ML (2003) Evolution of eusociality and the soldier caste in termites: influence of intraspecific competition and accelerated inheritance. Proc Natl Acad Sci USA 100:12808–12813
Thorne BL, Traniello JFA, Adams ES, Bulmer M (1999) Reproductive dynamics and colony structure of subterranean termites of the genus *Reticulitermes* (Isoptera Rhinotermitidae): a review of the evidence from behavioral, ecological and genetic studies. Ethol Ecol Evol 11:149–169
Tsutsui ND, Suarez AV, Grosberg RK (2003) Genetic diversity, asymmetrical aggression, and recognition in a widespread invasive species. Proc Natl Acad Sci USA 100:1078–1083
Uva P, Clément J-L, Austin JW et al (2004b) Origin of a new *Reticulitermes* termite (Isoptera, Rhinotermitidae) inferred from mitochondrial and nuclear DNA data. Mol Phylogenet Evol 30:344–353
Uva P, Clément J-L, Bagnères A-G (2004a) Colonial and geographic variations in agonistic behaviour, cuticular hydrocarbons and mtDNA of Italian populations of *Reticulitermes lucifugus* (Isoptera, Rhinotermitidae). Insectes Soc 51:163–170
Vargo EL (2000) Polymorphism at trinucleotide microsatellite loci in the subterranean termite *Reticulitermes flavipes*. Mol Ecol 9:817–820
Vargo EL (2003a) Genetic structure of *Reticulitermes flavipes* and *R. virginicus* (Isoptera: Rhinotermitidae) colonies in an urban habitat and tracking of colonies following treatment with hexaflumuron bait. Environ Entomol 32:1271–1282
Vargo EL (2003b) Hierarchical analysis of colony and population genetic structure in the eastern subterranean termite, *Reticulitermes flavipes*, using two classes of molecular markers. Evolution 57:2805–2818
Vargo EL, Carlson JC (2006) Comparative study of breeding systems of sympatric subterranean termites (*Reticulitermes flavipes* and *R. hageni*) in central North Carolina using two classes of molecular genetic markers. Environ Entomol 35:173–187
Vargo EL, Henderson G (2000) Identification of polymorphic microsatellite loci in the Formosan subterranean termite *Coptotermes formosanus* Shiraki. Mol Ecol 9:1935–1938
Vargo EL, Husseneder C (2009) Biology of subterranean termites: insights from molecular studies of *Reticulitermes* and *Coptotermes*. Annu Rev Entomol 54:379–403

Vargo EL, Husseneder C, Grace JK (2003) Colony and population genetic structure of the Formosan subterranean termite, *Coptotermes formosanus*, in Japan. Mol Ecol 12:2599–2608

Vargo EL, Husseneder C, Woodson D et al (2006a) Genetic analysis of colony and population genetic structure of three introduced populations of the Formosan subterranean termite (Isoptera: Rhinotermitidae) in the Continental United States. Environ Entomol 35: 151–166

Vargo EL, Juba TR, DeHeer CJ (2006b) Relative abundance and comparative breeding structure of subterranean termite colonies (*Reticulitermes flavipes*, *R. hageni*, *R. virginicus*, and *Coptotermes formosanus*) in a South Carolina lowcountry site as revealed by molecular markers. Ann Entomol Soc Am 99:1101–1109

Wang JS, Grace JK (2000) Genetic relationship of *Coptotermes formosanus* (Isoptera: Rhinotermitidae) populations from the United States and China. Sociobiology 36:7–19

Zhang DX, Hewitt GM (2003) Nuclear DNA analyses in genetic studies of populations: practice, problems and prospects. Mol Ecol 12:563–584

Chapter 13
Termite Mound Architecture, from Function to Construction

Judith Korb

> *Natural selection should operate on builders through the success of the structures that they build and also the success of the building process compared with rival methods of construction*
>
> (Hansell 2005)

Abstract Termite mounds function as nests for their inhabitants, which are colonies of small potentially vulnerable insects that are also susceptible to environmental fluctuations. Thus, the mound protects against enemies and hostile environmental condition. Mounds vary in shape, between and even within species. Yet, different species might also build similar mounds. Using current knowledge on fungus-growing termites and magnetic termites, I will show evidence that mound architecture is an adaptation to local environmental conditions. There is no single explanation for mound shape. The significance of different factors (e.g. thermoregulation, gas exchange) varies between species and between environments, and different mechanisms exist to achieve homeostatic nest conditions. In the second part of this review, I summarize what is known about the mechanisms involved in building such complex mounds. The construction of mounds that are adapted to local conditions does not imply a purposeful design. Rather, our current understanding suggests that it is the result of self-organisation, in interaction with stigmergy and templates. There remain many open questions in understanding termite mound architecture. To resolve them requires detailed studies and a closer cooperation among disciplines.

13.1 Introduction

Termite mounds have inspired the imagination of humans since the earliest journeys to the tropics (Smeathman 1781). They are conspicuous structural features, especially in tropical savannas and arid areas of South America, Africa and

J. Korb (✉)
Behavioural Biology, University of Osnabrueck, D-49076 Osnabrueck, Germany
e-mail: judith.korb@biologie.uni-osnabrueck.de

D.E. Bignell et al. (eds.), *Biology of Termites: A Modern Synthesis*,
DOI 10.1007/978-90-481-3977-4_13,

Australasia, where they can reach densities of more than 200 mounds per hectare (e.g. *Amitermes vitiosus* in Australia; Lepage and Darlington 2000). These buildings are constructed by thousands of termites, each less than one centimetre in length. They function as fortresses where a colony's nest (Box 13.1) is largely protected against enemies and hostile environmental conditions (Noirot and Darlington 2000). These defensible and controllable microcosms also allow the storage of food reserves, a selective advantage especially in areas with seasonal food shortage, for instance due to fire, flooding or ambient conditions restricting foraging activity.

Box 13.1 Glossary – General Terms

Colony: A social insect community. In termites, it generally is a family where the reproductives (generally one queen and one king) are the parents of all other castes (in termites: generally workers and soldiers).

Fungus chamber: The structure and location inside the nest where the fungus combs are located (see Fig. 13.2a, c).

Fungus comb: The fungus comb is an artificial substrate built by the termites using their partially digested faeces in which the fungus is cultivated (see Fig. 13.2a, c, d).

Hive: The proper nest location of a colony where the termites (and sometimes the fungus) are located (see Fig. 13.2).

Nest: The central location of a colony where all colony members come together.

Royal cell: Also often called royal chamber or queen cell/chamber. A special structure inside the hive where the queen and the king are located (see Figs. 13.2a, b; 13.5a).

Template: When using a template, the shape to be build already "exists" under/in the form of a pre-pattern in the environment. This pattern can result from natural gradients, fields or heterogeneities that are exploited (Bonabeau et al. 1997).

Self-organisation: Theories of self-organisation were originally developed in the context of physics and chemistry to describe the emergence of macroscopic patterns out of processes and interactions defined at the microscopic level (Bonabeau et al. 1997). In social organisms: the emergence of macroscopic patterns out of collective actions of relatively simple individuals.

Stigmergy: This concept was introduced by Grassé to explain task coordination and the regulation of building activity in Macrotermitinae. The template guiding activities progressively emerge through the sequence of numerous activities performed in space and time by workers.

Depending on the species, the environment and its function, termite mounds can range in form from simple earthen hillocks, pinnacles, or cones to complex mounds with many ridges that can reach a height of several metres (Grassé 1984; Korb 1997; Noirot and Darlington 2000) (Fig. 13.1). *Cubitermes* spp. mounds are

Fig. 13.1 Diversity in termite mound architecture. (**a**) *Cubitermes* in Côte d'Ivoire, (**b**) *Nasutitermes triodiae* in northern Australia, (**c**) dome-shaped *Macrotermes bellicosus* mound in the forest of northern Côte d'Ivoire, (**d**) cathedral-shaped *Macrotermes bellicosus* mound with ridges in the savanna of northern Côte d'Ivoire, (**e**) *Drepanotermes rubiceps* in northwestern Australia, and (**f**) *Trinervitermes oeconomus* in northern Benin

well-known for their mushroom-capped mounds (Fig. 13.1a). On West-African "bowal" plains, grassed areas on outcropping lateritic crust that rapidly dry out and are cleared by fire, *Cubitermes fungifaber* reaches impressive densities (Lepage 1984). By contrast, mounds of the grass-harvesting Australian cathedral termite *Nasutitermes triodiae* (Fig. 13.1b) are unmatched in their enormous size, up to eight metres, that is scarcely achieved by other termite species. There is also large intraspecific variation in mound structure. In the fungus-growing termite *Macrotermes bellicosus* mounds can be dome-shaped with thick walls and low structural diversity (Fig. 13.1c) to highly complex with many ridges or turrets (cathedral mound) (Fig. 13.1d). This structural diversity occurs between different regions, but also on a very small scale between habitats within a single area (Korb and Linsenmair 1999a). In contrast to this inter- and intraspecific variability, different termite species can also construct mounds with striking resemblances, apparently often under similar environmental conditions. The mounds of the Australian termite *Drepanotermes rubriceps* (Fig. 13.1e) closely resemble *Trinervitermes* mounds in Africa (Fig. 13.1f). Both species are grass-harvesters that occur in open grass savannas. Similarly, in open bush/tree savannas the cathedral mounds of the African fungus grower *Macrotermes bellicosus* (Fig 13.1d) resemble Australian *Nasutitermes triodiae* mounds (Fig. 13.1b), when judged from outer appearance only. This convergence in mound architecture implies that there is some adaptation to environmental conditions.

Explaining variability in mound architecture, its functional significance and the nuances of construction was and still is a fascinating topic for scientists and laymen alike. The aim of this review is twofold. First, I will summarize what is known about the function and functional significance of variation in mound architecture. Former comprehensive reviews on the architecture are provided by (Grassé 1937a; Harris 1956; Noirot 1970; Grassé 1984; Darlington 1989; Noirot and Darlington 2000). I update the state of the art and concentrate on two groups of termites that have been high on the research agenda, fungus growing Macrotermitinae and the so-called magnetic termites. Where appropriate, I will refer to other species, but our knowledge there is still very fragmentary. Second, I will summarize current concepts of the mechanisms whereby such complex structures can be built by many small insects with little brains.

13.2 Function and Functional Significance of Termite Mound Architecture

13.2.1 Fungus Growing Termites

The fungus growing termites, Macrotermitinae (Termitidae), are agriculturalists that have evolved an obligate ectosymbiotic relationship with basiomycete fungi of the genus *Termitomyces* (Batra and Batra 1966; Johnson et al. 1981; Wood and Thomas 1989). Although this association probably evolved in African forests

(Aanen and Eggleton 2005), nowadays their centre of diversity and ecological importance are African and Asian savannas (Grassé 1945; Ruelle 1970). Inside their mounds the termites cultivate the fungus within their nest (Box 13.1) in convoluted, greyish-brown to yellow combs that consist of dead plant material (for details see Traniello and Leuthold 2000) (Fig. 13.2). The fungus comb is an artificial substrate built by the termites using their partially digested faeces (Box 13.1) (Fig. 13.2). The combs are housed either inside the hive (Box 13.1) (as in all very young colonies and in mature ones in a few species), or more commonly in spe-

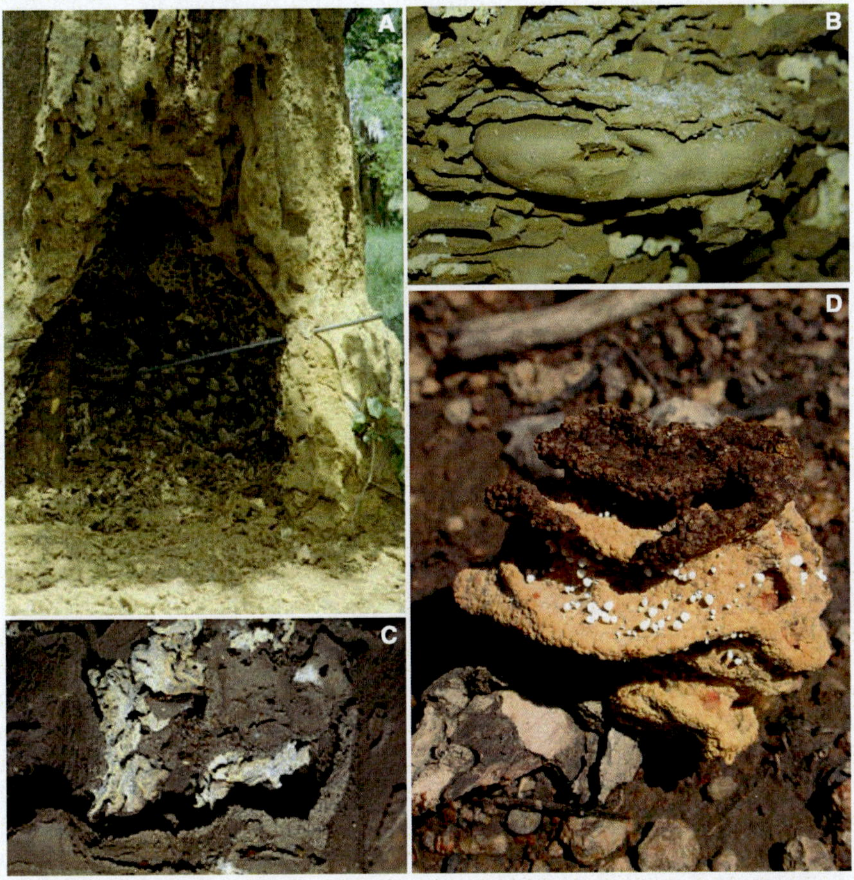

Fig. 13.2 Internal mound structures of fungus-growing termites (here: *Macrotermes bellicosus*). (**a**) An opened mound, view inside the hive: The *greyish brown* structures are fungus combs located in their fungus chambers. In the *middle* the central shaft rises to the *top* of the mound; at the *upper right*, a large air channel runs upward inside a ridge. A measuring device (*grey tube*) is installed inside the royal cell. (**b**) The royal cell (*enlarged*) that harbours the king and queen. (**c**) A fungus comb inside its fungus chamber. (**d**) A fungus comb (removed from its fungus chamber)

cially built chambers outside the hive. The termites provide a unique habitat for the fungus, then profit from improved nutrition. Wood and Thomas (1989), Veivers et al. (1991) and Darlington (1994) list the benefits of fungus cultivation. For the termites, these are (i) the fungi degrade chemically complex substances (e.g. cellulose, lignin) into substances that can be used by the termites, those substances include the tissues of the fungus itself, as mycelium and also as vegetative fruiting bodies, and (ii) the fungi increase the N:C ratio of the termites' diet, and allow them to exploit more diverse cellulose sources than most other termites are able to use. The fungi gain advantages in (i) access to plant material that can easily be penetrated and has an increased surface area, (ii) the provision of an optimal microclimate, and (iii) the selective inhibition of other fungi that are competitively superior and the prevention of microbial infection, probably by termite salivary secretions.

These examples illustrate that the association provides essential benefits for both partners, but it also comes with costs. For the termites, these include the huge amount of extra work required to build, maintain and house the fungus combs, and adjusting nest conditions to fungal requirements. The latter must always be kept in mind in discussing the functional significance of mound architecture of Macrotermitinae compared to other termite species (see below). Considerations of special relevance here are the provision of the fungus chambers (nest) with (i) constant high humidity, (ii) constant nest temperatures (laboratory experiments have shown that *Termitomyces* has a very narrow temperature tolerance; Lüscher 1961; Wood and Thomas 1989), and (iii) rather low CO_2 concentrations (the metabolism of the fungus is highly CO_2 dependent; Sands 1969). Fourthly, the high metabolic rate of the fungus demands efficient gas exchange mechanisms (Lüscher 1955).

13.2.1.1 *Macrotermes bellicosus*

One of the most intensively studied fungus-growing termites is *M. bellicosus* (Ruelle 1970, previously *Macrotermes natalensis*). Many papers, especially reviews (Heinrich 1993; Turner 2000b; Camazine et al. 2001), did not adopt the taxonomic revision, so that *M. bellicosus* is still commonly referred to as *M. natalensis*. Yet, these are two distinct species and studies done in West-Africa are therefore on *M. bellicosus*. *M. bellicosus* has a wide geographical distribution occurring in savannas ranging from West- to East-Africa, but is absent from rain forests (Ruelle 1970; Pomeroy 1977, 1978). In many areas it is the dominant decomposer with high colony densities and large colony sizes (up to a few million individuals) resulting in high biomasses (Collins 1977, 1981a, 1981b; Korb and Linsenmair 2001). In addition to a pair of reproductives, there are four castes with a strict division of labour (Leuthold et al. 1990; Lys and Leuthold 1991; Traniello and Leuthold 2000). The minor workers care for the brood and fungus and are the main builders in *M. bellicosus*. Major workers forage for dead plant material outside the nest. They are accompanied by minor soldiers that build a first line of defence, alarming colony members by banging their head on the ground (Röhrig et al. 1999). This "alarm call" recruits major soldiers that furiously defend their nestmates using their strong mandibles (Traniello and Leuthold 2000).

Pioneering work on the functioning of *M. bellicosus* mounds was done by Lüscher (1955, 1956, 1961) and Ruelle (1962, 1964). They were the first to monitor temperatures in- and outside a mound. They found that the nest temperatures are rather constant at around 30°C, fluctuating less than 3°C daily, with a mean annual fluctuation of less than 1°C, while outside temperatures fluctuate widely (Lüscher 1961; Ruelle 1962, 1964). These results quickly became textbook examples for thermoregulation in the animal kingdom (e.g. Seeley and Heinrich 1981; Heinrich 1993). Lüscher (1955) also was the first to recognize that the metabolism of the termites and the fungi requires efficient mechanisms to exchange respiratory gases. He calculated that a mound with about 2 million termites, amounting to 20 kg of insects and consuming about 500 mL oxygen per gram per hour, needs to exchange 1,200 L of air per day (Lüscher 1955, 1961). Although these calculations might be slightly exaggerated (Grassé 1984; Darlington et al. 1997), they illustrate a general problem for the termites. As *M. bellicosus* mounds are closed in these habitats (i.e. they lack open chimneys; Box 13.1), gases must be exchanged through the mound walls. Lüscher proposed that mound ridges (Box 13.2) serve as lungs (Lüscher 1956, 1961). According to his "thermosiphon hypothesis", air rises inside the central shaft (Box 13.2) from the nest to the top of the mound driven by buoyant forces imparted by the nest's metabolism (convective heat transfer) inside the mound (Fig. 13.3a). Then it descends inside the air channels (Box 13.2) of the ridges where respiratory gases are exchanged. Lüscher was cautious when proposing his hypothesis and clearly stated that more experiments were needed. Later this ventilation mode was questioned by Loos (1964) and indeed new data now demand new models of ventilation, which need to differ between different mound types and ambient conditions (Korb and Linsenmair 2000b). In cathedral mounds (Fig. 13.1d), like those studied by Lüscher, the main driving mechanisms seem to be temperature gradients within the mound and between the mound and its exterior environment, which lead to heat convection (convective and conductive heat transfer). In northern Côte d'Ivoire, this was demonstrated by temperature recordings combined with CO_2 and flow measurements (Korb and Linsenmair 2000b). During the day, air rises inside the air channels of the ridges and falls inside the central shaft (Fig. 13.3b). Thus the flow is in the opposite direction to that proposed by Lüscher (Fig. 13.3a). Correspondingly, CO_2 concentrations are highest at the base of the ridges and decline towards the top. The ventilation of the ridges has a striking daily pattern tracking the changing position of the sun. During the course of a day, the ridges with the most intense air flow shift from east via north/south to west as the sun heats the different sides of a mound in sequence. This pattern, together with the recorded temperature profiles, suggests that temperature gradients are the driving force underlying ventilation of cathedral mounds in the savanna. The conclusion was further confirmed by measurements during the night when temperature gradients break down or reverse (Fig. 13.3c). Then, the flow inside the air channels of the ridges stops and CO_2 accumulates (Korb and Linsenmair 2000b; Korb 2003b).

Box 13.2 Glossary – Structural Mound Elements

There is a very wide and often inconsistent usage of terms to describe different structures within a mound. I applied the following usage:

(Tall vertical structures that extend vertically from a mound)
Chimney: A central structure that opens at the top to the outside.

Spire: A central structure that it closed and does not open at the top to the outside.

Turrets: Several smaller spires.

Ridges: Peripheral structures extending vertically starting near the soil surface.

(Internal structures)
Passages: unspecified internal structures.

Central shaft: a rather large central space/passage that extends vertically from the nest to the top of the mound, generally running inside the chimney or a spire (Fig. 13.2a).

Air channels: smaller spaces that run vertically in the peripheral region of a mound, for example, inside ridges (Fig. 13.2a).

Lateral tunnels: passages that run more horizontally and connect the central shaft with the (peripheral) air channels (Fig. 13.2a).

The ventilation pattern in cathedral mounds contrasts sharply with that found in the dome-shaped type of *M. bellicosus* mounds (Fig. 13.1c) that occur in the same area (Korb and Linsenmair 1999a, 2000b). There, as molecular genetic analyses confirmed, a single population of *M. bellicosus* builds cathedral mounds in the savanna and dome-shaped mounds in the neighbouring gallery forest. In contrast to the cathedral savanna mounds, forest mounds have thick walls and lack complex ridges (Fig. 13.1c). The only structures with thin walls suitable for gas exchange are small turrets (Box 13.2) and, in tall mounds, a central spire (Box 13.2) (Fig. 13.1c). These structures, combined with the lower ambient temperatures and the shelter from wind provided by the forest, result in a simple rise of air, driven through buoyant forces, from the central nest to the top of the mound where respiratory gases seem to be exchanged through the relatively small surface of the turrets and the spire (Fig. 13.3d). This non-circular ventilation is less efficient than the ventilation in cathedral mounds (Korb and Linsenmair 2000b): While cathedral mounds have CO_2-concentrations corresponding to the ambient air, increasing only slightly during the night, those in the dome-shaped mounds are permanently higher. These elevated CO_2 concentrations have negative consequences for the cultivated fungi (and thus the termites) which depend on low CO_2 concentrations for

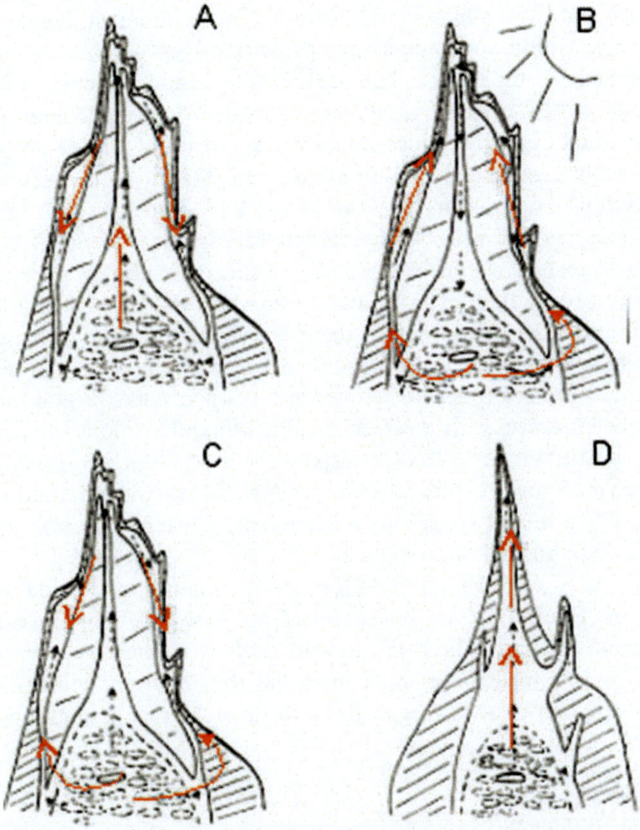

Fig. 13.3 Ventilation of *Macrotermes bellicosus* mounds. The *arrows* indicate the direction of air flow. (**a**) Thermosiphon ventilation mechanism as proposed by Lüscher. (**b**) Cathedral shaped termite mound in the savanna of northern Côte d'Ivoire during the day. (**c**) Cathedral shaped termite mound in the savanna of northern Côte d'Ivoire during the night. (**d**) Dome-shaped termite mound in the forest of northern Côte d'Ivoire. For more information see text

optimal growth (Sands 1969). So, why then do the termites in the forest not construct mounds with thin walls and a higher surface that would allow more efficient gas exchange? The answer seems to be the lower ambient temperature conditions in the forest. For an optimal growth *Termitomyces* fungi cultivated by *M. bellicosus* require constant temperatures of 30°C (Wood and Thomas 1989; Lüscher 1961). However, the mean ambient temperatures in the forest are lower than this. As in other termites (Holdaway and Gay 1948), the termites and their fungi produce heat inside the mound through their metabolism (Ruelle 1962, 1964; Korb and Linsenmair 2000a). Yet, to retain this heat, they must insulate their mounds to reduce heat loss to the outside. They do this by building dome-shaped mounds with thick walls and a relatively low surface area. That such an architecture is an adaptation to the cooler conditions in the forest was shown in an experiment where trees shading forest mounds were cut (Korb and Linsenmair 1998). After 1 year these forest mounds developed architecture more similar to savanna mounds, with ridges and thinner walls, and the CO_2 concentrations in the mounds had declined (Korb and Linsenmair 1999a). This shows that mound architecture is an adaptation to environmental conditions and suggests that the structure of the forest mounds is the result of a trade-off between the need to insulate the mound and to allow gas exchange (Korb and Linsenmair 1999a). Neither condition is optimal in forest mounds; CO_2 concentrations are increased and temperatures are below 30°C, which has fitness consequences. *M. bellicosus* colonies in the forest have a lower reproductive success with fewer sexual offspring per reproductive event and fewer reproductive events than colonies in the savanna (Korb and Linsenmair 1999b). Finally, the thermoregulation/gas exchange trade-off seems to limit the distribution of *M. bellicosus* at both the local and geographical scales. Colonies only occur at forest edges and in open stands inside forests, while they are absent in forests with a closed canopy which have low ambient temperatures (Korb and Linsenmair 1998). Correspondingly, *M. bellicosus* is geographically restricted to savanna regions and does not occur in West African rain forests (Ruelle 1970). However, in the latter regions it immediately establishes after the forest is cleared (Korb, unpublished observations).

There also exists a third type of *M. bellicosus* mound, in Uganda (Darlington, personal communication). It has holes around the base, opening into a low-roofed cellar underneath the hive. The cellar is walled off from the hive, so predators are excluded. Air seems to flow between the holes, but not very vigorously and the direction probably varies with the wind. Unfortunately, this mound type has never been studied in detail.

13.2.1.2 *Macrotermes michaelseni*

Macrotermes michaelseni (formerly *M. mossambicus*; Ruelle 1975) occurs in east Africa (Darlington 1985b) and southern Africa (Ruelle 1970) where it builds conspicuous closed mounds which can reach a height of several metres (Turner 2000a). Mounds studied in Kenya and Namibia differed in structure (Schuurman and Dangerfield 1996). In what follows, I refer to studies in Namibia focusing on ventilation mechanisms. Here, the hive is generally situated in the soil below the mound, and is about 1.5–2 m in diameter (Darlington, personal communication; Schuurman and Dangerfield 1996). Thus, in contrast to *M. bellicosus*, where about half of the nest extends into the mound, with the other half underground, only a small proportion of a *M. michaelseni* nest is epigeal. Hence, the *M. michaelseni* nest seems very well insulated within the soil, which might be an adaptation to arid conditions and low temperatures during winter in southern Africa. Large *M. michaelseni* mounds generally have a tall central spire with a broad outwash pediment, but in contrast to the central chimney in *M. jeanneli* (Box 13.2; see below), the mound is closed and does not open to the outside. Internally, the mound is permeated by a network of tunnels. As in *M. bellicosus* mounds (Fig. 13.3) (i) a central shaft extends upward through the centre of the mound from the nest to the top, (ii) mainly vertically running air channels lie a few centimetres below the surface, and (iii) passages connect the central shaft to the peripheral channels. In contrast to cathedral mounds of *M. bellicosus*, the latter do not run inside ridges and *M. michaelseni* has a more extended interweaving network of lateral tunnels (Box 13.2) that connects the central shaft with the surface air channels. Mound function of *M. michaelseni* has been extensively studied (Soar and Turner, personal communication), and ventilation is suggested to be mainly wind-driven (Turner 2000b, 2001). According to the Bernoulli principle, a positive pressure occurs when the outside air is slowed down and this pushes air within the mound, while an acceleration of outside air results in a suction pressure. As mounds extend upward through a wind boundary layer, they are surrounded by pressure fields. Positive pressures develop at the mound's upwind side and they push fresh air into the mound. Negative pressures prevail at the mound's trailing and lateral faces and they suck air out of the mound. Simply speaking, these pressures increase with increasing mound height as wind speed in the boundary layers intensifies. This leads to an exchange of air in the air channels through the porous mound walls. In contrast to the circulatory ventilation mechanism of cathedral shaped *M. bellicosus* mounds during the day (Korb and Linsenmair 2000b), this ventilation is highly variable with a strong dependence on wind speed (Turner 2000b, 2001). Inside the mound, air moves from the central nest upward in the central shaft, driven by buoyant forces imparted by the nest's metabolism. This mechanism is similar to Lüscher's thermosiphon hypothesis and to that proposed for dome-shaped *M. bellicosus* mounds from the forest (see above) (Korb and Linsenmair 2000b). Air from the central shaft mixes with other internal air in a mixing zone which comprises the lateral tunnels. Thus, *M. michaelseni* nests are ventilated by a tidal movement of air driven by temporal variation in wind speed and wind direction and its effectiveness depends upon the wind energy admitted to the mounds across its surface channels and the energy generated by the colony's metabolism.

This wind-driven tidal mechanism is unlikely to apply for *M. bellicosus* in northern Côte d'Ivoire. Here, there is no association between ventilation and wind conditions; although it can be windy, a strong wind is not reliable especially during the rainy season when significant air movements occur only before and during irregular storms (Korb, unpublished observations). By contrast, there is a clear association between ventilation and temperature: (i) within single mounds the ventilation of ridges is correlated with their heating up by the sun during the course of a day; (ii) ventilation of mounds slows down during the night when temperature gradients disappear. Thus, although wind might assist ventilation on occasion, the normal driving forces for ventilation of *M. bellicosus* mounds in the savannas of northern Côte d'Ivoire (and probably climatically comparable areas) are temperature associated convection currents. This shows that even for closed mounds, the ventilation mechanism differs between different areas.

13.2.1.3 Macrotermitinae with Open Mounds

So far, only fungus-growing termites with enclosed nests have been considered. However, there are also species which construct mounds with openings. For example, *Macrotermes jeanneli, Macrotermes subhyalinus* and some species in the genus *Odontotermes*. Some taxonomic clarification is required: *Macrotermes jeanneli* was formerly *Macrotermes subhyalinus* Marigat type, but recent data confirm the original description by Grassé (1937b) which recognised it as a distinct species (Darlington 1984; Bagine et al. 1989; Darlington et al. 1997). *Macrotermes subhyalinus* (Bissel type, found in East-Africa) is distinct from *M. subhyalinus* in West-Africa (Glover 1967; Pomeroy 1977; Darlington 1984; Brandl et al. 2007). *Odontotermes* species with open mounds are described by Coaton (1961), Grassé (1984) and Darlington (1997).

Here, mound structure can vary from relatively flat hillocks with several openings, e.g. *M. subhyalinus* (Darlington 1984), *Odontotermes latericius* and *Odontotermes transvaalensis* in South Africa (Coaton 1961, Grassé 1984) to a more or less huge mound with a central chimney in *M. jeanneli* (Grassé 1937b; Darlington 1984; Darlington et al. 1997) and *O. transvaalensis* (Ruelle 1985; Turner 1994). To date, systematic research on mound function is largely restricted to *M. subhyalinus*, *M. jeanneli*, and *O. transvaalensis*. Generally in mounds with several openings, differences in wind speed at different heights (Geiger 1965) and the shape of the openings (Weir 1972) promote wind-induced ventilation, also called the Venturi effect (Vogel and Bretz 1972; Vogel 1978). For a Venturi mechanism to work, at least two openings are required, one raised above the surface and one close to the ground. As these openings experience different wind speeds, a negative pressure is induced so that air is drawn out of the higher opening and into the lower opening. In *M. subhyalinus*, ambient air seems to enter the mound through peripheral, rimless, basal openings. It flows through the mound and leaves it through several exit tunnels with well-defined rims which are usually central and on top of the mound (Weir 1972). However, during this ventilation process the ambient air does not pass directly through the termite's nest, which is separated from the ventilation channels by a wall of soil (Darlington 1984, 1989). Strikingly, a similar ventilation system has been described for some fungus-growing ants (Kleineidam and Roces 2000; Kleineidam et al. 2001). In contrasts to this, *M. jeanneli* has no openings at the peripheral nest base but has only a single central chimney, which can reach a height of several metres and through which air leaves the mound (Darlington et al. 1992, 1997). Air probably enters the mound mainly via relatively distant foraging holes. Extensive studies in Kenya (Darlington et al. 1997) have shown that the gas outflow of large nests is huge, ranging from 100,000 to 400,000 L/day, including 800 to 1,500 L of CO_2. This comes at a cost: the gross water loss amounts to a rate of up to 13 L/day; when ambient humidity is subtracted, the net water loss is up to 5 L/day. A large proportion of this water comes from the soil, the rest is metabolic water from the termites and the fungus combs. *O. transvaalensis* studied by Turner (1994) in South Africa has an intermediate structure with a single central turret but with entrance holes in the ground close to the mound. However, the principal ventilation mechanism seems to be the same as those found for *M. subhyalinus* and *M. jeanneli*, namely wind-induced ventilation driven through Venturi effects, although the detailed mechanisms differ between species. In addition, in *M. jeanneli* a second mechanism based on buoyancy was found that works when the wind is not blowing (Darlington et al. 1997). This illustrates that there can be several mechanisms working in the same mound, with some as backup systems.

13.2.2 Magnetic Termites

The compass, "meridional" or "magnetic" termites are unique among termite mound builders in that they build epigeal nests that are elongated in plan with the long axis oriented along a north-south meridian (Jack 1896, Fig. 13.4). The mounds are wedge-shaped, tapering towards the top, and of a size ranging to 4 m in height and to 3 m in length (Grigg 1973; Duelli and Duelli-Klein 1978). Such mounds often occur in clusters (up to several 100 mounds; Jacklyn 1991) on flat plains, creating a striking, cemetery-like landscape. Compass mounds are built by only three tropical Australian species: *Amitermes meridionalis* (Frogatt) is an obligate compass mound builder in the Northern Territory, while *Amitermes vitiosus* (Hill) in North Queensland and *Amitermes laurensis* (Mjöberg) in North Queensland and the Northern Territory construct compass mounds only under certain conditions, namely when they occur in low lying, ill-drained areas (Gay and Calaby 1970). Although the exact orientation of the mounds varies in different areas (Grigg and Underwood 1977; Duelli and Duelli-Klein 1978; Spain et al. 1983; Jacklyn 1991), their shape considerably differs from mounds built by any other termite as their basic structure is a vertical plate not a cone or dome.

Currently this unique architecture is interpreted as a thermoregulatory adaptation. A north-south orientation provides more stable conditions than any other orientation. Yet, if thermostability is the crucial factor, why are the mounds not spheroidal in shape like all other termite mounds as such a shape would provide even more thermostable conditions? Thus the question of magnetic mound architecture must

Fig. 13.4 A compass mound of the magnetic termite *Amitermes meridionalis* in northern Australia. The broad eastern face is presented to the front

actually be split into three questions: (1) why are these mounds north-south orientated? (2) how can the variation in exact orientation be explained between different areas? (3) why do magnetic termites build plate-like elongated mounds? So far studies have mainly addressed the first and second questions, often not differentiating between the first (i.e. orientation) and the third question (i.e. shape). As a consequence, current explanations for the north-south orientation are often mistaken as explanations for their unique shape. At the proximate mechanistic level, the termites seem to use the earth's geomagnetic field to achieve the north-south orientation (Grigg et al. 1988; Jacklyn and Munro 2002).

13.2.2.1 Why Are Magnetic Termite Mounds Orientated North–South?

The first studies concentrated on the importance of north-south orientation (Hill 1942; Gay and Calaby 1970; Grigg 1973). It was soon suggested that the north-south orientation of the mounds is likely to be an adaptation to achieve thermostability, as elongated mounds of this type are more influenced by ambient air temperatures and solar irradiation than spheroidal mounds due to their high surface to volume ratio (Grigg 1973). North-south orientation seems to stabilise internal core temperature as demonstrated by an experiment in which a mound was rotated by 90° to an east–west orientation (Grigg 1973). Detailed investigations refined these ideas (Jacklyn 1991, 1992) and according to the currently accepted eastern-face-plateau hypothesis, face temperature gradients based on eastern face heating are maintained during the dry season with a thermal plateau relatively independent of the variations in ambient temperature (Jacklyn 1992; Schmidt et al., unpublished observations). This was confirmed by experimental rotation of mounds by 20°: it was found that face temperatures were sensitive to rotation whereas mound core temperatures were not. Thus, the eastern face of the mound warms up quickly in the early morning and maintains a relatively stable temperature. This is especially crucial during the dry season when temperatures are low during the night. In agreement with this, termites are known to concentrate towards the eastern face of mounds during cool morning hours (Korb 2003a).

13.2.2.2 How Can the Variation in Exact Orientation Be Explained Between Different Areas?

The exact orientation of *A. meridionalis* mounds differs with location, with population means ranging from 1 to 10°E (deviations from exact North/South; Jacklyn 1991, 1992). Grigg and Underwood (1977) and Spain et al. (1983) hypothesised that the exact orientation might reflect local adaptations to environmental conditions, or that variation in orientation is only a non-adaptive phenomenon caused by genetic drift due to the island like distribution of the mound populations. The influence of genetic drift has not been studied so far. However, Jacklyn (1992) suggested that the local and geographic variation in the orientation of the mounds is at least partially an adaptation to local shade and wind speed to ensure appropriate thermal conditions at the eastern face. He found that despite high variability within populations, a negative correlation exists between mean mound orientation at a site and its average wind exposure. Additionally, shade sites had larger deviations from 0° than open sites (Jacklyn 1991, 1992)

13.2.2.3 The Open Question: Why Are Mounds Elongated?

Despite considerable progress in understanding compass termite mounds, the basic question still lacks a convincing answer: why do compass mounds have an elongated axis (see also Grigg 1973). The significance of this striking shape needs to be addressed as it distinguishes compass termites from all other mound building termites.

In contrast to the majority of termites that build spheroidal mounds (Grassé 1984; Korb and Linsenmair 2000a; Noirot and Darlington 2000), with a high thermal inertia and stable core temperatures (see above and Lüscher 1955, 1961; Korb and Linsenmair 1998; Noirot and Darlington 2000), much of the northern Australian bush is dominated by the small cone-shaped mounds of other Australian *Amitermes* species, notably *A. vitiosus* (Gay and Calaby 1970). Like the compass mounds, these are also characterised by a relatively high surface to volume ratio and thus a low thermal inertia. Thus *Amitermes* in Australia might have been selected to have elongated mounds with an increased thermostable eastern face area (Jacklyn 1991, 1992). However, elongated mounds are only build by the three *Amitermes* species occurring in low-lying, ill-drained areas that are flooded during the rainy season for several months (Gay and Calaby 1970) and here this "compass mound" architecture seems to have evolved several times independently (Ozeki et al. 2007). This suggests that the seasonal flooding of the habitat might be crucial for selecting the elongated mound shape (see also Grigg 1973). Compared to domed or rounded mounds, elongated mounds with a high surface to volume ratio and thin walls offer more favourable conditions (a) to increase mound stability by drying the mounds quickly during the rainy season, (b) to dry the stored grass, on which the termites depend during the floods, and (c) to facilitate the exchange of respiratory gases when dampness decreases the porosity of the walls (Korb 2003a). These hypotheses suggest that the elongated, wedge shape of the compass mounds is an adaptation to the rainy season and flooding of the mounds. Thus, it would be no coincidence that the only termites that occur on seasonally flooded plains are also unique in their mound architecture. The second and third hypotheses are currently tested in long-term experiments where *A. meridionalis* mounds are manipulated to have a rounded shape similar to termite mounds of other species and compared to similar sized control mounds in the same area. First results show that nest temperatures are indeed more stable in manipulated rounded mounds than in elongated compass mounds (Schmidt et al., unpublished observations). Thus temperature stability does not explain the elongated mound shape, although it accounts for the north-south orientation of the mounds (see above). Also gas exchange seems not to be improved in elongated mounds compared to rounded mounds. Monthly CO_2-measurements covering 24 h during the rainy season did not reveal significant differences (ibidem). However, preliminary results in the same experiments also suggest that the microbial load of the grass storage during the rainy season was higher on manipulated rounded mounds compared to elongated control mounds, probably resulting in less forage at the end of the rainy season. According to these results, the elongated mound shape of *A. meridionalis* would be an adaptation to store food appropriately during the flooded period when the termites cannot forage in the immediate surroundings of their nest.

13.2.3 Conclusion on Mound Function

The main function of termite mounds certainly is to provide a home for a large colony. Compared to other nesting types (subterranean, wood-dwelling), epigeal nests are more exposed to ambient fluctuations as the protective layer between the nest and the environment is reduced. Interestingly, this contrasts with the mounds of many ant species where epigeal nests seem to be better insulated than subterranean nests (Bollazzi et al. 2008). This difference seems to be caused by different building materials in each group of social insects: soil particles in the case of termites and plant material for thatched ant mounds. As termites have thin cuticles and are rather prone to desiccation and sensitive to temperature fluctuations, there should be strong selection to build mounds which provide homeostatic conditions within the nest. As Hansell (2005) noted "Natural selection should operate on builders through the success of the structures that they build and also the success of the building process compared with rival methods of construction". The huge variety of mound types, differing intra- and interspecifically, is the outcome of this natural selection process.

There has been much controversy about the function of termite mound architecture, e.g. whether it mainly serves as a ventilation or as a thermoregulation device. A major flaw of these debates rests in the assumption that there is only one general function of termite mound architecture, yet this is certainly not true. At least three considerations must be taken into account when discussing the functional significance of termite mound architecture. Firstly, who are the "inhabitants" of the mound and how many are they? It makes a large difference whether the mound is only the home for the termites or whether it also harbours cultivated *Termitomyces* fungi or functions as storage for dead plant material. If there are many metabolising inhabitants, gas exchange and thus ventilation may play a crucial role. Secondly, where is the nest located? If the nest itself is below ground, underneath the proper mound, thermoregulation may be less important than if it is located above ground in the mound. Thirdly, what are the environmental conditions for the termites? Here it is important to consider the annual environmental cycle, as extreme events may occur only during a short period but may be crucial. In an arid environment, humidity and water storage might limit optimal ventilation in the same way as suboptimal cold ambient temperatures constrains the mound architecture and thus the ventilation of *M. bellicosus* mounds in the forests of northern Côte d'Ivoire. A mechanism to overcome the water problem in arid areas, and still guarantee sufficient gas exchange, might be to locate the nest below ground as *M. michaelseni* does in Namibia (Turner 2000a). This last example illustrates that the three listed questions must be considered together as they are not mutually independent (see below).

Additionally, there must be a clear distinction between the function of mound architecture and the mechanism of how this function is achieved. All fungus-growing termites are selected to have an efficient gas exchange (= function), yet the mechanism might differ and which mechanism works also depends on environmental conditions. An effective tidal movement of air through the mound, as shown in the studies on *M. michaelseni* in Namibia, only works in a windy area whereas warm ambient temperatures are necessary for an efficient temperature-driven ventilation for *M. bellicosus* in northern Côte d'Ivoire. Thus, there are several mechanisms with varying importance in different areas. Even within an area the mechanisms might vary depending on season and also on short-term fluctuations over a day; so that there might be backup systems when one mechanism fails to work. These considerations can seem trivial, but they have largely been overlooked in the past, resulting in often fruitless debates. Thus, useful future comparisons of studies will depend on a correct description of the species, a record of the exact location of the nest and documentation of its crucial environmental factors.

Thus, functional significance of mound shape depends on species and especially their environment. This can be most easily demonstrated by a comparison between *M. bellicosus* and *M. michaelseni,* which both have closed mounds. For *M. michaelseni* in Namibia the thermoregulatory function might be less important because the hives are mainly located below ground, probably reflecting an adaptation to the arid and cold winter conditions as it allows an effective insulation of the brood and the cultured fungi. As a consequence, gas exchange through the soil would be very inefficient and more efficient ventilation mechanisms would be strongly selected. The *M. michaelseni* mounds function exactly as such ventilation devices. As the nest is below ground, buoyant forces imparted by the metabolism of the termites and their fungi can only be of minor importance as driving mechanisms for the ventilation, but a wind-driven tidal movement of air can function efficiently in the typically open habitats where *M. michaelseni* occurs. Selection therefore favours mounds with a tall central spire. In contrast, for *M. bellicosus* in northern Côte d'Ivoire, ambient temperature is less of a problem in the savanna so that nests can be located above ground. Thus, buoyant forces imparted by temperature gradients function as efficient ventilation mechanisms, while wind-driven ventilation can only play a minor role as wind is not a reliable factor in the bushy habitat of West-African savannas. This selects for building mounds with a high surface that heats up quickly, whereas mounds with a single central spire (like those build by *M. michaelseni*) are less advantageous. In gallery forest, *M. bellicosus* has to cope with suboptimal low ambient temperatures and as a consequence it has to insulate its nest and build more dome-shaped mounds with thick walls. These mounds partially resemble those of *M. michaelseni,* probably reflecting similar selection pressures through cold ambient conditions, but the nest is partly located above ground and wind-driven ventilation is less of an option for *M. bellicosus* in a closed habitat like the forest. At the same time temperature-driven ventilation is also suboptimal, in contrast to the savanna setting. Thus, forest mounds at suboptimal conditions probably compromise with an intermediate ventilation systems: less efficient temperature-driven and less forceful wind-driven ventilation, the latter supported by tall central spires in large mounds.

13.3 Proximate Mechanisms of Mound Building

The construction of mounds that are adapted to local conditions does not imply a purposeful design. Rather, our current understanding suggests that it is the result of self-organisation, in interaction with stigmergy and templates (Box 13.1; Bonabeau et al. 1997; Theraulaz et al. 1999; Camazine et al. 2001). Contemporary models emphasize that the structure of social insect nests results from a succession of stimulus-response steps involving the workers, the environment, and the by-product of their activities. Although these do not yet explain complex features of termite mounds, several models have been developed to account for the construction of pillars and walls, the building elements of mounds (Grassé 1959; Deneubourg 1977; Bonabeau et al. 1998a, 1998b; O'Toole et al. 1999; Camazine et al. 2001) (Fig. 13.5).

For termites, the best analysis of building behaviour concerns the royal cell in fungus-growing species (Fig. 13.5a). The process of (re-) constructing a royal cell was first described quantitatively by Grassé (1939, 1959, 1967). However, the meticulous observations and experiments by Bruinsma (1979) on *M. subhyalinus* from Kenya have provided the most complete basis for our current understanding of this construction process. Royal cell re-construction has been observed several times in nature, for instance it occurs as a response to doryline army ant attacks (Darlington 1985a). Few other experimental studies have been done so far. Nest expansion and gallery construction were described by Jones (1979, 1980) in arboreal nesting *Nasutitermes costalis* and these studies served as a basis for a model by O'Toole et al. (1999). These examples best illustrate the general principles that seem to organise building behaviour in termites.

13.3.1 Description of Royal Cell Re-construction

To observe the building behaviour of *M. subhyalinus* workers, one large physogastric queen was removed from her royal cell and placed with different numbers of workers, ranging from 20 to 640, in an experimental observation dish without surrounding walls (Bruinsma 1979). The behaviour of individual workers appears to be made up of a few, relatively simple activities: workers pick up a soil pellet near the queen, transport it to a deposition zone 2–5 cm away from the queen, and deposit and cement the pellet in that zone. During the transport the worker mixes the soil pellet with saliva and produces a paste that is moulded and glued into place. At the beginning of the building process, the walk to the deposition zone can be described as a random walk during which workers seem to deposit the first pellets at surface irregularities. Gradually, the walk between the queen and the deposition zone becomes more directed and soil pellets are preferentially deposited where others already exist. This leads to the emergence of pillars up to 1 cm in height a few centimetres around the queen. These pillars are then laterally extended into lamellae and additional wall. By standing on the queen, they are finally connected to one another to form a roof over her.

Fig. 13.5 Termite building activity. (**a**) An opened royal cell with the queen (*large whitish* individual in the centre) and surrounding workers. The king is hidden beneath the queen. (**b**) A newly constructed wall composed of laterally extended pillars build by *Macrotermes bellicosus*. (**c**) Nest extension in *Trinervitermes geminatus*. Enlargement: The termites build pillars which are only extended at the *top*. For further information see text

13.3.2 Proposed Building Mechanisms

For these behavioural data it was postulated that the re-construction of the royal cell is the result of the use of (i) a template (the queen), (ii) self-organisation, and (iii) stigmergy mediated through three proposed pheromones: a building-, a trail- and a cement-pheromone, along with tactile stimuli (Camazine et al. 2001). In greater detail: (i) the queen functions as a template by producing a building pheromone. The concentration of this queen pheromone decreases with increasing distance from the queen. Assuming that the probability of workers to pick up soil particles increases with increasing queen pheromone concentration, while the probability to deposit soil particles decreases, workers that are closer to the queen are more likely to pick up soil and to deposit it further away from the queen. Experiments by Bruinsma (1979) suggest that the queen's building pheromone is produced by the so-called royal fat-bodies in the queen's abdomen. These fat bodies are located around the tracheal system, so that the volatile building pheromone might be carried through the tracheal system to the air surrounding the queen or the pheromone might leak out through the abdominal cuticle. (ii) Self-organisational processes occur through three positive feedback mechanisms mediated by the trail- and cement-pheromones and the spatial heterogeneity of the surface. Surface irregularities lead to the initial deposition of soil pellets in the deposition zone. This has also been shown by Stuart (1967) for a species of *Nasutitermes*, but where the principles of building behaviour differ from the Macrotermitinae. The deposited pellets are proposed to trigger further depositions by emitting an attractive cement pheromone. This substance orients workers from a distance of 1–2 cm to a deposition site and it induces workers to deposit new pellets onto recently deposited ones, however the pheromone seems to lose its biological activity within a few minutes of deposition. Workers walking to the deposition zone lay a trail pheromone which acts as long-range positive feedback and recruits workers to building sites. Thus the initial random walks gradually become replaced by straighter paths from the queen to the deposition sites. In line with these positive feedback mechanisms, Bruinsma found a pronounced effect of worker group size on pellet deposition rate during the first 60 min of building: (a) groups of 20 individuals were unable to reconstruct walls; (b) the total number of deposits increased non-linearly with time from the start of building and with group size and (c) the mean number of deposits per worker increased disproportionately with group size, but then reached a plateau when more than 80 workers were present. (iii) All these building processes are coordinated by stigmergy, a concept pioneered by Grassé (op. cit.) who argued that the regulation and coordination of the building activity do not depend on the workers themselves but are mainly achieved by the nest: i.e. a given configuration triggers a response by a termite worker, transforming the configuration into another configuration that may in turn trigger another, possibly different, action performed by the same termite or any other worker in the colony. The product of the behaviour of single individuals is the emerging pillars and wall.

13.4 Concluding Remarks

The re-construction of the royal cell of *Macrotermes* spp. is certainly the best understood building behaviour in termites and the general mechanisms of interactions between the use of a template, self-organisation processes and stigmergy probably applies to all building behaviours. Yet, there are some caveats and there remain many open questions for future studies. (i) The proposed pheromones during the royal cell re-construction process have not been identified so far. (ii) There are certainly differences in the details of the building process between species so that the detailed rules revealed from the re-construction of the royal cell do not apply generally. This is most obvious when comparing them with a second well-studied species, *N. costalis* (Jones 1979, 1980). There are at least two main differences: First, the building material differs between the two species. While *Macrotermes* uses soil particles with saliva, *N. costalis* builds with faecal pellets, which have different constructional properties (Korb 2003b). Second, the basic building elements also differ. In *Macrotermes*, walls are constructed by extending pillars (Fig. 13.5b), whereas during the nest expansion process in *N. costalis* pillars are extended by building a roof disk atop each pillar (Jones 1979) (Fig. 13.5c: here *Trinervitermes geminatus*). These species-specific differences could account for the debate among modellers whether an inhibition mechanism is necessary to get a smooth surface of joining tips of growing pillars and a lateral extension of the pillars into walls. Modellers working with *N. costalis* postulated that such a process would be necessary, while those working with *Macrotermes* did not need to postulate such an inhibition to account for royal cell re-construction. (iii) So far, we still only understand the basic processes of building behaviour. We do not know how they are "coordinated" at a larger scale to result in mounds with complex architectures consisting of variable structures within a single mound (e.g. outer wall, royal cell, structures surrounding the fungus garden). Neither do we understand how these basic behavioural elements are modified to result in mounds adapted to local conditions. The experiments with *M. bellicosus* in northern Côte d'Ivoire (see above) showed that the termites indeed adjust their building behaviour when shading trees are cut in the forest (Korb and Linsenmair 1998). Thus there remain many interesting questions for future study that require a close cooperation between modellers and empiricists.

Acknowledgments Joanna Darlington made very helpful comments on the manuscript.

References

Aanen DK, Eggleton P (2005) Fungus-growing termites originated in African rain forest. Curr Biol 15:851–855

Bagine RKN, Darlington JPEC, Kat P, Ritchie JM (1989) Nest structure, population structure and genetic differentiation of some morphologically similar species of *Macrotermes* in Kenya. Sociobiology 15:125–132

Batra LR, Batra SWT (1966) Fungus-growing termites of tropical India and associated fungi. J Kans Entomol Soc 39:725–738

Bollazzi M, Kronenbitter J, Roces F (2008) Soil temperature, digging behaviour, and the adaptive value of nest depth in South American species of *Acromyrmex* leaf-cutting ants. Oecologia 158:165–175

Bonabeau E, Theraulaz G, Deneubourg JL (1998a) Latency, time and absence of group effect. Insect Soc 45:191–195

Bonabeau E, Theraulaz G, Deneubourg JL et al (1998b) A model of emergence of pillars, walls and royal chambers in termite nests. Phil Trans R Soc Lond B 353:1561–1576

Bonabeau E, Theraulaz G, Deneubourg JL et al (1997) Self-organization in social insects. Trends Ecol Evol 12:188–193

Brandl R, Hyodo F, vonKorff-Schmising M (2007) Divergence times in the termite *Macrotermes* (Isoptera: Termitidae). Mol Phylog Evol 45:239–250

Bruinsma OH (1979) An analysis of building behaviour of the termite *Macrotermes subhyalinus*. PhD thesis. Wageningen Agricultural University

Camazine S, Deneubourg JL, Franks N et al (2001) Self-organization in biological systems. Princeton University Press, Princeton, NJ

Coaton WGH (1961) Association of termites and fungi. Afr Wildlife 15:39–54

Collins NM (1977) The population ecology and energetics of *Macrotermes bellicosus* (Smeathman) Isoptera. PhD thesis. University of London

Collins NM (1981a) Consumption of wood by artificially isolated colonies of the fungus-growing termite *Macrotermes bellicosus*. Entomol Exp Appl 29:313–320

Collins NM (1981b) Populations, age structure and survivorship of colonies of *Macrotermes bellicosus* (Isoptera: Macrotermitinae). J Anim Ecol 50:293–311

Darlington JPEC (1984) Two types of mound built by the termite *Macrotermes subhyalinus* in Kenya. Insect Sci Appl 5:481–492

Darlington JPEC (1985a) Attacks by doryline ants and termite nest defences (Hymenoptera; Formicidae; Isoptera; Termitidae). Sociobiology 11:189–200

Darlington JPEC (1985b) The structure of mature mounds of the termite *Macrotermes michaelseni* in Kenya. Insect Sci Appl 6:149–156

Darlington JPEC (1989) Ventilation systems in termite nests in Kenya. Sociobiology 15: 263–264

Darlington JPEC (1994) Nutrition and evolution in fungus-growing ants. In: Hunt JH, Nalepa CA (eds) Nourishment and evolution in insect societies. Westview Press, Boulder, CO, pp 105–130

Darlington JPEC (1997) Comparison of nest structure and caste parameters of sympatric species of *Odontotermes* (Termitidae, Macrotermitinae) in Kenya. Insect Soc 44:393–408

Darlington JPEC, Zimmerman PR, Greenberg J et al (1997) Production of metabolic gases by nests of the termite *Macrotermes jeanneli* in Kenya. J Trop Ecology 13:491–510

Darlington JPEC, Zimmerman PR, Wandiga SO (1992) Populations in nests of the termite *Macrotermes jeanneli* in Kenya. J Trop Ecol 8:73–85

Deneubourg JL (1977) Application de l'ordre par fluctuations a la description de certaines etapes de la construction du nid chez les termites. Insect Soc 24:117–130

Duelli P, Duelli-Klein R (1978) Die magnetische Nestausrichtung der australischen Kompasstermite *Amitermes meridionalis* Mitteilungen der Schweizerischen Entomologischen Gesellschaft. Bull Soc Entomol Suisse 51:337–342

Gay FJ, Calaby JH (1970) Termites of the Australian region. In: Krishna K, Weesner FM (eds) Biology of termites II. Academic Press, New York, NY, pp 393–448

Geiger R (1965) The climate near the ground. Harvard University Press, Cambridge

Glover R (1967) Further notes on some termites of Northern Somalia. East Afr Wildlife J 5: 121–132

Grassé P-P (1937a) Ecologies animale et microclimat. C R Trav Bull Admin Assoc Fr Avan Sci 383:383–390

Grassé P-P (1937b) Le *Bellicositermes jeanneli* n.sp. constructeur de grandes termitières a cheminée. Bull Soc Entomol Fr 42:71–73

Grassé P-P (1939) La reconstruction du nid et le travail collectif chez les termites supérieurs. J Psych Norm Pathol 30:370–396

Grassé P-P (1945) Recherches sur la biologie des termites champignonnistes (Macrotermitinae). Ann Sci Nat Zool Biol Anim 11:97–171

Grassé P-P (1959) La reconstruction du nid et les coordinations inter-individuelles chez *Bellicositermes natalensis* et *Cubitermes* sp. La théorie de la stigmergie: essai d'interprétation du comportement des termites constructeurs. Insect Soc 6:41–81

Grassé P-P (1967) Nouvelles expériences sur le termite de Müller (*Macrotermes mulleri*) et considération sur la théorie de la stigmergie. Insect Soc 14:73–102

Grassé P-P (1984) Termitologia Tome II. Fondation des societes-construction. Masson, Paris

Grigg GC (1973) Some consequences of the shape and orientation of "magnetic" termite mounds. Aust J Zool 21:231–237

Grigg G, Jacklyn P, Taplin L (1988) The effects of buried magnets on colonies of *Amitermes* spp building magnetic mounds in Northern Australia. Physiol Entomol 13:285–289

Grigg GC, Underwood AJ (1977) An analysis of orientation of magnetic termite mounds. Aust J Zool 25:87–94

Hansell MH (2005) Animal architecture. Oxford University Press, Oxford

Harris WV (1956) Termite mound-building. Insect Soc 3:261–268

Heinrich B (1993) The hot-blooded insects. Strategies and mechanisms of thermoregulation. Harvard University Press, Cambridge

Hill GF (1942) Termites (Isoptera) from the Australian region. CSIRO, Melbourne

Holdaway FG, Gay FJ (1948) Temperature studies of the habitat of *Eutermes exitiosus* with special reference to the temperature within the mound. Aust J Sci Res 1:464–493

Jack RL (1896) Note on the "meridional anthills" of the Cape York Peninsula. Proc R Soc Queensland 12:99–100

Jacklyn PM (1991) Evidence for adaptive variation in the orientation of *Amitermes* (Isoptera, Termitinae) mounds from northern Australia. Aust J Zool 39:569–578

Jacklyn PM (1992) "Magnetic" termite mound surfaces are oriented to suit wind and shade conditions. Oecologia 91:385–395

Jacklyn PM, Munro U (2002) Evidence for the use of magnetic cues in mound construction by the termite *Amitermes meridionalis* (Isoptera: Termitinae). Aust J Zool 50:357–368

Johnson RA, Thomas RJ, Wood TG, Swift MJ (1981) The inoculation of the fungus comb in newly founded colonies of some species of the Macrotermitinae (Isoptera) from Nigeria. J Nat Hist 15:751–756

Jones RJ (1979) Expansion of the nest of *Nasutitermes costalis*. Insect Soc 26:322–342

Jones RJ (1980) Gallery construction by *Nasutitermes costalis*: polyethism and the behaviour of individuals. Insect Soc 27:5–28

Kleineidam C, Ernst R, Roces F (2001) Wind-induced ventilation of the giant nests of the leaf-cutting ant *Atta vollenweideri*. Naturwissenschaften 88:301–305

Kleineidam C, Roces F (2000) Carbon dioxide concentrations and nest ventilation in nests of the leaf-cutting ant *Atta vollenweideri*. Insect Soc 47:241–248

Korb J (1997) Lokale und regionale Verbreitung von *Macrotermes bellicosus* (Isoptera; Macrotermitinae): Stochastik oder Deterministik. Wissenschaft und Technik, Berlin

Korb J (2003a) The shape of compass termite mounds and its biological significance. Insect Soc 50:218–221

Korb J (2003b) Thermoregulation and ventilation of termite mounds. Naturwissenschaften 90: 212–219

Korb J, Linsenmair K (1998) The effects of temperature on the architecture and distribution of *Macrotermes bellicosus* (Isoptera: Macrotermitinae) mounds in different habitats of a West African Guinea savanna. Insect Soc 45:51–65

Korb J, Linsenmair KE (1999a) The architecture of termite mounds: a result of a trade-off between thermoregulation and gas exchange? Behav Ecol 10:312–316

Korb J, Linsenmair KE (1999b) Reproductive success of *Macrotermes bellicosus* (Isoptera, Macrotermitinae) in two neighbouring habitats. Oecologia 118:183–191

Korb J, Linsenmair KE (2000a) Thermoregulation of termite mounds: what role does ambient temperature and metabolism of the colony play? Insect Soc 47:357–363

Korb J, Linsenmair KE (2000b) Ventilation of termite mounds: new results require a new model. Behav Ecol 11:486–494

Korb J, Linsenmair KE (2001) Resource availability and distribution patterns, indicators of competition between *Macrotermes bellicosus* and other macro-detritivores in the Comoé National Park, Côte d'Ivoire. Afr J Ecol 39:257–265

Lepage M (1984) Distribution, density and evolution of *Macrotermes bellicosus* nests (Isoptera: Macrotermitinae) in the north-east of Ivory Coast. J Anim Ecol 53:107–117

Lepage M, Darlington JPEC (2000) Population dynamics of termites. In: Abe T, Bignell DE, Higashi M (eds) Termites: evolution, sociality, symbiosis, ecology. Kluwer Academic Publishers, Dordrecht, pp 333–361

Leuthold RH, Barella L, Deneubourg JL, Goss S (1990) The formation of polyethic structures in the termite *Macrotermes bellicosus*. In: Veeresh GK, Mallik B, Viraktamath CA (eds) Social insects and the environment. Oxford & IBH, New Delhi, pp 389–390

Loos R (1964) A sensitive anemometer and its use for the measurement of air currents in nests of *Macrotermes natalensis* (Haviland). In: Bouillon A (ed) Etudes sur les termites Africains. Masson, Paris, pp 364–372

Lys JA, Leuthold RH (1991) Task-specific distribution of the worker castes in extranidal activities in *Macrotermes bellicosus* (Smeathman): observations of behaviour during food acquisition. Insect Soc 38:161–170

Lüscher M (1955) Der Sauerstoffverbrauch bei Termiten und die Ventilation des Nestes bei *Macrotermes natalensis* (Haviland). Acta Trop 12:289–307

Lüscher M (1956) Die Lufterneuerung im Nest der Termite *Macrotermes natalensis* (Haviland). Insect Soc 3:273–276

Lüscher M (1961) Air-conditioned termite nests. Sci Am 205:138–145

Noirot C (1970) The nests of termites. In: Krishna K, Weesner FM (eds) Biology of termites, vol 2. Academic Press, New York, NY, pp 311–350

Noirot C, Darlington J (2000) Termite nests: architecture, regulation and defence. In: Abe T, Bignell DE, Higashi M (eds) Termites: evolution, sociality, symbioses, ecology. Kluwer Academic Press, Dordrecht, pp 121–140

Ozeki M, Isagi Y, Tsubota H et al (2007) Phylogeography of an Australian termite, *Amitermes laurensis* (Isoptera, Termitidae), with special reference to the variety of mound shapes. Mol Phylogenet Evol 42:236–247

O'Toole DV, Robinson PA, Myerscough MR (1999) Self-organized criticality in termite architecture: a role for crowding in ensuring ordered nest expansion. J Theor Biol 198:305–327

Pomeroy DE (1977) The distribution and abundance of large termite mounds in Uganda. J Appl Ecol 14:465–475

Pomeroy DE (1978) The abundance of large termite mounds in Uganda in relation to their environment. J Appl Ecol 15:51–63

Rohrig A, Kirchner WH, Leuthold RH (1999) Vibrational alarm communication in the African fungus-growing termite genus *Macrotermes* (Isoptera, Termitidae). Insect Soc 46:71–77

Ruelle JE (1962) Etudes de quelques variables du microclimat du nid de *Macrotermes natalensis* (Hav.) (Isoptera, Termitidae) en rapport avec le declenchement de l'essaimage. PhD Thesis, Lovanium University, Leopoldville

Ruelle JE (1964) L'architecture du nid de *Macrotermes natalensis* et sons fonctionnel. In: Bouillon A (ed) Etudes sur les termites Africains. Masson, Paris, pp 327–362

Ruelle JE (1970) A revision of the termites of the genus *Macrotermes* from the Ethiopian Region (Isoptera: Termitidae). Bull Br Mus Nat Hist (Ent) 24:365–444

Ruelle JE (1975) *Macrotermes michaelseni* (Sjöstedt): a new name for *Macrotermes mossambicus* (Hagen) (Isoptera: Termitidae). J Entomol Soc S Afr 40:119

Ruelle JE (1985) Order Isoptera. In: Scholtz CH, Holm E (eds) Insects of southern Africa. Butterworth, Durban, p 502

Sands WA (1969) The association of termites and fungi. In: Krishna K, Weesner FM (eds) Biology of termites, vol 1. Academic Press, New York, NY, pp 495–524

Schuurman G, Dangerfield JM (1996) Mound dimensions, internal structure and potential colony size in the fungus growing termite *Macrotermes michaelseni* (Isoptera: Macrotermitinae). Sociobiology 27:29–38

Seeley T, Heinrich B (1981) Regulation of temperature in the nests of social insects. In: Heinrich B (ed) Insect thermoregulation. Wiley, New York, NY, pp 160–234

Smeathman H (1781) Some account of the termites which are found in Africa and other hot climates. Phil Trans R Soc Lond 71:139–192

Spain AV, Okello-Oloya T, John RD (1983) Orientation of the termitaria of two species of *Amitermes* (Isoptera: Termitinae) from Northern Queensland. Aust J Zool 31:167–177

Stuart AM (1967) Alarm, defense and construction behavior relationships in termites (Isoptera). Science 156:1123–1125

Theraulaz G, Bonabeau E, Deneubourg JL (1999) The mechanisms and rules of coordinated building in social insects. In: Detrain C, Deneubourg JL, Pasteels J (eds) Information processing in social insects. Birkhauser, Basel, pp 309–330

Traniello J, Leuthold R (2000) Behavior and ecology of foraging termites. In: Abe T Bignell DE, Higashi M (eds) Termites: evolution, sociality, symbioses, ecology. Kluwer Academic Publishers, Dordrecht, pp 141–168

Turner JS (1994) Ventilation and thermal constancy of a colony of a southern African termite (*Odontotermes transvaalensis*, Macrotermitinae). J Arid Environ 28:231–248

Turner JS (2000a) Architecture and morphogenesis in the mound of *Macrotermes michaelseni* (Sjöstedt) (Isoptera: Termitidae, Macrotermitinae) in northern Namibia. Cimbebasia 16: 143–175

Turner JS (2000b) The extended organism: the physiology of animal-built structures. Harvard University Press, Cambridge

Turner JS (2001) On the mound of *Macrotermes michaelseni* as an organ of respiratory gas exchange. Physiol Biochem Zool 74:798–822

Veivers PC, Mühlemann R, Slaytor M et al (1991) Digestion, diet and polyethism in two fungus-growing termites *Macrotermes subhyalinus* Rambur and *M. michaelseni* Sjöstedt. J Insect Physiol 37:675–682

Vogel S (1978) Organisms that capture currents. Sci Am 239:108–117

Vogel S, Bretz WL (1972) Interfacial organisms; passive ventilation in the velocity gradients near surface. Science 175:210–211

Weir JS (1972) Air flow, evaporation and mineral accumulation in mounds of *Macrotermes subhyalinus* (Rambur). J Anim Ecol 42:509–520

Wood TG, Thomas RJ (1989) The mutualistic association between Macrotermitinae and *Termitomyces*. In: Wilding N, Collins NM, Hammond PM, Webber JF (eds) Insect–fungus interactions. Academic Press, New York, NY, pp 69–92

Chapter 14
Morphology, Physiology, Biochemistry and Functional Design of the Termite Gut: An Evolutionary Wonderland

David E. Bignell

Abstract The chapter reviews termite gut structure and associations with mutualists, now informed by a great increase of data on intestinal microbial diversity made possible in the last decade by molecular genomics, and in the light of contemporary theories on the origin, evolution and trophic diversification of the Isoptera. Detailed morphological descriptions are not given, but the more modern synoptic literature on anatomy, histology and in situ coiling is listed and discussed in relation to current concepts of the termite gut as a bioreactor system. Knowledge of intestinal microbiology, and of microbial physiology and metabolism, has outstripped progress in understanding secretory and absorptive processes by the gut wall and associated structures, such that the primary substrates fermented in the hindgut and the end products utilised by the termite host are still not precisely identified in many cases. Current perceptions of the specialised digestive processes of fungus-growing and soil-feeding termites are summarised, and an overarching evolutionary thesis is proposed, arguing that social organisation in termites has developed primarily to safeguard the fidelity of symbiont transmission between individuals and generations.

14.1 Introduction

Although the presence of dense microbial populations in termite guts has been known for 80 years, and termite-microbe interactions are often used in textbooks to illustrate biological symbiosis, the role of the microbiota and their interactions with the intestine that contains them are yet to be specified exactly. Taxonomic characterization of termite-associated microbes is still a developing science, but the available evidence suggests that some are quite commonplace organisms, while others have co-evolved with their termite hosts and are not found elsewhere in nature.

D.E. Bignell (✉)
School of Biological and Chemical Sciences, Queen Mary, University of London, London E1 4NS, UK; Institute for Tropical Biology and Conservation, Universiti Malaysia Sabah, Kota Kinabalu, Sabah, Malaysia
e-mail: d.bignell@qmul.ac.uk

D.E. Bignell et al. (eds.), *Biology of Termites: A Modern Synthesis*,
DOI 10.1007/978-90-481-3977-4_14,

We are familiar with the many parallels between the Isoptera and the social Hymenoptera. The most notable differences concern nutrition, no social hymenopteran feeds in the same manner as termites, and the comprehensive symbioses between termites and almost all the major taxa of microorganisms is something unknown, rare or limited in ants, bees and wasps (Bignell 2000). Most explanations for the evolution of social organization in insects have been developed for the social Hymenoptera and centre around haplodiploid genetics, altruistic behaviour and the concept of inclusive fitness (Korb 2007, 2008; Hölldobler and Wilson 2009). With a little effort and mathematical initiative they can be applied to termites (Higashi et al. 2000), but it is only rarely suggested (as for example by Nalepa et al. 2001) that an adaptive microbiology had any part, or could even have been the main driver for the development of sociality in the Isoptera. Rather, it has been assumed the microorganisms simply appeared from some greater natural provenance, but as modern molecular methods define the unique character of the termite gut microbiota with ever greater clarity, such notions must be revised.

Rouland-Lefèvre and Bignell (2001) and Bignell (2006) defined at least 7 major functions of the gut microbiota in the Isoptera as a whole. These can be updated as follows:

a. dissimilatory carbohydrate metabolism (including cross-feeding reactions), either from plant cell wall polysaccharides or their depolymerization products, or from the products or derivatives of host glycolysis, yielding short chain fatty acids which are energy sources for other intestinal microbes and/or the termite host (e.g. Pasti and Belli 1985; Hethener et al. 1992).
b. oxygen consumption (as an electron acceptor), generally at the periphery of the gut lumen, rendering the centre of the lumen microaerobic or anaerobic (e.g. Slaytor et al. 1997). This permits fermentation of polymerised and labile carbohydrates, and also possibly proteins and amino acids, again yielding short chain fatty acids. Low redox may also facilitate symbiotic N_2 fixation, where this is advantageous to meet dietary nitrogen deficiencies.
c. dissimilatory and assimilatory N metabolism, providing for the conservation of excretory N (produced as uric acid) by the termite host as new microbial biomass, plus assimilation of the primary products of N_2 fixation in organic form and (probably) transamination, balancing the amino acid spectrum available to the host and upgrading poor sources of N to more nutritional ones (e.g. Potrikus and Breznak 1981). A strict N economy may be unnecessary where the C:N ratio of the food is favourable, for example in soil-feeding termites (Tayasu et al. 1997) and in Mactrotermitinae after composting of forage (and enrichment in N) by the fungal mutualist (Rouland-Lefèvre and Bignell 2001).
d. hydrogenesis (e.g. Taguchi et al. 1993) and hydrogen consumption by reductive acetogenesis (e.g. Brauman et al. 1992; Leadbetter et al. 1999) or methanogenesis (e.g. Ohkuma et al. 1995), assisting energy conservation by the system as a whole and preserving redox balances.
e. N_2 fixation on a facultative basis (e.g. Ohkuma et al. 1996; Lilburn et al. 2001). This may only be needed in wood-feeding termites lacking fungal symbionts.

f. demethylation, deacetylation and decarboxylation of aromatic polymers, possibly accompanied by limited aromatic ring cleavage (lignin degradation, e.g. Ohkuma 2003; Katsumata et al. 2007; Geib et al. 2008).
g. humification, or further humification, of complex organic material passing through the alimentary canal. Available evidence suggests that lignin is not modified during gut passage (Hopkins et al. 1998; Hyodo et al. 1999), but significant mineralization of carbon from the proteinaceous component of a model humic acid was demonstrated in a soil-feeding species (Ji et al. 2000). This accords with previous evidence from pyrolysis/mass spectrometry and other analyses that peptides are depleted in soil-feeder faeces, compared with their parent soil (Bignell 1994; Garnier-Sillam and Harry 1995). Physical changes to soil polysaccharides during gut passage by humivores, affecting structural stability, are described by Garnier-Sillam and Toutain (1995).

Other functions are possible, as the full diversity of the gut microbiota is not yet fully explored; for example there could be semiochemical and antibiotic roles (Dillon and Dillon 2004), or ingested microbes could be digested as food (Fujita et al. 2002). Sulphate reduction has been proposed as another microbial function (Brauman et al. 1990; Trinkerl et al. 1990; Kuhnigk et al. 1996; Droge et al. 2004), generating hydrogen, hydrogen sulphide or acetate from the degradation of organic material or from cross-feeding, the first two products poising the intestine to a low redox potential and the third product assisting energy conservation by the host.

At least two, possibly three, of the listed functions have been demonstrated in every termite species where the symbiosis with microbes has been explored, but not necessarily the same ones. It is unclear whether this reflects the inadequacy of the investigations or the diversity of symbiotic relationships. Every model of the relationship so far developed shows a level of complexity and sophistication equal to, if not surpassing that of comparable intestinal microbial communities in vertebrates, for example grass-eating fish, ungulates and leaf-eating monkeys. How and why did this evolve?

It remains a major issue whether some or all of the organisms mediating the above functions are specialist forms confined to termites, or are recruited from inocula available in their immediate environment (e.g. Donovan et al. 2004). The current plethora of molecular phylogenetic researches, based on nucleic acid sequencing, will probably confirm the unique character of termite gut microbiotas, but the concept is not entirely new. Even with traditional culture-based methods, Bignell et al. (1991) were able to show that viable (i.e. platable) actinomycetes within the gut lumen of two soil-feeding termites were not the same populations as those found in the food-soils or the mound material. This conclusion is now supported by studies based purely on molecular methods (e.g. Lefebvre et al. 2009), as well as plating (Watanabe et al. 2003). A similar result, also based on molecular methods, is reported for ascomycetes by Roose-Amsaleg et al. (2004).

There is a complexity of gut structure: examples from the higher termites typically show a long highly differentiated hindgut and unusual excretory arrangements, not seen elsewhere in insects (Bignell 1994). Extension of the intestine creates several different microbial communities, most of them retained and semi-permanent, suggesting that the processing of food and excretory products is sequential, and sophisticated. Where did this come from? The nutritional range within modern termites is from primitive wood-feeders living almost entirely within the substrate they consume and making minimal constructions with limited amounts of faecal material, through to soil-feeders (seen as highly derived forms, at least in early phylogenetic interpretations), consuming surface mineral horizons more or less without discrimination, and passing the soil through for the construction of complex mounds and gallery systems with the same material that they feed on. Many species are intermediate, either in terms of the type of organic material ingested (especially the degree of humification) or the use of soil for construction, although there is not complete consistency between construction with soil and other indicators of phylogenetic position. What is the significance of this trend? Finally, there is now compelling evidence that some of the most abundant and successful wood-feeders among higher termites (for example in the Nasutitermitinae and the Termitinae) have evolved from soil-feeding ancestors and acquired a secondary focus on lignocellulose digestion. What has selected for such a reversion?

This review examines issues in termite intestinal biology where progress has been made in the last 10 years, but also highlights the many questions that remain to be answered. The idiom is deliberately pre-genomic, but it is not possible to divorce any discussion of gut structure and function from the huge advances that have been made in the detailed documentation of termite intestinal microbes over the past decade, most of which are based on culture-independent molecular sequencing methods. Despite the more detailed and expert treatments to be found elsewhere in the book, an overview of these advances is therefore presented here, focusing on those that most clearly inform our understanding of gut functions. In that sense the chapter is intended to be free-standing, however the digestion of cellulose by termites in general, and the detailed mutualistic interactions of the Macrotermitinae with their fungal partners are covered in other chapters, and not included here.

14.2 Structure and Design: New Insights

14.2.1 Noirot's Synoptic Reviews

Structural diversity of the intestine and the attendant phylogenetic implications are considered by Noirot (1995, 2001), Donovan et al. (2000), and Bitsch and Noirot (2002). Of these the two single-authored reviews of Noirot (1995, on the lower termites; 2001 on higher termites) are major compilations of morphological information, comparable with Sands (1998), focusing particularly on coiling, curvitures and torsions, the gizzard, the midgut/hindgut junction, the Malpighian tubule system, the enteric valve and the posterior hindgut, but informed by new physiological, ultrastructural and microbiological data to the extent that functional implications of structure can be considered in the context of the broad themes of termite evolution.

These include paraphyly in lower termites and the evolutionary repositioning from wood-feeding to soil-feeding (and perhaps vice versa), and the acquisition of fungal mutualists, which are evident in higher termites (Noirot 1992; Bignell 1994). The seven modern works cited above list and summarise more than 100 years of previous studies of termite gut morphology, much of it exquisitely detailed observation never likely to be revisited, but not appropriate to inclusion in a general survey such as the present chapter. As noted by Bignell (2006), with gut morphology as well as with many other aspects of termite biology, the literature still shows bias towards the African fauna as a legacy of the colonial era, but this is changing as contributions on structure by Czolij et al. (1984), Fontes (1987), Landim and Costa-Leonardo (1990), Costa-Leonardo (1995), Costa-Leonardo and Landim (1997), Tokuda et al. (2001) and Godoy (2004) amply illustrate. Of the major reviews of gut structure, those of Noirot (1995, 2001) are probably the most catholic in terms of biogeography, although the cladistic phylogeny produced by Donovan et al. (2000) includes a very significant number of gut characters and representative species from Africa, South East Asia and South America. Noirot's intention to construct a phylogeny based on the intestine reaches fruition in Bitsch and Noirot (2002), discussed elsewhere in this book, but the synoptic reviews are also worthy of close inspection for other reasons, including comments on the terminology used to describe diverticula and section demarcation in the hindgut, the evolution of the mixed segment, and the numbers, configurations and partial enclosure of the Malpighian tubules. The last topic refers to a morphological feature which shows great variation within the Isoptera, but is of wholly unknown significance. The main conclusions of Bitsch and Noirot (2002) support other studies which show that the sub-families Macrotermitinae and Apicotermitinae are monophyletic within a monphyletic family, but the rest of the Termitidae (comprising the sub-families Nasutitermitinae and Termitinae) are an unresolved assemblage (see Engel et al. 2009).

14.2.2 Donovan's Feeding Group Classification

Heuristic classifications of termite trophic groups appear frequently in reports of field sampling, and in biogeographical and phylogenetic analyses (e.g. De Souza and Brown 1994; Eggleton et al. 1996; Donovan et al. 2000; Davies et al. 2003; Inward et al. 2007; Yamada et al. 2007). A related concept of lifetypes and lifeways was proposed by Eggleton and Tayasu (2001), which integrates feeding habits with some aspects of behaviour and with evolutionary history. Termites observed in the field can be allocated to trophic group by inspection of abdominal colour, behaviour, the microsites in which they are found and the character and contents of their nests or colony centres (Table 14.1). The advantage of such "functional" classifications is that termite assemblages can be compared one with another without the need for every specimen to be identified to species, or for all sampling to be exhaustive (Bignell et al. 2008), but it is clearly only an expedient to designate feeding habits without viewing the gut contents directly and no indication is given of the true diversity of intestinal physiologies and digestive mechanisms.

Table 14.1 Definitions of putative (heuristic) termite feeding groups, based on inspection and knowledge of natural history (after Eggleton et al. 1997). From Bignell (2006)

Group	Descriptions
Soil	Termites distributed in the soil profile, surface litter (leaves and twigs), and/or epigeal mounds, feeding on mineral soil; workers dark-bodied
Soil/wood	Termites feeding only or predominantly within soil under or plastered within logs, or feeding within highly decayed wood that has become friable and soil-like; workers dark-bodied (= intermediate feeders sensu De Souza and Brown 1994)
Wood	Termites feeding on wood and excavating galleries in large items of woody litter, which in some cases become colony centres. This group may also include species with arboreal (carton) nests, epigeal (soil or carton) nests or subterranean nests; also includes some Macrotermitinae cultivating fungus gardens
Litter	Termites that forage for leaf litter and small woody litter; includes some mound-building and subterranean Macrotermitinae, also epigeal mound builders and arboreal nesters of the Nasutitermitinae (nasute soldiers), which forage on the surface of the litter layer
Lichen	Termites that forage for lichen, mosses and algae on the bark of trees

While microscopic examination of termite gut contents to characterise symbionts has been reported for more than a century, Sleaford et al. (1996) pioneered the systematic analysis of the food consumed by classifying the items seen in suspensions of fresh whole hindgut lumen material after gentle maceration and staining with vital dyes. Nine termite species were examined, representing four heuristic feeding groups, from a humid forest assemblage in equatorial Africa. Overall, seven components could be identified by morphology: silica, plant roots, arthropod parts, fungal mycelium, plant tissue fragments, macerated organic material and lignified tissue (tracheids), with a further three item categories (humus, safranin +ve and gentian violet +ve characterised by staining or natural colour). The number of items seen in individual species varied from just four in the wood-feeding lower termite *Schedorhinotermes putorius* to 9 or 10 in soil-feeding termites from the Apicotermitinae and Termitinae, and (surprisingly) 10 in the fungus-growing *Pseudacanthotermes militaris*. Ordinations, based on proportionate content (% of all items) and on Pianka's proportional utilisation function suggested that 5 or 7 feeding groups, respectively, were distinguishable in the subset of the assemblage examined, forming a gradient of increasing humification of the material with which each termite is principally associated. Unfortunately, the procedure is not suitable for preserved specimens, but Donovan et al. (2001a) were able to modify the analysis for museum collections, using the P3 contents (only) diluted 40 times and viewed under bright field microscopy. In their analysis of 46 species covering the entire range of taxonomic and heuristic feeding forms in the Order, gut content items were scored for silica, plant root hairs, arthropod parts, fungal mycelium, plant tissue fragments and recognizable microbial cells. Principal component analysis showed that the 46 species could (once again) be placed along a humification gradient, using variations in the amount of silica and plant fragments in the gut. In a redundancy analysis, 22 characters of gut morphology were found to be significantly correlated with gut content variations, of which three (right mandible morphology, the number of Malpighian tubules and the degree of sclerotisation of the enteric valve) unambiguously defined four feeding groups (I – IV) in increasing order of humification (Table 14.2). By mapping the four groups onto the existing phylogenetic tree of the Isoptera (Donovan et al. 2000), it was possible to show that the more derived clades are feeding on the most humified material and confirming a broad evolutionary trend from wood-feeding to soil-feeding, as proposed by Noirot (1992). However, wood-feeding, grass-feeding and epiphyte-feeding Termitidae appear to have reversed the trend and reverted to consuming less humified material (see also Eggleton 2006; Inward et al. 2007).

The quantitative studies described above confirm the more intuitive conclusions of Noirot (1992), Bignell (1994) and Bignell and Eggleton (1995) concerning the directions of termite trophic diversification, with the additional twist that there may be a selective advantage conferred on taxa that have the gut morphology of a higher termite but feed on less decomposed plant material. Field biologists will recognise the success (if not dominance) of the genera *Amitermes*, *Microcerotermes*, *Trinervitermes*, *Nasutitermes* and *Hospitalitermes* in many tropical ecosystems, which illustrates the point. Soil-feeding may have first arisen from the behaviour of the Macrotermitinae, or a basal group from which the present Macrotermitinae are derived. Lower termites phylogenetically basal to the Macrotermitinae are wood- or grass-feeders, but use faeces as a construction material (this is true even in *Coptotermes acinaciformis*, where the interior of the mound-nest is a softer, organic rich substance). In the Macrotermitinae the faeces are instead used to grow the fungal symbiont and therefore soil is the building material, which then necessarily gets into the gut, pre-adapting the termite to soil-feeding, destroying the hindgut flagellates and providing a novel inoculum of humus-digesting gut bacteria to replace the protists at the same time. Current phylogenies of the Termitidae are consistent in showing the Macrotermitinae as basal and the Group III soil-feeders as the next most basal clade (Donovan et al. 2000; Inward et al. 2007). It is unlikely that the Group IV soil-feeders, which ingest highly humified mineral soil, could have evolved de novo without first passing through a Group III intestinal morphology.

14.2.3 Gut Configurations, the Enteric Valve and Particle Separation: Some Progress

Two relatively recent studies which strikingly illustrate the variable and complex configurations of the enteric valve in termites are well worth examining. Sands (1998) offers more than 1500 line drawings in his monumental work on the identification of African termites from the worker caste, of which more than half, approximately 57%, are of gut structure (dissected and in situ), Malpighian tubule attachment and enteric valve seating, This probably represents the largest single

Table 14.2 Feeding group classification of Donovan et al. (2001a). The gut contents of worker caste termites representing 46 species or morphospecies, taxonomically diverse and collected throughout the tropics, were analysed microscopically for the relative proportions of silica, plant tissue fragments, fungal mycelium, arthropod parts, plant root hairs and non-plant cells. Principal component analysis showed that an axis (*axis 1*) mainly derived from plant tissue fragments and silica, and defining a humification gradient, could explain the largest proportion of the variation in the data. A redundancy analysis, with Monte-Carlo permutation tests, showed that three gut morphological characters (out of 22 significantly associated with gut content variation) permitted an unambiguous classification of the specimens into four feeding groups, with generally consistent placement along the humification axis (except group II). This allows unknown specimens to be allocated to an objectively defined feeding group without resort to gut content analysis and without subjective assessment of natural history and abdominal colour. Clades follow Kambhampati and Eggleton (2000). From Bignell (2006)

	Diagnostic gut morphological characters			Typical gut contents		
Group	Malpighian tubules	Rt. mandible molar plate ridges	Enteric valve	Silica	Plant fragments	Examples
I	8 or more			Low	High	Lower termites
II	4 or less, attached at mesenteron/P1 junction only	Prominent	No armature between ridges	Low or moderate	High, but with variable degrees of humification	Macrotermitinae, and representatives of: *Amitermes*-clade *Cornitermes*-clade *Nasutitermes*-clade
III	4 or less, attached as above or otherwise	Vestigial	Unsclerotised ridges	Moderate or high	Moderate	
		Absent	Some sclerotisation	Moderate or high	Moderate	Representatives of: *Anoplotermes*-clade *Termes*-clade *Cornitermes*-clade *Nasutitermes*-clade
IV	4 or less, attached as above or otherwise	Absent	At least half the ridges 50% sclerotised	High	Low	*Apicotermes*-clade *Cubitermes*-clade and representatives of: *Anoplotermes*-clade *Amitermes*-clade *Termes*-clade *Cornitermes*-clade *Nasutitermes*-clade

body of information on termite gut morphology ever assembled, surpassing in quantity (though perhaps not quite in elegance) even the older French literature. In addition, there are 162 photographic plates showing dissected enteric valves, and (on p17) a detailed explanation of how preserved specimens can be dissected to show this and other key taxonomic features including posterior mesenteron appearance, Malpighian tubule insertion, the midgut/hindgut or midgut/mixed segment conjunctions and the enteric valve seating. Donovan (2002) presents detailed accounts of enteric valve armature in eleven African taxa of higher termites where the cuticle is highly sclerotized, illustrated by 40 spectacular scanning electron micrographs.

Sands (1972) and Johnson (1979) drew attention to the usefulness of the in situ gut configuration as a taxonomic character and as a guide to identification in the worker caste, and gut folding has duly assumed prominence in some modern keys (e.g. Sands 1992, 1995, 1998). Characteristically, in the absence of any substantive evidence Sands is reluctant to speculate in print on the functional significance of the coils which are consistent within taxa but different between them. However, it is intuitive that the coiling is not simply a means of compressing an extended alimentary system into a relatively short abdomen, but must have some further physiological significance (Sands, personal communication). Bignell et al. (1983) pointed out that in Group IV soil-feeding Termitinae (*Cubitermes* clade), the overall effect of coiling was to bring the posterior hindgut (especially the P4) into proximity with the mixed segment and anterior P1 (essentially the former overlays the latter, in dorsal view), which could facilitate transepithelial fluid transfer between the two regions, which in this clade are axially separated by the voluminous P1 and P3. Thus a circuit of fluid movement could be created flushing the larger hindgut compartments with labile nutrients to prime the diverse microbial community, which then degrade the primary resource (lignocellulose, or humified soil organic matter). With a somewhat different perspective and different data, Schmitt-Wagner and Brune (1999) demonstrated in the same clade that hydrogen production (a pre-requisite for methanogenesis) was characteristic of the anterior hindgut, while methane efflux was more characteristic of the posterior region, but also suggested that a transepithelial transfer of hydrogen or a highly reduced small organic molecule might link the two microbially-mediated processes, allowing the reductive production of both methane and acetate in the P3/P4 compartment. Both the mixed segment and the P4 epithelia have ultrastructure consistent with active transport and K^+ is acquired by the P1, presumably accounting for its very high pH (Bignell et al. 1983), but fluxes have not been investigated in vivo. By contrast the P1 and P3 epithelia are of a squamous character, but with enhanced cuticle. This does not rule out solute effluxes, but they would be passive flows along diffusion gradients. Many different gut designs can be seen in the three higher termite sub-families which are not Macrotermitinae: in many the P1 is a long tubular structure and in others the posterior hindgut consists only of a large P3, with relatively poor development of the P4 region (Bignell 1994; Sands 1998). These may represent a variety of digestive and symbiont-housing strategies the rationale of which is as yet unknown.

Donovan (2002) considered the aspect of the enteric valve in situ (something which is not obvious from flattened preparations on microscope slides) in basal and derived soil-feeding Foraminitermitinae and Apicotermitinae (sensu Inward et al. 2007) where the valve complex is everted posteriorly into the P3 from the P1/P3 junction. In these genera the armature is external to the valve ridges, i.e. it faces the periphery of the anterior P3, in most cases developing an elaborate but genus-typical array of spines. Sclerotized enteric valves are also a feature of the soil-feeding Termitinae, but here the ridges are generally carried on the wall of the extreme posterior P1, which approximates to the apex of a cone, and the armature therefore faces the lumen of the P1 rather than the periphery of the anterior P3 (numerous illustrations in Sands 1998). The Apicotermitinae are the only clade in which the armature has passed onto the opposite face of the valve, but apart from this basic observation it is not yet possible to assign homologies to the enteric structure because of their wide variation and the incomplete phylogeny of the subfamily (Donovan 2002).

Enteric valves have six longitudinal ridges where the muscle layers between the cuticle and the basal sheath are thickened, and which carry the main sclerotizations. These are most extensively developed in soil-feeders, but less so in wood-feeders whatever their phylogenetic position. However wood-feeders triturate food material using the gizzard (which also shows a hexagonal arrangement of ridges and armature) and the gizzard is consequently a prominent feature. Hexagonal radial symmetry is a feature of the rectum, where the epithelium (but not so much the musculature) is thickened into six longitudinal ridges where fluid and ion exchange processes are concentrated (Noirot and Noirot-Timothée 1977), but not the midgut (mesenteron) where the number of anterior caeca (where present) is 8 and the number of Malpighian tubules is 16, 10, 8 or 4, and rarely 2. These varying symmetries are a puzzle in termite embryological developmental, which like so much else in intestinal morphology remains to be fully explained.

There is subjective evidence from microscopy that coarse and fine fractions of ingested soil are separated in the posterior hindgut of Group IV Termitinae (see Fig. 6 in Bignell 2000, also Bignell et al. 1983), although the quantitative analysis to prove this has yet to be done. The coarse material, which looks identical to the contents of the P1 and which therefore presumably corresponds to what is ingested, forms a central core and is passed quite rapidly to the P4 and P5, from which it can be refluxed in an anterior direction by alternating waves of antiperistalsis and peristalsis. Finer material accumulates closer to the gut wall in the P3 and anterior P4, where large numbers of prokaryotic microorganisms are also present. It is not known how the finer material is separated from the coarse, but one function of the enteric valve may be to perform this fractionation, thus concentrating more digestible organic components of the soil (Donovan 2002).

14.2.4 Processing and Roles of Faecal Material

The available fossil evidence suggests the main adaptive radiation of termites (and their trajectory towards ecological dominance in the lowland tropics) was a mid-Tertiary event, and occurred after the start of the general continental break-up in the Cretaceous (Thorne et al. 2000). A major event in the evolution of eusocial termites seems to have been the use of soil, initially for construction of mounds and galleries, and subsequently as a food (Donovan et al. 2000). This apparently coincides with the loss of protist intestinal symbionts and the acquisition (or co-evolution), in some clades, of gut bacteria which can degrade soil organic matter sufficiently to provide the termite host with adequate nutrition. The advantage of soil-feeding is (1) that this resource is available in infinite amounts, although quality probably diminishes with increasing depth and (2) that it can be easily accessed by the construction of relatively simple subterranean gallery systems around the colony centre. The energetic cost of feeding is therefore low, competition for resources with other colonies or other types of animal is minimal and defence of the colony can be achieved largely by the avoidance or exclusion of predators from mounds and tunnels, rather than the production of large numbers of specialised soldiers and/or foraging workers protected by thick cuticles (Bignell 1994).

Faeces seem to be a major construction material employed by termites and are used to line existing galleries, but further metabolism occurs by microbes and the linings are re-ingested, more in the fashion of an external rumen (Brauman 2000; Fall et al. 2001, 2004). Recent evidence suggests that some soil-feeders can lyse and assimilate ingested microorganisms (Fujita et al. 2001), as is thought to be the mode for litter transformers (e.g. Bignell 1989), so perhaps there are different versions of soil-feeding or else soil-engineers (sensu Lavelle et al. 1997) are really only litter transformers under another guise. Assays of soil-feeder guts for cellulase and xylanase have proved negative (Rouland et al. 1986, 1989), however a more recent report of the isolation of an alkaliphilic xylanolytic bacterium from the *Termes*-clade member *Sinocapritermes mushae* (Ohkuma et al. 2003) suggest that residual polysaccharides might still be a resource, in addition to humic material.

Studies of the effects of transit of soil through the guts of Group IV soil-feeders by Fall et al. (2001, 2004, 2007) and Donovan et al. (2001b) have provided useful data to inform the question whether termites promote the fertility of tropical soils, in the absence of fertilisers applied by humans. Brauman (2000) pointed out that the mounds of the soil-feeding *Cubitermes niokolensis* are built from a combination of faeces and soil, resulting in significant increases in clay, inorganic nutrients (P, Ca and NH_4) and organically combined C and N, compared with the adjacent parent soil. This implies either selective feeding by the termites, perhaps excluding more of the coarse silica, or fractionation of ingested soil within the gut accompanied by selective defaecation of finer material within the mound as it is constructed. Unsurprisingly, an analysis of the bacterial communities in parent soil, gut contents and whole mounds by DGGE DNA fingerprinting showed that they were all different, the mound being dominated by actinobacteria in contrast to the Firmicutes and Proteobacteria phyla, whose profiles were most commonly seen in the gut contents and parent soil (Fall et al. 2007). Donovan et al. (2001b) conducted a field experiment in which soils collected from five sites along a forest disturbance gradient in Cameroon were worked by colonies of *Cubitermes fungifaber* in temporary mesocosms. After 2 weeks pH, organic C, water content and the proportional kaolinite content were increased relative to unworked controls, while quartz was reduced. Much more work of this kind is needed so that the contemporary explosion of information about microbial diversity in termite guts is balanced by a better grasp of the consequences of termite feeding activities at the microsite level in natural environments. For reviews of the broad interactions between termites and soils, see Holt and Lepage (2000) and Bignell (2006).

14.2.5 Accommodation of Microorganisms

The microbiology of the termite gut is reviewed by Breznak and Brune (1994), Breznak (2000), Inoue et al. (2000), König et al. (2002, 2006) and Brune (2006). Interpretations of the available data and hypotheses in the context (loosely) of bioreactor theory appear in Brune (1998, 2007), Brune and Friedrich (2000) and Ohkuma (2003, 2008). These short papers can be recommended as providing a simple introduction for those not intimately schooled in biochemistry and microbiology. Spirochaetes have excited special attention because of their abundance and diversity in (most) termites, and because culture-independent molecular methods have now been able to complement the data obtained with some difficulty from in vitro culture; these advances are reported by Breznak (2002, 2006), Breznak and Leadbetter (2006) and Berlanga et al. (2007) Intestinal protists in lower termites are addressed by Inoue et al. (2000) and Brugerolle and Radek (2006), while the relationships between Macrotermitinae and fungi are explored in depth by Rouland–Lefèvre and Bignell (2001) and Rouland-Lefèvre et al. (2006). Useful and straightforward schemata of the dissimilation process are given in Tholen and Brune (2000) and in Warnecke et al. (2007), based on lower termites (those with flagellated protist symbionts included in the gut microbiota) and higher termites (those without protists), respectively.

Details of new descriptions and the discoveries of novel types amongst termite associated prokaryotic microbiotas are given by Brauman et al. (2000), Breznak (2000), Lilburn et al. (2001), Wenzel et al. (2002); König et al. (2002), Ohkuma et al. (2002, 2003), Thongaram et al. (2003), Schmitt-Wagner et al. (2003a, 2003b), Donovan et al. (2004), Adams and Boopathy (2005), König (2006), König et al. (2006) and Herlemann et al. (2007). Purdy (2007) reviews the distribution and diversity of methanogens (Euyarchaeota) in termites. A long series of elegant papers from Japanese laboratories describe the application of modern molecular biological techniques to assess the heterogeneity of termite microbiotas and their possible phylogenies (Ohkuma and Kudo 1996, 1998; Ohkuma et al. 1995, 1996, 2002, 2005, 2007; Hongoh et al. 2003, 2005, 2006a; Thongaram et al. 2003). There is perhaps some danger of being overwhelmed by phylogenies, and the almost Gadarene rush into molecular sequencing does little to specify the actual roles (if any) of the individual clones identified, but the current evidence suggests that termite intestinal microbiotas evolved with their hosts and are therefore different communities in different termite taxa. In termite science a mild controversy has existed for years whether gut microbiotas are generically similar, but marshalled physiologically by the host in different ways in different taxa or functional groups, or whether termite evolution and diversification has been driven by the acquisition of different key microbial symbionts in different clades. Although the argument may now appear to be settled in favour of the latter view, this is only because sequencing methodologies focus attention on diversity, while the question of whether the termite host can regulate its microbiota physiologically is scarcely ever asked, yet alone investigated. Occasional reports suggest that gut microbiotas may change with age (that is to say with changing physicochemical conditions) or with the food ingested (Tanaka et al. 2006; Miyata et al. 2007), and in the Macrotermitinae there is even a clear-cut intra-colonial variation between castes, as well as a discernible variation between colonies (Hongoh et al. 2006b). Microbiological evidence seems to have settled the matter in favour of taxon-specific microbiotas, but this does not necessarily explain the great diversity of intestinal morphologies (and by implication, physiologies) found in the Isoptera, which in some respects exceeds the diversity of external morphologies, at least in the worker caste, suggesting that natural selection has acted more on internal intestinal interactions than on integumentary structure.

The termite hindgut contains the great majority of the microorganisms present in the intestinal system as a whole and which may contribute as much as 40% of the animal's weight (Shulz et al. 1986; Slaytor et al. 1997). A large proportion of the microorganisms are semi-permanent and attached to either the cuticular wall of the hindgut or to chitinous spines protruding in the lumen (Berchtold et al. 1999; Bignell 2000; Noirot 2001). The remainder are suspended in the gut contents, or internal symbionts of flagellates (in lower termites), but a series of valves and muscular constrictions between the main gut components regulate the movements of the contents so that each is, in effect, a continuous culture chemostat. Archaea, eubacteria, actinobacteria and spirochaetes have all been identified as elements of the permanent prokaryotic microbiota, although it is still unclear whether representatives of all four groups are present in every case or whether the balance changes with the food consumed (see Miyata et al. 2007). There is dense packing of cells in some places, especially adjacent to the gut wall, and bacterial population densities up to 10^{12} organisms mL^{-1} have been described (Bignell et al. 1980; Bignell 2006), close to the theoretical maximum (i.e. close packing with all available lumen space occupied). Where flagellate protists are present (in lower termites), they may reach a population of 10^5 per individual, representing about 60% of total hindgut weight. At such density they are also closely packed in the lumen, but still motile to some extent. It now seems clear that the flagellates are differentiated into cellulose and xylan utilisers (Inoue et al. 1997) and that much of their digestion involves the internalization of fragmented food. Bacteria can also be consumed and in both cases the process involved is endocytosis (Inoue et al. 2000). Since both the flagellates and the termite hosts produce polysaccharidases, the digestive process is shared (Watanabe and Tokuda 2001). Earlier views that cellulolytic bacteria are not present in termite guts (O'Brien and Slaytor 1982; Slaytor 2000) must now be revised in the light of new evidences (e.g. Rouland and Lenoir-Labé 1998; Wenzel et al. 2002; Ohkuma 2003), but the principle that termites do not have an absolute dependence on their microbial associates for cellulose and hemicellulose digestion still seems secure. It remains a major issue whether some or all of the organisms mediating the above functions are specialist forms confined to termites, or are recruited from inocula available in their immediate environments. The list of novel organisms isolated from termites, many of which are undescribed and/or uncultivable, continues to grow (op. cit.), while molecular genetic studies suggest that mutualist communities evolve with their hosts and are therefore different from one species (or functional group) to another (Ohkuma et al. 1995, 1996, 2002; cf. Aanen et al. 2002).

Parts of the intestine in higher termite (but excluding the Macrotermitinae) have a high or very high pH, reaching as much as pH >12 in some cases (Bignell and Eggleton 1995; Brune and Kühl 1996), occasioned by (or accompanied by) secretion of K^+ into the gut lumen (Bignell et al. 1983). In the Termitinae, the upward step in pH is clearly associated with the mixed segment and the highest values are found in the P1, with a subsequent neutralisation in the P4 (P3b, terminology of Noirot 2001). In the two other subfamilies of higher termites, where gut homologies are somewhat distant from the Termitinae, the elevations are generally less marked and the P3 is the most alkaline region, although the definitive microelectrode method of measurement devised by Brune and Kühl has been used in just a handful of species, all Termitinae. High pH was initially thought to suppress microorganisms (e.g. Bignell et al. 1980; Ji et al. 2000), but it is now clear that the niche supports a number of novel alkaliphiles including clostridia, bacilli, and bacteria from the phylum Planctomycetes (Ohkuma et al. 2003; Thongaram et al. 2003, 2005; Köhler et al. 2008). Planctomycetes in the P3 of *Cubitermes* spp. appeared microscopically as rods or curved rods with a population density estimated at 2.6×10^9 mL^{-1} and were said to comprise about one third of all bacterial cells detectable by FISH and 17% of the cells stained by DAPI (Köhler et al. 2008). While these numbers are not close to the estimated 1×10^{12} cells mL^{-1} in the P3 of the related soil-feeder *Procubitermes* sp., based on staining with phenolic aniline blue, they are about the same as the totals viable obtained on general growth media with either aerobic or anaerobic incubation (Bignell et al. 1980).

Planctomycetes are widespread in aquatic environments but have not been cultured from termites. In other settings they can show chemoheterotrophic metabolism with an ability to degrade a wide range of biopolymers (Ward et al. 2006) and can therefore be considered candidates for the degradation of humic substances which is presumably the basis of soil-feeder digestion. Of particular interest in soil-feeders is the high density of filamentous microorganisms associated with the lumen of the mixed segment and the posterior P4b (Kovoor 1968; Potts and Hewitt 1973; Bignell et al. 1979, 1983; P4b, terminology of Bignell et al. 1980). The filaments, which are clearly prokaryotic, have never been identified or allocated taxonomically, although the suggestion of Margulis et al. (1998) that they might be the so-called "Arthromitus" stage of *Bacillus cereus*, a common soil organism, is worth investigating. Tokuda et al. (2001) made a detailed examination of the mixed segment of *Nasutitermes takasagoensis*, the histology and ultrastructure of which broadly (but not exactly) matches that of *Procubitermes* and *Cubitermes* (cf. Bignell et al. 1983), and also found a characteristic bacterial microbiota, but this was identified by 16s rDNA sequencing as two species (clone groups) within the genus *Clostridium* (Tokuda et al. 2000). An in situ hydridisation procedure with clone-specific primers showed that the clostridia are strictly associated with the mesenteric epithelium of the mixed segment, either interdigitating with the microvilli or attached to the epithelium (as also described for *Procubitermes* and *Cubitermes,* op. cit.), but the report does not mention filaments in this location or in the gut lumen. The clostridia are also found in the (rudimentary) mixed segments of the soldiers, but nowhere else in the intestines of either caste. Iron-reducing bacteria have been identified in termites (Trinkerl et al. 1990; Vu et al. 2004) and provisionally allocated to the genera *Clostridium* and *Desulphovibrio.* Vu et al. (2004) speculate that iron and sulphate reductions may be major electron sinks in termite guts, rivaling or possibly complementary to methanogenesis and reductive acetogenesis, and also linked such that toxic sulphide is removed as FeS. A characteristic of soil-feeding Termitinae freshly dissected in the field is (often, though not invariably) a prominent black deposit within the anterior P1, co-located with the apparent dense bacterial monocultures which begin in the mixed segment.

14.3 Physiology

The radial and axial redox profiles of the intestine have now been established by separate probing with hydrogen and oxygen sensitive microelectrodes (Brune et al. 1995; Schmitt-Wagner and Brune 1999) following an earlier (but cruder) study of redox potential by a platinum electrode method (Bignell 1984), but as with pH these profiles have been systematically documented in only a few species. However, they include both wood-feeding and soil-feeding forms and so embrace the two principal feeding modes to be found in termites. Profiles for hydrogen and oxygen partial pressures within the gut lumen can be clearly established, but interpretation of net redox gradients (and therefore the prediction of possible modes of microbial metabolism) is more difficult because E_h (apparent redox potential) is strongly pH dependent. However it seems possible to conclude that in the main hindgut compartments (P1 and/or P3), soil-feeders are less electronegative overall than wood-feeders (Kappler and Brune 2002).

Knowledge of the physiological environment within the termite gut, advanced by the use of oxygen- and hydrogen-sensitive microelectrodes, is now such that the availability (and unavailability) of oxygen and electron donors, and the consequent diversity of metabolic niches for prokaryotes are much better understood (Leadbetter et al. 1999; Tholen and Brune 1999, 2000). New data permit some hypotheses to be advanced concerning the difference in intestinal processes between wood-feeding and soil-feeding termites (e.g. Kappler and Brune 1999; Ji et al. 2000; Brune and Friedrich 2000), but many unanswered questions remain, notably the nature of the carbon sources utilised by chemoorganotrophic bacteria in the midgut and the hindgut (including those that fix atmospheric nitrogen). Our ignorance on this matter is surprising, not only because of the long-established textbook dogma that termites can degrade cellulose only by virtue of their symbioses with mutualistic microbes, but also because of the overwhelming evidence now available that all termites feeding on predominantly cellulosic foods can produce their own cellulases (reviewed by Slaytor 2000; Watanabe and Tokuda 2001, see also Tokuda et al. 1997; Lo et al. 2000; Hyodo et al. 2000, 2003; Rouland-Lefèvre and Bignell 2001). It is by no means clear that degradation of lignocellulose is the raison d'être of the mutualisms in any termite; nevertheless it is still an unchallenged dogma that no termite can exist without its microbial symbionts (Bignell 2000). Further, some theoretical arguments concerning the evolution of the termite-symbiont system, and its reinforcement by social organisation, still turn on the advantages to be gained by harnessing microbes to digest recalcitrant food components (e.g. Nalepa et al. 2001).

Insect physiologists have barely noticed termite guts, perhaps because their relatively small size and delicacy makes experimental procedures such as cannulation difficult, whereas the retrieval of samples for the detection and amplification of genomes or parts of genomes is easy by comparison. Thus even 30 years ago more was known about the basis of secretion, absorption, motility and coordination of the cockroach alimentary canal (e.g. Bignell 1981) than can be claimed today about any termite. Of particular interest would be to identify any active absorptive processes associated with the midgut, to pinpoint the site or sites of absorption of acetate and other short chain fatty acids in termites with extended hindguts, and to characterize the products and fluids secreted by the Malpighian tubules. Bignell (1994) proposed that the mixed segment of higher termites developed as a substitute for Malpighian tubules that would permit a much greater rate of fluid secretion into the hindgut and thus perhaps form the basis of a mechanism regulating microbial metabolism, but the idea has never been pursued, despite the known existence of neuroendocrine systems in insects controlling diuresis, peristalisis and many aspects of metabolism, including that of lipids (e.g. Capinera 2008). Although the elevation of pH in the hindgut is widespread in higher termites and includes many wood-feeding as well as soil-feeding forms, the only published structural studies which confirm a secretory role for the mixed segment are in three species of Termitinae (Kovoor 1968; Bignell et al. 1983) and one species of Nasutitermitinae (Tokuda et al. 2001). In the Apicotermitinae, the midgut/hindgut junction is as morphologically complex and variable as in the Termitinae and Nasutitermitinae, and the four Malpighian tubules equally diminutive, but no histological or ultrastructural studies appear to exist. The agent raising pH in the anterior hindgut is K_2CO_3 (Bignell et al. 1983; Kappler and Brune 1999).

14.4 Biochemistry

14.4.1 Overview of Termite Digestion

Termites are aggressive detritivores and impact C mineralisation directly by consuming organic detritus in all stages of presentation and decay, ranging from (in different termite species) sound living wood, sound dead wood in all size categories (in situ or fallen), dry standing grass, decaying wood and leaf litter, highly friable decayed wood and tree roots (often mixed with mineral material) and mineral soil itself (in which the digestible material is predominantly soil organic matter).

Although this diversity of feeding has been rationalised into four basic modes (Eggleton and Tayasu 2001; Donovan et al. 2001a), it still represents a remarkable adaptive evolution as it covers decaying plant material in all stages at which it could represent a feasible food, including the refractory long-lived polyaromatic residues formed at the end of the humification process. Further, the efficiency of digestion of woody (lignocellulosic) materials by termites is high; as much as 90% of the polysaccharide components of organic detritus (cellulose and hemicellulose) may be degraded in a single transit of the gut (Breznak and Brune 1994; Hopkins et al. 1998; Bignell 2000). While the efficiency of cellulose and hemicellulose digestion in some termite guts is very high, the extent of lignin breakdown is uncertain and processing by termites may be confined to alterations in chemical structure, accompanied by some demethylation and decarboxylation, rather than ring cleavage (Breznak and Brune 1994). Consequently the faeces of wood-feeding species are characteristically enriched in lignin (Hopkins et al. 1998), and in this sense one of the major global impacts of termites is to accelerate humification. However, where Macrotermitinae are present, some faeces are incorporated into symbiotic fungus combs, where in addition to extensive polysaccharide degradation, some lignin is mineralised directly as CO_2 (Mora et al. 1998; Hyodo et al. 2003). Therefore plant material foraged by Macrotermitinae is so efficiently mineralized as to preclude the development of lower-order food webs based on the residues. A smaller proportion of the organic material collected by fungus-growing termites is actually ingested and consequently exposed to a possible fermentative digestion. Amongst other consequences, the potential methane production per unit of organic biovolume consumed by the colonies is therefore less than for non fungus-growing forms.

Those species feeding on more humified food (extensively decayed wood and organic-rich soil) seem to have a more limited digestive capacity of their own, so the intestinal microbes here are, presumably, the primary agents of nutrient acquisition by degrading lignocellulose and components of soil organic matter in a manner yet to be fully elucidated. In soil-feeding species, recent studies have indicated that in addition to polyphenols, peptide fractions of soil organic matter may be targets, much of this matter being solubilised in highly alkaline regions of the alimentary canal before being fermented (Ji and Brune 2001, 2005). Hence, the potential influence of termites on decomposition processes is large, at least in the tropics, impacting both the C and the N cycles. At higher latitudes, termite diversity and abundance both rapidly diminish (Eggleton 2000), and impacts are (presumably) correspondingly reduced.

Although the broad basis of lignocellulose digestion by termites has been known for more than 60 years (cf. Hungate 1946), numerous refinements continue to be added to the picture. Improvement in our understanding of the roles of individual microbial taxa has been especially rapid in the past decade as high-resolution tools have become available, especially the use of microsensors and metabolic profiling based on molecular sequences. Schemata of the dissimilation process are given in Tholen and Brune (2000) and in Warnecke et al. (2007), based on lower termites (those with flagellated protist symbionts included in the gut microbiota) and higher termites (those without protists), respectively.

A general equation for anoxic polysaccharide degradation by termites is given by Tholen and Brune (2000). Although specifically referring to *Reticulitermes flavipes* (a lower termite) it illustrates the overall process likely to be found in most wood-feeding types, whether the microbial agents are protists or bacteria, or a combination of both:

$$(C_6H_{12}O_6) + 2H_2O \rightarrow 2CH_3COO^- + 2H^+ + 2CO_2 + 4H_2$$

This meets the criteria of a fermentation because it is a chemotrophic process in which energy, electrons and carbon are all derived from organic sources and reduced gases and/or organic solutes are released as end products without respiration (i.e. without the use of an external electron acceptor), but with a net gain of ATP. In reality the process is highly variable in its details between different fermenting microorganisms but always takes place in a series of steps controlled by enzymes, some of which are inhibited if the end products (or some of the reduced intermediates) accumulate. Acetate, the major solute accumulated in termites, is absorbed and utilised by the termite host to support its own respiration (which is aerobic and conventional), other than in the use of acetate as the principal substrate and the related suppression of pyruvate decarboxylation (see Odelson and Breznak 1983; Slaytor et al. 1997). The reducing equivalents (as hydrogen) are removed by secondary symbioses between the fermenting microbes and others which conserve their own energy, either by oxidising hydrogen to water (Tholen and Brune 2000) or, more significantly, by reducing CO_2 with hydrogen to form either additional soluble acetate (e.g. Leadbetter et al. 1999) or methane gas, which is released from the system as an efflux (e.g. Schmitt-Wagner and Brune 1999).

Such intimate co-operations between microbes with differing trophic styles are common in nature, and especially in anoxic habitats (for example animal intestines, aquatic sediments and waterlogged soils) where they arise spontaneously and are generally dominated by methanogenesis as the main mechanism for maintaining redox balance. The arrangements in termites are different in a number of important ways and almost certainly represent a long co-evolution between the insects and their gut microbiotas. In termites, the consortium of organisms present and the prevailing physiological conditions maximise the yield of acetate, with much reduced accumulations of other solutes such as lactate, formate, succinate and propionate, even though such additional end products would be predicted and organisms that can produce them from carbohydrate substrates when in single culture are well represented in the gut microbial community (Leadbetter et al. 1999; Tholen and Brune 2000; Schmitt-Wagner et al. 2003a). These additional solutes may be present as intermediates with high turnover rates, although the primary material fermented in wood-feeding termites lacking protists has not been clearly identified because termite polysaccharidases secreted into the (aerobic) midgut play a significant role in at least the early stages of food dissolution and digestion (Tokuda et al. 1997; Slaytor 2000). In lower termites with protist symbionts, it is clear that the flagellates dominate the dissimilation of lignocellulose to acetate (Breznak and Brune 1994), even though some endogenous cellulases from the termite's salivary glands are active in the system (Tokuda et al. 2004; but see Tokuda and Watanabe 2007 and Zhou et al. 2007).

Hydrogen gas itself is rarely evolved in large amounts, but a high partial pressure exists in some or all of the (central) part of the hindgut (Ebert and Brune 1997; Schmitt-Wagner and Brune 1999). In wood-feeding species, homoacetogens (bacteria reducing CO_2 to acetate) and methanogens are found in the same regions of the gut (Tholen and Brune 1999) but are spatially separated with the methanogens confined to the peripheral areas of the intestinal adjacent to, and in many cases attached to, the epithelial cuticle (Leadbetter and Breznak 1996). This is paradoxical, not only because the periphery of the intestinal lumen is oxic or microaerobic (Brune et al. 1995; Ebert and Brune 1997) but also because methanogens have a much greater affinity for hydrogen and would be expected to dominate in a mixed population (Leadbetter et al. 1999). Estimates of the contribution made by homoacetogens and methanogens to overall hindgut carbon fluxes vary. It has been argued that acetogenesis may account for as much as 30% of the termite's respiratory requirements and that methanogenesis may represent up to 10% of electron flow (Tholen and Brune 1999, 2000). The overall effect is an enhancement of respirable carbon benefitting the termite host, the advantage of which has presumably driven the evolution of the system and mitigated the overall efflux of methane. Where protists are present, there is a further twist to the story: the symbiotic flagellate communities themselves vary from one termite taxon to another, and individual flagellate species maintain their own private redox balances by employing endosymbiotic (internalised) or epizootic bacteria (attached to the external surface). These symbionts can be acetogens or methanogens, depending on the flagellate species and its phylogenetic lineage (Tokura et al. 2000). Thus the redox networks of the intestine vary in different termite species and the net methane efflux is not always predictable from first principles or reference to taxonomic position.

14.4.2 Degradation of Cellulose, Non-cellulosic Polysaccharides and Lignin

A number of recent studies have re-examined the question of whether polyaromatic compounds or lignin monomers can be digested by termites. A useful review, with an unusually clear presentation of the relevant chemistry, is given by Katsumata et al. (2007) who investigated the ^{1}H and ^{13}C NMR spectra of Björkman lignin preparations from tropical hardwoods ingested by *Cryptotermes brevis*. As in previous studies which compare food and faeces in wood-feeding termites by NMR spectroscopy (for example Hopkins et al. 1998; Hyodo et al. 1999), they concluded that only minor changes occurred to lignin, notably an increased formation of C–C linkages in guaiacyl nuclei and a significant decrease in aliphatic hydroxyl groups. Lignin content of the faeces increased (also as reported in many previous studies) but remained able to recomplex with remaining polysaccharide, further evidence that major digestion had not occurred. A simple explanation is that oxygen is a critical co-substrate for biologically mediated aromatic ring cleavage, and that the broadly anaerobic conditions in the termite intestine therefore exclude any degradations more extensive than alterations to side chains. The same spectral data indicated that about 85% and 83%, respectively, of glucosyl and xylosyl residues (the main neutral sugars from cellulose and heteroxylan in woody angiosperms) were digested. Katsumata et al. demonstrate how this value (>80%) is consistent across many independent studies of termite polysaccharide digestion, as far back as Holdaway (1933), and offer an intriguing comparison with ruminant mammals, where only 30–40% of cell wall polysaccharides are digested. The explanation may lie in the combination of the efficient grinding of ingested materials by the mandibular molar surfaces and the gizzard, with the dual secretion of polysaccharidases by both the salivary glands and the midgut epithelium. In vivo and in vitro degradation of lignin monomers and related aromatic compounds has been demonstrated in termites as phylogenetically well separated as *Mastotermes darwiniensis*, *Reticulitermes santonensis* and *Nasutitermes nigriceps* (Katsumata et al. 2007). However most of this observed metabolism was aerobic and the extent of lignin breakdown under the natural micro-oxic conditions presumed to prevail in intact guts is unclear.

While it is tempting to conclude that modern technology has merely confirmed what was known about termite digestion 80 years ago, there remains a parallel set of published studies which indicate that at least some of the carbon contained in lignin is mineralised by transit through the termite, and a further set of reports of the isolation of prokaryotic organisms from termites which show laccase activity (the oxidation of phenolic substrates with the attendant reduction of oxygen to water) or can utilise a variety of aromatic monomers to support growth, including some which are incorporated into lignin (for recent reviews see Ohkuma 2003; Brune et al. 2006, and for critical comments on much of the existing literature see Geib et al. 2008). For example, the work of Butler and Buckerfield (1979) and a series of papers by Cookson (e.g. 1987, 1988) demonstrated that when synthetic lignin analogues labeled with ^{14}C or lignin fractions extracted from plants grown in $^{14}CO_2$ were consumed by termites, as much as 10% or more of the label was mineralised. Degradation of lignin can include demethylation and hydrogenation, as well as cleavage of some inter-monomer linkages, which may explain the apparent mineralisation of carbon, but does not necessarily imply disruption of the aromatic ring (Kuhnigk et al. 1994).

Microorganisms able to degrade lignin preparations or lignin analogues in vitro continue to be isolated from both lower and higher termites, and remain of interest in the context of bioprospecting (Harazano et al. 2007; Ngugi et al. 2007), however the reactions involved are peripheral, for example oxidations of aromatic aldehydes and alcohols to the corresponding acids. Harazano et al. were able to make use of 16s rDNA sequencing to identify the bacteria involved as novel strains of *Burkholderia* and *Citrobacter*, and made the useful suggestion that such reactions were basically for detoxification purposes within the intestinal ecosystem, and not major digestive pathways. Geib et al. (2008) produced mass spectra of lignin in undegraded softwood and faecal frass of *Zootermopsis angusticollis*, following tetramethylammonium hydroxide (TMAH) thermochemolysis, a method which they claim accounts for all components of native lignin, and showed that gut passage caused side chain oxidation, ring demethylation and ring hydroxylation. The changes observed were interpreted as sufficient to expose the polysaccharide components of wood to an efficient digestion by gut microorganisms, but uncertainty remains whether they are consistent with the overall anaerobic state of the intestine. It was also thought possible that fungal or actinobacterial enzymes ingested with the wood or even secreted within the gut may have a synergistic role in lignin modification. In a new development, Taprab et al. (2006) showed that laccase was detectable in the fungus combs of *Macrotermes gilvus*, *Odontotermes* sp. and *Microtermes* sp. from Thailand, and coded by a gene whose sequences indicated a similarity with enzymes from white-rot fungi known to have strong lignin-degrading abilities. However, the synergistic peroxidise characteristic of free-living lignin degrading fungi could not be detected.

Substantial evidence exists for hemicellulose digestion in termites, both by symbionts and by endogenous enzymes the secretion of which presumably accompanies that of cellulase in the salivary glands and/or midgut (reviewed in Slaytor 2000; see also Schafer et al. 1996; Ohkuma et al. 2003).

14.4.3 Digestion in Fungus-Growers

In the higher termite subfamily Macrotermitinae there is a quite different basis for symbiosis (Rouland-Lefèvre and Bignell 2001). Basidiomycete fungi in a taxonomic grouping uniquely associated with termites are cultivated within the nest where they compost foraged woody material, leaf litter or dry grass, prior to digestion by the termites. Mycelium of the basidiomycete genus *Termitomyces* is grown in purpose-built chambers (within the mound or entirely subterranean), then harvested and consumed by the termite host. This termite subfamily is distributed in the tropics from Africa (where diversity is highest) through the Middle East, South and South East Asia, but is absent from Central and South America, and from Australia (Eggleton 2000). The fungal mycelium is grown on a structurally complex substrate, derived from primary forage (mostly wood and leaf litter) and constituting a matrix known as fungus-comb. The termites manipulate the culture to exclude competing fungi and then consume, at different stages of their development, the spores and the senescent mycelium. The composting process is strongly aerobic and results in some lignin breakdown, although it is unclear whether this benefits the termites directly (Hyodo et al. 2000). Hyodo et al. (2003) analysed the chemical composition of fungus combs of four species of *Macrotermes*, and in addition *Odontotermes* sp., *Hypotermes makhamensis*, *Ancistrotermes pakistanicus*, and *Pseudacanthotermes militaris*, but found that lignin was preferentially degraded relative to carbohydrates only in *Macrotermes*, which typically uses leaf litter rather than dead wood as its principal resource. Termites ingest (at different times in their life cycle) fungal conidia which are rich in cellulase and in some cases other polysaccharidases, and ageing fungal mycelium which has a C/N ratio much more favourable to insect growth than raw forage (Rouland-Lefèvre and Bignell 2001). As in other termites the fungus-growers produce their own cellulolytic and xylanolytic digestive enzymes and the intestinal system accumulates short chain fatty acids (Anklin-Mühlemann et al. 1995).

Like other higher termites (family Termitidae), Macrotermitinae lack intestinal flagellates, but in contrast to the remaining three subfamilies of higher termites, the Macrotermitinae have not developed soil-feeding, sensu stricto. This is surprising, as the group is noted for the complex nature of their mound/nest constructions and for the ability to manipulate and translocate large quantities of soil (especially subsoil) during mound building and repair, gallery maintenance and foraging (Wood 1996). Macrotermitinae workers do accumulate soil in their guts, though it is clearly not a nutrient, presumably as a consequence of the working and carrying of mineral materials, in which the mandibles and buccal cavity are principally involved (Badertscher et al. 1983). Soil is a good source of fungal inoculum, so the specific association with basidiomycete fungi, and the complex behaviour patterns needed to support extensive construction activities and to propagate the fungus are, presumably, derived characters whereas the consumption of soil, even accidentally, is basal. Modern Macrotermitinae are physiologically advanced, showing a notable efficiency in moisture acquisition and conservation, which often allows them to dominate the termite faunas of arid and semi-arid environments (Desmukh 1989; Noirot and Darlington 2000; Traniello and Leuthold 2000), but gut structure is more akin to that of lower termites (Bignell 1994). Foraging by Macrotermitinae is very intensive per unit of termite biomass, presumably as the strongly aerobic metabolism of the fungus, less constrained by physical space than intestinal mutualists, is correspondingly vigorous (Wood and Sands 1978). Consumption of litter and C fluxes mediated by termites appear to increase when Macrotermitinae are well represented in the assemblage. Estimates of litter consumption by Macrotermitinae range up to 1,500 kg (dry weight)/ha/year in moist savannas, giving them a role in C mineralization comparable to or exceeding that of bush fires and/or mammalian herbivores (see Buxton 1981, Desmukh 1989). In a moist West African savanna, Konate et al. (2003) showed that soil respiration was greatly enhanced wherever fungus-comb chambers were present below in the soil column; further, they estimated that the two species of Macrotermitinae present were responsible for mineralizing 4.9% of the above-ground primary production and 11.3% of the carbon not mineralized by annual fires. Organic matter turnover by other types of termite in such dry-land systems would be less than 10% of this amount. Visual evidence of foraging by Macrotermitinae is provided by red soil sheeting covering stems, lying dead wood of all descriptions and the surface of the ground. Soil utilised by Macrotermitinae for building seems to come from deeper layers with a relatively low inherent organic content. Most other termites collect soil from the top 15 cm of the mineral profile.

Although short chain fatty acids can be detected in gut homogenates of worker caste Macrotermitinae (Anklin-Mühlemann et al. 1995), it is unclear whether fermentation is a significant nutritional process. The relationship between termite and fungus seems to cater to different needs in different circumstances: in some species fungal products assist with digestion, while in others the fungus accumulates nitrogenous compounds which are eventually passed to the termite when the mycelium is consumed (Rouland-Lefèvre and Bignell 2001; Aanen et al. 2002). In *Macrotermes subhyalinus*, both methane and hydrogen are evolved but in small amounts on a weight-specific basis (Anklin-Mühlemann et al. 1995), and this might suggest that fermentative metabolism is of minor significance in fungus-growers. However the definitive test is an assay of reductive acetogenesis and this does not seem to have been carried out in any representative of the Macrotermitinae. The gut microbiotas of *Odontotermes formosanus* and *Macrotermes subhyalinus* were characterised by Yara et al. (1989), Paul et al. (1993) and Anklin-Mühlemann et al. (1995), respectively. These studies indicated that intestinal bacteria were as abundant and diverse as in other termites, but variations with age and termite morphotype were observed within each species. Hongoh et al. (2006b) defined the bacterial gut microbiota of major and minor workers of *Macrotermes gilvus*, based on the sequencing of 16s rRNA in 179 clones defined by T-RFLP. Community profiles clustered by termite morphotype and age, rather than by colony, and were also quite distinct from those of bacteria associated with the fungus comb, the adjacent soil and the sympatric wood-feeding *Microcerotermes* sp. This again suggests that fungus-growing termites have a co-evolved gut microbiota, dominated by termite-specific clades of Bacteroidales and Clostridiales, which are typically fermenting bacteria. A similar result using the same methodology was obtained in *Odontotermes formosanus* by Shinzato et al. (2007), who were able to assign most of the gut bacteria to the phyla Firmicutes (includes clostridia), Bacteroidetes, delta-Proteobacteria and Actinobacteria. Most reports suggest that spirochaetes are absent or have low abundance in Macrotermitinae; this may be significant as spirochaetes have been identified as agents of both acetogenesis and nitrogen fixation in other termites (op. cit.). Potentially fermenting bacteria affiliated to the phyla Cytophaga-Flexibacter-Bacteroides, Proteobacteria, Clostridiales and Planctomycetes were identified by molecular methods in *Macrotermes michaelseni* by MacKenzie et al. (2007). The clones obtained resembled previously uncultured bacteria, while aerobic plating of gut homogenates produced mainly actinobacteria and bacilli typical of soils. The broad conclusion is that the Macrotermitinae, like other termites, harbour many bacterial phylotypes which are (a) as yet uncultured, (b) typically cluster by termite genus and by termite feeding type and (c) within their phylum more closely resemble other strains specifically associated with termites than strains associated with other environmental niches. Since the nutritional biology of the Macrotermitinae has always been discussed purely in terms of the interaction between the termite and it fungal partner, and with the assumption that the strongly aerobic mycobiont removes the need for fermentative metabolism of plant tissue, there is now an obvious need to define the nutritional role (if any) of gut bacteria in fungus-growing termites. The success and dominance of Macrotermitinae in the savannas of Africa and Asia is obvious by inspection, and the key distinction between these termites and others is the mutualism with *Termitomyces,* whose scenescent tissues appear to provide completely adequate nourishment for termites. Intuition therefore suggests that intestinal bacterial metabolism should not be critical, but the matter remains open to investigation.

14.4.4 Lysozyme and Proteases

Breznak (2000) suggested that termites digest symbiont bacteria transferred by trophallaxis, thus assimilating a portion of the nitrogen fixed in the hindgut and effecting an overall nitrogen economy. The idea was supported by Fujita et al. (2001), who showed that lysozyme was secreted by the salivary glands of *Reticulitermes speratus* (a wood-feeding lower termite) and had a measurable activity in the foregut. The enzyme, known to be synthesized in insects either to digest bacteria or (more commonly) as part of reactive immunity, disrupts the backbone polymers of bacterial cell wall peptidoglycan and thus causes lysis. In addition, protease activity and a high concentration of free amino acids (47 nmol/gut) were demonstrated in the contents of the midgut. Bearing in mind how little nitrogen the woody diet contains, one explanation for these observations would be a bacteria-digesting mechanism of the kind proposed. However, it is also possible that a high concentration of amino acids in the midgut lumen merely reflects the flux of fluid derived from haemolymph through the ectoperitrophic space, which is proposed as the basis of nutrient uptake in insects (Dow 1987; but see Wolfersberger 2000). In elegant molecular and histological work, Fujita et al. (2002) showed that insect-typical lysozyme was synthesized in the salivary glands from endogenous genes and could not be the product of any microbial symbiont. However, when a number of higher termite species in different trophic groups were assayed for lysozyme and protease, lysozyme did not appear to be detectable in the salivary glands and foregut in every case, although protease was present in all midguts (Fujita and Abe 2002).

Lysozyme is widely produced in insect haemolymph to reduce bacterial infections, so its secretion as a digestive enzyme is not as remarkable as the secretion of endogenous cellulase (see Chapter 3 by Lo et al., this volume), but perhaps deserves more attention. It is regrettable that, aside from a focus on the production of cellulases, so little work has been done on the physiology (and microbiology) of the termite midgut. The midgut is a substantial structure in lower termites and in the Macrotermitinae (Bignell 1994), and even in other higher termites where it is relatively diminutive, what little evidence there is suggests there is an active secretory process (Bignell et al. 1982).

14.4.5 How Do Soil-Feeders Work?

Soil-feeding termites constitute about 60% of all termite genera (Noirot 1992; Bignell 1994) and are especially abundant in humid tropical forests and wetter savannas (Bignell and Eggleton 2000). Although soil-feeding has evolved in several lineages (Eggleton 2006), the overall common features of such termites are a high degree of differentiation of the intestine (especially a trend for the hindgut to become compartmentalised, see Bignell 1994), a marked elevation of the pH in the anterior hindgut (Bignell and Eggleton 1995; Brune and Kühl 1996) and a higher weight-specific rate of methane efflux (Nunes et al. 1997). A decade of elegant laboratory work by Brune and co-workers on four species of the Group

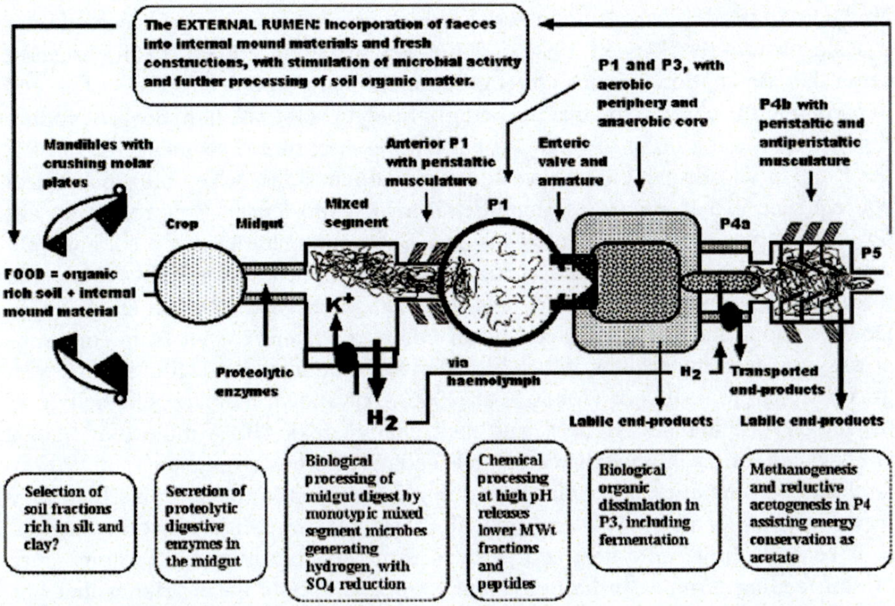

Fig. 14.1 Hypothesis of gut organization and sequential processing in soil-feeder termites of the *Cubitermes* clade (sensu Kambhampati and Eggleton 2000) and modified from the original proposed in Bignell (2006). The model interprets new information on the spatial separation of identified different elements of the microbial symbiont community and the sequential degradation of components of soil organic matter, from the recent publications of the Brune group (Brune et al. 2006)

IV African soil-feeder *Cubitermes* (*orthognathus*, *speciosus*, *umbratus* and *ugandensis*), with some additional study of the related *Thoracotermes macrothorax*, has enabled a fairly coherent and complete picture to be drawn of how the intestinal system works overall (Fig. 14.1). *Cubitermes* is chosen for study because it is a large termite, abundant and easy to find in forests or the wetter savannas, relatively robust after collection and transportation from the field, and shows a very high degree of hindgut differentiation into six distinct sections. Whether the same principles apply in other soil-feeders, especially the more trophically specialised Group III forms where hindgut differentiation is less pronounced than in *Cubitermes*, remains to be seen. Soil organic matter, much of which is polyaromatic in nature but with a higher nitrogen content than freshly dead wood, is in general regarded as an abundant resource of poor quality (i.e. recalcitrant to digestion), but one that can drive a corresponding evolutionary specialisation in ecosystems where competition for better quality food such as undecomposed or lightly degraded woody items is intense (Bignell 1994). However, mechanisms for the efficient degradation of polyaromatic substrates under completely anoxic conditions have not been previously known, so that models of the organisation of the soil-feeder gut have assumed a sequential process in which a primary chemical alkaline hydrolysis in the anterior hindgut releases labile organic materials which can be secondarily fermented after passing posteriorly (Bignell 1994; Schmitt-Wagner et al. 2003a). In the broadest sense this assumption remains correct, but additional properties of the system and in particular the observation that the anterior hindgut is the site of hydrogen production (in decreasing order of recorded partial pressure: mixed segment, P3 and P1, see Kappler and Brune 2002), have added new dimensions. A key biogeochemical consequence is that termite communities in which soil-feeders are prominent will be higher gross producers of methane gas. Such communities are predominantly equatorial and are rarely found at higher latitudes, but are especially vulnerable to disturbance and the effects of land-use change (Bignell and Eggleton 2000). Deforestation (and the associated loss of complex organic matter from soil) suppresses soil-feeding termites, which are often replaced by wood-feeding forms with greater apparent resilience (Jones et al. 2003). All known methanogens belong to the prokaryotic domain Archaea, and their characteristic rRNA molecular signals are stronger in gut homogenates of soil-feeding termites (Brauman et al. 2001). Archaeal communities in termite guts differ from those in the soil on which they feed (Friedrich et al. 2001; Donovan et al. 2004), making it likely that they are co-evolved with the termite host and perhaps vertically transmitted. The dominance of soil-feeding termites in humid primary tropical forests is so striking that one intuitively assumes they must represent some dramatic evolutionary leap forward, based on the efficient digestion of humic matter, but their gut contents also contain a great deal of dispersed, but recognisable plant tissue fragments (Sleaford et al. 1996; Donovan et al. 2001a) and the fate of this potential resource, once ingested, has never been investigated.

To identify the organic substrates digested by *Cubitermes*, Kappler and Brune (1999) extracted and fractionated organic matter from ingested and parent soils, and found that the alkaline environment of the P1 and P3 shifted the spectrum of humic acids towards smaller molecules and released them from their intimate association with inorganic soil constituents by a variety of possible mechanisms including deprotonation. Investigating the physiological conditions within the gut with microelectrodes, they showed that both the P1 and the P3 were significant oxygen sinks and were anoxic, excepting the most peripheral zone within 100 μm of the gut wall. However, the anoxic state of the P1 was generated chemically (autoxidation of soil organic matter), whereas that of the P3 appeared to be the consequence of chemoheterotrophic microbial metabolism. This is consistent with the estimated overall densities of microbial populations in the two compartments. In addition, conditions in the anterior hindgut reduce up to 80% of the iron in ingested soil (Kappler and Brune 2002), a process which is strongly stimulated by the presence of humic acids and which might therefore promote dissimilation of complex organic molecules (polysaccharides, polyphenolics or peptidoglycan) in the P3, with iron (III) acting as an electron sink within the greater part of the lumen and perhaps regenerated at the periphery where oxygen is available, or when faeces are exposed to air. Further experimental work by Ji et al. (2000) and Ji and Brune (2005) showed that 30% of the peptide component of a synthetic ^{14}C-humic acid was mineralised over a 10 day period when the construct was fed to *Cubitermes orthognathus*, compared to just 3% in a soil mesocosm without termites. Assays for proteolysis showed the highest activity in the midgut and mixed segment, and that the enzymes concerned were stable at the high pH prevailing in the anterior hindgut. The digestive strategy implied by these results explains some earlier observations, for example the ultrastructural evidence of intensive protein secretion by the midgut of *Cubitermes severus* (Bignell et al. 1982), the moderately negative apparent redox potential of the P1 (Bignell 1984) and the preliminary report of peptide depletion in pyrolysis-MS difference spectra between the foregut contents and faeces of *Cubitermes ugandensis* (Bignell 1994). Ji and Brune (2006) report a further consequence of the presumed peptide-based digestion of *Cubitermes*, which is the accumulation of large amounts of dissolved ammonia in the posterior hindgut contents (185 μmol/g dry weight), mound materials (25 μmol/g dry weight) and gaseous ammonia in the nest atmosphere (three orders of magnitude above ambient).

Although the substrates degraded by soil-feeders are still not precisely identified, micro-scale experiments by Schmitt-Wagner and Brune (1999), Tholen and Brune (1999), and Schmitt-Wagner et al. (2003b) have shown that the sites of hydrogen production and of methanogenesis (accompanied by acetogenesis) are axially separated, with the former anterior (mixed segment, P3) and the latter posterior (P4b) in the hindgut, the transfer of reducing equivalents between the two possibly taking place via the host haemolymph. It is not clear whether the separation of sites is responsible for the higher methane production, or whether this reflects some other feature of the overall digestive process, yet unidentified.

The overall picture that emerges from the *Cubitermes* model is of a five-part process (Fig. 14.1):

a. secretion of proteolytic digestive enzymes in the midgut.
b. a preliminary processing of the midgut digest by the (possibly monotypic) mixed segment microbial community, generating hydrogen.
c. chemical processing under anoxic, strongly alkaline conditions in the P1, releasing lower molecular weight fractions of soil organic matter into solution, with some biologically-mediated hydrogen generation.
d. microbially mediated organic dissimilation in the P3, including fermentation and generation of hydrogen, with end products absorbed by the host.
e. methanogenesis and reductive acetogenesis in the P4 (especially the P4b), utilising the hydrogen generated in b, c and d above, assisting energy conservation as acetate (absorbed by the host) and restoring overall redox balance.

This probably represents the most fully developed strategy for exploiting soil organic matter as a resource, and has without doubt been ecologically successful within the biogeographical domain of the *Cubitermes* clade (Eggleton et al. 1998; Eggleton 2000). However the clade is, in phylogenetic terms, quite highly derived (Inward et al. 2007) and it is necessary to explain why soil-feeding should have an evolutionary advantage at the earlier stages of its development, when gut structural differentiation would have been less marked.

14.5 An Overarching Hypothesis of Evolution

Insect evolution is closely linked to the appearance of vascular plants on land. Most schemes of Mesozoic adaptive radiation show an early line of detritivores as the common ancestors of modern-day cockroaches and termites, and there is growing evidence that these insects were dominant macroinvertebrates in Carboniferous swamp forests, where large quantities of decaying organic matter were available. However these distant ancestors were not social, nor is there any evidence of mutualism with microorganisms.

An argument can be made that the transition to subsocial organization in the ancestors of modern termites was a consequence of the employment of the external rumen principle and the associated behaviour of coprophagy, the habit of feeding on the faeces of other individuals, whether conspecific or otherwise (Nalepa et al. 2001). In many modern-day detritivorous invertebrates unrelated to cockroaches and termites but showing coprophagy, for example millipedes and woodlice, masticated litter serves as a good substrate for microbial growth, but changes little during passage through the gut (Bignell 1989). The faecal pellets consist of many fragments of relatively undecayed plant tissue compacted together, but are heavily colonised by bacteria when voided. The value of re-ingesting the pellet is self-evident, assuming the microbial cells can be lysed and assimilated, for which some evidence exists (Bignell 1989). There is a relatively homogeneous gut flora in the hindgut, colonizing the cuticular lining and which appears to be the main source of inoculum for bacterial growth on the voided pellet. The whole phenomenon is characterized as the external rumen and coprophagy can be seen as an early and basic adaptation of detritivores.

If faeces are ingested, microbes can be exploited selectively. They can be digested (there is evidence for this in modern termites as well as in modern millipedes), they can be exploited for any secretions, metabolites or exoenzymes of possible benefit to the animal host, they can be propagated in the faeces to amplify the external rumen effect, or they can be retained as mutualists, especially if the structure and physiology of the gut is modified to promote microbial diversity and co-evolution takes place. Coprophagy of conspecific material may have become a mechanism for ensuring that neonates received an inoculum of gut biota efficiently, as well as being a rich source of nutrients in its own right. The process of evolving a sophisticated community of hindgut microorganisms, such as that of termites, can be seen as a process of internalising the consortia comprising the external rumen, i.e. plant litter rots primarily in the hindgut after ingestion. An additional advantage of this system is that freshly dead plant material can be consumed before it has been significantly degraded or conditioned by microorganisms, a phenomenon seen in the more advanced grass-cutting and wood-feeding termites today. The internalised rumen is more productive and the turnover of nutrients from the substrate more predictable. It also allows for sustained interactions between microbes under redox conditions unlikely to be sustained in ageing faecal pellets.

Dependence on established gut consortia therefore grew, but these organisms became less predictably present in the external environment; consequently assured transmission came to require both gregarious behaviour and intraspecific coprophagy. There was then a transition from temporary aggregations to long-term membership of groups and then to sub-sociality by brooding with the mother and then to permanent family life. In the case where the gut biota included flagellates, which are sensitive to desiccation and oxygen (at least in the unencysted state), intergenerational transfer of mutualists had to occur via proctodaeal trophallaxis, the direct transfer of fluids from the rectal pouch of a donor to the mouth of a recipient. Trophallaxis also allows some of the tansferred flagellates to be used as food.

At some point in the evolutionary process termite faeces stopped being purely organic in nature and began to contain mineral soil. This was the end for the flagellates and the starting point for an impressive trophic and mutualistic diversification, with host and gut biota co-evolving, including several forms of soil-feeding. The rationale of soil-feeding seems to be that it is an infinitely abundant resource that does not need to be protected from competing colonies or species. Further, the principle of the external rumen appears to be re-evolved in soil-feeders, as the walls of galleries are lined with soil in which bacterial activity is high, and may be routinely consumed and re-lined (Fall et al. 2007). At the apex of termite evolution, faeces have decreased in importance as sources of food and mutualist inoculum, but have gained importance in other ways, notably as building materials.

References

Aanen DK, Eggleton P, Rouland-Lefèvre C et al (2002) The evolution of fungus-growing termites and their mutualistic symbionts. Proc Natl Acad Sci USA 99:14887–14892

Adams L, Boopathy R (2005) Isolation and characterization of enteric bacteria from the hindgut of Formosan termite. Bioresource Technol 96:1592–1598

Anklin-Mühlemann R, Bignell DE, Veivers PC et al (1995) Morphological, microbiological and biochemical studies of the gut flora in the fungus-growing termite *Macrotermes subhyalinus*. J Insect Physiol 41:929–940

Badertscher S, Gerber C, Leuthold RH (1983) Polyethism in food supply and processing in termite colonies of *Macrotermes subhyalinus* (Isoptera). Behav Ecol Sociobiol 12:115–119

Berchtold M, Chatzinotas A, Schönhuber W et al (1999) Differential enumeration and in situ localization of microorganisms in the hindgut of the lower termite *Mastotermes darwiniensis* by hybridization with rRNA-targeted probes. Archiv Microbiol 172:407–416

Berlanga M, Paster BJ, Guerrero R (2007) Coevolution of symbiotic spirochete diversity in lower termites. Int Microbiol 10:133–139

Bignell DE (1981) Nutrition and digestion. In: Bell WJ, Adiyodi KG (eds) The American cockroach. Chapman and Hall, London, pp 57–86

Bignell DE (1984) Direct potentiometric determination of redox potentials of the gut contents in the termite *Zootermopsis nevadensis* and *Cubitermes severus* and in three other arthropods. J Insect Physiol 30:169–174

Bignell DE (1989) Relative assimilations of ^{14}C-labelled microbial tissues and ^{14}C-plant fibre ingested with leaf litter by the millipede *Glomeris marginata* under experimental conditions. Soil Biol Biochem 21:819–827

Bignell DE (1994) Soil-feeding and gut morphology in higher termites. In: Hunt JH, Nalepa CA (eds) Nourishment and evolution in insect societies. Westview Press, Boulder, CO, pp 131–159

Bignell DE (2000) Introduction to symbiosis. In: Abe T, Bignell DE, Higashi M (eds) Termites: evolution, sociality, symbioses, ecology. Kluwer Academic Publishers, Dordrecht, pp 189–208

Bignell DE (2006) Termites as soil engineers and soil processors. In: König H and Varma, A (eds) Intestinal microorganisms of termites and other invertebrates. Springer, Berlin, pp 183–220

Bignell DE, Anderson JM, Crosse R (1991) Isolation of facultatively aerobic actinomycetes from the gut, parent soil and mound materials of the termites *Procubitermes aburiensis* and *Cubitermes severus*. FEMS Microbiol Ecol 85:151–160

Bignell DE, Constantino R, Czudi C et al (2008) Macrofauna. In: Moreira FMS, Huising J, Bignell DE (eds) A handbook of tropical soil biology: sampling and characterization of below-ground biodiversity. Earthscan, London, pp 43–75

Bignell DE, Eggleton P (1995) On the elevated intestinal pH of higher termites (Isoptera: Termitidae). Insect Soc 42:57–69

Bignell DE, Eggleton P (2000) Termites in ecosystems. In: Abe T, Bignell DE, Higashi M (eds) Termites: evolution, sociality, symbioses, ecology. Kluwer Academic Publishers, Dordrecht, pp 363–387

Bignell DE, Oskarsson H, Anderson JM (1979) Association of actinomycete-like bacteria with soil-feeding termites. Appl Environ Microbiol 37:339–342

Bignell DE, Oskarsson H, Anderson JM (1980) Distribution and abundance of bacteria in the gut of a soil-feeding termite *Procubitermes aburiensis* (Termitidae, Termitinae). J Gen Microbiol 117:393–403

Bignell DE, Oskarsson H, Anderson JM (1982) Formation of membrane-bounded secretory granules in the midgut epithelium of a termite, *Cubitermes severus*, and a possible intercellular route of discharge. Cell Tissue Res 222:187–200

Bignell DE, Oskarsson H, Anderson JM et al (1983) Structure, microbial associations and functions of the so-called "mixed segment" of the gut in two soil-feeding termites *Procubitermes aburiensis* Sjöstedt and *Cubitermes severus* Silvestri (Termitidae, Termitinae). J Zool 201: 445–480

Bitsch C, Noirot C (2002) Gut characters and phylogeny of the higher termites (Isoptera: Termitidae). A cladistic analysis. Ann Soc Entomol Fr 38:201–210

Brauman A (2000) Effect of gut transit and mound deposit on soil organic matter transformation in the soil feeding termite: a review. Eur J Soil Biol 36:117–125

Brauman A, Bignell DE, Tayasu I (2000) Soil-feeding termites: biology, microbial associations and digestive mechanisms. In Abe T, Bignell DE, Higashi M (eds) Termites: evolution, sociality, symbioses, ecology. Kluwer Academic Publishers, Dordrecht, pp 233–259

Brauman A, Doré J, Bignell DE et al (2001) Molecular phylogenetic profiling of microbial communities in guts of termites with different feeding habits. FEMS Microbiol Ecol 35:27–36

Brauman A, Kane MD, Labat M et al (1992) Genesis of acetate and methane by gut bacteria of nutritionally diverse termites. Science 257:1384–1387

Brauman A, Labat M, Garcia JL (1990) Preliminary studies on the gut microbiota of the soil feeding termite *Cubitermes speciosus*. In: Lesel R (ed) Microbiology of poikilotherms. Elsevier Science, Amsterdam, pp 73–77

Breznak JA (2000) Ecology of prokaryotic microbes in the guts of wood- and litter-feeding termites. In: Abe T, Bignell DE, Higashi M (eds) Termites: evolution, sociality, symbioses, ecology. Kluwer Academic Publishers, Dordrecht, pp 209–231

Breznak JA (2002) Phylogenetic diversity and physiology of termite gut spirochetes. Integr Comp Biol 42:313–318

Breznak JA (2006) Termite gut spirochetes. In: Radolph JD, Lukehart SA (eds) Pathogenic *Treponema:* molecular and cellular biology. Horizon Scientific Press, Norwich, pp 421–444

Breznak JA, Brune A (1994) Role of microorganisms in the digestion of lignocellulose by termites. Annu Rev Entomol 39:453–487

Breznak JA, Leadbetter J (2006) Termite gut spirochetes. In: Dworkin M, Falkow S, Rosenberg E et al (eds) The prokaryotes, vol 7. Springer, New York, NY, pp 318–329

Brugerolle G, Radek R (2006) Symbiotic protozoa of termites. In: König H, Varma A (eds) Intestinal microorganisms of soil invertebrates. Springer, Berlin, pp 243–269

Brune A (1998) Termite guts: the world's smallest bioreactors. Trends Biotechnol 16:16–21

Brune A (2006) Symbiotic associations between termites and prokaryotes. In: Dworkin M, Falkow S, Rosenberg E et al (eds) The prokaryotes, vol 7. Springer, New York, NY, pp 439–474

Brune A (2007) Woodworker's digest. Nature 450:487–488

Brune A, Emerson D, Breznak JA (1995) The termite gut microflora as an oxygen sink: microelectrode determination of oxygen and pH gradients in guts of lower and higher termites. Appl Environ Microbiol 61:2681–2687

Brune A, Friedrich M (2000) Microecology of the termite gut: structure and function on a microscale. Curr Opin Microbiol 3:263–269

Brune A, Kühl M (1996) pH profiles of the extremely alkaline hindguts of soil-feeding termites (Isoptera: Termitidae) determined with microelectrodes. J Insect Physiol 42:1121–1127

Butler JHA, Buckerfield JC (1979) Digestion of lignin by termites. Soil Biol Biochem 11: 507–513

Buxton RD (1981) Changes in the composition and activities of termite communities in relation to changing rainfall. Oecologia 51:371–378

Capinera JL (ed) (2008) Encyclopedia of entomology. Springer, London

Cookson LJ (1987) ^{14}C-lignin degradation by three Australian termite species. Wood Sci Technol 21:11–25

Cookson LJ (1988) The site and mechanisms of ^{14}C-lignin degradation by *Nasutitermes exitiosus*. J Insect Physiol 34:409–414

Costa-Leonardo AM (1995) Morphology of the digestive tube in the termite *Serritermes serrifer* (Isoptera, Serritermitidae). Naturalia 20:31–44

Costa-Leonardo AM, Landim CC (1997) Microorganisms of the digestive tract of Brazilian termites (Isoptera, Termitidae). Ciencia Cult J Braz Assoc Adv Sci 49:266–268

Czolij R, Slaytor M, Veivers PC, O'Brien R (1984) Gut morphology of *Mastotermes darwiniensis* Froggatt (Isoptera: Mastotermitidae). Int J Insect Morphol Embryol 13:337–355

Davies RG, Eggleton P, Dubbin WE et al (2003) Evolution of termite functional diversity: analysis and synthesis of local ecological and regional influences on local species richness. J Biogeogr 30:847–877

De Souza OFF, Brown VK (1994) Effect of habitat fragmentation on Amazonian termite communities. J Trop Ecol 10:197–206

Desmukh I (1989) How important are termites in the production ecology of African savannas? Sociobiology 15:155–168

Dillon RJ, Dillon VM (2004) The gut bacteria of insects: nonpathogenic interactions. Annu Rev Entomol 49:71–92

Donovan SE (2002) A morphological study of the enteric valves of the Afrotropical Apicotermitinae (Isoptera: Termitidae). J Nat Hist 36:1823–1840

Donovan SE, Eggleton P, Bignell DE (2001a) Gut content analysis and a new feeding group classification of termites. Ecol Entomol 26:356–366

Donovan SE, Eggleton P, Dubbin WE et al (2001b) The effect of a soil-feeding termite, *Cubitermes fungifaber* (Isoptera: Termitidae) on soil properties: termites may be an important source of soil microhabital heterogeneity in tropical forests. Pedobiologia 45:1–11

Donovan SE, Jones DT, Sands WA, Eggleton P (2000) The morphological phylogenetics of termites (Isoptera). Biol J Linn Soc 70:467–513

Donovan SE, Purdy KJ, Kane MD, Eggleton P (2004) Comparison of Euryarchaea strains in the guts and food-soil of the soil-feeding termite *Cubitermes fungifaber* across different soil types. Appl Environ Microbiol 70:3884–3892

Dow JAT (1987) Insect midgut function. Advs Insect Physiol 19:187–328

Droge S, Limper U, Emtiazi F et al (2004) In vitro and in vivo sulfate reduction in the gut contents of the termite *Mastotermes darwiniensis* and the rose-chafer *Pachnoda marginata*. J Gen Appl Microbiol 51:57–64

Ebert A, Brune A (1997) Hydrogen concentration profiles at the oxic-anoxic interface: a microsensor study the hindgut of the wood-feeding lower termite *Reticulitermes flavipes* (Kollar). Appl Environ Microbiol 63:4039–4046

Eggleton P (2000) Global patterns of termite diversity. In: Abe T, Bignell DE, Higashi M (eds) Termites: evolution, sociality, symbioses, ecology. Kluwer Academic Publishers, Dordrecht, pp 25–51

Eggleton P (2006) The termite gut habitat: its evolution and co-evolution. In: König H, Varma A (eds) Intestinal microorganisms of termites and other invertebrates. Springer, Berlin, pp 373–404

Eggleton P, Bignell DE, Sands WA et al (1996) The diversity, abundance and biomass of termites under differing levels of disturbance in the Mbalmayo Forest Reserve, southern Cameroon. Phil Trans R Soc Lond B 351:51–68

Eggleton P, Davies RG, Bignell DE (1998) Body size and energy use in termites (Isoptera): the responses of soil-feeders and wood-feeders differ in a tropical forest assemblage. Oikos 81:525–530

Eggleton P, Homathevi R, Jeeva D et al (1997) The species richness and composition of termites (Isoptera) in primary and regenerating lowland dipterocarp forest in Sabah, East Malaysia. Ecotropica 3:119–128

Eggleton P, Tayasu I (2001) Feeding groups, lifetypes, and the global ecology of termites. Ecol Res 16:941–960

Engel MS, Grimaldi DA, Krishna K (2009) Termites (Isoptera): their phylogeny, classification and rise to ecological dominance. Am Mus Novit 3650:1–27

Fall S, Brauman A, Chotte JL (2001) Comparative distribution of organic matter in particle and aggregate size fraction in the mounds of termites with different feeding habits in Senegal: *Cubitermes nikoloensis* and *Macrotermes bellicosus*. Appl Soil Ecol 17:131–140

Fall S, Hamelin J, Ndiaye F et al (2007) Differences between bacterial communities in the gut of a soil-feeding termite (*Cubitermes nikoloensis*) and its mound. Appl Environ Microbiol 73: 5199–5208

Fall S, Nazaret S, Chotte JL, Brauman A (2004) Bacterial density and community structure associated with aggregate size fractions of soil-feeding termite mounds. Microb Ecol 28:191–199

Fontes LR (1987) Morphology of the worker digestive tube of the soil-feeding nasute termites (Isoptera, Termitidae, Nasutitermitinae) from the Neotropical region. Rev Bras Zool 3(475-):501

Friedrich MW, Schmitt-Wagner D, Leuders T, Brune A (2001) Axial differences in community structure of *Crenarchaeota* and *Euryarchaeota* in the highly compartmentalized gut of the soil-feeding termite *Cubitermes orthognathus*. Appl Environ Microbiol 67:4880–4890

Fujita A, Abe T (2002) Amino acid concentration and distribution of lysozyme and protease activities in the guts of higher termites. Physiol Entomol 27:76–78

Fujita A, Minamoto T, Shimizu I, Abe T (2002) Molecular cloning of lysozyme-encoding cDNAs expressed in the salivary gland of a wood-feeding termite, *Reticulitermes speratus*. Insect Biochem Mol 32:1615–1624

Fujita A, Shimizu I, Abe T (2001) Distribution of lysozyme and protease, and amino acid concentration in the guts of a wood-feeding termite, *Reticulitermes speratus* (Kolbe): possible digestion of symbiont bacteria transferred by trophallaxis. Physiol Entomol 26:116–123

Garnier-Sillam E, Harry M (1995) Distribution of humic compounds in mounds of some soil-feeding termite species of tropical rainforests: its influence on soil structural stability. Insect Soc 42:167–185

Garnier-Sillam E, Toutain F (1995) Distribution of polysaccharides within the humic compounds of soil subjected to a humivorous termite *Thoracotermes macrothorax* Sjöstedt. Pedobiologia 39:462–469

Geib SM, Filley TR, Hatcher PG et al (2008) Lignin degradation in wood-feeding insects. Proc Natl Acad Sci USA 105:12932–12937

Godoy MC (2004) Gut structure of two species of the Neotropical genus *Tauritermes* Krishna (Isoptera: Kalotermitidae). Neotrop Entomol 33:163–167

Harazano K, Yamashita M, Shinzato N et al (2007) Isolation and characterization of aromatics-degrading microorganisms from the gut of the lower termite *Coptotermes formosanus*. Biosci Biotechnol Biochem 67:889–892

Herlemann DPR, Geissinger O, Brune A (2007) The termite group I phylum is highly diverse and widespread in the environment. Appl Environ Microbiol 73:6682–6685

Hethener P, Brauman A, Garcia J-L (1992) *Clostridium termitidis* sp. nov., a cellulolytic bacterium from the gut a wood-feeding termite, *Nasutitermes lujae*. Syst Appl Micobiol 15: 52–58

Higashi M, Yamamura N, Abe T (2000) Theories on the sociality of termites. In: Abe T, Bignell DE, Higashi M (eds) Termites: evolution, sociality, symbioses, ecology. Kluwer Academic Publishers, Dordrecht, pp 169–187

Holdaway FG (1933) The composition of different regions of mounds of *Eutermes exitiosus* Hill. Aust J Council Sci Indust Res 6:160–165

Hölldobler B, Wilson EO (2009) The superorganism. W.W. Norton and Company, New York, NY

Holt JA, Lepage M (2000) Termites and soil properties. In: Abe T, Bignell DE, Higashi M (eds) Termites: evolution, sociality, symbioses, ecology. Kluwer Academic Publishers, Dordrecht, pp 389–407

Hongoh Y, Deevong P, Hattori S et al (2006a) Phylogenetic diversity, localization and cell morphologies of members of the candidate phylum TG3 and a subphylum in the phylum Fibrobacteres, recently discovered bacterial groups in termite guts. Appl Environ Microbiol 72:6780–6788

Hongoh Y, Deevong P, Inoue T et al (2005) Intra-and interspecific comparisons of bacterial diversity and community structure support coevolution of gut microbiota and termite host. Appl Environ Microbiol 71:6590–6599

Hongoh Y, Ekpornprasit L, Inoue T et al (2006b) Intracolony variation of bacterial gut microbiota among castes and ages in the fungus-growing termite *Macrotermes gilvus*. Mol Ecol 15: 505–516

Hongoh Y, Ohkuma M, Kudo T (2003) Molecular analysis of bacterial microbiota in the gut of the termite *Reticulitermes speratus* (Isoptera: Rhinotermitidae). FEMS Microbiol Ecol 44: 231–242

Hopkins DW, Chudek JA, Bignell DE et al (1998) Application of ^{13}C NMR to investigate the transformations and biodegradation of organic materials by some soil- and litter-dwelling insects. Biodegradation 9:423–431

Hungate RE (1946) The symbiotic utilization of cellulose. J Elisha Mitch Sci S 62:9–24

Hyodo F, Azuma J-I, Abe T (1999) Estimation of effect of passage through the gut of a lower termite *Coptotermes formosanus* Shiraki, on lignin by solid-state CP/MAS ^{13}C NMR. Holzforschung 53:244–246

Hyodo F, Inoue T, Azuma J-J, Abe T (2000) Role of the mutualistic fungus in lignin degradation in the fungus-growing termite *Macrotermes gilvus* (Isoptera; Macrotermitinae). Soil Biol Biochem 32:653–658

Hyodo F, Tayasu I, Inoue T et al (2003) Differential role of symbiotic fungi in lignin degradation and food provision for fungus-growing termites (Macrotermitinae: Isoptera). Funct Ecol 17:186–193

Inoue T, Kitade O, Yoshimura T, Yamaoka I (2000) Symbiotic associations with protists. In: Abe T, Bignell DE, Higashi M (eds) Termites: evolution, sociality, symbioses, ecology. Kluwer Academic Publishers, Dordrecht, pp 235–242

Inoue T, Murashima K, Azuma J-I et al (1997) Cellulose and xylan utilisation in the lower termite *Reticulitermes speratus*. J Insect Physiol 43:235–242

Inward DJG, Vogler AP, Eggleton P (2007) A comprehensive phylogenetic analysis of termites (Isoptera) illuminates key aspects of their evolutionary biology. Mol Phylogenet Evol 44: 953–967

Ji R, Brune A (2001) Tansformation and mineralization of ^{14}C-labelled bacterial cells, protein, peptidoglycan, and cellulose by soil-feeding termites. Biol Fert Soils 33:166–174

Ji R, Brune A (2005) Digestion of peptidic residues in humic substances by an alkali-stable and humic-acid tolerant proteolytic activity in the gut of soil-feeding termites. Soil Biol Biochem 37:1648–1655

Ji R, Brune A (2006) Nitrogen mineralization, ammonia accumulation, and emission of gaseous NH_3 by soil-feeding termites. Biogeochemistry 78:267–283

Ji R, Kappler A, Brune A (2000) Transformation and mineralization of synthetic ^{14}C-labelled humic model compounds by soil-feeding termites. Soil Biol Biochem 32:1281–1291

Johnson RA (1979) Configuration of the digestive tube as an aid to identification of worker Termitidae (Isoptera). Syst Entomol 4:31–38

Jones DT, Susilo F-X, Bignell DE et al (2003) Termite assemblage collapses along a land-use intensification gradient in lowland central Sumatra, Indonesia. J Appl Ecol 40:380–391

Kambhampati S, Eggelton P (2000) Taxonomy and phylogeny of termites. In: Abe T, Bignell DE, Higashi M (eds) Termites: evolution, sociality, symbioses, ecology. Kluwer Academic Publishers, Dordrecht, pp 1–23

Kappler A, Brune A (1999) Influence of gut alkalinity and oxygen status on mobilization and size-class distribution of humic acids in the hindgut of soil-feeding termites. Appl Soil Ecol 13:219–229

Kappler A, Brune A (2002) Dynamics of redox potential and changes in redox state of iron and humic acids during gut passage in soil-feeding termites (*Cubitermes* spp.). Soil Biol Biochem 34:221–227

Katsumata KS, Jin Z, Hori K, Iiyama K (2007) Structural changes in lignin of tropical woods during digestion by termite, *Cryptotermes brevis*. J Wood Sci 53:419–426

Köhler T, Stingl U, Meuser K, Brune A (2008) Novel lineages of Planctomycetes densely colonize the alkaline gut of soil-feeding termites (*Cubitermes* spp.). Environ Microbiol 10:1260–1270

Konate S, Le Roux X, Verdier B, Lepage M (2003) Effect of underground fungus-growing termites on carbon dioxide emission at the point- and landscape-scales in an African savanna. Funct Ecol 17:305–314

König H (2006) *Bacillus* species in the intestine of termites and other soil invertebrates. J Appl Microbiol 101:620–627

König H, Fröhlich A, Berchtold M, Wenzel M (2002) Diversity and microhabitats of the hindgut flora of termites. Recent Res Devel Microbiol 6:125–156

König H, Fröhlich J, Hertel H (2006) Diversity and lignocellulolytic activities of cultured microorganisms. In: König H, Varma A (eds) Intestinal microorganisms of termites and other invertebrates. Springer, Berlin, pp 271–301

Korb J (2007) Termites. Curr Biol 17:R995–R999

Korb J (2008) Termites, hemimetabolous diploid white ants? Front Zool 5:15

Kovoor J (1968) L'intestin d'un termite supérieur (*Microcerotermes edentatus* Wasman, Amitermitinae). Histophysiologie, et flore bacterienne symbiotique. Bull Biol Fr Belg 102: 45–58

Kuhnigk T, Borst E-M, Ritter A et al (1994) Degradation of lignin monomers by the hindgut flora of xylophagous termites. Syst Appl Microbiol 17:76–85

Kuhnigk T, Branks J, Krekeler D et al (1996) A feasible role of sulfate-reducing bacteria in the termite gut. Syst Appl Microbiol 19:139–149

Landim CC, Costa-Leonardo AM (1990) Functional adaptations of the epithelium from different gut segments of *Grigiotermes bequaerti* (Isoptera, Termitidae, Apicotermitinae). Rev Bras Entomol 34:669–678

Lavelle P, Bignell DE, Lepage M et al (1997) Soil function in a changing world: the role of invertebrate ecosystem engineers. Eur J Soil Biol 33:59–193

Leadbetter JR, Breznak JA (1996) Physiological ecology of *Methanobrevibacter filiformis* sp. nov. and *Methanobrevibacter curvatus* sp. nov. isolated from the hindgut of the termite *Reticulitermes flavipes*. Appl Environ Microbiol 62:3620–3631

Leadbetter JR, Schmidt TM, Graber JR, Breznak JA (1999) Acetogenesis from H_2 plus CO_2 by spirochetes from termite guts. Science 283:686–689

Lefebvre T, Miambi E, Pando A et al (2009) Gut-specific actinobacterial community structure and diversity associated with the wood-feeding termite species, *Nasutitermes corniger* (Motschulsky) described by nested PCR-DGGE analysis. Insect Soc 56:269–276

Lilburn TG, Kim KS, Ostrom NE et al (2001) Nitrogen fixation by symbiotic and free-living spirochetes. Science 292:2495–2498

Lo N, Tokuda G, Watanabe H et al (2000) Evidence from multiple gene sequences indicates that termites evolved from wood-feeding cockroaches. Curr Biol 10:801–804

Mackenzie LM, Muigai AT, Osir EO et al (2007) Bacterial diversity in the intestinal tract of the fungus-cultivating termite *Macrotermes michaelseni* (Sjöstedt). Afr J Biotechnol 6: 658–667

Margulis L, Jorgensen JZ, Dolan S et al (1998) The *Arthromitus* stage of *Bacillus cereus*: intestinal symbionts of animals. Proc Natl Acad Sci USA 95:1236–1241

Miyata R, Noda N, Tamaki H et al (2007) Influence of feed components on symbiotic bacterial community structure in the gut of the wood-feeding higher termite *Nasutitermes takasagoensis*. Biosci Biotechnol Biochem 71:1244–1251

Mora P, Lattaud C, Rouland C (1998) Enzymes involved in lignin degradation among termites with various feeding habits. Actes des Colloques Insectes Soc 11:77–80

Nalepa C, Bignell DE, Bandi C (2001) Detritivory, coprophagy and the evolution of digestive mutualisms in Dictyoptera. Insect Soc 48:194–201

Ngugi DK, Tsanuo MK, Boga HI (2007) Benzoic acid-degrading bacteria from the intestinal tract of *Macrotermes michaelseni* Sjöstedt. J Basic Microbiol 47:87–92

Noirot C (1992) From wood-feeding to soil-feeding: an important trend in termite evolution. In: Billen J (ed) Biology and evolution of social insects. Leuven University Press, Leuven, pp 107–119

Noirot C (1995) The gut of termites (Isoptera). Comparative anatomy, systematics, phylogeny. I. Lower termites. Ann Soc Entomol Fr 31:197–226

Noirot C (2001) The gut of termites (Isoptera). Comparative anatomy, systematics, phylogeny. II. higher termites (Termitidae). Ann Soc Entomol Fr 37:431–471

Noirot C, Darlington JPEC (2000) Termite nests: architecture, regulation and defence. In: Abe T, Bignell DE, Higashi M (eds) Termites: evolution, sociality, symbioses, ecology. Kluwer Academic Publishers, Dordrecht, pp 121–139

Noirot C, Noirot-Timothée C (1977) Fine structure of the rectum in termites (Isoptera): a comparative study. Tissue Cell 9:693–710

Nunes L, Bignell DE, Lo N, Eggleton P (1997) On the respiratory quotient (RQ) of termites (Insecta: Isoptera). J Insect Physiol 43:749–758

O'Brien RW, Slaytor M (1982) Role of microorganisms in the metabolism of termites. Aust J Biol Sci 35:239–262

Odelson DA, Breznak JA (1983) Volatile fatty acid production by the hindgut microbiota of xylophagous termites. Appl Environ Microbiol 45:1602–1613

Ohkuma M (2003) Termite symbiotic systems: efficient biorecycling of lignocellulose. Appl Microb Biot 61:1–9

Ohkuma M (2008) Symbioses of flagellates and prokaryotes in the gut of lower termites. Trends Microbiol 16:345–352

Ohkuma M, Iida T, Ohtoko K et al (2005) Molecular phylogeny of parabasalids inferred from small subunit rRNA sequences, with emphasis on the Hypermastigea. Mol Phylogenet Evol 35:646–655

Ohkuma M, Kudo T (1996) Phylogenetic diversity of the intestinal bacterial community in the termite *Reticulitermes speratus*. Appl Environ Microbiol 62:461–468

Ohkuma M, Kudo T (1998) Phylogenetic analysis of the symbiotic intestinal microflora of the termite *Cryptotermes domesticus*. FEMS Microbiol Lett 164:389–395

Ohkuma M, Noda S, Hongoh Y, Kudo T (2002) Diverse bacteria related to the *Bacteroides* subgroup of the CFB phylum within the gut communities of various termites. Biosci Biotechnol Biochem 66:87–84

Ohkuma M, Noda S, Horikoshi K, Kudo T (1995) Phylogeny of symbiotic methanogens in the gut of the termite *Reticulitermes speratus*. FEMS Microbiol Lett 134:45–50

Ohkuma M, Noda S, Usami R et al (1996) Diversity of nitrogen fixing genes in the symbiotic intestinal microflora of the termite *Reticulitermes speratus*. Appl Environ Microbiol 62: 2747–2752

Ohkuma M, Sato T, Noda S et al (2007) The candidate phylum 'termite group 1' of bacteria: phylogenetic diversity, distribution, and endosymbiont members of various gut flagellated protists. FEMS Microbiol Ecol 60:467–476

Ohkuma M, Shimizu H, Thongaram T et al (2003) An alkaliphilic and xylanolytic *Paenibacillus* species isolated from the gut of a soil-feeding termite. Microbes Environ 18: 145–151

Pasti MB, Belli ML (1985) Cellulolytic activity of actinomycetes isolated from termites (Termitidae) guts. FEMS Microbiol Lett 26:107–112

Paul J, Saxena S, Varma A (1993) Ultrastructural studies of the termite (*Odontotermes obesus*) gut microflora and its cellulolytic properties. World J Microbiol Biotechnol 9:108–112

Potrikus CJ, Breznak JA (1981) Gut bacteria recycle uric acid nitrogen in termites: a strategy for nutrient conservation. Proc Natl Acad Sci USA 78:4601–4605

Potts RC, Hewitt PH (1973) The distribution of intestinal bacteria and cellulose activity in the harvester termite *Trinervitermes trinervoides* (Nasutitermitinae). Insect Soc 20: 215–220

Purdy KJ (2007) The distribution and diversity of *Euryarchaeota* in termite guts. Adv Appl Microbiol 62:63–80

Roose-Amsaleg C, Brygoo Y, Harry M (2004) Ascomycete diversity in soil-feeding termite nests and soils from a tropical rainforest. Environ Microbiol 6:462–469

Rouland C, Chararas C, Renoux J (1986) Etudes comparées des osidases de trois espèces de termites Africain à regime alimentaire différent. CR Acad Sci III Vie 302:341–345

Rouland C, Chararas C, Renoux J (1989) Les osidases digestives présente dans l'intestin moyen, l'intestin postérieur et les glandes salivaires du termite humivore *Crenetermes albotarsalis*. CR Acad Sci III Vie 308:281–285

Rouland C, Lenoir-Labé F (1998) Microflore intestinale symbiotique des insectes xylophages: mythe ou réalité? Cah Agric 7:37–47

Rouland-Lefèvre C, Bignell DE (2001) Cultivation of symbiotic fungi by termites of the subfamily Macrotermitinae. In: Seckbach J (ed) Symbiosis: mechanisms and model systems. Kluwer Academic Publishers, Dordrecht, pp 731–756

Rouland-Lefèvre C, Inoue T, Johima T (2006) *Termitomyces*/termite interactions. In: König H, Varma A (eds) Intestinal microorganisms of termites and other invertebrates. Springer, Berlin, pp 335–350

Sands WA (1972) The soldierless termites of Africa (Isoptera: Termitidae). Bull Br Mus Nat Hist (Ent) Suppl 18:3–244

Sands WA (1992) The termite genus *Amitermes* in Africa and the Middle East. Nat Res Inst Bull 51:1–140

Sands WA (1995) New species and genera of soil-feeding termites (Isoptera: Termitidae) from African savannas. J Nat Hist 29:1483–1515

Sands WA (1998) The identification of worker castes of termite genera from soils of Africa and the Middle East. CAB International, Wallingford

Schafer A, Konrad R, Kuhnigk T et al (1996) Hemicellulose-degrading bacteria and yeasts from the termite gut. J Appl Bacteriol 80:471–478

Schmitt-Wagner D, Brune A (1999) Hydrogen profiles and localization of methanogenic activities in the highly compartmentalized hindgut of soil-feeding higher termites (*Cubitermes* spp.). Appl Environ Microbiol 65:4490–4496

Schmitt-Wagner D, Friedrich MW, Wagner B, Brune A (2003a) Phylogenetic diversity, abundance and axial distribution of bacteria in the intestinal tract of two soil-feeding termites (*Cubitermes* spp.). Appl Environ Microbiol 69:6007–6017

Schmitt-Wagner D, Friedrich MW, Wagner B, Brune A (2003b) Axial dynamics, and interspecies similarlity of bacterial community structure in the highly compartmentalized gut of soil-feeding termites (*Cubitermes* spp.). Appl Environ Microbiol 69:6018–6024

Shinzato N, Matsumoto T, Yamaoka I et al (2007) Phylogenetic diversity of symbiotic methanogens living in the hindgut of the lower termite *Reticulitermes speratus* analysed by PCR and in situ hybridization. Appl Environ Microbiol 65:837–840

Shulz MW, Slaytor M, Hogan ME, O'Brien RW (1986) Components of cellulose from the higher termite, *Nasutitermes walkeri* (Hill). Insect Biochem 16:929–932

Slaytor M (2000) Energy metabolism in the termite and its gut microbiota. In: Abe T, Bignell DE, Higashi M (eds) Termites: evolution, sociality, symbioses, ecology. Kluwer Academic Publishers, Dordrecht, pp 307–332

Slaytor M, Veivers PC, Lo N (1997) Aerobic and anaerobic metabolism in the higher termite *Nasutitermes walkeri* (Hill). Insect Biochem Mol Biol 27:291–303

Sleaford F, Bignell DE, Eggleton P (1996) A pilot analysis of gut contents in termites from the Mbalmayo Forest Reserve, Cameroon. Ecol Entomol 21:279–288

Taguchi F, Chang JD, Mizukami N et al (1993) Isolation of a hydrogen-producing bacterium, *Clostridium beijerinckii* strain AM21B, from termites. Can J Microbiol 39:726–730

Tanaka H, Aoyagi H, Shina S et al (2006) Influence of the diet components on the symbiotic microorganisms community in hindgut of *Coptotermes formosanus* Shiraki. Appl Microbiol Biotechnol 71:907–917

Taprab Y, Johjima T, Maeda Y et al (2006) Symbiotic fungi produce laccases potentially involved in phenol degradation in fungus combs of fungus-growing termites in Thailand. Appl Environ Microbiol 71:7696–7704

Tayasu I, Abe T, Eggleton P, Bignell DE (1997) Nitrogen and carbon isotope ratios in termites (Isoptera): an indicator of trophic habit along the gradient from wood-feeding to soil-feeding. Ecol Entomol 22:343–351

Tholen A, Brune A (1999) Localization and in situ activities of homoacetogenic bacteria in the highly compartmentalized hindgut of soil-feeding higher termites (*Cubitermes* spp.). Appl Environ Microbiol 65:4497–4505

Tholen A, Brune A (2000) Impact of oxygen on the metabolic fluxes and in situ rates of reductive acetogenesis in the hindgut of the wood-feeding termite *Reticulitermes flavipes*. Environ Microbiol 2:436–449

Thongaram T, Hongoh Y, Kosono S et al (2005) Comparison of bacterial communities in the alkaline gut segment among various species of higher termites. Extremophiles 9: 229–238

Thongaram T, Kosono S, Ohkuma M et al (2003) Gut of higher termites as a niche for alkaliphiles as shown by culture-based and culture-independent studies. Microbes Environ 18: 152–159

Thorne BT, Grimaldi DA, Krishna K (2000) Early fossil history of the termites. In: Abe T, Bignell DE, Higashi M (eds) Termites: evolution, sociality, symbioses, ecology. Kluwer Academic Publishers, Dordrecht, pp 77–93

Tokuda G, Lo N, Watanabe H et al (2004) Major alteration of the expression site of endogenous cellulases in members of an apical termite lineage. Mol Ecol 13:3219–3228

Tokuda G, Nakamura T, Murakami R, Yamaoka I (2001) Morphology of the digestive system in the wood-feeding termite *Nasutitermes takasagoensis* (Shiraki) [Isoptera: Termitidae]. Zool Sci 18:869–877

Tokuda G, Watanabe H (2007) Hidden cellulases in termites: revision of an old hypothesis. Biol Lett 3:336–339

Tokuda G, Watanabe H, Matsumoto T, Noda H (1997) Cellulose digestion in the wood-eating higher termite, *Nasutitermes takasagoensis* (Shiraki): distribution of cellulases and properties of Endo-ß-1, 4-glucanase. Zool Sci 14:83–93

Tokuda G, Yamaoka I, Noda H (2000) Localization of symbiotic clostridia in the mixed segment of the termite *Nasutitermes takasagoensis* (Shiraki). Appl Environ Microbiol 66:2199–2207

Tokura M, Ohkuma M, Kudo T (2000) Molecular phylogeny of methanogens associated with flagellated protists in the gut and with the gut epithelium of termites. FEMS Microbiol Ecol 33:233–240

Traniello JFA, Leuthold RH (2000) Behaviour and ecology of foraging in termites. In: Abe T, Bignell DE, Higashi M (eds) Termites: evolution, sociality, symbioses, ecology. Kluwer Academic Publishers, Dordrecht, pp 141–168

Trinkerl M, Breunig A, Schauder R, König H (1990) *Desulphovibrio termitidis* sp. nov., a carbohydrate-degrading sulfate-reducing bacterium from the hindgut of a termite. Syst Appl Microbiol 13:372–377

Vu AT, Nguyen NC, Leadbetter JR (2004) Iron reduction in the metal-rich guts of wood-feeding termites. Geobiology 2:239–247

Ward N, Staley JT, Fuerst JA et al (2006) The order Planctomycetales, including the genera *Planctomyces, Pirellula, Gemmata* and *Isosphaera* and the Candidatus genera *Bracadia, Kuenenia* and *Scalindua*. In: Dworkin M, Falkow S, Rosenberg E et al (eds) The prokaryotes, vol 7. Springer, New York, NY, pp 757–793

Warnecke F, Luginbühl P, Ivanova N et al (2007) Metagenomic and functional analysis of hindgut microbiota of a wood-feeding higher termite. Nature 450:560–565

Watanabe Y, Shinzate N, Fukatsu T (2003) Isolation of actinomycetes from termites' guts. Biosci Biotechnol Biochem 67:1797–1801

Watanabe H, Tokuda G (2001) Animal cellulases. Cell Mol Life Sci 58:1167–1178

Wenzel M, Schönig I, Berchtold M et al (2002) Aerobic and facultatively anaerobic cellulolytic bacteria from the gut of the termite *Zootermopsis angusticollis*. J Appl Microbiol 92:32–40

Wolfersberger MG (2000) Amino acid transport in insects. Annu Rev Entomol 45:111–120

Wood TG (1996) The agricultural importance of termites in the tropics. Agr Zool Rev 7:117–155

Wood TG, Sands WA (1978) The role of termites in ecosystems. In: Brian MV (ed) Production ecology of ants and termite. Cambridge University Press, Cambridge, pp 245–292

Yamada A, Inoue T, Wjwatwitaya D, Ohkuma M (2007) A new concept of the feeding group composition of termites (Isoptera) in tropical ecosystems: carbon source competitions among fungus-growing termites, soil-feeding termites, litter-layer microbes, and fire. Sociobiology 50:135–153

Yara K, Jahana K, Hayashi H (1989) In situ morphology of the gut microbiota of the fungus-growing termite *Odontotermes formosanus* (Termitidae: Macrotermitinae). Sociobiology 15:2247–2260

Zhou XG, Smith JA, Koehler PG et al (2007) Correlation of cellulase gene expression and cellulolytic activity throughout the gut of the termite *Reticulitermes flavipes*. Gene 395:29–39

Chapter 15
Diversity, Structure, and Evolution of the Termite Gut Microbial Community

Moriya Ohkuma and Andreas Brune

Abstract The gut of termites is densely populated with microbial symbionts that aid in the efficient digestion of recalcitrant lignocellulose. Despite the formidable unculturability of the resident members, ribosomal RNA-based molecular analyses and other comprehensive and elaborate culture-independent studies of molecular microbial ecology in the past decade have gradually unveiled the complex nature of the intestinal microbiota. The microbial community, whose structure and spatial distribution seems to be characteristic for a termite species (but may differ between genera), consists of mostly novel lineages that seem to have co-evolved or converged with their particular host. A prominent feature of lower termites is the tripartite symbiosis with a variety of flagellated protists that are themselves associated with diverse prokaryotes. Here, the complete genome sequences of several bacterial endosymbionts have disclosed their functional interactions with their host flagellates, but the highly structured and coevolving nature of these associations requires more emphasis in future studies.

15.1 Introduction

Termites harbor diverse and unique microbial populations in their hindgut. These microorganisms form a complex community, with densities reaching up to 10^{11} cells/mL. Termites largely depend on the gut microbial symbionts for the digestion and utilization of their food, especially highly recalcitrant lignocellulose. Morphological observations, assessments of overall gut metabolism, and measurements of physicochemical conditions in the gut environment have all contributed to significant progress in understanding the roles of the gut microbial community in host nutrition, as have many studies involving pure cultures of the symbionts. Nevertheless, our understanding of the diversity and biology of the symbionts is

M. Ohkuma (✉)
Microbe Division (Japan Collection of Microorganisms), RIKEN BioResource Center, Saitama 351-0198, Japan
e-mail: mohkuma@riken.jp

D.E. Bignell et al. (eds.), *Biology of Termites: A Modern Synthesis*,
DOI 10.1007/978-90-481-3977-4_15,

still poor because most community members are difficult or impossible to culture, a situation not unusual in microbial ecological studies of natural environments.

Culture-independent molecular approaches in the past two decades have enhanced our ability to assess natural microbial communities and have been successfully applied for understanding the diversity, structure, and evolution of the microbiota in termite guts, as well as the roles of individual symbionts. These approaches, usually based on small-subunit (SSU) rRNA gene sequences, facilitate phylogenetic placement of the individual members, evaluation of their diversity, profiling of the community, and comparisons among communities. Specific probes or PCR primers for the sequences obtained can be applied to further studies, such as the in situ identification of the cells and their quantification.

Existing reviews of molecular studies on the intestinal microbial community of termites include Abe et al. (2000), Brune and Friedrich (2000), Ohkuma (2002, 2003, 2008), Brune and Stingl (2005), Brune (2006) and Ohkuma et al. (2006). Here, we will outline the cataloguing of the termite gut microbial community by culture-independent molecular studies with focus on the most recent advances. Functions and roles of the microbiota of the termite gut are addressed in Chapter 16.

15.2 Molecular Phylogeny and Evolution of Protists

15.2.1 Trends in Molecular Studies

Lower termites harbor unique anaerobic protists that are required for the host termites to feed on cellulose. The gut protists belong to either the phylum Parabasalia or the order Oxymonadida (phylum Preaxostyla). Early molecular phylogenetic studies of cultured representatives of the Parabasalia, i.e. trichomonads, and the observed lack of typical mitochondria in parabasalids and oxymonads, led to the hypothesis that these flagellates represent one of the earliest branching groups of eukaryotes before eukaryotes acquired mitochondria in the form of endosymbiotic bacteria. Instead of mitochondria, parabasalids have anaerobic energy-producing organelles, the hydrogenosomes, which are definitively characterized by hydrogen-generating activity. Oxymonads have neither mitochondria nor hydrogenosomes. Recent studies, however, have rejected this hypothesis because hydrogenosomes and mitochondria were found to share a common ancestor and remnant mitochondria-like organelles are found in many of such presumed primitive eukaryotes (Embley and Martin 2006). In addition, their deep-branching positions in eukaryote phylogenetic trees are now regarded as artefacts resulting from their abnormally divergent sequences. Consequently, both Parabasalia and Oxymonadida are now assigned to a loose assemblage of eukaryotes, the supergroup "Excavata" (Simpson 2003). The monophyly of this group has recently been demonstrated by a rigorous, genome-wide multi-gene analysis (Hampl et al. 2009).

SSU rRNA gene sequences from the gut protists have been obtained by PCR amplification of the mixed population of the gut without cultivation, and the corresponding organisms from which the SSU rRNA gene sequences were derived have been identified by whole-cell in situ hybridizations with sequence-specific probes (Berchtold and König 1995; Ohkuma et al. 1998, 2000, 2005, 2009; Dacks et al. 2001; Gerbod et al. 2002; Moriya et al. 2003; Stingl and Brune 2003; Noël et al. 2007). The sequences have also been obtained by PCR using cells manually isolated under microscopy (Dacks and Redfield 1998; Keeling et al. 1998; Fröhlich and König 1999; Dacks et al. 2001; Gerbod et al. 2002; Keeling 2002; Ohkuma et al. 2005, 2009; Heiss and Keeling 2006; Carpenter and Keeling 2007; Carpenter et al. 2009, 2010; Ikeda-Ohtsubo et al. 2007; Noël et al. 2007; Ikeda-Ohtsubo and Brune 2009; Noda et al. 2009a; Strassert et al. 2009; Desai et al. 2010). These extensive studies provide some important and sometimes unexpected insights into the classification and evolution within Parabasalia and Oxymonadida. Because the molecular phylogenetic studies globally conflict many traditional systematics, a taxonomic reclassification has been proposed. The rRNA gene sequences of these organisms are extremely divergent and only poorly resolve the relationships among higher taxonomic levels, such as families and orders (Hampl et al. 2004; Ohkuma et al. 2005). To improve the resolution, protein-encoding multi-gene analyses have already begun (Moriya et al. 1998, 2001; Gerbod et al. 2004; Keeling 2004; Ohkuma et al. 2007a, 2009; de Koning et al. 2008; Noda et al. 2009b). Environmentally expressed sequence tag (EST) analyses or related approaches have been applied to survey functionally important genes of gut protists, such as those encoding cellulases and hydrogenase as well as phylogenetic markers (Ohtoko et al. 2000; Nakashima et al. 2002; Watanabe et al. 2002; Li et al. 2003; Inoue et al. 2005; Slamovits and Keeling 2006; Inoue et al. 2007; Ohkuma et al. 2007a; Todaka et al. 2007, 2010; Zhou et al. 2007).

Termite guts harbor a great diversity of protist species. The morphologically described species amount to more than 400 parabasalids and 70 oxymonads, but a major obstacle in understanding their phylogeny is the inadequacy of morphology for complete taxonomical delineation of these protist (Harper et al. 2009; Strassert et al. 2009). Further, many taxonomically important species have not yet been subject of any molecular study and a multi-gene analysis is necessary for at least representative species of taxonomically important groups to establish a reliable phylogenetic framework.

15.2.2 Parabasalid Symbionts

The phylum Parabasalia is traditionally divided into two orders: Hypermastigida and Trichomonadida (Fig. 15.1). Hypermastigida and the families Devescovinidae and Calonymphidae in Trichomonadida are unique to lower termites and wood-feeding cockroaches of the genus *Cryptocercus*. The cells of the Hypermastigida are typically large and morphologically complex, with numerous flagella. One of the remarkable findings of molecular studies is the multiple origins of hypermastigids (Ohkuma et al. 2005, 2007a). Hypermastigida were traditionally divided into three groups based mainly on the organization of the flagellar apparatus: Trichonymphida, Spirotrichonymphida, and Lophomonadida. These groups

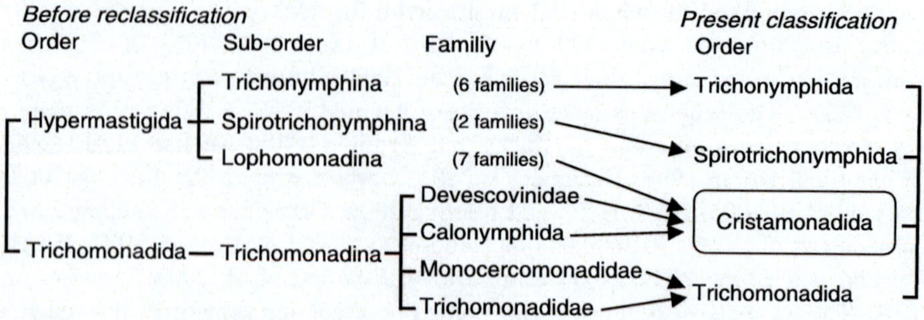

Fig. 15.1 Recent reclassification of the Parabasalia. Orders, sub-orders, and families before reclassification (*left*) are linked to the four new orders by *arrows* (*right*). Note the newly created order Cristamonadida (*encircled*). The previous three sub-orders of Hypermastigida are divided into three orders in the reclassification

have long been considered to have arisen once from a simple trichomonadid lineage. Parabasalia are now divided into four orders: Trichonymphida, Spirotrichonymphida, Cristamonadida, and Trichomonadida (Fig. 15.1) (Brugerolle and Patterson 2001; Adl et al. 2005). However, these revisions have not solved all the problems of parabasalian systematics (Hampl et al. 2004; Noël et al. 2007). Although the monophyly of each of the former three orders is well supported, the order Trichomonadida still remains as a paraphyletic or even polyphyletic group. Furthermore, a great deal of uncertainty remains with respect to phylogenetic relationships among and within these orders.

In these revisions, members of the newly created order Cristamonadida were extensively reorganized by uniting the former hypermastigid order Lophomonadida, the multi-nucleated polymonad Calonymphidae, and the well-developed trichomonad Devescovinidae. As a result, the order Cristamonadida comprises a diverse assemblage of species; the degree of complexity of the flagellar apparatus and associated cytoskeleton varies among taxa. Molecular phylogenetic studies have demonstrated that Cristamonadida descends from one lineage of rudimentary trichomonads and that most of the families and subfamilies comprising this order are polyphyletic, with lophomonads having the most basal position (Noda et al. 2009b). Therefore, the diversity and complexity of cristamonads have likely arisen from several successive and independent morphological transitions of the development and reduction of the flagellar apparatus and associated cytoskeleton, as well as transitions to a multi-nucleated status.

15.2.3 Oxymonad Symbionts

The order Oxymonadida is comprised of five families, four of which occur exclusively in the gut of termites and of cockroaches of the genus *Cryptocercus*. These gut oxmonads apparently include cellulolytic species, but their roles in the gut remain largely uncertain. A sister-group relationship between oxymonads and the genus *Trimastix*, a free-living heterotrophic flagellate inhabiting anoxic water, has been shown based on molecular phylogenetic analyses (Dacks et al. 2001). To date, phylogenetic markers have been examined for at least one taxon in each of the five families, but the exact relationships among the families are not yet fully resolved (Heiss and Keeling 2006; de Koning et al. 2008). From the inferred phylogenetic relationships, hypotheses on the evolution of the morphological characters have been discussed (Carpenter et al. 2008). For example, the holdfast, an anterior structure specialized for attachment to the gut wall, appears to have arisen independently at least twice, once in the family Streblomastigidae and once in the common ancestor of Oxymonadidae and Pyrsonymphidae. One genus in the Pyrsonymphidae family, *Pyrsonympha,* has a typical holdfast structure, and another genus in the family, *Dinenympha,* has lost the structure (Moriya et al. 2003). Molecular studies also provide clear evidence that *Dinenympha* and *Pyrsonympha* are separate genera, and not, as previously proposed, morphotypes of the same organism that differ in developmental stages (Moriya et al. 2003; Stingl and Brune 2003). The Streblomastigidae use a non-canonical genetic code (TAA and TGA for glutamine instead of for stop codons; Keeling and Leander 2003; de Koning et al. 2008) and spliceosomal introns, which are rare in Excavata (Slamovits and Keeling 2006).

15.3 Bacterial Diversity

15.3.1 Diversity Estimation

Compared with gut protists, the morphology of bacteria provides even less information on their taxonomy and diversity, although early studies documented dozens of distinct morphotypes. New light on the diversity of bacteria in the termite gut communities was shed by culture-independent molecular studies, usually based on 16S rRNA gene sequences. In these studies, bacterial 16S rRNA genes from the DNA extracted from the gut homogenates are amplified by PCR usually with universal bacteria primers; the amplified products are cloned, sequenced, and sorted into groups. The representative phylotypes of the groups are then analyzed. In most cases, phylotypes are defined as sequences representing clones of >97% sequence identity. Such molecular studies began in the mid-1990s, but initially dealt with only a limited number of clones or the sequences covered only a short region, or both (Berchtold et al. 1994; Berchtold and König 1996; Ohkuma and Kudo 1996, 1998; Paster et al. 1996; Ohkuma et al. 1999a, 2002).

One of the most comprehensive studies on the diversity of gut bacteria was reported by Hongoh et al. (2003a, 2003b) for the lower termite *Reticulitermes speratus*. The analysis of 1,923 clones of bacterial 16S rRNA genes resulted in the identification of 312 phylotypes. The bacterial diversity of a termite gut was estimated to be up to 740 phylotypes, which corresponds to approximately 6,000 phylotypes/mL. Later, analyses of nearly 100 or more clones of 16S rRNA gene sequences were reported not only in the lower termite genera *Reticulitermes* (Hongoh et al. 2005; Yang et al. 2005), *Coptotermes* (Shinzato et al. 2005), and *Cryptotermes* (Hongoh et al. 2007a), but also in the higher termite genera *Macrotermes* (Hongoh et al. 2006b), *Odontotermes* (Shinzato et al. 2007), *Cubitermes* (Schmitt-Wagner et al. 2003a), *Microcerotermes* (Hongoh et al. 2005; Deevong et al. 2006), and *Nasutitermes* (Hongoh et al. 2006a; Warnecke et al. 2007). In these studies, the estimated bacterial diversity was usually several hundred phylotypes per gut, the estimation depending on the number of analyzed clones as well as the methods applied. Community structure is typically uneven: while only a few phylotypes are numerically predominant, the other phylotypes present are generally quite rare. However, rarefaction analyses, when applied, have revealed that in most studies the sampling was not sufficient to cover the entire bacterial diversity in the gut community concerned.

15.3.2 Novelty of Gut Bacteria

A remarkable finding of molecular-based studies is that most gut bacteria have not yet been cultivated. In most cases, more than 90% of the phylotypes are novel, having no close relatives represented in the database sequences. Many phylotypes are clustered with each other and/or with sequences derived from other termite species in the trees constructed, which implies the existence of bacterial lineages unique to termite guts. These lineages are usually only distantly related to any cultured bacterial species and often represent novel genera or families. The novelty of gut bacteria is recognized even at the highest level of classification, and at least three candidate phyla have been described. They are designated as Termite Group (TG) 1, TG2, and TG3 (Ohkuma and Kudo 1996; Hongoh et al. 2003a, 2005, 2006a; Yang et al. 2005). Bacteria in the TG1 phylum are particularly abundant in lower termites, whereas bacteria in the TG3 phylum are abundant in wood-feeding higher termites, and may each account for more than 10% of the bacterial cells within the gut.

15.3.3 Composition of Bacterial Groups

The species richness of bacteria in the termite gut is similar to that observed in human and animal intestines, but at the phylum level, the bacterial community in termite guts is much more diverse. In a comprehensive survey of bacterial diversity in the human gut microbiome, only seven phyla of bacteria were represented (Ley et al. 2006), whereas the bacterial diversity in the gut of *Reticulitermes* species is distributed over more than 15 phyla. Among these phyla, Spirochaetes are the most dominant, and Bacteroidetes, Firmicutes, and the TG1 phylum are the second most dominant groups, together comprising approximately 80% of the bacteria in the gut community. Spirochetes are also abundant in more than 50% of the clones of 16S rRNA gene sequences from wood-feeding higher termites of the genera *Microcerotermes* and *Nasutitermes* (Hongoh et al. 2005, 2006a) as well as from many lower termites (Ohkuma, unpublished observations). However, an abundance of spirochetes is not always a general rule for wood-feeding termites. For example, in the lower termite *Coptotermes*, 70% of the analyzed clones belong to Bacteroidetes (Shinzato et al. 2005). Spirochetes are also infrequent in the fungus-growing genera *Macrotermes* and *Odontotermes* (Hongoh et al. 2006b; Shinzato et al. 2007) and in the soil-feeding genus *Cubitermes* (Schmitt-Wagner et al. 2003a, 2003b). In *Microcerotermes* and *Nasutitermes* species, the TG3 phylum and Fibrobacteres are the second most abundant bacterial groups (Hongoh et al. 2005, 2006a; Warnecke et al. 2007). In the fungus growers, Bacteroidetes and Firmicutes (particularly Clostridiales) are the most abundant, and in the soil feeders, Clostridiales are the most abundant. These observations indicate that the feeding habits of the termite host affect the bacterial composition; however, more comprehensive investigations with diverse termite species are necessary to understand the exact relationships.

The diversity of a specific phylogenetic groups of bacteria in various termites have also been examined with group-specific PCR primers, e.g. Spirochaetes (Lilburn et al. 1999; Noda et al. 2003; Berlanga et al. 2007), TG1 (Stingl et al. 2005; Herlemann et al. 2007; Ikeda-Ohtsubo et al. 2007; Ohkuma et al. 2007b), Synergistetes (Godon et al. 2005), TG3 and Fibrobacteres (Hongoh et al. 2006a), and Planctomycetes (Köhler et al. 2008). Sequences of many of these groups have also been found in other environments, but they often form lineages distinct from those found in termite guts, again indicating the uniqueness of termite-gut bacteria. Sometimes sequences from termite guts cluster with those of uncultivated bacteria found in the intestines of other insects or animals. Recently, the first representative of the candidate phylum TG1, *Elusimicrobium minutum*, has been isolated from the hindgut of a scarab beetle larva, and the TG1 has been classified as phylum Elusimicrobia (Geissinger et al. 2009).

15.4 Archaeal Diversity

Methanogenic archaea in the termite gut represent a relatively minor population and have been characterized by their 16S rRNA gene sequences (Ohkuma et al. 1995, 1998, 1999b; Fröhlich and König 1999; Shinzato et al. 1999, 2001; Tokura et al. 2000; Brauman et al. 2001; Friedrich et al. 2001; Hara et al. 2004; Donovan et al. 2005; Miyata et al. 2007a, see Purdy 2007 for a detailed review). Methanogens in the gut of lower termites are exclusively related to the genus *Methanobrevibacter* in the family Methanobacteriaceae, whereas those in the gut of higher termites belong to the family Methanosarcinaceae (related to the genus *Methanomicrococcus*), the order Methanomicrobiales, and the genus *Methanobrevibacter*. In general, soil-feeding higher termites emit more methane than wood feeders, and the relative archaeal abundance in the gut is correlated with the methane emission rates of the termites (Brauman et al. 2001). Most of the methanogens present in termite guts are indigenous to termites, although the soil around the mounds of *Cubitermes fungifaber* contained *Methanomicrococcus* and *Methanobrevibacter* species that were related to certain symbionts within the gut (Donovan et al. 2005). Related lineages of methanogens are also present in wood-feeding cockroaches of the genera *Panesthia* and *Salganea* (Hara et al. 2002). In lower termites, the presence of certain lineages of *Methanobrevibacter* may also be influenced by biogeographic factors (Shinzato et al. 2001). Also archaeal 16S rRNA gene sequences that do not correspond to methanogens have been identified; these are affiliated with the Thermoplasmales in lower termites (Shinzato et al. 1999, 2001) and with the genus *Natronococcus* (Donovan et al. 2005) and the Crenarchaeota (Friedrich et al. 2001).

15.5 Comparisons Among Host Termites

Termites practice proctodeal trophallaxis, which is the direct transfer of hindgut fluid from the rectal pouch of a donor to the mouth of a receiver. This behavior allows termites to exchange nutrients with nest mates and to transfer gut symbionts from a termite with symbionts to a termite that discarded the symbionts during the molting process. Gut symbionts are also carried over from the mother nest to a newly founded termite colony by the alates (king and queen). This mode of symbiont transfer assures vertical transmission of the gut microbial community from generation to generation. Therefore, the gut microbial community is likely to be conserved within a termite species. This implies that the gut microbial community has been inherited from the termite ancestors and that the termites and the gut microbial community co-evolved with each other. Indeed, the composition of gut protists is stable and almost identical in individuals of a termite species, and is apparently similar among species within any termite genus. The gut protist composition also has some relationships with the host termite phylogeny (Kitade et al. 2004). A clear co-speciation of termites of the Rhinotermitidae and their unique protist species of the genus *Pseudotrichonympha* is inferred based on molecular phylogenetic analyses (Noda et al. 2007).

Cockroaches of the genus *Cryptocercus* harbor cellulolytic gut protists related to the species found in termite guts. Robust sister-group relationships have been found within each of the monophyletic lineages that represent family-level groups of the gut protists of the order Trichonymphida in both *Cryptocercus* and termites (Ohkuma et al. 2009). This observation together with the evidence that termites evolved from wood-feeding cockroaches (Lo et al. 2000) strongly suggest that a set of gut protists essentially representative of the present-day diversity was already established in a common ancestor of these host insects, was inherited vertically to their offspring, and then diversified. The *Cryptocercus* cockroaches also practice proctodeal trophallaxis, which probably developed in the common ancestor of termites and *Cryptocercus*. It is becoming increasingly clear that the transfer of gut microbiota among individuals is tightly linked to the evolution of sociality of host insects (Nalepa et al. 2001).

Comparisons of the bacterial constituents within species, among species, and between genera of termites reveal that the gut bacterial community, even though it contains diverse species, is significantly conserved among congeneric termite species, irrespective of the individual, colony, and location. The consistent presence of certain elements of the bacterial microbiota in homologous gut compartments of three allopatric *Cubitermes* species has been roughly demonstrated

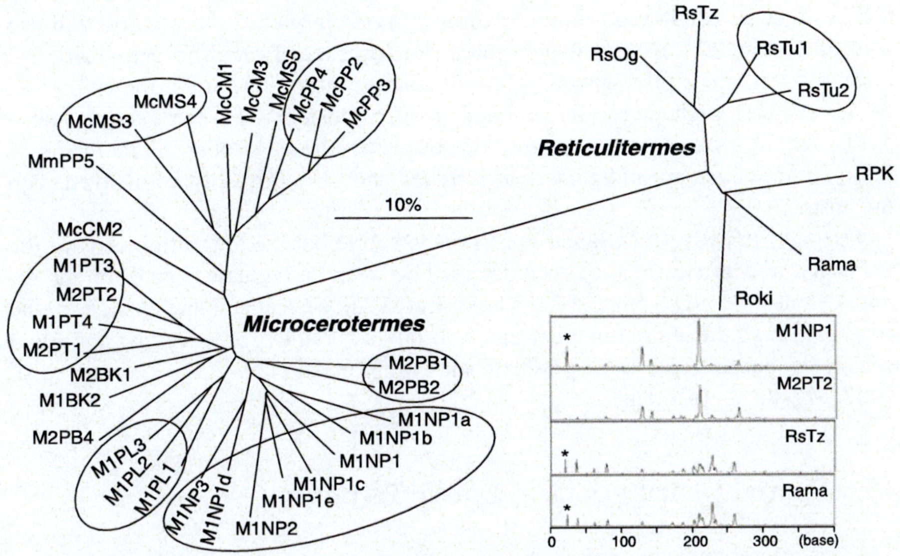

Fig. 15.2 Dendrogram showing the relationship of T-RFLP profiles of gut bacterial communities of termites of the genera *Reticulitermes* and *Microcerotermes*. T-RFLP profiles of 16S rRNA gene sequences amplified by PCR from termite guts were clustered based on their dissimilarity for 32 termite colonies and 5 individual termites from colony M1NP1 (M1NP1a–e). The first and second letters of each sample name indicate the species of host termite, and the third and fourth letters are abbreviations of their sampling locations. Four colonies of *Reticulitermes speratus* (Rs-) from three locations are clustered together, and this cluster is closely related to the other three *Reticulitermes* species. Four species of *Microcerotermes* termites (M1, M2, Mc, and Mm) form a large cluster. *Circles* indicate clustered samples collected in the same areas. The locality sometimes affects clustering of the profiles more significantly than the host taxonomy (e.g., samples M1PT and M2PT). The *bottom panel* on the *right* shows representative T-RFLP profiles. *Asterisks* in the T-RFLP profiles indicate the peaks from primer dimers. The images were reproduced from Hongoh et al. (2005)

using terminal-restriction fragment polymorphism (T-RFLP) analysis of the 16S rRNA gene sequences (Schmitt-Wagner et al. 2003b). In the wood-feeding genera *Reticulitermes* and *Microcerotermes*, the high similarity of the gut bacterial community within each genus was documented in more detail by a combining T-RFLP analyses with extensive clone analyses of 16S rRNA genes (Hongoh et al. 2005, Fig. 15.2), which revealed that congeneric species shared a substantial portion of bacterial phylotypes. Also in other studies of bacterial diversity in termite guts, the closest relatives of most bacterial phylotypes were typically found in the most closely related termite species (Shinzato et al. 2005, 2007; Yang et al. 2005; Hongoh et al. 2006a, 2006b, 2007a).

On the other hand, there is little overlap between the gut bacterial communities of termite higher taxa. This is not only the case at the family level, but has been reported also for members of those termite genera that belong to the same family or subfamily, such as *Reticulitermes* and *Coptotermes* of the Rhinotermitidae

(Shinzato et al. 2005) and *Macrotermes* and *Odontotermes* of the Macrotermitinae (Shinzato et al. 2007). Nevertheless, bacterial sequences from different genera frequently form larger phylogenetic clusters unique to termites, and a major portion of the bacterial sequences are included in such clusters. The consistency across gut communities and the uniqueness of gut bacteria indicate that the members of these communities are indigenous to termites and have probably co-evolved with the hosts.

Despite the above-mentioned consistency, subtle but distinct differences in the gut bacterial community have been detected between individuals from different termite colonies (Hongoh et al. 2005; Minkley et al. 2006). Remarkable changes in the composition of the gut community can be brought about by force feeding abnormal substrates, but this also greatly affects survival (Tanaka et al. 2006; Hayashi et al. 2007; Miyata et al. 2007b).

15.6 Spatial Distributions in Lower Termites

The gut microbiota of termites is not evenly dispersed; instead, the community is spatially highly structured. A remarkable feature of lower termites is the various associations of bacteria or archaea with gut protists (Brune and Stingl 2005; Ohkuma 2008). Typically, each flagellate harbors a dense population of intracellular endosymbionts or surface-attached ectosymbionts, or both. Gut protists are relatively large, and their enormous surface areas and their voluminous intracellular compartments provide an abundance of prokaryotic habitats. As a result, protist-associated prokaryotes constitute a large proportion of the gut bacterial population. In the case of *Mastotermes darwiniensis*, the protist cells take up ca. 95% of the volume of the hindgut paunch, and protist-associated prokaryotes account for ca. 90% of the prokaryotic cells (Berchtold et al. 1999). The gut wall is also colonized by prokaryotes, which account for 85% of the prokaryotic cells in the posterior hindgut of *M. darwiniensis* (Berchtold et al. 1999) and 3–19% of the prokaryotic cells in the total hindgut of some other lower termites (Nakajima et al. 2005, 2006). Analyses of 16S rRNA gene sequences after the fractionation of prokaryotic populations in the gut reveal that prokaryotic species associated with gut protist cells and the gut wall are distinct from each other and from those living freely in the gut fluid (Tokura et al. 2000; Nakajima et al. 2005, 2006; Yang et al. 2005). Indeed, FISH using a group- or species-specific probe detects unique populations, such as Actinobacteria or Bacteroidetes (Berchtold et al. 1999; Nakajima et al. 2005, 2006).

The protist-prokaryote associations have long been known and are often used in descriptions of protist species as cryptic morphological features (Dolan 2001, and references therein). Many of these associations have been documented by electron microscopy. However, the identity of the associated prokaryotes and the symbiotic relationship with host protists remained obscure until they were investigated by molecular ecological means and genome approaches. Protist-associated prokaryotes are typically identified based on their 16S rRNA gene sequences using manually isolated protist cells as PCR templates, and localized in situ using FISH and sequence-specific probes.

15.7 Protist-Prokaryote Associations

15.7.1 Methanogenic Archaea

Methanogens are easily detected by epifluorescence microscopy owing to the specific autofluorescence of their cofactor F_{420}. Endosymbiotic methanogens belonging to the genus *Methanobrevibacter* have been identified in several protist species (Fig. 15.3; Fröhlich and König 1999; Tokura et al. 2000; Hara et al. 2004; Inoue et al. 2008, see Hongoh and Ohkuma 2010 for a detailed review). The situation has been studied in detail in *Reticulitermes speratus* and *Hodotermopsis sjoestedti*. In both termites, the *Methanobrevibacter* phylotypes associated with protists are a different species from those associated with the gut wall (Tokura et al. 2000). Interestingly, the phylotypes of the endosymbiotic methanogens of the oxymonadid *Dinenympha* spp. and the spirotrichonymphid *Microjoenia* spp., two completely unrelated protists present within both termite species, are virtually identical among the protists of the same termite but clearly different between the two termite species.

15.7.2 Spirochetal Ectosymbionts

Ectosymbiotic spirochetes are frequently observed and have been phylogenetically identified as species of the genus *Treponema* in various protist species (Iida et al. 2000; Noda et al. 2003; Wenzel et al. 2003; Inoue et al. 2008). They often cover

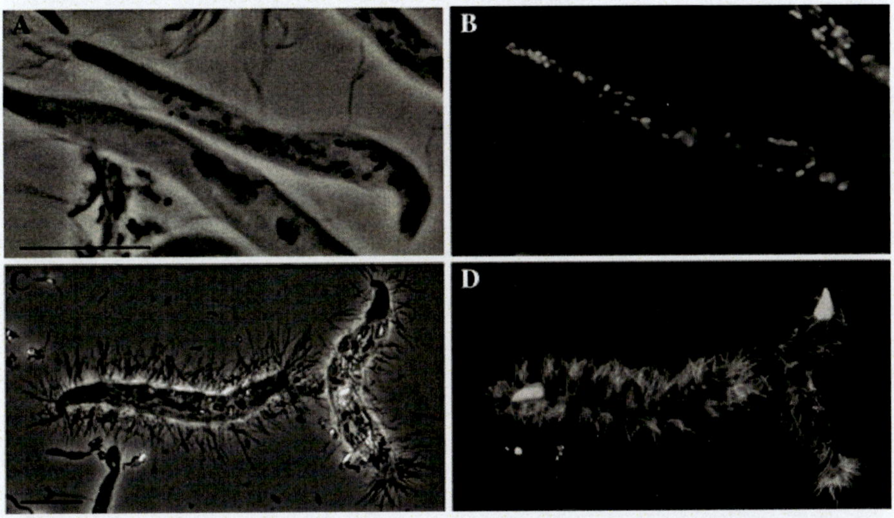

Fig. 15.3 Phase-contrast (**a**) and epifluorescence (**b**) images of cells of the oxymonad protist *Dinenympha parva*. Endosymbiotic methanogens are visible owing to autofluorescence of F_{420}. Phase-contrast (**c**) and epifluorescence (**d**) images of cells of the oxymonad protist *Dinenympha porteri*. The cells of ectosymbiotic spirochetes were visualized by staining with 4,6-diamidino-2-phenyl-indole-HCl. The scale bar in **a** and **b** is 10 μm. The images were reproduced from Ohkuma (2002)

almost the entire surface of the protist cell but sometimes only densely colonize the anterior or posterior end (Fig. 15.3). Two or more distinct spirochete species are usually associated with a single protist cell, and, as in the cases of endosymbiotic methanogens, a common spirochete species is shared by different protist species in the respective host termite (Iida et al. 2000; Noda et al. 2003). Ectosymbiotic spirochetes are often dominant in the gut. Also free-living spirochetes are abundant in the gut fluid. Although spirochete sequences are frequently represented also in clone libraries of the gut wall fraction, these spirochetes are probably ectosymbionts of protist species (e.g. *Pyrsonympha* and *Oxymonas*) attached to the gut wall through their holdfast apparatus (Nakajima et al. 2005, 2006; Yang et al. 2005).

15.7.3 Motility Symbiosis

A spectacular result of a spirochete association is motility symbiosis, in which the host protist cell is propelled solely by the synchronous waving motion of their ectosymbiotic spirochetes. Motility symbiosis has been observed in the protist *Mixotricha paradoxa* in *Mastotermes darwiniensis*, and at least three distinct spirochete species are localized to different areas of the protist cell surface (Wenzel et al. 2003). Removal of the ectobionts by antibiotic treatment leads to a disintegration of the cortical attachment systems (Radek and Nitsch 2007). The other example of motility symbiosis is the protist *Caduceia versatilis* in *Cryptotermes cavifrons*. Flagella of the rod-shaped, adjacently aligned ectosymbionts together form a bundle and create a synchronous helical wave that propels the host protist cell. This rod-shaped ectosymbiont species designated as "*Candidatus* Tammella caduceiae" has been phylogenetically identified as a member of the phylum Synergistetes (Hongoh et al. 2007a). The involvement of ectosymbionts of two distinct bacterial phyla in the motility symbioses implies that these symbiotic relationships are the result of evolutionary convergence.

15.7.4 Bacteroidales Ectosymbionts

Ectosymbiotic bacteria of the order Bacteroidales are also frequently observed and have been identified in various protist genera (Wenzel et al. 2003; Stingl et al. 2004; Noda et al. 2006a, 2006b, 2009b; Hongoh et al. 2007a, 2006b; Strassert et al. 2010). They exhibit a variety of morphological features. Bacteroidales ectosymbionts of *Devescovina* protists (designated as "*Candidatus* Armantifilum devescovinae") are long, fusiform rods that are aligned in longitudinal rows and cover almost the entire surface of the host cell (Noda et al. 2006a; Desai et al. 2010). Bacteroidales ectosymbionts of similar morphology are aligned on the surface of *Caduceia versatilis* between the rod-shaped "*Tammella*" ectosymbionts that are involved in motility (see above).

Filamentous ectosymbionts of similar shape but different phylogenetic position among the Bacteroidales have been identified in *Streblomastix* and *Hoplonympha* (Noda et al. 2006a). In these two cases, the ectosymbionts drastically affect the morphology of the host protist cell by forming deep furrows and thin, outwardly

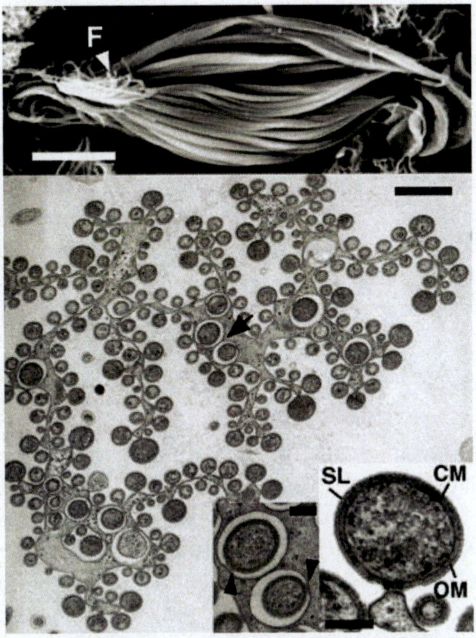

Fig. 15.4 Scanning electron micrograph (*upper*) and transmission electron micrograph of the transverse section (*lower*) of the parabasalid protist *Hoplonympha* sp. The scale bars in the *upper* and *lower panels* are 10 and 1.0 μm, respectively. In the *lower panel*, the vane-like cell body of the protist harbors numerous ectosymbiotic bacteria that have been identified as Bacteroidales (Noda et al. 2006a). The *arrow* points to phagocytosed bacteria, which are magnified in the *left inset*. The *arrowheads* point to attachment structures still present within the phagocytic vacuoles. The *right inset* shows the magnified image of the ectosymbionts, in which the S-layer (SL), outer membrane (OM), and cytoplasmic membrane (CM) of an ectosymbiont are indicated. Scale bars in the insets are 0.2 μm. The images, originally described by Noda et al. (2006a), were reproduced from Ohkuma (2008)

extending, vane-like structures (Fig. 15.4). Such structures apparently increase the surface area, allowing them to support a larger number of ectosymbionts. The ectosymbionts are necessary to maintain the slender cell form of the host protist, and thus have been proposed to act as an exo-cytoskeleton (Leander and Keeling 2004). These impressive structures are probably the result of evolutionary convergence because the host protists *Streblomastix* and *Hoplonympha* are only distantly related to each other, as are their Bacteroidales ectosymbionts (Noda et al. 2006b). The ectosymbiont lineage of *Hoplonympha* is clustered together with those of the other protists in the same protist family Hoplonymphidae and the related family Staurojoeninidae (both belonging to the order Trichonymphida), which suggests that they have acquired ectosymbionts once in their common ancestor (Noda et al. 2006a). However, all the ectosymbionts except those of *Hoplonympha* are short rods and do not form extensive furrows on the protist cell surface. The Bacteroidales ectosymbionts identified from the protists of Oxymonadida are usually only distantly related (Noda et al. 2006b, 2009a; Desai et al. 2010).

Bacteroidales ectosymbionts morphologically similar to ectosymbiotic spirochetes have been recently discovered; they appear as bristle-like appendages of the protist cell and are attached by their tip (Hongoh et al. 2007b). These bristle-like Bacteroidales ectosymbionts are usually scattered among ectosymbiotic spirochetes on the same protist cells. Therefore, their mode of attachment by the tip of the cell is likely more advantageous than a lateral attachment, which would require more space. One species attached to *Dinenympha* protists is designated as "*Candidatus* Symbiothrix dinenymphae" (Hongoh et al. 2007b); however, the bristle-like Bacteroidales ectosymbionts are phylogenetically dispersed, thus suggesting evolutionary convergence from different origins (Noda et al. 2009a).

15.7.5 Endosymbiotic Bacteroidales

Bacteria belonging to the Bacteroidales also occur as rod-shaped endosymbionts of gut protists of the genus *Pseudotrichonympha*. A single cell of *Pseudotrichonympha grassii* in *Coptotermes formosanus* harbors 10^5 cells of Bacteroidales endosymbionts and, surprisingly, this single endosymbiont species accounts for 70–80% of the gut bacteria (Noda et al. 2005). Indeed, its 16S rRNA gene sequence corresponds to the most abundant clones from the whole gut community (Shinzato et al. 2005). This endosymbiont species of *P. grassii* is designated as "*Candidatus* Azobacteroides pseudotrichonymphae" (Hongoh et al. 2008b). With one exception, all endosymbionts in the *Pseudotrichonympha* protists investigated (14 samples) form a monophyletic cluster (Noda et al. 2007). Phagocytosis of Bacteroidales-like ectosymbionts in vesicles of the host protist cells is often observed, and such phagocytosis has been proposed as an origin of bacterial endosymbionts; however, the endosymbiotic Bacteroidales of *Pseudotrichonympha* are distantly related to any ectosymbionts described to date (Noda et al. 2009a).

15.7.6 Endosymbiotic Endomicrobia

The phylum Elusimicrobia (formerly candidate phylum TG1) is a deep-branching clade of uncultivated bacteria from various environments (Herlemann et al. 2007). The first members of this phylum were originally discovered by Ohkuma and Kudo in 1996 and later recognized as endosymbionts of parabasalid and oxymonadid protists (Stingl et al. 2005; Ikeda-Ohtsubo et al. 2007; Ohkuma et al. 2007b; Ikeda-Ohtsubo and Brune 2009; Desai et al. 2010). In all phylogenetic analyses, these endosymbionts formed a distinct cluster of uncultivated bacteria designated as Endomicrobia (Stingl et al. 2005). Although related sequences have been obtained from higher termites, a cockroach, and bovine rumen, which typically lack the protists unique to termites (Ohkuma et al. 2007b; Ikeda-Ohtsubo et al. 2010), the diversity of sequences from lower termites likely reflects the specific associations with their protist hosts. Indeed, sequences of distinct endosymbiont lineages have been obtained from different protist genera examined to date (Ohkuma et al. 2007a; Ikeda-Ohtsubo et al. 2007; Desai et al. 2010). They are usually independent of the higher-level classifications of their host protist and are dispersed in the phylogenetic tree. In a single cell of a large protist species, such as *Trichonympha*, there may be several thousands of Endomicrobia cells, and in view of the single copy of rRNA genes in the Endomicrobia symbiont of a *Trichonympha* species (Hongoh et al. 2008a), Endomicrobia may make up an even larger fraction of the bacterial community of the hindgut than suggested by their proportion in bacterial SSU rRNA gene libraries (up to 10% of the clones for *Reticulitermes speratus*).

15.7.7 Associations of Multiple Species

A single protist cell often harbors multiple species of bacteria as either endosymbionts or ectosymbionts. As mentioned above, two or more species of spirochetes are usually attached to a single protist cell (Iida et al. 2000; Noda et al. 2003; Wenzel et al. 2003). *Caduceia versatilis* harbors both "*Tammella caduceiae*" and "*Armantifilum*"-like Bacteroidales as ectosymbionts (Hongoh et al. 2007a). Many of the *Devescovina* species harbor both Spirochaetes and Bacteroidales as ectosymbionts (Noda et al. 2003, 2006a). The cells of *Trichonympha agilis* harbor both Endomicrobia ("*Candidatus* Endomicrobium trichonymphae") and a member of the genus *Desulfovibrio* designated as "*Candidatus* Desulfovibrio trichonymphae" (Sato et al. 2009), and also the simultaneous association with endosymbiotic Endomicrobia and ectosymbiotic Bacteroidales is not uncommon among several protistan lineages (Desai et al. 2010; Strassert et al. 2010). A number of bacteria associated with gut protists have been described only in ultrastructural studies and remain to be identified, such as the symbionts in the host nucleus (Brugerolle and Radek 2006). These examples illustrate that the protist-prokaryote associations in the termite gut are not simply one-to-one relationships but are rather complex and involve multiple species.

Interestingly, when multiple symbiotic species simultaneously occur within a single protist cell, each species seems to locate to a specific part of the cell, probably because of specific interactions with a local activity in the protist. For instance, most cells of "*Desulfovibrio trichonymphae*" in *T. agilis* localize to the flagella- and hydrogenosome-enriched anterior portion, whereas Endomicrobia mainly occur at the posterior, dilated part of the cell (Sato et al. 2009). This particular location of "*D. trichonymphae*" implies that they utilize the molecular hydrogen produced by the hydrogenosomes of the host, but the electron acceptor of their hydrogen metabolism is unknown. *Desulfovibrio* species are well known for their ability to reduce sulfate with hydrogen as electron donor, but because the termite gut contains sulfate at low concentrations (as low as 0.3 mM), it is not clear whether sulfate in the gut fluid can be effectively utilized by these symbionts (see also Chapter 16 by Brune and Ohkuma, this volume).

15.7.8 Coevolution of Protists and Their Symbionts

Many of the bacterial clusters present in termite gut harbor several distinct lineages of protist-associated symbionts, indicating multiple independent acquisitions of the symbionts during evolution. It appears as if the gut protists have acquired their ectosymbionts or endosymbionts from a pool of free-living populations in the hindgut. For example, the ectosymbiotic *Treponema*, "*Tammella*", and Bacteroidales species form phylogenetic clusters with related species living freely in the hindgut fluid (Noda et al. 2003, 2006a, 2009a; Hongoh et al. 2007a). Evolutionary relationships are complicated by host switches and/or symbiont replacements, which (as examples) have been documented for endosymbiotic methanogens (Tokura et al. 2000) and ectosymbiotic spirochetes (Iida et al. 2000; Noda et al. 2003), or by convergence of certain morphologies, as described above.

Nevertheless, despite these complex evolutionary scenarios, the associations of symbionts with gut protists are species specific and apparently stable. The symbionts of congeneric protist species often form a monophyletic lineage and appear to have originated from a common ancestor. It can be expected that once the protists acquired a stable relationship with their bacterial symbionts, probably by gaining an advantage through adaptation and specialization, the protists and bacteria likely cospeciated with each other.

There are three cases where co-speciation of the partners has been clearly documented. One is the symbiosis between *Pseudotrichonympha* protists and their "*Azobacteroides*" endosymbionts. These flagellates occur exclusively in termites of the family Rhinotermitidae, and typically only a single *Pseudotrichonympha* species is found in each termite species. The flagellates are not only cospeciating with their endosymbionts, but also with their host termites, creating a unique case of co-speciation in a tripartite symbiosis (Noda et al. 2007). The second example is the symbiosis between *Trichonympha* protists and their "*Endomicrobium*" endosymbionts, which occur in the termite families Rhinotermitidae and Termopsidae (Ikeda-Ohtsubo and Brune 2009). The situation in the *Trichonympha*-"*Endomicrobium*" symbiosis is different from that in the *Pseudotrichonympha*-"*Azobacteroides*" symbiosis, because there is no co-speciation between the termite and their *Trichonympha* flagellates, and multiple *Trichonympha* species are simultaneously present in several termite species. This particular situation allows the conclusion that co-speciation of the symbiotic pair cannot be simply the result of a spatial separation of the host protist in different termite guts but must rather be based on the vertical transmission of the endosymbionts. Interestingly, the endosymbionts seem to be absent in *Trichonympha* species found in kalotermitid termites and cockroaches of the genus *Cryptocercus*, which form phylogenetic sister groups to the *Trichonympha* species in Rhinotermitidae and Termopsidae, indicating that the *Trichonympha* host acquired the endosymbionts after these termite lineages separated. Very recently, a third example of the co-speciation has been reported in the symbiosis between *Devescovina* protists and their "*Armantifilum*" ectosymbionts. As in the case of the *Trichonympha*-"*Endomicrobium*" symbiosis, there is no co-speciation between the *Devescovina* flagellates and their termite host, leaving vertical transmission of the symbionts as the most likely explanation for co-speciation also in this ectosymbiotic relationship (Desai et al. 2010).

15.7.9 Complete Genome of the Endosymbionts

Recently, the complete genomes of two endosymbiont species have been reported: one for "*Candidatus* Endomicrobium trichonymphae", the endosymbiont of *Trichonympha agilis* from the Elusimicrobia phylum (Hongoh et al. 2008a), and the other for "*Candidatus* Azobacteroides pseudotrichonymphae", the endosymbiont of *Pseudotrichonympha grassii* from the order Bacteroidales (Hongoh et al. 2008b). Both genome sequences were determined after whole-genome amplification using several hundred manually collected endosymbiont cells from a single host protist cell. It is crucial to start from a single host protist cell because the endosymbionts of different cells could have variations in the genome or polymorphisms.

The genomes of these two endosymbionts are thought to be dynamically adapting as intracellular symbionts because of their small genome size (both ca. 1.1 Mbp) and the loss of many gene functions. These small genomes contain numerous pseudogenes, particularly the "*E. trichonymphae*" genome, which contains 121 pseudogenes. The genome of the cultured representative (*Elusimicrobium minutum*) of Elusimicrobia, to which "*E. trichonymphae*" belongs, is 1.64 Mbp in size and contains almost no pseudogenes (Herlemann et al. 2009). Despite the reductive evolution of the genomes of the two endosymbionts, they retain the ability to synthesize most amino acids and various cofactors. Some of the genes responsible are duplicated, thereby reinforcing the synthesizing abilities. The genome of "*A. pseudotrichonymphae*" encodes for similar pathways but in addition contains a complete set of genes required for nitrogen fixation. Therefore, the endosymbionts may play important roles in supplying the essential nitrogenous nutrients deficient in lignocellulose to the host protists as well as the host termites (Hongoh et al. 2008a, 2008b; Ohkuma 2008; Hongoh and Ohkuma 2010). Details of the metabolic functions of these endosymbionts are described in Chapter 16 by Brune and Ohkuma, this volume.

15.8 Features of Microbial Communities in Higher Termites

15.8.1 Structure of the Microbial Community

Higher termites typically lack unique flagellated protists in their gut, and contain only prokaryotes. Higher termites exhibit an elaborate social system of caste development and division of labor. The fungus-growing termites of the subfamily Macrotermitinae employ the most complex polyethism, organized by different castes and ages. The gut microbial community varies among castes and ages, and is clearly more related to a difference in age than in caste (Hongoh et al. 2006a). This suggests that the variations are crucially affected by their food, which can comprise dead grass and leaves, composted forage, small dead wood, and the conidia and mycelia of the symbiotic fungus *Termitomyces*. The most conspicuous difference in the microbial community among castes and ages is the abundance of bacteria of the family Streptococcaceae (phylum Firmicutes); these bacteria are abundant particularly in newly molted major workers. Despite these intracolony variations, the gut microbial community is clearly distinct from those of the fungus garden and adjacent soil.

A prominent feature of the intestinal tract of higher termites (except fungus growers) is its highly compartmentalized structure, which provides a series of complex and distinct microbial habitats with pronounced axial dynamics of physicochemical parameters (Brune 1998, 2006; see also Chapter 14 by Bignell and Chapter 16 by Brune and Ohkuma, this volume). In particular the first proctodeal segment (P1) is highly alkaline, with pH values up to 12. The bacterial 16S rRNA gene sequences from the alkaline P1 segment of the soil-feeding *Cubitermes orthognathus* (Schmitt-Wagner et al. 2003a) and four other termite species, including soil, wood, and soil/wood interface feeders (Thongaram et al. 2005) are dominated by Firmicutes (clostridia and bacilli), but the phylogenetic compositions vary among the termite species. Closely related bacterial species (members of the class Clostridia) also occur in the alkaline mixed segment, specifically in the space between the mesentric midgut epithelium and the peritrophic membrane (Tokuda et al. 2000). In the soil-feeding *Cubitermes orthognathus*, the proportion of Firmicutes decreases progressively in the posterior gut segments, and the proportion of members of the Bacteroidetes, Proteobacteria, and Spirochaetes increase (Schmitt-Wagner et al. 2003a). In the P3 compartment, members of the phylum Planctomycetes constitute up to one-third of the bacteria detected by FISH, and indeed the P3 and other gut segments contain diverse lineages, including a major one unique to termites (Köhler et al. 2008). Also members of the archaeal community in *Cubitermes orthognathus* differ among the gut compartments; Methanosarcinaceae members dominate in the P1 segment, whereas Methanobacteriaceae and Methanomicrobiales predominate in the posterior segments (Friedrich et al. 2001).

The bacterial population in the mound of the soil-feeding termite *Cubitermes niokoloensis* drastically differs from those in their gut and the surrounding soil (Fall et al. 2004, 2007). The microbial community rapidly changes even in fresh feces, which is dominated by members of the phylum Actinobacteria, in contrast to the gut and surrounding soil, which are dominated by Firmicutes and Proteobacteria, respectively. Phylogenetic analyses have revealed the actinobacterial lineages in the mound differ from those in the surrounding soil and are more diverse. The composition of Ascomycetes also differs between nests of soil-feeding termites and surrounding soils (Roose-Amsaleg et al. 2004). A specific actinobacterial community seems to be present also in the gut of the wood-feeding higher termite *Nasutitermes corniger* (Lefebvre et al. 2009).

15.8.2 Metagenomic Analysis of a Higher Termite

Metagenomic analysis allows an investigation of the entire genetic material in a particular environment, and has been applied to the microbial community in the P3 segment of the gut of the wood-feeding higher termite *Nasutitermes* sp. (Warnecke et al. 2007). Approximately 71 Mbp of metagenome sequence data have been generated, but the assembly of the sequences was difficult, and only highly fragmented sequences have been obtained. In view of the complexity of the bacterial community and the fact that the largest contiguous fragment (contig) assembled had a length of only 15 kbp, it becomes apparent that much more sequence data will be required for a reasonable coverage of the microbial metagenome in any termite gut. Another problem that remains to be solved is the abundant presence of bacterial groups for which no reference genomes are available (in *Nasutitermes* species, e.g. the TG3 phylum and a deep-branching lineage of Fibrobacteres; see Section 15.3.3 above), which makes it difficult to assign the assembled contigs to phylogenetic groups. Although the authors tried to improve phylogenetic assignments by sequencing additional fosmid clones after the initial end-sequencing screen, only 9% of all contigs could be bioinformatically classified beyond the phylum level. Nevertheless, the first metagenome sequences from a termite gut offer a wealth of information on important functions of the gut community, e.g. cellulose and hemicellulose metabolism, hydrogen formation, nitrogen fixation, and reductive acetogenesis (Brune 2007, see also Chapter 16 by Brune and Ohkuma, this volume).

15.9 Conclusions and Perspective

Although the termite gut provides only a tiny, microliter-scale habitat, it is a reservoir of novel and complex microbial diversity. Molecular studies have disclosed that the diversity, especially that of bacteria, is far greater than previously expected and that the gut community consists of mostly novel and unique microbial species. Considerable differences in the community between termite genera are now recognized. Given that there are more than 200 termite genera and that only a small subset of them has been investigated, the termites as a whole should harbor an enormous, novel microbial diversity in their guts. Molecular phylogenetic studies of gut protists also provide unexpected implications of the evolutionary complexity of their cellular organizations.

The gut microbial community is a highly structured symbiotic complex. The microbial species usually show species-specific spatial distributions, as exemplified by a variety of protist-prokaryote associations in lower termites. Future studies should take this highly structured nature into consideration and should link the localization of species to the microbial activities there. For this purpose, genome analyses of individual microbial species are beneficial, as shown in the studies of two endosymbiont species of gut protists. Metagenomics is also advantageous, but the organisms need to be identified and sophisticated statistics and/or advanced experimental technology are required.

The gut microbial community is also a conserved symbiotic complex, at least among individuals of the same host termite species. Because of this consistency, researchers can easily and reproducibly address a stable microbial community in any given termite species. As a consequence, the termite gut provides an excellent and attractive model system to study microbial diversity, ecology, symbiosis, and evolution.

Acknowledgments We thank Karen A. Brune for editing an earlier version of the manuscript.

References

Abe T, Bignell DE, Higashi M (eds) (2000) Termites: evolution, sociality, symbiosis, ecology. Kluwer Academic Publishers, Dordrecht

Adl SM, Simpson AGB, Farmer MA et al (2005) The new higher level classification of eukaryotes with emphasis on the taxonomy of protists. J Eukaryot Microbiol 52:399–451

Berchtold M, Chatzinotas A, Schönhuber W et al (1999) Differential enumeration and in situ localization of micro-organisms in the hindgut of the lower termite *Mastotermes darwiniensis* by hybridization with rRNA-targeted probes. Arch Microbiol 172:407–416

Berchtold M, König H (1995) Phylogenetic position of the two uncultivated trichomonads *Pentatrichomonoides scroa* Kirby and *Metadevescovina extranea* Kirby from the hindgut of the termite *Mastotermes darwiniensis* Froggatt. Syst Appl Microbiol 18:567–573

Berchtold M, König H (1996) Phylogenetic analysis and in situ identification of uncultivated spirochetes from the hindgut of the termite *Mastotermes darwiniensis*. Syst Appl Microbiol 19:66–73

Berchtold M, Ludwig W, König H (1994) 16S rDNA sequence and phylogenetic position of an uncultivated spirochete from the hindgut of the termite *Mastotermes darwiniensis* Froggatt. FEMS Microbiol Lett 123:269–273

Berlanga M, Paster BJ, Guerrero R (2007) Coevolution of symbiotic spirochete diversity in lower termites. Int Microbiol 10:133–139

Brauman A, Dore J, Eggleton P et al (2001) Molecular phylogenetic profiling of prokaryotic communities in guts of termites with different feeding habits. FEMS Microbiol Ecol 35:27–36

Brugerolle G, Patterson DJ (2001) Ultrastructure of *Joenina pulchella* Grassi, 1917 (Protista, Parabasalia), a reassessment of evolutionary trends in the parabasalids, and a new order Cristamonadida for devescovinid, calonymphid and lophomonad flagellates. Org Divers Evol 1:147–160

Brugerolle G, Radek R (2006) Symbiotic protozoa of termites In: König H, Varma A (eds) Intestinal microorganisms of termites and other invertebrates. Springer, Berlin, pp 243–269

Brune A (1998) Termite guts: the world's smallest bioreactors. Trends Biotechnol 16:16–21

Brune A (2006) Symbiotic associations between termites and prokaryotes. In: Dworkin M, Falkow S, Rosenberg E et al (eds) The prokaryotes, vol 1, 3rd edn. Symbiotic associations, biotechnology, applied microbiology, Springer, New York, NY, pp 439–474

Brune A (2007) Woodworker's digest. Nature 450:487–488

Brune A, Friedrich M (2000) Microecology of the termite gut: structure and function on a microscale. Curr Opin Microbiol 3:263–269

Brune A, Stingl U (2005) Prokaryotic symbionts of termite gut flagellates: phylogenetic and metabolic implications of a tripartite symbiosis. In: Overmann J (ed) Molecular basis of symbiosis. Springer, Berlin, pp 39–60

Carpenter KJ, Chow L, Keeling PJ (2009) Morphology, phylogeny, and diversity of *Trichonympha* (Parabasalia: Hypermastigida) of the wood-feeding cockroach *Cryptocercus punctulatus*. J Eukaryot Microbiol 56:305–313

Carpenter KJ, Horak A, Keeling PJ (2010) Phylogenetic position and morphology of Spirotrichosomidae (Parabasalia): new evidence from *Leptospironympha* of *Cryptocercus punctulatus*. Protist 161:122–132

Carpenter KJ, Keeling PJ (2007) Morphology and phylogenetic position of *Eucomonympha imla* (Parabasalia: Hypermastigida). J Eukaryot Microbiol 54:325–332

Carpenter KJ, Waller RF, Keeling PJ (2008) Surface morphology of *Saccinobaculus* (Oxymonadida): implications for character evolution and function in oxymonads. Protist 159:209–221

Dacks JB, Redfield RJ (1998) Phylogenetic placement of *Trichonympha*. J Eukaryot Microbiol 45:445–447

Dacks JB, Silberman JD, Simpson AGB et al (2001) Oxymonads are closely related to the excavate taxon *Trimastix*. Mol Biol Evol 18:1034–1044

Deevong P, Hongoh Y, Inoue T et al (2006) Effect of temporal sample preservation on the molecular study of a complex microbial community in the gut of the termite *Microcerotermes* sp. Microbes Environ 21:78–85

de Koning AP, Noble GP, Heiss AA et al (2008) Environmental PCR survey to determine the distribution of a non-canonical genetic code in uncultivable oxymonads. Environ Microbiol 10:65–74

Desai MS, Strassert JFH, Meuser K et al (2010) Strict cospeciation of devescovinid flagellates and *Bacteroidales* ectosymbionts in the gut of dry-wood termites (*Kalotermitidae*). Environ Microbiol 12:2120–2132

Dolan MF (2001) Speciation of termite gut protists: the role of bacterial symbionts. Int Microbiol 4:203–208

Donovan SE, Purdy KJ, Kane MD, Eggleton P (2005) Comparison of *Euryarchaea* strains in the guts and food-soil of the soil-feeding termite *Cubitermes fungifaber* across different soil types. Appl Environ Microbiol 70:3884–3892

Embley TM, Martin W (2006) Eukaryotic evolution, changes and challenges. Nature 440: 623–630

Fall S, Hamelin J, Ndiaye F et al (2007) Differences between bacterial communities in the gut of a soil-feeding termite (*Cubitermes niokoloensis*) and its mounds. Appl Environ Microbiol 73:5199–5208

Fall S, Nazaret S, Chotte JL, Brauman A (2004) Bacterial density and community structure associated with aggregate size fractions of soil-feeding termite mounds. Microb Ecol 28:191–199

Friedrich MW, Schmitt-Wagner D, Lueders T, Brune A (2001) Axial differences in community structure of *Crenarchaeota* and *Euryarchaeota* in the highly compartmentalized gut of the soil-feeding termite *Cubitermes orthognathus*. Appl Environ Microbiol 67:4880–4890

Fröhlich J, König H (1999) Rapid isolation of single microbial cells from mixed natural and laboratory populations with the aid of a micromanipulator. Syst Appl Microbiol 22: 249–257

Geissinger O, Herlemann DPR, Mörschel E et al (2009) The ultramicrobacterium "*Elusimicrobium minutum*" gen. nov., sp. nov., the first cultivated representative of the termite group I phylum. Appl Environ Microbiol 75:2831–2840

Gerbod D, Noël C, Dolan MF et al (2002) Molecular phylogeny of parabasalids inferred from small subunit rRNA sequences, with emphasis on the Devescovinidae and Calonymphidae (Trichomonadea) Mol Phylogenet Evol 25:545–556

Gerbod D, Sanders E, Moriya S et al (2004) Molecular phylogenies of Parabasalia inferred from four protein genes and comparison with rRNA trees. Mol Phylogenet Evol 31:572–580

Godon JJ, Morinière J, Moletta M et al (2005) Rarity associated with specific ecological niches in the bacterial world: the '*Synergistes*' example. Environ Microbiol 7:213–224

Hampl V, Cepicka I, Flegr J et al (2004) Critical analysis of the topology and rooting of the parabasalian 16S rRNA tree. Mol Phylogenet Evol 32:711–723

Hampl V, Hug L, Leigh JW et al (2009) Phylogenomic analyses support the monophyly of Excavata and resolve relationships among eukaryotic "supergroups". Proc Natl Acad Sci USA 106: 3859–3864

Hara K, Shinzato N, Ohshima T, Yamagishi A (2004) Endosymbiotic *Methanobrevibacter* species living in symbiotic protists of the termite *Reticulitermes speratus* detected by fluorescent in situ hybridization. Microbes Environ 19:120–127

Hara K, Shinzato N, Seo M et al (2002) Phylogenetic analysis of symbiotic Archaea living in the gut of xylophagous cockroaches. Microbes Environ 17:185–190

Harper JT, Gile GH, James ER et al (2009) The inadequacy of morphology for species and genus delineation in microbial eukaryotes: an example from the parabasalian termite symbiont *Coronympha*. PLoS One 4:e6577

Hayashi A, Aoyagi H, Yoshimura T, Tanaka H (2007) Development of novel method for screening microorganisms using symbiotic association between insect (*Coptotermes formosanus* Shiraki) and intestinal microorganisms. J Biosci Bioeng 103:358–367

Heiss AA, Keeling PJ (2006) The phylogenetic position of the oxymonad *Saccinobaculus* based on SSU rRNA. Protist 157:335–344

Herlemann DPR, Geissinger O, Brune A (2007) The termite group I phylum is highly diverse and widespread in the environment. Appl Environ Microbiol 73:6682–6685

Herlemann DPR, Geissinger O, Ikeda-Ohtsubo W et al (2009) Genome analysis of "*Elusimicrobium minutum*," the first cultivated representative of the phylum "*Elusimicrobia*" (formerly Termite Group 1). Appl Environ Microbiol 75:2841–2849

Hongoh Y, Deevong P, Hattori S et al (2006a) Phylogenetic diversity, localization, and cell morphologies of members of the candidate phylum TG3 and a subphylum in the phylum *Fibrobacteres*, recently discovered bacterial groups dominant in termite guts. Appl Environ Microbiol 72:6780–6788

Hongoh Y, Deevong P, Inoue T et al (2005) Intra- and interspecific comparisons of bacterial diversity and community structure support coevolution of gut microbiota and termite host. Appl Environ Microbiol 71:6590–6599

Hongoh Y, Ekpornprasit L, Inoue T et al (2006b) Intracolony variation of bacterial gut microbiota among castes and ages in the fungus-growing termite *Macrotermes gilvus*. Mol Ecol 15: 505–516

Hongoh Y, Ohkuma M (2010) Termite gut flagellates and their methanogenic and eubacterial symbionts. In: Hackstein JHP (ed) Microbiology monographs: (endo) symbiotic methanogenic archaea. Springer, Berlin, Heidelberg, in press

Hongoh Y, Ohkuma M, Kudo T (2003a) Molecular analysis of bacterial microbiota in the gut of the termite *Reticulitermes speratus* (Isoptera; Rhinotermitidae). FEMS Microbiol Ecol 44: 231–242

Hongoh Y, Sato T, Dolan MF et al (2007a) The motility symbiont of the termite gut flagellate *Caduceia versatilis* is a member of the "Synergistes" group. Appl Environ Microbiol 73: 6270–6276

Hongoh Y, Sato T, Noda S et al (2007b) *Candidatus* Symbiothrix dinenymphae: bristle-like Bacteroidales ectosymbionts of termite gut protists. Environ Microbiol 9: 2631–2635

Hongoh Y, Sharma VK, Prakash T et al (2008a) Complete genome of the uncultured Termite Group 1 bacteria in a single host protist cell. Proc Natl Acad Sci USA 105:5555–5560

Hongoh Y, Sharma VK, Prakash T et al (2008b) Genome of an endosymbiont coupling N_2 fixation to cellulolysis within protist cells in termite gut. Science 322:1108–1109

Hongoh Y, Yuzawa H, Ohkuma M, Kudo T (2003b) Evaluation of primers and PCR conditions for the analysis of 16S rRNA genes from a natural environment. FEMS Microbiol Lett 221: 299–304

Iida T, Ohkuma M, Ohtoko K, Kudo T (2000) Symbiotic spirochetes in the termite hindgut: phylogenetic identification of ectosymbiotic spirochetes of oxymonad protists. FEMS Microbiol Ecol 34:17–26

Ikeda-Ohtsubo W, Brune A (2009) Cospeciation of termite gut flagellates and their bacterial endosymbionts: *Trichonympha* species and '*Candidatus* Endomicrobium trichonymphae'. Mol Ecol 18:332–342

Ikeda-Ohtsubo W, Desai M, Stingl U, Brune A (2007) Phylogenetic diversity of '*Endomicrobia*' and their specific affiliation with termite gut flagellates. Microbiology 153:3458–3465

Ikeda-Ohtsubo W, Faivre N, Brune A (2010) Putatively free-living "Endomicrobia" – ancestors of the intracellular symbionts of termite gut flagellates? Environ Microbiol Rep. 2:554–559

Inoue T, Moriya S, Ohkuma M, Kudo T (2005) Molecular cloning and characterization of a cellulase gene from a symbiotic protist of the lower termite, *Coptotermes formosanus*. Gene 349:67–75

Inoue JI, Noda S, Hongoh Y et al (2008) Identification of endosymbiotic methanogen and ectosymbiotic spirochetes of gut protists of the termite *Coptotermes formosanus*. Microbes Environ 23:94–97

Inoue JI, Saita K, Kudo T et al (2007) Hydrogen production by termite gut protists: characterization of iron hydrogenases of parabasalian symbionts of the termite *Coptotermes formosanus*. Eukaryot Cell 6:1925–1932

Keeling PJ (2002) Molecular phylogenetic position of *Trichomitopsis termopsidis* (Parabasalia) and evidence for the Trichomitopsiinae. Eur J Protistol 38:279–286

Keeling PJ (2004) Polymorphic insertions and deletions in parabasalian enolase genes. J Mol Evol 58:550–556

Keeling PJ, Leander BS (2003) Characterisation of a non-canonical genetic code in the oxymonad *Streblomastix strix*. J Mol Biol 326:1337–1349

Keeling PJ, Poulsen N, Mcfadden GI (1998) Phylogenetic diversity of parabasalian symbionts from termites, including the phylogenetic position of *Pseudotrypanosoma* and *Trichonympha*. J Eukaryot Microbiol 45:643–650

Kitade O (2004) Comparison of symbiotic flagellate faunae between termites and a wood-feeding cockroach of the genus *Cryptocercus*. Microbes Environ 19:215–220

Köhler T, Stingl U, Meuser K, Brune A (2008) Novel lineages of *Planctomycetes* densely colonize the alkaline gut of soil-feeding termites (*Cubitermes* spp). Environ Microbiol 10: 1260–1270

Leander BS, Keeling PJ (2004) Symbiotic innovation in the oxymonad *Streblomastix strix*. J Eukaryot Microbiol 51:291–300

Lefebvre T, Miambi E, Pando A et al (2009) Gut-specific actinobacterial community structure and diversity associated with the wood-feeding termite species, *Nasutitermes corniger* (Motschulsky) described by nested PCR-DGGE analysis. Insect Soc 56: 269–276

Ley RE, Peterson DA, Gordon JI (2006) Ecological and evolutionary forces shaping microbial diversity in the human intestine. Cell 124:837–848

Li L, Fröhlich J, Pfeiffer P, König H (2003) Termite gut symbiotic Archaezoa are becoming living metabolic fossils. Eukaryot Cell 2:1091–1098

Lilburn TG, Schmidt TM, Breznak JA (1999) Phylogenetic diversity of termite gut spirochaetes. Environ Microbiol 1:331–345

Lo N, Tokuda G, Watanabe H et al (2000) Evidence from multiple gene sequences indicates that termites evolved from wood-feeding cockroaches. Curr Biol 10:801–804

Minkley N, Fujita A, Brune A, Kirchner WH (2006) Nest specificity of the bacterial community in termite guts (*Hodotermes mossambicus*). Insect Soc 53:339–344

Miyata R, Noda N, Tamaki H et al (2007a) Phylogenetic relationship of symbiotic Archaea in the gut of the higher termite *Nasutitermes takasagoensis* fed with various carbon sources. Microbes Environ 22:157–164

Miyata R, Noda N, Tamaki H et al (2007b) Influence of feed components on symbiotic bacterial community structure in the gut of the wood-feeding higher termite *Nasutitermes takasagoensis*. Biosci Biotechnol Biochem 71:1244–1251

Moriya S, Dacks JB, Takagi A et al (2003) Molecular phylogeny of three oxymonad genera: *Pyrsonympha*, *Dinenympha* and *Oxymonas*. J Eukaryot Microbiol 50:190–197

Moriya S, Ohkuma M, Kudo T (1998) Phylogenetic position of symbiotic protist *Dinenympha exilis* in the hindgut of the termite *Reticulitermes speratus* inferred from the protein phylogeny of elongation factor 1 alpha. Gene 210:221–227

Moriya S, Tanaka K, Ohkuma M et al (2001) Diversification of the microtubule system in the early stage of eukaryote evolution: elongation factor 1 alpha and alpha-tubulin protein phylogeny of termite symbiotic oxymonad and hypermastigote protists. J Mol Evol 52:6–16

Nakajima H, Hongoh Y, Noda S et al (2006) Phylogenetic and morphological diversity of Bacteroidales members associated with the gut wall of termites. Biosci Biotech Biochem 70:211–218

Nakajima H, Hongoh Y, Usami R et al (2005) Spatial distribution of bacterial phylotypes in the gut of the termite *Reticulitermes speratus* and the bacterial community colonizing the gut epithelium. FEMS Microbiol Ecol 54:247–255

Nakashima KI, Watanabe H, Azuma JI (2002) Cellulase genes from the parabasalian symbiont *Pseudotrichonympha grassii* in the hindgut of the wood-feeding termite *Coptotermes formosanus*. Cell Mol Life Sci 59:1554–1560

Nalepa CA, Bignell DE, Bandi C (2001) Detritivory, coprophagy, and the evolution of digestive mutualisms in Dictyoptera. Insect Soc 48:194–201

Noda S, Hongoh Y, Sato T, Ohkuma M (2009a) Complex coevolutionary history of symbiotic Bacteroidales bacteria of various protists in the gut of termites. BMC Evol Biol 9:158

Noda S, Iida T, Kitade S et al (2005) Endosymbiotic Bacteroidales bacteria of the flagellated protist *Pseudotrichonympha grassii* in the gut of the termite *Coptotermes formosanus*. Appl Environ Microbiol 71:8811–8817

Noda S, Inoue T, Hongoh Y et al (2006a) Identification and characterization of ectosymbionts of distinct lineages in Bacteroidales attached to flagellated protists in the gut of termites and a wood-feeding cockroach. Environ Microbiol 8:11–20

Noda S, Kawai M, Nakajima H et al (2006b) Identification and in situ detection of two lineages of *Bacteroidales* ectosymbionts associated with a termite gut protist, *Oxymonas* sp. Microbes Environ 21:16–22

Noda S, Kitade O, Inoue T et al (2007) Cospeciation in the triplex symbiosis of termite gut protists (*Pseudotrichonympha* spp.), their hosts, and their bacterial endosymbionts. Mol Ecol 16: 1257–1266

Noda S, Mantini C, Bordereau C et al (2009b) Molecular phylogeny of parabasalids with emphasis on the order Cristamonadida and its complex morphological evolution. Mol Phylogenet Evol 52:217–224

Noda S, Ohkuma M, Yamada A et al (2003) Phylogenetic position and in situ identification of ectosymbiotic spirochetes on protists in the termite gut. Appl Environ Microbiol 69: 625–633

Noël C, Noda S, Mantini C et al (2007) Molecular phylogenetic position of the genera *Stephanonympha* and *Caduceia* (Parabasalia) inferred from nuclear small subunit rRNA gene sequences. J Eukaryot Microbiol 54:93–99

Ohkuma M (2002) Symbiosis in the termite gut: culture-independent molecular approaches. In: Seckbach J (ed) Symbiosis: mechanisms and model systems. Kluwer Academic Publishers, Dordrecht, pp 715–730

Ohkuma M (2003) Termite symbiotic systems: efficient bio-recycling of lignocellulose. Appl Microbiol Biotechnol 61:1–9

Ohkuma M (2008) Symbioses of flagellates and prokaryotes in the gut of lower termites. Trends Microbiol 16:345–352

Ohkuma M, Hongoh Y, Kudo T (2006) Diversity and molecular analyses of yet-uncultivated microorganisms. In: König H, Varma A (eds) Intestinal microorganisms of termites and other invertebrates. Springer, Heidelberg, pp 303–317

Ohkuma M, Iida T, Kudo T (1999a) Phylogenetic relationships of symbiotic spirochetes in the gut of diverse termites. FEMS Microbiol Lett 181:123–129

Ohkuma M, Iida T, Ohtoko K et al (2005) Molecular phylogeny of parabasalids inferred from small subunit rRNA sequences, with emphasis on the Hypermastigea. Mol Phylogenet Evol 35:646–655

Ohkuma M, Kudo T (1996) Phylogenetic diversity of the intestinal bacterial community in the termite *Reticulitermes speratus*. Appl Environ Microbiol 62:461–468

Ohkuma M, Kudo T (1998) Phylogenetic analysis of the symbiotic intestinal microflora of the termite *Cryptotermes domesticus*. FEMS Microbiol Lett 164:389–395

Ohkuma M, Noda S, Hongoh Y et al (2009) Inheritance and diversification of symbiotic trichonymphid flagellates from a common ancestor of termites and the cockroach *Cryptocercus*. Proc R Soc B 276:239–245

Ohkuma M, Noda S, Hongoh Y, Kudo T (2002) Diverse bacteria related to the bacteroides subgroup of the CFB phylum within the gut symbiotic communities of various termites. Biosci Biotech Biochem 66:78–84

Ohkuma M, Noda S, Horikoshi K, Kudo T (1995) Phylogeny of symbiotic methanogens in the gut of the termite *Reticulitermes speratus*. FEMS Microbiol Lett 134:45–50

Ohkuma M, Noda S, Kudo T (1999b) Phylogenetic relationships of symbiotic methanogens in diverse termites. FEMS Microbiol Lett 171:147–153

Ohkuma M, Ohtoko K, Grunau C et al (1998) Phylogenetic identification of the symbiotic hypermastigote *Trichonympha agilis* in the hindgut of the termite *Reticulitermes speratus* based on small-subunit rRNA sequence. J Eukaryot Microbiol 45:439–444

Ohkuma M, Ohtoko K, Iida T et al (2000) Phylogenetic identification of hypermastigotes, *Pseudotrichonympha*, *Spirotrichonympha*, *Holomastigotoides*, and parabasalian symbionts in the hindgut of termites. J Eukaryot Microbiol 47:249–259

Ohkuma M, Saita K, Inoue T, Kudo T (2007a) Comparison of four protein phylogeny of parabasalian symbionts in termite guts. Mol Phylogenet Evol 42:847–853

Ohkuma M, Sato T, Noda S et al (2007b) The candidate phylum 'Termite Group 1' of bacteria: phylogenetic diversity, distribution, and endosymbiont members of various gut flagellated protists. FEMS Microbiol Ecol 60:467–476

Ohtoko K, Ohkuma M, Moriya S et al (2000) Diverse genes of cellulase homologues of glycosyl hydrolase family 45 from the symbiotic protists in the hindgut of the termite *Reticulitermes speratus*. Extremophiles 4:343–349

Paster BJ, Dewhirst FE, Cooke SM et al (1996) Phylogeny of not-yet-cultured spirochetes from termite guts. Appl Environ Microbiol 62:347–352

Purdy KJ (2007) The distribution and diversity of Euryarchaeota in termite guts. Adv Appl Microbiol 62:63–80

Radek R, Nitsch G (2007) Ectobiotic spirochetes of flagellates from the termite *Mastotermes darwiniensis*: attachment and cyst formation. Eur J Protistol 43:281–294

Roose-Amsaleg C, Brygoo Y, Harry M (2004) Ascomycete diversity in soil-feeding termite nests and soils from a tropical rainforest. Environ Microbiol 6:462–469

Sato T, Hongoh Y, Noda S et al (2009) *Candidatus* Desulfovibrio trichonymphae, a novel intracellular symbiont of the flagellate *Trichonympha agilis* in termite gut. Environ Microbiol 11:1007–1015

Schmitt-Wagner D, Friedrich MW, Wagner B, Brune A (2003a) Phylogenetic diversity, abundance, and axial distribution of bacteria in the intestinal tract of two soil-feeding termites (*Cubitermes* spp). Appl Environ Microbiol 69:6007–6017

Schmitt-Wagner D, Friedrich MW, Wagner B, Brune A (2003b) Axial dynamics, stability, and interspecies similarity of bacterial community structure in the highly compartmentalized gut of soil-feeding termites (*Cubitermes* spp). Appl Environ Microbiol 69:6018–6024

Shinzato N, Matsumoto T, Yamaoka I et al (1999) Phylogenetic diversity of symbiotic methanogens living in the hindgut of the lower termite *Reticulitermes speratus* analyzed by PCR and in situ hybridization. Appl Environ Microbiol 65:837–840

Shinzato N, Matsumoto T, Yamaoka I et al (2001) Methanogenic symbionts and the locality of their host lower termites. Microbes Environ 16:43–47

Shinzato N, Muramatsu M, Matsui T, Watanabe Y (2005) Molecular phylogenetic diversity of the bacterial community in the gut of the termite *Coptotermes formosanus*. Biosci Biotechnol Biochem 69:1145–1155

Shinzato N, Muramatsu M, Matsui T, Watanabe Y (2007) Phylogenetic analysis of the gut bacterial microflora of the fungus-growing termite *Odontotermes formosanus*. Biosci Biotechnol Biochem 71:906–915

Simpson AGB (2003) Cytoskeletal organization, phylogenetic affinities and systematics in the contentious taxon Excavata (Eukaryota). Int J Syst Evol Microbiol 53:1759–1777

Slamovits CH, Keeling PJ (2006) A high density of ancient spliceosomal introns in oxymonad excavates. BMC Evol Biol 6:34

Stingl U, Brune A (2003) Phylogenetic diversity and whole-cell hybridization of oxymonad flagellates from the hindgut of the wood-feeding lower termite *Reticulitermes flavipes*. Protist 154:147–155

Stingl U, Maass A, Radek R, Brune A (2004) Symbionts of the gut flagellate *Staurojoenina* sp. from *Neotermes cubanus* represent a novel, termite-associated lineage of Bacteroidales: description of '*Candidatus* Vestibaculum illigatum'. Microbiology 150:2229–2235

Stingl U, Radek R, Yang H, Brune A (2005) "Endomicrobia": cytoplasmic symbionts of termite gut protozoa form a separate phylum of prokaryotes. Appl Environ Microbiol 71:1473–1479

Strassert JFH, Desai MS, Brune A, Radek R (2009) The true diversity of devescovinid flagellates in the termite *Incisitermes marginipennis*. Protist 160:522–535

Strassert JFH, Desai MS, Radek R, Brune A (2010) Identification and localization of the multiple bacterial symbionts of the termite gut flagellate *Joenia annectens*. Microbiology 156:2068–2079

Tanaka H, Aoyagi H, Shina S et al (2006) Influence of the diet components on the symbiotic microorganisms community in hindgut of *Coptotermes formosanus* Shiraki. Appl Microbiol Biotechnol 71:907–917

Thongaram T, Hongoh Y, Kosono S et al (2005) Comparison of bacterial communities in the alkaline gut segment among various species of higher termites. Extremophiles 9:229–238

Todaka N, Inoue T, Saita K et al (2010) Phylogenetic analysis of cellulolytic enzyme genes from representative lineages of termites and a related cockroach. PLoS One 5:e8636

Todaka N, Moriya S, Saita K et al (2007) Environmental cDNA analysis of the genes involved in lignocellulose digestion in the symbiotic protist community of *Reticulitermes speratus*. FEMS Microbiol Ecol 59:592–599

Tokuda G, Yamaoka I, Noda H (2000) Localization of symbiotic clostridia in the mixed segment of the termite *Nasutitermes takasagoensis* (Shiraki). Appl Environ Microbiol 66:2199–2207

Tokura M, Ohkuma M, Kudo T (2000) Molecular phylogeny of methanogens associated with flagellated protists in the gut and with the gut epithelium of termites. FEMS Microbiol Ecol 33:233–240

Warnecke F, Luginbühl P, Ivanova N et al (2007) Metagenomic and functional analysis of hindgut microbiota of a wood-feeding higher termite. Nature 450:560–565

Watanabe H, Nakashima K, Saito H, Slaytor M (2002) New endo-β-1,4-glucanases from the parabasalian symbionts, *Pseudotrichonympha grassii* and *Holomastigotoides mirabile* of *Coptotermes* termites. Cell Mol Life Sci 59:1983–1992

Wenzel M, Radek R, Brugerolle G, König H (2003) Identification of the ectosymbiotic bacteria of *Mixotricha paradoxa* involved in movement symbiosis. Eur J Protistol 39:11–23

Yang H, Schmitt-Wagner D, Stingl U, Brune A (2005) Niche heterogeneity determines bacterial community structure in the termite gut (*Reticulitermes santonensis*). Environ Microbiol 7: 916–932

Zhou X, Smith JA, Oi FM et al (2007) Correlation of cellulase gene expression and cellulolytic activity throughout the gut of the termite *Reticulitermes flavipes*. Gene 395:29–39

Chapter 16
Role of the Termite Gut Microbiota in Symbiotic Digestion

Andreas Brune and Moriya Ohkuma

Abstract The symbiotic gut microbiota of termites plays important roles in lignocellulose digestion and nitrogen metabolism. Termites possess a dual cellulolytic system: in lower termites the cellulases are contributed by both the insect and its gut flagellates, whereas in higher termites, host cellulases and hindgut bacteria participate in fiber digestion. Commonly, the microbial feeding chain is driven by the primary fermentations of carbohydrates. However in soil-feeding taxa, which exploit the peptidic component of soil organic matter as a dietary resource and show pronounced differences in physiochemical conditions along their highly compartmented intestinal tract, amino acids are an important substrate for the microbiota. Hydrogen appears to be the central intermediate in the hindgut fermentations in all termites. In wood-feeding taxa, it is efficiently recycled by homoacetogenic spirochetes, which prevail over methanogenic archaea probably because of their ability to colonize the bulk volume of the hindgut, whereas methanogens are restricted to particular microniches at the hindgut wall or within the gut flagellates. As a general rule, the spatial separation of microbial populations and metabolic activities gives rise to steep gradients of metabolites. The continuous influx of oxygen into the hindgut affects microbial metabolism in the microoxic periphery, and the anoxic status of the gut center is maintained only by the rapid reduction of oxygen by both aerobic and anaerobic microorganisms. Lignin is not significantly mineralized during gut passage, but modification of polyphenols by processes yet uncharacterized may increase the digestibility of both lignocellulose and humic substances. In wood-feeding termites, gut microbiota fix and upgrade nitrogen and recycle nitrogenous waste products. The microorganisms responsible for these reactions are mostly unknown, but recent studies have indicated that bacterial ectosymbionts and endosymbionts of the gut flagellates play a major role in the nitrogen metabolism of lower termites.

A. Brune (✉)
Department of Biogeochemistry, Max Planck Institute for Terrestrial Microbiology, 35043 Marburg, Germany
e-mail: brune@mpi-marburg.mpg.de

D.E. Bignell et al. (eds.), *Biology of Termites: A Modern Synthesis*,
DOI 10.1007/978-90-481-3977-4_16, © Springer Science+Business Media B.V. 2011

16.1 Introduction

Termites are among the few animals capable of degrading lignocellulose. As in other instances where fiber-rich but nutrient-poor material is being digested, this capacity is based largely on microbial symbionts housed in specialized regions of the intestinal tract (Brune 2009a, Fig. 16.1). This chapter reviews the metabolic processes catalyzed by the gut microbiota of termites and the functions of the different microbial guilds. Phylogenetic diversity, community structure, and evolution of the microbial communities of the termite gut are treated separately (Chapter 15 by Ohkuma and Brune, this volume).

The older literature on this subject has already been addressed in several comprehensive reviews, including two excellent chapters in the predecessor volume (Breznak 2000; Slaytor 2000). Here, the focus will be on conceptual advances made in the past decade, to a large extent based on the development of novel microanalytical techniques, such as microsensor measurements and the microinjection of radiotracers, and the introduction of culture-independent molecular approaches to the characterization of the intestinal microbiota and its activities (Ohkuma 2002; Brune and Pester 2005; Schramm 2006). Detailed microbiological perspectives of the symbiotic associations between termites and prokaryotes are given elsewhere (Brune and Stingl 2005; Brune et al. 2006; Ohkuma 2008). The digestion of lignocellulose has been reviewed also by Ohkuma (2003).

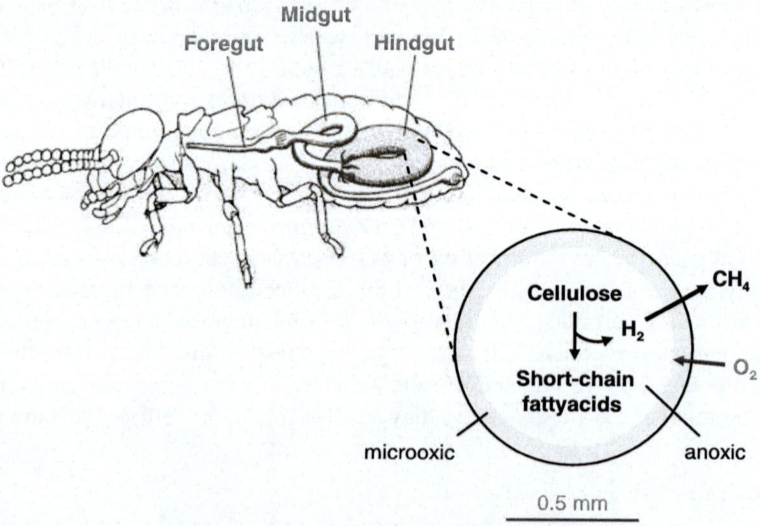

Fig. 16.1 The hindgut of wood-feeding termites is a microbial bioreactor that ferments lignocellulose to acetate and some methane. The fermentation products of the anaerobic gut microbiota are the major carbon and energy sources for the host. However, a termite gut is not a purely anoxic fermenter. The constant influx of oxygen across the gut epithelium strongly affects the microbial processes within the gut

16.2 Digestion of Wood Polysaccharides

A fundamental function of the termite–symbiont system in the digestion of lignocellulose is the hydrolysis of cellulose and hemicelluloses. The symbiotic digestion of lignocellulose is a highly efficient process, resulting in assimilation rates of wood glucan by termites that are far higher than those of grasses by ruminants (Breznak and Brune 1994). While cows exploit only 30–40% of the polysaccharides in their forage, the passage of lignocellulose through the gut of lower termites removes most neutral polysaccharides and more than half of the acidic sugars (Hyodo et al. 1999; Katsumata et al. 2007, and references therein).

Although host-derived cellulases are secreted in the foregut or midgut region and increasingly appear to be of importance in fiber digestion in higher termites (Slaytor 2000; Tokuda et al. 2004), the hindgut remains unarguably the most important site for this process (Tokuda et al. 2005). The considerable dilation of this gut region slows down gut transit and prolongs the exposure of the digesta to the activities of the intestinal microbiota: this is the key to the breakdown of the recalcitrant diet.

16.2.1 The Dual Cellulolytic System of Lower Termites

Although the capacity to digest cellulose is present also in other Dictyoptera, termites became efficient wood feeders only after their ancestor had established an association with cellulolytic gut flagellates. Termite gut flagellates belong to two separate lineages of unicellular eukaryotes, parabasalids and oxymonadids, and occur only in the hindguts of lower termites and of the closely related wood-feeding cockroaches of the genus *Cryptocercus*, the sister group of the termite lineage (Chapter 2 by Lo and Eggleton, this volume). The diversity and phylogeny of termite gut flagellates are reviewed in Chapter 15 (by Ohkuma and Brune, this volume).

The partners in the termite–flagellate symbiosis are obligately dependent on each other. The vital role of the flagellates in cellulose digestion, first documented by Cleveland (1923) in a series of elegant experiments, is illustrated by the fact that termites die of starvation within a few weeks after removal of their flagellate symbionts even though they continue to feed on wood, whereas they remain viable if maintained on a starch diet or re-inoculated with gut contents of normally faunated nestmates. The termites comminute the wood through the action of their mandibles and proventriculus, a process that forms microscopic particles (ca. 20–100 μm) that can be phagocytosed by the flagellates. Incorporation into the digestive vacuoles of the flagellates also slows down gut passage of the digesta and increases exposure time to digestive enzymes. Each termite species typically harbors several flagellate species that seem to be nutritionally specialized, each filling a specific niche in lignocellulose digestion. The composition of the flagellate community is affected when termites are fed with artificial diets (e.g. pure cellulose or xylan), and selective defaunation changes the polysaccharidase activities in the hindgut. This has allowed the identification of the role of certain flagellates in xylan degradation, or the specialization of others on cellulose fractions of different molecular weights.

Many studies of this kind date far back into the twentieth century and have been comprehensively reviewed (e.g. Yoshimura 1995; Inoue et al. 2000; Brugerolle and Radek 2006). A more recent study of *Coptotermes formosanus* showed that such dietary shifts also affect the bacterial community (Tanaka et al. 2006), although it remains to be investigated whether polymer-degrading bacteria are affected directly or whether changes in community structure reflect changes in the microfauna and their associated prokaryotic symbionts.

In addition to the cellulolytic activity of their symbionts, termites also produce their own "host cellulases". The production of endogenous cellulases had been postulated already by several authors in the 1970s (see Slaytor 1992, 2000) and was unequivocally proven with Watanabe and Tokuda's demonstration of cellulase genes and their expression in the salivary glands or the midgut epithelium (see Chapter 3 by Lo et al., this volume). The host cellulases are endoglucanases exclusively of the glycosyl hydrolase family (GHF) 9 (see Li et al. 2006; Watanabe and Tokuda 2009), which degrade amorphous substrates well, but unlike exo-cellobiohydrolases depolymerize crystalline celluloses inefficiently. Many existing reports of high symbiont-independent levels of cellulase activity in termites and cockroaches have to be interpreted with caution since they were mostly determined with carboxymethyl cellulose (CMC), a chemically modified substrate that is not representative of the native polymer in wood. Differences between the relative degradation rates of CMC and microcrystalline cellulose underline the importance of using the appropriate substrates when assessing cellulolytic capacities of different gut regions (Tokuda et al. 2005).

The cellulases produced by flagellates, on the other hand, comprise both exo-cellobiohydrolases and endo-β-1,4-glucanases. The enzymes characterized to date stem from flagellates in the hindguts of *Reticulitermes speratus*, *Coptotermes formosanus*, and *Mastotermes darwiniensis*. In all cases, the origin of the flagellate cellulases has been verified either by using capillary-picked flagellate suspensions or by fluorescence in situ hybridization with mRNA-specific primers. By cloning, heterologous expression and subsequent biochemical characterization of the properties of the recombinant proteins, the cellulases have been identified as members of GHF 5, GHF 7, and GHF 45 (Ohtoko et al. 2000; Nakashima et al. 2002; Watanabe et al. 2002; Li et al. 2003; Inoue et al. 2005; Todaka et al. 2010b). A comparative cDNA analysis of glycosyl hydrolases of the flagellates from several species of lower termites and from the closely related wood-feeding cockroach *Cryptocercus punctulatus* revealed that many of the genes are ancestral to the termite–flagellate symbiosis (Todaka et al. 2010a). Phylogenetic analysis indicated that the GHF 7 enzymes are innate to the flagellates, whereas GHF 10 and 11 enzymes are most likely of bacterial origin and must have been acquired by the protists via lateral gene transfer.

Since hemicelluloses protect the cellulose fibers from enzymatic attack, the simultaneous hydrolysis of both polymers is a prerequisite for efficient lignocellulose digestion. Although it was recognized that both cellulose and hemicelluloses disappear during gut passage, little is known about the enzymes responsible for the degradation of pentosans. *Trichomitopsis termopsidis*, the only termite gut flagellate that has been studied in this regard in pure culture, digests both microcrystalline cellulose and xylan, one of the major hemicelluloses of wood (Odelson and Breznak 1985b). Most of the endoxylanase and β-xylosidase activities of *Reticulitermes speratus* are in the anterior hindgut, and the loss of xylanolytic activities upon selective defaunation (removal of certain protozoa from the hindgut) indicates a major role of gut flagellates in the process (Inoue et al. 1997).

Analysis of cDNA revealed that the scope of glycosyl hydrolase genes expressed by the flagellates in the hindgut of *Reticulitermes speratus* comprises not only cellulases but also several putative xylanases, arabinosidase, mannosidase, and arabinofuranosidase (Todaka et al. 2007). Also in *Reticulitermes flavipes*, xylanolytic activities are restricted to the hindgut, but the presumed role in xylan degradation of cellulase orthologs originating from the flagellates remains to be substantiated (Zhou et al. 2007). Xylanase activity is present also in the hindgut of *Coptotermes formosanus* (Azuma et al. 1984), and functional xylanases originating from the flagellate *Holomastigotoides mirabile* have been characterized (Arakawa et al. 2009). Recently, a comprehensive cDNA analysis documented that putative xylanases of several GHFs are present in all termite species investigated and also in the closely related wood-feeding cockroach *C. punctulatus* (Todaka et al. 2010a)

Although the "dual cellulolytic system" of lower termites (Nakashima et al. 2002) opens the possibility of cooperative effects among the polymer-degrading activities of host and symbionts, it is not clear to what extent the pretreatment of the digesta with host enzymes in fore- and midgut facilitates their final digestion by the flagellate community in the hindgut. In addition to the rapid passage of the digesta through the tubular midgut, it has to be considered that different glycosyl hydrolases can cooperate only if they are simultaneously present within the same digestive vacuole of a protist, rendering synergistic effects of spatially separated activities unlikely (Tokuda et al. 2007). Li et al. (2003) reported that the gut flagellates of *Mastotermes darwiniensis* do not express their own cellulases but incorporate host cellulases produced in the salivary glands into their digestive vacuoles, however this hypothesis was challenged by Watanabe et al (2006) who demonstrated that the protists in the gut of *M. darwiniensis* indeed produce their own cellulases, which are important in cellulose digestion.

Very little is known about the cellulolytic or hemicellulolytic activities of oxymonad flagellates. Since only the larger species seem to phagocytose wood (Kiuchi et al. 2004), others may thrive on bacteria or soluble substrates, in a similar fashion to some smaller parabasalid species (Inoue et al. 2000; Brugerolle and Radek 2006).

16.2.2 Role of Bacteria in Fiber Digestion

Although numerous bacteria catalyzing individual stages of lignocellulose degradation have been cultured by enrichment from the hindgut of termites (see Brune et al. 2006; König et al. 2006), none of the isolates seems to play an essential role in fiber digestion. This does not represent a conceptual problem in the lower termites, where lignocellulose digestion in the hindgut may be exclusively catalyzed by the gut flagellates, which efficiently incorporate any wood particles entering the hindgut into their digestive vacuoles, rendering them inaccessible to the attack by prokaryotes in the gut fluid. However, the situation differs in higher termites (family Termitidae). All members of this family, which are considered evolutionarily advanced and make up the majority of all extant termite species, have lost the gut flagellates characteristic of lower termites and typically harbor an essentially prokaryotic microbial community in their hindguts. A preliminary report that the amoebae found in the guts of some higher termites play an important role in cellulose digestion (cited by Slaytor 2000) remains to be substantiated.

An involvement of bacteria in lignocellulose digestion in higher termites has long been controversial (see Slaytor 2000). The debate was fueled by the reports of host cellulases produced by the midgut epithelium, and of apparently low endoglucanase activities in the hindgut contents of higher termites, which together suggested a greater importance in cellulose digestion for the termite itself (or in the fungus-cultivating termites, the symbiotic fungus: see Chapter 3 by Lo et al. and Chapter 8 by Nobre et al., this volume). However, new findings have rebalanced the argument. The first evidence for an involvement of hindgut bacteria in cellulose digestion was provided by the observation that significant cellulase activities acting on crystalline cellulose are present not only in cell-free supernatants of centrifuged midgut contents, but also in the pellets of centrifuged hindgut contents of *Nasutitermes takasagoensis* (Tokuda et al. 2005). The activity in the hindgut had been missed in previous studies, which had employed only the clarified supernatant of the hindgut contents for enzyme assays, thus excluding the activities of particle-associated enzymes or bacteria (Tokuda and Watanabe 2007).

Characterization of the hindgut microbiota of wood-feeding higher termites with cultivation-independent methods led to the discovery of several lineages of uncultivated bacteria, which were classified as Termite Group 3 (TG3; Hongoh et al. 2005; Hongoh et al. 2006). Such bacteria are abundant in *Microcerotermes* and *Nasutitermes* species and appear to be auchthonous to termite guts. Interestingly, the members of these lineages are distantly related to Fibrobacteres, which comprise the most important cellulose-degrading bacteria in the rumen.

Finally, the emerging role of uncultivated bacteria in fiber digestion was substantiated by a metagenomic analysis of the hindgut microbiota of a different *Nasutitermes* species, yielding numerous endoglucanase genes which phylogenetic binning suggested were encoded by members of Fibrobacteres and Spirochaetes (Warnecke et al. 2007). It is tempting to assign the putative Fibrobacteres sequences to deep-branching Fibrobacteres or the TG3, but in view of the complete absence of genome information for members of these bacterial clades further studies are required to substantiate the case. Many of the glycosyl hydrolases identified in the metagenomic dataset are major proteins secreted into the gut fluid, and the catalytic properties of recombinant enzymes encoded by these genes have been confirmed. Genes encoding various enzymes depolymerizing and debranching hemicellulose (xylanase, arabinofuranosidase, and others) were also identified and were attributed to Spirochaetes (treponemes). A bacterial xylanase gene was previously found to be expressed in the gut of an unidentified member of Nasutitermitinae (Brennan et al. 2004). These and other aspects of the findings are discussed by Brune (2007).

16.3 The Anaerobic Food Web

The low concentrations of soluble sugars and the accumulation of their metabolites in the hindgut fluid of termites indicate that lignocellulose depolymerization is coupled to the fermentative degradation of its hydrolysis products. Also, other dietary components, endogenous substrates, and certain products of the primary fermentations are metabolized. The final products of microbial metabolism are short-chain fatty acids, which are subsequently resorbed by the host and form the basis for its energy metabolism (Fig. 16.2).

16.3.1 Fermentative Degradation of Carbohydrates

The bulk of the fermentative activities in lower termites should be catalyzed by the flagellates that occupy most of the hindgut volume and efficiently sequester any wood particles entering the hindgut. However, despite their sheer abundance and the fact that their central role in the fermentative metabolism of lignocellulose in the gut of *Zootermopsis* spp. was established more than 60 years ago by the pioneering work of Robert E. Hungate (see Breznak and Brune 1994), our knowledge of the metabolism of termite gut flagellates is quite limited (see Brune and Stingl 2005; Ohkuma 2008). So far, the only termite gut flagellate whose metabolism of cellulose

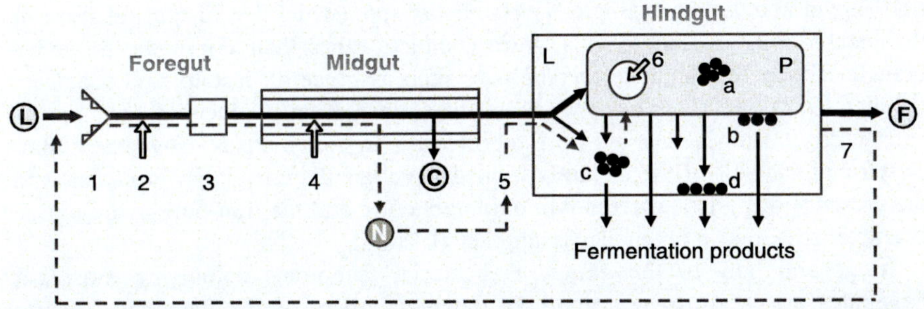

Fig. 16.2 Symbiotic digestion in a wood-feeding lower termite. The *bold lines* show the path of the insoluble material, the lignin-rich residues of which are released as feces, whereas the thinner lines represent soluble degradation products that are eventually resorbed by the host. The *dashed lines* indicate the cycling of nitrogenous compounds. Hollow *arrows* mark the sites where cellulolytic enzymes are secreted. Lower-case letters refer to the different groups of prokaryotes in the hindgut, which are either endosymbionts (*a*) or ectosymbionts (*b*) of the gut flagellates, suspended in the gut lumen (*c*), or attached to the gut wall (*d*). The scheme has been simplified for the sake of clarity; not all possible interactions are shown. L, lignocellulose; C, soluble carbohydrates; N, nitrogenous compounds; F, fecal matter; L, gut lumen; P, protozoa; 1, mandibles; 2, salivary glands; 3, proventriculus; 4, midgut epithelium; 5, Malpighian tubules; 6, digestive vacuoles; 7, proctodeal feeding. (Scheme from Brune 2009a, modified.)

has been reasonably studied is *Trichomitopsis termopsidis* (Yamin 1978; Yamin and Trager 1979; Odelson and Breznak 1985a, 1985b). It ferments cellulose following the equation

$$[C_6H_{12}O_6] + 2H_2O \rightarrow 2CH_3COO^- + 2H^+ + 2CO_2 + 4H_2$$

(Odelson and Breznak 1985a). Since the fermentation balance is closed and no other reduced fermentation products were found in the culture fluid, part of the H_2 must be derived from the oxidation of reduced pyridine nucleotides formed in glycolysis, a reaction that is endergonic under standard conditions. Cocultivation with a methanogen leads to a higher flagellate cell yield, which indicates that flagellates may benefit from hydrogen removal by other members of the gut microbiota (Odelson and Breznak 1985a).

The formation of hydrogen seems to be a common feature of all parabasalid flagellates in the termite guts so far investigated. Hydrogen formation has been identified directly in temporary cultures of *Trichonympha* spp. (Yamin 1981) and indirectly by the effect of selective defaunation on methane emission in *Zootermopsis* (Messer and Lee 1989). Moreover, ultrastructural studies indicate that the presence of hydrogenosomes is a common feature of parabasalids (Brugerolle and Radek 2006). The only hydrogenases of a termite gut flagellate characterized to date are those of the parabasalid flagellate *Pseudotrichonympha grassii* from the termite *Coptotermes formosanus* (Inoue et al. 2007). They are of the iron-only type; one of them is related to the hydrogenases of *Trichomonas* spp. and (in agreement with their hydrogenosomal location) preferentially catalyze hydrogen evolution rather than uptake. However, the presence of hydrogenosomes does not mean that all parabasalids necessarily ferment carbohydrates according to the equation shown above. For example, *Trichomonas* spp., which are distant relatives of the parabasalids in termite guts, form products other than H_2 in the reductive branch of their fermentative metabolism, such as glycerol, lactate, and succinate (Müller 1988). Nothing is known about the fermentation products of oxymonads, and hydrogenosomes have not been reported for any oxymonads. However, a transcriptional analysis of the distantly related flagellate *Trimastix pyriformis* showed the presence of pyruvate:ferredoxin oxidoreductase and the iron-only hydrogenase characteristic of such organelles (Hampl et al. 2008).

In general, any of the numerous bacterial symbionts colonizing most gut flagellates are likely to contribute to the metabolism of the flagellate–bacteria system (Brune and Stingl 2005; Ohkuma 2008). The genome sequences of "*Candidatus* Endomicrobium trichonymphae" and "*Candidatus* Azobacteroides pseudotrichonymphae", the abundant endosymbionts of *Trichonympha agilis* and *Pseudotrichonympha grassii*, are strict anaerobes that produce their energy by fermentation of sugars to acetate (Hongoh et al. 2008a, 2008b). They are very restricted in their potential substrate spectrum, the former utilizing solely glucose-6-phosphate and the latter only glucose, xylose, or hexuronates; most likely, these substrates are abundantly available in the host cells during cellulose and hemicellulose degradation. The hydrogenase of "E. trichonymphae" is probably NADH-dependent (Hongoh et al. 2008a), which makes it unlikely that the endosymbiont can produce hydrogen against the high hydrogen partial pressure in the gut lumen of *Reticulitermes* species (see below). Also "A. pseudotrichonymphae" possesses putative hydrogenase genes (Hongoh et al. 2008b) and exhibits a strong hydrogen uptake activity (Inoue et al. 2007). The electron acceptor for hydrogen oxidation is not clear; it is possible that the enzyme reoxidizes the H_2 developed by the nitrogenase of this endosymbiont (see below). Neither endosymbiont genome has homologous sequences to known genes for enzymes involved in cellulose hydrolysis, which clarifies the issue concerning a possible contribution of cellulolytic activities to the flagellate host (see Brune and Stingl 2005).

The contribution of the diverse bacterial community in the hindgut of lower termites to the fermentative processes is far from clear. The bacteria associated with the flagellates clearly represent the majority of the prokaryotes in the system, but there are diverse populations that are either free-swimming (mostly spirochetal forms) or attached to the gut epithelium. The lack of representative isolates makes it difficult to predict their metabolic activities, and the microbial products accumulating in the hindgut fluid are not necessarily indicative of the nature of the primary fermentations. It is likely that, as in other anoxic systems, many of the reduced fermentation products of the primary fermenters are consumed in secondary fermentations. Also, the removal of H_2 or the influx of oxygen into the gut should affect the carbon and electron fluxes in the anaerobic feeding chains (see below).

The classical example of a metabolic link between primary and secondary fermentations in a feeding chain is lactate. Although lactate does not accumulate in the hindgut fluid to high concentrations, it has been suspected that any lactate produced by primary fermenters is rapidly turned over by secondary fermenters (Schultz and Breznak 1979). Microinjection of radioactive lactate into intact guts of *Reticulitermes* spp. revealed a rapid removal of label from the lactate pool, demonstrating that lactate is indeed formed at high rates but immediately converted to acetate (Tholen and Brune 2000; Pester and Brune 2007). The identity of the lactate-producing microorganisms is not clear. A part of the lactate may be formed by the intestinal protozoa (see above), but the lactococci and enterococci colonizing the gut of most termites are also candidates for lactate production based on their ability to ferment cellobiose and xylose (Bauer et al. 2000, and references therein). In contrast, the turnover of free glucose has been shown to contribute only marginally to metabolic fluxes in *R. flavipes* hindguts (Tholen and Brune 2000). The microorganisms responsible for lactate consumption seem to be propionigenic bacteria located in the hindgut periphery because the intestinal fluxes of lactate carbon shift from acetate to propionate when oxygen is excluded (Tholen and Brune 2000, see below).

While lactate fermentation to propionate and acetate is thermodynamically feasible and becomes even more exergonic when the oxidative branch is promoted by the transfer of reducing equivalents to oxygen, most secondary fermentations can take place only at extremely low H_2 partial pressure (Schink 1997). However, in view of the high hydrogen partial pressures observed in the hindgut of lower termites (see below), a syntrophic oxidation of short-chain fatty acids is highly unlikely, and small amounts of propionate and butyrate accumulating in the hindgut fluid are probably indicative of low formation rates of these compounds.

In view of the compelling evidence for the absorption of cellulose-derived glucose already in the midgut (see above) and the apparent absence of pyruvate dehydrogenase activity in termite tissues, it had been hypothesized that the gut microbiota of termites is essential for the terminal oxidation of carbohydrates, converting pyruvate released into the hindgut to acetate, which is then reabsorbed by the host (see Slaytor 2000). However, more recent findings documented that the mitochondria of both lower and higher termites possess an active pyruvate dehydrogenase complex (Itakura et al. 1999, 2003). Rather, the absorption of glucose in the midgut may serve to meet the need of the termite for hexoses, since the end products of the hindgut metabolism contain no gluconeogenic substrates. Just like other animals, insects can use acetate and butyrate only for energy metabolism and lipid synthesis (formate, hydrogen, or methane cannot be metabolized at all), and the concentrations of propionate, the major substrate of gluconeogenesis in ruminants, are typically low, an issue that has been flagged already by Slaytor (1992).

16.3.2 Hydrogen as Central Intermediate

The major products of primary fermentations in lower termites appear to be acetate, CO_2 and H_2 (Fig. 16.3), which would be in agreement with the predominance of parabasalid flagellates and their putative activities (see above). Hydrogen production by these protists leads to an enormous accumulation of hydrogen at the gut center and steep, decreasing radial hydrogen gradients toward the gut periphery of *Reticulitermes* and *Zootermopsis* spp. (Ebert and Brune 1997; Pester and Brune 2007). In cases where hydrogen does not accumulate to high concentrations, it nonetheless remains a central metabolite in the microbial food chain, surpassing volume-specific hydrogen production rates in the bovine rumen by a factor of three (Pester and Brune 2007). This corroborates the hypothesis that differences in hydrogen emission by termites merely reflects the efficiency of hydrogen transfer between its sources and sinks (Sugimoto et al. 1998).

Higher termites also emit hydrogen (Sugimoto et al. 1998; Schmitt-Wagner and Brune 1999). Since they lack gut flagellates, H_2 must be formed by bacterial fermentations. A number of 16S rRNA genes obtained from *Cubitermes orthognathus* guts (Schmitt-Wagner et al. 2003a) are closely related to genes of the cellulolytic, hydrogen-producing bacterium *Clostridium termitidis* isolated from the higher termite *Nasutitermes lujae* (Hethener et al. 1992), but the roles of the bacteria represented by these and many other uncultivated lineages in cellulose degradation and hydrogen production remain to be established.

16.3.3 Methanogenesis

Termites, cockroaches, and scarab beetles are the only insects known to emit methane, and they do so in impressive amounts (Hackstein et al. 2006). Methane formation in the guts of termites and the closely related wood-feeding cockroach

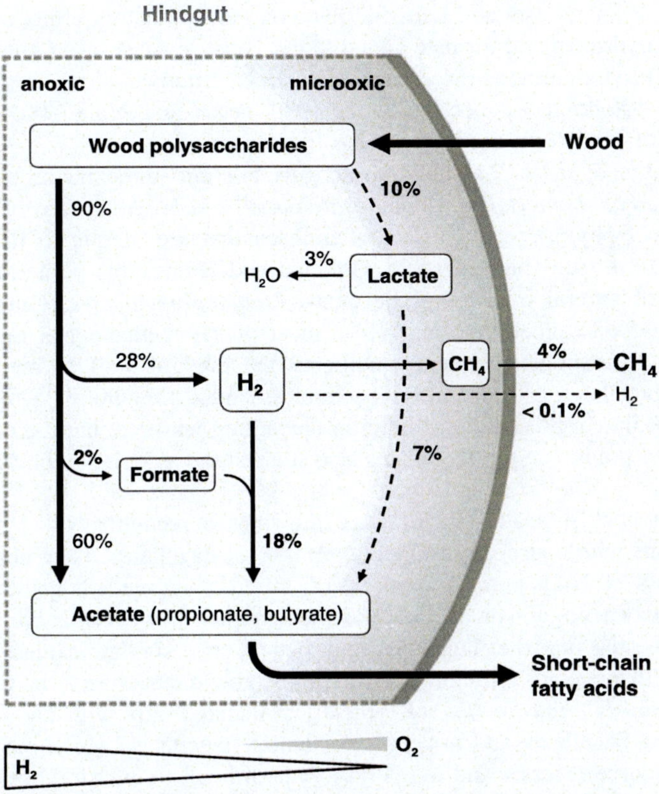

Fig. 16.3 Major metabolic fluxes in the hindgut of the wood-feeding termite *Reticulitermes santonensis* relative to the respiratory electron flow (the electrons necessary to account for O_2 consumption and its reduction to H_2O by the entire system, i.e. the intact whole termite). The model is based on radiotracer studies with intact guts and the respiratory rate measurements of living termites; the fluxes determined by direct measurements are given in *bold* (Pester and Brune 2007). The *dashed lines* indicate metabolic fluxes that seem to be influenced by the continuous influx of oxygen into the gut periphery

Cryptocercus punctulatus had been suspected already in the 1930s and was finally proven by Breznak and coworkers (Breznak et al. 1974; for historical details, see Brune 2009b). Wood-feeding and fungus-cultivating termite species typically emit less methane than wood–soil interface feeders and true soil feeders (Brauman et al. 1992; Sugimoto et al. 1998), whose rates of methane emission (based on fresh weight) may surpass that of ruminants. For a comprehensive review of literature on methanogenesis in insects, see Brune (2009b).

In all termites, methanogenesis is restricted to the enlarged hindgut compartments. It seems to be fueled by hydrogen and reduced one-carbon compounds formed during the fermentative breakdown of plant fiber and humus, as indicated by the strict dependence of methane emission by lower termites on the presence of (hydrogen-producing) gut flagellates (Odelson and Breznak 1983; Rasmussen and Khalil 1983; Messer and Lee 1989) and on the stimulatory effect on methanogenesis of hydrogen and formate added to gut homogenates, intact guts, or (in the case of H_2) living termites (Messer and Lee 1989; Brauman et al. 1992; Ebert and Brune 1997; Schmitt-Wagner and Brune 1999; Pester and Brune 2007). There are no reports of aceticlastic methanogenesis in termites.

The methanogens colonizing termite guts fall into three major phylogenetic groups (*Methanobacteriales*, *Methanosarcinales*, and *Methanomicrobiales*). They form distinct phylogenetic clusters of archaea that are unique to the intestinal tract of insects (see Chapter 15 by Ohkuma and Brune, this volume). With the exception of several isolates of the genus *Methanobrevibacter* (Leadbetter and Breznak 1996; Leadbetter et al. 1998), most of the methanogens remain to be cultivated. Methanogens typically colonize the hindgut wall or are associated with gut flagellates or filamentous bacteria as endosymbionts or ectosymbionts. Attachment may represent an adaptation preventing washout, but the colonization of hydrogen-producing gut flagellates also suggest the association has a metabolic function.

Odelson and Breznak (1985a) were the first to recognize the symbiosis of methanogens with hydrogen-producing termite gut flagellates. They noted that the growth yield of *Trichomitopsis termopsidis* cured of its methanogenic symbionts increased when co-cultivated with the methanogen *Methanospirillum hungatei*, which suggested that the flagellates may benefit in a similar manner from their normal methanogenic symbiont. Surprisingly, the methanogens in the hindgut of *Zootermopsis angusticollis* are typically associated only with the smaller, trichomonadid flagellates and not with the larger flagellates, which are the major hydrogen source (Messer and Lee 1989). It is difficult to envision how the larger flagellates would benefit from the removal of hydrogen by methanogens associated with the smaller species, considering the high hydrogen concentrations throughout the gut lumen of most lower termites investigated (Pester and Brune 2007). Odelson and Breznak (1983) also realized that methane production in the hindgut of lower termites is much lower than one would expect based on the large amount of hydrogen formed by the protozoa. From the strong increase in both hydrogen and methane emission rates after feeding with antibacterial drugs, they concluded that methanogenic archaea apparently compete with bacteria for the hydrogen formed by the flagellates.

16.3.3.1 Reductive Acetogenesis

The presence of reductive acetogenesis from H_2 and CO_2 in termite guts was discovered by Breznak and Switzer (1986) and later confirmed for species from all major feeding guilds (Brauman et al. 1992). Interestingly, reductive acetogenesis dominates over methanogenesis as the hydrogenotrophic process in most wood-feeding termites, whereas the opposite is true for most fungus-cultivating and soil-feeding termite species (both for gut homogenates and for intact guts: Breznak and Switzer 1986; Brauman et al. 1992; Tholen and Brune 1999, 2000; Pester and Brune 2007).

The identity of the bacteria responsible for reductive acetogenesis in termite guts was long unclear, because the numerical significance of the homoacetogenic firmicutes initially isolated from termite guts was either low or had not been established (see Brune et al. 2006). The situation changed when Leadbetter et al. (1999) isolated *Treponema primitia*, which was not only the first representative of the spirochetes that numerically predominate in the hindgut of most termite species to be cultured, but also the first spirochete identified to be capable of reductive acetogenesis. Analysis of the diversity of the genes encoding formyl tetrahydrofolate synthetase (FTHFS), a key enzyme in reductive acetogenesis, revealed that most homologs of this gene in the guts of lower termites fall into a cluster comprising the FTHFS gene of *Treponema primitia* and other termite gut spirochetes, which indicated that the majority of the homoacetogens in termite guts belong to this phylum (Salmassi and Leadbetter 2003; Pester and Brune 2006). The key role of spirochetes in reductive acetogenesis in termite guts was further corroborated by expression profiles of the FTHFS genes in several lower termites (Pester and Brune 2006) and a metagenomic analysis of the functional genes of the microbial community of a higher termite (*Nasutitermes* sp.) by Warnecke et al. (2007).

Although it seems safe to conclude that spirochetes dominate reductive acetogenesis in termite guts, one should not assume that all spirochetes in termite guts are homoacetogens. Several other spirochetes that have been isolated from termite guts are not capable of reductive acetogenesis but rather ferment various mono-, di-, and oligosaccharides, forming formate, acetate, and ethanol as the main fermentation end products (Graber et al. 2004; Dröge et al. 2006, 2008). Microfluidic digital PCR of individual gut bacteria confirmed that all FTHFS genes among the numerically predominant bacteria in the gut of *Zootermopsis nevadensis* belong to the uncultivated *Treponema* species typical of termite guts, but also indicated that not all treponemes in the community carry an FTHFS homolog (Ottesen et al. 2006). Moreover, cDNA analysis of the gut contents of several lower termites revealed that not all FTHFS genes present in the community are being expressed (Pester and Brune 2006).

16.3.4 Anaerobic Respirations

Sulfate-reducing bacteria have been isolated from the guts of lower and higher termites (see Brune et al. 2006). However, the sulfate-reducing activities in the gut contents of *Mastotermes darwiniensis* are relatively low (Dröge et al. 2005). Feasible roles of sulfate-reducing bacteria in hindgut metabolisms other than sulfate reduction have been proposed (Kuhnigk et al. 1996), and the identification of a hitherto uncultivated *Desulfovibrio* species as a specific endosymbiont of the flagellate *Trichonympha agilis* (Sato et al. 2008) adds another facet to their so far enigmatic function.

Nitrate is an excellent electron acceptor for many bacteria, and it is unsurprising that all nitrate contained in the soil consumed by soil-feeding *Cubitermes* species disappears in the anterior gut regions (Ji and Brune 2006). Also the nitrate accumulating in the posterior hindgut may be subject to a microbial turnover (Ji and Brune 2006, Ngugi and Brune, unpublished observations). The soils consumed by soil-feeding termites are typically characterized by an enormous content of ferric iron, which is reduced during gut passage (Kappler and Brune 2002). However, the processes and organisms responsible for these reactions have not been studied. Substantial iron concentrations have been reported also for the gut of lower termites, and several bacterial strains with iron-reducing capacities have been isolated (Vu et al. 2004).

16.4 Termite Guts as Gradient Systems

The small dimensions of termite guts make diffusion an efficient means of metabolite transport. No mixing is necessary, and gut peristalsis seems to serve mainly in axial transport. Microsensors studies revealed steep gradients of oxygen and hydrogen, and radial or axial gradients should be present also for any other metabolite where sources and sinks are spatially separated (Brune et al. 2000).

The introduction of microsensor techniques into termite gut research (Brune et al. 1995a; Ebert and Brune 1997) has fundamentally changed our concept of the metabolic processes in the termite–symbiont system. Originally considered as purely anoxic fermenters, termite guts are now regarded as gradient systems, where different microbial populations and their activities are distributed on a microscale (see Brune 1998; Brune and Friedrich 2000). This not only reconciled the presence of both aerobic and anaerobic bacteria within the same microliter environment, but it also helped explain the enigmatic predominance of reductive acetogenesis over methanogenesis, an unusual characteristic of termite guts that had puzzled microbiologists for many years.

16.4.1 Spatial Separation of Microbial Processes

For thermodynamic reasons, methanogens should outcompete homoacetogens for hydrogen, their common substrate. However, this scenario is relevant only if the methanogens can keep the hydrogen partial pressures of the environment at, or below, the hydrogen threshold concentrations of the homoacetogens (i.e. <35–50 Pa; Breznak 1994; Graber and Breznak 2004). However, at the hydrogen partial pressures observed in the hindgut proper of *Zootermopsis* or *Reticulitermes* spp. (1–100 kPa: Ebert and Brune 1997; Pester and Brune 2007), both processes would operate at substrate saturation. This means that at least in these termites, methanogens and homoacetogens are not competing for hydrogen, and it is no longer necessary to seek explanations for the predominance of reductive acetogenesis over methanogenesis in termite guts on the basis of an increased competitiveness of homoacetogens, e.g. in their ability to grow mixotrophically on H_2 and other substrates (Breznak and Switzer 1991; Breznak 1994).

Moreover, homoacetogens and methanogens cannot compete for hydrogen if they are spatially separated within the gut. The considerable hydrogen emissions of many termites indicate that production and consumption of hydrogen are not tightly coupled (Zimmerman et al. 1982; Odelson and Breznak 1983; Ebert and Brune 1997; Sugimoto et al. 1998; Schmitt-Wagner and Brune 1999; Pester and Brune 2007). The strong hydrogen production of the large parabasalid flagellates in the gut lumen of *Reticulitermes flavipes* and *Zootermopsis nevadensis* leads to steep hydrogen gradients towards the gut periphery, with a strong hydrogen sink directly at the gut wall (Ebert and Brune 1997; Pester and Brune 2007, Fig. 16.4). For *R. flavipes*, it has been shown that this hydrogen sink represents an anaerobic process (Ebert and Brune 1997), most likely caused by the methanogens that densely colonize the hindgut cuticle of these and many other termites (for references, see Brune 2009b). Methanogens located in the gut periphery are clearly hydrogen-limited (see above), and the differences in the rates of hydrogen and methane emission may depend on the specific location of methanogens relative to the hydrogen source in the gut of a termite species (Sugimoto et al. 1998). The location of the methanogens at the gut wall also explains the apparent hydrogen limitation of methanogenesis in intact guts, which is strongly stimulated by hydrogen supplied from the outside, whereas reductive acetogenesis, which is presumably catalyzed by the spirochetes located within the gut lumen, is not (Tholen and Brune 2000).

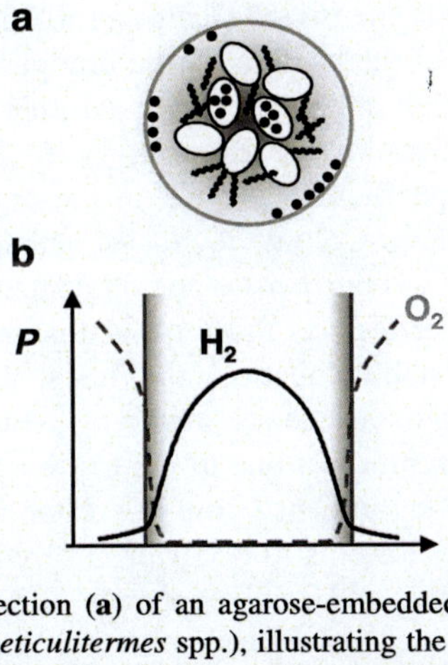

Fig. 16.4 Schematic cross section (**a**) of an agarose-embedded hindgut (*paunch region*) of a wood-feeding lower termite (*Reticulitermes* spp.), illustrating the location of methanogens (*filled circles*) attached to the hindgut wall and homoacetogenic spirochetes (*spirals*) within the gut proper. In some species, methanogens are also associated with the hydrogen-producing gut flagellates (*white ovals*). Radial profiles (**b**) of oxygen and hydrogen partial pressure show that the respiratory activity of the gut microbiota maintains steep oxygen gradients (*dashed line*) within the gut periphery, rendering the center anoxic. Hydrogen (*solid line*) accumulates at the gut center but is consumed throughout the entire gut. The strong hydrogen sink below the gut wall is probably caused by methanogens, which prevent larger amounts of H_2 escaping from the gut. (Scheme based on data from Ebert and Brune 1997; Pester and Brune 2007.)

In soil-feeding *Cubitermes* species, hydrogen production and consumption are separated in different compartments, and both methanogenesis and reductive acetogenesis are strongly hydrogen-limited (Schmitt-Wagner and Brune 1999; Tholen and Brune 1999). The strong stimulation of methanogenesis by external hydrogen and formate added to intact gut compartments led to the hypothesis that hydrogen diffuses across the gut epithelia between hydrogen-producing and hydrogen-consuming gut regions, which are in close contact in situ. Such cross-epithelial transfer of reducing equivalents has been documented in detail in cockroaches and scarab beetle larvae (Lemke et al. 2001, 2003) and would explain the low hydrogen and high methane emissions of *Cubitermes* species. Inspired by the situation in insects, it has been discovered that that cross-compartmental transfer of H_2 occurs also in mammals. H_2 produced by the colonic microbiota of mice is transported via the bloodstream and forms an important factor for colonization and virulence of *Helicobacter pylori* in the stomach (Olson and Maier 2002).

16.4.2 Effects of Oxygen on Hindgut Metabolism

The steep oxygen gradients in the hindgut periphery drive a continuous flux of oxygen. Although the gut microbiota efficiently consumes inflowing oxygen within fractions of a millimeter, the microoxic part makes up a significant fraction of the hindgut volume. All microorganisms residing in this region, and those occasionally crossing the oxic–anoxic boundary, are inevitably exposed to oxygen, either continuously or for a limited time period (Brune et al. 2000). It is therefore not surprising that a large portion of the bacteria in the hindgut of *Reticulitermes flavipes* that are cultivable on complex media are either aerotolerant, facultatively aerobic, or obligately aerobic (see Schultz and Breznak 1978; Tholen et al. 1997). Other populations appear to be specifically adapted to the oxygen-limited environment presented by the gut periphery. Colony counts on agar plates inoculated with gut homogenates of *Zootermopsis angusticollis* and incubated under hypoxic conditions (2% O_2) were about twice as high as those on plates incubated under air (Graber and Breznak 2005). Using such incubation conditions, Wertz and Breznak isolated *Stenoxybacter acetivorans*, a novel acetate-oxidizing β-Proteobacterium that is obligately microaerophilic and represents one of the few bacteria isolated from termite guts for which a numerical significance could be established (Wertz and Breznak 2007a, 2007b). *S. acetivorans* seems to occur primarily in the peripheral, hypoxic region of hindguts of *R. flavipes* and oxidizes acetate using a high-affinity, *cbb*$_3$-type cytochrome oxidase.

Even the anaerobic bacteria isolated from termite guts are capable of reducing oxygen at astonishingly high rates. In fermenting bacteria, such as lactic acid bacteria (Tholen et al. 1997; Bauer et al. 2000) or the hydrogen-oxidizing propionigenic *Sporotalea propionica* (Boga et al. 2007), oxygen reduction shifts the fermentation pathways from lactate or propionate to acetate production. The relevance of these phenomena was tested by microinjection of radiotracers into intact guts of *Reticulitermes flavipes*, which confirmed that carbon fluxes through the lactate pool are shifted from propionate to acetate production when oxygen is present (Tholen and Brune 2000). The bacteria responsible for these activities are not known, but uncultivated members of Bacteroidales and Propionibacteriaceae specifically colonizing the gut wall of *Reticulitermes* species are feasible candidates (see Chapter 15 by Ohkuma and Brune, this volume).

Especially intriguing is the capacity for hydrogen-dependent oxygen reduction in notoriously oxygen-sensitive microorganisms, such as homoacetogens and methanogens. Oxygen reduction by *Sporomusa aerivorans*, an oxygen-reducing homoacetogenic bacterium from the hindgut of a soil-feeding termite (Boga et al. 2003), is inhibited by KCN, suggesting an involvement of an electron transport chain (Boga and Brune 2003). While reductive acetogenesis of *Sporomusa* species is inhibited even at the lowest oxygen fluxes, the methanogenic archaeon *Methanobrevibacter cuticularis,* although even more sensitive to oxygen accumulation than the homoacetogens, appears to be well adapted to its life in the microoxic gut periphery of lower termites. *Methanobrevibacter* species remain metabolically active in dense cell suspensions that are exposed to controlled oxygen fluxes as long as the influx of oxygen does not exceed their capacity for oxygen removal (Tholen et al. 2007). It has been proposed that the redirection of electron flow from methanogenesis towards oxygen reduction enables *Methanobrevibacter* species to colonize the hindgut periphery of termites. The mechanisms of tolerance to reactive oxygen species and the biochemistry of oxygen reduction in *Methanobrevibacter* species have been discussed elsewhere in detail (Brune 2009b), taking into account also the implication of spiracular control of tracheal gas exchange for the oxygen status of the gut microenvironment (Lighton and Ottesen 2005).

Nevertheless, the location of methanogens at the gut wall of lower termites, at the unfavorable end of the outwardly directed hydrogen gradient, remains enigmatic. It has been suggested that an attachment to the hindgut cuticle may protect against predation or prevent washout from the gut, which may compensate methanogens for the negative effects of hydrogen limitation and exposure to inflowing oxygen (Breznak 2000). From that perspective, the association of *Methanobrevibacter* species with the parabasalid gut flagellates, which allowed the methanogens to maintain a stable position in the anoxic and hydrogen-rich hindgut lumen, should provide a considerable advantage, not only for methanogens but also for other obligately anaerobic bacteria. Microbial cells also are attached to cuticular structures projecting from the hindgut wall into the lumen in soil-feeding higher termites (Bignell et al. 1980), many of which are putative methanogens based on their characteristic UV-fluorescence (Schmitt-Wagner and Brune 1999).

Although estimates of oxygen consumption rates determined with intact guts under ex situ conditions have to be treated with caution (see discussion by Brune et al. 1995a), both oxygen concentration profiles of agarose-embedded guts and respiratory oxygen consumption rates of guts immersed in Ringer's solution indicated that a substantial fraction of the respiratory activity of *R. flavipes* worker larvae is attributable to the microbiota in their intestinal tract (Brune et al. 1995a; Wertz and Breznak 2007b). The relative contributions of aerobic respiration and anaerobic (non-respiratory) processes to the oxygen-reducing activity of termite guts are not clear. Since the estimated oxygen reduction potential of specific bacterial or archaeal populations in the hindgut of lower termites appears to be rather small (Tholen et al. 1997, 2007; Wertz and Breznak 2007b), the protists have also to be considered as a potential oxygen sink. When *R. flavipes* was fed artificial diets that selectively affect the abundance of the bacterial and protozoan populations in the hindgut, it turned out that a decimation of the gut flagellates had a much stronger impact on O_2 consumption rates of whole guts than an almost complete elimination of gut bacteria by antibiotic treatment (Wertz and Breznak 2007b). The metabolic effects of oxygen on termite gut flagellates remain to be investigated, but it has been suggested that the continuous influx of oxygen into the gut might shift their fermentations towards more oxidized products (Ebert and Brune 1997, see also Lloyd 2004 for a critical commentary on the emerging oxygen relationships of "amitochondriate" protists in general).

16.4.3 Absence of Methane Oxidation in Termites

The abundance of methane and the constant influx of oxygen in the gut periphery resemble the situation that gives rise to aerobic methane oxidation in many other habitats. However, there is no report of methane oxidation or evidence for the presence of methanotrophic bacteria in termites. In contrast, methane emissions of several termite species from different families are not, or only slightly, affected by the absence of oxygen (Messer and Lee 1989; Pester et al. 2007). The addition of $^{14}CH_4$ to the headspace of vials containing living termites (*Reticulitermes flavipes* or *Cubitermes orthognathus*) did not yield any $^{14}CO_2$, and attempts to amplify the *pmoA* gene, a marker for particulate methane monooxygenase, from hindgut DNA extracts of these termites were unsuccessful (Pester et al. 2007). This means that there are no processes reducing the emission of methane in the gut itself, which indicates that the differences in methane emission between termites of different feeding guilds directly reflect differences in methane production. However, it has been shown that methane oxidation in mound material and surrounding soil is a potentially important factor mitigating methane production by termites at the environmental level (Sugimoto et al. 2000; see also Brune 2009b).

16.5 Role of the Gut Microbiota in Nitrogen Metabolism

Wood-feeding termites are strongly nitrogen-limited. The termite–symbiont system employs several strategies compensating for the strong C/N imbalance of the diet (Higashi et al. 1992). They are found both within the termite–symbiont system, such as the fixation of atmospheric nitrogen and an efficient nitrogen recycling by the hindgut microbiota, and in extracorporeal associations that add another trophic level, such as the elimination of carbon in fungus gardens (see Chapter 8 by Nobre et al., this volume). There should be a strong link between cellulose decomposition and N metabolism because an improved nitrogen supply in the hindgut of wood-feeding termites should promote the growth of the cellulose-decomposing members of the gut microbiota, thus also ensuring high rates of carbon mineralization (Breznak 2000).

16.5.1 Nitrogen Recycling in Wood-Feeding Termites

In wood-feeding termites, the growth of the colony is limited by the low nitrogen content of wood. Therefore, many wood-feeding termites show a preference for lignocellulosic substrates that are colonized by fungi and therefore have a decreased C-to-N ratio (Rouland-Lefèvre 2000; Cornelius et al. 2002). Another option is to complement the diet by taking up the hindgut content of nestmates. This special behavioral trait is called proctodeal trophallaxis and increases in frequency with the nitrogen deficiency of the diet (Machida et al. 2001). Digestion of hindgut contents and nitrogenous dietary components (microbial biomass) and resorption of the products probably take place in the foregut and midgut, as indicated by lysozyme and protease activities in these gut regions (Fujita et al. 2001; Fujita and Abe 2002; Fujita 2004), providing amino acids and vitamins to the insect. In freshly molted individuals, proctodeal trophallaxis also serves to inoculate the gut with a representative set of symbionts, which explains the high similarity of the gut microbiota in individuals of the same colony (Minkley et al. 2006).

A peculiar feature of the nitrogen metabolism in termites is the fate of uric acid, the major waste product of their nucleic acid and protein metabolism (reviewed in detail by Breznak 2000). Terrestrial insects are primarily uricotelic, i.e. uric acid is secreted via the Malpighian tubules, which merge into the intestinal tract at the midgut–hindgut junction (Fig. 16.2), and is typically voided with the feces. In wood-feeding termites and cockroaches, however, uric acid is stored in the uricocytes of the fat body, where it may accumulate to considerable concentrations. While the related cockroaches may remobilize uric acid during times of high nitrogen demand, termites lack uricase activity and recycle uric acid with the help of the gut microbiota, which hydrolyzes any uric acid entering the hindgut to acetate, CO_2, and ammonia (see Potrikus and Breznak 1981, and references therein).

In the case of cockroaches, it has been postulated that intracellular bacteria (*Blattabacterium* sp.) housed in the bacteriocytes of the fat body are involved in the mobilization of stored nitrogen (see Chapter 2 by Lo and Eggleton, this volume). Genome sequencing of the *Blattabacterium* symbiont of *Periplaneta americana* indicates that it lacks uricase but can use (and upgrade) urea N, formed from uric acid N, once uric acid is mobilized, presumably by the cockroach (see Sabree et al. 2009). However termites, with the exception of the primitive *Mastotermes darwiniensis,* lack these symbionts. It is not clear whether this is the reason why termites keep accumulating uric acid during their entire life cycle. In *Reticulitermes flavipes* kept in the laboratory for 18 months, uric acid makes up almost half of the body weight (Potrikus and Breznak 1981), and it has been questioned whether termites possess any means of remobilizing uric acid once it has been deposited in the uricocytes (see Slaytor and Chappell 1994). However, this would make uric acid storage in the fat body only an apparent dead end. Since termites are both necrophagous and cannibalistic, any uric acid accumulated in the bodies will eventually be reingested by nestmates. Mineralization of uric acid and uptake of ammonia by the gut microbiota would lead to the formation of fresh biomass and close the nitrogen cycle.

16.5.2 Nitrogen Fixation by Gut Bacteria

An efficient recycling may serve to avoid nitrogen loss via the feces, but a termite colony living on sound wood is still severely nitrogen limited. In dry-wood termites, colony growth depends to a substantial part on the capacity of the prokaryotic gut microbiota to fix atmospheric nitrogen. The nitrogen-fixing activity of the termite gut microbiota had been discovered independently by Benemann (1973) and Breznak et al. (1973). Since then a large number of papers have been published on this subject, demonstrating the presence of dinitrogen-fixing bacteria, the incorporation of fixed N into termite biomass, and the widespread distribution of nitrogen fixation among termites of all families (see Breznak 2000 for a detailed review of the older literature).

Depending on the nitrogen content of the diet, nitrogen-fixation rates of termites may differ considerably, and intraspecific variations, interspecific differences, seasonal patterns, and effects of laboratory maintenance make extrapolation of activities to the levels of colony and ecosystem a difficult task. Nevertheless, the importance of nitrogen fixation for termites feeding on sound wood has been underscored by stable isotope analysis, which revealed that 30–60% of the nitrogen in *Neotermes koshunensis* workers is derived via this pathway (Tayasu 1998). Nitrogenase activity in certain *Nasutitermes* spp. would be sufficient to double the nitrogen content of a colony within a few years (Breznak 2000).

Stable isotope analyses of fungus-growing and soil-feeding higher termites indicate that nitrogen fixation is much less important to their nitrogen economy than in the case of wood-feeding termites (see Eggleton and Tayasu 2001). In the fungus-cultivating termites, the symbiotic fungi seem to play different roles in digestion and nutrition. While the fungus in *Macrotermes* spp. serves mainly in lignin degradation, enabling the termites and their gut microbiota to utilize the polysaccharides more efficiently, it serves directly as a nutritious food source in several representatives of other genera (Hyodo et al. 2003). In soil-feeding termites, nitrogen fixation is not required (and cannot be demonstrated) because adequate nitrogen is available in the food (see below).

The capacity to fix dinitrogen is the exclusive domain of prokaryotes, and nitrogen-fixing bacteria have been isolated from various termites (see Brune et al. 2006). However, cultivation-independent studies of the diversity of the *nifH* gene in termite guts have revealed a spectrum of potentially N_2-fixing microorganisms that is much wider than that represented by the dinitrogen-fixing bacteria isolated from the same environments (Ohkuma et al. 1996, 1999; Noda et al. 1999, 2002; Lilburn et al. 2001). There is actually little overlap between the results provided by these strategies, indicating a severe cultivation bias also in the nitrogen-fixing gut bacteria. The *nifH* gene sequences obtained from termite guts form several unique clusters in the phylogenetic tree of *nifH* and its paralogs (Noda et al. 1999, 2002; Ohkuma et al. 1999; Lilburn et al. 2001; Yamada et al. 2007). Among the *nifH* sequences clustering with those of well-characterized nitrogenases, most sequences fall into Clusters II and III (nomenclature of Zehr et al. 2003). Sequences from Cluster I are rare and not closely related to the *nifH* genes of the nitrogen-fixing enterobacteria; although the latter are readily isolated from termite guts, they apparently do not represent dominant nitrogen-fixing populations. Numerous sequences in Clusters II and III, however, clustered with the *nifH* homologs of cultured *Treponema* species from the gut of *Zootermopsis* spp. (Lilburn et al. 2001; Graber et al. 2004), underlining that related spirochetes may play a significant role in nitrogen fixation in a variety of lower termites. Other sequences in Cluster III were affiliated with those of cellulolytic clostridia and may belong to hitherto unidentified firmicutes represented in 16S rRNA-based analyses. Many sequences in Clusters II (the "anf-methano" group; nomenclature of Ohkuma et al. 1999), which harbors the so-called alternative nitrogenases, and III (the "anaerobe group"), which comprises the sequences from diverse lineages of strictly anaerobic microorganisms (e.g. clostridia, sulfate reducers, spirochetes, and methanogens), formed termite gut clusters that were not close to any sequences of known origin. Also the origin of the *nifH* homologues identified in a metagenomic analysis of the hindgut contents (P3) of a *Nasutitermes* species is not clear, although phylogenetic binning of other genes from the *nif* operon indicated that both Spirochaetes and Fibrobacteres are potential nitrogen fixers in the gut of this higher termite (Warnecke et al. 2007).

Since a wide variety of termites have been sampled with respect to the community diversity of *nifH* genes, it became possible to test whether the composition of the nitrogen-fixing community depends on phylogeny of the termite. Interestingly, *nifH* diversity corresponded better with the lifestyles than the phylogenetic relationships of the respective hosts (Yamada et al. 2007). Close relationships among congeneric termites, however, indicate the overall similarities of the bacterial community in each genus of termites examined so far, and most of the potential nitrogen-fixing organisms are probably indigenous to termite guts and have evolved together with their hosts (see Chapter 15 by Ohkuma and Brune, this volume). At this point, it may be prudent to consider that the presence of a *nifH* homolog is not a failsafe predictor for the capacity of a microorganism to reduce N_2. For instance, several species of spirochetes that possess *nifH* genes gave no evidence of N_2 fixation when cultivated in nitrogen-deficient media (Lilburn et al. 2001). Since the *nifH* gene encodes dinitrogenase reductase, i.e. the electron-delivering component but not the N_2-reducing components of the dinitrogenase complex, it is conceivable that some NifH-like proteins might serve in reductive processes unrelated to N_2 fixation. Lastly, even if a microorganism has the capacity to fix nitrogen, it may not do so under the conditions present within the gut. In the dry-wood termite *Neotermes koshunensis*, which feeds on sound wood and has high nitrogenase activity, only a few homologs of the diverse *nifH* gene sequences present in the gut microbial community were transcribed (Noda et al. 1999). They represented alternative nitrogenases (encoded by the *anf* gene) in Cluster II, which comprises paralogs of *nif* genes encoding alternative nitrogenases that contain neither molybdenum nor vanadium cofactors. Their expression levels were positively correlated with the nitrogen fixation activity displayed by the termites under different feeding regimens. Interestingly, *anf* gene expression in *Neotermes koshunensis* was not affected by addition of molybdenum to the diet, although alternative nitrogenases in microorganisms that possess both conventional and alternative nitrogenases are typically expressed only if molybdenum (and vanadium) availability is low.

16.5.3 Upgrading of Nitrogen Quality

Like other animals, most insects lack the ability to assimilate ammonia and require a dietary supply of essential amino acids and vitamins that is quite similar to that of humans (Dadd 1973). Since most of these components are only scarcely represented in sound wood, an upgrading of any low-quality nitrogen component becomes necessary. Fungus-cultivating higher termites receive high-quality nutrients in the form of ingested fungal biomass, and also in lower termites a preference for fungus-infested wood is well established and even exploited to make poisonous baits in termite control. If the food does not meet the dietary requirements of the termite, it is supplemented with microbial biomass acquired from nestmates by proctodeal trophallaxis (see above), similar to the strategies of other hindgut fermenters, such as rodents and lagomorphs, which consume caecum content rich in microbial biomass to upgrade their diet. In this context, it is important to consider also the dietary requirements of and nutritional dependencies among the gut microbiota, where our knowledge is severely limited by the lack of pure cultures. Homoacetogenic spirochetes isolated from the hindgut of *Zootermopsis* require folate to catalyze a key step in the acetyl-CoA pathway (Graber and Breznak 2004). They cannot synthesize this coenzyme but take up a precursor (folinate; 5-formyl-tetrahydrofolate) that is excreted by other members of the gut microbiota (e.g. *Serratia* and *Lactococcus* species: Graber and Breznak 2005). Identification of other components of termite gut fluid that stimulate bacterial growth (analogous to the strategy successfully applied to cultivate fastidious rumen bacteria) may be a promising approach to isolating bacteria from termite guts that so far have resisted cultivation.

16.5.4 Nitrogen Metabolism of the Flagellate Symbionts

Not much is known about the dietary requirements of the gut flagellates. In principle, they can complement their diet of wood particles by phagocytosis of hindgut bacteria. Although digestion of their own ectosymbionts has also been observed (Brugerolle and Radek 2006), it has not been reported that the endosymbionts are harvested in the same manner. In contrast, recent genome sequencing efforts have revealed important insights into the metabolic capacities of the uncultivated endosymbionts of several gut flagellates, which suggest that the endosymbionts themselves may be involved in the fixation, assimilation, and nutritional upgrading of nitrogen. The genome sequence of "*Candidatus* Endomicrobium trichonymphae" revealed that these uncultured endosymbionts of the Elusimicrobia phylum possess a hitherto unappreciated capacity for assimilation of ammonia and upgrading of nitrogen (Hongoh et al. 2008a). Despite a considerable reduction in their genome size that is typical of obligate endosymbionts, Endomicrobia seem to have retained a diversity of pathways for the biosynthesis of amino acids and vitamins that are lacking in a free-living member of the *Elusimicrobia* phylum (Geissinger et al. 2009; Herlemann et al. 2009).

Another example of endosymbiont involvement in nitrogen metabolism is "*Candidatus* Azobacteroides pseudotrichonymphae", which colonizes the cytoplasm of flagellates of the genus *Pseudotrichonympha* and is the most abundant bacterium in the hindgut of *Coptotermes formosanus* (Noda et al. 2007). Genome sequencing revealed that this uncultivated representative of the phylum Bacteroidetes not only possesses diverse pathways for the biosynthesis of amino acids and cofactors, but also has a complete set of genes necessary for nitrogen fixation and utilization of ammonium and urea (Hongoh et al. 2008b, Fig. 16.5); both of these functions were unprecedented among members of this phylum. Considering the enormous abundance of such endosymbionts in many of the large gut flagellates, they may play important roles in supplying their protist hosts, and thereby indirectly

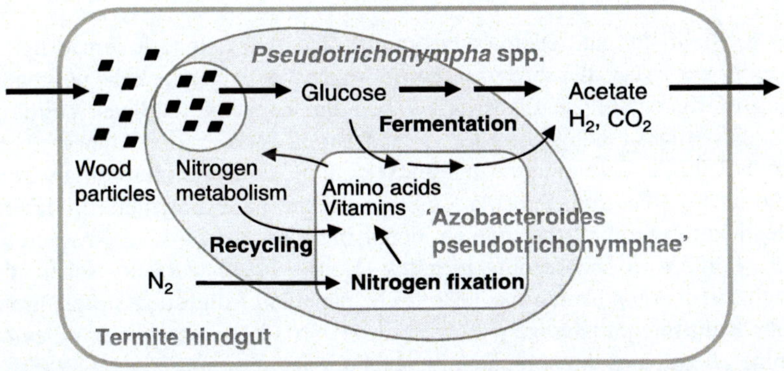

Fig. 16.5 Schematic view of the proposed symbiotic relationship between the endosymbiotic '*Candidatus* Azobacteroides pseudotrichonymphae' and its host flagellates (*Pseudotrichonympha* spp.). Wood particles or cellulose are endocytosed by the flagellate, degraded into monomeric sugars, and then fermented to acetate, carbon dioxide, and molecular hydrogen. The endosymbiont utilizes glucose (or xylose or uronate) supplied by the host and produces acetate as a fermentation product. The host termite utilizes acetate as its major energy and carbon source. The capacity to fix nitrogen and to synthesize a variety of amino acids and cofactors enables the endosymbiont to provide essential nitrogenous nutrients to the host flagellate. A recycling of nitrogenous wastes (ammonia or urea) of the protists by the endosymbiont is also predicted. (Scheme modified after Hongoh et al. 2008b.)

also the termite, with essential nutrients lacking in the lignocellulosic diet. Such a dependency may also be part of the selective forces behind the cospeciation of the symbiotic pairs (Noda et al. 2007; Ikeda-Ohtsubo and Brune 2009, see Chapter 15 by Ohkuma and Brune, this volume).

16.6 Digestion of Soil Organic Matter

The majority of termite species do not feed on sound wood, but rather on lignocellulose in an advanced state of humification. This change in diet is most obvious in the true soil feeders, which are an important component of the soil fauna in tropical ecosystems (Bignell 2006, see Chapter 14 by Bignell and Chapter 17 by Jones and Eggleton, this volume). Since the C:N ratio of organic matter decreases with the degree of humification, soil-feeding termites should not be limited by nitrogen but by the availability of energy in their highly refractory diet (Eggleton and Tayasu 2001). The following section focuses on the findings concerning the nutritional basis of soil-feeding termites and the putative roles in the digestive process of the microorganisms inhabiting their highly structured intestinal tracts. For other aspects of the biology of soil-feeding termites and their role in biogeochemical processes, the reader is referred to other treatises on this subject (Brauman et al. 2000; Eggleton 2006).

16.6.1 Nutritional Basis of Soil-Feeding Termites

Humus is a complex and chemically diverse diet. It was long assumed that soil-feeding termites either digest residual carbohydrates or even attack the polyphenolic products of lignocellulose digestion. While the capacity to digest cellulose is present also in soil-feeding termites (Ji and Brune 2001, and references therein), the residual lignin–carbohydrate complexes should become increasingly recalcitrant during humification. Speculations that the aromatic polymers of lignin and their degradation products became an increasingly important dietary resource during the evolution of soil-feeding termites (Noirot 1992) were not substantiated. Although soil-feeding Termitinae seem to be capable of digesting simple aromatic compounds during gut passage (Brune et al. 1995b), there is currently no evidence for a mineralization of the polyphenolic constituents of humus (Ji et al. 2000).

To understand the nutritional basis of soil-feeding termites, it is important to realize that most of the nitrogen in humus is present in amide or peptide structures or other nitrogen-rich residues of microbial biomass, which are structurally stabilized by their intimate association with clay mineral surfaces and covalent linkage to polyphenols (see Vairavamurthy and Wang 2002; Eusterhues et al. 2003, and references therein). Feeding experiments with artificial humic substances showed that soil-feeding *Cubitermes* spp. are able to mobilize and digest the peptidic components of soil organic matter (Ji et al. 2000; Ji and Brune 2001), which renders peptides, amino sugars, and microbial biomass an abundant dietary resource for humivorous termites. The mineralization of nitrogenous soil components during gut passage is evidenced by the enormous ammonium concentrations accumulating in the feces and nest material (constructed from feces) of all humivorous termites species investigated (Ndiaye et al. 2004; Ji and Brune 2006). Ammonia concentrations within the intestinal tract (up to 130 mM in the posterior hindgut) are among the highest values ever reported for soil macroinvertebrates and are matched only by insects feeding on an extremely protein-rich diet, e.g. the sarcophagous larvae of blowflies (Ji and Brune 2006). Nitrogenous soil components have been estimated to contribute substantially (20–40%) to the dietary carbon oxidized by *Cubitermes* species (Ngugi and Brune 2010).

16.6.2 Functional Compartmentalization of the Gut

In lower termites, the bulk of the microbiota is located in a single dilation of the anterior hindgut, called the paunch. In all higher termites except the Macrotermitinae, there is an evolutionary trend towards an increasing gut compartmentalization (see Chapter 14 by Bignell, this volume). The situation is most complex in the true soil feeders (Group IV, sensu Donovan et al. 2001) of the *Cubitermes* clade, which possess several consecutive hindgut dilations, separated by muscular sphincters or enteric valves (Fig. 16.6).

The fundamental differences in the physicochemical conditions and microbial activities in these compartments have been studied using microsensor and radiotracer techniques (Brune and Kühl 1996; Schmitt-Wagner and Brune 1999; Tholen

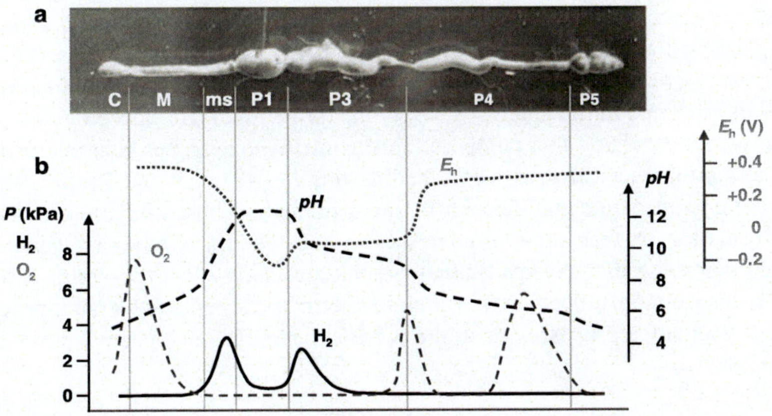

Fig. 16.6 Gut morphology (**a**) and axial profiles (**b**) of different physicochemical parameters along the gut axis of a soil-feeding termite (*Cubitermes* spp.). Oxygen and hydrogen partial pressures, intestinal pH, and apparent redox potential (against a standard hydrogen electrode) were measured with microsensors. The gut was stretched out and embedded in agarose-solidified Ringer's solution. The borders between the different gut regions are indicated by the *vertical lines*. (Scheme based on data compiled by Brune et al. 2006.)

and Brune 1999; Kappler and Brune 2002). They are reflected also in the structure of the microbiological communities in the respective compartments (see Chapter 15 by Ohkuma and Brune, this volume). There are striking differences among the bacterial fingerprint patterns of the different gut segments (Schmitt-Wagner et al. 2003b) and also between those of gut and fresh mound material, which is composed of feces (Fall et al. 2007). Our understanding of the processes occurring in the individual gut compartments of *Cubitermes* spp. is only fragmentary, and the functional role of many microbial populations, e.g. of the uncultivated Planctomycetes abundantly colonizing certain gut regions (Köhler et al. 2008), is totally unclear. The most prominent feature of the anterior hindgut compartments is the pronounced alkalinity of the P1 and sometimes also the P3 segment (Bignell and Eggleton 1995; Brune and Kühl 1996), which clearly increases the solubility of humic substances and, together with autoxidative processes possibly catalyzed by the presence of iron, may effect changes in the size distribution of humic acids (Kappler and Brune 1999). The midgut and the alkaline gut regions also possess alkali-stable and humic acid-tolerant proteolytic activities of so far unknown origin that can digest peptides, both free and copolymerized with humic acids (Ji and Brune 2005).

Little is known about the microbial fermentations in the different gut regions of *Cubitermes* spp. All gut regions accumulate microbial fermentation products and are colonized by unique bacterial communities (Schmitt-Wagner et al. 2003a). High similarity between community profiles of the same gut region in allopatric species indicates a similar function (Schmitt-Wagner et al. 2003b). Hydrogen, presumably of bacterial origin since gut flagellates are absent, accumulates only in the anterior, alkaline compartments (Schmitt-Wagner and Brune 1999), which are dominated by diverse clostridial lineages that seem to be specific for this region and are represented also in the alkaline gut segment of other species of higher termites (Schmitt-Wagner et al. 2003a; Thongaram et al. 2005). Methanogenic capacities are highest in the posterior, less alkaline to slightly acidic gut regions (Schmitt-Wagner and Brune 1999; Kappler and Brune 2002) and may depend in part on hydrogen produced in the anterior gut regions (see above). Homoacetogenic bacteria are restricted to the same compartments as methanogens, but their activities are much less pronounced that in wood-feeding termites (Tholen and Brune 1999). It is tempting to speculate that these differences in methane emission are diet related, but information on fermentative processes in soil-feeding termites is sparse. It is possible that the rather diverse methanogenic community in *Cubitermes* spp. and its axial distribution reflects differences in the methanogenic substrates produced by the fermentative processes or in the pH tolerance of the respective populations (Brune 2009b).

16.7 Do Termites Degrade Lignin?

The possibility of finding novel organisms involved in, or even new principles of lignin degradation in termites has stimulated researchers for several decades. However, many older reports of lignin degradation in termites have to be interpreted with caution, and a comprehensive review of the literature up to the early 1990s (Breznak and Brune 1994) came to the conclusion that there was no unambiguous evidence for the degradation of core lignin during passage through the termite gut. Moreover, no mechanism is known that would explain the cleavage of the non-hydrolyzable bonds in the lignin polymer in the absence of oxygen.

Aerobic lignin degradation by the symbiotic *Termitomyces* species cultivated by the fungus-growing termites (Macrotermitinae) is well established (see Rouland-Lefèvre et al. 2006). These termites circumvent the lignin barrier by employing the lignin-degrading capacities of their basidiomycete partner and, depending on the termite species, either utilize the delignified wood or the fungus itself as food source (Hyodo et al. 2003). A peculiar feature of lignin degradation by *Termitomyces* spp. sets them apart from most other basidiomycete fungi. Biochemical and genetic data indicate that laccases are the only lignolytic enzymes produced by the symbiotic *Termitomyces* spp. of several fungus-cultivating termites; lignin- and Mn-peroxidases, the major catalysts of lignin degradation in most other basidiomycetes, are not expressed (Taprab et al. 2005; Johjima et al. 2006). Delignification, together with an improved C-to-N ratio and a higher nitrogen quality, may also explain why many wood-feeding termites are attracted to wood that has undergone fungal decay (Rouland-Lefèvre 2000; Cornelius et al. 2002, and references therein).

While lignin degradation in the fungal gardens is rather orthodox, as it does not require any properties unusual for a basidiomycete fungus, lignin degradation during gut passage is a much more controversial subject. In view of the lack of any mechanism for the cleavage of carbon–carbon and ether bonds that would operate under anoxic conditions, and the absence of any significant decrease in lignin content of the feces versus the parent material, there was little evidence for the existence of such a process (Breznak and Brune 1994). Although termite guts may contain small numbers of anaerobic microorganisms that degrade simple aromatic compounds (e.g. Brauman et al. 1998), the significance of such isolates is questionable; lignin-derived phenylpropanoids and other monoaromatic compounds are not mineralized by gut homogenates of wood-feeding termites incubated under anoxic conditions. The problem gained a somewhat different perspective when it was recognized that termite guts are not anoxic fermenters, but gradient systems characterized by the continuous influx of oxygen (see Brune 1998). Termite guts were found to contain considerable numbers of bacteria that rapidly mineralize monoaromatic compounds and dimeric lignin model compounds during gut passage and in gut homogenates incubated in air (Kuhnigk et al. 1994; Brune et al. 1995b; Kuhnigk and König 1997; Ngugi et al. 2007). Nevertheless, there is still no evidence for a mineralization of core lignin itself under oxic conditions.

Yet the question of whether or not lignin is mineralized does not touch the heart of the matter. Any modification of lignin that improves the accessibility of the polysaccharides to glycosyl hydrolases would enhance the efficiency of lignocellulose digestion in termite guts. Application of spectroscopic techniques that allow detection of structural changes in the lignocellulose polymer following gut passage provided deeper insights into the problem. Several studies comparing ^{13}C-NMR spectra of wood substrate and feces of wood-feeding lower (*Coptotermes formosanus*; Hyodo et al. 1999) and higher termites (*Microcerotermes parvus*; Hopkins et al. 1998) came to the conclusion that the wood-feeding termites studied possessed little or no ability to degrade lignin. The structural features of lignin in wood and feces changed only a little during digestion by *Cryptotermes brevis* (Katsumata et al. 2007), confirming earlier, wet chemical analyses reporting that lignin was not significantly degraded by wood-feeding termites (see Breznak and Brune 1994).

Solid-state NMR has revealed that lignin is not completely inert during gut passage but is subject to structural changes. In *Cryptotermes brevis*, these changes are considered to arise from polymerization of guaiacyl units and cleavage of ether linkages in side chains rather than from anaerobic demethylation or aerobic demethoxylation during ring cleavage (Katsumata et al. 2007). In *Zootermopsis angusticollis*, the lignin in the feces is oxidized in the side-chains, along with demethylation and (most notably) also some ring hydroxylation (Geib et al. 2008). However, the nature of the processes responsible for such changes remains unclear. Recently, a metatranscriptome analysis of *Reticulitermes flavipes* gut tissue documented the expression of host genes that encode for putative laccases and peroxidases (Tartar et al. 2009). Although laccase-like multicopper oxidases are also present in other insects and are involved in cuticle sclerotization, there are other putative functions for such enzymes (see Dittmer et al. 2004 for a discussion). The location of the activity in the foregut of *R. flavipes* and its catalytic properties provide good arguments for following up on their potential role in lignin transformation (Tartar et al. 2009).

It should also be considered that modifications of polyphenolic dietary components may be brought about by abiotic processes. In soil-feeding termites, there are marked changes in the molecular weight of humic substances between different gut compartments (Kappler and Brune 1999). The high concentrations of ferrous iron and the steep oxygen gradients in the alkaline gut compartments (Kappler and Brune 2002) are prerequisites for iron-mediated hydroxyl radical formation (Fenton reaction), which may initiate the chemical oxidation of lignin or polyphenols in humic substances. In this context, it may be of importance that guts of wood-feeding termites contain astonishingly high concentrations of iron (Vu et al. 2004). Hence, it seems likely that the relatively high digestibility of lignocellulose in termite guts is based on more than just mechanical grinding.

16.8 Conclusions

The symbiotic gut microbiota enables termites to thrive on a lignocellulosic diet. The roles of the microorganisms in C and N metabolism and the relative importance of individual processes depend on the lifetype of the host and its particular diet. Cellulose and hemicelluloses are depolymerized by both lower and higher termites, but the strategies employed by the two groups of termites differ. While lignocellulose digestion in the hindgut of lower termites depends largely on symbiotic protists, bacteria are emerging as important players in higher termites. However, the relative importance of host enzymes in the foregut and midgut to the overall process remains to be established. Nitrogenous soil components seem to be an important dietary resource for soil-feeding higher termites, but also in this case the contribution of bacteria to the hydrolysis of peptides and possibly other humus components requires further investigation. Lignin is not mineralized to a significant extent, but modification of the polyphenolic fraction of lignocellulose by so far uncharacterized processes may increase the digestibility of lignocellulose and humic substances.

In all termites, the depolymerization products are the starting point for a microbial feeding chain that leads to the production of short-chain fatty acids, mostly acetate. Little is known about the nature of the primary fermentations. Hydrogen is an important intermediate, but most of it is converted to additional acetate or methane. The relative importance of methanogenesis and reductive acetogenesis differs between the feeding guilds and seems to be based on the spatial organization of the microbial communities. Metabolic fluxes in the gut microbial communities are affected by the influx of oxygen, and its removal by the gut microbiota is a prerequisite for an overall fermentative metabolism and thus the basis for the digestive symbiosis. Although oxygen is reduced probably by both aerobic and anaerobic bacteria, their relative importance in this process is not clear. The pronounced axial gut compartmentalization in higher termites creates several consecutive microbial bioreactors, but the exact nature of the processes catalyzed in each gut compartment requires further studies.

Nitrogen metabolism in termite guts depends strongly on the feeding guild. In wood-feeding termites, nitrogen fixation is an important process, but the identity of the microorganisms responsible for this process and the emerging role of bacterial symbionts of gut flagellates in nitrogen metabolism need to be explored in greater detail. In fungus-cultivating and soil-feeding termites, the dietary nitrogen supply exceeds the demand of the insect and leads to high ammonia content in the feces. In soil-feeding termites, nitrogen transformations during gut passage need to be better understood, especially in terms of their potential impact on nitrogen metabolism in tropical soils.

Acknowledgments We thank John A. Breznak for helpful comments and Karen A. Brune for editing an earlier version of the manuscript.

References

Arakawa G, Watanabe H, Yamasaki H et al (2009) Purification and molecular cloning of xylanases from the wood-feeding termite, *Coptotermes Formosanus* Shiraki. Biosci Biotechnol Biochem 73:710–718

Azuma J, Nishimoto K, Koshijima T (1984) Studies on digestive system of termites. II. Properties of carbohydrolases of termite *Coptotermes formosanus* Shiraki. Wood Res 70:1–16

Bauer S, Tholen A, Overmann J, Brune A (2000) Characterization of abundance and diversity of lactic acid bacteria in the hindgut of wood- and soil-feeding termites by molecular and culture-dependent techniques. Arch Microbiol 173:126–173

Benemann JR (1973) Nitrogen fixation in termites. Science 181:164–165

Bignell DE (2006) Termites as soil engineers and soil processors. In: König H, Varma A (eds) Intestinal microorganisms of termites and other invertebrates. Springer, Berlin, pp 183–220

Bignell DE, Eggleton P (1995) On the elevated intestinal pH of higher termites (Isoptera: Termitidae). Insect Soc 42:57–69

Bignell DE, Oskarsson H, Anderson JM (1980) Specialization of the hindgut wall for the attachment of symbiotic microorganisms in a termite *Procubitermes Aburiensis*. Zoomorphology 96:103–112

Boga HI, Brune A (2003) Hydrogen-dependent oxygen reduction by homoacetogenic bacteria isolated from termite guts. Appl Environ Microbiol 69:779–786

Boga HI, Ji R, Ludwig W, Brune A (2007) *Sporotalea Propionica* gen. nov. sp. nov., a hydrogen-oxidizing, oxygen-reducing, propionigenic firmicute from the intestinal tract of a soil-feeding termite. Arch Microbiol 187:15–27

Boga HI, Ludwig W, Brune A (2003) *Sporomusa Aerivorans* sp. nov., an oxygen-reducing homoacetogenic bacterium from the gut of a soil-feeding termite. Int J Syst Evol Microbiol 53:1397–1404

Brauman A, Bignell DE, Tayasu I (2000) Soil-feeding termites: biology, microbial associations and digestive mechanisms. In: Abe T, Bignell DE, Higashi M (eds) Termites: evolution, sociality, symbiosis, ecology. Kluwer Academic Publishers, Dordrecht, pp 233–259

Brauman A, Kane MD, Labat M, Breznak JA (1992) Genesis of acetate and methane by gut bacteria of nutritionally diverse termites. Science 257:1384–1387

Brauman A, Müller JA, Garcia J-L et al (1998) Fermentative degradation of 3-hydroxybenzoate in pure culture by a novel strictly anaerobic bacterium, *Sporotomaculum hydroxybenzoicum* gen. nov., sp. nov. Int J Syst Bacteriol 48:215–221

Brennan Y, Callen WN, Christoffersen L et al (2004) Unusual microbial xylanases from insect guts. Appl Environ Microbiol 70:3609–3617

Breznak JA (1994) Acetogenesis from carbon dioxide in termite guts. In: Drake HL (ed) Acetogenesis. Chapman and Hall, New York, NY, pp 303–330

Breznak JA (2000) Ecology of prokaryotic microbes in the guts of wood- and litter-feeding termites. In: Abe T, Bignell DE, Higashi M (eds) Termites: evolution, sociality, symbiosis, ecology. Kluwer Academic Publishers, Dordrecht, pp 209–231

Breznak JA, Blum JS (1991) Mixotrophy in the termite acetogen gut acetogen, *Sporomusa termitida*. Arch Microbiol 156:105–110

Breznak JA, Brill WJ, Mertins JW, Coppel HC (1973) Nitrogen fixation in termites. Nature 244:577–580

Breznak JA, Brune A (1994) Role of microorganisms in the digestion of lignocellulose by termites. Annu Rev Entomol 39:453–487

Breznak JA, Mertins JW, Coppel HC (1974) Nitrogen fixation and methane production in a wood-eating cockroach, *Cryptocercus Punctulatus* Scudder (Orthoptera: Blattidae). Univ Wisc Forest Res Notes 184:1–2

Breznak JA, Switzer JM (1986) Acetate synthesis from H_2 plus CO_2 by termite gut microbes. Appl Environ Microbiol 52:623–630

Brugerolle G, Radek R (2006) Symbiotic protozoa of termites. In: König H, Varma A (eds) Intestinal microorganisms of termites and other invertebrates. Springer, Berlin, pp 243–269

Brune A (1998) Termite guts: the world's smallest bioreactors. Trends Biotechnol 16:16–21

Brune A (2006) Symbiotic associations between termites and prokaryotes. In: Dworkin M, Falkow S, Rosenberg E et al (eds) The prokaryotes. 3rd edn, vol 1. Symbiotic associations, biotechnology, applied microbiology, Springer, New York, NY, pp 439–474

Brune A (2007) Woodworker's digest. Nature 450:487–488

Brune A (2009a) Symbionts aiding digestion. In: Cardé RT, Resh VH (eds) Encyclopedia of insects, 2nd edn. Academic Press, New York, NY, pp 978–983

Brune A (2009b) Methanogenesis in the digestive tracts of insects. In: Timmis KN (ed) Handbook of hydrocarbon and lipid microbiology, vol 1. Springer, Heidelberg, pp 707–728

Brune A, Emerson D, Breznak JA (1995a) The termite gut microflora as an oxygen sink: microelectrode determination of oxygen and pH gradients in guts of lower and higher termites. Appl Environ Microbiol 61:2681–2687

Brune A, Frenzel P, Cypionka H (2000) Life at the oxic-anoxic interface: microbial activities and adaptations. FEMS Microbiol Rev 24:691–710

Brune A, Friedrich M (2000) Microecology of the termite gut: structure and function on a microscale. Curr Opin Microbiol 3:263–269

Brune A, Kühl M (1996) pH profiles of the extremely alkaline hindguts of soil-feeding termites (Isoptera: Termitidae) determined with microelectrodes. J Insect Physiol 42: 1121–1127

Brune A, Miambi E, Breznak JA (1995b) Roles of oxygen and the intestinal microflora in the metabolism of lignin-derived phenylpropanoids and other monoaromatic compounds by termites. Appl Environ Microbiol 61:2688–2695

Brune A, Pester M (2005) *In situ* measurements of metabolite fluxes: microinjection of radiotracers into insect guts and other small compartments. In: Leadbetter JR (ed) Methods in enzymology, vol 397. Elsevier, London, pp 200–212

Brune A, Stingl U (2005) Prokaryotic symbionts of termite gut flagellates: phylogenetic and metabolic implications of a tripartite symbiosis. In: Overmann J (ed) Molecular basis of symbiosis. Springer, Berlin, pp 39–60

Cleveland LR (1923) Symbiosis between termites and their intestinal protozoa. Proc Natl Acad Sci USA 9:424–428

Cornelius ML, Daigle DJ, Connick WJ Jr et al (2002) Responses of *Coptotermes Formosanus* and *Reticulitermes Flavipes* (Isoptera: Rhinotermitidae) to three types of wood rot fungi cultured on different substrates. J Econ Entomol 95:121–128

Dadd RH (1973) Insect nutrition; current developments and metabolic implications. Annu Rev Entomol 18:381–420

Dittmer NT, Suderman RJ, Jiang H et al (2004) Characterization of cDNAs encoding putative laccase-like multicopper oxidases and developmental expression in the tobacco hornworm, *Manduca Sexta*, and the malaria mosquito, *Anopheles Gambiae*. Insect Biochem Molec Biol 34:29–34

Donovan SE, Eggleton P, Bignell DE (2001) Gut content analysis and a new feeding group classification of termites. Ecol Entomol 26:356–366

Dröge S, Fröhlich J, Radek R, König H (2006) *Spirochaeta Coccoides* sp. nov., a novel coccoid spirochete from the hindgut of the termite *Neotermes Castaneus*. Appl Environ Microbiol 72:392–397

Dröge S, Limper U, Emtiazi F et al (2005) In vitro and in vivo sulfate reduction in the gut contents of the termite *Mastotermes Darwiniensis* and the rose-chafer *Pachnoda Marginata*. J Gen Appl Microbiol 51:57–64

Dröge S, Rachel R, Radek R, König H (2008) *Treponema Isoptericolens* sp. nov., a novel spirochaete from the hindgut of the termite *Incisitermes Tabogae*. Int J Syst Evol Microbiol 58:1079–1083

Ebert A, Brune A (1997) Hydrogen concentration profiles at the oxic-anoxic interface: a microsensor study of the hindgut of the wood-feeding lower termite *Reticulitermes Flavipes* (Kollar) . Appl Environ Microbiol 63:4039–4046

Eggleton P (2006) The termite gut habitat: its evolution and co-evolution. In: König H, Varma A (eds) Intestinal microorganisms of termites and other invertebrates. Springer, Berlin, pp 373–404

Eggleton P, Tayasu I (2001) Feeding groups, lifetypes and the global ecology of termites. Ecol Res 16:941–960

Eusterhues K, Rumpel C, Kleber M, Kögel-Knabner I (2003) Stabilisation of soil organic matter by interactions with minerals as revealed by mineral dissolution and oxidative degradation. Org Geochem 34:1591–1600

Fall S, Hamelin J, Ndiaye F et al (2007) Differences between bacterial communities in the gut of a soil-feeding termite (*Cubitermes Niokoloensis*) and its mounds. Appl Environ Microbiol 73:5199–5208

Fujita A (2004) Lysozymes in insects: what role do they play in nitrogen metabolism?. Physiol Entomol 299:305–310

Fujita A, Abe T (2002) Amino acid concentration and distribution of lysozyme and protease activities in the guts of higher termites. Physiol Entomol 27:76–78

Fujita A, Shimizu I, Abe T (2001) Distribution of lysozyme and protease, and amino acid concentration in the guts of a wood-feeding termite, *ReTiculitermes Speratus* (Kolbe): possible digestion of symbiont bacteria transferred by trophallaxis. Physiol Entomol 26: 116–123

Geib SM, Filley TR, Hatcher PG et al (2008) Lignin degradation in wood-feeding insects. Proc Natl Acad Sci USA 105:12932–12937

Geissinger O, Herlemann DPR, Mörschel E et al (2009) The ultramicrobacterium "*Elusimicrobium Minutum*" gen. nov., sp. nov., the first cultivated representative of the Termite Group 1 phylum. Appl Environ Microbiol 75:2831–2840

Graber JR, Breznak JA (2004) Physiology and nutrition of *Treponema primitia*, an H_2/CO_2-acetogenic spirochete from termite hindguts. Appl Environ Microbiol 70:1307–1314

Graber JR, Breznak JA (2005) Folate cross-feeding supports symbiotic homoacetogenic spirochetes. Appl Environ Microbiol 71:1883–1889

Graber JR, Leadbetter JR, Breznak JA (2004) Description of *Treponema Azotonutricium* sp. nov. and *Treponema Primitia* sp. nov., the first spirochetes isolated from termite guts. Appl Environ Microbiol 70:1315–1320

Hackstein JHP, van Alen TA, Rosenberg J (2006) Methane production by terrestrial arthropods. In: König H, Varma A (eds) Intestinal microorganisms of termites and other invertebrates. Springer, Berlin, pp 155–180

Hampl V, Silberman JD, Stechmann A et al (2008) Genetic evidence for a mitochondriate ancestry in the 'amitochondriate' Flagellate *Trimastix pyriformis*. PLoS ONE 3:e1383. doi:10.1371/journal.pone.0001383

Herlemann DPR, Geissinger O, Ikeda-Ohtsubo W et al (2009) Genomic analysis of "*Elusimicrobium Minutum*," the first cultivated representative of the phylum "*Elusimicrobia*" (formerly Termite Group 1). Appl Environ Microbiol 75:2841–2849

Hethener P, Brauman A, Garcia JL (1992) *Clostridium Termitidis* sp. nov., a cellulolytic bacterium from the gut of the wood-feeding termite, *Nasutitermes Lujae*. Syst Appl Microbiol 15:52–58

Higashi M, Abe T, Burns TP (1992) Carbon-nitrogen balance and termite ecology. Proc R Soc Lond B 249:303–308

Hongoh Y, Deevong P, Hattori S et al (2006) Phylogenetic diversity, localization and cell morphologies of the candidate phylum TG3 and a subphylum in the phylum *Fibrobacteres*, recently found bacterial groups dominant in termite guts. Appl Environ Microbiol 72:6780–6788

Hongoh Y, Deevong P, Inoue T et al (2005) Intra- and interspecific comparisons of bacterial diversity and community structure support coevolution of gut microbiota and termite host. Appl Environ Microbiol 71:6590–6599

Hongoh Y, Sharma VK, Prakash T et al (2008a) Complete genome of the uncultured Termite Group 1 bacteria in a single host protist cell. Proc Natl Acad Sci USA 105:5555–5560

Hongoh Y, Sharma VK, Prakash T et al (2008b) Genome of an endosymbiont coupling N_2 fixation to cellulolysis within protist cells in termite gut. Science 322:1108–1109

Hopkins DW, Chudek JA, Bignell DE et al (1998) Application of ^{13}C NMR to investigate the transformations and biodegradation of organic materials by wood- and soil-feeding termites, and a coprophagous litter-dwelling dipteran larva. Biodegradation 9:423–431

Hyodo F, Azuma J, Abe T (1999) Estimation of effect of passage through the gut of a lower termite, *Coptotermes Formosanus* Shiraki, on lignin by solid-state CP MAS C-13 NMR. Holzforschung 53:244–246

Hyodo F, Tayasu I, Inoue T et al (2003) Differential role of symbiotic fungi in lignin degradation and food provision for fungus-growing termites (Macrotermitinae: Isoptera). Funct Ecol 17:186–193

Ikeda-Ohtsubo W, Brune A (2009) Cospeciation of termite gut flagellates and their bacterial endosymbionts: *Trichonympha* species and '*Candidatus* Endomicrobium trichonymphae'. Mol Ecol 18:332–342

Inoue T, Kitade O, Yoshimura T, Yamaoka I (2000) Symbiotic associations with protists. In: Abe T, Bignell DE, Higashi M (eds) Termites: evolution, sociality, symbiosis, ecology. Kluwer Academic Publishers, Dordrecht, pp 275–288

Inoue T, Moriya S, Ohkuma M, Kudo T (2005) Molecular cloning and characterization of a cellulase gene from a symbiotic protist of the lower termite, *Coptotermes formosanus*. Gene 349:67–75

Inoue T, Murashima K, Azuma J-I et al (1997) Cellulose and xylan utilization in the lower termite *Reticulitermes Speratus*. J Insect Physiol 43:235–242

Inoue JI, Saita K, Kudo T et al (2007) Hydrogen production by termite gut protists: characterization of iron hydrogenases of parabasalian symbionts of the termite *Coptotermes Formosanus*. Eukaryot Cell 6:1925–1932

Itakura S, Tanaka H, Enoki A (1999) Occurrence and metabolic role of the pyruvate dehydrogenase complex in the lower termite *Coptotermes formosanus* (Shiraki). Insect Biochem Molec Biol 29:625–633

Itakura S, Tanaka H, Enoki A et al (2003) Pyruvate and acetate metabolism in termite mitochondria. J Insect Physiol 49:917–926

Ji R, Brune A (2001) Transformation and mineralization of ^{14}C-labeled cellulose, peptidoglycan, and protein by the soil-feeding termite *Cubitermes Orthognathus*. Biol Fertil Soils 33:166–174

Ji R, Brune A (2005) Digestion of peptidic residues in humic substances by an alkali-stable and humic-acid-tolerant proteolytic activity in the gut of soil-feeding termites. Soil Biol Biochem 37:1648–1655

Ji R, Brune A (2006) Nitrogen mineralization, ammonia accumulation, and emission of gaseous NH_3 by soil-feeding termites. Biogeochemistry 78:267–283

Ji R, Kappler A, Brune A (2000) Transformation and mineralization of synthetic ^{14}C-labeled humic model compounds by soil-feeding termites. Soil Biol Biochem 32:1281–1291

Johjima T, Taprab Y, Noparatnaraporn N et al (2006) Large-scale identification of transcripts expressed in a symbiotic fungus (*Termitomyces*) during plant biomass degradation. Appl Microbiol Biotechnol 73:195–203

Kappler A, Brune A (1999) Influence of gut alkalinity and oxygen status on mobilization and size-class distribution of humic acids in the hindgut of soil-feeding termites. Appl Soil Ecol 13:219–229

Kappler A, Brune A (2002) Dynamics of redox potential and changes in redox state of iron and humic acids during gut passage in soil-feeding termites (*Cubitermes* spp). Soil Biol Biochem 34:221–227

Katsumata KS, Jin Z, Hori K, Iiyama K (2007) Structural changes in lignin of tropical woods during digestion by termite, *Cryptotermes Brevis*. J Wood Sci 53:419–426

Kiuchi I, Moriya S, Kudo T (2004) Two different size-distributions of engulfment-related vesicles among symbiotic protists of the lower termites, *Reticulitermes Speratus*. Microb Environ 19:211–214

Kuhnigk T, Borst E-M, Ritter A et al (1994) Degradation of lignin monomers by the hindgut flora of xylophagous termites. System Appl Microbiol 17:76–85

Kuhnigk T, Branke J, Krekeler D et al (1996) A feasible role of sulfate-reducing bacteria in the termite gut. System Appl Microbiol 19:139–149

Kuhnigk T, König H (1997) Degradation of dimeric lignin model compounds by aerobic bacteria isolated from the hindgut of xylophagous termites. J Basic Microbiol 37:205–211

Köhler T, Stingl U, Meuser K, Brune A (2008) Novel lineages of *Planctomycetes* densely colonize the alkaline gut of soil-feeding termites (*Cubitermes* spp). Environ Microbiol 10:1260–1270

König H, Fröhlich J, Hertel H (2006) Diversity and lignocellulolytic activities of cultured microorgansims. In: König H, Varma A (eds) Intestinal microorganisms of termites and other invertebrates. Springer, Berlin, pp 271–301

Leadbetter JR, Breznak JA (1996) Physiological ecology of *Methanobrevibacter Cuticularis* sp. nov. and *MethanoBrevibacter Curvatus* sp. nov., isolated from the hindgut of the termite *Reticulitermes Flavipes*. Appl Environ Microbiol 62:3620–3631

Leadbetter JR, Crosby LD, Breznak JA (1998) *Methanobrevibacter Filiformis* sp. nov., a filamentous methanogen from termite hindguts. Arch Microbiol 169:287–292

Leadbetter JR, Schmidt TM, Graber JR, Breznak JA (1999) Acetogenesis from H_2 plus CO_2 by spirochetes from termite guts. Science 283:686–689

Lemke T, Stingl U, Egert M et al (2003) Physicochemical conditions and microbial activities in the highly alkaline gut of the humus-feeding larva of *Pachnoda Ephippiata* (Coleoptera: Scarabaeidae). Appl Environ Microbiol 69:6650–6658

Lemke T, van Alen T, Hackstein JHP, Brune A (2001) Cross-epithelial hydrogen transfer from the midgut compartment drives methanogenesis in the hindgut of cockroaches. Appl Environ Microbiol 67:4657–4661

Li L, Fröhlich J, König H (2006) Cellulose digestion in the termite gut. In: König H, Varma A (eds) Intestinal microorganisms of termites and other invertebrates. Springer, Berlin, pp 221–241

Li L, Fröhlich J, Pfeiffer P, König H (2003) Termite gut symbiotic Archaezoa are becoming living metabolic fossils. Eukaryot Cell 2:1091–1098

Lighton JRB, Ottesen EA (2005) To DGC or not to DGC: oxygen guarding in the termite *Zootermopsis Nevadensis* (Isoptera: Termopsidae). J Exp Biol 208:4671–4678

Lilburn TG, Kim KS, Ostrom NE et al (2001) Nitrogen fixation by symbiotic and free-living spirochetes. Science 292:2495–2498

Lloyd D (2004) 'Anaerobic protists': some misconceptions and confusions. Microbiology 150:1115–1116

Machida M, Kitade O, Miura T, Matsumoto T (2001) Nitrogen recycling through proctodeal trophallaxis in the Japanese damp-wood termite *Hodotermopsis japonica* (Isoptera, Termopsidae). Insect Soc 48:52–56

Messer AC, Lee MJ (1989) Effect of chemical treatments on methane emission by the hindgut microbiota in the termite *Zootermopsis Angusticollis*. Microb Ecol 18:275–284

Minkley N, Fujita A, Brune A, Kirchner WH (2006) Nest specificity of the bacterial community in termite guts (*Hodotermes mossambicus*). Insect Soc 53:339–344

Müller M (1988) Energy metabolism of protozoa without mitochondria. Annu Rev Microbiol 42:465–488

Nakashima K, Watanabe H, Saitoh H (2002) Dual cellulose-digesting system of the wood-feeding termite, *Coptotermes Formosanus* Shiraki. Insect Biochem Molec Biol 32:777–784

Ndiaye D, Lensi R, Lepage M, Brauman A (2004) The effect of the soil-feeding termite *Cubitermes Niokoloensis* on soil microbial activity in a semi-arid savanna in West Africa. Plant Soil 259:277–286

Ngugi DK, Ji R, Brune A (2010) Nitrogen mineralization, denitrification, and nitrate ammonification by soil-feeding termites – a 15 N-based approach. Biogeochemistry (DOI:10.1007/s10533-010-9478-6)

Ngugi DK, Tsanuo MK, Boga HI (2007) Benzoic acid-degrading bacteria from the intestinal tract of *Macrotermes michaelseni* Sjostedt. J Basic Microbiol 47:87–92

Noda S, Kitade O, Inoue T et al (2007) Cospeciation in the triplex symbiosis of termite gut protists (*Pseudotrichonympha* spp.), their hosts, and their bacterial endosymbionts. Mol Ecol 16: 1257–1266

Noda S, Ohkuma M, Kudo T (2002) Nitrogen fixation genes expressed in the symbiotic microbial community in the gut of the termite *Coptotermes Formosanus*. Microbes Environ 17:139–143

Noda S, Ohkuma M, Usami R et al (1999) Culture-independent characterization of a gene responsible for nitrogen fixation in the symbiotic microbial community in the gut of the termite *Neotermes Koshunensis*. Appl Environ Microbiol 65:4935–4942

Noirot C (1992) From wood- to humus-feeding: an important trend in termite evolution. In: Billen J (eds) Biology and evolution of social insects. Leuven University Press, Leuven, Belgium, pp 107–119

Odelson DA, Breznak JA (1983) Volatile fatty acid production by the hindgut microbiota of xylophagous termites. Appl Environ Microbiol 45:1602–1613

Odelson DA, Breznak JA (1985a) Nutrition and growth characteristics of *Trichomitopsis Termopsidis*, a cellulolytic protozoan from termites. Appl Environ Microbiol 49:614–621

Odelson DA, Breznak JA (1985b) Cellulase and other polymer-hydrolyzing activities of *Trichomitopsis Termopsidis*, a symbiotic protozoan from termites. Appl Environ Microbiol 49:622–626

Ohkuma M (2002) Symbiosis in the termite gut: culture-independent molecular approaches. In: Seckbach J (ed) Symbiosis: mechanisms and model systems. Kluwer Academic Publishers, Dordrecht, pp 717–730

Ohkuma M (2003) Termite symbiotic systems: efficient bio-recycling of lignocellulose. Appl Microbiol Biotechnol 61:1–9

Ohkuma M (2008) Symbioses of flagellates and prokaryotes in the gut of lower termites. Trends Microbiol 16:345–352

Ohkuma M, Noda S, Kudo T (1999) Phylogenetic diversity of nitrogen fixation genes in the symbiotic microbial community in the gut of diverse termites. Appl Environ Microbiol 65:4926–4934

Ohkuma M, Noda S, Usami R et al (1996) Diversity of nitrogen-fixation genes in the symbiotic intestinal microflora of the termite *ReticuLitermes Speratus*. Appl Environ Microbiol 62: 2747–2752

Ohtoko K, Ohkuma M, Moriya S et al (2000) Diverse genes of cellulase homologues of glycosyl hydrolase family 45 from the symbiotic protists in the hindgut of the termite *Reticulitermes Speratus*. Extremophiles 4:343–349

Olson JW, Maier RJ (2002) Molecular hydrogen as an energy source for *Helicobacter pylori*. Science 298:1788–1790

Ottesen EA, Hong JW, Quake SR, Leadbetter JR (2006) Microfluidic digital PCR enables multigene analysis of individual environmental bacteria. Science 314:1464–1467

Pester M, Brune A (2006) Expression profiles of *fhs* (FTHFS) genes support the hypothesis that spirochaetes dominate reductive acetogenesis in the hindgut of lower termites. Environ Microbiol 8:1261–1270

Pester M, Brune A (2007) Hydrogen is the central free intermediate during lignocellulose degradation by termite gut symbionts. ISME J 1:551–565

Pester M, Tholen A, Friedrich MW, Brune A (2007) Methane oxidation in termite hindguts: absence of evidence and evidence of absence. Appl Environ Microbiol 73:2024–2028

Potrikus CJ, Breznak JA (1981) Gut bacteria recycle uric acid nitrogen in termites: a strategy for nutrient conservation. Proc Natl Acad Sci USA 78:4601–4605

Rasmussen RA, Khalil MAK (1983) Global production of methane by termites. Nature 301: 700–702

Rouland-Lefèvre C (2000) Symbiosis with fungi. In: Abe T, Bignell DE, Higashi M (eds) Termites: evolution, sociality, symbiosis, ecology. Kluwer Academic Publishers, Dordrecht, pp 289–306

Rouland-Lefèvre C, Inoue T, Johjima T (2006) *Termitomyces*/termite interactions. In: König H, Varma A (eds) Intestinal microorganisms of termites and other invertebrates. Springer, Berlin, pp 335–350

Sabree ZL, Kambhampati S, Moran NA (2009) Nitrogen recycling and nutritional provisioning by *Blattabacterium*, the cockroach endosymbiont. Proc Natl Acad Sci USA 106:19521–19526

Salmassi TM, Leadbetter JR (2003) Molecular aspects of CO_2-reductive acetogenesis in cultivated spirochetes and the gut community of the termite *Zootermopsis Angusticollis*. Microbiology 149:2529–2537

Sato T, Hongoh Y, Noda S et al (2008) *Candidatus* Desulfovibrio Trichonymphae, a novel intracellular symbiont of the flagellate *Trichonympha Agilis* in termite gut. Environ Microbiol 11:1007–1015

Schink B (1997) Energetics of syntrophic cooperation in methanogenic degradation. Microbiol Mol Biol Rev 61:262–280

Schmitt-Wagner D, Brune A (1999) Hydrogen profiles and localization of methanogenic activities in the highly compartmentalized hindgut of soil-feeding higher termites (*Cubitermes* spp). Appl Environ Microbiol 65:4490–4496

Schmitt-Wagner D, Friedrich MW, Wagner B, Brune A (2003a) Phylogenetic diversity, abundance, and axial distribution of bacteria in the intestinal tract of two soil-feeding termites (*Cubitermes* spp). Appl Environ Microbiol 69:6007–6017

Schmitt-Wagner D, Friedrich MW, Wagner B, Brune A (2003b) Axial dynamics, stability, and interspecies similarity of bacterial community structure in the highly compartmentalized gut of soil-feeding termites (*Cubitermes* spp). Appl Environ Microbiol 69:6018–6024

Schramm A (2006) Microsensors for the study of microenvironments and processes in the intestine of invertebrates. In: König H, Varma A (eds) Intestinal microorganisms of termites and other invertebrates. Springer, Berlin, pp 463–473

Schultz JE, Breznak JA (1978) Heterotrophic bacteria present in hindguts of wood-eating termites [*Reticulitermes Flavipes*] (Kollar). Appl Environ Microbiol 35:930–936

Schultz JE, Breznak JA (1979) Cross-feeding of lactate between *Streptococcus Lactis* and *Bacteroides* sp. isolated from termite hindguts. Appl Environ Microbiol 37:1206–1210

Slaytor M (1992) Cellulose digestion in termites and cockroaches: what role do symbionts play?. Comp Biochem Physiol 103B:775–784

Slaytor M (2000) Energy metabolism in the termite gut and its gut microbiota. In: Abe T, Bignell DE, Higashi M (eds) Termites: evolution, sociality, symbiosis, ecology. Kluwer Academic Publishers, Dordrecht, pp 307–332

Slaytor M, Chappell DJ (1994) Nitrogen metabolism in termites. Comp Biochem Physiol 107B:1–10

Sugimoto A, Bignell DE, MacDonald JA (2000) Global impact of termites on the carbon cycle and atmospheric trace gases. In: Abe T, Bignell DE, Higashi M (eds) Termites: evolution, sociality, symbioses, ecology. Kluwer Academic Publishers, Dordrecht, pp 409–435

Sugimoto A, Inoue T, Tayasu I et al (1998) Methane and hydrogen production in a termite-symbiont system. Ecol Res 13:241–257

Tanaka H, Aoyagi H, Shina S et al (2006) Influence of the diet components on the symbiotic microorganisms community in hindgut of *Coptotermes formosanus* Shiraki. Appl Microbiol Biotechnol 71:907–917

Taprab Y, Johjima T, Maeda Y et al (2005) Symbiotic fungi produce laccases potentially involved in phenol degradation in fungus combs of fungus-growing termites in Thailand. Appl Environ Microbiol 71:7696–7704

Tartar A, Wheeler MM, Zhou X et al (2009) Parallel metatranscriptome analyses of host and symbiont gene expression in the gut of the termite *Reticulitermes flavipes*. Biotechnol Biofuels 2:25

Tayasu I (1998) The use of carbon and nitrogen isotope ratios in termite research. Ecol Res 13: 377–387

Tholen A, Brune A (1999) Localization and in situ activities of homoacetogenic bacteria in the highly compartmentalized hindgut of soil-feeding higher termites (*Cubitermes* spp). Appl Environ Microbiol 65:4497–4505

Tholen A, Brune A (2000) Impact of oxygen on metabolic fluxes and in situ rates of reductive acetogenesis in the hindgut of the wood-feeding termite *Reticulitermes Flavipes*. Environ Microbiol 2:436–449

Tholen A, Pester M, Brune A (2007) Simultaneous methanogenesis and oxygen reduction by *Methanobrevibacter Cuticularis* at low oxygen fluxes. FEMS Microbiol Ecol 62:303–312

Tholen A, Schink B, Brune A (1997) The gut microflora of *Reticulitermes Flavipes*, its relation to oxygen, and evidence for oxygen-dependent acetogenesis by the most abundant *Enterococcus* sp. FEMS Microbiol Ecol 24:137–149

Thongaram T, Hongoh Y, Kosono S et al (2005) Comparison of bacterial communities in the alkaline gut segment among various species of higher termites. Extremophiles 9:229–238

Todaka N, Inoue T, Saita K et al (2010a) Phylogenetic analysis of cellulolytic enzyme genes from representative lineages of termites and a related cockroach. PLoS ONE 5:e8636

Todaka N, Lopez CM, Inoue T et al (2010b) Heterologous expression and characterization of an endoglucanase from a symbiotic protist of the lower termite, *Reticulitermes speratus*. Appl Microbiol Biotechnol 160:1168–1178

Todaka N, Moriya S, Saita K et al (2007) Environmental cDNA analysis of the genes involved in lignocellulose digestion in the symbiotic protist community of *Reticulitermes Speratus*. FEMS Microbiol Ecol 59:592–599

Tokuda G, Lo N, Watanabe H et al (2004) Major alteration of the expression site of endogenous cellulases in members of an apical termite lineage. Mol Ecol 13:3219–3228

Tokuda G, Lo N, Watanabe H (2005) Marked variations in patterns of cellulase activity against crystalline- vs. carboxymethyl-cellulose in the digestive systems of diverse, wood-feeding termites. Physiol Entomol 30:372–380

Tokuda G, Watanabe H (2007) Hidden cellulases in termites: revision of an old hypothesis. Biol Lett 3:336–339

Tokuda G, Watanabe H, Lo N (2007) Does correlation of cellulase gene expression and cellulolytic activity in the gut of termite suggest synergistic collaboration of cellulases? Gene 401:131–134

Vairavamurthy A, Wang S (2002) Organic nitrogen in geomacromolecules: insights on speciation and transformation with K-edge XANES spectroscopy. Environ Sci Technol 36:3050–3056

Vu AT, Nguyen NC, Leadbetter JR (2004) Iron reduction in the metal-rich guts of wood-feeding termites. Geobiology 2:239–247

Warnecke F, Luginbühl P, Ivanova N et al (2007) Metagenomic and functional analysis of hindgut microbiota of a wood-feeding higher termite. Nature 450:560–565

Watanabe H, Nakashima K, Saito H, Slaytor M (2002) New endo-beta-1,4-glucanases from the parabasalian symbionts, *PseuDotrichonympha Grassii* and *Holomastigotoides Mirabile* of *Coptotermes* termites. Cell Mol Life Sci 59:1983–1992

Watanabe H, Takase A, Tokuda G et al (2006) Symbiotic "Archaezoa" of the primitive termite *Mastotermes darwiniensis* still play a role in cellulase production. Eukaryot Cell 5:1571–1576

Watanabe H, Tokuda G (2009) Cellulolytic systems in insects. Annu Rev Entomol 55:609–632

Wertz JT, Breznak JA (2007a) *Stenoxybacter Acetivorans* gen. nov., sp. nov., an acetate-oxidizing obligate microaerophile among diverse O_2-consuming bacteria from termite guts. Appl Environ Microbiol 73:6819–6828

Wertz JT, Breznak JA (2007b) Physiological ecology of *Stenoxybacter Acetivorans*, an obligate microaerophile in termite guts. Appl Environ Microbiol 73:6829–6841

Yamada A, Inoue T, Noda Y et al (2007) Evolutionary trend of phylogenetic diversity of nitrogen fixation genes in the gut community of wood-feeding termites. Mol Ecol 16:3768–3777

Yamin MA (1978) Axenic cultivation of the cellulolytic flagellate *Trichomitopsis termopsidis* (Cleveland) from the termite *Zootermopsis*. J Protozool 25:535–538

Yamin MA (1981) Cellulose metabolism by the flagellate *Trichonympha* from a termite is independent of endosymbiotic bacteria. Science 211:58–59

Yamin MA, Trager W (1979) Cellulolytic activity of an axenically-cultivated termite flagellate, *Trichomitopsis Termopsidis*. J Gen Microbiol 113:417–420

Yoshimura T (1995) Contribution of the protozoan fauna to nutritional physiology of the lower termite, *Coptotermes formosanus* Shiraki (Isoptera: Rhinotermitidae). Wood Res 82:68–129

Zehr JP, Jenkins BD, Short SM, Steward GF (2003) Nitrogenase gene diversity and microbial community structure: a cross-system comparison. Environ Microbiol 5:539–554

Zhou X, Smith JA, Oi FM et al (2007) Correlation of cellulase gene expression and cellulolytic activity throughout the gut of the termite *Reticulitermes Flavipes*. Gene 395:29–39

Zimmerman PR, Greenberg JP, Wandiga SO, Crutzen PJ (1982) Termites: A potentially large source of atmospheric methane, carbon dioxide, and molecular hydrogen. Science 218:563–565

Chapter 17
Global Biogeography of Termites: A Compilation of Sources

David T. Jones and Paul Eggleton

Abstract Present day local termite assemblages, and the distribution patterns of the different evolutionary clades, result from long-term processes of diversification, radiation and dispersal, as well as the influence of contemporary environments. In the past four decades ecological studies of termites have been focussed on tropical forests and forest margins, and a corresponding theory of regional and global biogeography across the different biogeographical realms has emerged. What is now lacking is a comparison of the taxonomic and functional assemblage structures in habitats other than closed canopy tropical forest. To address this, we have compiled the generic assemblages of the five major biomes where termites occur (tropical rain forest, tropical savanna woodland, semi-desert, temperate woodland and temperate rain forest), designating and comparing 23 exemplar ecosystems around the world. The highest generic richness, with 62 genera, occurs in African Congolese rain forest, followed by Neotropical Guianan rain forest (55 genera), and then the Asian rain forest of northern Borneo (44 genera). The other three tropical rain forest sites, Madagascar, Australia and New Guinea, all have less than one third of the number of genera that occur in the Bornean rain forest. Of the six savanna woodland ecosystems, the African representative is the richest with 37 genera, followed by those of the Neotropics (25 genera) and Australia (24 genera). Temperate woodland and temperate rain forest ecosystems have the lowest richness, with three genera or fewer. The tropical semi-deserts have more genera than the temperate ecosystems. Soil- and humus-feeding termites have their highest generic richness in the African, Neotropical and Asian tropical rain forests, and decline across the other biomes. In contrast, wood-feeders are more evenly distributed across all the biomes. The evolution of different clades and their dispersal to other ecosystems are briefly discussed, and the implications for termite mediated decomposition are emphasised.

D.T. Jones (✉)
Soil Biodiversity Group, Department of Entomology, Natural History Museum, London, SW7 5BD, UK
e-mail: david.jones@nhm.ac.uk

D.E. Bignell et al. (eds.), *Biology of Termites: A Modern Synthesis*,
DOI 10.1007/978-90-481-3977-4_17, © Springer Science+Business Media B.V. 2011

17.1 Introduction

The ecological importance of termites is well recognised and documented (Wood and Sands 1978; Bignell and Eggleton 2000; Holt and Lepage 2000; Ji and Brune 2006). In contrast, their biogeography is much less well-studied. And yet, the ecological impact of termites in any location depends in large part on the taxonomic and functional composition of the local assemblage. Present day local assemblages and the distribution patterns of the different evolutionary clades are a result of the diversification, radiation and dispersal of termites, as well as the influence of contemporary environmental forces (Eggleton 2000; Eggleton and Tayasu 2001; Davies et al. 2003; Chapter 2 by Lo and Eggleton, this volume).

Termites are tractable model organisms for biogeographical studies for several reasons. First, the group is relatively small with only about 200 genera, less than 2,700 described species and perhaps a further 500 novel species (mostly in tropical rain forests) remaining to be described (Kambhampati and Eggleton 2000; Constantino 2010). We therefore have a reasonably well informed view of termite diversity. Second, recent morphological and molecular studies have significantly improved our understanding of termite phylogenetics (Donovan et al. 2000; Inward et al. 2007) and have provided a well supported tree for testing biogeographical hypotheses. Third, there is now a huge number of geo-referenced species records that can be used for investigating global and regional distributional patterns.

Many of the raw data needed for detailed biogeographical analyses are scattered throughout the literature in numerous ecological studies and site checklists. These offer very uneven geographical coverage, and inevitably there are still large parts of the world from which we have little reliable data. Notable contributions to our understanding of the termite faunas of individual biogeographical realms were provided by chapters in volume II of *Biology of Termites* (Krishna and Weesner 1970). Furthermore, our knowledge of termite distribution in several parts of the world has been greatly enhanced by more systematic projects to map termite species across broad geographical areas such as Southern Africa (Coaton and Sheasby 1972), Australia (Watson and Abbey 1993), and the Neotropical region (listed in Bignell et al. 2008; see also Constantino 2010). Despite existing limitations in the available data, there is still sufficient coverage across most of the world to allow meaningful analyses to be attempted.

Biogeographical studies of termites began with Emerson's pioneering treatise on the geographical origins of the group (Emerson 1955). With hindsight, Emerson's synthesis was compromised by the erroneous arrangement of the higher taxa of the day, and was without the benefit of our current understanding of plate tectonics (Eggleton 2000). The first modern attempts to analyse global patterns of distribution were those of Eggleton et al. (1994) and Eggleton (2000). These studies presented generic richness in large grid-squares (equal area cells based on 10° intervals of latitude and variable degrees of longitude), compiled from the available raw data. The generic richness was implicitly treated as being representative of the dominant biome in the grid-square. However, one drawback of this approach is that the terrain in such large grid-squares can be highly variable in elevation, climate and vegetation type, and as a result will often include more than one biome. Genera found in different biomes were therefore often conflated to give increased and non-habitat specific generic richness figures, thus confounding the comparative analyses. However, the same genera are commonly found across a number of adjacent biomes, and so this effect was less pronounced at the generic taxonomic level than it would be at the species level.

Bignell and Eggleton (2000) listed estimates for the abundance and biomass of termites from all ecological studies available up to that time. While this covered studies from different habitats in the African, Oriental and Neotropical regions, it gave no details of the taxonomic or functional structure of these assemblages. Collins (1989) had offered the first detailed comparison of generic termite assemblages of tropical rain forests in Africa, Asia and South America. However, differences in sampling design (especially sampling effort, sampling area, season and replication) used in the previous studies reviewed meant that reliable direct comparisons of assemblage structure among different sites was problematic. In an attempt to eliminate this problem and to build on previous experience, Jones and Eggleton (2000) developed a standardised rapid sampling protocol for assessing termite species density and assemblage structure in tropical rain forests. An extended campaign using this protocol has allowed the comparison of tropical forest assemblages at the local (for example Eggleton et al. 1997, 1999; Davies 2002), regional (Gathorne-Hardy et al. 2002) and global (Davies et al. 2003) levels. However, what is still lacking is a comparison of the assemblage structure in different habitat types in different biogeographical realms, and especially those outside closed canopy tropical forests.

The aim of this chapter is to set out the generic assemblage structure within and across biomes (major habitat types defined by vegetation structure and precipitation) in the different biogeographical realms of the world, with an initial examination of differences in the functional structure of those assemblages. To achieve this, the generic pools of termites in all the five major biomes where termites occur (tropical rain forest, tropical savanna woodland, semi-desert, temperate woodland and temperate rain forest) have been compared in 23 exemplar ecosystems around the world. Regional and global patterns are briefly described, and some of the origins and consequences of those patterns are discussed. This is the first modern summary of the structure of assemblages in all major biomes and is intended to define a reliable dataset on which further analyses of distributional patterns may be based.

17.2 Termite Functional and Taxonomic Classification

For the purposes of analyses, two distinct classification systems have been adopted. The first is based on the feeding preferences of termites, while the second focuses on our current understanding of termite phylogenetics as its framework (see Chapter 2 by Lo and Eggleton, this volume). Termites were first assigned to one of four feeding groups following the modern classification of Donovan et al. (2001; see Chapter 14 by Bignell, this volume). The feeding group of each genus is listed in Table 17.2. The groupings reflect feeding preferences along the humification gradient of the dietary substrates used by termites, and correlate broadly with the variations in the amount of plant tissue fragments (from organic detritus) and silica (from the parent soil) in the gut (Donovan et al. 2001). The groups are:

Group I: All lower termites. The Mastotermitidae, Kalotermitidae, Rhinotermitidae and Serritermitidae. Almost all wood-feeders, except the Hodotermitidae which also include grass-feeders.

Group II: Some Termitidae species. These have a range of feeding habits, including wood-feeders, litter-feeders, and micro-epiphyte-feeders. The fungus-growing Macrotermitinae are notionally included, but can also be treated as a separate feeding group (as in Fig. 17.2).

Group III: Species of Termitidae that feed on highly decayed wood that has lost its structure and become friable and soil-like, and those that feed on soil with a high organic content. These termites can be considered "organic-rich soil-feeders" or "humus-feeders", or "soil/wood interface feeders", sensu De Souza and Brown (1994).

Group IV: Species of Termitidae that feed on soil with a lower organic content. These termites can be considered "true soil-feeders".

The second classification involved subdividing the feeding groups of Donovan et al. (2001) into Functional Taxonomic Groups (FTG) by assigning species to putative clades, based on the phylogenetic framework of Inward et al. (2007) and representing a refinement of the FTG system proposed by Davies et al. (2003). Genera from 26 FTGs are included in Table 17.2, although there are a few others which do not occur in the exemplar habitats considered in this chapter. The use of FTGs allows greater functional and taxonomic resolution than simply using feeding groups, and it retains a significant phylogenetic signal.

17.3 Exemplar Assemblages

Twenty-three ecosystems have been selected to give a broad representation of termite assemblages across all biogeographical realms (Fig. 17.1). They encompass a wide range of termite diversities, from the ecosystem with the highest known termite species richness (the Northwestern Congolese rain forest) to ecosystems at the latitudinal limits of termite occurrence where only one or two species are found (for example, the temperate rain forests of Northern Japan). For comparative purposes, and where possible, the same subset of biomes were selected in each biogeographical realm. The 23 ecosystems represent five distinct biomes: tropical rain forest, tropical savannah woodland, semi-desert, temperate woodland and temperate rain forest. The vegetation structure and environmental conditions were more similar among ecosystems within the same biome than among ecosystems in different biomes. For instance, the ecosystems in the three forest biomes (tropical rain forest, temperate forest and temperate rain forest) all have a closed canopy, unlike the savanna woodland ecosystems which have a lower density of trees and an open canopy. The three forest biomes differ in their temperature and rainfall regimes. Table 17.1 lists and characterises the 23 sites and the ecosystems they represent.

The publications used for the compilation of the termite genera found in the 23 exemplar ecosystems are given in Table 17.1. This is modified from the dataset used in Eggleton (2000) to make it ecosystem specific, and up-dated with additional

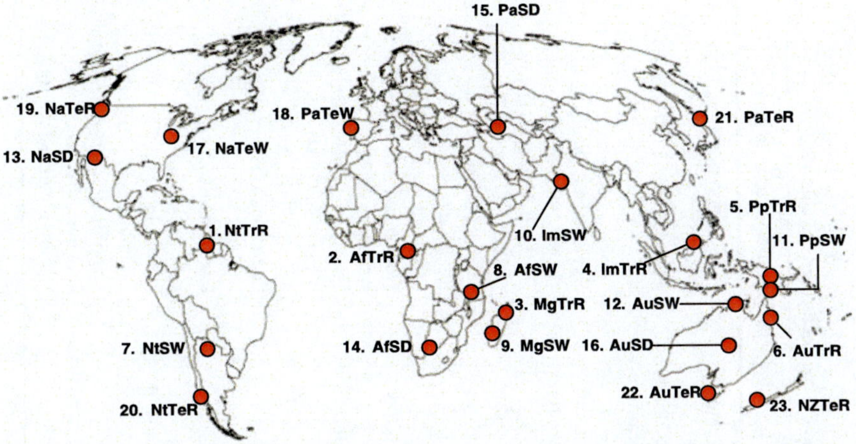

Fig. 17.1 Map showing the location of the 23 exemplar ecosystems. See Table 17.2 for descriptions of each site. Biogeographical codes are: Nt = Neotropical; Na = Nearctic; Af = Afrotropical; Mg = Malagasy; Im = Indomalayan; Pa = Palearctic; Pp = Papuan; Au = Australian; NZ = New Zealand. Biome codes are: TrR = tropical rain forest; SW = savanna woodland; SD = semi-desert; TeW = temperate woodland; TeR = temperate rain forest

records published in the last ten years. Table 17.2 lists the genera recorded in each ecosystem. Each list represents the available pool of genera recorded from sites across the whole ecosystem. As a result, the generic pool is often larger than the generic diversity found at any one site within the ecosystem.

17.4 Taxonomic Richness

Figure 17.2 shows the number of genera in each of the 23 exemplar ecosystems. As expected, there is a general trend of declining generic richness with increasing latitude (Eggleton et al. 1994; Eggleton 2000). The highest richness is found in the African rain forest of the Congolese forest block with 62 genera, followed by Neotropical rain forest (55 genera), and then the Asian rain forest of northern Borneo (44 genera). The other three tropical rain forest sites, Madagascar, Australia and New Guinea, all have less than one third of the number of genera that occur in the Bornean rain forest. The African savanna woodland is the richest (37 genera) of the six exemplars of this biome, again closely followed by the Neotropical ecosystem (25 genera) and Australia (24 genera). Australia is unusual because its savanna woodland and semi-desert are both richer in termite genera (for example see Abensperg-Traun and de Boer 1990; Abersperg-Traun and Steven 1997) than its tropical rain forest (in northern Queensland, see Gay and Calaby 1970). The semi-deserts of Africa, Australia and Nearctic have generic richness comparable to that of some savanna woodland systems. In contrast, the Palearctic semi-desert, which is at a higher latitude than the other semi-deserts, contains only two termite genera. The temperate woodland and temperate rain forest ecosystems have the lowest richness, with three or fewer genera.

Table 17.1 The 23 exemplar sites giving their region, biome, location and vegetation cover, and the site codes used in Table 17.2

No. (code)	Region and biome	Location	Vegetation cover	Source of termite genera
1 (NtTrR)	Neotropical rain forest	French Guiana, Guiana Shield	Lowland moist broadleaf forest	Davies (2002), Constantino (2010)
2 (AfTrR)	Afrotropical rain forest	Southern Cameroon, Northwestern Congolian forest	Lowland and premontane moist semi-deciduous broadleaf forest	Eggleton et al. (1996, 2002)
3 (MgTrR)	Malagasy rain forest	Northeastern Madagascar	Lowland moist broadleaf forest	Eggleton and Davies (2003)
4 (ImTrR)	Indomalayan rain forest	Northern Borneo	Lowland mixed dipterocarp forest	Thapa (1981), Eggleton et al. (1999), Gathorne-Hardy (2004), Gathorne-Hardy et al. (2002)
5 (PpTrR)	Papuan rain forest	East Sepik, northern Papua New Guinea	Lowland and hill forest	Roisin and Pasteels (1996), Bourguignon et al. (2008)
6 (AuTrR)	Australian rain forest	Wet Tropics of Queensland, Cairns	Tropical moist broadleaf forest	Watson and Abbey (1993), Miller (1991)
7 (NtSW)	Neotropical savanna woodland	Northern Argentina	Southern humid Gran Chaco	Constantino (2010), Roisin and Leponce (2004), Torales et al. (2005, 2007)
8 (AfSW)	Afrotropical savanna woodland	Southern Tanzania	Miombo woodland	Eggleton (2000)
9 (MgSW)	Malagasy savanna woodland	Southern Madagascar	Dry deciduous woodland	Eggleton and Davies (2003)
10 (ImSW)	Indomalayan savanna woodland	Gujurat, northwest India	Dry deciduous woodland and thorn scrub	Roonwal and Chhotani (1989), Chhotani (1997), Rathore and Bhattacharyya (2004)
11 (PpSW)	Papuan savanna woodland	Western Province, southern Papua New Guinea	Savanna and open dry sclerophyllous woodland	Roisin and Pasteels (1996), Bourguignon et al. (2008)

Table 17.1 (continued)

No. (code)	Region and biome	Location	Vegetation cover	Source of termite genera
12 (AuSW)	Australian savanna woodland	Arnhem Land, Northern Territory	Tropical open *Eucalyptus* woodland	Braithwaite et al. (1988), Watson and Abbey (1993), Miller (1991), Dawes-Gromadzki (2005, 2008)
13 (NaSD)	Nearctic semi-desert	Sonoran desert, southern Arizona, USA	Temperate semi-desert and semi-arid scrub	Weesner (1970)
14 (AfSD)	Afrotropical arid-desert	Kalahari Desert, Botswana	Tropical semi-desert	Uys (2002)
15 (PaSD)	Palearctic semi-desert	Western Turkmenistan, Central Asian Southern Desert	Temperate desert and xeric shrubland	Eggleton (2000)
16 (AuSD)	Australian semi desert	Central Australia, Northern Territory	Tropical semi-desert and thorn scrub	Watson and Abbey (1993), Miller (1991)
17 (NaTeW)	Nearctic temperate woodland	Eastern USA	Broadleaf deciduous woodland	Weesner (1970)
18 (PaTeW)	Palearctic temperate woodland	Portugal and western Spain	Broadleaf deciduous and mixed woodland	Eggleton (2000)
19 (NaTeR)	Nearctic temperate rain forest	Olympic Peninsula, Washington State, USA	Coniferous forest	Weesner (1970)
20 (NtTeR)	Neotropical temperate rain forest	Valdivian Forest, Southern Chile	Broadleaf evergreen *Nothofagus* forest	Eggleton (2000)
21 (PaTeR)	Palearctic temperate rain forest	Southern Hokkaido, Japan	Broadleaf and mixed forest	Eggleton (2000)
22 (AuTeR)	Australian temperate rain forest	Western Tasmania, Australia	Broadleaf evergreen *Nothofagus* forest	Watson and Abbey (1993)
23 (NZTeR)	New Zealand temperate rain forest	South Island, New Zealand	Broadleaf evergreen *Nothofagus* forest	Eggleton (2000)

Table 17.2 List of termite genera with their distribution across five biomes. This table does not include all extant termite genera but only those that occur in the 23 exemplar assemblages. (gp = group)

		Site No.	1	2	3	4	5	6	7	8	9	10	11	12	13	14	15	16	17	18	19	20	21	22	23
			Tropical rain forest						Savanna woodland						Semi-desert				Temperate woodland		Temperate rain forest				
	Food type	FTG	NtTrR	AfTrR	MgTrR	ImTrR	PpTrR	AuTrR	NtSW	AfSW	MgSW	ImSW	PpSW	AuSW	NaSD	AfSD	PaSD	AuSD	NaTeW	PaTeW	NaTeR	NtTeR	PaTeR	AuTeR	NZTeR
Mastotermitidae																									
Mastotermes	Wood	Masto-I												+				+							
Termopsidae																									
Zootermopsis	Wood-nesting	Termop-I													+						+				
Porotermes	Wood-nesting	Termop-I																				+		+	
Stolotermes	Wood-nesting	Termop-I						+																+	+
Hodotermitidae																									
Anacanthotermes	Grass	Hodo-I															+								
Hodotermes	Grass	Hodo-I								+						+									
Kalotermitidae																									
Kalotermes	Wood-nesting	Kalo-I						+												+				+	+
Epicalotermes	Wood-nesting	Kalo-I			+					+						+									
Pterotermes	Wood-nesting	Kalo-I													+										
Paraneotermes	Wood-nesting	Kalo-I													+										
Incisitermes	Wood-nesting	Kalo-I	+												+										
Tauritermes	Wood-nesting	Kalo-I							+																
Rugitermes	Wood-nesting	Kalo-I	+						+																
Bifiditermes	Wood-nesting	Kalo-I								+	+														
Postelectrotermes	Wood-nesting	Kalo-I			+																				
Calcaritermes	Wood-nesting	Kalo-I	+																						
Neotermes	Wood-nesting	Kalo-I	+	+	+	+	+		+	+	+	+	+	+											
Allotermes	Wood-nesting	Kalo-I									+														
Procryptotermes	Wood-nesting	Kalo-I												+											
Bicornitermes	Wood-nesting	Kalo-I		+						+															
Glyptotermes	Wood-nesting	Kalo-I	+			+	+						+												
Marginitermes	Wood-nesting	Kalo-I													+										
Eucryptotermes	Wood-nesting	Kalo-I	+																						
Cryptotermes	Wood-nesting	Kalo-I	+	+		+	+	+	+	+		+	+	+				+							
Serritermitinae																									
Glossotermes	Wood	Serri-I	+																						
Rhinotermitidae																									
Psammotermes	Detritus	Rhino-I														+									
Termitogeton	Wood	Rhino-I				+																			
Stylotermes	Wood	Rhino-I				+																			
Dolichorhinotermes	Wood	Rhino-I	+																						
Acorhinotermes	Wood	Rhino-I	+																						

Table 17.2 (continued)

		Site No.	1	2	3	4	5	6	7	8	9	10	11	12	13	14	15	16	17	18	19	20	21	22	23
			Tropical rain forest						Savanna woodland						Semi-desert				Temperate woodland		Temperate rain forest				
	Food type	FTG	NtTrR	AfTrR	MgTrR	ImTrR	PpTrR	AuTrR	NtSW	AfSW	MgSW	ImSW	PpSW	AuSW	NaSD	AfSD	PaSD	AuSD	NaTeW	PaTeW	NaTeR	NtTeR	PaTeR	AuTeR	NZTeR
Rhinotermes	Wood	Rhino-I	+																						
Parrhinotermes	Wood	Rhino-I				+		+																	
Schedorhinotermes	Wood	Rhino-I		+		+	+	+		+			+	+				+							
Coptotermes	Wood	Rhino-I	+	+	+	+	+	+		+	+	+	+	+				+							
Heterotermes	Wood	Rhino-I	+			+	+		+			+	+	+	+			+							
Reticulitermes	Wood	Rhino-I													+				+	+			+		
Macrotermitinae																									
Acanthotermes	Fungus	Macro-II		+																					
Pseudacanthotermes	Fungus	Macro-II		+						+															
Macrotermes	Fungus	Macro-II		+		+				+		+				+									
Ancistrotermes	Fungus	Macro-II								+															
Allodontermes	Fungus	Macro-II								+						+									
Synacanthotermes	Fungus	Macro-II		+																					
Microtermes	Fungus	Macro-II		+	+					+	+	+				+									
Protermes	Fungus	Macro-II		+																					
Odontotermes	Fungus	Macro-II		+		+				+		+				+									
Hypotermes	Fungus	Macro-II				+																			
						4	5	6	7	8															
Sphaerotermitinae																									
Sphaerotermes	Wood	Sphae-II		+																					
Foraminitermitinae																									
Foraminitermes	Humus	Fora-III		+																					
Labritermes	Humus	Fora-III				+																			
Apicotermitinae																									
Eburnitermes	Soil	Apico-IV		+																					
Labidotermes	Soil	Apico-IV		+																					
Phoxotermes	Soil	Apico-IV		+																					
Machadotermes	Soil	Apico-IV		+																					
Duplidentitermes	Soil	Apico-IV		+																					
Apicotermes	Soil	Apico-IV		+						+															
Coxotermes	Soil	Apico-IV		+																					
Heimitermes	Soil	Apico-IV		+																					
Jugositermes	Soil	Apico-IV		+																					
Amalotermes	Humus	Anop-III		+																					
Astalotermes	Humus	Anop-III		+						+						+									
Astratotermes	Humus	Anop-III		+						+															
Alyscotermes	Humus	Anop-III		+						+						+									

Table 17.2 (continued)

		Site No.	1	2	3	4	5	6	7	8	9	10	11	12	13	14	15	16	17	18	19	20	21	22	23
			Tropical rain forest						Savanna woodland						Semi-desert				Temperate woodland		Temperate rain forest				
	Food type	FTG	NtTrR	AfTrR	MgTrR	ImTrR	PpTrR	AuTrR	NtSW	AfSW	MgSW	ImSW	PpSW	AuSW	NaSD	AfSD	PaSD	AuSD	NaTeW	PaTeW	NaTeR	NtTeR	PaTeR	AuTeR	NZTeR
Acholotermes	Humus	Anop-III		+																					
Aderitotermes	Humus	Anop-III		+						+															
Adaiphrotermes	Humus	Anop-III		+						+															
Speculitermes	Humus	Anop-III										+													
Euhamitermes	Humus	Anop-III				+						+													
Eurytermes	Humus	Anop-III										+													
Skatitermes	Humus	Anop-III														+									
Anoplotermes	Humus	Anop-III	+						+																
Aparatermes	Humus	Anop-III							+																
Ruptitermes	Humus	Anop-III	+						+																
Guiana Gen. nov. 1	Humus	Anop-III	+																						
Guiana Gen. nov. 2	Humus	Anop-III	+																						
Guiana Gen. nov. 3	Humus	Anop-III	+																						
Guiana Gen. nov. 4	Humus	Anop-III	+																						
Guiana Gen. nov. 5	Humus	Anop-III	+																						
Guiana Gen. nov. 6	Humus	Anop-III	+																						
Guiana Gen. nov. 7	Humus	Anop-III	+																						
Guiana Gen. nov. 8	Humus	Anop-III	+																						
Amicotermes	Soil	Anop-IV		+																					
Ateuchotermes	Soil	Anop-IV		+						+															
Anenteotermes	Soil	Anop-IV		+						+															
Apagotermes	Soil	Anop-IV		+																					
Grigiotermes	Soil	Anop-IV							+																
Guiana Gen. nov. 9	Soil	Anop-IV	+																						
Guiana Gen. nov. 10	Soil	Anop-IV	+																						
Guiana Gen. nov. 11	Soil	Anop-IV	+																						
Guiana Gen. nov. 12	Soil	Anop-IV	+																						
															13	14	15	16	17	18	19	20	21	22	23
Amitermes-gp																									
Microcerotermes	Wood	Ami-II	+	+	+	+	+	+	+	+	+	+	+	+		+		+							
Pseudomicrotermes	Wood	Ami-II		+																					
Eremotermes	Detritus	Ami-II										+													
Amitermes	Wood	Ami-II							+	+	+		+	+	+	+	+	+							
Drepanotermes	Grass	Ami-II												+				+							
Ahamitermes	Carton	Ami-II												+											
Prohamitermes	Wood	Ami-II				+																			
Globitermes	Wood	Ami-II				+																			

Table 17.2 (continued)

		Site No.	1	2	3	4	5	6	7	8	9	10	11	12	13	14	15	16	17	18	19	20	21	22	23
			Tropical rain forest						Savanna woodland						Semi-desert				Temperate woodland		Temperate rain forest				
	Food type	FTG	NtTrR	AfTrR	MgTrR	ImTrR	PpTrR	AuTrR	NtSW	AfSW	MgSW	ImSW	PpSW	AuSW	NaSD	AfSD	PaSD	AuSD	NaTeW	PaTeW	NaTeR	NtTeR	PaTeR	AuTeR	NZTeR
Gnathamitermes	Detritus	Ami-II													+										
Synhamitermes	Wood	Ami-II										+													
Syntermes	Litter	Ami-II	+						+																
Cornitermes	Wood	Ami-II	+						+																
Rhynchotermes	Litter	Ami-II							+																
Angulitermes	Humus	Ami-III										+				+									
Promirotermes	Humus	Ami-III		+						+						+									
Onkotermes	Humus	Ami-III							+																
Neocapritermes	Humus	Ami-III	+						+																
Planicapritermes	Humus	Ami-III	+																						
Armitermes	Humus	Ami-III	+																						
Procornitermes	Humus	Ami-III							+																
Embiratermes	Humus	Ami-III	+																						
Cyrilliotermes	Humus	Ami-III	+																						
Invasitermes	Mound	Ami-IV												+											
Orthognathotermes	Mound	Ami-IV	+																						
Dentispicotermes	Mound	Ami-IV	+																						
Labiotermes	Soil	Ami-IV	+						+																
Cylindrotermes-gp																									
Cephalotermes	Wood	Cylin-II		+																					
Cylindrotermes	Wood	Cylin-II	+																						
Protohamitermes-gp																									
Protohamitermes	Wood	Proto-II				+																			
Orientotermes	Wood	Proto-II				+																			
Cubitermes-gp																									
Tuberculitermes	Soil	Cubi-IV		+																					
Crenetermes	Soil	Cubi-IV		+																					
Thoracotermes	Soil	Cubi-IV		+						+															
Apilitermes	Soil	Cubi-IV		+																					
Cubitermes	Soil	Cubi-IV		+						+															
Procubitermes	Soil	Cubi-IV		+																					
Lepidotermes	Soil	Cubi-IV								+						+									
Noditermes	Soil	Cubi-IV		+						+															
Unguitermes	Soil	Cubi-IV		+																					
Mucrotermes	Soil	Cubi-IV		+																					
Proboscitermes	Soil	Cubi-IV		+																					
Fastigitermes	Soil	Cubi-IV		+																					

Table 17.2 (continued)

		Site No.	1	2	3	4	5	6	7	8	9	10	11	12	13	14	15	16	17	18	19	20	21	22	23
			Tropical rain forest						Savanna woodland						Semi-desert				Temperate woodland		Temperate rain forest				
	Food type	FTG	NtTrR	AfTrR	MgTrR	ImTrR	PpTrR	AuTrR	NtSW	AfSW	MgSW	ImSW	PpSW	AuSW	NaSD	AfSD	PaSD	AuSD	NaTeW	PaTeW	NaTeR	NtTeR	PaTeR	AuTeR	NZTeR
Basidentitermes	Soil	Cubi-IV		+						+															
Orthotermes	Soil	Cubi-IV		+																					
Profastigitermes	Soil	Cubi-IV		+																					
Ophiotermes	Mound	Cubi-IV		+						+															
Euchilotermes	Mound	Cubi-IV		+						+															
Furculitermes	Mound	Cubi-IV		+																					
Pericapritermes-gp																									
Pericapritermes	Humus	Peri-III		+		+	+			+		+	+												
Homallotermes	Humus	Peri-III				+																			
Procapritermes	Humus	Peri-III				+																			
Coxocapritermes	Humus	Peri-III				+																			
Dicuspiditermes	Humus	Peri-III				+						+													
Malaysiocapritermes	Humus	Peri-III				+																			
Mirocapritermes	Humus	Peri-III				+																			
Pseudocapritermes	Humus	Peri-III				+																			
Kemneritermes	Soil	Peri-IV				+																			
Oriencapritermes	Soil	Peri-IV				+																			
Syncapritermes	Soil	Peri-IV				+																			
							5	6	7	8	9	10	11	12											
Termes-gp																									
Termes	Humus	Term-III	+	+		+			+	+						+									
Capritermes	Humus	Term-III			+																				
Quasitermes	Humus	Term-III			+																				
Cavitermes	Humus	Term-III	+																						
Cornicapritermes	Humus	Term-III	+																						
Crepititermes	Humus	Term-III	+																						
Dihoplotermes	Humus	Term-III							+																
Inquilinitermes	Humus	Term-III	+																						
Spinitermes	Humus	Term-III	+																						
Xylochomitermes	Humus	Term-III												+											
Aspenterotermes	Humus	Term-III												+											
Ephelotermes	Humus	Term-III											+	+				+							
Ekphysotermes	Humus	Term-III																+							
Cristatitermes	Humus	Term-III												+											
Macrognathotermes	Humus	Term-III											+	+											
Paracapritermes	Humus	Term-III						+																	
Protocapritermes	Humus	Term-III					+																		

Table 17.2 (continued)

		Site No.	1	2	3	4	5	6	7	8	9	10	11	12	13	14	15	16	17	18	19	20	21	22	23
			Tropical rain forest						Savanna woodland						Semi-desert				Temperate woodland		Temperate rain forest				
	Food type	FTG	NtTrR	AfTrR	MgTrR	ImTrR	PpTrR	AuTrR	NtSW	AfSW	MgSW	ImSW	PpSW	AuSW	NaSD	AfSD	PaSD	AuSD	NaTeW	PaTeW	NaTeR	NtTeR	PaTeR	AuTeR	NZTeR
Lophotermes	Soil	Term-IV						+					+	+											
Saxatilitermes	Soil	Term-IV												+											
Hapsidotermes	Soil	Term-IV												+											
Nasutitermitinae																									
Australitermes	Humus	Subu-III												+				+							
Macrosublitermes	Humus	Subu-III						+																	
Subulitermes	Humus	Subu-III	+																						
Atlantitermes	Humus	Subu-III	+																						
Caetetermes	Humus	Subu-III	+																						
Angularitermes	Humus	Subu-III	+																						
Coatitermes	Humus	Subu-III	+																						
Araujotermes	Humus	Subu-III	+						+																
Cyranotermes	Soil	Subu-IV																							
Verrucositermes	Soil	Subu-IV		+																					
Eutermellus	Soil	Subu-IV		+																					
Postsubulitermes	Soil	Subu-IV		+																					
Rhadinotermes	Litter	Nas-II								+															
Grallatotermes	Microepiphytes	Nas-II					+																		
Nasutitermes	Wood	Nas-II	+	+	+	+	+	+	+	+	+	+	+	+				+							
Trinervitermes	Grass	Nas-II								+		+				+									
Coarctotermes	Grass	Nas-II									+														
Occultitermes	Wood	Nas-II												+				+							
Tumulitermes	Grass	Nas-II											+	+				+							
Diwaitermes	Wood	Nas-II					+						+												
Niuginitermes	Wood	Nas-II					+						+												
Tenuirostritermes	Detritus	Nas-II													+										
Velocitermes	Litter	Nas-II	+																						
Constrictotermes	see footnote	Nas-II	+						+																
Diversitermes	Wood	Nas-II							+																
Cortaritermes	Litter?	Nas-II							+																
Bulbitermes	Wood	Nas-II				+																			
Havilanditermes	Wood	Nas-II				+																			
Hirtitermes	Wood	Nas-II				+																			
Lacessititermes	Litter	Nas-II				+																			
Longipeditermes	Litter	Nas-II				+																			
Hospitalitermes	Microepiphytes	Nas-II				+	+						+												
Leptomyxotermes	Humus	Nas-III		+																					

Table 17.2 (continued)

		Site No.	1	2	3	4	5	6	7	8	9	10	11	12	13	14	15	16	17	18	19	20	21	22	23
			Tropical rain forest						Savanna woodland						Semi-desert				Temperate woodland		Temperate rain forest				
	Food type	FTG	NtTrR	AfTrR	MgTrR	ImTrR	PpTrR	AuTrR	NtSW	AfSW	MgSW	ImSW	PpSW	AuSW	NaSD	AfSD	PaSD	AuSD	NaTeW	PaTeW	NaTeR	NtTeR	PaTeR	AuTeR	NZTeR
Malagasitermes	Humus	Nas-III			+																				
Gen. nov. 1 (Mad)	Humus	Nas-III			+																				
Oriensublitermes	Humus	Nas-III				+																			
Malaysiotermes	Humus	Nas-III				+																			
Subulioiditermes	Humus	Nas-III				+																			
Leucopitermes	Humus	Nas-III				+																			
Eleanoritermes	Humus	Nas-III				+																			
Aciculitermes	Humus	Nas-III				+																			
Convexitermes	Humus	Nas-III	+																						
		Total	55	62	11	44	14	11	25	37	9	18	17	24	10	17	2	14	1	2	1	1	1	3	2

In rain forest, *Constrictotermes* feeds on microephytes (Martius et al. 2000) but feeds on dead wood in savanna woodland (Moura et al. 2008)

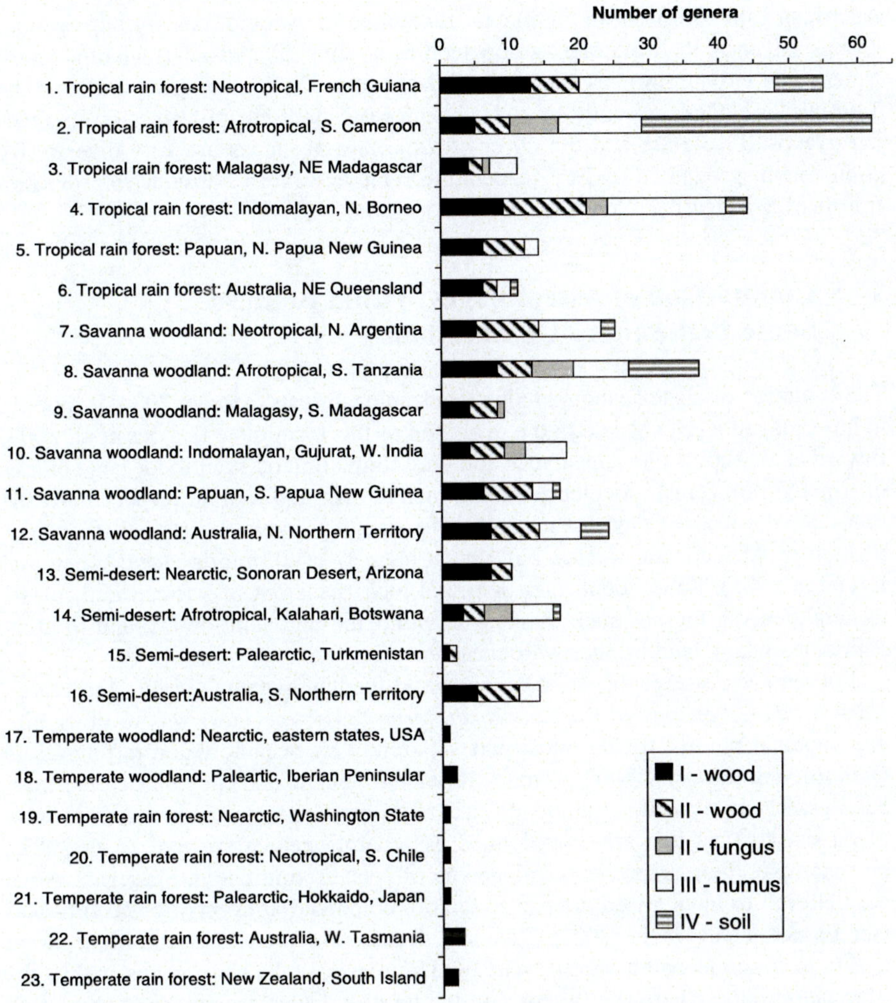

Fig. 17.2 Number of genera and feeding groups in termite assemblages at the 23 exemplar sites

Group IV soil-feeders are the dominant feeding group in the African rain forest and savanna woodland ecosystems, whereas group III humus-feeders dominate in the tropical rain forests of the Neotropics and Borneo (Fig. 17.2). Although Group II wood-feeders (excluding fungus-growers) have their highest richness in Borneo, the number of group II genera across all the tropical rain forests and savanna woodlands is similar. Only group I wood-feeders occur in the temperate woodlands and temperate rain forests. Compared with other biomes, it is the group III humus-feeding and the group IV soil-feeding genera that account for most of the greater richness in the rain forests of Africa, South America and Southeast Asia (Davies et al. 2003). In African rain forests, the *Cubitermes* clade and the *Apicotermes*-IV and *Anoplotermes*-III clades are significant in terms of numbers of genera. South

American rain forests have significant radiations of Anoplotermes-group feeding groups III and IV, Amitermes-group feeding groups III and IV (including those genera formerly known as the mandibulate nasutes), Termes-group III and the *Sublitermes* clades. In Southeast Asian rain forests the Pericapritermes-group feeding groups III and IV, and the *Oriensubulitermes*-III clades are very diverse. By comparison, groups III and IV feeders are relatively rare or absent from the rain forests of Madagascar, New Guinea and Australia.

17.5 Comparison of Assemblages Within Biomes: Some Preliminary Observations

The evolution of closed canopy multistrata tropical forests (Morley 2000) is thought to have played a critical role in the evolution of the Termitidae (Davies et al. 2003; Inward et al. 2007). The appearance and long-term climatic stability of large blocks of tropical rain forest, coupled with continental drift and termite dispersal, set the scene for the present day distributions. The rise of the rain forests led to diversification in African rain forests, but almost entirely of Termitidae (see Aanen and Eggleton 2005). Subsequent filter dispersal took place out of African rain forests to neighbouring biomes such as savanna woodland and semi-desert, and to other continents where land bridges were present.

Further sweepstakes dispersal by key wood-feeding genera (*Termes*, *Amitermes*, *Nasutitermes*, *Pericapritermes*, *Microcerotermes*, *Microtermes*) was to other biogeographical realms, for the most part apparently by oceanic wood rafting (as in Gathorne-Hardy et al. 2000). These dispersive genera (except *Microcerotermes*) have given rise to new radiations on different continents, and in some cases have given rise to taxa that are now defined as separate genera (Inward et al. 2007). For example, there are six endemic genera of Nasutitermitinae in Madagascar that are believed to have arisen after one or more colonization events by *Nasutitermes* (see Eggleton and Davies 2003). Diversification of Macrotermitinae in Madagascar is due to a single colonisation event by *Microtermes*, the only genus within the Macrotermitinae where the fungus and the termite disperse together (Nobre et al. 2010). However, drier or colder climates and oceans act as filters to prevent many clades from dispersing, and some remain endemic, such as the Cubitermes-group. The energetic constraints inherent in feeding lower down the humification gradient (groups III and IV) and their nesting habits may have prevented these groups from effectively dispersing much beyond rain forests. In contrast, group II wood-feeders are energetically much more resilient and often nest in dead wood.

The broadleaf temperate woodlands of the Nearctic (eastern USA) and western Palearctic (Iberian Peninsular) have very few termite genera. Only species of *Reticulitermes* occur across most of the eastern forests of North America, and much of the western Palearctic. *Kalotermes* also occurs in the Iberian Peninsular and other parts of southern Palearctic. Temperate woodlands in the southern hemisphere at a similar latitude, such as coastal southern Western Australia have many more genera (Abensperg-Traun and Perry 1998).

As with the temperate woodlands, the five temperate rain forests also have very low termite richness, and all consist of group I wood-feeders. All genera are single-piece wood-nesters (feeding and nesting inside the same piece of wood). The exception is the temperate rain forest in Hokkaido, northern Japan, where only *Reticulitermes* occurs, a genus that forages away from its nest site. Termopsids are widely scattered at the margins of the termites' global range, with *Porotermes* occurring in Southern Chile and Tasmania, *Stolotermes* in Tasmania and South Island New Zealand, and *Zootermopsis* in Washington State, North America. In addition, *Kalotermes* is also recorded from the rain forests of Tasmania and New Zealand. Elsewhere, drywood termites (Kalotermitidae) are found in very dry wood (for example in the canopies of tropical forests), or on isolated islands (including Madagascar).

17.6 Implications of Varying Assemblage Structures for Termite Mediated Decomposition in Different Biomes

The seminal works by Wood and Sands (1978) and Wood (1978) in *Production Ecology of Ants and Termites* (Brian 1978) established a framework for quantitative studies of feeding and rates of consumption by termites. They compiled, analysed and evaluated all the available literature on consumption by individual species and gave the earliest calculated estimates of consumption by local guilds and whole assemblages (reviewed in Bignell and Eggleton 2000). Since that pioneering work, assemblage-level consumption by termites have been estimated at only nine new sites (Abe 1978; Abe and Matsumoto 1979; Buxton 1981a; Whitford et al. 1982; Collins 1984; Jones 1996; Salick et al. 1983; Martius 1994; Lepage 1981; Bagine 1989).

As Wood and Sands (1978) and Wood (1978) make clear, estimates of consumption are fraught with problems. Consumption by termites in the field calculated from average weight-specific rates of consumption are dependent on accurate estimates of population densities. Rates of consumption measured in the field using baiting techniques can produce overestimates if untypically palatable food is offered (Wood 1978), while extrapolations from laboratory feeding trials can also lead to errors as they may not be representative of field conditions. Furthermore, laboratory measurements show that daily food consumption by some termites, for example *Trinervitermes geminatus*, can double if alate nymphs are in the nest (Josens 1983).

It has been understood for decades that the ability of some groups of termites to feed on fresh litter opens up a completely new pathway in tropical decomposition processes (Wood 1976). In savanna regions, where fresh-litter-feeding termites are usually dominant, considerable quantities of plant material may be consumed before being attacked by saprotrophic micro-organisms. In Africa, Asia and to a lesser extent Madagascar, Macrotermitinae usually dominate this niche, with grass-harvesters prominent in many parts of Australia.

Schuurman (2005) shows that significant variation in decomposition can occur in similar habitats depending on the presence and relative abundance of fungus-growers. Controlled field experiments are needed to compare the impact in different continents of the presence or absence of the Macrotermitinae. However, there are suggestions that consumption may still be considerable even in ecosystems where fungus-growers are absent. For example, in Brazilian Caatinga (semi-arid thorn scrubland) *Constrictotermes cyphergaster* consumes 33.6 g m^{-2} yr^{-1} of dead plant material, or approximately 13% of annual wood litter production (Moura 2008). This compares with Wood and Sand's (1978) estimate that termites consumed 55% of surface litter in the Southern Guinea savanna of Nigeria (where fungus-growers are present), about 20% of production, and Buxton (1981b) who suggested that 90% of lying dead wood was taken in arid Kenya.

Estimates are available from Australia on the incidence of termite attack on standing trees. Elliott and Bashford (1984) reported that in two forest blocks of lowland dry sclerophyll eucalypt forest in north eastern Tasmania, 43 and 60% of commercially logged *Eucalyptus* species were found to have been piped by *Porotermes adamsoni*. In the montane forest of *Eucalyptus delegatensis* and *E. dalrympleana* in New South Wales, 56–81% of the volume of commercial timber was rejected due to damage by *P. adamsoni* (Greaves et al. 1965). Drawing on the work of Stamm (2006), Cornwell et al. (2009) suggested that termites in the temperate forests of Tasmania may be responsible for between 20–30% of the loss of volume of *Eucalyptus obliqua* logs. However, M Stamm (personal communication) states that this is a gross overestimate of the losses due to termites.

17.7 Conclusions

Termites are generally tropical animals, but their distributions do range significantly into colder (temperate forests) and drier environments (deserts). However, the global ecological impact of termites is very strongly biased towards tropical rain forest and tropical savannas, both wet and dry. Within tropical systems, the taxonomic richness and functional structure of termites is strongly constrained by biogeography. This is due to the particular biogeographical history of the Termitidae, which make up more than 90% of all tropical termites.

The Termitidae clearly evolved in Africa and have penetrated other tropical regions, to a greater or lesser extent, by species dispersing across land bridges or by wood-rafting. This is especially true of soil-feeding termites (feeding groups III and IV), which are not generally very dispersive at all, and each biogeographical region appears to have its own endogenous soil feeding clades. These have probably evolved from groups that originally colonised from Africa, many of them ancestrally wood-feeders. A full description of these repeated waves of dispersal and local diversification remains to be made, but will be vital if we are to understand the evolutionary context of termite distributions and their ecological consequences.

References

Aanen DK, Eggleton P (2005) Fungus-growing termites in African rain forest. Curr Biol 15: 851–855

Abe T (1978) Studies on the distribution and ecological role of termites in a lowland rain forest of west Malaysia. 1. Faunal composition, size, colouration and nest of termites in Pasoh Forest Reserve. Kontyu 46:273–290

Abe T, Matsumoto T (1979) Studies on the distribution and ecological role of termites in a lowland rain forest of West Malaysia (3). Distribution and abundance of termites in Pasoh Forest Reserve. Jpn J Ecol 29:337–351

Abensperg-Traun M, de Boer ES (1990) Species abundance and habitat differences in biomass of subterranean termites (Isoptera) in the wheatbelt of Western Australia. Aust J Ecol 15: 219–226

Abensperg-Traun M, Perry DH (1998) Distribution and characteristics of mound-building termites (Isoptera) in Western Australia. J R Soc West Aust 81:191–200

Abersperg-Traun M, Steven D (1997) Ant- and termite-eating in Australian mammals and lizards: a comparison. Aust Ecol 22:9–17

Bagine RKN (1989) Plant litter consumption by termites in an arid environment of northern Kenya. Sociobiology 15:114–124

Bignell DE, Constantino R, Czudi C et al (2008) Macrofauna. In: Moreira FMS, Huising J, Bignell DE (eds) A handbook of tropical soil biology: sampling and characterization of below-ground biodiversity. Earthscan, London, pp 43–75

Bignell DE and Eggleton P (2000) Termites in ecosystems. In: Abe T, Bignell DE, Higashi M (eds) Termites: evolution, sociality, symbioses, ecology. Kluwer Academic Publishers, Dordrecht, pp 363–387

Bourguignon T, Leponce M, Roisin Y (2008) Revision of the Termitinae with snapping soldiers (Isoptera: Termitidae) from New Guinea. Zootaxa 1769:1–34

Braithwaite RW, Miller L, Wood JT (1988) The structure of termite communities in the Australian tropics. Aust Ecol 13:375–391

Brian MV (ed) (1978) Production ecology of ants and termites. Cambridge University Press, Cambridge, 409p

Buxton RD (1981a) Termites and the turnover of dead wood in an arid tropical environment. Oecologia 51:379–384

Buxton RD (1981b) Changes in the composition and activities of termite communities in relation to changing rainfall. Oecologia 51:371–378

Chhotani OB (1997) Fauna of India. Isoptera (Termites), vol 2. Zoological Survey of India, Calcutta, p 801

Coaton WGH, Sheasby JL (1972) Preliminary report on a survey of termites (Isoptera) of South West Africa. Cimbebasia Mem 2:1–128

Collins NM (1984) The termites (Isoptera) of the Gunung Mulu National Park, with a key to the genera known from Sarawak. Sarawak Mus J 30:65–87

Collins NM (1989) Termites. In: Leith H, Werger MJA (eds) Tropical rain forest ecosystems. Elsevier Science Publishers, Amsterdam, pp 455–471

Constantino R (2010) On-line termite database. http://vsites.unb.br/ib/zoo/docente/constant/catalog/ and http://vsites.unb.br/ib/zoo/catalog.html

Cornwell WK, Cornelissen JHC, Allison A et al (2009) Plant traits and wood fates across the globe: rotted, burned, or consumed? Glob Change Biol 15:2341–2449

Davies RG (2002) Feeding group responses of a Neotropical termite assemblage to rain forest fragmentation. Oecologia 133:233–242

Davies RG, Eggleton P, Jones DT et al (2003) Evolution of termite functional diversity: analysis and synthesis of local ecological and regional influences on local species richness. J Biogeogr 30:847–877

Dawes-Gromadzki TZ (2005) The termite (Isoptera) fauna of a monsoonal rain forest near Darwin, northern Australia. Aust J Entomol 44:152–157

Dawes-Gromadzki TZ (2008) Abundance and diversity of termites in a savanna woodland reserve in tropical Australia. Aust J Entomol 47:307–314

De Souza OFF, Brown VK (1994) Effect of habitat fragmentation on Amazonian termite communities. J Trop Ecol 10:197–206

Donovan SE, Eggleton P, Bignell DE (2001) Gut content analysis and a new feeding group classification of termites. Ecol Entomol 26:356–366

Donovan SE, Jones DT, Sands WA, Eggleton P (2000) The morphological phylogenetics of termites (Isoptera). Biol J Linn Soc 70:467–513

Eggleton P (2000) Global patterns of termite diversity. In: Abe T, Bignell DE, Higashi M (eds) Termites: evolution, sociality, symbioses, ecology. Kluwer Academic Publishers, Dordrecht, pp 25–51

Eggleton P, Bignell DE, Sands WA et al (1996) The diversity, abundance and biomass of termites under differing levels of disturbance in the Mbalmayo Forest Reserve, southern Cameroon. Phil Trans R Soc Lond B 351:51–68

Eggleton P Davies RG (2003) Isoptera, termites. In: Goodman SM, Benstead JP (eds) The natural history of Madagascar. University of Chicago Press, Chicago, IL, pp 654–660

Eggleton P, Davies RG, Connetable S et al (2002) The termites of the Mayombe Forest Reserve, Congo (Brazzaville): transect sampling reveals an extremely high diversity of ground-nesting soil feeders. J Nat Hist 36:1239–1246

Eggleton P, Homathevi R, Jeeva D et al (1997) The species richness and composition of termites (Isoptera) in primary and regenerating lowland dipterocarp forest in Sabah, east Malaysia. Ecotropica 3:119–128

Eggleton P, Homathevi R, Jones DT, et al (1999) Termite assemblages, forest disturbance, and greenhouse gas fluxes in Sabah, East Malaysia. Phil Trans R Soc B 354:1791–1802

Eggleton P, Tayasu I (2001) Feeding groups, lifetypes, and the global ecology of termites. Ecol Res 16:941–960

Eggleton P, Williams PH, Gaston KJ (1994) Explaining termite global richness: productivity or history? Biodivers Conserv 3:318–330

Elliott HJ, Bashford R (1984) Incidence and effects of the dampwood termite, *Porotermes adamsoni*, in two Tasmanian east coast eucalypt forests. Aust Forest 4:11–15

Emerson AE (1955) Geographical origins and dispersion of termite genera. Fieldiana Zool 37: 465–521

Gathorne-Hardy FJ (2004) The termites of Sundaland: a taxonomic review. Sarawak Mus J 60: 89–133

Gathorne-Hardy F, Collins NM, Buxton RD, Eggleton P (2000) A faunistic review of the termites (Insecta: Isoptera) of Sulawesi, including an updated checklist of the species. Malay Nat J 54:347–353

Gathorne-Hardy FJ, Syaukani, Davies RG et al (2002) Quaternary rain forest refugia in Southeast Asia: using termites (Isoptera) as indicators. Biol J Linn Soc 75:453–466

Gay FJ, Calaby JE (1970) Termites of the Australian region. In: Krishna K, Weesner FM (eds) Biology of termites, vol II. Academic Press, New York, NY, pp 393–447

Greaves T, McInnes RS, Dowse JE (1965) Timber losses caused by termites, decay and fire in an alpine forest in New South Wales. Aust Forest 29:161–174

Holt JA, Lepage M (2000) Termites and soil properties. In: Abe T, Bignell DE, Higashi M (eds) Termites: evolution, sociality, symbioses, ecology. Kluwer Academic Publishers, Dordrecht, pp 389–407

Inward DJG, Vogler AP, Eggleton P (2007) A comprehensive phylogenetic analysis of termites (Isoptera) illuminates key aspects of their evolutionary biology. Mol Phylogenet Evol 44: 953–967

Ji R, Brune A (2006) Nitrogen mineralization, ammonia accumulation, and emission of gaseous NH_3 by soil-feeding termites. Biogeochemistry 78:267–283

Jones DT (1996) A quantitative survey of the termite assemblage and its consumption of food in lowland mixed dipterocarp forest in Brunei Darussalam. In: Edwards DS, Booth WE,

Choy S (eds) Tropical rain forest research – current issues. Kluwer Academic Publishers, Dordrecht, pp 297–305
Jones DT, Eggleton P (2000) Sampling termite assemblages in tropical forests: testing a rapid biodiversity assessment protocol. J Appl Ecol 37:191–203
Josens G (1983) The soil fauna of tropical savannas. III The termites. In: Boulière F (ed) Ecosystems of the world, vol 13. Tropical savannas. Elsevier, Amsterdam, pp 505–524
Kambhampati S, Eggleton P (2000) Taxonomy and phylogeny of termites. In: Abe T, Bignell DE, Higashi M (eds) Termites: evolution, sociality, symbioses, ecology. Kluwer Academic Publishers, Dordrecht, pp 1–23
Krishna K, Weesner FM (eds) (1970) Biology of termites, vol II. Academic Press, New York, NY, 643p
Lepage M (1981) L'impact des populations récoltantes de *Macrotermes michaelseni* (Sjöstedt) (Isoptera: Macrotermitinae) dans un écosystème semi-aride (Kajiado – Kenya). 2 – La nourriture récoltantée, comparison avec les grands herbivores. Insect Soc 28:309–319
Martius C (1994) Diversity and ecology of termites in Amazonian forests. Pedobiologia 38: 407–428
Martius C, Amelung W, Garcia MVB (2000) The Amazonian forest termite (Isoptera: Termitidae) *Constrictotermes cavifrons* feeds on microepiphytes. Sociobiology 35:379–383
Miller LR (1991) A revision of the *Termes/Capritermes* branch of the Termitinae in Australia (Isoptera, Termitidae). Invertebr Taxon 4:1147–1282
Morley J (2000) Origin and evolution of tropical rain forests. Wiley, Chichester
Moura FMS, Vasconcellos A, de Araújo VFP, Bandeira AG (2008) Consumption of vegetal organic matter by *Constrictotermes cyphergaster* (Isoptera, Termitidae, Nasutitermitinae) in an Area of Caatinga, Northeastern Brazil. Sociobiology 51:181–189
Nobre T, Eggleton P, Aanen DK (2010) Vertical transmission as the key to the colonization of Madagascar by fungus-growing termites? Proc R Soc Lond B 277:359–365
Rathore NS, Bhattacharyya AK (2004) Termite (Insecta: Isoptera) fauna of Gujarat and Rajasthan – present state of knowledge. Records of the Zoological Survey of India: occasional Paper No. 223, vi, 94p
Roisin Y, Leponce M (2004) Characterizing termite assemblages in fragmented forests: a test case in the Argentinian Chaco. Austr Ecol 29:637–646
Roisin Y, Pasteels JM (1996) The nasute termites (Isoptera: Nasutitermitinae) of Papua New Guinea. Invertebr Taxon 10:507–616
Roonwal ML, Chhotani OB (1989) Fauna of India. Isoptera (Termites), vol 1. Zoological Survey of India, Calcutta, p 672
Salick J, Herrera R, Jordan CF (1983) Termitaria: nutrient patchiness in nutrient-deficient rain forests. Biotropica 15:1–7
Schuurman G (2005) Decomposition rates and termite assemblage composition in semiarid Africa. Ecology 86:1236–1249
Stamm L (2006) Measuring and modelling coarse woody debris in tall wet eucalypt forests in southern Tasmania. Honours thesis, University of Tasmania, Hobart, Tasmania
Thapa RS (1981) Termites of Sabah. Sabah Forest Rec 12:1–374
Torales GJ, Coronel JM, Laffont ER et al (2007) Termite (Insecta, Isoptera) faunal composition in natural forests of the humid Chaco (Argentina). Sociobiology 50:419–433
Torales GJ, Laffont ER, Godoy MC et al (2005) Update on taxonomy and distribution of Isoptera from Argentina. Sociobiology 45:853–886
Uys V (2002) A guide to the termite genera of Southern Africa. Plant Protection Research Handbook No.15. Agricultural Research Council, Pretoria
Watson JAL, Abbey HA (1993) Atlas of Australian termites. CSIRO, Canberra, 155p
Weesner FM (1970) Termites of the Nearctic region. In: Krishna K, Weesner FM (eds) Biology of termites, vol II. Academic Press, New York, NY, pp 477–525
Whitford WG, Steinberger Y, Ettershank G (1982) Contributions of subterranean termites to the "economy" of Chihuahuan desert ecosystems. Oecologia 55:298–302

Wood TG (1976) The role of termites (Isoptera) in decomposition processes. In: Anderson JM, MacFadyen A (eds) The role of terrestrial and aquatic organisms in decomposition processes. Blackwell Scientific, Oxford, pp 145–168

Wood TG (1978) Food and feeding habits of termites. In: Brian MV (ed) Production ecology of ants and termites. Cambridge University Press, Cambridge, pp 55–80

Wood TG, Sands WA (1978) The role of termites in ecosystems. In: Brian MV (ed) Production ecology of ants and termite. Cambridge University Press, Cambridge, pp 245–292

Chapter 18
Termites as Pests of Agriculture

Corinne Rouland-Lefèvre

Abstract While pest species of termites are thought to belong predominantly to four families comprising the lower termites, the species causing the most damage to tropical agriculture belong to three of the four subfamilies of higher termites, lacking protist symbionts. A wide range of crops are affected, including trees in plantations and orchards, coconuts, palms, sugar cane, rice, maize, wheat, sorghum, groundnuts, coffee, tea, cocoa, yam, cassava and cotton. Market gardens and pastures can also be damaged. Organochlorine and organophosphate insecticides, once very effective agents of chemical control, are now prohibited or restricted, owing to their several toxicities, but alternatives of comparable efficacy in cropping systems are yet to be found. Integrated control strategies, combining improved cultural practices, attempts to increase plant vigour, and the introduction of biological agents to reduce termite populations may hold some hope for the future. Among the biological agents, entomopathogenic fungi and nematodes seem the most suitable for research and development.

18.1 Introduction

By reason of their feeding habits and preferences for relatively undecayed living and dead plant material, about 10% of the approximately 2,700 described species of termites have been reported as pests (Wood 1996). The damage caused can be to cultivated plants, buildings, pastures and forests, and in addition even to non-cellulosic materials such as electric cables. Further, the larger epigeal mounds can hamper harvesting, whether this is mechanical or by hand (Renoux et al. 1991; Mora 1992). The species recorded as pests from the lower termites generally belong to four families, namely:

C. Rouland-Lefèvre (✉)
Unité de recherche Biodiversité et fonctionnement du sol, Institut de recherche pour le développement, 93143 Bondy Cedex, France
e-mail: Corinne.Rouland-lefevre@bondy.ird.fr

D.E. Bignell et al. (eds.), *Biology of Termites: A Modern Synthesis*,
DOI 10.1007/978-90-481-3977-4_18, © Springer Science+Business Media B.V. 2011

- the *Mastotermitidae*: this monotypic family is represented by a single species *Mastotermes darwiniensis*, found only in northern Australia and New Guinea, and principally attacks forest trees (Wood 1996), but also damages sugar cane plantations and fruit orchards (Ratcliffe et al. 1952).
- the *Kalotermitidae*: the approximately 350 species in this family are known as "drywood termites" (Sands 1973; Cowie et al. 1989) and can cause damage in long-term plantations such as tea, cocoa and a variety of fruits and coconut (Sands 1973) as well as in forestry (Harris 1969). Termites recorded as agricultural pests are listed by Harris (1971) and include drywood species in West Africa, South Asia, many Pacific Islands, as well as Central and South America. Eleven genera can include pests: *Kalotermes, Postelectrotermes, Neotermes, Rugitermes, Comatermes, Paraneotermes, Glyptotermes, Epicalotermes, Incisitermes, Bifiditermes* and *Cryptotermes* (Sands 1973).
- the *Hodotermitidae*: encountered mostly in semi-arid zones and prairies at temperate or sub-tropical latitudes (Wood 1996). The genera *Hodotermes, Microhodotermes* and *Anacanthotermes* have about 20 species.
- the *Rhinotermitidae*: this group of about 200 species are principally wood-feeders. The genus *Reticulitermes*, found at temperate latitudes in Europe, North Africa and North America, causes widespread damage to wood in service (Pearce 1997). The genus *Coptotermes* is the principal pest in this family in Asia, Australia and the United States, where damage can include plantations of eucalyptus and coconut palms (Harris 1954; French 1986; Mariau et al. 1992)

However, the termites which cause the most damage to tropical agriculture belong to three of the four subfamilies of higher termites (Family *Termitidae*):

- The *Termitinae*: a single pest genus, *Amitermes*, is particularly active in West Africa where *A. evuncifer* attacks crops such as yam, groundnut, manioc (cassava) and sugar cane, as well as young trees (Harris 1969; Sands 1977; Wood et al. 1980; Han et al. 1998). In India and Malaysia *Termes*, *Capritermes* and *Pericapritermes* attack the roots of crop plants and *Microcerotermes* the roots of tea and eucalyptus (Krishna and Weesner 1970; Srivastava and Butani 1987)
- The *Nasutitermitinae*: their distribution is very wide, both pantropical and subtropical, although they are perhaps best represented in the Neotropical fauna. The genera *Syntermes*, *Cornitermes*, and *Procornitermes*, encountered in South America, attack rice, sugar cane and young trees. Their nests can be subterranean or arboreal (Costa Lima 1939; Mathews 1977; Mill 1992).
- The *Macrotermitinae*: this group is particularly active in Africa, Southeastern Arabia, India and Southeast Asia (Wood 1996). The genera considered pests (of arable crops, trees and wood in service) are *Odontotermes*, *Microtermes*, *Macrotermes*, *Pseudacanthotermes* and *Ancistrotermes* (Harris 1969; Wood et al. 1980, 1987; Wood and Cowie 1988; Tiben et al. 1990; Renoux et al. 1991; Logan 1992; Rouland et al. 1993; Mora et al. 1996).

18.2 Damage to Tropical Crops

Most tropical crops are susceptible to termite attack, but only those situations where damage has a strong economic impact are cited here.

18.2.1 Forestry

18.2.1.1 Eucalyptus

In Brazil, the first reports of attack surfaced in the 1940s in the state of St. Paulo (Fonseca 1949). Numerous studies showed that Kalotermitidae, Rhinotermitidae and Termitidae were responsible (Berti-Filho 1993; Resende et al. 1995). The genera of Termitidae identified by Dietrich (1989) in eucalyptus plantations include *Aparatermes, Cylindrotermes, Embiratermes, Rhynchotermes, Obtusitermes* and *Subulitermes*, to which may be added *Heterotermes tenuis* and *Nasutitermes aquilicus,* which consume living trees (Torales 1998; Constantino 2002), while *Armitermes euamignathus*, *Procornitermes araujoi*, and *Neocapritermes opacus* attack the roots (Silva et al. 1968). Wilcken (1992) described attacks by *Cornitermes* sp. on *Eucalyptus grandis*, reporting that young trees are susceptible 34–76 days after planting, causing 18% mortality. Otherwise, *Syntermes molestus* is a principal pest species of eucalyptus plantations, aggressively stripping the base of the trunk and responsible for 70% of losses (Alves et al. 1997). *Cornitermes bequaerti* is reported as a principal factor limiting eucalyptus afforestation of the Cerrado, due to its destruction of roots (Wilcken and Raetano 1998).

Numerous studies in Africa have shown that termites have a significant impact on tree nurseries: three species, *Mactrotermes bellicosus*, *M. natalensis* and *Pseudacanthotermes militaris* are responsible for the majority of attacks in West Africa (Sands 1960; Brown 1965; Chilima 1991). Other species cited as causing up to 80% mortality in the 4–6 months following seedling planting are *Ancistrotermes amphidon, Microtermes* sp. and *Amitermes evuncifer* (Wardell 1987; Wood and Pearce 1991). In Zimbabwe, termite-related mortality of eucalyptus (*E. grandis* and *E. camaldulensis*) can even reach 100% (Harris 1971). Each termite species has its own mode of attacking woody tissue: *Macrotermes* spp., *Pseudacanthotermes* and *Ancistrotermes* spp. eat the outside of the roots, removing the cortex and the cambium, and from time to time destroying the belowground plant entirely (Mitchell 1989; Logan and El Bakri 1990). Some species of *Coptotermes* can also produce significant losses, but more readily in older plantations (Wardell 1987). Even in Australia, losses of eucalyptus to termites have been equally marked (Fox and Clark 1972; Andersen et al. 2005; Werner et al. 2008). *Odontotermes gurdaspurensis* and *O. obesus* are frequently cited as pests of eucalypts in India (Parihar 1981; Rajagopal 1982).

18.2.1.2 Coconuts and Palms

In Africa Aisagbonhi (1985) noted that 8–10% of coconut seednuts in a Nigerian agricultural institution nursery had been attacked by *Macrotermes bellicosus*. In

India, Abushama and Kambal (1977) reported attacks by *Microtermes traegardhi* and Reddy (1983) by *Odontotermes obsesus* and *O. bellahunisensis*. In Sumatra *Coptotermes curvignathus* is described as the principal pest of coconut planted in peat (Mariau et al. 1992), while in China unidentified species have a similar role (Tang et al. 2006). Date palms are also attacked, for example in India by *O. obesus* (Kranz et al. 1978), the termites consuming roots, the stems of young plants and even the trunks of older plants (Wood and Kambal 1984). In the Dongola region of Northern Sudan, Logan and El Bakri (1990) found that *Odontotermes smeathmani* was the sole termite associated with palms, but they attacked after 10–15 years when the plants had reached about 4 m in height, causing 70% losses. In Côte d'Ivoire young oil palms were heavily damaged by *Amitermes* and *Odontotermes* (Han et al. 1998).

18.2.1.3 Fruit Trees

Fruit trees are susceptible to termite attack, notably young mangos. Damage to bark, cambium and wood by *Ancistrotermes cavithorax* in Côte d'Ivoire was reported by Han and Ndiaye (1996), and attacks by 10 species of termites on mangos and lemons in Sénégal are described by Ndiaye and Han (2006). In a recent study of mangos close to the Thiés region of Sénégal, Assié (2008) found that termite attack by *Odontotermes* spp., *Microtermes* or *Coptotermes* was evident in all sites investigated. Mangos are also attacked in the south of China and in Taiwan by *Odontotermes formosanus* (Kranz et al. 1981), and by *O. wallonensis*, *O. obesus* and *O. horni* in India (Parihar 1981). In South America, the genera most frequently encountered damaging fruit trees are *Neotermes*, *Coptotermes* and *Nasutitermes* (Constantino 2002).

18.2.2 Sugar Crops

Sugar cane in Brazil has 200 species of known pests (Melo-Filho and Veiga 1997), amongst which the termites are *Cornitermes ovatus, Heterotermes tenuis, H. crinitus, Procornitermes* sp., *Syntermes molestus, S. grandis, Nasutitermes costalis, Coptotermes testaceus* and *Neocapritermes* sp. (Sands 1973; Novaretti and Carderan 1988; Berti-Filho 1993), attacking cuttings and roots of seedlings, or even penetrating the stem and killing the plant. Losses of yield can reach 10–0 t/ha/year (Novaretti and Fontes 1998). *Anoplotermes pacificus* devours growing roots; *Heterotermes* spp. attacks cane already infested with other species, excavating stems and roots in the manner of *Microcerotermes strunkii* and *Termes nigritus* (Mill 1992; Haifig et al. 2008). Harris (1969) states that 53 species of termites are noted as pests of sugar cane in Africa, with the earliest report (from Natal) that of Dick (1951). A study in Sudan by Abushama and Kambal (1977) reported that workers of *Microtermes traegardhi* attacked cane at three different stages of its development: as a cutting just after planting (found inside the plant tissues), at maturation (penetrating into the stem) and after harvesting (consuming the roots). Collins

(1984) examined the source of yield losses to termites in Nigeria and identified *Macrotermes subhyalinus* and *Odontotermes smeathmani* as the culprits. In Central Africa *Microtermes subhyalinus* was responsible for a 10% yield loss annually by attacking cane roots (Renoux et al. 1991; Mora et al. 1996), while in Tchad *Ancistrotermes guineensis* is known to build its nest at the base of cane stems, then consume the sweet parenchyma (Rouland et al. 1993). In India, termites are considered the most serious pests of sugar cane, for example in Bihar 60% of cuttings were destroyed by *O. obesus* (Parihar 1985). In Sri Lanka the principal pest is *O. redemanni* (Khumasinghe and Ranasinghe 1988), in China *O. formosanus* and in SE Asia *Macrotermes gilvus* (Kranz et al. 1981; Johnson et al. 1981).

18.2.3 Cereal Crops

18.2.3.1 Rice

Hempel in 1920 was the first to document termite attacks in rice plantations. The genera consistently mentioned in the Brazilian literature are *Syntermes*, *Cornitermes* and *Procornitermes,* attacking both foliage and roots of cereal crops generally, but especially rice, and *Anoplotermes*, *Grigiotermes*, *Heterotermes, Aparatermes* and *Procornitermes* are specifically mentioned in the context of rice (Sands 1973; Czepak et al. 1993). *Syntermes molestus* and *S. nanus* can destroy as much as 65% or even the whole crop in small plantations through the destruction of roots, so this genus is probably the major pest overall (Fonseca 1949; Mill 1992; Dario and Villela-Filho 1998)

18.2.3.2 Maize

In South America, cereal crops in the south of Brazil are threatened by *Syntermes molestus, Nasutitermes surinamensis, Anoplotermes pacificus* and *Procornitermes triacifer*, which attack the seeds and seedlings of maize (Mill 1992; Berti-Filho 1993; Figueira et al. 1998). *Cornitermes cumulans* is another pest of this crop (Cancello 1989) and on occasions can cause a 50% loss of yield (Fernandes and Alves 1992). In Ethiopia, maize is attacked by several species of *Microtermes* and *Ancistrotermes*, and occasionally *Macrotermes subhyalinus* will cut the stems (Cowie and Wood 1989). In Nigeria, maize is also attacked by *Microtermes*, which enters via the root and bores up into the stem (Wood et al. 1980). Population densities can reach to 4,000 m^{-2}, resulting in crop losses of up to 10%. In Zambia, maize is attacked by *Ancistrotermes latinosus*, *Macrotermes falciger*, *Pseudacanthotermes spiniger* and species of *Odontotermes*, causing up to 20% crop losses (Nkunika 1989, 1994).

18.2.3.3 Wheat

Wheat is attacked by *Microtermes thoracalis* in Northeast and Central Africa (Kranz et al. 1981), however in Ethiopia Cowie and Wood (1989) considered the damage to be minor, except during extended dry periods. Ripe wheat is attacked by *Microtermes* sp., but again with only small losses. A study in Pakistan identified *Microtermes obesi* and *Odontotermes obesus* as pests of wheat (Hashimi et al. 1983).

18.2.3.4 Sorghum

Sorghum is an important source of energy and protein for millions of people in Africa and Asia. While generally little attacked by termites, damage has been reported in Malawi with as much as 25% of the crop lost to *Odontotermes* sp., *Microtermes* sp. and *Macrotermes michaelseni* (Logan 1991).

18.2.4 Oil Crops

18.2.4.1 Groundnut

Sudan is the largest producer of groundnuts in the world, but the plant is attacked by *Microtermes thoracalis* and *M. lepidus,* and by *Macrotermes bellicosus*, and *M. natalensis* (El Amin et al. 1983). Losses were estimated as 20% of seed nuts and 10% each of mature plants and the nut harvest (Wood and Pearce 1991). *M. lepidus* is particularly voracious in Darfur, eating roots and holing the husks of the nuts to reach the interior. Other termites reported to attack the roots are *Amitermes evuncifer* and *Eremotermes* sp. (Pearce et al. 1995); however *Odontotermes obesus*, uniquely, consumes the foliage of the plant (Logan 1992). In Ethiopia, *Microtermes* causes a characteristic pathology: wilted plants with hollowed out roots and stems (Cowie and Wood 1989). In Nigeria as much as 88% of the pods may be lost following hollowing of the roots and scarification of the internal tissues by *M. lepidus* (Johnson and Gumel 1981; Johnson et al. 1981).

18.2.4.2 Castor Oil

Termite damage has been observed in India, where roots and stems are attacked by *Microtermes mycophagus*. Generally, the plants are most susceptible at the time of fruiting and least vulnerable during germination and growth (Parihar 1985).

18.2.5 Beverage Crops

18.2.5.1 Coffee

Roots of coffee bushes are vulnerable to termites. *Cornitermes cumulans*, *Procornitermes triacifer* and *P. striatus* are identified as pests in Brazil (Berti-Filho 1993; Neves and Alves 1999), *Macrotermes*, *Odontotermes* and *Microtermes* in Ethiopia (Cowie and Wood 1989), *Odontotermes obesus* in India and *O. formosanus* in China (Kranz et al. 1981).

18.2.5.2 Tea

Ten species of termites are noted as occurring in tea plantations in India, but only *Microtermes* seems to attack the living plants (Parihar 1981). In Tanzania, there are periodic attacks by *Pseudacanthotermes militaris*, and in Uganda by *Macrotermes natalensis*.

18.2.5.3 Cocoa

West Africa and Brazil are the centres of cocoa production. In Africa *Macrotermes bellicosus* is the principal pest (Cowie and Wood 1989), and in Brazil seven species are implicated, these being *Neotermes gestroi*, *Coptotermes testaceus*, *Cryptotermes cubioceps*, *Neotermes wagneri* and *Nasutitermes rippertii* (Mill 1992; Berti-Filho 1993).

18.2.6 Pastures

In Brazil *Cornitermes bequaerti*, *C. snyderi* and above all *C. cumulans* are serious pests of pastures (Furquim et al. 1968; Biondo et al. 1988; Mariconi et al. 1994; Fernandes et al. 1998; Cruvinel et al. 1998). The *Brachiaria decumbens* pasture is infested by *Syntermes wheeleri*; *Cornitermes* and *Procornitermes* may also be damaging (Sands 1973; Moura et al. 1998). In Africa, the large pasturing zones in the west of Ethiopia are subject to degradation by *Macrotermes*, *Odontotermes* and *Pseudacanthotermes* which take away both standing grass and lying litter in the wake of cattle grazing (Cowie and Wood 1989). In Nigeria a similar role is attributed to *Trinervitermes geminatus* (Ohiagu 1979) and in South Africa to *Hodotermes mossambicus* (Coaton 1954; Nel and Hewitt 1969).

18.2.7 Tuber Crops

18.2.7.1 Yam

Yam belongs to the genus *Dioscorea* (Coursey 1967), which includes about 600 species. Edible species are cultivated in West Africa, the Caribbean and SE Asia. In Africa, it is regularly attacked at the cutting stage by *Ancistrotermes* and *Microtermes*, with up to 70% losses (Sands 1973; Atu 1993).

18.2.7.2 Manioc (Cassava)

This perennial herbaceous plant grows all over the tropics, but especially in the warmer lowland plains, where it is a starchy staple for subsistence farmers, who may on average eat more than 100 kg of the tuber per year (Phillips 1983). Manioc is attacked at all stages of its development, as seedlings when planted in wet soil and as a mature plant in the dry season (Bandial 1992). *Ancistrotermes*, *Microtermes*, *Odontotermes* and *Pseudacanthotermes* are all implicated in damage to mature plants by excavating the stems at ground level (Sands 1973).

18.2.8 Market Gardens

Tomatoes are attacked by *Microtermes thoracalis* in Northeast and Central Africa (Kranz et al. 1981). In Yemen, Wood et al. (1987) identified pepper, okra and tomato as susceptible to *Microtermes naidensis*, with losses up to 20%, though less so than cotton and maize. In Sénégal, aubergine, tomato and okra were principally attacked by *Microtermes* spp., but also by *Psammotermes hybostoma*, *Microcerotermes* sp. and *Odontotermes nilensis* (Han and Ndiaye 1998).

18.2.9 Fibre Crops

Cotton is a notable crop in the Delta Tokar region of Sudan, but is damaged by *Microtermes naidensis* and *M. thoracalis*, which attack both roots and stems (Tiben et al. 1990), with losses to eventual yield as much as 25% (Harris 1968; Schmutterer 1969). *Microtermes* excavates the stems of older cotton, filling them with earth which then leads to the death of the plants. In Tanzania, the young flush of leaves can be eaten by *Hodotermes mossambicus*, and mature plants are threatened by *Microtermes, Amitermes* and *Microcerotermes,* which excavate the stem, causing the plant to wilt (Harris 1954). In Central Africa *Ancistrotermes latinotus*, *Odontotermes badius* and *Allodontermes* sp. can destroy as many as 3,000 plants/ha (Sands 1960), while in SE Asia the notable pest is *Odontotermes obesus* (Kranz et al. 1981).

18.3 Chemical Control

18.3.1 Use of Insecticides

In the tropics, control of pest termites has usually occasioned the use of powerful persistent organochlorine insecticides such as dieldrin and aldrin (Harris 1969; Wood and Pearce 1991). These were effective, for example the treatment of cotton with aldrin to protect the crop in the Sudan against *Microtermes naidensis* increased yield by 132% (Tiben et al. 1990) and in Tanzania the treatment of soy grains with dieldrin reduced attacks by *Microtermes albopartitus* and *Ancistrotermes latinotus* (Bigger 1966). However the use of these compounds has been restricted in the USA since the 1970s, and the evidence of toxicity to mammals and damaging environmental effects resulted in a prohibition everywhere in 1985 (Wood and Pearce 1991). To replace them, organophosphates (malathion, chlorpyriphos, dichlorvos), carbamates (aldicarb, carbosulfan) and chlorinated phenols (pentachlorophenol) have been brought forward to the market (Wardell 1987) and successes are noted. For instance, chlorpyriphos is more effective than aldrin for protecting groundnuts against *Odontotermes obesus* and doubles the yield (Logan 1992). In Malawi, carbosulfan reduced mortality due to *Macrotermes*, *Odontotermes* and *Ancistrotermes* by 34% in a eucalyptus plantation (Chilima 1991).

In South America, aldrin, dieldrin and telodrine have been used to treat the nests of *Cornitermes* and *Syntermes* (Sands 1973; Amante 1962; Furquim et al. 1968; Nakano and Prospero 1969). Since the restriction of these three products the pyrethroids and organophosphates have made their appearance (Loek and Nakano 1988). Wilcken (1992) showed the efficacy of a synthetic pyrethroid, tefluthrin, against *Cornitermes cumulans*, and application in a *Eucalyptus grandis* plantation destroyed all termite colonies (Mariconi et al. 1994). Heptachlor and thiodan protected sugar cane against termites (Novaretti and Carderan 1988), while benfuracarb eradicated most of the populations of *Syntermes molestus* and *Procornitermes triacifer* in a millet plantation (Figueira et al. 1998). However the new insecticides do not possess the persistence that is important in tropical conditions (Mauldin et al. 1987; Wood et al. 1987; Mampe 1988; Horwood 2007).

Other persistent insecticides such as the phenyl pyrazole fipronil (Bobe et al. 1998; Sharma et al. 2008) or thiamethoxam (Maienfisch et al. 2001) have come on stream during the 1990s. These two compounds were tested in the laboratory against several African termites including *Odontotermes nilensis, Macrotermes subhyalinus, O. smeathmani* (fungus-growers) *Trinervitermes trinervius* and *Amitermes evuncifer, Nasutitermes costalis, Syntermes molestus* and *Cornitermes snyderi* (wood-feeders). The compounds were very toxic for a group of species and not restricted biogeographically, furthermore their action is quite slow, permitting transmission of the toxin between individuals, but in certain formulations (Termidor or Regent for fipronil, Actara for thiamethoxam) there is some evidence of repulsion from baits or a phagodepressant effect in some species (Delgarde and Rouland 2002a, 2002b). However, for treating seedlings in sugar cane plantations, fipronil is effective (Mampouya 1996; Delgarde and Rouland 2001).

18.3.2 Chitin Synthesis Inhibitor

The efficacy of hexaflumuron, a chitin inhibitor insecticide, has been studied on the fungus-growing termite *Pseudacanthotermes spiniger*, a pest of sugar cane plantations in the Congo. It showed potent larvicidal action in the termite, but did not act as a fungicide and growth of *Termitomyces* was not inhibited (Peppuy et al. 1998). Hexaflumuron baits were tested against *Reticulitermes flavipes*, a pest of young citrus orchards in Florida, by Stansly et al. (2001), and termite activity was eliminated within 2–3 months.

18.3.3 Fungicides

Fungicides targeted at the symbionts of Macrotermitinae (the basidiomycete fungus *Termitomyces)* were tested by El Bakri et al. (1989) and Wardell (1990). Death of the symbiont blocks the assimilation of foraged food by the termite and the whole colony dies. Erpacide® 450T and 490T were effective against *Pseudacanthotermes spiniger* and *Microtermes subhyalinus* (Mora 1992), while Bavastin and Sumi 8, when trialed against the *Termitomyces* symbiont of *Ancistrotermes guineensis*, completely inhibited its growth. A fungicidal control measure based on broadcasting these two products was subsequently put in place in the sugar cane plantations of SONASUT (Societé Nationale Sucrière du Tchad) against the same termite, and destroyed about 20% of colonies (Rouland et al. 1993; Rouland-Lefèvre and Mora 2002)

18.4 Control by Non-Chemical Means

18.4.1 Cultural Methods

Healthy plants can sometimes be damaged by termites, but those that are infected or stressed are generally more susceptible (Cowie and Wood 1989). Drought, diseases, wounding, weeds, lack of fertilizer and poor husbandry can all assist the attacks of termites (Logan and El Bakri 1990). Further, cropping and tree plantation increasingly extends into marginal zones, making use of unsuitable ground (Black and Okwakol 1997). Cultural control comprises everything that contributes to acquiring or maintaining plant vigour, but can also include reducing the number of termites or modifying their behaviour.

18.4.1.1 Practices Improving Plant Vigour

Good seed quality, planting methods and plantation conditions all affect plant health. For example, in Uganda seedlings not planted upright are more likely to be attacked by termites (Logan et al. 1990). In forestry, the use of polythene tubes to protect the root bunches and keep them moist until planting out also reduces subsequent stress and increases resistance to termites. Rates of attack can depend on hydrological conditions; high rainfall favours termite attack on tea and groundnuts through increased foraging, and also promotes mildew, producing even greater losses (Reddy and Ghewande 1986). On the other hand, evidence from cotton, groundnut and eucalyptus plantations shows that vulnerability to termites is increased in dry zones, where drought conditions prevail or between irrigations (Cowie and Wood 1989). Lack of irrigation was also suggested as a cause of new termite attacks on date palms in the Sudan (Logan and El Bakri 1990) and on cane sugar in Sri Lanka (Khumasinghe and Ranasinghe 1988). Attacks on *Eucalyptus* spp. in Sénégal were negatively correlated with the sap content of the trees (Lepage 1983). In low-rainfall regions, irrigation can reduce water stress and thereby lessen termite attack. In India, such irrigations have reduced damage to wheat and rice (Verma 1980; Sharma et al. 2004).

Intensive cultivation, and in particular monoculture, reduces fertility and modifies soil structure (Black and Okwakol 1997). Under these conditions studies have shown that plants are less vigorous and more susceptible to termite attack, for example maize in Ethiopia and trees in Zambia (Cowie and Wood 1989; Sileshi et al. 2005). Rotation of crops, including a period of fallow to remedy changes to soil structure, is therefore recommended (Sekamatte et al. 2003). Competition from weeds, low light levels and low water availability are all stress factors for crops and young trees, which equally increase susceptibility to termite attack (Black and Okwakol 1997). Fertiliser can restore vigour and reduce vulnerability (Wardell 1990).

18.4.1.2 Practices to Reduce Termite Numbers

Some cover crops such as *Pueraria* seem to eliminate termites after use as a fallow (Renoux et al. 1991). Plant extracts or mineral preparations may also be useful; adding the chopped leaves of *Euphorbia tirucalli, Aloe graminicola, Melia azedarach,* and *Lippia javanica* to the planting holes, and mulching with *Cassia siamea* or *Azadirachta indica* can all be recommended (Wardell 1990; Verma et al. 2009).

18.5 Biological Control

Biological control comprises the manipulation of predators, pathogens or parasites with the objective of reducing a pest population to an economically acceptable level. The natural predators of termites are generally larger animals, both invertebrate and vertebrate, for example alates are taken by birds (Thiollay 1970) and foraging workers by scorpions, spiders, cockroaches, crickets and bees (Wood 1978). However the principal predators of termites are ants, and on this basis Sekamatte et al. (2001) baited ants in Uganda with sugar and meat to encourage their populations and thereby reduced both termite populations and subsequent damage to maize. Despite this diversity of predators, most work on biological control has focused on microbial pathogens: viruses, bacteria, fungi, protists and nematodes.

18.5.1 Viruses

There are few studies on the use of termite viruses in control, however Ahlam et al. (1988) found that a virus which can be isolated from the cotton leafworm *Spodoptera littoralis* can also infect termites. This may be an avenue for future research, especially since infection with viruses is easier to achieve than for other biological agents (Pearce 1987).

18.5.1.1 Bacteria

The pathological character of the spore of *Bacillus thuringiensis* has been much studied (Khan et al. 1985; Pearce 1997), but there have been few trials with termites under natural conditions. In fact the spores of the bacillus show poor survival in the soil and seem unsuitable as agents of biological control (Burges 1981). Other bacteria, notably *Serratia marcescens* are pathogenic for many termites (Khan et al. 1985), but trials have not been effective. Three different species of HCN-producing rhizobacteria *Rhizobacterium radiobacter, Alcaligenes latus* and *Aeromonas caviae* were effective in killing *Odontotermes obesus* under in vitro conditions (Devi et al. 2006).

18.5.2 Fungi

Metarhizium anisopliae and *Beauveria bassiana* seem quite effective against termites. The spores can be introduced into nests placed in baits or applied as a powder either to the nests or to the termites directly (Starnes et al. 1993; Sun et al. 2003). Many studies have been conducted: for example *Beauveria bassiana* and *Metarhizium anisopliae* have been very effective against *Cornitermes cumulans,* the principal termite pest of agriculture in Brazil (Berti-Filho 1993; Neves and Alves 1999). The pathogenicity of *M. anisopliae* was also tested on seven termite pest species: *Mastotermes darwiniensis, Hodotermopsis sjoestedti, Hodotermes mossambicus, Kalotermes flavicollis, Reticulitermes flavipes, Prorhinotermes canalifrons* and *Nasutitermes voeltzkowi*. The results showed that susceptibility to *M. anisopliae* is influenced by the termite's nesting ecology (Chouvenc et al. 2009). Ivermectin, a metabolite produced by *Streptomyces avermitilis,* decreases the food consumption of *C. formosanus* workers (Mo et al. 2006)

18.5.3 Protists

Protists pathogenic in termites are few in number and are not generally considered suitable agents for biological control (Jafri et al. 1976).

18.5.4 Nematodes

Entomopathogenic nematodes are also utilised for biological control. *Nasutitermes costalis*, infested by *Steinernema feltiae*, showed 100% mortality after only 3 days (Laumond et al. 1979). The strains *Steinernema carpocapsae* K27 and *S. kushidai* E2 have clearly shown their potential against the reproductives of several species of Macrotermitinae: *Ancistrotermes guineensis*, *Pseudacanthotermes spiniger*, *Odontotermes* sp. and *Macrotermes bellicosus* (Rouland et al. 1996). All castes are susceptible to infection, but the life cycle of the nematode can only be completed in the alates (Benmoussa-Haïchour et al. 1998).

18.6 Conclusions

Attacks by termites on crops are a continuing problem in the tropics and not least in countries where agriculture is the principal economic resource. There is an urgency to find effective methods of control which do not involve the use of organochlorine or organophosphate insecticides, since these are either prohibited or have severe restrictions, owing to their various toxicities. But currently there is no adequate substitute. Amongst non-chemical approaches, cultural methods such as short cropping cycles, early harvesting, deep tillage and the application of fertilizer can all be utilized, but none offers a guarantee of lasting protection against attack, so recourse to a chemical application carefully selected to target the foraging behaviours of the termites in question remains necessary.

Development of a termite-specific control which takes advantage of their biology and accommodates the many concerns over the use of chemicals in the natural environment (persistence, collateral selection for resistance) should nevertheless be possible (Fig. 18.1). Beyond this, integrated and interdisciplinary control strategies would include the development of resistant crop cultivars, the encouragement of termitophagous faunas, preventative cultural practices and the deployment of repulsive or phagodeterrent products perhaps based on termite-specific pheromones or plant extracts.

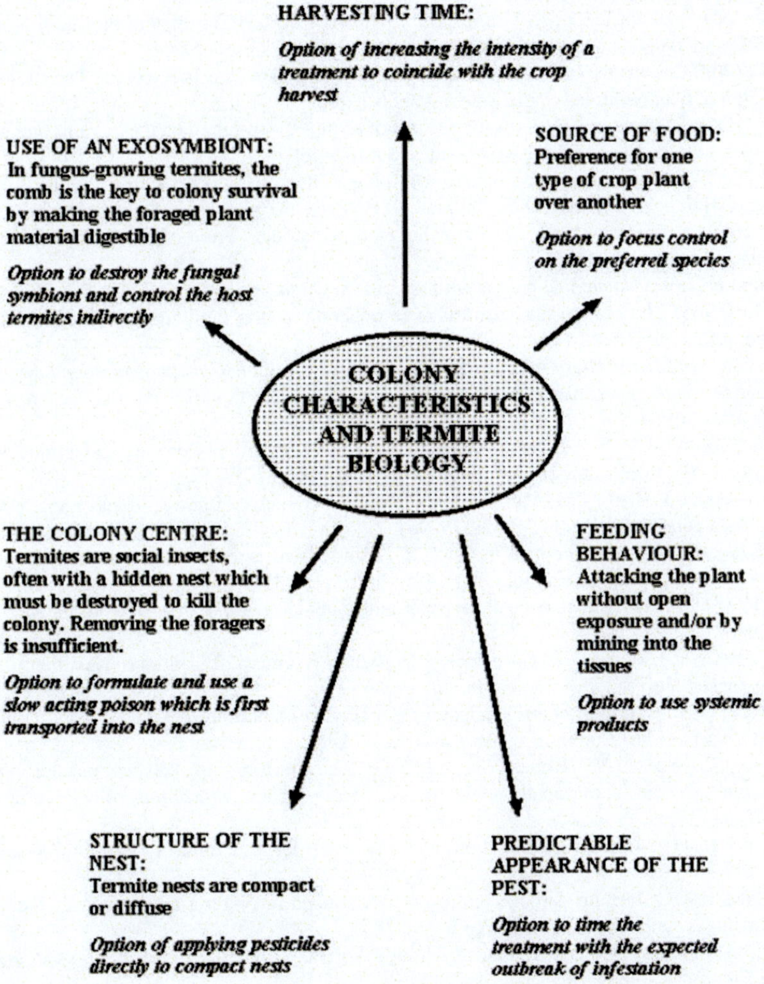

Fig. 18.1 Factors to be taken into account for termite control in tropical agriculture

References

Abushama FT, Kambal MA (1977) The role of sugars in the food-selection of the termite *Microtermes traegardhi* (Sjost). Z Angew Entomol 84:250–255

Ahlam A, Fazairy AL, Hassan FA (1988) Infection of termites by *Spodoptera littoralis* nuclear polyhedrosis virus. Insect Sci Appl 9:37–39

Aisagbonhi CI (1985) A survey of the destructive effect of *Macrotermes bellicosus* Smeathman (Isoptera: Termitidae – Macrotermitinae) on coconut seednuts at NIFOR, Benin, Nigeria. Int J Pest Manage 35:380–381

Alves AN, Wilcken CF, Raetano CG (1997) Contrôle de cupins subterrâneos (Isopter) em plantios de eucalipto com imidacloprida e dissulfoton mais triadimenol. Congresso Brasileiro de Entomologia 16, Salvador, BA, p 253

Amante E (1962) Ensãio de campo para controlar o cupim de monticule *Cornitermes cumulans*. Archos Ins Biol São Paulo 29:133–138

Andersen AN, Jacklyn P, Dawes-Gromadzki T, Morris I (2005) Termites of Northern Australia. CSIRO and Tropical Savannas CRC. Barker, Alice Springs

Assié B (2008) Les termites ravageurs des vergers de manguiers dans la région de Thiès (Sénégal). Report M2 Bioressources (Paris XII University), 38p

Atu UG (1993) Cultural practices for the control of termite (Isoptera) damage to yams and cassava in South-eastern Nigeria. Int J Pest Manage 39:462–466

Bandial (1992) Rapport de mission IRD (juin) Lobaye-Ombella-m'poko-Kémo, 26 pp

Benmoussa-Haîchour D, Reversat G, Rouland C (1998) Sensitivity of the different castes of higher termites to an infection with entomopathogenic nematodes – role of the lipids composition. Actes Coll Insect Soc 11:45–52

Berti-Filho E (1993) Manual de pragas em florestas. 3. Cupins ou térmitas. IPEF/SIF, Viçosa

Bigger M (1966) The biology and control of termites damaging field crops in Tanganyika. Bull Entomol Res 56:417–444

Biondo CJ, Geraldi FI, Clari AI et al (1988) Cupim de monte *Cornitermes cumulans* (Kollar 1832): combate experimental com formulações inseticidas líquidae. An Esc Sup Agr "Luiz de Queiroz"45:91–97

Black HIJ, Okwakol MJN (1997) Agricultural intensification, soil biodiversity and agroecosystem function in the tropics: the role of termites. Appl Soil Ecol 6:37–53

Bobe A, Cooper J, Coste CM, Muller M-A (1998) Behaviour of fipronil in soil under Sahelian plain field conditions. Pestic Sci 52:275–281

Brown KW (1965) Termite control research in Uganda (with special reference to the control of attacks in *Eucalyptus* plantations). East Afr Agr Forest J 31:218–223

Burges HD (1981) Microbial control of pests and plant diseases 1970–1980. Academic Press, London

Cancello E (1989) Revisão de *Cornitermes* Wassman (Isoptera, Termitidae, Nasutitermitinae). Ph.D. thesis, University of São Paulo, São Paulo

Chilima CZ (1991) Termite control in young *Eucalyptus* plantation in Malawi using controlled release insecticides. Common Forest Rev 70:237–247

Chouvenc T, Su N-Y, Robert A (2009) Susceptibility of seven termite species (Isoptera) to the entomopathogenic fungus *Metarhizium anisopliae*. Sociobiology 54: 723–748

Coaton WGH (1954) Veld reclamation and harvester termite control. Farming S Afr 29: 243–248

Collins NM (1984) Termite damage and crop loss studies in Nigeria-assessment of damage to upland sugar cane. Trop Pest Manage 30:26–28

Constantino R (2002) The pest termites of South America: taxonomy, distribution and status. J Appl Entomol 126:355–365

Costa Lima AM (1939) Termites. In: Insetos do Brasil, vol 1. Escola Nacional de Agronomia, Rio de Janeiro, pp 263–327

Coursey DG (1967) Yams. An account of the nature, origins, cultivation and utilisation of the useful members of the *Dioscoreaceae*. Tropical Agricultural Series Longmans, Green and Co Ltd, London

Cowie RH, Wood TG (1989) Damage to crops, forestry and rangeland by fungus-growing termites (Termitidae: Macrotermitinae) in Ethiopia. Sociobiology 15:139–153

Cowie RH, Logan JMW, Wood TG (1989) Termite (Isoptera) damage and control in tropical forestry with special reference to Africa and Indo-Malaysia: a review. Bull Entomol Res 79:173–184

Cruvinel IR, Leandro Junior G de M, Alves GMR et al (1998) Comparação de doses e formas de aplicação de gastoxin (fosfeto de alumínio) em pastilhas de 0,6 g, no controle de colônias de *Cornitermes snyderi* em pastagens. XVII Congresso Brasileiro de Entomologia -Resumos- 9 a 14 de agosto de 1998, Rio de Janeiro. p 236

Czepak C, Ferreira E, Nogueira SB (1993) Identifição e quantificação de cupins rizófagos na cultura do arroz de sequeiro. Pesq Agropec Bras 28:871–875

Dario GJA, Villela-Filho RS (1998) Imidacloprid – none insecticida para o controle do cupim (*Syntermes molestus*) na cultura do arroz (*Oryza sativa* L) In: XVII Congresso Barieiro de Entomologia Resumos, pp 206–222

Delgarde S, Rouland-Lefèvre C (2001) Effects of fipronil on *Odontotermes nilensis* (Emerson) (Termitidae-Macrotermitinae) pest of sugarcanes in Chad. Sociobiology 37:539–550

Delgarde S, Rouland-Lefèvre C (2002a) Evaluation of the effects of thiamethoxam on the three species of African termite (Isoptera: Termitidae) crop pests. J Econ Entomol 95:531–536

Delgarde S, Rouland-Lefèvre C (2002b) Efficacy of Actara 25WG on three species of Brazilian termite (Isoptera: Nasutiterminae) crop pests. Sociobiology 40:669–710

Devi KK, Seth N, Kothamasi S, Kothamasi D (2006) Hydrogen cyanide-producing rhizobacteria kill subterranean termite *Odontotermes obesus* (Rambur) by cyanide poisoning under in vitro conditions. Curr Microbiol 54:74–78

Dick J (1951) Insect pests in Natal cane. S Afr Sug J 26:99–100

Dietrich CRRC (1989) Ocorrência de cupins (Insecta: Isoptera) em reflorestamento de *Eucalyptus* spp. M.Sc. thesis, Universidade de São Paulo, Piracicaba

El Bakri A, Eldein N, Kambal MA et al (1989) Effect of fungicide impregnated food on the viability of fungus combs and colonies of *Microtermes* sp. nr *albopartitus* (Isoptera: Macrotermitinae). Sociobiology 15:175–180

El Amin EM, Ishag HM, Burhan HO (1983) Important factors affecting the yield of groundnuts (*Arachis hypogaea* L) in the Sudan. Z Angew Zool 70:29–55

Fernandes PM, Alves SB (1992) Preferência alimentare e danos de *Cornitermes cumulans* (Kollar 1832) (Isoptera: Termitidae) às plantas cultivadas em labaoratório. An Soc Entomol Brasil 21:125–132

Fernandes PM, Czepak C, Veloso RS et al (1998) Cupins de montículos em pastagens: prejuizo real ou praga estética? In: Fontes LR, Berti-Filho E (eds) Cupins: desafio do conhecimento. FEALQ, Picacicaba, pp 187–210

Figueira FAV, Shitara T, da Silva AL (1998) Controle químico dos cupins *Syntermes molestus* e *Procornitermes triacifer* via tratamento de sementes na cultura do milho. XVII Congresso Brasileiro de Entomologia -Resumos- 9 a 14 de agosto de 1998, Rio de Janeiro, p 101

Fonseca (1949) Experiências de combate químico a cupins subterrâneos no Horta Floresta de Guarani. Acq Inst Biol Brasil 19:57–88

Fox RE, Clark NB (1972) The incidence of termites in eucalypts of the Darwin area. Aust Forest Res 5:29–36

French JRJ (1986) Termites and their economic importance in Australia. In: Vinson SB (ed) Economic impact and control of social insects. Praeger, New York, NY, pp 103–129

Furquim MR, Kamizono Y, Andrade SC et al (1968) Combate experimental ao cupim *Cornitermes cumulans* (Kolar). O Solo 60:57–62

Haifig I, Costa-Leonardo AM, Marchetti FF (2008) Effects of nutrients on feeding activities of the pest termite *Heterotermes tenuis* (Isoptera: Rhinotermitidae). J Appl Entomol 132:497–501

Han SH, Ndiaye AB (1996) Dégâts causés par les termites (Isoptera) sur les arbres fruitiers dans la region de Dakar (Sénégal). Actes Coll Insect Soc 10:111–117

Han SH, Ndiaye AB (1998) L'attaque des cultures maraîchères par les termites (Isoptera) dans la région de Dakar (Sénégal). Actes Coll Insect Soc 11:37–43

Han SH, Tokro GP, Tano Y, Lepage M (1998) Dégâts des termites dans les plantations de palmiers à huile en Côte D''Ivoire: evaluation et méthode de lutte. Plantations Rech Développe 5:119–126

Harris WV (1954) Termites and tropical agriculture. Trop Agric Trin 31:11–18

Harris WV (1968) Termites of the Sudan. Sudan Nat Hist Bull 4:1–29

Harris WV (1969) Termites as pests of crops and trees. Commonwealth Agricultural Bureau, London

Harris WV (1971) Termites, their recognition and control. Longman, London

Hashmi AA, Hussain MM, Itaf M (1983) Insect pest complex of wheat crop. Pak J Zool 15: 169–176

Horwood MA (2007) Rapid degradation of termiticides under field conditions. Aust J Entomol 46:75–78

Jafri RH, Ahmad M, Idrees K (1976) Microsporidian infection in the workers of termite *Microcerotermes championi*. Pak J Zool 8:234–236

Johnson RA, Gumel MH (1981) Termite damage and crop loss studies in Nigeria – the incidence of termite-scarified groundnut pods and resulting kernel contamination in field and market samples. Int J Pest Manage 27:343–350

Johnson RA, Lamb WA, Wood TG (1981) Termite damage and crop loss studies in Nigeria – a survey of damage to groundnuts. Trop Pest Manage 27:325–342

Khan K, Jafri RH, Ahmad M (1985) The pathogenicity and development of *Bacillus thuringiensis* in termites. Pak J Zool 17:201–209

Khumasinghe NC, Ranasinghe MASK (1988) Incidence of termite damage in sugar cane grown in Sri Lanka. Beitr Trop Landwirt 26:303–307

Kranz J, Schmutterer H, Koch W (1978) Diseases, pests and weeds in tropical crops. Wiley, London

Kranz J, Schmutterer H, Koch W (1981) Maladie, ravageurs et mauvaises herbes des cultures tropicales. Termites, fourmis blanches. Verlag Paul Parey, Berlin et Hambourg pp 304–306

Krishna K, Weesner FM (eds) (1970) Biology of termites (Vol 2). Academic Press, New York, NY

Laumond C, Mauleon H, Kermarrec A (1979) New data on the host spectrum and the parasitism of the entomophagous nematode, *Neoplectana carpocapsae*. Entomophaga 24:13–27

Lepage M (1983) Structure et dynamique des peuplements de termites tropicaux. Acta Oecol 4: 65–87

Loek AE, Nakano O (1988) Persistência de dois insecticidas piretroides num solo barro-areno-argiloso em condições de campo. Pesqui Agropecu Bras 23:709–715

Logan JMW (1991) Damage to sorghum by termites (Isoptera: Macrotermitinae) in the Lower Shire Valley, Malawi. Sociobiology 19:305–307

Logan JWM (1992) Termites (Isoptera): a pest or resource for small farmers in Africa?. Trop Sci 32:71–79

Logan JWM, Cowie RH, Wood TG (1990) Termite (isopteran) control in agriculture and forestry by non-chemical methods: a review. Bull Entomol Res 80:309–333

Logan JMW, El Bakri A (1990) Termite damage to date palms (*Phoenix dactylifera* L) in Northern Sudan with particular reference to the Dongola District. Trop Sci 99:1363–1368

Maienfisch P, Angst M, Brandl F et al (2001) Chemistry and biology of thiamethoxam: a second generation neonicotinoid. Pest Manage Sci 57:906–913

Mampe CD (1988) Making the right choice. Pest Control 56:34–36

Mampouya D (1996) Les populations de termites d'une culture de canne à sucre irriguée dans un écosystème sahélien; le fipronil dans une lutte ciblée contre les termites. University thesis, Université Paris XII Val de Marne, Créteil

Mariau D, Renoux J, de Chenon RD (1992) *Coptotermes curvignathus* Holmgren, Rhinotermitidae, main pest of coconut planted on peat in Sumatera. Oleagineux 47: 562–568

Mariconi FAM, Galan VB, Rocha MT (1994) Ensaios de combate ao cupim de monte *Cornitermes cumulans* (Kollar 1832) (Isoptera, Termitidae). Sci Agricola 51:505–508

Mathews AGA (1977) Studies on termites from the Mato Grosso. Ph.D. thesis, University of Edinburgh, Edinburgh

Mauldin JK, Jones SC, Beal RH (1987). Soil termiticides: a review of efficacy data from field tests. Int Res Group Wood Preservation, Doc. No. IRG/WP/1323. Secretariat, Stockholm, Sweden.

Melo-Filho RM, Veiga AF de SL(1997) Eficiência do fipronil no controle do cupim de montículo, *Nasutitermes sp.* (Isoptera: Termitidae), em cana-de- açúcar. Anais do 16° Congresso Brasileiro de Entomologia -Resumos- 02 a 07 de março de 1997-Salvador-BA, p 156

Mill AE (1992) Termites as agricultural pests in Amazonas, Brazil. Outlook Agr 21:41–46

Mitchell MR (1989) Susceptibility to termite attack of various tree species planted in Zimbabwe. In: Roland DJ (ed) Trees for the tropics, vol 10. Australian Centre for International Agricultural Research, Canberra, Monograph, pp 215–226

Mo J, Wang Z, Song X et al (2006) Effects of sublethal concentrations of ivermectin on behaviors of *Coptotermes formosanus* (Isoptera: Rhinotermitidae). Sociobiology 47:687–696

Mora P (1992) Dégâts des termites champignonnistes (Macrotermitinae) *Pseudacanthotermes spiniger* et *Microtermes subhyalinus* dans les plantations de canne à sucre. Ph.D. thesis. University of Paris XII, Val-de-Marne

Mora P, Rouland C, Renoux J (1996) Foraging and nesting damage caused by *Microtermes subhyalinus* (Isoptera: Termitidae) in a sugar cane plantation in the Central African Republic. Bull Entomol Res 86:387–395

Moura E, Fernandes PM, Czepak C et al (1998) Eficiência do inseticida Regent 20G (fipronil) em diferentes doses e formas de aplicação, no controle de colônias de *Syntermes wheeleri* em pastagens. XVII Congresso Brasileiro de Entomologia -Resumos- 9 a 14 de agosto de 1998, Rio de Janeiro, p 486

Nakano O, Prospero AO (1969) Efeito comparativo de diversos inseticidas no controle do cupim de montículo *Cornitermes cumulans* (Kollar, 1832). Revi Agricultura Brasil 44:98–102

Ndiaye B, Han SH (2006) L'attaque des arbres fruitiers par les termites dans la region de Thiès (Sénégal). Bull Soc Entomol Fr 111:59–64

Nel JJC, Hewitt PH (1969) A study of the food eaten by a field population of the harvester termite *Hodotermes mossambicus* (Hagen) and its relation to population density. J Entomol Soc South Afr 32:123–131

Neves POJ, Alves SB (1999) Controle associado de *Cornitermes cumulans* (Kollar 1832) (Isoptera: Termitidae) com *Metarhizium anisopliae, Beauveria bassiana* e imidacloprid. Sci Agric 56:313–319

Nkunika POY (1989) Termites: a dilemma for the small scale farmer. Prod Farming (August), Zambia National Farmers Union, Lusaka, pp 7–10

Nkunika POY (1994) Control of termites in Zambia: practical realities. Insect Sci Appl 15:241–245

Novaretti WRT, Carderan JO (1988) Controle químico de cupins em canade-açúcar. An III Seminario Technologia Agronomica, pp 111–114

Novaretti WRT, Fontes LR (1998) Cupins: uma grave ameaça à cana-de-açúcar no nordeste do Brasil. In: Fontes LR, Berti-Filho E (eds) Cupins: odesafio do conhecimento. FEALQ, Piracicaba, pp 163–172

Ohiagu CE (1979) A quantitative study of seasonal foraging by the grass harvesting termite *Trinervitermes geminatus* (Wasmann) (Isoptera, Nasutitermitinae), in Southern Guinea Savanna, Mokwa, Nigeria. Oecologia 40:179–188

Parihar DR (1981) Termites affecting eucalyptus plantations and their control in the arid régions of India. Z Angew Entomol 92:106–111

Parihar DR (1985) Crop termite damage in Indian desert and its control in castor. Z Angew Zool 72:309–315

Pearce MJ (1987) Seals, tombs, mummies and tunneling in the drywood termite *Cryptotermes* (isopteran: Kalotermitidae). Sociobiology 13:217–226

Pearce MJ (1997) Termites, biology and pest management. CAB International, Wallingford

Pearce MJ, Logan JWM, Tiben A (1995) Termites (Isoptera) from the Darfur region of the Sudan with comments on their pest status. J Arid Environ 30:197–206
Peppuy A, Robert A, Delbecque JP et al (1998) Efficacy of hexaflumuron against the fungus-growing termite *Pseudacanthotermes spiniger* (Sjöstedt) (Isoptera, Macrotermitinae). Pestic Sci 54:22–26
Phillips TP (1983). An overview of cassava consumption and production. In: Delange F, Ahluwalia R (eds) Cassava toxicity and thyroid research. Proceedings of a Workshop, Ottawa, 1982 (International Development Research Centre Monograph 207e), Ottawa, Canada, International Development Research Centre, pp 83–88
Rajagopal D (1982) Relative incidence of termites on exotic species of *Eucalyptus* in India and their control. Z Angew Entomol 98:225–230
Ratcliffe FN, Gay FJ, Greaves T (1952) Australian termites. CSIRO, Melbourne
Reddy MV (1983) First record of *Odontotermes bellahunisensis* Holmg. feeding on *Cocos lucifera* Linn. J Bomb Nat Hist Soc 80:653–656
Reddy PS, Ghewande MP (1986) Major insect pests of groundnut and their management. Pesticides 20:52–56
Renoux J, Rouland C, Mora P, Dibangou V (1991) Les constructions du termite *P. spiniger* dans les champs de canne à sucre. Essai de lutte spécifique. AFCAS R Int Can Suc 1:124–130
Resende VF, Zanuncio JC, Buedes NC et al (1995) Efeitos comparativos do carbosulfan e do aldrin na proteção de mudas de eucaliptos contra cupins subterrâneos. An Soc Entomol Brasil Londrina 24:645–648
Rouland C, Benmoussa D, Reversat G, Laumond C (1996) Étude de la sensibilité de sexués de termites Macrotermitinae à une infestation par des nématodes entomopathogènes des genres *Heterorhabditis* et *Steinernema*. C R Acad Sci III Vie 319:997–1001
Rouland C, Ikhouane A, Nayalta N (1993) Etude biologique des populations d'*Ancistrotermes guineensis* présentes dans les plantations de la SONASUT. Act Coll IUSSI 8:79–87
Rouland-Lefèvre C, Mora P (2002) Control of *Ancistrotermes guineensis* Silvestri (Termitidae: Macrotermitinae), a pest of sugarcane in Chad. Int J Pest Manage 48:81–86
Sands WA (1960). Termite control in West African afforestation. Rep 7th Commonw Ent Conf, pp 91–95.
Sands WA (1973) Termites as pests of tropical food crops. Pest Artic News Summ 19:167–177
Sands WA (1977) The role of termites in tropical agriculture. Outlook Agr 9:136–143
Sekamatte MB, Latigo MWO, Smith AR (2001) The effect of maize stover used as mulch on termite damage to maize and activity of predatory ants. Afr Crop Sci J 9:411–419
Sekamatte BM, Latigo MO, Smith MRA (2003) Effects of maize-legume intercrops on termite damage to maize, activity of predatory ants and maize yields in Uganda. Crop Protect 22:87–93
Sharma RK, Babu KS, Chhokar RS, Sharma AK (2004) Effect of tillage on termites, weed incidence and productivity of spring wheat in rice-wheat system of North Western Indian plains. Crop Prot 23:1049–1054
Sharma KK, Kalpana V, Sharma P et al (2008) Persistence and vertical distribution of termiticide fipronil in modified ground board test. Environ Monit Assess 137:179–184
Sileshi G, Mafongoya PL, Kwesiga F, Nkunika P (2005) Termite damage to maize grown in agroforestry systems, traditional fallows and monoculture on nitrogen-limited soils in eastern Zambia. Agric Forest Entomol 7:61–69
Silva AG, Gonçalves CR, Galvão DM et al (1968) Quarto catálogo dos insetos que vivem nas plantas do Brasil, seus parasitas e predadores. Ministério da Agricultura, Rio de Janeiro
Smutterer H (1969) Pests of crops in Northeast and Central Africa with particular reference to the Sudan. Gustav Fischer, Stuttgart
Srivastava KP, Butani DK (1987) Insect pests of tea in India and their control. Pesticides 21:16–21
Stansly PA, Su N-Y, Conner JM (2001) Management of subterranean termites, *Reticulitermes* spp. (Isoptera: Rhinotermitidae) in a citrus orchard with hexaflumuron bait. Crop Prot 20:199–206
Starnes RL, Lui CL, Marone PG (1993) History, use and future of microbial insecticides. Am Entomol 39:83–91

Sun J, Fuxa JR, Henderson G (2003) Virulence and in vitro characteristics of pathogenic fungi isolated from soil by baiting with *Coptotermes formosanus* (isopteran: Rhinotermitidae). J Entomol Sci 38:342–358

Tang B, Tang M, Chen C et al (2006) Characteristics of soil fauna community in the Dongjiao coconut plantation ecosystem in Hainan, China. Acta Ecol Sinica 26:26–32

Thiollay JM (1970) L'exploitation par les oiseaux des essaimages de fourmis et termites dans une zone de contact savane-forêt en Côte D'Ivoire. Alauda 38:255–273

Tiben A, Pearce MJ, Wood TG et al (1990) Damage to crop by *Microtermes anjdensis* (Isoptera: Macrotermitinae) in irrigated semi-desert areas of the Red Sea Coast. 2. Cotton in the Tokar Delta region of Sudan. Trop Pest Manage 36:296–304

Torales GJ (1998) Isoptera. In: Morrone JJ, Coscarón S (eds) Biodiversidad de artrópodos argentinos: una perspectiva biotaxonómica. Ediciones Sur, La Plata, pp 48–66

Verma AN (1980) Effect of number of irrigations on termite damage in wheat crop. Harvana Agricultural. Univ J Res (India) 10:564–565

Verma M, Sharma S, Prasad R (2009) Biological alternatives for termite control: a review. Int Biodeterior 63:959–972

Wardell DA (1987) Control of termites in nurseries and young plantations in Africa: established practices and alternative courses of action. Common Forest Rev 66:77–89

Wardell DA (1990) The African termite: peaceful coexistence or total war? Agroforest Today 2:4–6

Werner PA, Prior LD, Forner J (2008) Growth and survival of termite-piped *Eucalyptus tetrodonta* and *E. miniatra* in Northern Australia: implications for harvest of trees for didgeridoos. Forest Ecol Manage 25:328–334

Wilcken CF (1992) Danos de cupins subterrâneos *Cornitermes* sp. (Isoptera: Termitidae) em plantios de *Eucalyptus grandis* e contrôle com insecticidas no solo. An Soc Entomol Brasil 21:329–338

Wilcken CF, Raetano CG (1998) Atualidades no contrôle de cupins em florestas de eucalipto. In: Fontes LR, Berti-Filho E (eds) Cupins: o desafio do conhecimento. FEALQ, Piracicaba, pp 173–185

Wood TG (1978) Food and feeding habits of termites. In: Brian MV (ed) Production ecology of ants and termites. University Press, Cambridge, pp 55–80

Wood TG (1996) The agricultural importance of termites in the tropics. Agric Zool Rev 7:117–155

Wood TG, Bednarzik M, Aden H (1987) Damage to crops by *Microtermes najdensis* (Isoptera, Macrotermitinae) in irrigated semi-desert areas of the Red Sea coast. 1. The Tihama region of the Yemen Arab Republic. Trop Pest Manage 33:142–150

Wood TG, Cowie RH (1988) Assessment of on-farm losses in cereals in Africa due to soil insects. Insect Sci Appl 9:709–716

Wood TG, Johnson RA, Ohiagu CE (1980) Termite damage and crop loss studies in Nigeria – a review of termite damage to maize and estimation of damage, loss of yield and *Microtermes* abundance at Mokwa. Trop Pest Manage 26:241–253

Wood TG, Kambal MA (1984) Damage to date palms in northern Sudan by *Odontotermes* (Isoptera). Int J Pest Manage 30:469–470

Wood TG, Pearce MJ (1991) Termites in Africa: the environmental impact of control measures and damage to crops, trees, rangeland and rural buildings. Sociobiology 19:221–234

Chapter 19
Invasive Termites

Theodore A. Evans

Abstract When the first comprehensive review of introduced termites was made in 1967, 17 species could be considered invasive using modern definitions. The current number is 26 species: 11 are new, 10 have larger distributions, four have no change in distribution, and one species is uncertain – possibly eradicated. Invasive species share three characteristics: wood for food, nesting in food, and a high capacity to generate secondary reproductives. These characteristics combine to increase probability of transportation of viable propagules and are most prevalent in two families, the Kalotermitidae and Rhinotermitidae, which comprise 85% of invasive species. Three genera, *Cryptotermes*, *Heterotermes* and *Coptotermes*, comprise over half (58%) of all invasive species. The largest termite family, the Termitidae, has relatively few species with these characteristics, and is underrepresented. Islands have double the invasive species typical of continents, with islands in the South Pacific the most invaded geographical region. In contrast to earlier findings and in part due to better identification methods and greater elapsed time since introduction, invasive species are now confirmed to colonise natural habitats and have inland distributions.

19.1 Introduction

Frank Gay wrote the first world wide review of termites introduced to new countries (Gay 1967) and distilled this into a chapter (Gay 1969) in Volume 1 of Krishna and Weesner's milestone *Biology of Termites* (Krishna and Weesner 1969). Almost all the termite species listed by Gay (1967, 1969) have received attention during the ensuing four decades, and two termite species were listed among the 100 worst invasive species (Lowe et al. 2000). A similar scale of review has not been attempted since, although there have been synoptic papers published on particular species or country situations. This chapter provides a review of all termite invasions covering the 40 years since Gay's seminal work, and is focussed on: publications since

T.A. Evans (✉)
CSIRO Division of Entomology, Canberra, ACT 2600, Australia
e-mail: theo.evans@csiro.au

D.E. Bignell et al. (eds.), *Biology of Termites: A Modern Synthesis*,
DOI 10.1007/978-90-481-3977-4_19, © CSIRO Australia 2011

1969, earlier publications that were not included by Gay (1967, 1969), and some contributions that are re-interpreted in a modern context.

Invasion biology has grown as a discipline in the past 40 years. Gay reported on species and distributions, but he did not attempt to place his information into a theoretical context, primarily because no established body of theory existed at the time. Therefore this chapter attempts to place termite invasions into the contexts of invasive biology, life history and ecological theory. In particular, standard invasion biology definitions will be applied and Gay's (1967, 1969) lists revised and updated with these definitions. Also, life history and ecological traits of invasive species are compared and contrasted with non-invasive species at meaningful cladistic levels (family for lower termites, subfamily or other group level for higher termites). Characteristics of invaded regions and habitats, and those of the endemic regions of the invasive species, are compared and patterns considered. Finally, the limitations of current knowledge and future possibilities are considered.

19.2 Definitions

Gay's seminal paper (Gay 1967) was entitled "A world review of introduced species of termites" and his chapter (Gay 1969) "Species introduced by man". These were written a decade after Elton (1958) had defined invasion ecology as a new field of research, yet Gay did not use the terms "invasion" or "invasive" in his works, but instead used "introduction" and "introduced". It is possible that Gay did not consider "introduced" termites to be "invasive" because the majority of cases he documented were of introduced species found in or around human habitations and structures in urban settings, not natural environments. He stated, "Introduced species frequently become of considerable economic importance, although they are rarely able to invade native habitats and are almost always restricted to the man-modified environments of houses, building, and cultivated crops". This distinction was used by Gay in his two works in order to distinguish native from introduced/invaded ranges, and has been used by others to help separate the endemic range from the invaded range for various termite species, for example, *Cryptotermes brevis* (Scheffrahn et al. 2009).

Definitions are important because they create and limit a field of study. This is perhaps of increased importance since Gay's contributions, because invasion biology has grown as a discipline, even spawning its own journal (*Biological Invasions*). The field is jargon-rich and Davis and Thompson (2000) even suggested that the many unclear definitions had slowed progress. This has been remedied by Falk-Petersen et al. (2006), and their definitions are used here, principally the following:

- *Native* is a synonym of indigenous and of original, and means: "an organism occurring within its natural, past or present range and dispersal potential (organisms whose dispersal is independent of human intervention)".
- *Endemic* means: "organism restricted to a specified region or locality"; i.e. native species with a restricted distribution.

- *Introduction* means "direct or indirect movement by human agency of an organism from its native, past or present range to a range outside its distribution potential".
- *Introduced* is a synonym of alien, adventive, exotic, foreign, non-indigenous, non-native and novel, and means: "an organism occurring outside its natural, past or present range and dispersal potential including any parts of the organism that might survive and subsequently reproduce (organisms whose dispersal is caused by human action)".
- *Established* means: "native or non-native organism that has obtained a self-sustaining population in an area where it previously did not occur".
- *Transfer* is a synonym of translocation and transplantation and means "human mediated movement of an organism within its past or present range and dispersal potential".
- *Invasive* means: "alien organisms that have established in a new area and are expanding their range".
- *Naturalized* means: a non-native organism that has obtained a self-sustaining population.
- *Pest* means: "organisms considered harmful to human activities".

These definitions are illustrated in Fig. 19.1. Native, introduced and naturalized species can all be pests by the above definition. Note that the list does not include a distinction between "established and non-invasive", meaning the introduced species has a self-perpetuating population that remains localized around the point of introduction, and "invasive". This is because the difference between the two definitions is one of the amount of spread, which could be arbitrary. Spread depends on the elapsed time since introduction and establishment, and these dates are rarely known for unintentionally introduced species.

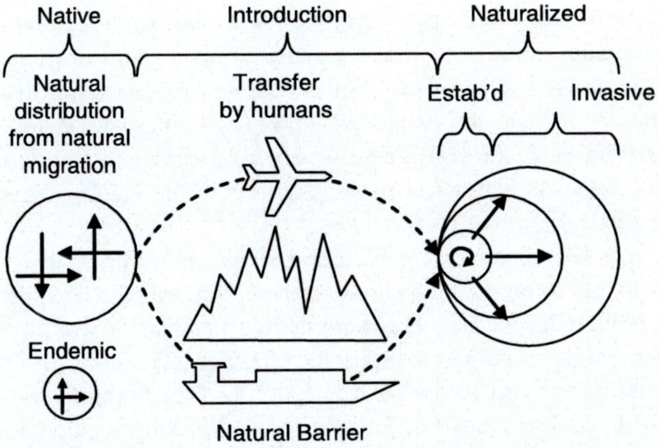

Fig. 19.1 Schematic showing the definitions of invasion biology after Falk-Petersen et al. (2006)

Falk-Petersen et al. (2006) give definitions of "introduced" plus "established" and "invasive" which more or less match those of Gay's (1967, 1969) status categories of "intercepted" and "established". This suggests that Gay had a good understanding of differences now considered important 40 years later, and the problem of variation in terminology.

19.3 List of Invasive Species

Gay (1967, 1969) listed 42 identified and 8 unidentified species that had been introduced to new areas or countries. The origins of some species were uncertain and other species failed to establish; excluding these Gay reported "there are at least 17 species for which there is unequivocal evidence of their establishment in new habitats." He also stated, "Introduced species frequently become of considerable economic importance, although they are rarely able to invade native habitats and are almost always restricted to the man-modified environment." In fact it is because the 17 invasive species listed by Gay were always pests in human habitation that they have been considered important enough for study and to be specifically targeted for control. In addition, Gay and other authors cited by him use the presence of a termite species in a native habitat as evidence that it is not introduced, which would now be considered a dubious assumption.

Accurate identification of native and invaded distributions was, and in some cases still is, unresolved for some species. This was particularly true of drywood species, as these appear to have been distributed as tramps by human activity, perhaps during European colonial times, as much as four centuries previously. Even if termites were noticed then, taxonomic expertise was not sufficient for the description of species and by the time identification was possible, the tramp termites had been distributed pan-globally and thus their origins remained obscure.

Since 1969 the origin of some species has become more certain. For example, *Cryptotermes brevis* was considered to have Chinese origin by Light (1931), but Emerson (1936) considered the Caribbean to be the natural range. Gay (1969) agreed with Emerson, in part due to the paucity of historical information in China, but he also noted that the "Central and South American records for this species suggest that in some areas it is an introduced rather than a native species" (Gay 1969: p. 474). Scheffrahn et al. (2009) recently provided evidence (outdoor and native habitats) that *C. brevis* is native to riparian habitats in the Pacific coastal deserts of Peru and Chile.

The situation is less clear for *Cryptotermes havilandi*. Gay (1967, 1969) stated this species "appears to be native to the eastern countries of Africa and to the rain forests of the Congo Republic." He thought this because it was found in native habitats but only rarely in human habitations. However, Williams (1976) stated that the species is native to West Africa as it is found in forests there, and therefore he considered that it was introduced to East Africa. Other authors have found this termite in both forest and urban areas in West and East Africa (Forsyth 1966; Wagner et al. 1991; Schabel 2006), thus tropical forests across Africa are now considered the native range. However, it is possible that this species originated in one part of Africa but was introduced to the other, and has since invaded and naturalised, perhaps many centuries ago. Such a situation is demonstrated by *Coptotermes formosanus*. Gay (1967, 1969) considered this species to be native to Japan, Taiwan and southern China because the species "attacks not only buildings and furniture but living trees as well." However, Mori (1987) found that *C. formosanus* was introduced to Japan perhaps 400 years ago, and therefore has had time to spread beyond human urban areas and become naturalised. Therefore Taiwan and Southern China are now considered the native range.

Better taxonomic identifications have identified two rhinotermitid invaders more clearly. *Reticulitermes flavipes* from the Western USA and *R. santonensis* from Western France have been found definitively to be the same species (Ye et al. 2004; Austin et al. 2004, 2005), providing evidence that the species was introduced to France in the eighteenth or nineteenth century (Bagnères et al. 1990; Vieau 2001). *Reticulitermes arenincola* has also been found to be *R. flavipes* (Ye et al. 2004; Austin et al. 2007), and therefore removed from the current list (Table 19.1). Similarly, *Coptotermes gestroi* and *C. havilandi* from South East Asia, and *C. vastator* from the Philippines are the same species (Kirton and Brown 2003; Yeap et al. 2007). This taxonomic discovery raises the question of how much of the range of *C. gestroi*, from India in the west to the Philippines in the east, is native and how much is invaded. The same question can be posed for two other invaders found in this area: *Cryptotermes domesticus* and *Cryptotermes dudleyi*. Perhaps human shipping and other transport moved these three species many centuries ago, so identification of the original native habitat will require detailed population level genetics. Also, three *Nasutitermes* species, *N. corniger, N. polygynus* and *N. costalis*, once believed to be native as they were found in the natural forests of Panama, New Guinea and the West Indies, respectively (Roisin and Pasteels 1996), have all been found to be *N. corniger* (Scheffrahn et al. 2005a, 2005b), and now recognised as the most widespread invasive termitid species, but of undetermined origin.

In total, by the definitions of Falk-Petersen et al. (2006) and with the above comments in mind, 26 termite species can be classified as invaders. These are listed in Table 19.1, which also gives the currently understood native range, the invaded range as reported by Gay (1967, 1969), and the currently understood invaded range. The details are listed below, categorised according to change or otherwise in status.

19.3.1 New Species

Eleven species are new to the list of invasive species since 1969. Three are kalotermitid species, six are rhinotermid species, and two are termitid species.

19.3.1.1 Kalotermitidae

Kalotermes banksiae. This termite originates on the south east coasts of Australia (Hill 1942), and has invaded New Zealand in one location (Bain and Jenkin 1983).

Table 19.1 The 25 species of invasive termites (sensu Falk-Petersen et al. 2006), their natural range, invasion status in 1969 and in 2009)

Species	Natural range	1969 Invaded regions	1969 Invaded localities	2009 Invaded regions	2009 Invaded localities	2009 References
Mastotermes darwiniensis	Australia	S Pacific	PNG (Lae)	S Pacific	PNG (Lae)	Gray (1968), Thistleton et al. (2007)
Porotermes adamsoni	Australia	S Pacific	New Zealand	S Pacific	New Zealand	Walker (2007), Philip et al. (2008)
Incisitermes immigrans[a]	? Pacific N and S America	N Pacific	Polynesia (Kiribati [Fanning, Jarvis Is])			Grace (2009)
Incisitermes minor	N America (Mexico, USA[CA])			N Pacific	Japan, Ryukyu Is, Hawaii	Light (1934), Mori (1976), Ichiko et al. (2000), Indrayani et al. 2004, 2005), Grace (2009)
				E Asia	China, Zhejiang	Xie et al. (2001)
				N America	USA (TX, AK, LA, OK, SC)	Hathorne et al. (2000), Scheffrahn et al. (2001), Messenger et al. (2002)
Cryptotermes brevis[b]	S America (Peru, Chile)	(Caribbean)		Caribbean	N Is (Bahamas, Gd Turk, Gt Inagua)	Scheffrahn and Su (1990), Scheffrahn and Krecek (1999), Scheffrahn et al. (2009)
					Gtr Antilles (Cuba, Jamaica, Hispaniola, Puerto Rico)	Scheffrahn and Krecek (1999), Scheffrahn et al. (1994, 2003)
					Lsr Antilles (Antigua, Barbados, Dominica, Guadeloupe, Nevis, Martinique)	Scheffrahn and Krecek (1999), Scheffrahn et al. (1994, 2003)

Table 19.1 (continued)

Species	Natural range	1969 Invaded regions	1969 Invaded localities	2009 Invaded regions	2009 Invaded localities	2009 References
					S Is (Trinidad)	Scheffrahn and Krecek (1999), Scheffrahn et al. (1994, 2003)
		N Atlantic	Canary Islands	N Atlantic	Azores	Borges and Myles (2007)
		S Atlantic	Saint Helena Island			
		N Pacific	Hawaii, Midway			
		S Pacific	Melanesia (New Caledonia, Fiji)	S Pacific	Melanesia (Tonga)	Peters and Fitzgerald (2008)
			Polynesia (Marquesas, Easter)			
		N America	USA (NY-FL, WI-TN, CA)		Panama to Mexico	Scheffrahn et al. (2009)
			El Salvador			
		S America	Venezuela (Caracus, Cua), Guiana (Georgetown), Brazil (Sao Paulo)	S America	Ecuador to Uruguay	Constantino (2002)
			Peru (Lima)	Europe	Portugal (Lisbon)	Nunes (2008)
		Africa	Congo (Boma), Sierra Leone (Njala)			
			S Africa (Cape Prov, Durban, Natal, Zululand)			
		Australia	SE Qld	Australia	SE Qld	Peters and Fitzgerald (2008)
Cryptotermes cynocephalus	SE Asia (Philippines)	S Pacific	PNG, New Britain (Rabaul)	N Pacific	Hawaii	Scheffrahn et al. (2000), Grace (2009)
		Indian	Sri Lanka			
		Australia	Qld	Australia	Torres Strait Islands	Gay and Watson (1982)

Table 19.1 (continued)

Species	Natural range	1969 Invaded regions	1969 Invaded localities	2009 Invaded regions	2009 Invaded localities	2009 References
Cryptotermes domesticus	? SE Asia	N Pacific	Guam			
		S Pacific	Melanesia (New Britain, Solomon Is, Fiji)	S Pacific	Melanesia (New Guinea [Port Moresby], Fiji, Samoa)	Gray (1968), Lal and Tuvou (2003), Iosua and Peseta (2003)
			Polynesia (Samoa, Marquesas, Pitcairn [Oeno Is], Tahiti [Austral Is, Society Is])			
		N Pacific	Kiribati (Fanning and Flint Is)			
		N America	Panama			
		E Asia	China (Hainan)			
		Australia	Qld, Torres Strait Is	Australia	N Coast of Australia	Gay and Watson (1982)
Cryptotermes dudleyi	? S and SE Asia	Caribbean	Gtr Antilles (Jamaica)		Gtr Antilles (Jamaica)	Scheffrahn and Krecek (1999)
			S Is (Trinidad)		S Is (Trinidad)	Scheffrahn and Krecek (1999)
		N Pacific	Marshall Islands	N Pacific	Guam	Scheffrahn, unpublished observations
		Indian	Madagascar, Mauritius Cocos/Keeling Islands			
		N America	Panama, Costa Rica	N America	Nicaragua	Scheffrahn, unpublished observations
				S America	Brazil, Colombia	Constantino (2002)

Table 19.1 (continued)

		1969		2009		
Species	Natural range	Invaded regions	Invaded localities	Invaded regions	Invaded localities	2009 References
		S Asia	India (Orissa, Lwr Bengal); Bangladesh, Khulna			
		Africa	Kenya, Tanzania, Somalia, Uganda	Africa	Tanzania	Williams (1976), Schabel (2006)
		Australia	Darwin, Torres Strait Islands		Darwin, Torres Strait Islands	Gay and Watson (1982)
Cryptotermes havilandi[c]	Africa (tropical west and east)[c]	Caribbean	S Is (Trinidad, Tobago)	Caribbean	Lsr Antilles (Antigua, Barbados, Guadeloupe, St Kitts, Nevis, St. Lucia, St Vincent, Little Tobago, St Croix, St Thomas)	Forsyth (1966), Scheffrahn and Krecek (1999), Wagner et al. (1991), Schabel (2006), Scheffrahn, unpublished observations
		S Atlantic	African Is (Fernando Póo Is)			
		Indian	Madagascar, Comoro, Europa Is	Indian	Madagascar	Constantino (2002)
		S America	Guiana (Kartabo), Surinam (Paramaribo)	S America	Brazil, Guianas	Constantino (2002)
			Brazil (Santos and Guaratuba)			
		S Asia	India	S Asia	India and Bangaldesh	Chhotani (1963), Maiti (1983), Bose (1984)
		Africa	Tanzania			
Kalotermes banksiae	SE Australia			S Pacific	New Zealand	Hill (1942), Bain and Jenkin (1983)

Table 19.1 (continued)

		1969		2009		
Species	Natural range	Invaded regions	Invaded localities	Invaded regions	Invaded localities	2009 References
Glyptotermes breviconis				S Pacific	New Zealand	Bain and Jenkin (1983)
				S Pacific	Fiji	Hill (1942), Evenhuis (2007)
Heterotermes convexinotatus	S America			Caribbean	Gtr Antilles (Hispaniola, Puerto Rico)	Araujo (1977), Szalanski et al. (2004)
					Lsr Antilles (Antigua, Barbados, Curacao, Martinique, St. Kitts, St. Maarten)	Szalanski et al. (2004)
				N America?	USA (FL)?	Scheffrahn and Su (1995)
				S Pacific	Galapagos?	Emerson (1971), Peck (2001)
Heterotermes perfidus[d]	?	S Atlantic	Saint Helena Island			
Heterotermes philippinensis	Philippines	Indian	Madagascar, Mauritius	Indian	Madagascar	Cachan (1950)
Heterotermes tenuis	S America			Caribbean	Lesser Antilles	Szalanski et al. (2004)
Heterotermes n. sp.	Caribbean			N America	USA (FL, Miami)	Szalanski et al. (2004)
Reticulitermes flavipes	N America	Europe	France, Germany	Europe	France	Becker (1970), Jenkins et al. (2001), Austin et al. 2004, 2005)
		N America	Canada			
				Caribbean	N Is (Gd Bahama)	Scheffrahn et al. (1999)

Table 19.1 (continued)

Species	Natural range	1969 Invaded regions	1969 Invaded localities	2009 Invaded regions	2009 Invaded localities	2009 References
				S Pacific	Galapagos	Austin et al. (2005)
				S America	Uruguay, Chile	Aber and Fontes (1993), Aber and Beltrami (2002), Austin et al. (2005), Su et al. (2006)
Reticulitermes grassei	Europe (France)			N Atlantic	England	Jenkins et al. (2001), Verkerk and Bravery (2001), Bravery and Verkerk (2010)
					Azores	Borges and Myles (2007)
Coptotermes acinaciformis	Australia	S Pacific	New Zealand	S Pacific	New Zealand	Miller (1941), Bain and Jenkin (1983), Ross (2005a, 2005b), Philip et al. (2008)
			Fiji		Fiji	Lal and Tuvou (2003)
Coptotermes curvignathus	SE Asia			East Asia	China	Xie et al. (2001)
Coptotermes formosanus[e]	E Asia (China, Taiwan)	N Pacific	Hawaii, Guam, Midway, Marshall Is, Japan		Japan	Mori (1987)
		Indian	Sri Lanka			
		N America	USA (LA, TX?)	N America	USA (AB, CA, FL, GA, LA, MS, NC, SC, TN, TX)	Scheffrahn et al. (1988), Atkinson et al. (1993), Woodson et al. (2001), Messenger et al. (2002), Austin et al. (2006), Sun et al. (2007)

Table 19.1 (continued)

Species	Natural range	1969 Invaded regions	1969 Invaded localities	2009 Invaded regions	2009 Invaded localities	2009 References
		Africa	S Africa	(Africa?)	(S Africa?)	Coaton and Sheasby (1976), Mitchell (2002)
Coptotermes frenchi[f]	Australia	S Pacific	New Zealand	(S Pacific?)	(New Zealand?)	Bain and Jenkin (1982)
Coptotermes gestroi[g]	SE Asia,	Caribbean	Barbados, Jamaica	Caribbean	Gtr Antilles (Cuba, Haiti, Grand Cayman, Little Cayman, Puerto Rico)	Scheffrahn et al. (1994), Darlington, unpublished observations
					Lsr Antilles (Antigua, Barbuda, Guadeloupe, St. Kitts, Nevis, Virgin Gorda, Monserrat)	Hernández (1994); Scheffrahn et al. (1994, 2003) and Jenkins et al. (2007)
					N Is (Grand Turk, Providenciales)	Scheffrahn and Su (1990)
		N Pacific	Hawaii	N Pacific	Taiwan	Tsai and Chen (2003); Li et al. (2009)
					Yap, Guam, Midway, Marshalls	Scheffrahn, unpublished observations
					Hawaii	Woodrow et al. (2001)
		S Pacific	Marquesas			
		Indian	Mauritius, Reunion			
				N America	USA (FL)	Su et al. (1997), Scheffrahn and Su (2005)
					Mexico	Ferraz and Montiel (2004)
		S America	Brazil (Sao Paulo, Rio de Janeiro, Santos)	S America	Brazil	Constantino (2002)

Table 19.1 (continued)

Species	Natural range	1969 Invaded regions	1969 Invaded localities	2009 Invaded regions	2009 Invaded localities	2009 References
Coptotermes sjostedti	Africa			Caribbean	Guadeloupe	Harris (1966), Scheffrahn et al. (2004)
Termes hispaniolae[h]	S America and Antilles	(Caribbean)	(Lsr Antilles (Barbados))	Caribbean	Bahamas (New Providence, S. Andros, Paradise)	Scheffrahn, unpublished observations
Nasutitermes corniger	Neotropics			Caribbean	Bahamas (Abaco Island)	Constantino (1998), Scheffrahn et al. (2002, 2005a)
				S Pacific	Papua New Guinea	Scheffahn et al. (2005b)
				N America	USA (FL)	Scheffrahn et al. (2002)

Regions (and localities, if present) were defined as follows: Caribbean (Greater Antilles, Lesser Antilles, North group, South group), North Atlantic (British Is., Islands off North Africa, Islands off West Africa), South Atlantic, North Pacific (North West Pacific, Micronesia, Hawaii), South Pacific (Melanesia, Polynesia, New Zealand, Galapagos), Indian Ocean (East islands, South West islands, Sri Lanka), North America ([temperate] northern North America, [subtropical] southern North America), South America ([sub/tropical] northern South America, [temperate] southern South America), Europe, Africa (West Africa, East Africa, Southern Africa), Arabia, Asia ([tropical] South and South East Asia, [temperate] East Asia), Australia. N = north, S = south, E = east, W = west, Is = island or isle (incl. plurals), Gd = Grand, Gtr = Greater, Lsr = Lesser, ? = unknown, uncertain or disputed. States in the USA indicated with standard two letter abbreviations.

[a]Unknown, probably Central and South America (Constantino 1998). Pacific Is distribution (Galapagos, Hawaii, Marquesas) may predate European discovery (Grace et al. 2002).

[b]Gay (1967, 1969) considered *Cryptotermes brevis* to be native to the Caribbean, South and Central America.

[c]Gay (1967, 1969) considered *Cryptotermes havilandi* to be native to east Africa and Belgian Congo therefore all west African countries considered to be invaded.

[d]Gay (1967, 1969) believed this to be a valid species, but possibly *Heterotermes convexinotatus* or *H. tenuis*; see text.

[e]*Coptotermes formosanus* possibly eradicated from South Africa.

[f]*Coptotermes frenchi* possibly eradicated from New Zealand.

[g]Includes information for *Coptotermes havilandi* and *C. vastator*, as these names have been synonymised under *C. gestroi* (Kirton and Brown 2003; Yeap et al. 2007).

[h]Gay (1967, 1969) listed as introduced, hence parentheses.

Incisitermes minor. This termite originates on the Pacific coast on the Southern USA and Northern Mexico. It has invaded the Eastern USA (Hathorne et al. 2000; Scheffrahn et al. 2001; Messenger et al. 2002), Pacific Islands, including Hawaii and Japan (Indrayani et al. 2004, 2005; Grace 2009), and China (Xie et al. 2001).

Glyptotermes breviconis. This termite originates on the south east coasts of Australia (Hill 1942). It has invaded New Zealand (Bain and Jenkin 1983), and may have invaded Fiji as well (Hill 1942). There has been uncertainty about the exact *Glyptotermes* species in Fiji, designated *brevicornis* by Hill (1942), *taveuniensis* by Gray (1968) and by Lal and Tuvou (2003), or unspecified (Gross 1975). However as Evenhuis (2007) documents both species, Fiji is included in Table 19.1 as an invaded location.

19.3.1.2 Rhinotermitidae

Heterotermes convexinotatus. This termite originates in South America. It has invaded the Greater and Lesser Antilles in the Caribbean (Szalanski et al. 2004), possibly Florida in North America, and the Galapagos in the Pacific.

Heterotermes tenuis. This termite originates in South America. It has invaded the Lesser Antilles in the Caribbean.

Heterotermes sp. nov. This is an undescribed species suggested by Szalanski et al. (2004) from a study of *Heterotermes* in the Caribbean using a section of the 16S rRNA (mitochondrial) gene. The new species is based on their groups II and III. However there are reasons to consider that these do not define a new species: they were not a single clade and cluster at the base of *H. cardini*. Single gene analyses may have low resolution (e.g. *Coptotermes* phylogeny using COII by Gentz et al. 2008) and within termites using even five or six genes can still have low resolution or conflicting results (Inward et al. 2007; Legendre et al. 2008; see also Muirhead et al. 2008 for sampling issues in identification of source populations). In addition, no resolution could be obtained between *H. aureus* from western USA, one undescribed species from Mexico, one undescribed species from Belize, and *H. convexinotatus* from South America (Szalanski et al. 2004). Therefore it is possible that unidentifiable *Heterotermes* specimens from the neotropics will be reassigned to either *H. convexinotatus* or *H. cardini* should more data emerge. However, until this is resolved, the conclusions of Szalanski et al. (2004) are presented in Table 19.1.

Reticulitermes grassei. This species originates in Europe (France) and established in one location in Britain (Saunton, Devon). It was believed to have been eradicated (Jenkins et al. 2001; Verkerk and Bravery 2001; Bravery and Verkerk 2008) after a baiting program. However, 8 years later termites were found again adjacent to in ground hot water pipes (Bravery and Verkerk 2010), a similar situation is know for *Reticuliterme flavipes* in colder climates in Toronto, Canada and Hamburg, Germany (Becker 1970; Clément et al. 2001; Austin et al. 2005).

Coptotermes curvignathus. This species originates from South East Asia (Malaysia, Thailand) and has established in one location in China (Xie et al. 2001).

Coptotermes sjostedti. This species originates from Africa (in tropical rainforests from West to Central Africa) and has established in one location in the Lesser Antilles in the Caribbean (Guadeloupe) (Scheffrahn et al. 2004).

19.3.1.3 Termitidae

Termes hispaniolae. This species originates from the Caribbean shores of South and North America. It has been found on two islands in the Greater Antilles (Hispaniola and Puerto Rico), and three islands in the Lesser Antilles (Guadeloupe, St Croix and St Thomas), and has established on three islands in the Bahamas Group in the northern Caribbean (New Providence, S. Andros and Paradise). This species lives in carton arboreal nests, derived from digested wood, which are built in the branches of trees. It is uncertain how the termite was transported to the islands (Scheffrahn, unpublished observations), but some indication is given by Gay (1967, 1969) who reported that "in 1939–1940 it was intercepted in shipments of firewood entering Barbados from British Guyana".

Nasutitermes corniger. This is one of the more abundant and widespread termites in the tropical areas of North and South America and the Caribbean. It has invaded three locations, one each in the North Island group in the Caribbean, Florida, and Papua New Guinea in the South Pacific, where it was synonymised with *N. polygynus* (Scheffrahn et al. 2005a). Constantino (2002) considered *N. costalis*, *N. araujoi*, *N. globiceps* and *N. tatarendae*; to be the same species as *N. corniger*, while *N. costalis* was synonymised by Scheffrahn et al. (2005b), giving this species an even wider distribution. However, the extent of the native compared with the invaded habitat is unknown. Like *Termes hispaniolae*, *N. corniger* lives in arboreal carton nests derived from digested wood. Scheffrahn et al. (2005a) suggested that "a single early maritime introduction centuries ago from which the termites dispersed around New Guinea, possibly helped by human transportation" was the more likely explanation for how this neotropical species came to be found in New Guinea, very distant from its natural range. The Dutch had colonies in the natural range of *N. corniger* in the Caribbean (Dutch Guiana and the Netherlands Antilles: Aruba, Curacao, Bonaire) and in the Indonesian archipelago (centred on Java, then called Batavia, but including Western New Guinea); therefore Dutch colonial transport may have been the method by which this species was spread. Gay (1969) reported that it has been intercepted on numerous occasions in plants, mainly orchids, imported from Central America and in packaging material from Martinique.

19.3.2 Species with Larger Distributions

There are ten species that were listed in 1969, but now have larger distributions incorporating locations subsequently invaded. One is *Mastotermes*, five are kalotermitid species, and four are rhinotermid species.

19.3.2.1 Mastotermitidae

Mastotermes darwiniensis. This is a tropical Australian species, with a new location in the Gold Coast, Queensland (Peters and Fitzgerald 2008). This infestation is around 800 km south of the previous most southerly location (Emerald), which is north of the Tropic of Capricorn (Watson and Abbey 1993). The termites have been found in an area ca. one hectare, with damage to timber in service suggesting that the infestation is perhaps 10 years old (Brenton Peters, Queensland Department of Primary Industries, personal communication). The termites may survive the cooler (subtropical) winters due to artificial heating in houses. This range extension is equivalent to that of *Reticulitermes flavipes* in North America, which has extended into Canada. *M. darwiniensis* is also still present in the city of Lae, Papua New Guinea, despite ongoing eradication efforts. Thistleton et al. (2007) compiled an extensive report on the presence of *M. darwiniensis* in Papua New Guinea. The species was originally discovered in 1959, probably transported from Australia during or soon after World War 2. There were two sites in Lae city which were subjected to a 12 year eradication effort completed in 1973. This was believed to have been successful (e.g. Gray 1968) until new infestations in the same sites were discovered 20 years later. The infested area is now seven times larger than the area known from 1973, and termites have been found not only in buildings, but in 42 species of native and exotic trees.

19.3.2.2 Kalotermitidae

Cryptotermes brevis. Gay (1967, 1969) considered this species to be native to South America and the Caribbean. However, Scheffrahn et al. (2009), using Gay's own criteria of regenerating outdoor populations in native (i.e. non-human modified) habitat, have found that the native range of the species is riparian habitats in the coastal desert from south Peru into North Chile. Gay (1967, 1969) included China, but Scheffrahn et al. (2009) considered this locality record to be suspect. Newly invaded localities include: Azores in the North Atlantic, Portugal (Lisbon) in Europe, and Tonga in Melanesia.

Cryptotermes cynocephalus. This tropical South East Asian species has been found in new localities including Hawaii (Scheffrahn et al. 2000; Grace 2009) and the Torres Strait Islands between Australia and New Guinea (Gay and Watson 1982).

Cryptotermes domesticus. This tropical South East Asian species has been found in Samoa (Iosua and Peseta 2003, who thought this was a native species as it is found in native forests), and four sites in Australia: Darwin and three locations on the Queensland peninsula (Gay and Watson 1982). There has been discussion about whether the species is native or naturalised to Australia because it has been found in native trees (Yule and Watson 1976; Gay and Watson 1982). However, it is included as invasive in Table 19.1 because the sites are discontinuous and disjunct, and all discussions agree that the likely introduction was from human transportation (in fishing boats from Indonesia).

Cryptotermes dudleyi. Another tropical South East Asian species; new localities include Guam in Micronesia, North Pacific, Nicaragua in North America (Scheffrahn, unpublished observations) and Brazil and Colombia in South America (Constantino 2002).

Cryptotermes havilandi. This tropical African species has been found in new localities across 11 islands in the Lesser Antilles (Scheffrahn and Krck 1999; Scheffrahn, unpublished observations).

19.3.2.3 Rhinotermitidae

Reticulitermes flavipes. This temperate North American species has been found in several new localities: France, Galapagos in the Pacific, Bahamas in the Caribbean (Scheffrahn et al. 1999), Montevideo in Uruguay and Santiago in Chile (Austin et al. 2005; Su et al. 2006). As noted above, the population of *R. flavipes* in France was originally described as a new native species, *R. santonensis*, in 1924. Subsequent authors noted similarities to the America species, and the validity of the species was debated until quite recently, in part because it was found in natural forest and in part because of limited sampling for genotyping (Clément et al. 2001; Dronnet et al. 2005). The issue has been resolved with more extensive molecular studies (Jenkins et al. 2001; Ye et al. 2004; Austin et al. 2004, 2005). Taxonomic confusion has also occurred in Uruguay, as *R. flavipes* was originally designated as a *Heterotermes* species, then as *R. lucifugus* (Aber and Fontes 1993; Aber and Beltrami 2002), but has been declared *R. flavipes* using three haplotypes of 16S rRNA (Austin et al. 2005, however note that no phylogenetic tree is given or analyses performed, and note above comments on single gene analyses). The conclusions of Austin et al. (2005) are presented in Table 19.1.

Coptotermes acinaciformis. This Australian species has received continuous management in New Zealand (Pearson and Bennett 2008). In the 1930s, this species was found in two locations (Auckland and New Plymouth; Miller 1941) ca 350 km apart on the west coast of the North Island. Ten years later it was found in one location (near Gisborne) on the east coast of the North Island, 450 km from the two other sites. By the 1980s, *C. acinaciformis* was found in two more locations (Whakatane and Te Puka) on the North Island (Bain and Jenkin 1983), and by 2006, three more locations (Otorohanga, Matamata and Morrinsville) on the North Island (Ross 2005a) and one on the South Island at Nelson (Ross 2006). Some authors consider this species established, supported by the presence of the termite in native timbers (Miller 1941; Bain and Jenkin 1983), but others not (Ross 2005a). Philip et al. (2008) reported alate flights whereas earlier studies did not, and neotenics were implicated (Ross 2005b). Given the geographic spread of *C. acinaciformis* in New Zealand from telegraph poles and railway sleepers/ties imported from Australia decades ago, this species is designated as established in New Zealand in Table 19.1. The interception and infestation of wharf timbers in Suva, Fiji from 1939 was reported by Gay (1967, 1969). Forty years later, Lal and Tuvou (2003) report that *C. acinaciformis* is a major forestry pest in Fiji, indicating this species has expanded its range substantially over time.

Coptotermes formosanus. Gay (1967, 1969) considered Japan as endemic habitat, however Mori (1987) reported that Japan was invaded perhaps 400 years ago; therefore China and Taiwan are now considered to be the native range (Kistner 1985; Li et al. 2009). Fang et al. (2008) have found a low level of genetic diversity in *C. formosanus* populations in southern China (nine COII haplotypes compared with six from the USA), which raises questions about the natural distribution of this species in China. New localities to be invaded are all in the USA. Gay (1967, 1969) reported five sites in Louisiana, one in Texas and one in North Carolina (the latter from alates only). The following 40 years shows a continued spread from Louisiana and Texas to the states of Alabama, California, Florida, Georgia, Mississippi, North Carolina (all castes), South Carolina, and Tennessee (Su and Tamashiro 1987; Atkinson et al. 1993; Austin et al. 2006), and within these states. For example, in Louisiana in 1968, *C. formosanus* was found in five cities in four parishes, and 40 years later was found in 42 cities in 25 parishes (Messenger et al. 2002; Brown et al. 2007). Similar patterns have been reported in other US states. In Texas, *C. formosanus* was discovered in one city in 1958, but 40 years later it was found in 14 counties (Howell et al. 2001). In Florida, *C. formosanus* was discovered in 1980, and 25 years later was found in 40 cities in 20 counties (Scheffrahn and Su 2005). In Mississippi, *C. formosanus* was discovered in 1984, and 22 years later was found in 25 counties (Sun et al. 2007). It is now known that there were at least two independent introductions of *C. formosanus* to the USA (Haverty et al. 1990; Austin et al. 2006). Long distance spread of the termite within the USA has been by humans moving infested wood, especially railway sleepers/ties (Jenkins et al. 2002; Messenger et al. 2002), as alate flight distances are less than one kilometre (Messenger and Mullins 2005) and therefore insufficient. The termite has spread from urban areas to natural forests, with higher populations (as inferred from higher alate numbers) in those forests (Sun et al. 2007). This species was also introduced to South Africa, but may have been eliminated or at least confined to the very small area of introduction. Indirect evidence for this is the complete absence of publications on this species in South Africa for 34 years (since Coaton and Sheasby 1976, who reported eradication methods as well the occurrence). There is also no mention in the most recent review of pest termites in South Africa (Mitchell 2002; Mitchell, personal communication), and the absence of specimens submitted for identification by the national authorities for over 25 years (V. Uys, South African National Collection of Insects, Plant Protection Research Institute, South African Agricultural Research Council, personal communication). This compares favourably with the re-emergence of *Reticulitermes grassei* in the UK after 8 years (Bravery and Verkerk 2010).

Coptotermes gestroi. This South and Southeast Asian species is included in the information given for *C. havilandi* and *C. vastator* in Gay (1967, 1969), as the species have been synonymised (Kirton and Brown 2003; Yeap et al. 2007). New localities in to be invaded include fourteen islands in the Caribbean Sea, Taiwan (Tsai and Chen 2003; Li et al. 2009), four islands in the North Pacific (Scheffrahn, unpublished observations), Florida (Su et al. 1997; Scheffrahn and Su 2005; Li et al. 2009) and Mexico (Ferraz and Montiel 2004).

19.3.3 Species with No Change

There are four species that were listed in 1969 that have no change in their distributions since then. One is a termopsid species, one a kalotermitid species, and two are rhinotermid species.

19.3.3.1 Termopsidae

Porotermes adamsoni. This Australian species was listed as "established?" in New Zealand by Gay (1969), however, Bain and Jenkin (1983) did not consider *P. adamsoni* had in fact established despite infestations in five locations in the North Island, because it was found only in eucalyptus railway sleepers imported from Australia. Twenty-five years later, Philip et al. (2008) reported alates in private houses and in wharf timbers from three locations 500 km apart; one of which was reported by Bain and Jenkin (1983), suggesting at least one self supporting population. This species has been found in trees in New Zealand (C. Stowe, NZ Ministry of Agriculture and Forestry, personal communication), which suggests they have moved from imported railway sleepers, and thus indicates that this species is established.

19.3.3.2 Kalotermitidae

Incisitermes immigrans. No further information has been available since 1969; one extra information source has been added.

19.3.3.3 Rhinotermitidae

Heterotermes perfidus. Gay (1967, 1969) notes that this species was described from St Helena Island in the South Atlantic Ocean, but was an introduced termite of unknown origin. Gay notes that the species was originally considered to be *H. tenuis* from South America before it was given a new description (*Leucotermes perfidus*) by Silvestri in 1936, and then considered to be *H. platycephalus* from Australia by Emerson in 1951. Gay states definitively that "there is no doubt now, however, that *H. perfidus* is a valid species not referable to any other described species of *Heterotermes*", however he did not provide evidence for this statement. Snyder (1949) lists the original description of this species as the only reference in his catalogue and only one other study has been reported since the description, that of Bacchus (1979) who used long-established (possibly 30–40 years) lab colonies in the UK to study exocrine glands. The question arises as to whether *H. perfidus* really is a valid species, but the history of the termite on the island is less contentious. Gay (1967, 1969) states "this termite is believed to have been introduced into St Helena about 1840 when a captured slaver was brought into the port of Jamestown. . ." Given such an origin, it seems likely that the termite originated from either West Africa, South America, or the Caribbean. There are six pest species of *Heterotermes* in South America, two of which are known invaders, including islands in the Caribbean. There are several *Heterotermes* species in Africa, although less well known, and Australia as well. Of comparative interest, the four ant species on St Helena island are all invasive tramp species; two pantropical and two African (Taylor and Wilson 1961). Until the species status of *H. perfidus* is resolved, the species remains listed as invasive in Table 19.1.

Heterotermes philippinensis. There has been no change since 1969; one extra information source has been added.

19.3.4 Species No Longer Considered Invasive

There are three species that were listed in 1969 but have not been the subject of further reports of activity. Two species were listed as "established" by Gay (1969), but were probably interceptions and not invasive populations, as defined by Falk-Petersen et al. (2006). These have been removed from the list. One species has less clear status.

19.3.4.1 Kalotermitidae

Neotermes insularis. This Australian species was considered to be established in New Zealand by Gay (1967, 1969). However this status has been revised to introduced without establishment (Bain and Jenkin 1983; Philip et al. 2008), and therefore has been removed from the list.

19.3.4.2 Rhinotermitidae

Coptotermes frenchi. This Australian species has possibly been eliminated from New Zealand since 1969. There has been ongoing management of invasive termites in New Zealand, as noted for *C. acinaciformis* above. There have been scattered discoveries from the 1940s to the 1970s from three locations (Whangarei, Auckland and Wellington), ca 800 km apart on the north and south points of the North Island. The most recent report to mention active infestations of *C. frenchi*, then being treated, was that of Bain and Jenkin (1983). It is possible that treatment was effective as the four articles from New Zealand in the following 27 years mention only *C. acinaciformis*.

Coptotermes heimi. This species originates from the Indian subcontinent and Gay (1967, 1969) listed two locations of interceptions in South East Asia: Java (possibly established) and Singapore. However, since there have been no further reports of this species in these locations, it has been removed from the list.

19.4 Characteristics of Invasive Species

The number and relative abundance of invasive species in termite families and sub-families are listed in Table 19.2. Three broad patterns can be seen from this table: wood for food, nesting in food, and a high capacity to generate secondary reproductives. These three patterns are most prevalent in two families, the Kalotermitidae and Rhinotermitidae.

19.4.1 Wood-Feeding

All invasive termites are wood feeding species; there are no documented invasive species that eat litter, grass or any type of soil. Furthermore, these species do not eat every form of wood, but only sound heart wood; all 26 species in Table 19.1 are known to attack this form of wood. Sound heart wood is exactly the type of wood most valued by humans for housing, fencing, poles, ship building (historically), and

Table 19.2 Main termite clades showing total and invasive number of species, life way group, food, and capacity to produce secondary reproductives

	Species						Secondary reproductives[b]				
Clade (Family/subfamily/group)[a]	Total #	# Inv	%Clade Inv	%Inv	Life way type[a]	Food[a] Total	Inv	Type	# Spp (e n)	%Clade (e n)	%Inv (e n)
Mastotermitidae	1	1	100	3.8	Int I	dw	dw	e	1	100	100
Hodotermitidae	20				Sep I	g		n	4	21	
Termopsidae	20	1	5	19.2	Sin I	ww	dw	n s	14	70	100
Kalotermitidae	430	9	2.1	34.6	Sin I	dw	dw	n a	51	12	100
Serritermitidae	2				Sep I	w		e n			
Rhinotermitidae	330	13	3.9	50	Sin I, Int I, Sep I	vw	dw	e n	38	11	100
Termitidae											
Macrotermitinae	330				Sep II	f (w g l)	var	a			
Foraminitermes group	10				III	o		?			
Apicotermitinae, Anoplotermes group	160				III, IV	o s		a			
Apicotermitinae, Apicotermes group	35				IV	s		a			
Termitinae, Amitermes group[c]	305				Int II, Sep II, III	vw g l o		e n a	34	11	
Termitinae, Termes group[c,d]	285	1	0.4	3.8	Int II, Sep II, III, IV	vw o s	dw	n a	10	4	?

Table 19.2 (continued)

Clade (Family/subfamily/ group)[a]	Species: Total #	# Inv	%Clade Inv	%Inv	Life way type[a]	Food[a] Total	Secondary reproductives[b]: Inv	Type	# Spp (e n)	%Clade (e n)	%Inv (e n)
Cubitermes group	165				IV	s		n a	1	< 1	
Nasutitermitinae[e]	650	1	0.2	3.8	Int II, Sep II, III, IV	vw g l o s	dw	e n a	24	4	100

For species: Total # = in the clade # Inv = invasive species in clade. Life way types: Sin = single piece nester, Int = intermediate nesters, Sep = separate piece nester, I = non-humified plant food (w g l), with gut flagellates; II = non-humified plant food (w g l) without gut flagellates; III = degraded organic material–soil interface; IV = mineral soil; n.b. III and IV have no nest as they live in soil. Food: w = wood (dw = dry, sound wood, ww = wet, rotted wood, vw = wood in various states), g = grass, l = litter, f() = fungus culturing, o = (degraded) organic material–soil interface, s = mineral soil. Secondary reproductives, types: e = ergatoid (excludes any form with wingbuds), n = nymphoid (includes any form with wingbuds), a = adultoid (all forms, includes retained alates), s = soldier, ? = uncertain. Secondary reproductives: #Species (en) = number of species with ergatoids or nymphoids; %Clade (en) = percentage of species in clade with ergatoids or nymphoids %Inv (en) = percentage of species in clade with ergatoids or nymphoids.

[a]After Donovon et al. (2001), Eggleton and Tayasu (2001), and Inward et al. (2007).

[b]After Sieber (1985), Myles (1999), Roisin (1987, 2000) and Bourguignon et al. (2009).

[c]Paraphyletic group; Inward et al. (2007).

[d]Ergatoid neotenics reported for four species only: *Amitermes hastatus*, *Microcerotermes biroi*, *M. papuanus* and *M. strunkii*.

[e]Ergatoid neotenics reported for thirteen species: *Lacessititermes batavus, Leptomyxotermes doriae* and 13 *Nasutitermes* spp. including *corniger*.

still transported as logs, sawn squared standard lengths, or as manufactured objects. This pattern was noted by Gay (1967, 1969) who considered the wood trade as the primary cause of invasion, rather than wood feeding per se, viz: "it is obvious that the commonest method of transport from place to place is in timber"

Two species illustrate apparent exceptions. A single termite from the Termopsidae is included in the list of invasive species, having been reported in New Zealand (Philip et al. 2008): *Porotermes adamsoni*. This species would seem to be atypical because termopsids are also known as dampwood termites as they eat wet and rotted wood. Yet even this species attacks sound wood and so conforms to the pattern (Lenz et al. 1982b). *Porotermes adamsoni* normally attacks a tree from a dead and rotting part, such as a weathered branch stub, but then tunnels into sound heart wood in the trunk, spreading wood rotting fungus and thus expands the food resource and habitat available to it. The same pattern of attack is also effective for timber in service, such as telephone poles and bridge supports; consequently the species is considered a pest in forestry and urban areas (Hill 1942; Hadlington 1987). There is a single species from the Termitidae, Termitinae, Termes group: *Termes hispaniolae*. It is unusual in that it is a wood-feeding species from a group that predominantly consumes organic matter – soil/wood interface and mineral soil feeders (Roisin et al. 2006); note however that the Termes group is paraphyletic (Inward et al. 2007), so *T. hispaniolae* may be removed from this clade after further studies.

19.4.2 Nesting Habit

There have been various classifications of termites using "life types" or nesting behaviours (e.g. Abe 1987; Shellman-Reeve 1997). Table 19.2 uses the "life way" classification of Eggleton and Tayasu (2001), a more complete descriptive classification based on nesting behaviour modified from the life types and feeding substrate types (of Abe 1987) and feeding groups (following Donovan et al. 2001).

Single-piece nesters feed on and nest in the same single piece of substrate, which is always wood. Intermediate nesters are like single piece nesters, except they also forage away from their nest substrate to find other patches of feeding substrate nearby, which is always wood. Once the nest substrate is exhausted, the colony relocates the nest to a new substrate. Separate-piece nesters live a nest that is separate and distinct from their feeding substrate; they forage away from their nest to find patches of feeding substrate nearby. Separate piece nesters feed on wood, grass and litter. No distinction is made between the location of the separate nest i.e. whether hypogeal (below-ground), epigeal (above-ground), or arboreal.

Donovan et al. (2001) defined four feeding groups based on gut morphology and content analysis. Type I feeds on non-humified plant matter, easily recognised as wood, grass or litter, which is digested with the aid of gut flagellates. The type I feeding group is comprised of all lower termites (i.e. all families other than the Termitidae). Type II feeds on non-humified plant matter as for type I, but digested without the aid of gut flagellates. This group is comprised of higher termites that feed on plant matter and live in nests. Type III feeds on more degraded and thus humified plant matter at the organic matter–soil interface, and comprises only higher termites. Type IV feeds on highly degraded and humified plant matter in mineral soil. The last group are also known as the "true soil feeders" and comprises only the higher termites. Types III and IV frequently have no nest as they live in the soil on which they feed and move through the soil, although many prominent epigeal mound-builders are also affiliated to these groups.

It is clear from Table 19.2 that invasive species are either type I single piece nesters (termopsid and kalotermitid species: the ten species in *Porotermes*, *Cryptotermes*, *Incisitermes*, *Kalotermes* and *Glyptotermes*), type I intermediate piece nesters (*Mastotermes darwiniensis* and the twelve rhinotermitids species in *Reticulitermes*, *Heterotermes* and *Coptotermes*), or type II intermediate nesters (*Termes hispaniolae*, *Nasutitermes corniger*). There are neither separate type nesters nor any type III or IV invasive termite species.

19.4.3 Secondary Reproductives

Primary reproductives in termites are defined as those imago (i.e. the adult form of all hemimetabolous insects) individuals that depart from their natal nests, fly to a new location, find a mate and then found a new nest and colony independently (i.e. without assistance from helper castes). Secondary reproductives are those individuals that do not undertake this hazardous process independently, but remain in their natal nest or colony (or may bud from their natal colony) and so retain assistance from their nest mate helper castes (Thorne 1996; Myles 1999; Roisin 2000). There are many sub-definitions within this overarching definition of secondary reproductives (see Noirot 1985a, 1985b, 1985c; Lenz 1985, 1994; Lenz and Barrett 1982; Lenz et al. 1986, 1988; Sieber 1985; Roisin 1987; Thorne 1996; Myles 1999), based on temporal sequence (supplementary reproductives co-occur with primary reproductives whereas replacement reproductives appear once the primary reproductives die) and spatial location (assister reproductives are located in the main nest whereas budding reproductives are located away from the main nest, often at peripheral feeding sites). Also, the categories can overlap.

Secondary reproductives are also defined by their caste of origin. There are many precise definitions based on caste, instar, level of wing and eye development and other characters (see Thorne 1996; Myles 1999; Roisin 2000; Korb and Hartfelder 2008), however such exacting definitions hamper cross-taxon comparisons as particular types are often found in only one or two families. Therefore only three broad categories are recognised in this chapter: (1) *ergatoids* develop from wingless individuals (those usually called "workers", therefore excluding any form that produces wingbuds during the developmental moult(s) to reproductive status), (2) *nymphoids* which are those individuals developing from nymphs through the standard hemimetabolous pathway to adulthood (nymphoids can have or gain wingbuds during the moult(s) to reproductive status), (3) *adultoids* developing from alates or penultimate instars, similar to alates and not departed from the nest and (4) *soldier reproductives* developing from soldiers (known only from the Termopsidae).

The origin of secondary reproductives may be more important than their temporal and spatial locations for invasiveness. The crucial factor is the opportunity of creating a viable propagule (sensu Mack et al. 2000; Lockwood et al. 2005) for transport (in wood by humans). Species that form ergatoid reproductives have the highest potential to be transported as any nest or food resource containing foraging workers can become a viable propagule at any time of year. Species that form nymphoid reproductives have a medium to high potential to be transported, lower if the nymphs do not forage away from the nest, and higher if they do. This is moderated by availability of nymphs; they are present between 6 and 10 months in the year, depending on the species. Species that form adultoid reproductives have a low potential to be transported because alates (and penultimate instar nymphs) are only found in the natal nest. In addition, alates are present for a short period, perhaps 1 or 2 months in the year.

Approximately 200 species have described secondary reproductive (for review see Myles 1999, for Serritermitidae see Bourguignon et al. 2009). The list is far from complete as the great majority of species have not received sufficient study for secondary reproductives to be known; for comparison, even food type or preference is unknown or poorly known for the majority of termite species (Donovan et al. 2001; Eggleton and Tayasu 2001). However the 200 known examples are from all clades, except the Termitidae Foraminitermes group (Table 19.2). Therefore it is likely that the majority of termite genera, if not species, are capable of producing secondary reproductives. However, the caste from which they develop differs between clades: the lower termites can produce ergatoids or nymphoids or both, whereas the higher termites mostly produce nymphoids or adultoids; three termitid clades produce only adultoids.

All invasive species, including the two termitid species, are capable of forming ergatoid or nymphoid secondary reproductives. Gay (1967, 1969) did not mention the pattern of neotenic reproduction explicitly, but he did state that two Australian invasive *Coptotermes* species "appear to rely mainly on supplementary reproductives for colony foundation in New Zealand, whereas in Australia such forms are very uncommon and colony foundation is almost invariably by winged reproductives". This comment may serve as a general rule for invasive rhinotermitids, although there is considerable variation with latitude for *R. flavipes* and *C. formosanus* (see Table 19.2 in Vargo and Husseneder 2009); further studies are needed to determine a definitive pattern.

19.4.4 Kalotermitidae and Rhinotermitidae

These families account for 22 of the 26 (85%) invasive termite species: nine (35%) are in the Kalotermitidae and thirteen (50%) are in the Rhinotermitidae. The relative difference between kalotermitids and rhinotermitids is accentuated when the proportion of invasive species in each family is considered: 2.1% of the Kalotermitidae are invasive compared with 3.9% of the Rhinotermitidae (Table 19.2). The pattern has changed little since Gay (1969), who noted that invasive species "occur mainly in the Kalotermitidae and Rhinotermitidae, and particularly in the genera *Cryptotermes* and *Coptotermes*." These two genera account for five and six invasive species respectively. *Heterotermes* appears equally invasive with five species; however this number may decrease by two with further taxonomic study (see comments under *Heterotermes perfidus* and *H.* sp. nov. above).

There is a clear explanation for the over-representation of the Kalotermitidae and Rhinotermitidae and the perhaps even more remarkable under-representation of the Termitidae (two invasive species despite this family comprising the majority (71%;) of all termite species). The invasion-typical patterns of wood feeding, nest type and secondary reproductives are not independent, as almost all lower termites are wood feeders, whereas most higher termites are not (Donovan et al. 2001; Eggleton and Tayasu 2001). In addition, lower termites have more opportunity to be transported as they are either type I single-piece nesters (Kalotermitidae), or type I intermediate nesters (Rhinotermitidae), plus they have greater capacity to form ergatoid or nymphoid reproductives, often because they have more flexible developmental pathways.

Any piece of wood containing kalotermitids is a potential propagule because (a) as type I single piece nesters the wood is the nest (by definition as the wood substrate is the nest as well as the food), thus the reproductives are present. If the piece of wood is cut such that the nest is divided and the reproductives are not present in the transferred piece, this piece of wood is still a potential propagule because (b) secondary reproductives can develop from any individual (other than the soldiers). Kalotermitids do not have a wingless or worker developmental pathway, as all individuals are capable of maturing into alates (Noirot 1985a). Wingless individuals are simply juveniles, the younger instar nymphs, though note there is discussion about possible variation on a simple and direct development into alates on the pattern of other hemimetabolous insects. Whatever variation there may be appears to be a consequence of a strategy to delay maturation for an opportunity to inherit the natal nest, at least in some species of *Cryptotermes* (Korb and Katrantzis 2004; Korb and Lenz 2004; Korb and Schmidinger 2004). Although described for *Cryptotermes* species, the same situation exits for the other invasive kalotermitid species and the termopsid *Porotermes adamsoni*.

Similarly, many, perhaps most pieces of wood containing rhinotermitids are potential propagules also. As type I intermediate nesters the wood is often the nest, so argument (a) for kalotermitids applies for rhinotermitids. These termites also forage away from the nest; any such food-only wood transferred by humans may also become a propagule as for kalotermitids (argument b) if individuals capable of developing into secondary reproductives are present. This will be more likely for species capable of producing ergatoids, and less likely for those capable of producing nymphoids only, and then only for those species in which nymphs forage for themselves. *Heterotermes* and especially *Reticulitermes* species have very flexible developmental pathways and capacity to produce ergatoid and nymphoid secondary reproductives (e.g. Buchli 1956; Gay and Calaby 1970; Howard and Haverty 1980;

Jones 1990). In addition, *R. flavipes* carries larvae to food sources, allowing any form of cast to develop at that site, should it be cut off from the natal colony (Lenz et al. 2009).

The same situation exits for *Mastotermes darwiniensis* (Hill 1942; Thistleton et al. 2007), and the two termitid invasive species. *Mastotermes darwiniensis* is a type I intermediate nester and the two termitid species are type II intermediate nesters, which are functionally equivalent (the difference being in gut symbionts). Both *Termes hispaniolae* and *Nasutitermes corniger* live in tree canopies and either live in branches or build arboreal nests on branches (Thorne 1982; Thorne and Noirot 1982; Roisin et al. 2006), but as these nests are built on and from their food substrate, these species do not qualify as separate piece nesters.

The situation is less clear in *Coptotermes*; a few older references suggest that ergatoids do form in this genus (Kelsey 1945; Tang and Li 1960) though most references either give no description or state that nymphoids were found (see Myles 1999 for review). More recent studies have found that secondary reproductives are always nymphoid (Roy-Noël 1966; Lenz and Barrett 1982; Lenz et al. 1986, 1988; Lenz and Runko 1993; Costa-Leonardo et al. 2004; Raina et al. 2004), and groups of workers do not produce secondary reproductives (e.g. Raina et al. 2004; Morales-Ramos and Rojas 2005). *Coptotermes* nymphs forage away from the nest (e.g. Haverty et al. 1996; Evans et al. 1998; Raina et al. 2004), which increases the potential that human assisted transfer of wood will produce viable propagules.

These patterns have been demonstrated in field populations by population genetic data. These data indicate the presence of inbreeding with multiple related secondary reproductives in *Mastotermes darwiniensis* (Goodisman and Crozier 2002), *Reticulitermes flavipes* (Bulmer et al. 2001; Bulmer and Traniello 2002; DeHeer and Vargo 2004; Dronnet et al. 2005; Parman and Vargo 2008), *R. grassei* (DeHeer et al. 2005), *Coptotermes formosanus* (Husseneder et al. 2005; Vargo et al. 2006; Aluko and Husseneder 2007) and *C. gestroi* (Costa-Leonardo et al. 1999, 2004; Costa-Leonardo and Arab 2004). The pattern has been found in *Nasutitermes corniger* also (Atkinson and Adams 1997).

Coptotermes has the highest number of invasive species, but not all *Coptotermes* species are invasive. The effect of life way can be demonstrated by considering three mound-building species, which are all separate piece nesters: *Coptotermes acinaciformis*, *C. lacteus* and *C. brunneus*. These species have been not transported outside their natural ranges in Australia (note that the mound-building form of *C. acinaciformis* is likely to be a separate species from the invasive hypogeal and tree-trunk nesting form (Brown et al. 1990, 2004)). This is true even though incipient colonies can be described as type I intermediate nesters, as they start from dead wood litter. No separate-piece nester species are invasive.

It is important to note that only a small minority of kalotermitids and rhinotermitids are actually invasive. Therefore the life way type by itself is insufficient to explain invasiveness. Lenz et al. (1982a) and Lenz (1994) demonstrate the much higher and faster neotenic production in invasive species, compared with some non-invasive natives, and also between different invasive populations. More work will be required to elucidate these differences.

19.5 Invaded Habitats

Gay (1967, 1969) made several statements about invaded habitats, including broad biogeographic scale (island and coasts), type of habitat (human modified) and possible resilience. The information available at the time he wrote made these statements reasonable, even obvious. However in the forty intervening years, new and better identification methods, constant surveying of new introductions (especially of *Coptotermes formosanus*, a morphologically distinct species in its invaded habitat in the USA) and better theoretical understanding call these statements into doubt.

19.5.1 Islands and Coasts

Gay (1969) stated, "with remarkably few exceptions, all successful introductions and establishments have taken place in islands or on the coastal fringes of continental areas." This pattern is somewhat still true 40 years later, but with modification. Table 19.3 shows the invaded regions and localities simplified from Table 19.1. The South Pacific Ocean is the most invaded region, with 13 species, followed by North America (9), the Caribbean Sea (8) and the North Pacific Ocean (8), the Indian Ocean (6), South America (5), Australia (4), East Asia (3) Africa (3), North Atlantic Ocean (3), South Atlantic Ocean (2), South and South East Asia (2) and Europe (1). A simple t-test comparing the number of invasive species on islands in oceans (6.8 ± 1.8) and on continents (3.9 ± 2.6) shows there is no significant difference ($t_{11} = 1.679, p = 0.121$). This lack of difference is driven by North America having a high number, and the North and South Atlantic Oceans having low numbers of invasive species. This asymmetry results from several invasive species being found only in Florida in North America, which could be considered more part of the Caribbean than continental North America in biogeographical terms. The tropical islands of the Atlantic Ocean are mostly found in the Caribbean Sea. Removing Florida from North America and placing it into the Caribbean, and merging the Caribbean into the North Atlantic Ocean shows that islands (7.8 ± 3.7) have significantly more invasive species than continents (3.4 ± 1.7) ($t_{10} = 2.772$, $p = 0.020$).

Gay (1969) believed that only coastal fringes of continents were invaded, and examples of this remain in evidence. For instance, the four species of invasive *Cryptotermes* in Australia are limited to the humid coasts and are absent from the dry interior (Yule and Watson 1976; Gay and Watson 1982). But this pattern is due to physiological limitations of the termites; kalotermitids fulfil their water requirements from the metabolism of the food, which is possible only with high humidity. Thus two invasive *Cryptotermes* species are not limited to coastal habitats in South America and India, but are found hundreds of kilometres inland due to the much wetter climate (Chhotani 1963; Constantino 2002).

Table 19.3 The invaded regions of the world and the species that have invaded them

Invaded regions/localities	#	Invasive species (1 unless specified)	Natural range
Caribbean	9		
Greater Antilles	3	*Cryptotermes*, *Heterotermes*, *Coptotermes*	S America, SE Asia
Lesser Antilles	8	2 *Cryptotermes*, 2 *Coptotermes*, 2 *Heterotermes*, *Nasutitermes*, *Adaiphrotermes*	S America, Africa, SE Asia
N group[a]	4	*Cryptotermes*, *Coptotermes*, *Nasutitermes*, *Termes*	S America, SE Asia, Caribbean
S group[b]	3	3 *Cryptotermes*	S America, SE Asia, Africa
North Atlantic	3		
British Is.	1	*Reticulitermes*	Europe
Is off N Africa[c]	1	*Cryptotermes*	S America
Is off W Africa[d]	1	2 *Cryptotermes*	S America, Africa
South Atlantic	3		S America, Africa, unknown[e]
North Pacific	8		
NW Pacific[f]	3	*Incisitermes*, 2 *Coptotermes*	N America, E Asia, SE Asia
Micronesia[g]	6	3 *Cryptotermes*, *Incisitermes*, 2 *Coptotermes*	S America, SE Asia, E Asia
Hawaii	5	2 *Cryptotermes*, *Incisitermes*, 2 *Coptotermes*	S America, N America, SE Asia, E Asia
South Pacific	13		
Melanesia[h]	8	*Mastotermes*, 3 *Cryptotermes*, *Glyptotermes*, *Coptotermes*, *Nasutitermes*	S America, Australia, SE Asia
Polynesia[i]	3	2 *Cryptotermes*, *Coptotermes*	S America, SE Asia
New Zealand	5	*Porotermes*, *Kalotermes*, *Glyptotermes*, 2 *Coptotermes*	Australia
Galapagos	1	*Heterotermes*, *Reticulitermes*	S America
Indian Ocean	6		
E islands[j]	1	*Cryptotermes*	S and SE Asia
SW islands[k]	4	2 *Cryptotermes*, *Heterotermes*, *Coptotermes*	SE Asia, Africa
Sri Lanka	2	*Cryptotermes*, 2 *Coptotermes*	E Asia
North America	9		
Northern NA[l]	2	*Cryptotermes*, *Reticulitermes*	SE Asia N America
Southern NA[m]	8	3 *Cryptotermes*, *Incisitermes*, *Heterotermes*, 2 *Coptotermes*, *Nasutitermes*	N America, S America, SE Asia, E Asia
South America	5		
Northern SA[n]	4	3 *Cryptotermes*, *Coptotermes*	S America, SE Asia, Africa
Southern SA[o]	1	*Reticulitermes*	N America
Europe	1	*Reticulitermes*	N America
Africa	3		
W Africa	1	*Cryptotermes*	S America
E Africa	1	*Cryptotermes*	SE Asia
S Africa	2	*Cryptotermes*, *Coptotermes*	S America, E Asia
Arabia	1	*Coptotermes*	S Asia

Table 19.3 (continued)

Invaded regions/localities	# Invasive species (1 unless specified)		Natural range
Asia	5		
S and SE Asia	2	2 *Cryptotermes*	Africa, SE Asia
E Asia	3	*Cryptotermes, Incisitermes, Coptotermes*	S America, SE Asia, N Amercia
Australia	4	4 *Cryptotermes*	S America, SE Asia

Data from Table 19.1. NB, the number in front of the genera indicates the number of species if more than one.
[a]Bahamas and Turks.
[b]Is near Venezuela, including Trinidad and Tobago.
[c]Azores, Madeira, Canary, Cape Verde.
[d]Príncipe, São Tomé, Annobón.
[e]Species ID uncertain.
[f]Japan, Taiwan.
[g]From Yap and Guam in west to Midway and Kiribati in east.
[h]Not incl. New Zealand.
[i]From Samoa to Tahiti, incl. Easter, not incl. Hawaii.
[j]Cocos/Keeling islands.
[k]Madagascar, Mauritius, Rodrigues, Reunion.
[l]Temperate areas of Canada and Northern USA.
[m]Sub/tropical areas of USA, Mexico, Central America.
[n]Sub/tropical areas, including Venezuela, Guiana, Brazil.
[o]Temperate areas, including Uruguay, Chile.

With the continuation of invasions over time and the introduction of molecular identification methods, there are now many examples which undermine Gay's dogma. The factor of elapsed time is most clearly seen for *Coptotermes formosanus* in the USA, which has seen a spectacular spread from the initial sites of Lake Charles in Louisiana and Houston Ship Channel in Pasadena, Texas, across these two states and to eight others, with locations several hundred kilometres inland, mostly by human transport (Woodson et al. 2001; Jenkins et al. 2002; Messenger et al. 2002; Austin et al. 2006; Sun et al. 2007). The effect of better identification methods is most clearly seen for *Reticulitermes flavipes* in France, which has spread across southern France, including well into the interior, probably from alate flight as well as by human transport (see Clément et al. 2001; Austin et al. 2005). However, they may be an influence of apparency on records of termite occurrence: in low biodiversity habits (e.g. islands), termites may be noticed when they appear, and especially if they have alate swarming, or where damage to previously sound timber in service is noticed. Thus continental areas may be invaded at similar (or even higher) rates, but detection is more difficult against a background of higher biodiversity and the presence of non-invasive native species, or where all termites including the invaders have cryptic habits.

To illustrate this, we can cite the cases of (1) the USA, which has relatively low termite diversity and no *Coptotermes* species, where identification can easily be made from the soldiers with their distinctive defensive secretion and (2) Uruguay and Chile, which also have low termite diversity and where invading *Reticulitermes* are readily detected. On the other hand, species that are similar to native species are likely to remain undetected, for example the French *Reticulitermes santonensis* is really the invasive North American species *R. flavipes*, and the New Guinean *Nasutitermes polygynus* is really the South American *N. corniger*. Perceived patterns may be affected by the availability of taxonomic expertise. Perhaps it is not mere coincidence that the most invaded state in the USA also has the greatest concentration of termite specialists. If so, then invasions in much of Africa, Asia and South America will be more difficult to detect. Similarly, invasions in urban areas with immediate damage to houses may be detected more readily than invasions of natural or semi-natural habitats.

19.5.2 Habitat Type and Resilience

The patterns of increased spread question another existing dogma reflected by Gay (1969): "they [invasive termites] are rarely able to invade native habitats and are almost always restricted to the man-modified environment. This restriction seems to be more definite in continental areas than in islands or island-like areas and may be due to biotic barriers (resulting from the presence of closely related native species), which are strong in continental areas, but relatively weak in the simpler island communities."

As has been indicated for particular species previously, the statement about native habitats is not valid. Although still confined to Lae (almost certainly due to the extremely wet habitat and two huge rivers) *Mastotermes darwiniensis* has been found in more than 40 species of native trees including in the city's Botanic Gardens (Thistleton et al. 2007). *Porotermes adamsoni* has been found in native trees in New Zealand, and *Cryptotermes brevis* and *C. cynocephalus* have been found in native trees in Hawaii (Scheffrahn et al. 2000). *Coptotermes formosanus* is now found not only in trees in urban parklands (Guadalupe Rojas et al. 2001; Mankin et al. 2002; Brown et al. 2007), but in native forests in the USA (Sun et al. 2007), and in Japan (Mori 1987). As mentioned previously, invasive populations of *Reticulitermes flavipes* were described as a native species because of being found in native forests in France (Clément et al. 2001; Dronnet et al. 2005), as were those of *Nasutitermes corniger* in New Guinea (Roisin and Pasteels 1985; Scheffrahn et al. 2005b).

These discoveries question the use of the presence of a termite species in human modified environments and its absence from natural habitats as the defining criterion (after Gay 1967, 1969) of an introduction. Termite introductions occur through trade, which is focussed in ports and cities, which are human modified areas. Therefore it is expected for termites to be introduced to and be found in coastal urban areas in the first instance; the question is how long is required for the introduced species to become invasive (moving beyond the point of introduction, including other urban areas) and eventually become "naturalized", as defined by Falk-Petersen et al. (2006). Many introduced plant and animal species have become naturalised, and there is no obvious reason why introduced termites are incapable of doing so also. Wood eating insects are known to have appreciably and significantly longer life cycles than insects that eat other materials (Danks 1992), due in part to in digestibility and low nitrogen availability (LaFage and Nutting 1978; Mattson 1980; Nalepa 1994; Nalepa et al. 2001). Therefore it is likely that the progress from invasive to naturalised species would also take longer for wood eating insects, including termites. *Reticulitermes flavipes* was introduced into Western France around two centuries or more ago (Bagnères et al. 1990; Vieau 2001); perhaps this period of time is indicative of how long is required for this species, or even this termite life way (type I intermediate nester), to naturalise. A similar period of time or longer has elapsed for the drywood species *Cryptotermes brevis*; it too is found in native habitats (Scheffrahn et al. 2000) albeit rarely (Scheffrahn et al. 2009). The original native habitats for *Cryptotermes domesticus* and *Cryptotermes dudleyi* in Asia are unknown; perhaps the many hundreds or even thousands of years of trading in Asia was required for these species (or life type way) to naturalise.

Gay's comments about biotic barriers are antecedent to the modern concept of ecosystem resilience (Holling 1973; Peterson et al. 1998; Stachowicz et al. 2002; Elmqvist et al. 2003), although ecological function is now considered more important that phylogenetic relationships. There is no clear evidence to suggest resilience exists in termite communities, in part because no studies have asked this question explicitly. Global diversity of termites according to families and life types has been determined by Eggleton et al. (1994), Eggleton (2000) and Eggleton and Tayasu (2001). If resilience is conferred by either ecological functional groups (equivalent to life ways) or related species, then the regions with the greatest diversity should prove to be the most resilient. Evidence is mixed from the Kalotermitidae. *Cryptotermes brevis* is absent from South and South East Asia and East Asia (Gay 1967, 1969; Scheffrahn et al. 2009), which has a moderately high kalotermitid fauna. Yet *Cryptotermes brevis*, *C. dudleyi* and *C. havilandi* have invaded most of the Caribbean and South America and Madagascar even though these areas have the highest kalotermitid diversity on the planet. Evidence is also mixed in the Rhinotermitidae and type I and II intermediate nesters. The highest diversity of rhinotermitids and type I and type II intermediate nesters is in South East Asia, and no rhinotermitids have invaded this region. There is a lack of rhinotermitids in the temperate locations in South America where *Reticulitermes flavipes* has invaded, and a lack of functional ecological equivalents also, perhaps because the fauna is mostly affiliated to the Nasuititerminae, which is fundamentally a tropical termitid group. However, *R. flavipes* has invaded France, which has several native *Reticulitermes* species and *Coptotermes formosanus* has invaded Japan and the USA, which both have native *Reticulitermes* species, which are of the same life type (type I intermediate nesters). That said, there may still be a case for a niche separation. *Coptotermes* species attack live trees (Hill 1942; Roy-Noël 1966; Guadalupe Rojas et al. 2001; Werner and Prior 2007) and *Reticulitermes* species do not; thus live trees may represent otherwise untapped resources.

19.6 Source Habitats of Invasive Species

Table 19.4 re-organises the information from Table 19.1 according to the source region of invasive species. South and South East Asia is the source of the greatest number of invasive species (7), then South America (6), Australia (6), Africa (3), North America (2), Caribbean (1), East Asia (1) and Europe (1). All source areas are continental, except for the Caribbean. Note however, the concerns (above) about the identity of the invasive *Heterotermes* sp. nov. from the Caribbean. These patterns reflect diversity of the typical invasive termite species (type I single piece nester, and type I and type II intermediate nester) and trade routes. The highest diversities of type I single piece nester species are in (northern) South America, South Asia, (north) Australia, and Madagascar. The highest diversity of type I and type II intermediate nester species is in South and South East Asia, then (north) South America and (north) Australia. The north of South America and South and South East Asia were the most important trading destinations during European colonial expansion, when ocean voyaging used wooden ships; Scheffrahn et al. (2009) have suggested Spanish trading during the colonial period was the cause

Table 19.4 The regions that have produced invasive species, and the regions that have been invaded

Natural range	# Invasive species (1 unless specified)		Invaded island regions (# of 6 total)		Invaded continental regions (# of 7 total)	
Caribbean	1	*Heterotermes*	1	Caribbean	1	N America
N America	2	*Reticulitermes*, *Incisitermes*	1	N Pacific	4	N America, S America, Europe, E Asia
S America	6	*Cryptotermes*, *Incisitermes*, 2 *Heterotermes*, *Termes*, *Nasutitermes*	5	Caribbean, N Atlantic, S Atlantic, N Pacific, S Pacific	5	N America, S America, W Africa, S Africa, E Asia, Australia
Europe	1	*Reticulitermes*	1	N Atlantic	0	
Africa	2	*Cryptotermes*, *Coptotermes*	3	Caribbean, S Atlantic, Indian	2	S America, S and SE Asia
S and SE Asia	7	3 *Cryptotermes*, *Heterotermes*, 2 *Coptotermes*	4	Caribbean, N Pacific, S Pacific, Indian	6	N America, S America, E Africa, S and SE Asia, E Asia, Australia
E Asia	1	*Coptotermes*	2	N Pacific, S Pacific	2	N America, Africa
Australia	6	*Mastotermes*, *Glyptotermes*, *Kalotermes*, *Porotermes*, 2 *Coptotermes*	1	S Pacific	0	

Data from Table 19.1. NB, the number in front of the genus name indicates the number of species if more than one.

of the spread of *Cryptotermes brevis*. Australia has become a major trading partner of Pacific nations over the past century; raw logs and cut timber (especially railway sleepers or ties) have been implicated as the source of several Australian species in New Zealand (Bain and Jenkin 1983; Philip et al. 2008). In comparison, Madagascar has not been a major trading area; therefore its diverse type I single piece nester fauna has not had the same level of opportunity for human assisted transport.

Trade has been found to be more important than island status or country size for invasions (Westphal et al. 2008). Trade takes place over many distances; the information in Table 19.4 suggests that invasive species tend to originate from the closest continental regions. This is most obvious for Australian invasive species in the South Pacific, but is also clear for European species to North Atlantic (Britain), for South and South East Asian species to other parts of the same region, to the North and South Pacific, India, East Asia and Australia, and for South American species to other parts of South America the Caribbean, North and South Atlantic, and North America. The type of trade favouring the transportation of termites is unknown. Measuring movement of wood, either as ships or as a traded commodity over the millennia of human trade is not a feasible analysis, but it can be conjectured that perhaps four generic episodes of long distance oceanic travel may have transplanted termites to the greatest extent. (1) The initial human dispersal into and subsequent trade within the Pacific Ocean, as recently as 3000 years ago (Kayser et al. 2000). (2) Arab and Persian trade from east Africa and Arabia across the Indian Ocean to South and South East Asia using dhows, and Indian and East Asian trade in Asia and back to Arabia has occurred over the past 1,500 years. (3) European colonisation and trade of nearly 500 years, from the Spanish and Portuguese "Age of Discovery" (early 1400s) to the age of "New Imperialism" (1800). (4) Trade on a modern scale in the twentieth and twenty-first century. The first three episodes used wooden ships (although ships were made of iron and steel towards the end of European colonial period), and the ships themselves may have been an important vector. Watson and Gay (1982) commented that *Cryptotermes domesticus* and *C. dudleyi* may have been introduced to northern Australia by wooden Indonesian boats fishing in coastal Australian waters. However, in all periods, smaller wooden items were and are transported. Today, the main concern is solid wood packing (e.g. Haack and Cavey 1997, 2000; Haack 2006).

19.7 Future Invasions?

The most likely new invasive species will resemble the existing 26 invasive species: wood-eating, type I single piece nester or type I or II intermediate nester, with ergatoid or nymphoid or both secondary reproductives, from an area with a high volume of trade. If so, then more kalotermitids (especially *Cryptotermes* species), rhinotermitids (especially *Reticulitermes*, *Heterotermes* and *Coptotermes* species), and termitids (*Nasutitermes*, but perhaps also type II intermediate nester species from *Microcerotermes* and *Amitermes*) may join the list. Future new invasive termite species may be native pests that live in urban habitats. Such termites are pre-adapted to living with humans. For example, *Coptotermes gestroi* is the major pest species in urban areas in South East Asia; 85% of house infestations in urban areas of peninsula Malaysia are from this species, which becomes less common in suburban and semi rural areas, and is rare in natural forests (Kirton and Azmi 2005). The pattern is repeated for *C. formosanus* in China and Taiwan (note collecting sites of Fang et al. 2008 and Li et al. 2009: Chun-Chun Tsai, Tunghai University, Taiwan; Junhong Zhong Guangdong Entomological Institute, China; Jiangzhu Mo, Zhejiang University, China; personal communications). This pattern suggests that *C. gestroi* and *C. formosanus* are better adapted to survive in disturbed areas, such as human urban areas. If so, then these *Coptotermes* and perhaps other species experience competitive release in human disturbed lands, increase their populations, and so increase the potential for propagule formation.

Molecular studies at species and population level will give greater insights. At the species level, new invasive species or greater distributions of invasive species are likely to be discovered. For example, Kirton (2005) suggested that *C. vastator*, *C. heimi* and *C. ceylonicus*, the main pests in urban areas in the Philippines, Indian subcontinent and Sri Lanka respectively, may be *C. gestroi*. Using molecular markers, Yeap et al. (2007) have since demonstrated that *C. vastator* is *C. gestroi* so perhaps the other two species may also be synonymised. If so, then the reports of *C. heimi* in Singapore and Java (Gay 1967, 1969) will be infestation by a native species. The identity of the mysterious *Heterotermes perfidus* may also be resolved.

At the population level, the distribution routes may be ascertained, building on studies that have determined that *Coptotermes formosanus* was introduced at least twice to the USA (Haverty et al. 1990; Austin et al. 2006). Perhaps routes taken by drywood termitess such as *Cryptotermes* and *Incisitermes* can also be identified and *Cryptotermes brevis* is the ideal candidate, comparing Spanish colonial trading routes (as suggested by Scheffrahn et al. 2009) and population genetics. A similar, but less well documented example may be *Incisitermes immigrans*, perhaps brought to Hawaii and other Pacific islands by native Polynesian travel, before European colonial times (Grace et al. 2002), using a similar approach as Matisoo-Smith et al. (1998) for rats. As biological invasions appear to be increasing, there is likely to be plenty of scope for future research on existing and new invasions.

Acknowledgements I thank Rudi Scheffrahn (University of Florida) for suggesting the topic of this chapter and for providing notes from his extensive surveys of termites in the Americas. I thank many people for providing their published and unpublished information: Ana Maria Costa Leonardo (Universidade Estadual Paulista, Brazil); Laurence Kirton (Forest Research Institute Malaysia), Chow-Yang Lee (Universiti Sains Malaysia), Jannette Mitchell (Agricultural Research Council, South Africa), Jianchu Mo (Zhejiang University, China), Tim Myles (City of Guelph, Canada), Brenton Peters (Department of Primary Industry, Queensland, Australia), Chris Stowe (New Zealand Ministry of Agriculture and Forestry), Brian Thistleton (Northern Territory Department of Primary Industry, Australia), Vivience Uys (South African National Collection of Insects), Tsuyoshi Yoshimura (Kyoto University, Japan), and Junhong Zhong (Guangdong Entomological Institute, China). I thank Patrick Gleeson (CSIRO) for his assistance finding references, and the editors for their patience and assistance.

References

Abe T (1987) Evolution of life types in termites. In: Kawano S, Connell JH, Hidaka T (eds) Evolution and coadaptation in biotic communities. University of Tokyo Press, Tokyo, pp 26–148

Aber A, Beltrami M (2002) *Reticulitermes lucifugus* in urban households in Uruguay (Isoptera: Rhinotermitidae). Sociobiology 40:153–162

Aber A, Fontes LR (1993) *Reticulitermes lucifugus* (Isoptera, Rhinotermitidae) a pest of wooden structures, is introduced into the South American Continent. Sociobiology 21:335–339

Aluko GA, Husseneder C (2007) Colony dynamics of the Formosan subterranean termite in a frequently disturbed urban landscape. J Econ Entomol 100:1037–1046

Araujo RL (1977) Catálogo dos Isoptera do Novo Mundo. Acad Brasileira de Ciências, Rio de Janeiro

Atkinson L, Adams ES (1997) The origins and relatedness of multiple reproductives in colonies of the termite *Nasutitermes corniger*. Proc R Soc Lond B 264:1131–1136

Atkinson TH, Rust MK, Smith JL (1993) The Formosan subterranean termite, *Coptotermes formosanus* Shiraki (Isoptera, Rhinotermitidae), established in California. Pan Pacific Entomol 69:111–113

Austin JW, Bagnères AG, Szalanski AL et al (2007) *Reticulitermes malletei* (Isoptera: Rhinotermitidae): a valid Nearctic subterranean termite from eastern North America. Zootaxa 1554:1–26

Austin JW, Szalanski AL, Cabrera BJ (2004) A phylogenetic analysis of the subterranean termite family Rhinotermitidae (Isoptera) using the mitochondrial cytochrome oxidase (COII) gene. Ann Entomol Soc Am 97:548–555

Austin JW, Szalanski AL, Scheffrahn RH et al (2005) Genetic evidence for the synonymy of two *Reticulitermes* species: *Reticulitermes flavipes* and *Reticulitermes santonensis*. Ann Entomol Soc Am 98:395–401

Austin JW, Szalanski AL, Scheffrahn RH et al (2006) Genetic evidence for two introductions of the Formosan subterranean termite, *Coptotermes formosanus* (Isoptera: Rhinotermitidae), to the United States. Fla Entomol 89:183–193

Bacchus S (1979) New exocrine gland on the legs of some Rhinotermitidae (Isoptera). Int J Insect Morphol 8:135–142

Bagnères A-G, Clément J-L, Blum MS et al (1990) Cuticular hydrocarbons and defensive compounds of *Reticulitermes flavipes* (Kollar) and *R. santonensis* (Feytaud): polymorphism and chemotaxonomy. J Chem Ecol 16:3213–3244

Bain J, Jenkin MJ (1983) *Kalotermes banksiae*, *Glyptotermes brevicornis*, and other termites (Isoptera) in New Zealand. N Z Entomol 7:365–371

Becker G (1970) *Reticulitermes* in mittel und west Europa. Z Angew Entomol 65:268–278

Borges PAV, Myles T (eds) (2007) Térmitas dos Açores. Princípia, Lisboa

Bose G (1984) Termite fauna of southern India. Records of the Zoological Survey of India, Occasional Papers, 49

Bourguignon T, Šobotník J, Hanus R, Roisin Y (2009) Developmental pathways of *Glossotermes oculatus* (Isoptera, Serritermitidae): at the cross-roads of worker caste evolution in termites. Evol Dev 11:659–668

Bravery T, Verkerk R (2008) Termite eradication programme, Saunton, North Devon. Communities and Local Government (UK) Information Leaflet No 18. Reference number 08 BD 05496

Bravery T, Verkerk R (2010) Termite eradication programme, Saunton, North Devon. Communities and Local Government (UK) Information Leaflet No 19. Reference number 09 LGQ 06225

Brown WV, Lacey MJ, Lenz M (2004) Further examination of cuticular hydrocarbons of worker termites of Australian *Coptotermes* (Isoptera: Rhinotermitidae) reveals greater taxonomic complexity within species. Sociobiology 44:623–658

Brown WV, Watson JAL, Carter FL et al (1990) Preliminary examination of cuticular hydrocarbons of worker termites as chemotaxonomic characters for some Australian species of *Coptotermes* (Isoptera: Rhinotermitidae). Sociobiology 16:305–328

Brown KS, Yokum BP, Riegel C, Carroll MK (2007) New parish records of *Coptotermes formosanus* (Isoptera: Rhinotermitidae) in Louisiana. Fla Entomol 90:570–572

Buchli H (1956) Die Neotenie bei *Reticulitermes*. Insect Soc 3:132–143

Bulmer MS, Adams ES, Traniello JFA (2001) Variation in colony structure in the subterranean termite *Reticulitermes flavipes*. Behav Ecol Sociobiol 49:236–243

Bulmer MS, Traniello JFA (2002) Foraging range expansion and colony genetic organization in the subterranean termite *Reticulitermes flavipes* (Isoptera: Rhinotermitidae). Environ Entomol 31:293–298

Cachan P (1950) Les termites de Madagascar et leurs degats. Publications de l'Institut de Recherche Scientifique Tananarive-Tsimbazaza

Chhotani OB (1963) The termite *Cryptotermes havilandi* (Sjöstedt) from the interior of India. J Bombay Nat Hist Soc 60:287–288

Clément J-L, Bagnères AG, Uva P et al (2001) Biosystematics of *Reticulitermes* termites in Europe: morphological, chemical and molecular data. Insect Soc 48:202–215

Coaton WGH, Sheasby JL (1976) National survey of the Isoptera of southern Africa. II. The genus *Coptotermes* Wasmann (Rhinotermitidae: Coptotermitinae). Cimbebasia 3:139–172

Constantino R (1998) Catalog of the living termites of the New World (Insecta: Isoptera). Arq Zool S Paulo 325:135–230

Constantino R (2002) The pest termites of South America: taxonomy, distribution and status. J Appl Entomol 126:355–365

Costa-Leonardo AM, Arab A (2004) Reproductive strategy of *Coptotermes gestroi* (Isoptera: Rhinotermitidae) in Brazil. Sociobiology 44:123–125

Costa-Leonardo AM, Arab A, Casarin FE (2004) Neotenic formation in laboratory colonies of the termite *Coptotermes gestroi* after orphaning. J Insect Sci 4:10

Costa-Leonardo AM, Barsotti RC, Camargo-Dietrich CRR (1999) Review and update on the biology of *Coptotermes havilandi* (Isoptera, Rhinotermitidae). Sociobiology 33: 339–356

Danks HV (1992) Long life cycles in insects. Can Entomol 124:167–187

Davis MA, Thompson K (2000) Eight ways to be a colonizer; two ways to be an invader: a proposed nomenclature scheme for invasion ecology. Bull Ecol Soc Am 81:226–230

DeHeer CJ, Kutnik M, Vargo EL, Bagnères A-G (2005) The breeding system and population structure of the termite *Reticulitermes grassei* in southwestern France. Heredity 95:408–415

DeHeer CJ, Vargo EL (2004) Colony genetic organization and colony fusion in the termite *Reticulitermes flavipes* as revealed by foraging patterns over time and space. Mol Ecol 13:431–441

Donovan SE, Eggleton P, Bignell DE (2001) Gut content analysis and a new feeding group classification of termites (Isoptera). Ecol Entomol 26:356–366

Dronnet S, Chapuisat M, Vargo EL et al (2005) Genetic analysis of the breeding system of an invasive subterranean termite, *Reticulitermes santonensis*, in urban and natural habitats. Mol Ecol 14:1311–1320

Eggleton P (2000) Global patterns of termite diversity. In: Abe T, Bignell DE, Higashi M (eds) Termites: evolution, sociality, symbioses, ecology. Kluwer Academic Publishers, Dordrecht, pp 25–54

Eggleton P, Tayasu I (2001) Feeding groups, lifetypes and the global ecology of termites. Ecol Res 16:941–960

Eggleton P, Williams PH, Gaston KJ (1994) Explaining global termite diversity: productivity or history? Biodivers Conserv 3:318–330

Elmqvist T, Folke C, Nyström M et al (2003) Response diversity, ecosystem change, and resilience. Front Ecol Environ 1:488–494

Elton CS (1958) The ecology of invasions by animals and plants. Methuen, London

Emerson AE (1936) Distribution of termites. Science 83:410–411

Emerson AE (1971) Tertiary fossil species of the Rhinotermitidae (Isoptera), phylogeny of genera, and reciprocal phylogeny of associated Flagellata (Protozoa) and the Staphylinidae (Coleoptera). Bull Am Mus Nat Hist 146:243–304

Evans TA, Lenz M, Gleeson PV (1998) Testing assumptions of mark-recapture protocols for estimating population size using Australian, mound-building, subterranean termites. Ecol Entomol 23:101–121

Evenhuis NL (2007) Checklist of the Isoptera of Fiji. Bishop Mus Tech Rep 38(12), Honolulu, Hawaii

Falk-Petersen J, Bøhn T, Sandlund OT (2006) On the numerous concepts in invasion biology. Biol Invasions 8:1409–1424

Fang R, Huang L, Zhong JH (2008) Surprising low levels of genetic diversity of Formosan subterranean termites in South China as revealed by the COII gene (Isoptera: Rhinotermitidae). Sociobiol 51:1–20

Ferraz MV, Montiel JTM (2004) First record of a subterranean termite, *Coptotermes havilandi* Holgrem (Isoptera, Rhinotermitidae), on the west coast of North America (Mexico). Acta Zoológica Mexicana 20:39–43

Forsyth J (1966) Agricultural insects of Ghana: a list of insects recorded in Ghana on field crops and stored produce with their host plants, damage caused, locality, parasites and predators where known, together with references to the literature for Ghana and neighbouring countries covering the period 1910–1960. Ghana Universities Press, Accra

Gay FJ (1967) A world review of introduced species of termites. Bull CSIRO 286:1–88

Gay FJ (1969) Species introduced by man. In: Krishna K, Weesner FM (eds) Biology of termites, vol 1. Academic Press, New York, NY, pp 459–494

Gay FJ, Calaby JH (1970) Termites of the Australian region. In: Krishna K, Weesner FM (eds) Biology of termites, vol 2. Academic Press, New York, NY, pp 393–448

Gay F, Watson JAL (1982) The genus *Cryptotermes* in Australia (Isoptera: Kalotermitidae). Aust J Zool Suppl 30(88):1–64

Gentz MC, Rubinoff D, Grace JK (2008) Phylogenetic analysis of subterranean termites (*Coptotermes* spp., Isoptera: Rhinotermitidae) indicates the origins of Hawaiian and North American invasions: potential implications for invasion biology. Proc Hawaiian Entomol Soc 40:1–9

Goodisman MAD, Crozier RH (2002) Population and colony genetic structure of the primitive termite *Mastotermes darwiniensis*. Evolution 56:70–83

Grace JK (2009) What can fecal pellets tell us about cryptic drywood termites (Isoptera: Kalotermitidae)? The International Research Group on Wood Protection, IRG-WP doc 09–20407

Grace JK, Woodrow RP, Yates JRIII (2002) Distribution and management of termites in Hawaii. Sociobiology 40:87–93

Gray B (1968) Forest tree and timber insect pests in the territory of Papua and New Guinea. Pac Insects 10:301–323

Gross GF (1975) The land invertebrates of the New Hebrides and their relationships. Phil Trans R Soc Lond B 272:391–421

Guadalupe Rojas M, Morales-Ramos JA, Klich MA, Wright M (2001) Three fungal species isolated from *Coptotermes formosanus* (Isoptera: Rhinotermitidae) bodies, carton material, and infested wood. Fla Entomol 84:156–158

Haack RA (2006) Exotic bark- and wood-boring Coleoptera in the United States: recent establishments and interceptions. Can J For Res 36:269–288

Haack RA, Cavey JF (1997) Insects intercepted on wood articles at ports-of entry in the United States: 1985–1996. Newslett Mich Entomol Soc 42:1–7

Haack RA, Cavey JF (2000) Insects intercepted on solid wood packing materials at United States ports-of-entry: 1985–1998. In: Quarantine pests, risks for the forestry sector and their effects on foreign trade. Proc Silvotecna 14, 27–28 June 2000, Concepcion, Chile (CD, 16 pp)

Hadlington P (1987) Australian termites and other common timber pests, 2nd edn. UNSW Press, Sydney

Harris WV (1966) On the genus *Coptotermes* in Africa (Isoptera: Rhinotermitidae). Proc R Entomol Soc Lond B 35:161–171

Hathorne KT, Zungoli PA, Benson EP, Bridges WC (2000) The termite (Isoptera) fauna of South Carolina. J Agr Urban Entomol 17:219–229
Haverty MI, Grace JK, Nelson LJ, Yamamoto RT (1996) Intercaste, intercolony and temporal variation in cuticular hydrocarbons of *Coptotermes formosanus* Shiraki (Isoptera: Rhinotermitidae). J Chem Ecol 22:1813–1834
Haverty MI, Nelson LJ, Page M (1990) Cuticular hydrocarbons of four populations of *Coptotermes formosanus* Shiraki in the United States: similarities and origins of introductions. J Chem Ecol 16:1635–1647
Hernández LM (1994) Una nueva especie del género *Incisitermes* y dos nuevos registros de termites (Isoptera) para Cuba. Avicennia. 1:87–99
Hill GF (1942) Termites (Isoptera) from the Australian region. CSIR, Melbourne
Holling CS (1973) Resilience and stability of ecological systems. Annu Rev Ecol Syst 4:1–23
Howard RW, Haverty MI (1980) Reproductives in mature colonies of *Reticulitermes flavipes*: abundance, sex-ratio and association with soldiers. Environ Entomol 9:458–460
Howell HN, Gold RE, Glenn GJ (2001) *Coptotermes* distribution in Texas (Isoptera: Rhinotermitidae). Sociobiology 37:687–697
Husseneder C, Messenger MT, Su N-Y et al (2005) Colony social organization and population genetic structure of an introduced population of Formosan subterranean termite from New Orleans, Louisiana. J Econ Entomol 98:1421–1434
Ichiko Y, Yukio N, Kazuhiko K, Shiryo Y (2000) Morphology and distribution of termites in Ryukyu islands and North and South Daito islands. Jpn J Entomol New Series 3:139–156
Indrayani Y, Yoshimura T, Fujii T (2004) Survey on the infestation of houses by *Incisitermes minor* (Hagen) in Kansai and Hokuriku areas. Jpn J Environ Entomol Zool 15:261–268
Indrayani Y, Yoshimura T, Fujii T et al (2005) A case study of *Incisitermes minor* (Isoptera: Kaltermitidae) infestation in Wakayama Prefecture, Japan. Sociobiology 46:45–63
Inward DJG, Vogler AP, Eggleton P (2007) A comprehensive phylogenetic analysis of termites (Isoptera) illuminates key aspects of their evolutionary biology. Mol Phylogenet Evol 44: 953–967
Iosua F, Peseta O (2003) Development of forest health surveillance systems for South Pacific countries and Australia; country report – Samoa. Australian Government, Australian Centre for International Agricultural Research, Canberra
Jenkins TM, Dean RE, Forschler BT (2002) DNA technology, interstate commerce, and the likely origin of Formosan subterranean termite (Isoptera: Rhinotermitidae) infestation in Atlanta, Georgia. J Econ Entomol 95:381–389
Jenkins TM, Dean RE, Verkerk R, Forschler BT (2001) Phylogenetic analyses of two mitochondrial genes and one nuclear intron region illuminate European subterranean termite (Isoptera: Rhinotermitidae) gene flow, taxonomy, and introduction dynamics. Mol Phylogenet Evol 20:286–293
Jenkins TM, Jones SC, Lee C-Y et al (2007) Phylogeography illuminates maternal origins of exotic *Coptotermes gestroi* (Isoptera: Rhinotermitidae). Mol Phylogenet Evol 42:612–621
Jones SC (1990) Colony size of the desert subterranean termite *Heterotermes aureus* (Isoptera: Rhinotermitidae). Southwest Nat 35:285–291
Kayser M, Brauer S, Weiss G et al (2000) Melanesian origin of Polynesian Y chromosomes. Curr Biol 10:1237–1246
Kelsey JM (1945) The influence of termite reproductive castes on the colony and on control measures. N Z J Sci Technol 26:353–359
Kirton LG (2005) The importance of accurate termite taxonomy in the broader perspective of termite management. In: Lee C-Y, Robinson WH (eds) Proceedings of 5th international conference on urban pests. Perniagaan Ph'ng @ P&Y Design, Peneng, pp 1–7
Kirton LG, Azmi M (2005) Patterns in the relative incidence of subterranean termite species infesting buildings in peninsular Malaysia. Sociobiology 46:1–15
Kirton LG, Brown VK (2003) The taxonomic status of pest species of *Coptotermes* in southeast Asia: resolving the paradox in the pest status of the termites, *Coptotermes gestroi*, *C. havilandi* and *C. travians* (Isoptera: Rhinotermitidae). Sociobiology 42:43–63

Kistner DH (1985) A new genus and species of termitophilous Aleocharinae from mainland China associated with *Coptotermes formosanus* and its zoogeographic significance (Coleoptera: Staphylinidae). Sociobiology 10:93–104

Korb J, Hartfelder K (2008) Life history and development – a framework for understanding developmental plasticity in lower termites. Biol Rev 83:295–313

Korb J, Katrantzis S (2004) Influence of environmental conditions on the expression of the sexual dispersal phenotype in a lower termite: implications for the evolution of workers in termites. Evol Dev 6:342–352

Korb J, Lenz M (2004) Reproductive decision-making in the termite *Cryptotermes secundus* (Kalotermitidae) under variable food conditions. Behav Ecol 15:390–395

Korb J, Schmidinger S (2004) Help or disperse? Cooperation in termites influenced by food conditions. Behav Ecol Sociobiol 56:89–95

Krishna K, Weesner FM (eds) (1969) Biology of termites, vols 1 and 2. Academic Press, New York, NY

LaFage JP, Nutting WL (1978) Nutrient dynamics of termites. In: Brian MV (ed) Production ecology of ants and termites. Cambridge University Press, Cambridge, pp 165–232

Lal S, Tuvou L (2003) Development of forest health surveillance systems for South Pacific countries and Australia; country report – Fiji. Australian Government, Australian Centre for International Agricultural Research, Canberra

Legendre F, Whiting MF, Bordereau C et al (2008) The phylogeny of termites (Dictyoptera: Isoptera) based on mitochondrial and nuclear markers: implications for the evolution of the worker and pseudergate castes, and foraging behaviors. Mol Phylogenet Evol 48:615–627

Lenz M (1985) Is inter-and intra specific variability of lower termite neotenic numbers due to adaptive thresholds for neotenic elimination? – Considerations from studies on *Porotermes adamsoni* (Froggatt) (Isoptera: Termopsidae). In: Watson JAL, Okot-Kotber BM (eds) Caste differentiation in social insects. Pergamon Press, Oxford, pp 125–146

Lenz M (1994) Food resources, colony growth and caste development in wood-feeding termites. In: Hunt J, Nalepa CA (eds) Nourishment and evolution in insect societies. Westview Press, Boulder, CO, pp 159–209

Lenz M, Barrett RA (1982) Neotenic formation in field colonies of *Coptotermes lacteus* (Froggatt) in Australia, with comments on the roles of neotenics in the genus *Coptotermes* (Isoptera: Rhinotermitidae). Sociobiology 7:47–59

Lenz M, Barrett RA, Miller LR (1986) The capacity of colonies of *Coptotermes acinaciformis* from Australia to produce neotenics (Isoptera: Rhinotermitidae). Sociobiology 11:237–244

Lenz M, Barrett RA, Miller LR (1988) Mechanisms of colony reestablishment after orphaning in *Coptotermes lacteus* (Froggatt) (Isoptera; Rhinotermitidae). Sociobiology 14:245–268

Lenz M, Barrett RA, Williams ER (1982b) Influence of diet on the survival and wood consumption of *Porotermes adamsoni* (Froggatt) (Isoptera: Termopsidae) at different temperatures. Bull Entomol Res 72:423–435

Lenz M, Kard B, Evans TA et al (2009) Differential use of identical food resources by *Reticulitermes flavipes* (Isoptera: Rhinotermitidae) in two types of habitats. Environ Entomol 38:35–42

Lenz M, McMahan EA, Williams ER (1982a) Neotenic production in *Cryptotermes brevis* (Walker): influence of geographical origin, group composition, and maintenance conditions (Isoptera: Kalotermitidae). Insect Soc 29:148–163

Lenz M, Runko S (1993) Long-term impact of orphaning on field colonies of *Coptotermes lacteus* (Froggatt) (Isoptera: Rhinotermitidae). Insect Soc 40:439–456

Li HF, Ye W, Su N-Y, Kanzaki N (2009) Phylogeography of *Coptotermes gestroi* and *Coptotermes formosanus* (Isoptera: Rhinotermitidae) in Taiwan. Ann Entomol Soc Am 102:684–693

Light SF (1931) Present status of our knowledge of the termties of China. Lingnan Sci J 7:581–600

Light SF (1934) The termite fauna of North America with special reference to the United States. In: Kofoid CA, Light SF, Horner AC et al (eds) Termites and termite control, 2nd edn. University of California Press, Berkeley, CA, pp 127–135

Lockwood JL, Cassey P, Blackburn T (2005) The role of propagule pressure in explaining species invasions. Trends Ecol Evol 20:223–228

Lowe S, Browne M, Boudjelas S, De Poorter M (2000) 100 of the world's worst invasive alien species: a selection from the global invasive species database. World Conservation Union (Species Survival Commission) Aliens 12, 12 pp

Mack RN, Simberloff D, Lonsdale WM et al (2000) Biotic invasions: causes, epidemiology, global consequences, and control. Ecol Appl 10:689–710

Maiti PK (1983) Termite fauna (Isoptera) of west Bengal, India, their recognition, biology and ecology. Records of the Zoological Survey of India, Occasional Papers, 48

Mankin RW, Osbrink WL, Oi FM, Anderson JB (2002) Acoustic detection of termite infestations in urban trees. J Econ Entomol 95:981–988

Matisoo-Smith E, Roberts RM, Irwin GJ et al (1998) Patterns of prehistoric human mobility in Polynesia indicated by mtDNA from the Pacific rat. Proc Natl Acad Sci USA 95:15145–15150

Mattson WJ (1980) Herbivory in relation to plant nitrogen content. Annu Rev Ecol Syst 11: 119–161

Messenger MT, Mullins AJ (2005) New flight distance recorded for *Coptotermes formosanus* (Isoptera: Rhinotermitidae). Fla Entomol 88:99–100

Messenger MT, Su N-Y, Scheffrahn RH (2002) Current distribution of the Formosan subterranean termite and other termite species (Isoptera: Rhinotermitidae, Kalotermitidae) in Louisiana. Fla Entomol 85:580–587

Miller D (1941) The species of termites in New Zealand. N Z J For 4:333–334

Mitchell JD (2002) Termites as pests of crops, forestry, rangeland and structures in Southern Africa and their control. Sociobiology 40:47–69

Morales-Ramos JA, Rojas GM (2005) Wood consumption rates of *Coptotermes formosanus* (Isoptera: Rhinotermitidae): a three-year study using groups of workers and soldiers. Sociobiology 45:707–719

Mori H (1976) Establishment of the American common dry-wood termite *Incisitermes minor* (Hagen) in Tokyo. Shiroari (Termite) 27:45–47 (in Japanese)

Mori H (1987) The Formosan subterranean termite in Japan: distribution, damage, and current and potential control measures. In: Tamashiro M, Su N-Y (eds) Biology and control of the Formosan subterranean termite. Research Extension Series 083. University of Hawaii, Honolulu, pp 23–26

Muirhead JR, Gray DK, Kelly DW et al (2008) Identifying the source of species invasions: sampling intensity vs. genetic diversity. Mol Ecol 17:1020–1035

Myles TG (1999) Review of secondary reproduction in termites (Insecta: Isoptera) with comments on its role in termite ecology and social evolution. Sociobiology 33:1–91

Nalepa CA (1994) Nourishment and the origin of termite eusociality. In: Nalepa CA, Hunt JH (eds) Nourishment and evolution in insect societies. Westview Press, Boulder, CO, pp 57–104

Nalepa CA, Bignell DE, Bandi C (2001) Detritivory, coprophagy, and the evolution of digestive mutualisms in Dictyoptera. Insect Soc 48:194–201

Noirot C (1985a) Pathways of caste development in the lower termites. In: Watson JAL, Okot-Kotber BM, Noirot C (eds) Caste differentiation in social insects. Pergamon Press, Frankfurt, pp 41–57

Noirot C (1985b) The caste system in higher termites. In: Watson JAL, Oko-Kotber BM, Noirot C (eds) Caste differentiation in social insects. Pergamon Press, Frankfurt, pp 75–86

Noirot C (1985c) Differentiation of reproductives in higher termites. In: Watson JAL, Oko-Kotber BM, Noirot C (eds) Caste differentiation in social insects. Pergamon Press, Frankfurt, pp 177–186

Nunes L (2008) Termite infestation risk in Portuguese historic buildings. International Conference on Wood Science for Preservation of Cultural Heritage. Braga, Portugal, 5–7 Nov 2008

Parman V, Vargo EL (2008) Population density, species abundance, and breeding structure of subterranean termite colonies in and around infested houses in central North Carolina. J Econ Entomol 101:1349–1359

Pearson HG, Bennett SJ (2008) Australian subterranean termites (*Coptotermes acinaciformis*) in New Zealand. In: Froud KJ, Popay AI, Zydenbos SM (eds) Surveillance for biosecurity: pre-border to pest management. The New Zealand Plant Protection Society, Auckland, pp 167–174

Peck SB (2001) Smaller orders of insects of the Galápagos Islands, Ecuador: evolution, ecology and diversity. National Research Council Press, Ottawa

Peters BC, Fitzgerald CJ (2008) Giant northern termite: gold coast inspection report. Report of Department of Primary Industries and Fisheries, Queensland Government

Peterson G, Allen CR, Holling CS (1998) Ecological resilience, biodiversity and scale. Ecosystems 1:6–18

Philip B, Pearson H, Bennett S (2008) Australian termites in New Zealand. Biosecur Mag 82: 12–13

Raina A, Osbrink N, Park YI (2004) Nymphs of the Formosan subterranean termite (Isoptera: Rhinotermitidae): aspects of formation and transformation. Ann Entomol Soc Am 97: 757–764

Roisin Y (1987) Polygyny in *Nasutitermes* species: field data and theoretical approaches. Experientia Suppl 54:379–404

Roisin Y (2000) Diversity and evolution of caste patterns. In: Abe T, Bignell DE, Higashi M (eds) Termites: evolution, sociality, symbioses, ecology. Kluwer Academic Publishers, Dordrecht, pp 95–119

Roisin Y, Dejean A, Corbara B et al (2006) Vertical stratification of the termite assemblage in a neotropical rainforest. Oecologia 149:301–311

Roisin Y, Pasteels JM (1985) A new *Nasutitermes* species from New Guinea (Isoptera: Termitidae). Indo-Malayan Zool 2:325–330

Roisin Y, Pasteels JM (1996) The nasute termites (Isoptera: Nasutitermitinae) of Papua New Guinea. Invertebr Taxon 10:507–616

Ross M (2005a) Australian subterranean termites eradicated. Biosecur Mag 60:9

Ross MG (2005b) Responding to incursions of Australian subterranean termites in New Zealand. In: Lee C-Y, Robinson WH (eds) Proceedings of the 5th international conference on urban pests. Perniagaan Ph'ng @ P&Y Design, Peneng, pp 233–238

Ross M (2006) Subterranean termites found in Nelson. Biosecur Mag (New Zealand) 67:9

Roy-Noël J (1966) Description de l'imago et de quelques formes jeunes de néoténiques chez *Coptotermes intermedius* Silvestri (Isoptère). Insect Soc 13:217–224

Schabel HG (2006) Forest entomology in East Africa: forest insects of Tanzania. Springer, Dordrecht

Scheffrahn RH, Cabrera BJ, Kern WH Jr, Su N-Y (2002) *Nasutitermes costalis* (Isoptera: Termitidae) in Florida: first record of a non-endemic establishment by a higher termite. Fla Entomol 85:273–275

Scheffrahn RH, Chase JA, Mangold JR et al (1999) First record of *Reticulitermes* (Isoptera: Rhinotermitidae) from the West Indies: *R. flavipes* on Grand Bahama Island. Fla Entomol 82:480–482

Scheffrahn RH, Darlington JPEC, Collins MS et al (1994) Termites (Isoptera: Kalotermitidae, Rhinotermitidae, Termitidae) of the West Indies. Sociobiol 24:213–238

Scheffrahn RH, Jones SC, Krck J et al (2003) Taxonomy, distribution, and notes on the termites (Isoptera: Kalotermitidae, Rhinotermitidae, Termitidae) of Puerto Rico and the US Virgin Islands. Ann Entomol Soc Am 96:181–201

Scheffrahn RH, Krck J (1999) Termites of the Genus *Cryptotermes* Banks (Isoptera: Kalotermitidae) from the West Indies. Insecta Mundi 13:111–171

Scheffrahn RH, Krck J, Maharajh B et al (2004) Establishment of the African termite, *Coptotermes sjostedti* (Isoptera: Rhinotermitidae), on the Island of Guadeloupe, French West Indies. Ann Entomol Soc Am 97:872–876

Scheffrahn RH, Krck J, Szalanski AL et al (2005a) Synonymy of two arboreal termites (Isoptera: Termitidae: Nasutitermitinae): *Nasutitermes corniger* from the neotropics and *N. polygynus* from New Guinea. Fla Entomol 88:28–33

Scheffrahn RH, Křeček J, Ripa R, Luppichini P (2009) Endemic origin and vast anthropogenic dispersal of the West Indian drywood termite. Biol Invasions 11:787–799

Scheffrahn RH, Krck J, Szalanski AL, Austin JW (2005b) Synonymy of neotropical arboreal termites *Nasutitermes corniger* and *N. costalis* (Isoptera: Termitidae: Nasutitermitinae), with evidence from morphology, genetics, and biogeography. Ann Entomol Soc Am 98:273–281

Scheffrahn RH, Mangold JR, Su N-Y (1988) A survey of structure infesting termites in peninsula Florida. Fla Entomol 71:615–630

Scheffrahn RH, Su N-Y (1990) Native, introduced and structure-infesting termites of the Turks and Caicos Islands, BWI (Isoptera: Kalotermitidae, Rhinotermitidae, Termitidae). Fla Entomol 73:622–627

Scheffrahn RH, Su N-Y (1995) A new subterranean termite introduced to Florida: *Heterotermes froggatt* (Rhinotermitidae: Heterotermitinae) established in Miami. Fla Entomol 78:623–627

Scheffrahn RH, Su N-Y (2005) Distribution of the termite genus *Coptotermes* (Isoptera: Rhinotermitidae) in Florida. Fla Entomol 88:201–203

Scheffrahn RH, Su N-Y, Chase JA et al (2000) First record of *Cryptotermes cynocephalus* Light (Isoptera: Kalotermitidae) and natural woodland infestations of *C. brevis* (Walker) on Oahu, Hawaiian Islands. Proc Hawaiian Entomol Soc 34:121–125

Scheffrahn RH, Su N-Y, Chase JA, Forschler BT (2001) New termite (Isoptera: Kalotermitidae, Rhinotermitidae) records from Georgia. J Entomol Sci 36:109–113

Shellman-Reeve JS (1997) The spectrum of eusociality in termites. In: Choe JC, Crespi BJ (eds) The evolution of social behavior in insects and arachnids. Cambridge University Press, Cambridge, pp 52–93

Sieber R (1985) Replacement reproductives in Macrotermitinae (Isoptera: Termitidae). In: Watson JAL, Okot-Kotber BM (eds) Caste differentiation in social insects. Pergamon Press, Oxford, pp 201–208

Snyder TE (1949) Catalog of the termites (Isoptera) of the world. Smithson Misc Coll 112. The Smithsonian Institution, Washington, DC

Stachowicz JJ, Fried H, Osman RW, Whitlatch RB (2002) Biodiversity, invasion resistance, and marine ecosystem function: reconciling pattern and process. Ecology 83:2575–2590

Su N-Y, Scheffrahn RH, Weissling T (1997) A new introduction of a subterranean termite, *Coptotermes havilandi* Holmgren (Isoptera: Rhinotermitidae) in Miami, Florida. Fla Entomol 80:408–411

Su N-Y, Tamashiro M (1987) An overview of the Formosan subterranean termite in the world. In: Tamashiro M, Su N-Y (eds) Biology and control of the Formosan subterranean termite. Res Exten Ser 083. University of Hawaii, Honolulu, pp 3–15

Su N-Y, Ye W, Ripa R et al (2006) Identification of Chilean *Reticulitermes* (Isoptera: Rhinotermitidae) inferred from three mitochondrial gene DNA sequences and soldier morphology. Ann Entomol Soc Am 99:352–363

Sun J-Z, Lockwood ME, Etheridge JL et al (2007) Distribution of Formosan subterranean termite (Isoptera: Rhinotermitidae) in Mississippi. J Econ Entomol 100:1400–1408

Szalanski AL, Scheffrahn RH, Austin JW et al (2004) Molecular phylogeny and biogeography of *Heterotermes* (Isoptera: Rhinotermitidae) in the West Indies. Ann Entomol Soc Am 97: 574–585

Tang C, Li S (1960) Notes on the types of reproductive castes of subterranean termites *Coptotermes formosanus* Shirake and *Reticulitermes flaviceps* Oshima (Isoptera, Rhinotermitidae). Acta Entomol Sinica 10:302–306

Taylor RW, Wilson EO (1961) Ants from three remote oceanic islands. Psyche 68:134–144

Thistleton BM, Neal M, Peki M, Dobunaba J (2007) *Mastotermes darwiniensis* in the Lae area of PNG: an assessment of current and potential status of the incursion, and options for management, final report FST/2005/049. Australian Government, Australian Centre for International Agricultural Research, Canberra

Thorne BL (1982) Polygyny in termites: multiple primary queens in colonies of *Nasutitermes corniger* (Motschulsky) (Isoptera: Termitidae). Insect Soc 29:102–117
Thorne BL (1996) Termite terminology. Sociobiology 28:253–261
Thorne BL, Noirot C (1982) Ergatoid reproductives in *Nasutitermes corniger* (Motschulsky) (Isoptera: Termitidae). Int J Insect Morphol Embryol 11:213–226
Tsai C-C, Chen C-S (2003) First record of *Coptotermes gestroi* (Isoptera: Rhinotermitidae) from Taiwan. Formosan Entomol 23:157–161
Vargo EL, Husseneder C (2009) Biology of subterranean termites: insights from molecular studies of *Reticulitermes* and *Coptotermes*. Annu Rev Entomol 54:379–403
Vargo EL, Husseneder C, Woodson D et al (2006) Genetic analysis of colony and population genetic structure of three introduced populations of the Formosan subterranean termite (Isoptera: Rhinotermitidae) in the continental United States. Environ Entomol 35: 151–166
Verkerk RHJ, Bravery AF (2001) The UK termite eradication programme: justification and implementation. Sociobiology 37:351–360
Vieau F (2001) Comparison of the spatial distribution and reproductive cycle of *Reticulitermes santonensis* Feytaud and *R. lucifugus grassei* Clément (Isoptera, Rhinotermitidae) suggests that they represent introduced and native species, respectively. Insect Soc 48:57–62
Wagner MR, Atuahene SKN, Cobbinah JR (1991) Forest entomology in west tropical Africa: forest insects of Ghana. Kluwer Academic Publishers, Dordrecht
Walker J (2007) Pest watch: 12/02/2007 – 16/03/2007. Biosecur Mag (New Zealand) 75:27
Watson JAL, Abbey HE (1993) Atlas of Australian termites. CSIRO, Canberra
Werner PA, Prior LD (2007) Tree-piping termites and growth and survival of host trees in savanna woodland of north Australia. J Trop Ecol 23:611–622
Westphal MI, Browne M, MacKinnon K, Noble I (2008) The link between international trade and the global distribution of invasive alien species. Biol Invasions 10:391–398
Williams RMC (1976) Factors limiting the distribution of building-damaging dry-wood termites (Isoptera, *Cryptotermes* spp.) in Africa. Mater Org 3:393–406
Woodrow RJ, Grace JK, Higa SY (2001) Occurrence of *Coptotermes vastator* (Isoptera: Rhinotermitidae) on the island of Oahu, Hawaii. Sociobiology 38:667–673
Woodson WD, Wiltz BA, Lax AR (2001) Current distribution of the Formosan subterranean termite (Isoptera: Rhinotermitidae) in the United States. Sociobiology 37:661–671
Xie Y, Li Z, Gregg WI, Li D (2001) Invasive species in China – an overview. Biodivers Conserv 10:1317–1341
Ye W, Lee C-Y, Scheffrahn RH et al (2004) Phylogenetic relationships of nearctic *Reticulitermes* species (Isoptera: Rhinotermitidae) with particular reference to *Reticulitermes arenincola* Goellnerq. Mol Phylogenet Evol 30:815–822
Yeap BK, Othman AS, Lee VS, Lee CY (2007) Genetic relationship between *Coptotermes gestroi* and *Coptotermes vastator* (Isoptera: Rhinotermitidae). J Econ Entomol 100:467–474
Yule RA, Watson JAL (1976) Two further domestic species of *Cryptotermes* from the Australian mainland (Isoptera: Kalotermitidae). J Aust Ent Soc 15:349–352

Index

D.E. Bignell et al. (eds.), *Biology of Termites: A Modern Synthesis*,
DOI 10.1007/978-90-481-3977-4, © Springer Science+Business Media B.V. 2011

D

G

T